Transport Processes:
Momentum, Heat, and Mass

CHRISTIE J. GEANKOPLIS

The Ohio State University

Transport Processes: Momentum, Heat, and Mass

ALLYN AND BACON, INC.

Boston London Sydney Toronto

This book is part of the
ALLYN AND BACON SERIES IN ENGINEERING
Consulting Editor: Frank Kreith
University of Colorado

Portions of this book appear in *Transport Processes and Unit Operations*, second edition, by Christie J. Geankopolis, copyright © 1983, 1978 by Allyn and Bacon, Inc.

LIBRARY OF CONGRESS CATALOGING IN PUBLICATION DATA

Geankoplis, Christie J.
  Transport processes—momentum, heat, and mass.

  "Part I of the text, Transport processes and unit operations"—Pref.
  Bibliography: p.
  Includes index.
  1. Transport theory. I. Title.
TP156.T7G425 660.2'842 82-3928
ISBN 0-205-07787-0   AACR2
ISBN (international) 0-205-07942-3

Printed in the United States of America

10 9 8 7 6 5 4 3 2 1   87 86 85 84 83 82

Dedicated to
my father and mother
for their inspiration
and encouragement

Contents

Preface

The principles of momentum transfer and heat transfer have been taught historically in most engineering fields, such as chemical, mechanical, civil, agricultural, ceramic, and process metallurgy. The study of mass transfer has been limited primarily to chemical engineers. However, engineers in other fields have become more interested in mass transfer in gases, liquids, and solids. These subjects of momentum, heat, and mass transfer have often been taught as completely separate subjects and the similarities between these areas have not been emphasized.

The three transport processes of momentum, heat or energy, and mass transfer are considered in this text in a single volume using a more unified approach to a fundamental and systematic description of these transport processes. These three areas are considered as one general discipline and the similarities of the basic equations and calculational methods are used in increasing the understanding of the three individual transport processes. Also, the differences that occur in the actual physical processes which modify the general equations are pointed out.

Some engineering departments are gradually including in their curricula more of the basic core areas of the transport processes of momentum, heat, and mass. It is hoped that this textbook will enable and/or influence more engineering departments to include the transport processes in their curricula.

In this text, Chapters 2 and 3 deal with momentum transfer, Chapters 4 and 5 involve energy or heat transfer, and Chapters 6 and 7 consider mass transfer.

In Chapter 1 elementary principles of mathematical and graphical methods, laws of chemistry and physics, material balances, and heat balances are reviewed. Many, especially chemical engineers, may be familiar with most of these principles and may omit all or parts of this chapter.

A few topics, involved primarily with the processing of biological materials, may be omitted at the discretion of the reader or instructor: Sections 3.5, 4.12, 5.5 and 6.4. Over 150 example or sample problems and over 340 homework problems on all topics are included in the text. Some of the homework problems are concerned with biological systems, for those readers who are especially interested in that area.

A feature of this text is the division of each chapter into elementary sections followed by selected topic sections. In the elementary sections at the beginning of each chapter the basic fundamentals needed for a thorough understanding of the principles are covered. Then, depending on the needs of the reader or instructor, various selected topic sections can be studied. As an example, Section 2.11, (Selected Topic) Compressible Flow of Gases, is a topic usually omitted in elementary coverage of the basic principles. This arrangement is followed in every chapter of the book.

This text may be used for a course of study using any of the following four suggested plans. In all plans, Chapter 1 may or may not be included.

1. *Study of transport processes of momentum, heat, and mass*

In this plan, the complete text covering the principles of the transport processes are covered. Also, most of the selected topics could be studied. This plan could be applicable to all engineers at the junior and/or senior level in a two-semester or three-quarter course.

2. *Elementary study of transport processes of momentum, heat, and mass*

Only the elementary sections of the text are covered with Chapter 7 (Principles of Unsteady-State and Convective Mass Transfer) being omitted and the selected topic sections being omitted in a one-semester or two-quarter course.

3. *Study of transport processes of heat and mass*

For those such as chemical or mechanical engineers who have had a course in momentum transfer, Chapters 4, 5, 6, and 7 on heat and mass transfer would be covered in a one-semester course. In some cases, particularly for mechanical engineers who might desire less coverage of mass transfer, Chapter 7 could also be omitted in a one-semester course.

4. *Study of mass transfer*

For those such as chemical, mechanical, civil, or agricultural engineers who have had momentum and heat transfer, or those who desire only a background in mass transfer in a one-quarter course, all of Chapters 6 and 7 would be covered.

The SI (Système International d'Unités) system of units has been adopted by the scientific community. Because of this, the SI system of units has been adopted in this text for use in the equations, example problems, and homework problems. However, the most important equations derived in the text are also given in a dual set of units, SI and English, when different. Some example and homework problems are also given using English units for those desiring such coverage.

This text, published separately, is Part I of the text *Transport Processes and Unit Operations*. For those desiring to cover the applications of the transport processes of momentum, heat, and mass in the unit operations, the reader is referred to Part II of *Transport Processes and Unit Operations*.

C. J. G.

Transport Processes:
Momentum, Heat, and Mass

CHAPTER 1

Introduction to Engineering Principles and Units

1.1 CLASSIFICATION OF UNIT OPERATIONS AND TRANSPORT PROCESSES

1.1A Introduction

In the chemical and other physical processing industries and the food and biological processing industries, many similarities exist in the manner in which the entering feed materials are modified or processed into final materials of chemical and biological products. We can take these seemingly different chemical, physical, or biological processes and break them down into a series of separate and distinct steps called *unit operations*. These unit operations are common to all types of diverse process industries.

For example, the unit operation *distillation* is used to purify or separate alcohol in the beverage industry and hydrocarbons in the petroleum industry. Drying of grain and other foods is similar to drying of lumber, filtered precipitates, and rayon yarn. The unit operation *absorption* occurs in absorption of oxygen from air in a fermentation process or in a sewage treatment plant and in absorption of hydrogen gas in a process for liquid hydrogenation of oil. Evaporation of salt solutions in the chemical industry is similar to evaporation of sugar solutions in the food industry. Settling and sedimentation of suspended solids in the sewage and the mining industries are similar. Flow of liquid hydrocarbons in the petroleum refinery and flow of milk in a dairy plant are carried out in a similar fashion.

The unit operations deal mainly with the transfer and change of energy and the transfer and change of materials primarily by physical means but also by physical–chemical means. The important unit operations, which can be combined in various sequences in a process and which are covered in Part II of this text, are described next.

1.1B Classification of Unit Operations

1. *Fluid flow.* This concerns the principles that determine the flow or transportation of any fluid from one point to another.
2. *Heat transfer.* This unit operation deals with the principles that govern accumulation and transfer of heat and energy from one place to another.

3. *Evaporation.* This is a special case of heat transfer, which deals with the evaporation of a volatile solvent such as water from a nonvolatile solute such as salt or any other material in solution.
4. *Drying.* In this operation volatile liquids, usually water, are removed from solid materials.
5. *Distillation.* This is an operation whereby components of a liquid mixture are separated by boiling because of their differences in vapor pressure.
6. *Absorption.* In this process a component is removed from a gas stream by treatment with a liquid.
7. *Membrane separation.* This process involves the diffusion of a solute from a liquid or gas through a semipermeable membrane barrier to another fluid.
8. *Liquid–liquid extraction.* In this case a solute in a liquid solution is removed by contacting with another liquid solvent which is relatively immiscible with the solution.
9. *Liquid–solid leaching.* This involves treating a finely divided solid with a liquid that dissolves out and removes a solute contained in the solid.
10. *Crystallization.* This concerns the removal of a solute such as a salt from a solution by precipitating the solute from the solution.
11. *Mechanical–physical separations.* These involve separation of solids, liquids, or gases by mechanical means, such as filtration, settling, and size reduction, which are often classified as separate unit operations.

Many of these unit operations have certain fundamental and basic principles or mechanisms in common. For example, the mechanism of diffusion or mass transfer occurs in drying, absorption, distillation, and crystallization. Heat transfer occurs in drying, distillation, evaporation, and so on. Hence, the following classification of a more fundamental nature is often made into transfer or transport processes.

1.1C Fundamental Transport Processes

1. *Momentum transfer.* This is concerned with the transfer of momentum which occurs in moving media, such as in the unit operations of fluid flow, sedimentation, and mixing.
2. *Heat transfer.* In this fundamental process, we are concerned with the transfer of heat from one place to another; it occurs in the unit operations heat transfer, drying, evaporation, distillation, and others.
3. *Mass transfer.* Here mass is being transferred from one phase to another distinct phase; the basic mechanism is the same whether the phases are gas, solid, or liquid. This includes distillation, absorption, liquid–liquid extraction, and leaching.

1.1D Arrangement in Parts I and II

This text is arranged in two parts:

Part I: Transport Processes: Momentum, Heat, and Mass. These fundamental principles are covered extensively in Chapters 1 to 7 to provide the basis for study of unit operations.

Part II: Unit Operations. The various unit operations and their applications to process areas are studied in Part II of this text.

There are a number of elementary engineering principles, mathematical techniques, and laws of physics and chemistry which are basic to a study of the principles of momentum, heat, and mass transfer and the unit operations. These are reviewed for the reader in this first chapter. Some readers, especially chemical engineers, agricultural engineers, civil engineers, and chemists, may be familiar with many of these principles and techniques and may wish to omit all or parts of this chapter.

Homework problems at the end of each chapter are arranged in different sections, each corresponding to the number of a given section in the chapter. Each chapter is divided into an elementary section at the beginning and a selected topic section. The selected topic sections can be omitted or certain topics selected for study, depending on the needs of the reader.

1.2 SI SYSTEM OF BASIC UNITS USED IN THIS TEXT AND OTHER SYSTEMS

There are three main systems of basic units employed at present in engineering and science. The first and most important of these is the *SI* (Système International d'Unités) *system*, which has as its three basic units the meter (m), the kilogram (kg), and the second (s). The others are the English foot (ft)–pound (lb)–second (s), or *fps*, *system* and the centimeter (cm)–gram (g)–second (s), or *cgs*, *system*.

At present the SI system has been adopted officially for use exclusively in engineering and science, but the older English and cgs systems will still be used for some time. Much of the physical and chemical data and empirical equations are given in these latter two systems. Hence, the engineer should not only be proficient in the SI system but must also be able to use the other two systems to a limited extent.

1.2A SI System of Units

The basic quantities used in the SI system are as follows: the unit of length is the meter (m); the unit of time is the second (s); the unit of mass is the kilogram (kg); the unit of temperature is the degree Kelvin (K); and the unit of an element is the kilogram mole (kg mol). The other standard units are derived from these basic quantities.

The basic unit of force is the newton (N), defined as

$$1 \text{ newton (N)} = 1 \text{ kg} \cdot \text{m/s}^2$$

The basic unit of work, energy, or heat is the newton-meter, or joule (J).

$$1 \text{ joule (J)} = 1 \text{ newton} \cdot \text{m } (\text{N} \cdot \text{m}) = 1 \text{ kg} \cdot \text{m}^2/\text{s}^2$$

Power is measured in joules/s or watts (W).

$$1 \text{ joule/s (J/s)} = 1 \text{ watt (W)}$$

The unit of pressure is the newton/m^2 or pascal (Pa).

$$1 \text{ newton/m}^2 \text{ (N/m}^2) = 1 \text{ pascal (Pa)}$$

[Pressure in atmospheres (atm) is not a standard SI unit but is being used during the transition period.] The standard acceleration of gravity is defined as

$$1 \text{ } g = 9.80665 \text{ m/s}^2$$

A few of the standard prefixes for multiples of the basic units are as follows: giga (G) = 10^9, mega (M) = 10^6, kilo (k) = 10^3, centi (c) = 10^{-2}, milli (m) = 10^{-3}, micro (μ) = 10^{-6}, and nano (n) = 10^{-9}. The prefix c is not a preferred prefix. Temperatures are defined in degrees Kelvin (K) as the preferred unit in the SI system. However, in practice, wide use is made of the degree Celsius (°C) scale, which is defined by

$$t°C = T(K) - 273.15$$

Note that 1°C = 1 K and that in the case of temperature difference,

$$\Delta t°C = \Delta T \ K$$

The standard preferred unit of time is the second (s), but time can be in nondecimal units of minutes (min), hours (h), or days (d).

1.2B   CGS System of Units

The cgs system is related to the SI system as follows:

$$1 \ g \ mass \ (g) = 1 \times 10^{-3} \ kg \ mass \ (kg)$$

$$1 \ cm = 1 \times 10^{-2} \ m$$

$$1 \ dyne \ (dyn) = 1 \ g \cdot cm/s^2 = 1 \times 10^{-5} \ newton \ (N)$$

$$1 \ erg = 1 \ dyn \cdot cm = 1 \times 10^{-7} \ joule \ (J)$$

The standard acceleration of gravity is

$$g = 980.665 \ cm/s^2$$

1.2C   English fps System of Units

The English system is related to the SI system as follows:

$$1 \ lb \ mass \ (lb_m) = 0.45359 \ kg$$

$$1 \ ft = 0.30480 \ m$$

$$1 \ lb \ force \ (lb_f) = 4.4482 \ newtons \ (N)$$

$$1 \ ft \cdot lb_f = 1.35582 \ newton \cdot m \ (N \cdot m) = 1.35582 \ joules \ (J)$$

$$1 \ psia = 6.89476 \times 10^3 \ newton/m^2 \ (N/m^2)$$

$$1.8°F = 1 \ K = 1°C \ (centigrade \ or \ Celsius)$$

$$g = 32.174 \ ft/s^2$$

The proportionality factor for Newton's law is

$$g_c = 32.174 \ ft \cdot lb_m/lb_f \cdot s^2$$

The factor g_c in SI units and cgs units is 1.0 and is omitted.

In Appendix A.1, convenient conversion factors for all three systems are tabulated. Further discussions and use of these relationships are given in various sections of the text.

This text uses the SI system as the primary set of units in the equations, sample problems, and homework problems. However, the important equations derived in the text are given in a dual set of units, SI and English, when these equations differ. Some example problems and homework problems are also given using English units. In some cases, intermediate steps and/or answers in example problems are also stated in English units.

Transport Processes: Momentum, Heat, and Mass

1.2D Dimensionally Homogeneous Equations and Consistent Units

A dimensionally homogeneous equation is one in which all the terms have the same units. These units can be the base units or derived ones (for example, $kg/s^2 \cdot m$ or Pa). Such an equation can be used with any system of units provided that the same base or derived units are used throughout the equation. No conversion factors are needed when consistent units are used.

The reader should be careful in using any equation and always check it for dimensional homogeneity. To do this, a system of units (SI, English, etc.) is first selected. Then units are substituted for each term in the equation and like units in each term canceled out.

EXAMPLE 1.2-1. Checking Units in Dimensionally Homogeneous Equations

The equation for heat transfer from a fluid to a surface is

$$q = hA(T_f - T_w) \qquad (1.2\text{-}1)$$

where q is the rate of heat transfer (energy/time), h is the heat-transfer coefficient (energy/time–area–temperature), A is the surface area, and T is temperature. Use SI units and check if the equation is dimensionally homogeneous.

Solution: Using $kg \cdot m^2/s^2$ for energy and substituting the base SI units into Eq. (1.2-1),

$$q \frac{kg \cdot m^2/s^2}{s} = h \frac{kg \cdot m^2/s^2}{s \cdot m^2 \cdot K} A(m^2)(T_f - T_w)(K) \qquad (1.2\text{-}2)$$

Canceling like units, the left-hand term has the same units of $kg \cdot m^2/s^3$ as the right-hand term, and the equation is dimensionally homogeneous. If the derived unit J is used for energy, both terms will have units of J/s, or W, which is more commonly used.

1.3 METHODS OF EXPRESSING TEMPERATURES AND COMPOSITIONS

1.3A Temperature

There are two temperature scales in common use in the chemical and biological industries. These are degrees Fahrenheit (abbreviated °F) and Celsius (°C). It is often necessary to convert from one scale to the other. Both use the freezing point and boiling point of water at 1 atmosphere pressure as base points. Often temperatures are expressed as absolute degrees K (SI standard) or degrees Rankine (°R) instead of °C or °F. Table 1.3-1 shows the equivalences of the four temperature scales.

TABLE 1.3-1. *Temperature Scales and Equivalents*

	Centigrade	Fahrenheit	Kelvin	Rankine	Celsius
Boiling water	100°C	212°F	373.15 K	671.7°R	100°C
Melting ice	0°C	32°F	273.15 K	491.7°R	0°C
Absolute zero	−273.15°C	−459.7°F	0 K	0°R	−273.15°C

The difference between the boiling point of water and melting point of ice at 1 atm is 100°C or 180°F. Thus, a 1.8°F change is equal to a 1°C change. Usually, the value of $-273.15°C$ is rounded to $-273.2°C$ and $-459.7°F$ to $-460°F$. The following equations can be used to convert from one scale to another.

$$°F = 32 + 1.8(°C) \tag{1.3-1}$$

$$°C = \frac{1}{1.8}(°F - 32) \tag{1.3-2}$$

$$°R = °F + 460 \tag{1.3-3}$$

$$K = °C + 273.15 \tag{1.3-4}$$

EXAMPLE 1.3-1.   Sterilization Temperature in Fermentation
A medium in a fermenter is sterilized at 120°C. What is the temperature in °F, °R, and K?

Solution: Using Eq. (1.3-1),

$$°F = 32 + 1.8(°C) = 32 + 1.8(120) = 248°F$$

For calculation of °R and K, Eqs. (1.3-3) and (1.3-4) are used.

$$°R = °F + 460 = 248 + 460 = 708°R$$

$$K = °C + 273.2 = 120 + 273.2 = 393.2 \text{ K}$$

1.3B   Mole Units, and Weight or Mass Units

There are many methods used to express compositions in gases, liquids, and solids. One of the most useful is molar units, since chemical reactions and gas laws are simpler to express in terms of molar units. A mole (mol) of a pure substance is defined as the amount of that substance whose mass is numerically equal to its molecular weight. Hence, 1 kg mol of methane CH_4 contains 16.04 kg. Also, 1.0 lb mol contains 16.04 lb_m.

The mole fraction of a particular substance is simply the moles of this substance divided by the total number of moles. In like manner, the weight or mass fraction is the mass of the substance divided by the total mass. These two compositions, which hold for gases, liquids, and solids, can be expressed as follows for component A in a mixture:

$$x_A(\text{mole fraction of } A) = \frac{\text{moles of } A}{\text{total moles}} \tag{1.3-5}$$

$$w_A(\text{mass or wt fraction of } A) = \frac{\text{mass } A}{\text{total mass}} \tag{1.3-6}$$

EXAMPLE 1.3-2.   Mole and Mass or Weight Fraction of a Solution
A container holds 50 g of water (B) and 50 g of NaOH (A). Calculate the weight fraction and mole fraction of NaOH. Also, calculate the lb_m of NaOH (A) and H_2O (B).

Solution: Taking as a basis for calculation 50 + 50 or 100 g of solution, the following data are calculated:

Component	G	Wt Fraction	Mol Wt	G Moles	Mole Fraction
H_2O (B)	50.0	$\dfrac{50}{100} = 0.500$	18.02	$\dfrac{50.0}{18.02} = 2.78$	$\dfrac{2.78}{4.03} = 0.690$
NaOH (A)	50.0	$\dfrac{50}{100} = 0.500$	40.0	$\dfrac{50.0}{40.0} = 1.25$	$\dfrac{1.25}{4.03} = 0.310$
Total	100.0	1.00		4.03	1.000

Hence, $x_A = 0.310$ and $x_B = 0.690$ and $x_A + x_B = 0.310 + 0.690 = 1.00$. Also, $w_A + w_B = 0.500 + 0.500 = 1.00$. To calculate the lb_m of each component, Appendix A.1 gives the conversion factor of 453.6 g per $1 lb_m$. Using this,

$$\text{lb mass of } A = \frac{50 \text{ g } A}{453.6 \text{ g } A/lb_m A} = 0.1102 \text{ lb}_m A$$

Note that the g of A in the numerator cancels the g of A in the denominator, leaving lb_m of A in the numerator. The reader is cautioned to put all units down in an equation and cancel those appearing in the numerator and denominator. In a similar manner we obtain $0.1102 lb_m B$ (0.0500 kg B).

The analyses of solids and liquids are usually given as weight or mass fraction or weight percent, and gases as mole fraction or percent. Unless otherwise stated, analyses of solids and liquids will be assumed to be weight (mass) fraction or percent, and of gases to be mole fraction or percent.

1.3C Concentration Units for Liquids

In general, when one liquid is mixed with another miscible liquid, the volumes are not additive. Hence, compositions of liquids are usually not expressed as volume percent of a component but as weight or mole percent. Another convenient way to express concentrations of components in a solution is *molarity*, which is defined as g mol of a component per liter of solution. Other methods used are kg/m^3, g/liter, g/cm^3, lb mol/cu ft, lb_m/cu ft, and $lb_m/gallon$. All these concentrations depend on temperature, so the temperature must be specified.

The most common method of expressing total concentration per unit volume is density, kg/m^3, g/cm^3, or lb_m/ft^3. For example, the density of water at 277.2 K (4°C) is 1000 kg/m^3, or 62.43 lb_m/ft^3. Sometimes the density of a solution is expressed as *specific gravity*, which is defined as the density of the solution at its given temperature divided by the density of a reference substance at its temperature. If the reference substance is water at 277.2 K, the specific gravity and density of the substance are numerically equal.

EXAMPLE 1.3-3. Concentration of Albumin Solution
The density, ρ, of a 2.0 wt % bovine serum albumin (A) solution (O1) has been measured as 1.028 g/cm^3 at 298 K (25°C). The molecular weight of the albumin protein is 67 000 g/g mol (S1). Calculate the following:
  (a) The specific gravity of this solution referred to water at 277 K.
  (b) The mole fraction of albumin (A) in this solution.
  (c) The density in lb mass/gallon and kg/m^3.
  (d) The molarity.

Solution: For part (a), the density of water at 277 K is 1.000 g/cm³. Hence,

$$\text{specific gravity} = \frac{1.0028 \text{ g/cm}^3}{1.000 \text{ g/cm}^3} = 1.0028$$

In part (b), taking 1.00 cm³ of solution as a basis for calculation,

$$\text{g mass } A = \frac{2}{100}(1.0028) = 0.020056 \text{ g } A$$

$$\text{g mass } B = \frac{98}{100}(1.0028) = 0.9827 \text{ g } B$$

$$\text{g mole } A = \frac{0.020056}{67\,000} = 3.00 \times 10^{-7} \text{ g mol } A$$

$$\text{g mole } B = \frac{0.9827}{18.0} = 5.46 \times 10^{-2} \text{ g mol } B$$

$$\text{total g mol} = 3.00 \times 10^{-7} + 5.46 \times 10^{-2} = 5.46 \times 10^{-2}$$

$$\text{mole fraction, } x_A = \frac{3.00 \times 10^{-7}}{5.46 \times 10^{-2}} = 5.49 \times 10^{-6}$$

In part (c), the conversion factor in Appendix A.1 to convert gal to cm³ is 3785 cm³/gal, and for lb mass to g mass it is 453.6 g mass/lb mass. So the density is

$$\text{lb mass/gal} = \frac{(1.0028 \text{ g mass/cm}^3)(3785 \text{ cm}^3/\text{gal})}{453.6 \text{ g mass/lb mass}} = 8.37 \frac{\text{lb}_m}{\text{gal}}$$

$$\text{kg mass/m}^3 = \frac{(1.0028 \text{ g mass/cm}^3)(10^6 \text{ cm}^3/\text{m}^3)}{1000 \text{ g mass/kg mass}} = 1002.8 \text{ kg/m}^3$$

For part (d), the molarity is defined as g mol albumin per L (liter) of solution. Now, 1.00 cm³ of solution contains 3.00×10^{-7} g mol A; then,

$$\text{molarity} = 3.00 \times 10^{-7} \frac{\text{g mol } A}{\text{cm}^3}\left(1000 \frac{\text{cm}^3}{\text{L}}\right) = 3.00 \times 10^{-4} \frac{\text{g mol } A}{\text{L}}$$

In this section of the text, when specifying the weight of a substance we have used the terms lb mass (lb_m) or g mass instead of the more commonly used terms lb or g. This was done to emphasize the fact that in a discussion of mass units we are dealing with lb mass or g mass, not with lb force (lb_f) or g force. However, in the remainder of the text we shall generally use the shorter terms lb or g unless we wish to differentiate mass and force. In English units, lb force (lb_f) units are often used. In cgs units, g force units are rarely used.

1.4 GAS LAWS AND VAPOR PRESSURE

1.4A Pressure

There are numerous ways of expressing the pressure exerted by a fluid or system. An *absolute pressure* of 1.00 atm is equivalent to 760 mm Hg at 0°C, 29.921 in. Hg, 0.760 m Hg, 14.696 lb force per square inch (psia), or 33.90 ft of water at 4°C. *Gage pressure* is the

Transport Processes: Momentum, Heat, and Mass

pressure above the absolute pressure. Hence, a pressure of 21.5 lb per square inch gage (*psig*) is 21.5 + 14.7 (rounded off), or 36.2 psia. In SI units, 1 psia = 6.89476×10^3 pascal (Pa) = 6.89476×10^3 newtons/m². Also, 1 atm = 1.01325×10^5 Pa.

In some cases, particularly in evaporation, one may express the pressure as inches of mercury vacuum. This means the pressure as inches of mercury measured "below" the absolute barometric pressure. For example, a reading of 25.4 in. Hg vacuum is $29.92 - 25.4$, or 4.52 in. Hg absolute pressure. Pressure conversion units are given in Appendix A.1.

1.4B   Ideal Gas Law

An *ideal gas* is defined as one that obeys simple laws. Also, in an ideal gas the gas molecules are considered as rigid spheres which themselves occupy no volume and do not exert forces on one another. No real gases obey these laws exactly, but at ordinary temperatures and pressures of not more than several atmospheres, the ideal laws give answers within a few percent or less of the actual answers. Hence, these laws are sufficiently accurate for engineering calculations.

The *ideal gas law* of Boyle states that the volume of a gas is directly proportional to the absolute temperature and inversely proportional to the absolute pressure. This is expressed as

$$pV = nRT \qquad (1.4\text{-}1)$$

where p is the absolute pressure in N/m², V the volume of the gas in m³, n the kg mol of the gas, T the absolute temperature in K, and R the gas law constant of $8314.3 \, \text{kg} \cdot \text{m}^2/\text{kg mol} \cdot \text{s}^2 \cdot \text{K}$. When the volume is in ft³, n in lb moles, and T in °R, R has a value of $0.7302 \, \text{ft}^3 \cdot \text{atm/lb mol} \cdot °\text{R}$. For cgs units (see Appendix A.1), $V = \text{cm}^3$, $T = K$, $R = 82.057 \, \text{cm}^3 \cdot \text{atm/g mol} \cdot \text{K}$, and $n = \text{g mol}$.

In order that amounts of various gases may be compared, *standard conditions of temperature and pressure* (abbreviated STP or SC) are arbitrarily defined as 101.325 kPa (1.0 atm) abs and 273.15 K (0°C). Under these conditions the volumes are as follows:

$$\text{volume of 1.0 kg mol (SC)} = 22.414 \text{ m}^3$$

$$\text{volume of 1.0 g mol (SC)} = 22.414 \text{ L (liter)}$$

$$= 22\,414 \text{ cm}^3$$

$$\text{volume of 1.0 lb mol (SC)} = 359.05 \text{ ft}^3$$

EXAMPLE 1.4-1.   Gas-Law Constant
Calculate the value of the gas-law constant R when the pressure is in psia, moles in lb mol, volume in ft³, and temperature in °R. Repeat for SI units.

Solution: At standard conditions, $p = 14.7$ psia, $V = 359$ ft³, and $T = 460 + 32 = 492°\text{R}$ (273.15 K). Substituting into Eq. (1.4-1) for $n = 1.0$ lb mol and solving for R,

$$R = \frac{pV}{nT} = \frac{(14.7 \text{ psia})(359 \text{ ft}^3)}{(1.0 \text{ lb mol})(492°\text{R})} = 10.73 \, \frac{\text{ft}^3 \cdot \text{psia}}{\text{lb mol} \cdot °\text{R}}$$

$$R = \frac{pV}{nT} = \frac{(1.01325 \times 10^5 \text{ Pa})(22.414 \text{ m}^3)}{(1.0 \text{ kg mol})(273.15 \text{ K})} = 8314 \, \frac{\text{m}^3 \cdot \text{Pa}}{\text{kg mol} \cdot \text{K}}$$

A useful relation can be obtained from Eq. (1.4-1) for n moles of gas at conditions

p_1, V_1, T_1, and also at conditions p_2, V_2, T_2. Substituting into Eq. (1.4-1),

$$p_1 V_1 = nRT_1$$

$$p_2 V_2 = nRT_2$$

Combining gives

$$\frac{p_1 V_1}{p_2 V_2} = \frac{T_1}{T_2} \tag{1.4-2}$$

EXAMPLE 1.4-2. Compressing a Gas

A volume of 10.0 liters of N_2 gas at 1.20 atm pressure and 100°C is compressed to 3.0 atm and cooled to 50°C. Calculate the g mol N_2 and the final volume in liters and m³.

Solution: Substituting the original conditions of $p_1 = 1.20$ atm, $V_1 = 10.0$ liters, and $T_1 = 273.2 + 100 = 373.2$ K into Eq. (1.4-1) and solving for n,

$$n = \frac{p_1 V_1}{RT_1} = \frac{(1.20 \text{ atm})(10.0 \text{ L})}{0.08205 \dfrac{\text{L} \cdot \text{atm}}{\text{g mol} \cdot \text{K}} (373.2 \text{ K})} = 0.393 \text{ g mol}$$

To calculate the final volume, the final conditions are substituted into Eq. (1.4-1) and the equation solved for V_2.

$$V_2 = \frac{nRT_2}{p_2} = \frac{(0.393 \text{ g mol})\left(0.08205 \dfrac{\text{L} \cdot \text{atm}}{\text{g mol} \cdot \text{K}}\right)(323.2 \text{ K})}{3.0 \text{ atm}} = 3.46 \text{ L}$$

Alternatively, substituting into Eq. (1.4-2) and solving for V_2,

$$V_2 = \frac{T_2 p_1 V_1}{T_1 p_2} = \frac{(323.2 \text{ K})(1.20 \text{ atm})(10.0 \text{ L})}{(373.2 \text{ K})(3.0 \text{ atm})} = 3.46 \text{ L} \ (0.00346 \text{ m}^3)$$

1.4C Ideal Gas Mixtures

Dalton's law for mixtures of ideal gases states that the total pressure of a gas mixture is equal to the sum of the individual partial pressures:

$$P = p_A + p_B + p_C + \cdots \tag{1.4-3}$$

where P is total pressure and p_A, p_B, p_C, ... are the partial pressures of the components A, B, C, ... in the mixture.

Since the number of moles of a component is proportional to its partial pressure, the mole fraction of a component is

$$x_A = \frac{p_A}{P} = \frac{p_A}{p_A + p_B + p_C + \cdots} \tag{1.4-4}$$

The volume fraction is equal to the mole fraction. Gas mixtures are almost always represented in terms of mole fractions and not weight fractions. For engineering purposes, Dalton's law is sufficiently accurate to use for actual mixtures at total pressures of a few atmospheres or less.

EXAMPLE 1.4-3. Composition of a Gas Mixture

A gas mixture contains the following components and partial pressures: CO_2, 75 mm Hg; CO, 50 mm Hg; N_2, 595 mm Hg; O_2, 26 mm Hg. Calculate the total pressure and the composition in mole fraction.

Solution: Substituting into Eq. (1.4-3),

$$P = p_A + p_B + p_C + p_D = 75 + 50 + 595 + 26 = 746 \text{ mm Hg}$$

The mole fraction of CO_2 is obtained by using Eq. (1.4-4).

$$x_A(CO_2) = \frac{p_A}{P} = \frac{75}{746} = 0.101$$

In like manner, the mole fractions of CO, N_2, and O_2 are calculated as 0.067, 0.797, and 0.035, respectively.

1.4D Vapor Pressure and Boiling Point of Liquids

When a liquid is placed in a sealed container, molecules of liquid will evaporate into the space above the liquid and fill it completely. After a time, equilibrium is reached. This vapor will exert a pressure just like a gas and we call this pressure the *vapor pressure* of the liquid. The value of the vapor pressure is independent of the amount of liquid in the container as long as some is present.

If an inert gas such as air is also present in the vapor space, it will have very little effect on the vapor pressure. In general, the effect of total pressure on vapor pressure can be considered as negligible for pressures of a few atmospheres or less.

The vapor pressure of a liquid increases markedly with temperature. For example, from Appendix A.2 for water, the vapor pressure at 50°C is 12.333 kPa (92.51 mm Hg). At 100°C the vapor pressure has increased greatly to 101.325 kPa (760 mm Hg).

The *boiling point* of a liquid is defined as the temperature at which the vapor pressure of a liquid equals the total pressure. Hence, if the atmospheric total pressure is 760 mm Hg, water will boil at 100°C. On top of a high mountain, where the total pressure is considerably less, water will boil at temperatures below 100°C.

> *EXAMPLE 1.4-4. Evaporation of Fruit Juice Under Vacuum*
> A fruit juice solution containing 5% solids is being concentrated by evaporation of water. Since the juice is heat-sensitive, too high a temperature may impair its flavor. It is decided that 100°F is a maximum temperature to be used. Any effect of the solids on the vapor pressure of the water solution should not be too large and will be neglected in this problem. What total pressure should be used?
>
> *Solution:* For the water solution to boil, the vapor pressure of the solution at 100°F must equal the total pressure. First, we use Eq. (1.3-2) to convert 100°F to °C.
>
> $$°C = \frac{1}{1.8}(°F - 32) = \frac{1}{1.8}(100 - 32) = 37.8°C$$
>
> From Appendix A.2 the vapor pressure of water at 37.8°C is (by interpolation) 6.60 kPa (49.5 mm Hg). Hence, for the solution to boil or evaporate at 100°F, the total pressure must be equal to the vapor pressure of 6.60 kPa.

A plot of vapor pressure P_A of a liquid versus temperature does not yield a straight line but a curve. However, for moderate temperature ranges, a plot of $\log P_A$ versus $1/T$ is a reasonably straight line, as follows.

$$\log P_A = m\left(\frac{1}{T}\right) + b \tag{1.4-5}$$

where m is the slope, b is a constant for the liquid A, and T is the temperature in K.

1.5 CONSERVATION OF MASS AND MATERIAL BALANCES

1.5A Conservation of Mass

One of the basic laws of physical science is the *law of conservation of mass*. This law, stated simply, says that mass cannot be created or destroyed (excluding, of course, nuclear or atomic reactions). Hence, the total mass (or weight) of all materials entering any process must equal the total mass of all materials leaving plus the mass of any materials accumulating or left in the process.

$$\text{input} = \text{output} + \text{accumulation} \qquad (1.5\text{-}1)$$

In the majority of cases there will be no accumulation of materials in a process, and then the input will simply equal the output. Stated in other words, "what goes in must come out." We call this type of process a *steady-state process*.

$$\text{input} = \text{output (steady state)} \qquad (1.5\text{-}2)$$

1.5B Simple Material Balances

In this section we do simple material (weight or mass) balances in various processes at steady state with no chemical reaction occurring. We can use units of kg, lb_m, lb mol, g, kg mol, etc., in our balances. The reader is cautioned to be consistent and not to mix several units in a balance. When chemical reactions occur in the balances (as discussed in Section 1.5D), one should use kg mol units, since chemical equations relate moles reacting. In Section 2.6, overall mass balances will be covered in more detail and in Section 3.6, differential mass balances.

To solve a material-balance problem it is advisable to proceed by a series of definite steps, as listed below.

1. *Sketch a simple diagram of the process.* This can be a simple box diagram showing each stream entering by an arrow pointing in and each stream leaving by an arrow pointing out. Include on each arrow the compositions, amounts, temperatures, and so on, of that stream. All pertinent data should be on this diagram.
2. *Write the chemical equations involved (if any).*
3. *Select a basis for calculation.* In most cases the problem is concerned with a specific amount of one of the streams in the process, which is selected as the basis.
4. *Make a material balance.* The arrows into the process will be input items and the arrows going out output items. The balance can be a total material balance in Eq. (1.5-2) or a balance on each component present (if no chemical reaction occurs).

Typical processes that do not undergo chemical reactions are drying, evaporation, dilution of solutions, distillation, extraction, and so on. These can be solved by setting up material balances containing unknowns and solving these equations for the unknowns.

> **EXAMPLE 1.5-1. Concentration of Orange Juice**
> In the concentration of orange juice a fresh extracted and strained juice containing 7.08 wt % solids is fed to a vacuum evaporator. In the evaporator, water is removed and the solids content increased to 58 wt % solids. For 1000 kg/h entering, calculate the amounts of the outlet streams of concentrated juice and water.
>
> **Solution:** Following the four steps outlined, we make a process flow diagram (step 1) in Fig. 1.5-1. Note that the letter W represents the unknown

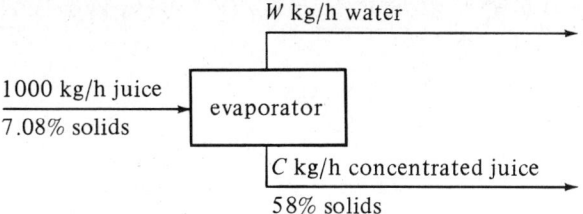

FIGURE 1.5-1. *Process flow diagram for Example 1.5-1.*

amount of water and C the amount of concentrated juice. No chemical reactions are given (step 2). Basis: 1000 kg/h entering juice (step 3).

To make the material balances (step 4), a total material balance will be made using Eq. (1.5-2).

$$1000 = W + C \qquad\qquad \textbf{(1.5-3)}$$

This gives one equation and two unknowns. Hence, a component balance on solids will be made.

$$1000\left(\frac{7.08}{100}\right) = W(0) + C\left(\frac{58}{100}\right) \qquad\qquad \textbf{(1.5-4)}$$

To solve these two equations, we solve Eq. (1.5-4) first for C since W drops out. We get $C = 122.1$ kg/h concentrated juice.

Substituting the value of C into Eq. (1.5-3),

$$1000 = W + 122.1$$

and we obtain $W = 877.9$ kg/h water.

As a check on our calculations, we can write a balance on the water component.

$$1000\left(\frac{100 - 7.08}{100}\right) = 877.9 + 122.1\left(\frac{100 - 58}{100}\right) \qquad\qquad \textbf{(1.5-5)}$$

Solving,

$$929.2 = 877.9 + 51.3 = 929.2$$

In Example 1.5-1 only one unit or separate process was involved. Often a number of processes in series are involved. Then we have a choice of making a separate balance over each separate process and/or a balance around the complete overall process.

EXAMPLE 1.5-2.   *Evaporation of NaOH Solution in Two Steps*
In a process for producing caustic (NaOH), 4000 kg/h of a solution containing 10 wt % NaOH is evaporated in the first evaporator, giving an 18 wt % NaOH solution. This is then fed to a second evaporator, which gives a product of 50 wt % NaOH. Calculate the water removed from each evaporator, the feed to the second evaporator, and the amount of product.

Solution: The process flow diagram is given in Fig. 1.5-2. Note that the letters W and V are used to indicate the unknown amounts of water per hour and B and P the unknown amounts of solution per hour.

No chemical reactions are involved, and the basis of 4000 kg/h of 10 wt % NaOH entering is used. We notice that we have four unknowns to solve for. Also we have a choice of making a component or total balance on

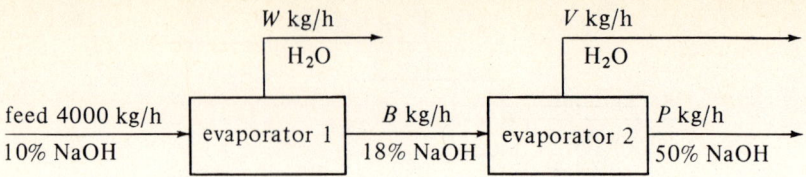

FIGURE 1.5-2. *Process flow diagram for Example 1.5-2.*

evaporator 1, evaporator 2, or on the entire process. We start by making a total balance on evaporator 1.

$$4000 = W + B \qquad (1.5\text{-}6)$$

Next, we make a balance for NaOH on evaporator 1.

$$4000\left(\frac{10}{100}\right) = W(0) + B\left(\frac{18}{100}\right) \qquad (1.5\text{-}7)$$

Solving this for B, $B = 2222$ kg/h solution. Substituting this value of B into Eq. (1.5-6), we obtain $W = 1778$ kg/h water.

Now, we repeat the process on evaporator 2 by writing a total balance and then a balance on NaOH.

$$2222 = V + P \qquad (1.5\text{-}8)$$

$$2222\left(\frac{18}{100}\right) = V(0) + P\left(\frac{50}{100}\right) \qquad (1.5\text{-}9)$$

Solving Eq. (1.5-9) first for P, we get $P = 800$. Then, from Eq. (1.5-8), we obtain $V = 1422$ kg/h water.

As a check, we will make an overall balance over the entire process and a water balance.

$$4000 = W + V + P = 1778 + 1422 + 800 = 4000 \quad (1.5\text{-}10)$$

$$4000\left(\frac{90}{100}\right) = 3600 = 1778 + 1422 + 800\left(\frac{50}{100}\right) = 3600 \quad (1.5\text{-}11)$$

Hence, the calculations check.

1.5C   Material Balances and Recycle

Processes that have a recycle or feedback of part of the product into the entering feed are sometimes encountered. For example, in a sewage treatment plant, part of the activated sludge from a sedimentation tank is recycled back to the aeration tank where the liquid is treated. In some food-drying operations, the humidity of the entering air is controlled by recirculating part of the hot wet air that leaves the dryer. In chemical reactions, the material that did not react in the reactor can be separated from the final product and fed back to the reactor.

EXAMPLE 1.5-3.   Crystallization of KNO₃ and Recycle

In a process producing KNO_3 salt, 1000 kg/h of a feed solution containing 20 wt % KNO_3 is fed to an evaporator, which evaporates some water at 422 K to produce a 50 wt % KNO_3 solution. This is then fed to a crystallizer at 311 K, where crystals containing 96 wt % KNO_3 are removed. The saturated solution containing 37.5 wt % KNO_3 is recycled to the evapor-

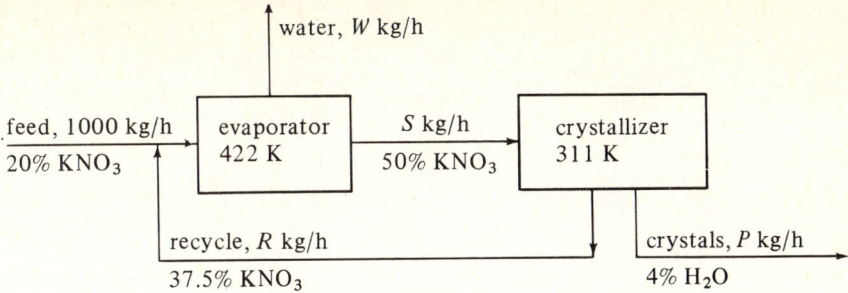

feed, 1000 kg/h
20% KNO_3

water, W kg/h

evaporator
422 K

S kg/h
50% KNO_3

crystallizer
311 K

recycle, R kg/h
37.5% KNO_3

crystals, P kg/h
4% H_2O

FIGURE 1.5-3.   *Process flow diagram for Example 1.5-3.*

ator. Calculate the amount of recycle stream R in kg/h and the product stream of crystals P in kg/h.

Solution: Figure 1.5-3 gives the process flow diagram. As a basis we shall use 1000 kg/h of fresh feed. No chemical reactions are occurring. We can make an overall balance on the entire process for KNO_3 and solve for P directly.

$$1000(0.20) = W(0) + P(0.96) \qquad \textbf{(1.5-12)}$$

$$P = 208.3 \text{ kg crystals/h}$$

To calculate the recycle stream, we can make a balance around the evaporator or the crystallizer. Using a balance on the crystallizer since it now includes only two unknowns, S and R, we get for a total balance,

$$S = R + 208.3 \qquad \textbf{(1.5-13)}$$

For a KNO_3 balance on the crystallizer,

$$S(0.50) = R(0.375) + 208.3(0.96) \qquad \textbf{(1.5-14)}$$

Substituting S from Eq. (1.5-13) into Eq. (1.5-14) and solving, $R = 766.6$ kg recycle/h and $S = 974.9$ kg/h.

1.5D   Material Balances and Chemical Reaction

In many cases the materials entering a process undergo chemical reactions in the process so that the materials leaving are different from those entering. In these cases it is usually convenient to make a molar and not a weight balance on an individual component such as kg mol H_2 or kg atom H, kg mol CO_3^- ion, kg mol $CaCO_3$, kg atom Na^+, kg mol N_2, and so on. For example, in the combustion of CH_4 with air, balances can be made on kg mol of H_2, C, O_2, or N_2.

EXAMPLE 1.5-4.   *Combustion of Fuel Gas*
A fuel gas containing 3.1 mol % H_2, 27.2% CO, 5.6% CO_2, 0.5% O_2, and 63.6% N_2 is burned with 20% excess air (i.e., the air over and above that necessary for complete combustion to CO_2 and H_2O). The combustion of CO is only 98% complete. For 100 kg mol of fuel gas, calculate the moles of each component in the exit flue gas.

Solution: First, the process flow diagram is drawn (Fig. 1.5-4). On the

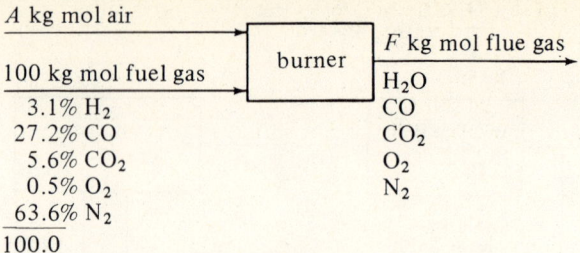

A kg mol air

100 kg mol fuel gas

burner

F kg mol flue gas

H₂O
CO
CO₂
O₂
N₂

100 kg mol fuel gas
3.1% H_2
27.2% CO
5.6% CO_2
0.5% O_2
63.6% N_2
100.0

FIGURE 1.5-4. *Process flow diagram for Example 1.5-4.*

diagram the components in the flue gas are shown. Let A be moles of air and F be moles of flue gas. Next the chemical reactions are given.

$$CO + \tfrac{1}{2}O_2 \rightarrow CO_2 \qquad\qquad \textbf{(1.5-15)}$$

$$H_2 + \tfrac{1}{2}O_2 \rightarrow H_2O \qquad\qquad \textbf{(1.5-16)}$$

An accounting of the total moles of O_2 in the fuel gas is as follows:

mol O_2 in fuel gas $= (\tfrac{1}{2})27.2(CO) + 5.6(CO_2) + 0.5(O_2) = 19.7$ mol O_2

For all the H_2 to be completely burned to H_2O, we need from Eq. (1.5-16) $\tfrac{1}{2}$ mol O_2 for 1 mol H_2 or $3.1(\tfrac{1}{2}) = 1.55$ total mol O_2. For completely burning the CO from Eq. (1.5-15), we need $27.2(\tfrac{1}{2}) = 13.6$ mol O_2. Hence, the amount of O_2 we must add is, theoretically, as follows:

mol O_2 theoretically needed $= 1.55 + 13.6 - 0.5$ (in fuel gas)

$$= 14.65 \text{ mol } O_2$$

For a 20% excess, we add 1.2(14.65), or 17.58 mol O_2. Since air contains 79 mol % N_2, the amount of N_2 added is (79/21)(17.58), or 66.1 mol N_2.

To calculate the moles in the final flue gas, all the H_2 gives H_2O, or 3.1 mol H_2O. For CO, 2.0% does not react. Hence, 0.02(27.2), or 0.54, mol CO will be unburned.

A total carbon balance is as follows: inlet moles $C = 27.2 + 5.6 = 32.8$ mol C. In the outlet flue gas, 0.54 mol will be as CO and the remainder of $32.8 - 0.54$, or 32.26, mol as CO_2.

For calculating the outlet mol O_2, we make an overall O_2 balance.

$$O_2 \text{ in} = 19.7 \text{ (in fuel gas)} + 17.58 \text{ (in air)} = 37.28 \text{ mol } O_2$$

$$O_2 \text{ out} = (3.1/2) \text{ (in } H_2O) + (0.54/2) \text{ (in CO)} + 32.26 \text{ (in } CO_2) + \text{free } O_2$$

Equating inlet O_2 to outlet, the free remaining $O_2 = 3.2$ mol O_2. For the N_2 balance, the outlet $= 63.6$ (in fuel gas) $+ 66.1$ (in air), or 129.70 mol N_2. The outlet flue gas contains 3.10 mol H_2O, 0.54 mol CO, 32.26 mol CO_2, 3.20 mol O_2, and 129.7 mol N_2.

In chemical reactions with several reactants, the limiting reactant component is defined as that compound which is present in an amount less than the amount necessary for it to react stoichiometrically with the other reactants. Then the percent completion of a reaction is the amount of this limiting reactant actually converted, divided by the amount originally present, times 100.

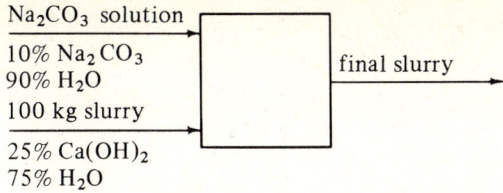

FIGURE 1.5-5. *Process flow diagram for Example 1.5-5.*

EXAMPLE 1.5-5. **Production of a Caustic (NaOH) Solution**
A caustic (NaOH) solution is produced by adding a solution containing 10 wt % Na_2CO_3 in the stoichiometric proportions to an inlet slurry containing 25 wt % $Ca(OH)_2$. What would be the composition of the final slurry if the reaction goes to only 99% completion? If the reaction is 100% complete, what would be the composition?

Solution: The process flow diagram is shown in Fig. 1.5-5. Note that a basis of 100 kg of inlet slurry is selected. Next, we write the chemical reaction to give $CaCO_3$ precipitate.

$$Na_2CO_3 + Ca(OH)_2 \rightarrow 2NaOH + CaCO_3 \downarrow \qquad (1.5\text{-}17)$$

Next we calculate the mol $Ca(OH)_2$ in the inlet slurry using a molecular weight of 74.1.

$$\text{mol } Ca(OH)_2 \text{ in slurry} = 100\left(\frac{0.25}{74.1}\right) = 0.338 \text{ kg mol } Ca(OH)_2$$

$$\text{kg } H_2O \text{ in slurry} = 100(0.75) = 75.0 \text{ kg } H_2O$$

Now we can calculate the mol Na_2CO_3 needed since, by Eq. (1.5-17), 1 mol Na_2CO_3 is needed for 1 mol $Ca(OH)_2$.

$$\text{mol } Na_2CO_3 \text{ needed} = \text{mol } Ca(OH)_2 = 0.338 \text{ kg mol } Na_2CO_3$$

This gives 0.338(mol wt Na_2CO_3) or 0.338(106.0) = 35.8 kg Na_2CO_3. Since the Na_2CO_3 solution contains 90 wt % H_2O,

$$\text{kg water} = \frac{0.90}{0.10}(35.8) = 322.2 \text{ kg } H_2O$$

$$\text{total kg } Na_2CO_3 \text{ solution} = 35.8 + 322.2 = 358.0 \text{ kg solution}$$

To calculate the final slurry, the reaction is 99% complete.

$$\text{mol } CaCO_3 \text{ produced} = 0.99(0.338) = 0.335 \text{ kg mol } CaCO_3$$

$$\text{mol } Ca(OH)_2 \text{ left} = 0.01(0.338) = 0.00338 \text{ kg mol } Ca(OH)_2$$

$$\text{mol NaOH produced} = 0.99(2)(0.338) = 0.669 \text{ kg mol NaOH}$$

$$\text{mol } Na_2CO_3 \text{ left} = 0.01(0.338) = 0.00338 \text{ kg mol } Na_2CO_3$$

$$\text{kg } H_2O \text{ (from inlet slurry plus } Na_2CO_3 \text{ solution)}$$

$$= 75.0 + 322.2 = 397.2 \text{ kg } H_2O$$

Tabulating the final slurry composition and converting kg mol to kg and then to wt %, we obtain the following:

	Kg Mol	Kg	Wt %
$CaCO_3$	0.335	33.53	7.32
$Ca(OH)_2$	0.00338	0.25	0.06
NaOH	0.669	26.76	5.84
Na_2CO_3	0.00338	0.36	0.08
H_2O		397.2	86.70
Total		458.10	100.00

As a check, an overall weight balance of 358.0 kg Na_2CO_3 solution plus 100.0 kg inlet slurry gives a final slurry of 458.0 kg, which is close to the value of 458.10 in the table. If the reaction goes to 100% completion, which is more likely with good contacting, no $Ca(OH)_2$ and Na_2CO_3 would be in the final slurry and the kg mol $CaCO_3$ would be 0.338 and kg mol NaOH 0.676.

1.6 ENERGY AND HEAT UNITS

1.6A Joule, Calorie, and Btu

In a manner similar to that used in making material balances on chemical and biological processes, we can also make energy balances on a process. Often a large portion of the energy entering or leaving a system is in the form of heat. Before such energy or heat balances are made, we must understand the various types of energy and heat units.

In the SI system energy is given in joules (J) or kilojoules (kJ). Energy is also expressed in btu (British thermal unit) or cal (calorie). The g calorie (abbreviated cal) is defined as the amount of heat needed to heat 1.0 g water 1.0°C (from 14.5°C to 15.5°C). Also, 1 kcal (kilocalorie) = 1000 cal. The btu is defined as the amount of heat needed to raise 1.0 lb water 1°F. Hence, from Appendix A.1,

$$1 \text{ btu} = 252.16 \text{ cal} = 1.05506 \text{ kJ} \tag{1.6-1}$$

1.6B Heat Capacity

The *heat capacity* of a substance is defined as the amount of heat necessary to increase the temperature by 1 degree. It can be expressed for 1 g, 1 lb, 1 g mol, 1 kg mol, or 1 lb mol of the substance. For example, a heat capacity is expressed in SI units as J/kg mol · K; in other units as cal/g · °C, cal/g mol · °C, kcal/kg mol · °C, btu/lb$_m$ · °F, or btu/lb mol · °F.

It can be shown that the actual numerical value of a heat capacity is the same in mass units or in molar units. That is,

$$1.0 \text{ cal/g} \cdot °C = 1.0 \text{ btu/lb}_m \cdot °F \tag{1.6-2}$$

$$1.0 \text{ cal/g mol} \cdot °C = 1.0 \text{ btu/lb mol} \cdot °F \tag{1.6-3}$$

For example, to prove this, suppose that a substance has a heat capacity of 0.8

$btu/lb_m \cdot °F$. The conversion is made using 1.8°F for 1°C or 1 K, 252.16 cal for 1 btu, and 453.6 g for 1 lb_m, as follows:

$$\text{heat capacity} \left(\frac{cal}{g \cdot °C} \right) = \left(0.8 \frac{btu}{lb_m \cdot °F} \right) \left(252.16 \frac{cal}{btu} \right) \left(\frac{1}{453.6 \text{ g/lb}_m} \right) \left(1.8 \frac{°F}{°C} \right)$$

$$= 0.8 \frac{cal}{g \cdot °C}$$

The heat capacities of gases (sometimes called *specific heat*) at constant pressure c_p are functions of temperature and for engineering purposes can be assumed to be independent of pressure up to several atmospheres. In most process engineering calculations, one is usually interested in the amount of heat needed to heat a gas from one temperature t_1 to another at t_2. Since the c_p varies with temperature, an integration must be performed or a suitable mean c_{pm} used. These mean values for gases have been obtained for T_1 of 298 K or 25°C (77°F) and various T_2 values, and are tabulated in Table 1.6-1 at 101.325 kPa pressure or less as c_{pm} in kJ/kg mol · K at various values of T_2 in K or °C.

EXAMPLE 1.6-1. Heating of N_2 Gas

The gas N_2 at 1 atm pressure absolute is being heated in a heat exchanger. Calculate the amount of heat needed in J to heat 3.0 g mol N_2 in the following temperature ranges:
  (a) 298–673 K (25–400°C)
  (b) 298–1123 K (25–850°C)
  (c) 673–1123 K (400–850°C)

Solution: For case (a), Table 1.6-1 gives c_{pm} values at 1 atm pressure or less and can be used up to several atm pressures. For N_2 at 673 K, $c_{pm} = 29.68$ kJ/kg mol · K or 29.68 J/g mol · K. This is the mean heat capacity for the range 298–673 K.

$$\text{heat required} = M \text{ g mol} \left(c_{pm} \frac{J}{\text{g mol} \cdot K} \right) (T_2 - T_1) K \qquad \textbf{(1.6-4)}$$

Substituting the known values,

$$\text{heat required} = (3.0)(29.68)(673 - 298) = 33\,390 \text{ J}$$

For case (b), the c_{pm} at 1123 K (obtained by linear interpolation between 1073 and 1173 K) is 31.00 J/g mol · K.

$$\text{heat required} = 3.0(31.00)(1123 - 298) = 76\,725 \text{ J}$$

For case (c), there is no mean heat capacity for the interval 673–1123 K. However, we can use the heat required to heat the gas from 298 to 673 K in case (a) and subtract it from case (b), which includes the heat to go from 298 to 673 K plus 673 to 1123 K.

$$\text{heat required (673–1123 K)} = \text{heat required (298–1123 K)}$$

$$- \text{heat required (298–673)} \qquad \textbf{(1.6-5)}$$

Substituting the proper values into Eq. (1.6-5),

$$\text{heat required} = 76\,725 - 33\,390 = 43\,335 \text{ J}$$

On heating a gas mixture, the total heat required is determined by first calculating the heat required for each individual component and then adding the results to obtain the total.

TABLE 1.6-1. *Mean Molar Heat Capacities of Gases Between 298 and $T°K$ (25 and $T°C$) at 101.325 kPa or Less (SI Units: $c_p = kJ/kg \, mol \cdot K$)*

$T(K)$	$T(°C)$	H_2	N_2	CO	Air	O_2	H_2O	CO_2	CH_4	SO_2
298	25	28.86	29.14	29.16	29.19	29.38	33.59	37.20	35.8	39.9
373	100	28.99	29.19	29.24	29.29	29.66	33.85	38.73	37.6	41.2
473	200	29.13	29.29	29.38	29.40	30.07	34.24	40.62	40.3	42.9
573	300	29.18	29.46	29.60	29.61	30.53	34.39	42.32	43.1	44.5
673	400	29.23	29.68	29.88	29.94	31.01	35.21	43.80	45.9	45.8
773	500	29.29	29.97	30.19	30.25	31.46	35.75	45.12	48.8	47.0
873	600	29.35	30.27	30.52	30.56	31.89	36.33	46.28	51.4	47.9
973	700	29.44	30.56	30.84	30.87	32.26	36.91	47.32	54.0	48.8
1073	800	29.56	30.85	31.16	31.18	32.62	37.53	48.27	56.4	49.6
1173	900	29.63	31.16	31.49	31.48	32.97	38.14	49.15	58.8	50.3
1273	1000	29.84	31.43	31.77	31.79	33.25	38.71	49.91	61.0	50.9
1473	1200	30.18	31.97	32.30	32.32	33.78	39.88	51.29	64.9	51.9
1673	1400	30.51	32.40	32.73	32.76	34.19	40.90	52.34		

Mean Molar Heat Capacities of Gases Between 25 and $T°C$ at 1 atm Pressure or Less (English Units: $c_p = btu/lb \, mol \cdot °F$)

$T(°C)$	H_2	N_2	CO	Air	O_2	NO	H_2O	CO_2	HCl	Cl_2	CH_4	SO_2	C_2H_4	SO_3	C_2H_6
25	6.894	6.961	6.965	6.972	7.017	7.134	8.024	8.884	6.96	8.12	8.55	9.54	10.45	12.11	12.63
100	6.924	6.972	6.983	6.996	7.083	7.144	8.084	9.251	6.97	8.24	8.98	9.85	11.35	12.84	13.76
200	6.957	6.996	7.017	7.021	7.181	7.224	8.177	9.701	6.98	8.37	9.62	10.25	12.53	13.74	15.27
300	6.970	7.036	7.070	7.073	7.293	7.252	8.215	10.108	7.00	8.48	10.29	10.62	13.65	14.54	16.72
400	6.982	7.089	7.136	7.152	7.406	7.301	8.409	10.462	7.02	8.55	10.97	10.94	14.67	15.22	18.11
500	6.995	7.159	7.210	7.225	7.515	7.389	8.539	10.776	7.06	8.61	11.65	11.22	15.60	15.82	19.39
600	7.011	7.229	7.289	7.299	7.616	7.470	8.678	11.053	7.10	8.66	12.27	11.45	16.45	16.33	20.58
700	7.032	7.298	7.365	7.374	7.706	7.549	8.816	11.303	7.15	8.70	12.90	11.66	17.22	16.77	21.68
800	7.060	7.369	7.443	7.447	7.792	7.630	8.963	11.53	7.21	8.73	13.48	11.84	17.95	17.17	22.72
900	7.076	7.443	7.521	7.520	7.874	7.708	9.109	11.74	7.27	8.77	14.04	12.01	18.63	17.52	23.69
1000	7.128	7.507	7.587	7.593	7.941	7.773	9.246	11.92	7.33	8.80	14.56	12.15	19.23	17.86	24.56
1100	7.169	7.574	7.653	7.660	8.009	7.839	9.389	12.10	7.39	8.82	15.04	12.28	19.81	18.17	25.40
1200	7.209	7.635	7.714	7.719	8.068	7.898	9.524	12.25	7.45	8.94	15.49	12.39	20.33	18.44	26.15
1300	7.252	7.692	7.772	7.778	8.123	7.952	9.66	12.39							
1400	7.288	7.738	7.818	7.824	8.166	7.994	9.77	12.50							
1500	7.326	7.786	7.866	7.873	8.203	8.039	9.89	12.69							
1600	7.386	7.844	7.922	7.929	8.269	8.092	9.95	12.75							
1700	7.421	7.879	7.958	7.965	8.305	8.124	10.13	12.70							
1800	7.467	7.924	8.001	8.010	8.349	8.164	10.24	12.94							
1900	7.505	7.957	8.033	8.043	8.383	8.192	10.34	13.01							
2000	7.548	7.994	8.069	8.081	8.423	8.225	10.43	13.10							
2100	7.588	8.028	8.101	8.115	8.460	8.255	10.52	13.17							
2200	7.624	8.054	8.127	8.144	8.491	8.277	10.61	13.24							

Source: O. A. Hougen, K. W. Watson, and R. A. Ragatz, *Chemical Process Principles*, Part I, 2nd ed. New York: John Wiley & Sons, Inc., 1954. With permission.

The heat capacities of solids and liquids are also functions of temperature and independent of pressure. Data are given in Appendix A.2, Physical Properties of Water; A.3, Physical Properties of Inorganic and Organic Compounds; and A.4, Physical Properties of Foods and Biological Materials. Additional data are available in reference (P1).

EXAMPLE 1.6-2. Heating of Milk

Rich cows' milk (4536 kg/h) at 4.4°C is being heated in a heat exchanger to 54.4°C by hot water. How much heat is needed?

Solution: From Appendix A.4 the average heat capacity of rich cows' milk is 3.85 kJ/kg·K. Temperature rise, $\Delta T = (54.4 - 4.4)°C = 50$ K.

heat required $= (4536 \text{ kg/h})(3.85 \text{ kJ/kg·K})(1/3600 \text{ h/s})(50 \text{ K}) = 242.5$ kW

The enthalpy, H, of a substance in J/kg represents the sum of the internal energy plus the pressure–volume term. For no reaction and a constant-pressure process with a change in temperature, the heat change as computed from Eq. (1.6-4) is the difference in enthalpy, ΔH, of the substance relative to a given temperature or base point. In other units, $H = \text{btu/lb}_m$ or cal/g.

1.6C Latent Heat and Steam Tables

Whenever a substance undergoes a change of phase, relatively large amounts of heat changes are involved at a constant temperature. For example, ice at 0°C and 1 atm pressure can absorb 6013.4 kJ/kg mol. This enthalpy change is called the *latent heat of fusion*. Data for other compounds are available in various handbooks (P1, W1).

When a liquid phase vaporizes to a vapor phase under its vapor pressure at constant temperature, an amount of heat called the *latent heat of vaporization* must be added. Tabulations of latent heats of vaporization are given in various handbooks. For water at 25°C and a pressure of 23.75 mm Hg, the latent heat is 44 020 kJ/kg mol, and at 25°C and 760 mm Hg, 44 045 kJ/kg mol. Hence, the effect of pressure can be neglected in engineering calculations. However, there is a large effect of temperature on the latent heat of water. Also, the effect of pressure on the heat capacity of liquid water is small and can be neglected.

Since water is a very common chemical, the thermodynamic properties of it have been compiled in steam tables and are given in Appendix A.2 in SI and in English units.

EXAMPLE 1.6-3. Use of Steam Tables

Find the enthalpy change (i.e., how much heat must be added) for each of the following cases using SI and English units.

    (a) Heating 1 kg (lb_m) water from 21.11°C (70°F) to 60°C (140°F) at 101.325 kPa (1 atm) pressure.

    (b) Heating 1 kg (lb_m) water from 21.11°C (70°F) to 115.6°C (240°F) and vaporizing at 172.2 kPa (24.97 psia).

    (c) Vaporizing 1 kg (lb_m) water at 115.6°C (240°F) and 172.2 kPa (24.97 psia).

Solution: For part (a), the effect of pressure on the enthalpy of liquid water is negligible. From Appendix A.2,

$$H \text{ at } 21.11°C = 88.60 \text{ kJ/kg} \quad \text{or} \quad \text{at } 70°F = 38.09 \text{ btu/lb}_m$$

$$H \text{ at } 60°C = 251.13 \text{ kJ/kg} \quad \text{or} \quad \text{at } 140°F = 107.96 \text{ btu/lb}_m$$

$$\text{change in } H = \Delta H = 251.13 - 88.60 = 162.53 \text{ kJ/kg}$$

$$= 107.96 - 38.09 = 69.87 \text{ btu/lb}_m$$

In part (b), the enthalpy at 115.6°C (240°F) and 172.2 kPa (24.97 psia) of the saturated vapor is 2699.9 kJ/kg or 1160.7 btu/lb$_m$.

$$\text{change in } H = \Delta H = 2699.9 - 88.60 = 2611.3 \text{ kJ/kg}$$

$$= 1160.7 - 38.09 = 1122.6 \text{ btu/lb}_m$$

The latent heat of water at 115.6°C (240°F) in part (c) is

$$2699.9 - 484.9 = 2215.0 \text{ kJ/kg}$$

$$1160.7 - 208.44 = 952.26 \text{ btu/lb}_m$$

1.6D Heat of Reaction

When chemical reactions occur, heat effects always accompany these reactions. This area where energy changes occur is often called *thermochemistry*. For example, when HCl is neutralized with NaOH, heat is given off and the reaction is exothermic. Heat is absorbed in an endothermic reaction. This heat of reaction is dependent on the chemical nature of each reacting material and product and on their physical states.

For purposes of organizing data we define a standard heat of reaction ΔH^0 as the change in enthalpy when 1 kg mol reacts under a pressure of 101.325 kPa at a temperature of 298 K (25°C). For example, for the reaction

$$H_2(g) + \tfrac{1}{2}O_2(g) \rightarrow H_2O(l) \tag{1.6-6}$$

the ΔH^0 is -285.840×10^3 kJ/kg mol or -68.317 kcal/g mol. The reaction is exothermic and the value is negative since the reaction loses enthalpy. In this case, the H_2 gas reacts with the O_2 gas to give liquid water, all at 298 K (25°C).

Special names are given to ΔH^0 depending upon the type of reaction. When the product is formed from the elements, as in Eq. (1.6-6), we call the ΔH^0, *heat of formation* of the product water, ΔH_f^0. For the combustion of CH_4 to form CO_2 and H_2O, we call it *heat of combustion*, ΔH_c^0. Data are given in Appendix A.3 for various values of ΔH_c^0.

> **EXAMPLE 1.6-4. Combustion of Carbon**
> A total of 10.0 g mol of carbon graphite is burned in a calorimeter held at 298 K and 1 atm. The combustion is incomplete and 90% of the C goes to CO_2 and 10% to CO. What is the total enthalpy change in kJ and kcal?
>
> **Solution:** From Appéndix A.3 the ΔH_c^0 for carbon going to CO_2 is -393.513×10^3 kJ/kg mol or -94.0518 kcal/g mol, and for carbon going to CO is -110.523×10^3 kJ/kg mol or -26.4157 kcal/g mol. Since 9 mol CO_2 and 1 mol CO are formed,
>
> $$\text{total } \Delta H = 9(-393.513) + 1(-110.523) = -3652 \text{ kJ}$$
>
> $$= 9(-94.0518) + 1(-26.4157) = -872.9 \text{ kcal}$$

If a table of heats of formation, ΔH_f^0, of compounds is available, the standard heat of the reaction, ΔH^0, can be calculated by

$$\Delta H^\circ = \sum \Delta H_{f \text{ (products)}}^0 - \sum \Delta H_{f \text{ (reactants)}}^0 \tag{1.6-7}$$

In Appendix A.3, a short table of some values of ΔH_f is given. Other data are also available (H1, P1, S1).

EXAMPLE 1.6-5. Reaction of Methane

For the following reaction of 1 kg mol of CH_4 at 101.32 kPa and 298 K,

$$CH_4(g) + H_2O(l) \rightarrow CO(g) + 3H_2(g)$$

calculate the standard heat of reaction ΔH^0 at 298 K in kJ.

Solution: From Appendix A.3, the following standard heats of formation are obtained at 298 K:

	ΔH^0_f (kJ/kg mol)
$CH_4(g)$	-74.848×10^3
$H_2O(l)$	-285.840×10^3
$CO(g)$	-110.523×10^3
$H_2(g)$	0

Note that the ΔH^0_f of all elements is, by definition, zero. Substituting into Eq. (1.6-7),

$$\Delta H^0 = [-110.523 \times 10^3 - 3(0)] - (-74.848 \times 10^3 - 285.840 \times 10^3)$$

$$= +250.165 \times 10^3 \text{ kJ/kg mol} \quad \text{(endothermic)}$$

1.7 CONSERVATION OF ENERGY AND HEAT BALANCES

1.7A Conservation of Energy

In making material balances we used the law of conservation of mass, which states that the mass entering is equal to the mass leaving plus the mass left in the process. In a similar manner, we can state the *law of conservation of energy*, which says that all energy entering a process is equal to that leaving plus that left in the process. In this section elementary heat balances will be made. More elaborate energy balances will be considered in Sections 2.7 and 5.6.

Energy can appear in many forms. Some of the common forms are enthalpy, electrical energy, chemical energy (in terms of ΔH reaction), kinetic energy, potential energy, work, and heat inflow.

In many cases in process engineering, which often takes place at constant pressure, electrical energy, kinetic energy, potential energy, and work either are not present or can be neglected. Then only the enthalpy of the materials (at constant pressure), the standard chemical reaction energy (ΔH^0) at 25°C, and the heat added or removed must be taken into account in the energy balance. This is generally called a *heat balance*.

1.7B Heat Balances

In making a heat balance at steady state we use methods similar to those used in making a material balance. The energy or heat coming into a process in the inlet materials plus any net energy added to the process is equal to the energy leaving in the materials. Expressed mathematically,

$$\sum H_R + (-\Delta H^0_{298}) + q = \sum H_p \qquad (1.7\text{-}1)$$

Introduction to Engineering Principles and Units

where $\sum H_R$ is the sum of enthalpies of all materials entering the reaction process relative to the reference state for the standard heat of reaction at 298 K and 101.32 kPa. If the inlet temperature is above 298 K, this sum will be positive. ΔH^0_{298} = standard heat of the reaction at 298 K and 101.32 kPa. The reaction contributes heat to the process, so the negative of ΔH^0_{298} is taken to be positive input heat for an exothermic reaction. q = net energy or heat added to the system. If heat leaves the system, this item will be negative. $\sum H_p$ = sum of enthalpies of all leaving materials referred to the standard reference state at 298 K (25°C).

Note that if the materials coming into a process are below 298 K, $\sum H_R$ will be negative. Care must be taken not to confuse the signs of the items in Eq. (1.7-1). If no chemical reaction occurs, then simple heating, cooling, or phase change is occurring. Use of Eq. (1.7-1) will be illustrated by several examples. For convenience it is common practice to call the terms on the left-hand side of Eq. (1.7-1) input items, and those on the right, output items.

EXAMPLE 1.7-1. *Heating of Fermentation Medium*

A liquid fermentation medium at 30°C is pumped at a rate of 2000 kg/h through a heater, where it is heated to 70°C under pressure. The waste heat water used to heat this medium enters at 95°C and leaves at 85°C. The average heat capacity of the fermentation medium is 4.06 kJ/kg · K, and that for water is 4.21 kJ/kg · K (Appendix A.2). The fermentation stream and the wastewater stream are separated by a metal surface through which heat is transferred and do not physically mix with each other. Make a complete heat balance on the system. Calculate the water flow and the amount of heat added to the fermentation medium assuming no heat losses. The process flow is given in Fig. 1.7-1.

Solution: It is convenient to use the standard reference state of 298 K (25°C) as the datum to calculate the various enthalpies. From Eq. (1.7-1) the input items are as follows.

Input items. $\sum H_R$ of the enthalpies of the two streams relative to 298 K (25°C) (note that $\Delta t = 30 - 25°C = 5°C = 5$ K):

$$H(\text{liquid}) = (2000 \text{ kg/h})(4.06 \text{ kJ/kg} \cdot \text{K})(5 \text{ K})$$

$$= 4.060 \times 10^4 \text{ kJ/h}$$

$$H(\text{water}) = W(4.21)(95 - 25) = 2.947 \times 10^2 \ W \text{ kJ/h} \qquad (W = \text{kg/h})$$

$$(-\Delta H^0_{298}) = 0 \qquad \text{(since there is no chemical reaction)}$$

$$q = 0 \qquad \text{(there are no heat losses or additions)}$$

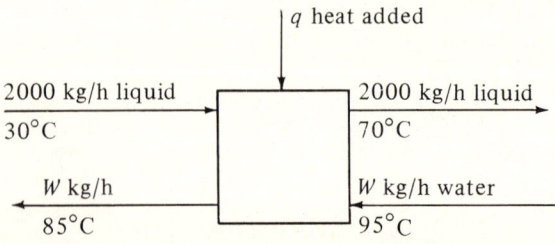

FIGURE 1.7-1. *Process flow diagram for Example 1.7-1.*

Output items. $\sum H_P$ of the two streams relative to 298 K (25°C):

$$H(\text{liquid}) = 2000(4.06)(70 - 25) = 3.65 \times 10^5 \text{ kJ/h}$$

$$H(\text{water}) = W(4.21)(85 - 25) = 2.526 \times 10^2 \, W \text{ kJ/h}$$

Equating input to output in Eq. (1.7-1) and solving for W,

$$4.060 \times 10^4 + 2.947 \times 10^2 \, W = 3.654 \times 10^5 + 2.526 \times 10^2 \, W$$

$$W = 7720 \text{ kg/h water flow}$$

The amount of heat added to the fermentation medium is simply the difference of the outlet and inlet liquid enthalpies.

$$H(\text{outlet liquid}) - H(\text{inlet liquid}) = 3.654 \times 10^5 - 4.060 \times 10^4$$

$$= 3.248 \times 10^5 \text{ kJ/h (90.25 kW)}$$

Note in this example that since the heat capacities were assumed constant, a simpler balance could have been written as follows:

$$\text{heat gained by liquid} = \text{heat lost by water}$$

$$2000(4.06)(70 - 30) = W(4.21)(95 - 85)$$

Then, solving, $W = 7720$ kg/h. This simple balance works well when c_p is constant. However, when the c_p varies with temperature and the material is a gas, c_{pm} values are only available between 298 K (25°C) and t K and the simple method cannot be used without obtaining new c_{pm} values over different temperature ranges.

EXAMPLE 1.7-2 Heat and Material Balance in Combustion

The waste gas from a process of 1000 g mol/h of CO at 473 K is burned at 1 atm pressure in a furnace using air at 373 K. The combustion is complete and 90% excess air is used. The flue gas leaves the furnace at 1273 K. Calculate the heat removed in the furnace.

Solution: First the process flow diagram is drawn in Fig. 1.7-2 and then a material balance is made.

$$CO(g) + \tfrac{1}{2}O_2(g) \rightarrow CO_2(g)$$

$$\Delta H_{298}^0 = -282.989 \times 10^3 \text{ kJ/kg mol}$$

(from Appendix A.3)

$$\text{mol CO} = 1000 \text{ g mol/h} = \text{moles CO}_2$$

$$= 1.00 \text{ kg mol/h}$$

$$\text{mol O}_2 \text{ theoretically required} = \tfrac{1}{2}(1.00) = 0.500 \text{ kg mol/h}$$

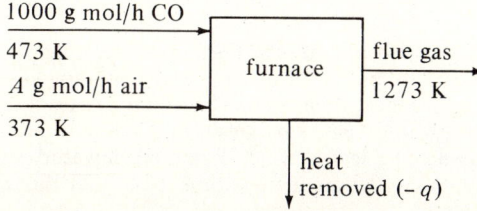

FIGURE 1.7-2. *Process flow diagram for Example 1.7-2.*

$$\text{mol } O_2 \text{ actually added} = 0.500(1.9) = 0.950 \text{ kg mol/h}$$

$$\text{mol } N_2 \text{ added} = 0.950 \, \frac{0.79}{0.21} = 3.570 \text{ kg mol/h}$$

$$\text{air added} = 0.950 + 3.570 = 4.520 \text{ kg mol/h} = A$$

$$O_2 \text{ in outlet flue gas} = \text{added} - \text{used}$$

$$= 0.950 - 0.500 = 0.450 \text{ kg mol/h}$$

$$CO_2 \text{ in outlet flue gas} = 1.00 \text{ kg mol/h}$$

$$N_2 \text{ in outlet flue gas} = 3.570 \text{ kg mol/h}$$

For the heat balance relative to the standard state at 298 K, we follow Eq. (1.7-1).

Input items

$$H \text{ (CO)} = 1.00(c_{pm})(473 - 298) = 1.00(29.38)(473 - 298) = 5142 \text{ kJ/h}$$

(The c_{pm} of CO of 29.38 kJ/kg mol $\cdot$ K between 298 and 473 K is obtained from Table 1.6-1.)

$$H \text{ (air)} = 4.520(c_{pm})(373 - 298) = 4.520(29.29)(373 - 298) = 9929 \text{ kJ/h}$$

$$q = \text{heat added, kJ/h}$$

(This will give a negative value here, indicating that heat was removed.)

$$-\Delta H^0_{298} = -(-282.989 \times 10^3 \text{ kJ/kg mol})(1.00 \text{ kg mol/h}) = 282\,990 \text{ kJ/h}$$

Output items

$$H(CO_2) = 1.00(c_{pm})(1273 - 298) = 1.00(49.91)(1273 - 298) = 48\,660 \text{ kJ/h}$$

$$H(O_2) = 0.450(c_{pm})(1273 - 298) = 0.450(33.25)(1273 - 298) = 14\,590 \text{ kJ/h}$$

$$H(N_2) = 3.570(c_{pm})(1273 - 298) = 3.570(31.43)(1273 - 298) = 109\,400 \text{ kJ/h}$$

Equating input to output and solving for q,

$$5142 + 9929 + q + 282\,990 = 48\,660 + 14\,590 + 109\,400$$
$$q = -125\,411 \text{ kJ/h}$$

Hence, heat is removed: $-34\,837$ W.

Often when chemical reactions occur in the process and the heat capacities vary with temperature, the solution in a heat balance can be trial and error if the final temperature is the unknown.

EXAMPLE 1.7-3. *Heating Air and Trial-and-Error Solution*
A flow rate of 5.55 lb mol of air at 85°F and 1 atm abs pressure is being heated by steam in a heat exchanger. The air and steam do not physically mix. A total of 5.0 lb of saturated steam at 330°F and 66.98 psia is condensed and liquid water leaves at 300°F and 66.98 psia. Assuming no heat losses, calculate the outlet air temperature. Note that this will be a trial-and-error solution.

Transport Processes: Momentum, Heat, and Mass

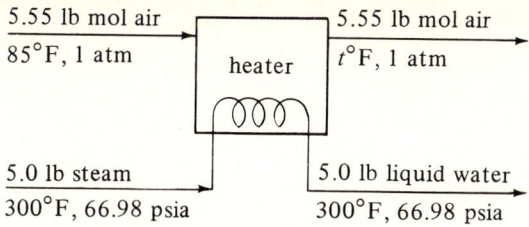

FIGURE 1.7-3. *Process flow diagram for Example 1.7-3.*

Solution: The process flow diagram is shown in Fig. 1.7-3. The datum temperature will be 77°F (25°C).

Input items

$$H \text{ (air)} = (5.55 \text{ lb mol})\left(c_{pm} \frac{\text{btu}}{\text{lb mol} \cdot {}^\circ\text{F}}\right)(85 - 77){}^\circ\text{F}$$

$$= (5.55)(6.98)(85 - 77) = 309.9 \text{ btu}$$

(The c_{pm} of the air is obtained from Table 1.6-1 at 85°F or 29.4°C.)

$$H \text{ (steam)} = 5[1180.2(\text{relative to } 32{}^\circ\text{F}) - 45.09 \text{ (relative to } 77{}^\circ\text{F)}]$$

(The value of 1180.2 btu/lb$_m$ is obtained from steam tables and is relative to 32°F. The value of 45.09 btu/lb$_m$ is the enthalpy of liquid water between 77°F and 32°F. This value corrects the data to a basis of 77°F.)

$$q = \text{heat added} = 0$$

Output items

$$H(\text{air}) = 5.55(c_{pm})(t - 77)$$

$$H(\text{water}) = 5[269.73 \text{ (relative to } 32{}^\circ\text{F)} - 45.09 \text{ (relative to } 77{}^\circ\text{F)}]$$

Setting input = output, and solving,

$$309.9 + 5(1180.2 - 45.09) = 5.55(c_{pm})(t - 77) + 5(269.73 - 45.09)$$

$$4863.4 = 5.55(c_{pm})(t - 77) \qquad \textbf{(1.7-2)}$$

Note that the correction of 45.09 btu/lb$_m$, which corrects the steam table to a basis of 77°F, cancels out on both sides of the equation and could have been omitted.

The solution is trial and error. Estimating a t of 250°F (121.1°C), $c_{pm} = 7.01$ (from the table). Substituting into Eq. (1.7-2) and solving, $t = 201.8$°F. For the second trial, using $t = 200$°F (93.3°C), $c_{pm} = 7.00$. Solving, $t = 202.0$°F (94.4°C). The c_{pm} value changes very little, hence only two trials were needed.

EXAMPLE 1.7-4. Oxidation of Lactose
In many biochemical processes, lactose is used as a nutrient, which is oxidized as follows:

$$C_{12}H_{22}O_{11}(s) + 12O_2(g) \rightarrow 12CO_2(g) + 11H_2O(l)$$

The heat of combustion ΔH_c^0 in Appendix A.3 at 25°C is -5648.8×10^3 J/g

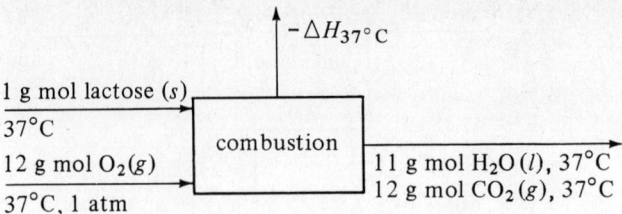

FIGURE 1.7-4. *Process flow diagram for Example 1.7-4.*

mol. Calculate the heat of complete oxidation (combustion) at 37°C, which is the temperature of many biochemical reactions. The c_{pm} of solid lactose is 1.20 J/g·K, and the molecular weight is 342.3 g mass/g mol.

Solution: This can be treated as an ordinary heat-balance problem. First, the process flow diagram is drawn in Fig. 1.7-4. Next, the datum temperature of 25°C is selected and the input and output enthalpies calculated. The temperature difference $\Delta t = (37 - 25)°C = (37 - 25)$ K.

Input items

$$H(\text{lactose}) = (342.3 \text{ g})\left(c_{pm} \frac{\text{J}}{\text{g} \cdot \text{K}}\right)(37 - 25) \text{ K} = 342.3(1.20)(37 - 25)$$

$$= 4929 \text{ J}$$

$$H(\text{O}_2 \text{ gas}) = (12 \text{ g mol})\left(c_{pm} \frac{\text{J}}{\text{g mol} \cdot \text{K}}\right)(37 - 25) \text{ K}$$

$$= 12(29.38)(37 - 25) = 4230 \text{ J}$$

(The c_{pm} of O_2 was obtained from Table 1.6-1.)

$$-\Delta H_{25}^0 = -(-5648.8 \times 10^3)$$

Output items

$$H(\text{H}_2\text{O liquid}) = 11(18.02 \text{ g})\left(c_{pm} \frac{\text{J}}{\text{g} \cdot \text{K}}\right)(37 - 25) \text{ K}$$

$$= 11(18.02)(4.18)(37 - 25) = 9943 \text{ J}$$

(The c_{pm} of liquid water was obtained from Appendix A.2.)

$$H(\text{CO}_2 \text{ gas}) = (12 \text{ g mol})\left(c_{pm} \frac{\text{J}}{\text{g mol} \cdot \text{K}}\right)(37 - 25) \text{ K}$$

$$= 12(37.45)(37 - 25) = 5393 \text{ J}$$

(The c_{pm} of CO_2 is obtained from Table 1.6-1.)

$\Delta H_{37°C}$:
Setting input = output and solving,

$$4929 + 4230 + 5648.8 \times 10^3 = 9943 + 5393 - \Delta H_{37°C}$$

$$\Delta H_{37°C} = -5642.6 \times 10^3 \text{ J/g mol} = \Delta H_{310 \text{ K}}$$

1.8 GRAPHICAL, NUMERICAL, AND MATHEMATICAL METHODS

1.8A Graphical Integration

Often the mathematical function $f(x)$ to be integrated is too complex and we are not able to integrate it analytically. Or in some cases the function is one that has been obtained from experimental data, and no mathematical equation is available to represent the data so that they can be integrated analytically. In these cases, we can use graphical integration.

Integration between the limits $x = a$ to $x = b$ can be represented graphically as shown in Fig. 1.8-1. Here a function $y = f(x)$ has been plotted versus x. The area under the curve $y = f(x)$ between the limits $x = a$ to $x = b$ is approximately equal to the sum of the rectangles, each with a base width of Δx and a height $y_1, y_2, \ldots, y_n$.

$$\text{area} \cong y_1\,\Delta x + y_2\,\Delta x + y_3\,\Delta x + \cdots + y_n\,\Delta x \tag{1.8-1}$$

As n approaches ∞, the Δx becomes a differential dx and the calculation of the area will be exact. This area is also equal to the integral, as follows:

$$\text{area} = \int_{x=a}^{x=b} y\,dx \tag{1.8-2}$$

Hence, evaluation of an integral can also be accomplished by graphically obtaining the area under the curve. The method of doing this will be shown in the following example.

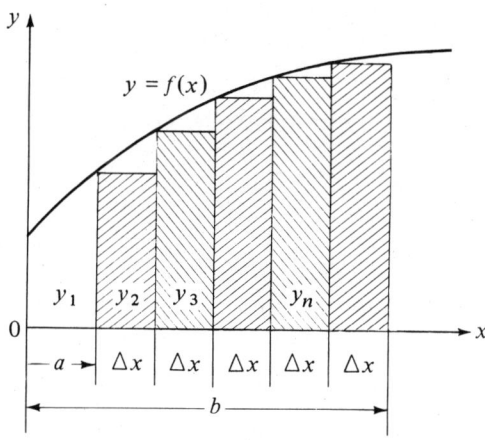

FIGURE 1.8-1. *Concept of integration graphically.*

EXAMPLE 1.8-1. Graphical Integration to Determine Total Flow in a Sewage Plant

The following data were obtained in measuring the flow rate $R\,(\text{m}^3/\text{min})$ at various times $x\,(\text{min})$ in a sewage treatment plant.

Flow, R (m^3/min)	Time, x (min)
79	0
72	10
62	20
54	30
51.5	40
55	50
61	60

Determine the total flow in m^3 for the entire 60 min.

Solution: First the data are plotted in Fig. 1.8-2 as R, which is $f(x)$ versus x. Then we draw a series of rectangles, as shown, with the height of the rectangle being determined by having the shaded area for a given rectangle, which is above the curve, equal to the shaded area below the curve for the same rectangle. From Eq. (1.8-2), we are integrating between the limits $x = a = 0$ min to $x = b = 60$ min. Then, $y = f(x) = R$. The total area under the curve is Eq. (1.8-2) with R substituted for y and is equal to the total flow in m^3. This area is the sum of the area of all the rectangles A through F. Since the area of each rectangle is height times width,

total area $= 75.8(10) + 67.5(10) + 57.7(10) + 52.4(10) + 53.0(10) + 57.8(10)$

$= 3642 \ m^3$

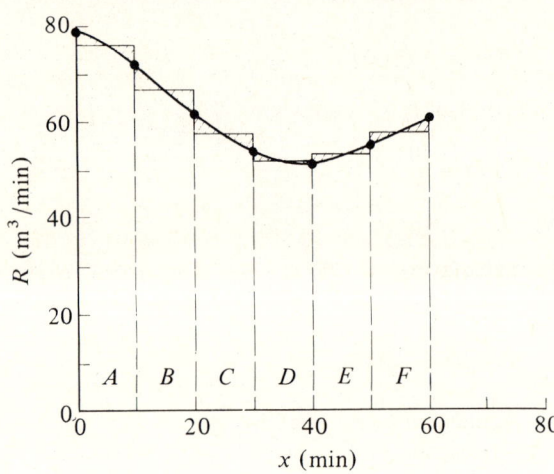

FIGURE 1.8-2. *Integration for total flow in Example 1.8-1.*

1.8B Numerical Integration and Simpson's Rule

Often it is desired or necessary to perform a numerical integration by computing the value of a definite integral from a set of numerical values of the integrand $f(x)$. This, of course, can be done graphically, but if data are available in large quantities, numerical methods suitable for the digital computer are desired.

The integral to be evaluated is as follows:

$$\int_{x=a}^{x=b} f(x) \ dx \qquad (1.8\text{-}3)$$

where the interval is $b - a$. The most generally used numerical method is the parabolic rule often called *Simpson's rule*. This method divides the total interval $b - a$ into an even number of subintervals m, where

$$m = \frac{b - a}{h} \qquad (1.8\text{-}4)$$

The value of h, a constant, is the spacing in x used. Then, approximating $f(x)$ by a parabola on each subinterval, Simpson's rule is

$$\int_{x=a}^{x=b} f(x)\, dx = \frac{h}{3} \left[f_0 + 4(f_1 + f_3 + f_5 + \cdots + f_{m-1}) \right.$$

$$\left. + 2(f_2 + f_4 + f_6 + \cdots + f_{m-2}) + f_m \right] \qquad (1.8\text{-}5)$$

where f_0 is the value of $f(x)$ at $x = a$, f_1 the value of $f(x)$ at $x = x_1, \ldots, f_m$ the value of $f(x)$ at $x = b$. The reader should note that m must be an even number and the increments evenly spaced. This method is well suited for digital computation.

EXAMPLE 1.8-2. Numerical Integration by Simpson's Method

Using the data from Example 1.8-1, perform the integration numerically by Simpson's method using six intervals.

Solution: The value of m by Eq. (1.8-4) is as follows for $h = 10$:

$$m = \frac{b - a}{h} = \frac{60 - 0}{10} = 6$$

The value of f_0 is 79, of f_1 is 72, and so on. Substituting into Eq. (1.8-5),

$$\int_{x=0}^{x=60} f(x)\, dx = \frac{10}{3} \left[79 + 4(72 + 54 + 55) + 2(62 + 51.5) + 61 \right]$$

$$= 3637$$

This compares favorably with the value of 3642 obtained graphically in Example 1.8-1.

1.8C Linear Plot

Experimental data are often presented in graphical form rather than in tabular form to give a picture as to the significance and correlation of the data. In many cases it is necessary or desirable to obtain a mathematical equation that will fit a given set of experimental data. An important method of doing this is to plot the data on graph paper and then draw a smooth curve or line through the points. In some cases, if the proper type of graph paper is selected, the data plotted lie on an approximately straight line. Any equation obtained from experimental data is called an *empirical equation*.

The simplest type of graph paper is linear or arithmetic. Experimental data involving the variables x and y lie on a straight line on linear graph paper if they follow the equation

$$y = mx + a \qquad (1.8\text{-}6)$$

where m = slope of the straight line and a = a constant (intercept on y axis).

EXAMPLE 1.8-3. *Empirical Equation for Drying Paper Pulp*
Experimental data obtained by drying a slab of compressed paper pulp are as follows:

Rate of Drying, R (kg $H_2O/h \cdot m^2$)	Free H_2O Content, W (kg H_2O)
0.287	1.51
0.279	1.47
0.248	1.30
0.224	1.12
0.192	0.91
0.150	0.67
0.138	0.55
0.105	0.40

Graphically determine the empirical equation fitting these data.

Solution: The data are plotted on linear graph paper in Fig. 1.8-3. The equation of the straight line is

$$R = mW + a$$

The slope m is the distance A over B or $m = 0.272/1.72 = 0.158$. The intercept a is 0.048. The final equation is

$$R = 0.158W + 0.048$$

1.8D   Semilog Plot

Many equations occurring in process engineering have the form

$$y = a \times 10^{bx} \qquad (1.8\text{-}7)$$

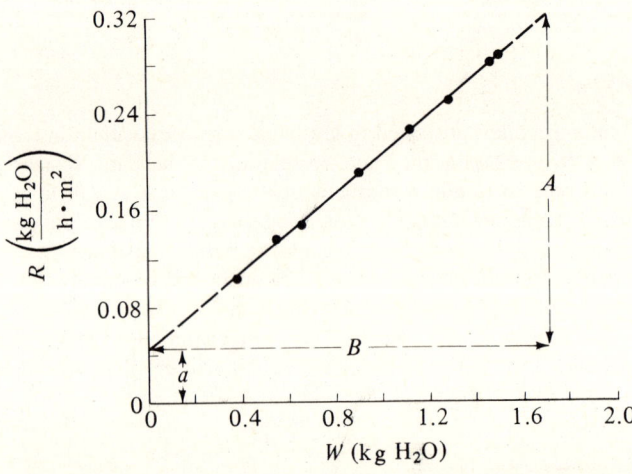

FIGURE 1.8-3.   *Determination of empirical equation for Example 1.8-3.*

Transport Processes: Momentum, Heat, and Mass

where a and b are constants. If we take the log of both sides, then

$$\log y = bx + \log a \qquad \text{(1.8-8)}$$

Hence, if we plot $\log y$ versus x on linear graph paper a straight line should be obtained with a slope of b and an intercept on the y axis of the constant $\log a$. A more convenient method is to use semilog graph paper, where one scale is linear and the other scale is a log scale and plot y on the log scale and x on the linear scale.

1.8E Log–Log Plot

Often data are represented by the equation

$$y = ax^b \qquad \text{(1.8-9)}$$

where a and b are constants. Taking the log of both sides,

$$\log y = b \log x + \log a \qquad \text{(1.8-10)}$$

Hence, if log–log graph paper is used and y is plotted versus x, a straight line should result where the slope of the line is b and a is the value of y when x equals 1. Another method to obtain a and b is to substitute two points from the line drawn into Eq. (1.8-10) and solve for a and b.

PROBLEMS

1.2-1. *Temperature of a Chemical Process.* The temperature of a chemical reaction was found to be 353.2 K. What is the temperature in °F, °C, and °R?
Ans.   176°F, 80°C, 636°R

1.2-2. *Temperature for Smokehouse Processing of Meat.* In smokehouse processing of sausage meat, a final temperature of 155°F inside the sausage is often used. Calculate this temperature in °C, K, and °R.

1.3-1. *Molecular Weight of Air.* For purposes of most engineering calculations, air is assumed to be composed of 21 mol % oxygen and 79 mol % nitrogen. Calculate the average molecular weight.
Ans.   28.9 g mass/g mol, lb mass/lb mol, or kg mass/kg mol

1.3-2. *Oxidation of CO and Mole Units.* The gas CO is being oxidized by O_2 to form CO_2. How many kg of CO_2 will be formed from 56 kg of CO? Also, calculate the kg of O_2 theoretically needed for this reaction. (*Hint:* First write the balanced chemical equation to obtain the mol O_2 needed for 1.0 kg mol CO. Then calculate the kg mol of CO in 56 kg CO.)
Ans.   88.0 kg CO_2, 32.0 kg O_2

1.3-3. *Composition of a Gas Mixture.* A gaseous mixture contains 20 g of N_2, 83 g of O_2, and 45 g of CO_2. Calculate the composition in mole fraction and the average molecular weight of the mixture.
Ans.   Average mol wt $= 34.2$ g mass/g mol, 34.2 kg mass/kg mol

1.3-4. *Composition of a Protein Solution.* A liquid solution contains 1.15 wt % of a protein, 0.27 wt % KCl, and the remainder water. The average molecular weight of the protein by gel permeation is 525 000 g mass/g mol. Calculate the mole fraction of each component in solution.

1.3-5. *Concentration of NaCl Solution.* An aqueous solution of NaCl has a concentration of 24.0 wt % NaCl with a density of 1.178 g/cm^3 at 25°C. Calculate the following.
(a) Mole fraction of NaCl and water.
(b) Concentration of NaCl as g mol/liter, lb_m/ft^3, lb_m/gal, and kg/m^3.

Introduction to Engineering Principles and Units   **33**

1.3-6. Concentration of Ammonium Nitrate Solution. An aqueous ammonium nitrate solution (molecular weight = 80.05 g mass/g mol) contains 1.807 lb_m salt per gallon of solution, which weighs 9.01 lb_m per gallon. Calculate the following.
(a) Wt % salt.
(b) Lb mass salt/ft^3.
(c) Molarity of solution.

Ans. (a) 20.1 wt %, (b) 13.52 lb_m/ft^3, (c) 2.70 g mol/liter

1.3-7. Fuel Oil Composition. The composition of a fuel oil is 83.6 wt % carbon and 16.4% hydrogen. What is its composition in mol % C and H_2?

Ans. 46.1 mol % C, 53.9 mol % H_2

1.4-1. Conversion of Pressure Measurements in Freeze Drying. In the experimental measurement of freeze drying of beef, an absolute pressure of 2.4 mm Hg was held in the chamber. Convert this pressure to atm, in. of water at 4°C, μm of Hg, and Pa. (*Hint:* See Appendix A.1 for conversion factors.)

Ans. 3.16×10^{-3} atm, 1.285 in. H_2O, 2400 μm Hg, 320 Pa

1.4-2. Compression and Cooling of Nitrogen Gas. A volume of 65.0 ft^3 of N_2 gas at 90°F and 29.0 psig is compressed to 75 psig and cooled to 65°F. Calculate the final volume in ft^3 and the final density in lb_m/ft^3. [*Hint:* Be sure to convert all pressures to psia first and then to atm. Substitute original conditions into Eq. (1.4-1) to obtain n, lb mol.]

1.4-3. Gas Composition and Volume. A gas mixture of 0.13 g mol NH_3, 1.27 g mol N_2, and 0.025 g mol H_2O vapor is contained at a total pressure of 830 mm Hg and 323 K. Calculate the following.
(a) Mole fraction of each component.
(b) Partial pressure of each component in mm Hg.
(c) Total volume of mixture in m^3 and ft^3.

1.4-4. Evaporation of a Heat-Sensitive Organic Liquid. An organic liquid is being evaporated from a liquid solution containing a few percent nonvolatile dissolved solids. Since it is heat-sensitive and may discolor at high temperatures, it will be evaporated under vacuum. If the lowest absolute pressure that can be obtained in the apparatus is 12.0 mm Hg, what will be the temperature of evaporation in K? It will be assumed that the small amount of solids does not affect the vapor pressure, which is given as follows:

$$\log P_A = -2250\left(\frac{1}{T}\right) + 9.05$$

where P_A is in mm Hg and T in K.

Ans. $T = 282.3$ K or 9.1°C

1.5-1. Evaporation of Cane Sugar Solutions. An evaporator is used to concentrate cane sugar solutions. A feed of 10 000 kg/d of a solution containing 38 wt % sugar is evaporated, producing a 74 wt % solution. Calculate the weight of solution produced and amount of water removed.

Ans. 5135 kg/d of 74 wt % solution, 4865 kg/d water

1.5-2. Processing of Fish Meal. Fish are processed into fish meal and used as a supplementary protein food. In the processing the oil is first extracted to produce wet fish cake containing 80 wt % water and 20 wt % bone-dry cake. This wet cake feed is dried in rotary drum dryers to give a "dry" fish cake product containing 40 wt % water. Finally, the product is finely ground and packed. Calculate the kg/h of wet cake feed needed to produce 1000 kg/h of "dry" fish cake product.

Ans. 3000 kg/h wet cake feed

1.5-3. Drying of Lumber. A batch of 100 kg of wet lumber containing 11 wt % moisture is dried to a water content of 6.38 kg water/1.0 kg bone-dry lumber. What is the weight of "dried" lumber and the amount of water removed?

1.5-4. Processing of Paper Pulp. A wet paper pulp contains 68 wt % water. After the pulp was dried, it was found that 55% of the original water in the wet pulp was removed. Calculate the composition of the "dried" pulp and its weight for a feed of 1000 kg/min of wet pulp.

1.5-5. Production of Jam from Crushed Fruit in Two Stages. In a process producing jam (C1), crushed fruit containing 14 wt% soluble solids is mixed in a mixer with sugar (1.22 kg sugar/1.00 kg crushed fruit) and pectin (0.0025 kg pectin/1.00 kg crushed fruit). The resultant mixture is then evaporated in a kettle to produce a jam containing 67 wt% soluble solids. For a feed of 1000 kg crushed fruit, calculate the kg mixture from the mixer, kg water evaporated, and kg jam produced.

> **Ans.**   2222.5 kg mixture, 189 kg water, 2033.5 kg jam

1.5-6. Drying of Cassava (Tapioca) Root. Tapioca flour is used in many countries for bread and similar products. The flour is made by drying coarse granules of the cassava root containing 66 wt % moisture to 5% moisture and then grinding to produce a flour. How many kg of granules must be dried and how much water removed to produce 5000 kg/h of flour?

1.5-7. Processing of Soybeans in Three Stages. A feed of 10 000 kg of soybeans is processed in a sequence of three stages or steps (E1). The feed contains 35 wt % protein, 27.1 wt % carbohydrate, 9.4 wt % fiber and ash, 10.5 wt % moisture, and 18.0 wt % oil. In the first stage the beans are crushed and pressed to remove oil, giving an expressed oil stream and a stream of pressed beans containing 6% oil. Assume no loss of other constituents with the oil stream. In the second step the pressed beans are extracted with hexane to produce an extracted meal stream containing 0.5 wt % oil and a hexane-oil stream. Assume no hexane in the extracted meal. Finally, in the last step the extracted meal is dried to give a dried meal of 8 wt % moisture. Calculate:
(a) Kg of pressed beans from the first stage.
(b) Kg of extracted meal from stage 2.
(c) Kg of final dried meal and the wt % protein in the dried meal.

> **Ans.**   (a) 8723 kg, (b) 8241 kg, (c) 7816 kg, 44.8 wt % protein

1.5-8. Recycle in a Dryer. A solid material containing 15.0 wt % moisture is dried so that it contains 7.0 wt % water by blowing fresh warm air mixed with recycled air over the solid in the dryer. The inlet fresh air has a humidity of 0.01 kg water/kg dry air, the air from the drier that is recycled has a humidity of 0.1 kg water/kg dry air, and the mixed air to the dryer, 0.03 kg water/kg dry air. For a feed of 100 kg solid/h fed to the dryer, calculate the kg dry air/h in the fresh air, the kg dry air/h in the recycle air, and the kg/h of "dried" product.

> **Ans.**   95.6 kg/h dry air in fresh air, 27.3 kg/h dry air in recycle air, and 91.4 kg/h "dried" product

1.5-9. Crystallization and Recycle. It is desired to produce 1000 kg/h of $Na_3PO_4 \cdot 12H_2O$ crystals from a feed solution containing 5.6 wt % Na_3PO_4 and traces of impurity. The original solution is first evaporated in an evaporator to a 35 wt% Na_3PO_4 solution and then cooled to 293 K in a crystallizer, where the hydrated crystals and a mother liquor solution are removed. One out of every 10 kg of mother liquor is discarded to waste to get rid of the impurities, and the remaining mother liquor is recycled to the evaporator. The solubility of Na_3PO_4 at 293 K is 9.91 wt %. Calculate the kg/h of feed solution and kg/h of water evaporated.

> **Ans.**   7771 kg/h feed, 6739 kg/h water

1.5-10. Evaporation and Bypass in Orange Juice Concentration. In a process for concentrating 1000 kg of freshly extracted orange juice (C1) containing 12.5 wt % solids, the juice is strained, yielding 800 kg of strained juice and 200 kg of pulpy juice. The strained juice is concentrated in a vacuum evaporator to give an evaporated juice of 58% solids. The 200 kg of pulpy juice is bypassed around the evaporator and mixed with the evaporated juice in a mixer to improve the flavor. This final concentrated juice contains 42 wt % solids. Calculate the concentration of solids in the strained juice, the kg of final concentrated juice, and the concentration of solids in the pulpy juice bypassed. (*Hint:* First, make a total balance and then a solids balance on the overall process. Next, make a balance on the evaporator. Finally, make a balance on the mixer.)

Ans. 34.2 wt % solids in pulpy juice

1.5-11. *Manufacture of Acetylene.* For the making of 6000 ft^3 of acetylene (CHCH) gas at 70°F and 750 mm Hg, solid calcium carbide (CaC$_2$) which contains 97 wt % CaC$_2$ and 3 wt % solid inerts is used along with water. The reaction is

$$CaC_2 + 2H_2O \rightarrow CHCH + Ca(OH)_2 \downarrow$$

The final lime slurry contains water, solid inerts, and Ca(OH)$_2$ lime. In this slurry the total wt % solids of inerts plus Ca(OH)$_2$ is 20%. How many lb of water must be added and how many lb of final lime slurry is produced? [*Hint:* Use a basis of 6000 ft^3 and convert to lb mol. This gives 15.30 lb mol C$_2$H$_2$, 15.30 lb mol Ca(OH)$_2$, and 15.30 lb mol CaC$_2$ added. Convert lb mol CaC$_2$ feed to lb and calculate lb inerts added. The total lb solids in the slurry is then the sum of the Ca(OH)$_2$ plus inerts. In calculating the water added, remember that some is consumed in the reaction.]

Ans. 5200 lb water added (2359 kg), 5815 lb lime slurry (2638 kg)

1.5-12. *Combustion of Solid Fuel.* A fuel analyzes 74.0 wt % C and 12.0% ash (inert). Air is added to burn the fuel, producing a flue gas of 12.4% CO$_2$, 1.2% CO, 5.7% O$_2$, and 80.7% N$_2$. Calculate the kg of fuel used for 100 kg mol of outlet flue gas and the kg mol of air used. (*Hint:* First calculate the mol O$_2$ added in the air, using the fact that the N$_2$ in the flue gas equals the N$_2$ added in the air. Then make a carbon balance to obtain the total moles of C added.)

1.5-13. *Burning of Coke.* A furnace burns a coke containing 81.0 wt % C, 0.8% H, and the rest inert ash. The furnace uses 60% excess air (air over and above that needed to burn all C to CO$_2$ and H to H$_2$O). Calculate the moles of all components in the flue gas if only 95% of the carbon goes to CO$_2$ and 5% to CO.

1.5-14. *Production of Formaldehyde.* Formaldehyde (CH$_2$O) is made by the catalytic oxidation of pure methanol vapor and air in a reactor. The moles from this reactor are 63.1 N$_2$, 13.4 O$_2$, 5.9 H$_2$O, 4.1 CH$_2$O, 12.3 CH$_3$OH, and 1.2 HCOOH. The reaction is

$$CH_3OH + \tfrac{1}{2}O_2 \rightarrow CH_2O + H_2O$$

A side reaction occurring is

$$CH_2O + \tfrac{1}{2}O_2 \rightarrow HCOOH$$

Calculate the mol methanol feed, mol air feed, and percent conversion of methanol to formaldehyde.

Ans. 17.6 mol CH$_3$OH, 79.8 mol air, 23.3% conversion

1.5-15. *Production of Methanol and Recycle.* Methanol is produced by passing H$_2$ and CO gases at high pressure and 573 K over a catalyst bed. The reaction is

$$CO + 2H_2 \rightarrow CH_3OH$$

In the reactor bed 20% of the total CO entering is converted to methanol. The exit gases from the reactor are cooled in a condenser and most (but not all) of the methanol goes out as liquid product. Some of the methanol as a vapor stays in the gas from the condenser. This gas stream contains 2 mol % methanol and is recycled to the reactor. The fresh feed to the reactor (2.64 mol) contains 2 mol H$_2$/mol CO and is mixed with the recycle stream on entering the reactor. Calculate the mol product and mol recycle. (*Hint:* Let R = total moles recycle stream containing 2% alcohol and the rest H_2 and CO. The ratio mol H_2/mol CO is still 2 : 1 in the recycle stream. Let P = total mol CH_3OH product. Make an overall balance first to solve for P. Add R mol recycle to 2.64 mol fresh feed to give the total feed to the reactor and convert 20% of the CO to give the total outlet to the condenser. Then make a balance on the condenser for each component.)

Ans. 0.880 mol product, 10.77 mol recycle

1.5-16. *Ammonia Synthesis and Recycle.* In a process for making NH_3 a stoichiometric molar mixture of N_2 to H_2 of 1 : 3 is fed to a catalytic reactor, where 10% conversion to NH_3 takes place. The NH_3 formed is removed as a product in a condenser and the unconverted gases recycled to the reactor. The fresh feed to the reactor contains 25 mol N_2, 75 mol H_2, and 0.304 mol inert argon. To avoid accumulation of the argon, part of the recycle stream is discarded as a bleed stream. The mol % argon in the total feed (fresh feed plus recycle) to the reactor is held at 0.50%. Calculate the total moles of bleed to hold this concentration at 0.50%.

1.6-1. *Heating of CO_2 Gas.* A total of 250 g of CO_2 gas at 373 K is heated to 623 K at 101.32 kPa total pressure. Calculate the amount of heat needed in cal, btu, and kJ.

Ans. 15 050 cal, 59.7 btu, 62.98 kJ

1.6-2. *Heating a Gas Mixture.* A mixture of 25 lb mol N_2 and 75 lb mol CH_4 is being heated from 400°F to 800°F at 1 atm pressure. Calculate the total amount of heat needed in but.

1.6-3. *Final Temperature in Heating Applesauce.* A mixture of 454 kg of applesauce at 10°C is heated in a heat exchanger by adding 121 300 kJ. Calculate the outlet temperature of the applesauce. (*Hint:* In Appendix A.4 a heat capacity for applesauce is given at 32.8°C. Assume that this is constant and use this as the average c_{pm}.)

Ans. 76.5°C

1.6-4. *Use of Steam Tables.* Using the steam tables, determine the enthalpy change for 1 lb water for each of the following cases.
   (a) Heating liquid water from 40°F to 240°F at 30 psia. (Note that the effect of total pressure on the enthalpy of liquid water can be neglected.)
   (b) Heating liquid water from 40°F to 240°F and vaporizing at 240°F and 24.97 psia.
   (c) Cooling and condensing a saturated vapor at 212°F and 1 atm abs to a liquid at 60°F.
   (d) Condensing a saturated vapor at 212°F and 1 atm abs.

Ans. (a) 200.42 btu/lb_m, (b) 1152.7 btu/lb_m, (c) -1122.4 btu/lb_m, (d) -970.3 btu/lb_m, -2256.9 kJ/kg

1.6-5. *Heating and Vaporization Using Steam Tables.* A flow rate of 1000 kg/h of water at 21.1°C is heated to 110°C when the total pressure is 244.2 kPa in the first stage of a process. In the second stage at the same pressure the water is heated further, until it is all vaporized at its boiling point. Calculate the total enthalpy change in the first stage and in both stages.

1.6-6. *Combustion of CH_4 and H_2.* For 100 g mol of a gas mixture of 75 mol % CH_4 and 25% H_2, calculate the total heat of combustion of the mixture at 298 K and 101.32 kPa, assuming that combustion is complete.

1.6-7. *Heat of Reaction from Heats of Formation.* For the reaction

$$4NH_3(g) + 5O_2(g) \rightarrow 4NO(g) + 6H_2O(g)$$

calculate the heat of reaction, ΔH, at 298 K and 101.32 kPa for 4 g mol of NH_3 reacting.

Ans. ΔH, heat of reaction = -904.7 kJ

1.6-8. *Standard Heat of Reaction.* Calculate the standard heat of reaction of the following at 298 K.

$$HCl(g) + NH_3(g) \rightarrow NH_4Cl(c)$$

Ans. $\Delta H° = -176\,887$ kJ/kg mol

1.7-1. *Heat Balance and Cooling of Milk.* In the processing of rich cows' milk, 4540 kg/h of milk is cooled from 60°C to 4.44°C by a refrigerant. Calculate the heat removed from the milk.

Ans. Heat removed = 269.6 kW

1.7-2. *Heating of Oil by Air.* A flow of 2200 lb_m/h of hydrocarbon oil at 100°F enters a heat exchanger, where it is heated to 150°F by hot air. The hot air enters at 300°F and is to leave at 200°F. Calculate the total lb mol air/h needed. The mean heat capacity of the oil is 0.45 btu/$lb_m \cdot$ °F.

Ans. 70.1 lb mol air/h, 31.8 kg mol/h

1.7-3. *Combustion of Methane in a Furnace.* A gas stream of 10 000 kg mol/h of CH_4 at 101.32 kPa and 373 K is burned in a furnace using air at 313 K. The combustion is complete and 50% excess air is used. The flue gas leaves the furnace at 673 K. Calculate the heat removed in the furnace. (*Hint:* Use a datum of 298 K and liquid water at 298 K. The input items will be the following: the enthalpy of CH_4 at 373 K referred to 298 K; the enthalpy of the air at 313 K referred to 298 K; $-\Delta H_c^0$, the heat of combustion of CH_4 at 298 K which is referred to liquid water; and q, the heat added. The output items will include: the enthalpies of CO_2, O_2, N_2, and H_2O gases at 673 K referred to 298 K; and the latent heat of H_2O vapor at 298 K and 101.32 kPa from Appendix A.2. It is necessary to include this latent heat since the basis of the calculation and of the ΔH_c^0 is liquid water.)

1.7-4. *Preheating Air by Steam for Use in a Dryer.* An air stream at 32.2°C is to be used in a dryer and is first preheated in a steam heater, where it is heated to 65.5°C. The air flow is 1000 kg mol/h. The steam enters the heater saturated at 148.9°C, is condensed and cooled, and leaves as a liquid at 137.8°C. Calculate the amount of steam used in kg/h.

Ans. 450 kg steam/h

1.7-5. *Cooling of Cans of Potato Soup After Thermal Processing.* A total of 1500 cans of potato soup undergo thermal processing in a retort at 240°F. The cans are then cooled to 100°F in the retort before being removed from the retort by cooling water, which enters at 75°F and leaves at 85°F. Calculate the lb of cooling water needed. Each can of soup contains 1.0 lb of liquid soup and the empty metal can weighs 0.16 lb. The mean heat capacity of the soup is 0.94 btu/$lb_m \cdot$ °F and that of the metal can is 0.12 btu/$lb_m \cdot$ °F. A metal rack or basket which is used to hold the cans in the retort weighs 350 lb and has a heat capacity of 0.12 btu/$lb_m \cdot$ °F. Assume that the metal rack is cooled from 240°F to 85°F, the temperature of the outlet water. The amount of heat removed from the retort walls in cooling from 240 to 100°F is 10 000 btu. Radiation loss from the retort during cooling is estimated as 5000 btu.

Ans. 21 320 lb water, 9670 kg

1.7-6. *Oxidation of SO₂ Gas in Converter.* The gas SO_2 is oxidized at 101.3 kPa in a catalytic converter to SO_3 using 100% excess air.

$$SO_2(g) + \tfrac{1}{2}O_2(g) \rightarrow SO_3(g)$$

Only 80% conversion to SO_3 is obtained. The SO_2 and air enter at 673 K and the exit gas mixture leaves at 723 K. There is a heat exchanger in the converter which removes heat. How many kJ are absorbed by the exchanger for 1 kg mol SO_2 added? The ΔH_c^0 for the reaction as written is -98.11×10^3 kJ/kg mol at 298 K.

1.7-7. Energy from Oxidation of Sucrose in a Fermentation Process. In a fermentation process sucrose, $C_{12}H_{22}O_{11}$, is used as a nutrient and is oxidized to CO_2 gas and H_2O liquid. Calculate the heat of oxidation at 45°C in this fermentation. The c_{pm} of solid sucrose is 1.25 kJ/kg · K.

Ans. $\Delta H_{45°} = -5633.8 \times 10^3$ kJ/kg mol

1.8-1. Graphical Integration and Numerical Integration Using Simpson's Method. The following experimental data of $y = f(x)$ were obtained.

x	$f(x)$	x	$f(x)$
0	100	0.4	53
0.1	75	0.5	60
0.2	60.5	0.6	72.5
0.3	53.5		

It is desired to determine the integral

$$A = \int_{x=0}^{x=0.6} f(x)\, dx$$

(a) Do this by a graphical integration.
(b) Repeat using Simpson's numerical method.

Ans. (a) $A = 38.55$; (b) $A = 38.45$

1.8-2. Graphical and Numerical Integration to Obtain Wastewater Flow. The rate of flow of wastewater in an open channel has been measured and the following data obtained:

Time (min)	Flow (m³/min)	Time (min)	Flow (m³/min)
0	655	70	800
10	705	80	725
20	780	90	670
30	830	100	640
40	870	110	620
50	890	120	610
60	870		

(a) Determine the total flow in m³ for the first 60 min and also the total for 120 min by graphical integration.
(b) Determine the flow for 120 min using Simpson's numerical method.

Ans. (a) 48 460 m³ for 60 min, 90 390 m³ for 120 m

1.8-3. Graphical and Numerical Integration to Determine Distance an Airplane Travels. The ground-speed indicator on an airplane gave the following velocities during a gradual slowing down of the plane as it approached a landing field and landed:

Time (s)	0	10	20	30	40	50	60	70	80
Speed (miles/h)	275	263	245	220	190	160	130	86	50 (landed)

(a) How far did the airplane travel in the 80 s to landing, and what was the average velocity? (*Hint*: First convert to a consistent set of units such as time in seconds and velocity in miles/s, or time in minutes and velocity in miles/min. Then graphically integrate to get total area or distance traveled. To calculate average velocity, divide total distance by total time.)

(b) Repeat using Simpson's numerical method.

1.8-4. Graphical Integration in Drying. A granular food product is dried in a pan to determine the rate of drying. The pan contains $M = 2.0$ kg of dry solid with an original moisture content of $X = 0.30$ kg water/kg dry solid. The following data were obtained, where R is the rate of drying as kg/h water evaporated in the pan:

X (kg water/kg solid)	R (kg water/h)	X (kg water/kg solid)	R (kg water/h)
0.30	0.111	0.12	0.084
0.25	0.109	0.090	0.075
0.20	0.110	0.075	0.062
0.175	0.106	0.050	0.050
0.15	0.095		

The equation for the rate of drying can be written

$$R\frac{\text{kg water}}{\text{h}} = -M \text{ kg solid} \frac{dX}{d\Theta h}(\text{kg water/kg solid})$$

Rearranging and integrating,

$$\int_{\Theta=0}^{\Theta=\Theta_F} d\Theta = \Theta_F = -M \int_{X=X_1}^{X=X_2} \frac{1}{R} dX = M \int_{X_2}^{X_1} \frac{1}{R} dX$$

It is desired to calculate the time Θ_F in hours to dry the solid from a moisture content of $X_1 = 0.25$ kg water/kg solid to $X_2 = 0.06$, using the same pan and drying conditions. (*Hint*: Make a plot of $1/R$ versus X. Then graphically integrate between $X_2 = 0.06$ and $X_1 = 0.25$.)

Ans. $\Theta_F = 4.22$ h

1.8-5. Empirical Equation for Strength of Concrete. Experimental data on the compressive strength of concrete T in psi as affected by the amount of water R in kg water/kg concrete are as follows:

T(psi)	R (kg H_2O/kg concrete)	T(psi)	R (kg H_2O/kg concrete)
5330	0.500	2450	0.900
4500	0.600	2000	1.000
3700	0.700	1400	1.200
3000	0.800	1000	1.400

Determine by graphical means an empirical equation to represent the data.

Ans. $T = 14\,000(10)^{-0.832R}$

1.8-6. *Empirical Equation for Flow in Pipes.* The following experimental data were obtained for fluids flowing in pipes:

Friction Factor, f	Reynolds Number, N_{Re}
0.144	110
0.053	298
0.0281	570
0.0222	705
0.0177	852
0.0144	1054
0.0135	1127

Determine by a graphical method an empirical equation to represent the data.

Ans. $f = 16.0/N_{Re}^{1.007}$

REFERENCES

(C1)   CHARM, S. E. *The Fundamentals of Food Engineering*, 2nd ed. Westport, Conn.: Avi Publishing Co., Inc., 1971.

(E1)   EARLE, R. L. *Unit Operations in Food Processing*. Oxford: Pergamon Press, Inc., 1966.

(H1)   HOUGEN, O. A., WATSON, K. M., and RAGATZ, R. A. *Chemical Process Principles*, Part I, 2nd ed. New York: John Wiley & Sons, Inc., 1954.

(O1)   OKOS, M. R. M.S. thesis. Ohio State University, Columbus, Ohio, 1972.

(P1)   PERRY, R. H., and CHILTON, C. H. *Chemical Engineers' Handbook*, 5th ed. New York: McGraw-Hill Book Company, 1973.

(S1)   SOBER, H. A. *Handbook of Biochemistry, Selected Data for Molecular Biology*, 2nd ed. Boca Raton, Fla.: Chemical Rubber Co., Inc., 1970.

(W1)   WEAST, R. C., and SELBY, S. M. *Handbook of Chemistry and Physics*, 48th ed. Boca Raton, Fla.: Chemical Rubber Co., Inc., 1967–1968.

CHAPTER 2

Principles of
Momentum Transfer
and Overall Balances

2.1 INTRODUCTION

The flow and behavior of fluids is important in many of the unit operations in process engineering. A fluid may be defined as a substance that does not permanently resist distortion and, hence, will change its shape. In this text gases, liquids, and vapors are considered to have the characteristics of fluids and to obey many of the same laws.

In the process industries, many of the materials are in fluid form and must be stored, handled, pumped, and processed, so it is necessary that we become familiar with the principles that govern the flow of fluids and also with the equipment used. Typical fluids encountered include water, air, CO_2, oil, slurries, and thick syrups.

If a fluid is inappreciably affected by changes in pressure, it is said to be *incompressible*. Most liquids are incompressible. Gases are considered to be *compressible* fluids. However, if gases are subjected to small percentage changes in pressure and temperature, their density changes will be small and they can be considered to be incompressible.

Like all physical matter, a fluid is composed of an extremely large number of molecules per unit volume. A theory such as the kinetic theory of gases or statistical mechanics treats the motions of molecules in terms of statistical groups and not in terms of individual molecules. In engineering we are mainly concerned with the bulk or macroscopic behavior of a fluid rather than with the individual molecular or microscopic behavior.

In momentum transfer we treat the fluid as a continuous distribution of matter or as a "continuum". This treatment as a continuum is valid when the smallest volume of fluid contains a large enough number of molecules so that a statistical average is meaningful and the macroscopic properties of the fluid such as density, pressure, and so on, vary smoothly or continuously from point to point.

The study of *momentum transfer*, or *fluid mechanics* as it is often called, can be divided into two branches: *fluid statics*, or fluids at rest, and *fluid dynamics*, or fluids in motion. In Section 2.2 we treat fluid statics; in the remaining sections of Chapter 2 and in Chapter 3, fluid dynamics. Since in fluid dynamics momentum is being transferred, the term "momentum transfer" or "transport" is usually used. In Section 2.3 momentum transfer is related to heat and mass transfer.

2.2 FLUID STATICS

2.2A Force, Units, and Dimensions

In a static fluid an important property is the pressure in the fluid. Pressure is familiar as a surface force exerted by a fluid against the walls of its container. Also, pressure exists at any point in a volume of a fluid.

In order to understand *pressure*, which is defined as force exerted per unit area, we must first discuss a basic law of Newton's. This equation for calculation of the force exerted by a mass under the influence of gravity is

$$F = mg \quad \text{(SI units)}$$

$$(2.2\text{-}1)$$

$$F = \frac{mg}{g_c} \quad \text{(English units)}$$

where in SI units F is the force exerted in newtons $N(kg \cdot m/s^2)$, m the mass in kg, and g the standard acceleration of gravity, $9.80665 \, m/s^2$.

In English units, F is in lb_f, m in lb_m, g is $32.1740 \, ft/s^2$, and g_c (a gravitational conversion factor) is $32.174 \, lb_m \cdot ft/lb_f \cdot s^2$. The use of the conversion factor g_c means that g/g_c has a value of $1.0 \, lb_f/lb_m$ and that $1 \, lb_m$ conveniently gives a force equal to $1 \, lb_f$. Often when units of pressure are given, the word "force" is omitted, such as in $lb/in.^2$ (psi) instead of $lb_f/in.^2$. When the mass m is given in g mass, F is g force, $g = 980.665$ cm/s^2, and $g_c = 980.665 \, g \, mass \cdot cm/g \, force \cdot s^2$. However, the units g force are seldom used.

Another system of units sometimes used in Eq. (2.2-1) is that where the g_c is omitted and the force ($F = mg$) is given as $lb_m \cdot ft/s^2$, which is called *poundals*. Then $1 \, lb_m$ acted on by gravity will give a force of 32.174 poundals $(lb_m \cdot ft/s^2)$. Or if 1 g mass is used, the force ($F = mg$) is expressed in terms of dynes $(g \cdot cm/s^2)$. This is the centimeter–gram–second (cgs) systems of units.

Conversion factors for different units of force and of force per unit area (pressure) are given in Appendix A.1. Note that always in the SI system, and usually in the cgs system, the term g_c is not used.

EXAMPLE 2.2-1. Units and Dimensions of Force
Calculate the force exerted by 3 lb mass in terms of the following.
   (a) Lb force (English units).
   (b) Dynes (cgs units).
   (c) Newtons (SI units).

Solution: For part (a), using Eq. (2.2-1),

$$F \, (\text{force}) = m \frac{g}{g_c} = (3 \, lb_m)\left(32.174 \, \frac{ft}{s^2}\right)\left(\frac{1}{32.174 \frac{lb_m \cdot ft}{lb_f \cdot s^2}}\right) = 3 \, \text{lb force} \, (lb_f)$$

For part (b),

$$F = mg = (3 \, lb_m)\left(453.59 \, \frac{g}{lb_m}\right)\left(980.665 \, \frac{cm}{s^2}\right)$$

$$= 1.332 \times 10^6 \, \frac{g \cdot cm}{s^2} = 1.332 \times 10^6 \, dyn$$

As an alternative method for part (b), from Appendix A.1,

$$1 \text{ dyn} = 2.2481 \times 10^{-6} \text{ lb}_f$$

$$F = (3 \text{ lb}_f)\left(\frac{1}{2.2481 \times 10^{-6}\text{lb}_f/\text{dyn}}\right) = 1.332 \times 10^6 \text{ dyn}$$

To calculate newtons in part (c),

$$F = mg = \left(3 \text{ lb}_m \times \frac{1 \text{ kg}}{2.2046 \text{ lb}_m}\right)\left(9.80665 \frac{\text{m}}{\text{s}^2}\right)$$

$$= 13.32 \frac{\text{kg}\cdot\text{m}}{\text{s}^2} = 13.32 \text{ N}$$

As an alternative method, using values from Appendix A.1,

$$1 \frac{\text{g}\cdot\text{cm}}{\text{s}^2}(\text{dyn}) = 10^{-5} \frac{\text{kg}\cdot\text{m}}{\text{s}^2}(\text{newton})$$

$$F = (1.332 \times 10^6 \text{ dyn})\left(10^{-5} \frac{\text{newton}}{\text{dyn}}\right) = 13.32 \text{ N}$$

2.2B Pressure in a Fluid

Since Eq. (2.2-1) gives the force exerted by a mass under the influence of gravity, the force exerted by a mass of fluid on a supporting area or force/unit area (pressure) also follows from this equation. In Fig. 2.2-1 a stationary column of fluid of height h_2 m and constant cross-sectional area A m^2, where $A = A_0 = A_1 = A_2$, is shown. The pressure above the fluid is P_0 N/m^2; that is, this could be the pressure of the atmosphere above the fluid. The fluid at any point, say h_1, must support all the fluid above it. It can be shown that the forces at any given point in a nonmoving or static fluid must be the same in all directions. Also, for a fluid at rest, the force/unit area or pressure is the same at all points with the same elevation. For example, at h_1 m from the top, the pressure is the same at all points shown on the cross-sectional area A_1.

The use of Eq. (2.2-1) will be shown in calculating the pressure at different vertical points in Fig. 2.2-1. The total mass of fluid for h_2 m height and density ρ kg/m^3 is

$$\text{total kg fluid} = (h_2 \text{ m})(A \text{ m}^2)\left(\rho \frac{\text{kg}}{\text{m}^3}\right) = h_2 A\rho \text{ kg} \qquad \text{(2.2-2)}$$

FIGURE 2.2-1. *Pressure in a static fluid.*

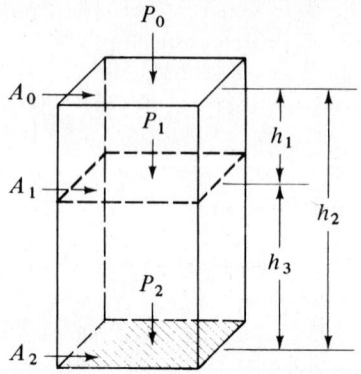

Substituting into Eq. (2.2-2), the total force F of the fluid on area A_1 due to the fluid only is

$$F = (h_2 A\rho \text{ kg})(g \text{ m/s}^2) = h_2 A\rho g \frac{\text{kg} \cdot \text{m}}{\text{s}^2} \text{ (N)} \qquad \text{(2.2-3)}$$

The pressure P is defined as force/unit area:

$$P = \frac{F}{A} = (h_2 A\rho g)\frac{1}{A} = h_2 \rho g \text{ N/m}^2 \quad \text{or} \quad \text{Pa} \qquad \text{(2.2-4)}$$

This is the pressure on A_2 due to the mass of the fluid above it. However, to get the total pressure P_2 on A_2, the pressure P_0 on the top of the fluid must be added.

$$P_2 = h_2 \rho g + P_0 \text{ N/m}^2 \quad \text{or} \quad \text{Pa} \qquad \text{(2.2-5)}$$

Equation (2.2-5) is the fundamental equation to calculate the pressure in a fluid at any depth. To calculate P_1,

$$P_1 = h_1 \rho g + P_0 \qquad \text{(2.2-6)}$$

The pressure difference between points 2 and 1 is

$$P_2 - P_1 = (h_2 \rho g + P_0) - (h_1 \rho g + P_0) = (h_2 - h_1)\rho g \qquad \text{(SI units)}$$

$$P_2 - P_1 = (h_2 - h_1)\rho \frac{g}{g_c} \qquad \text{(English units)}$$

(2.2-7)

Since it is the vertical height of a fluid that determines the pressure in a fluid, the shape of the vessel does not affect the pressure. For example, in Fig. 2.2-2, the pressure P_1 at the bottom of all three vessels is the same and equal to $h_1 \rho g + P_0$.

EXAMPLE 2.2-2. Pressure in Storage Tank

A large storage tank contains oil having a density of 917 kg/m³ (0.917 g/cm³). The tank is 3.66 m (12.0 ft) tall and is vented (open) to the atmosphere of 1 atm abs at the top. The tank is filled with oil to a depth of 3.05 m (10 ft) and also contains 0.61 m (2.0 ft) of water in the bottom of the tank. Calculate the pressure in Pa and psia 3.05 m from the top of the tank and at the bottom. Also calculate the gage pressure at the tank bottom.

Solution: First a sketch is made of the tank, as shown in Fig. 2.2-3. The pressure $P_0 = 1$ atm abs $= 14.696$ psia (from Appendix A.1). Also,

$$P_0 = 1.01325 \times 10^5 \text{ Pa}$$

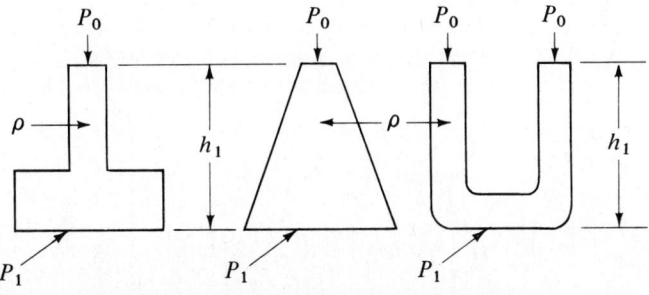

FIGURE 2.2-2. *Pressure in vessels of various shapes.*

Principles of Momentum Transfer and Overall Balances

FIGURE 2.2-3. *Storage tank in Example 2.2-2.*

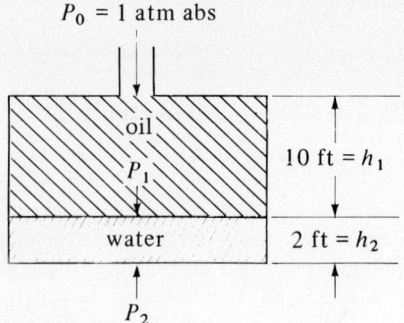

From Eq. (2.2-6) using English and then SI units,

$$P_1 = h_1 \rho_{\text{oil}} \frac{g}{g_c} + P_0 = (10 \text{ ft})\left(0.917 \times 62.43 \frac{\text{lb}_m}{\text{ft}^3}\right)\left(1.0 \frac{\text{lb}_f}{\text{lb}_m}\right)\left(\frac{1}{144 \text{ in.}^2/\text{ft}^2}\right)$$

$$+ 14.696 \text{ lb}_f/\text{in.}^2 = 18.68 \text{ psia}$$

$$P_1 = h_1 \rho_{\text{oil}} g + P_0 = (3.05 \text{ m})\left(1000 \frac{\text{kg}}{\text{m}^3}\right)\left(9.8066 \frac{\text{m}}{\text{s}^2}\right) + 1.0132 \times 10^5$$

$$= 1.287 \times 10^5 \text{ Pa}$$

To calculate P_2 at the bottom of the tank, $\rho_{\text{water}} = 1.00 \text{ g/cm}^3$ and

$$P_2 = h_2 \rho_{\text{water}} \frac{g}{g_c} + P_1 = (2.0)(1.00 \times 62.43)(1.0)(\tfrac{1}{144}) + 18.68$$

$$= 19.55 \text{ psia}$$

$$= h_2 \rho_{\text{water}} g + P_1 = (0.61)(1000)(9.8066) + 1.287 \times 10^5$$

$$= 1.347 \times 10^5 \text{ Pa}$$

The gage pressure at the bottom is equal to the absolute pressure P_2 minus 1 atm pressure.

$$P_{\text{gage}} = 19.55 \text{ psia} - 14.696 \text{ psia} = 4.85 \text{ psig}$$

2.2C Head of a Fluid

Pressures are given in many different sets of units, such as psia, dyn/cm², and newtons/m², as given in Appendix A.1. However, a common method of expressing pressures is in terms of head in m or feet of a particular fluid. This height or head in m or feet of the given fluid will exert the same pressure as the pressures it represents. Using Eq. (2.2-4), which relates pressure P and height h of a fluid and solving for h, which is the head in m,

$$h(\text{head}) = \frac{P}{\rho g} \text{ m} \quad \text{(SI)}$$

(2.2-8)

$$h = \frac{P g_c}{\rho g} \text{ ft} \quad \text{(English)}$$

46   *Transport Processes: Momentum, Heat, and Mass*

EXAMPLE 2.2-3. Conversion of Pressure to Head of a Fluid
Given the pressure of 1 standard atm as $101.325 \, kN/m^2$ (Appendix A.1), do
as follows.
    (a) Convert this pressure to head in m water at 4°C.
    (b) Convert this pressure to head in m Hg at 0°C.

Solution: For part (a), the density of water at 4°C in Appendix A.2 is 1.000
g/cm^3. From A.1, a density of 1.000 g/cm^3 equals 1000 kg/m^3. Substituting
these values into Eq. (2.2-8),

$$h(\text{head}) = \frac{P}{\rho g} = \frac{101.325 \times 10^3}{(1000)(9.80665)}$$

$$= 10.33 \text{ m of water at } 4°C$$

For part (b), the density of Hg in Appendix A.1 is 13.5955 g/cm^3. For
equal pressures P from different fluids, Eq. (2.2-8) can be rewritten as

$$P = \rho_{Hg} h_{Hg} g = \rho_{H_2O} h_{H_2O} g \qquad (2.2\text{-}9)$$

Solving for h_{Hg} in Eq. (2.2-9) and substituting known values,

$$h_{Hg}(\text{head}) = \frac{\rho_{H_2O}}{\rho_{Hg}} h_{H_2O} = \left(\frac{1.000}{13.5955}\right)(10.33) = 0.760 \text{ m Hg}$$

2.2D   Devices to Measure Pressure and Pressure Differences

In chemical and other industrial processing plants it is often important to measure and
control the pressure in a vessel or process and/or the liquid level in a vessel. Also, since
many fluids are flowing in a pipe or conduit, it is necessary to measure the rate at which
the fluid is flowing. Many of these flow meters depend upon devices to measure a
pressure or pressure difference. Some common devices are considered in the following
paragraphs.

1. Simple U-tube manometer.   The U-tube manometer is shown in Fig. 2.2-4a. The

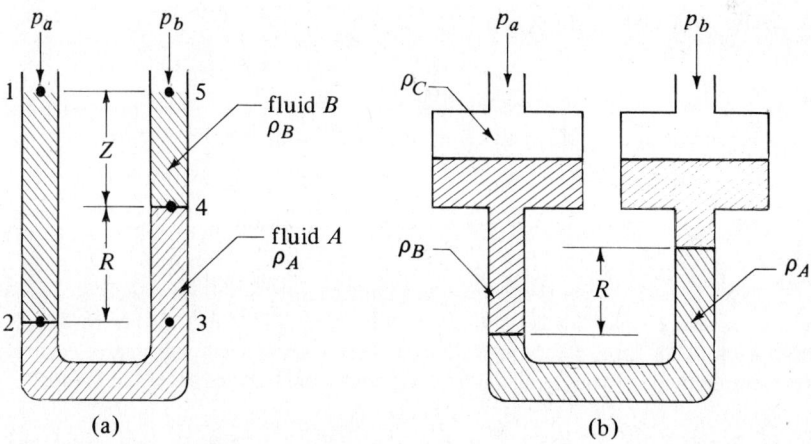

FIGURE 2.2-4.   *Manometers to measure pressure differences: (a) U tube, (b) two-fluid*
           U tube.

pressure p_a N/m^2 is exerted on one arm of the U tube and p_b on the other arm. Both pressures p_a and p_b could be pressure taps from a fluid meter, or p_a could be a pressure tap and p_b the atmospheric pressure. The top of the manometer is filled with liquid B, having a density of ρ_B kg/m^3, and the bottom with a more dense fluid A, having a density of ρ_A kg/m^3. Liquid A is immiscible with B. To derive the relationship between p_a and p_b, p_a is the pressure at point 1 and p_b at point 5. The pressure at point 2 is

$$p_2 = p_a + (Z + R)\rho_B g \text{ N/m}^2 \qquad \text{(2.2-10)}$$

where R is the reading of the manometer in m. The pressure at point 3 must be equal to that at 2 by the principles of hydrostatics.

$$p_3 = p_2 \qquad \text{(2.2-11)}$$

The pressure at point 3 also equals the following:

$$p_3 = p_b + Z\rho_B g + R\rho_A g \qquad \text{(2.2-12)}$$

Equating Eq. (2.2-10) to (2.2-12) and solving,

$$p_a + (Z + R)\rho_B g = p_b + Z\rho_B g + R\rho_A g \qquad \text{(2.2-13)}$$

$$p_a - p_b = R(\rho_A - \rho_B)g \qquad \text{(SI)}$$

$$\qquad \text{(2.2-14)}$$

$$p_a - p_b = R(\rho_A - \rho_B)\frac{g}{g_c} \qquad \text{(English)}$$

The reader should note that the distance Z does not enter into the final result nor do the tube dimensions, provided that p_a and p_b are measured in the same horizontal plane.

EXAMPLE 2.2-4. *Pressure Difference in a Manometer*
A manometer, as shown in Fig. 2.2-4a, is being used to measure the head or pressure drop across a flow meter. The heavier fluid is mercury, with a density of 13.6 g/cm^3, and the top fluid is water, with a density of 1.00 g/cm^3. The reading on the manometer is $R = 32.7$ cm. Calculate the pressure difference in N/m^2 using SI units.

Solution: Converting R to m,

$$R = \frac{32.7}{100} = 0.327 \text{ m}$$

Also converting ρ_A and ρ_B to kg/m^3 and substituting into Eq. (2.2-14),

$$p_a - p_b = R(\rho_A - \rho_B)g = (0.327 \text{ m})[(13.6 - 1.0)(1000 \text{ kg/m}^3)](9.8066 \text{ m/s}^2)$$

$$= 4.040 \times 10^4 \text{ N/m}^2 \text{ (5.85 psia)}$$

2. *Two-fluid U tube.* In Fig. 2.2-4b a two-fluid U tube is shown, which is a sensitive device to measure small heads or pressure differences. Let A m^2 be the cross-sectional area of each of the large reservoirs and a m^2 be the cross-sectional area of each of the tubes forming the U. Proceeding and making a pressure balance as for the U tube,

$$p_a - p_b = (R - R_0)\left(\rho_A - \rho_B + \frac{a}{A}\rho_B - \frac{a}{A}\rho_C\right)g \qquad \text{(2.2-15)}$$

where R_0 is the reading when $p_a = p_b$, R is the actual reading, ρ_A is the density of the heavier fluid, and ρ_B is the density of the lighter fluid. Usually, a/A is made sufficiently

Transport Processes: Momentum, Heat, and Mass

small to be negligible, and also R_0 is often adjusted to zero; then

$$p_a - p_b = R(\rho_A - \rho_B)g \qquad \text{(SI)}$$
$$p_a - p_b = R(\rho_A - \rho_B)\frac{g}{g_c} \qquad \text{(English)} \qquad \text{(2.2-16)}$$

If ρ_A and ρ_B are close to each other, the reading R is magnified.

EXAMPLE 2.2-5. Pressure Measurement in a Vessel
The U-tube manometer in Fig. 2.2-5a is used to measure the pressure p_A in a vessel containing a liquid with a density ρ_A. Derive the equation relating the pressure p_A and the reading on the manometer as shown.

Solution: At point 2 the pressure is

$$p_2 = p_{\text{atm}} + h_2 \rho_B g \ \text{N/m}^2 \qquad \text{(2.2-17)}$$

At point 1 the pressure is

$$p_1 = p_A + h_1 \rho_A g \qquad \text{(2.2-18)}$$

Equating $p_1 = p_2$ by the principles of hydrostatics and rearranging,

$$p_A = p_{\text{atm}} + h_2 \rho_B g - h_1 \rho_A g \qquad \text{(2.2-19)}$$

Another example of a U-tube manometer is shown in Fig. 2.2-5b. This device is used in this case to measure the pressure difference between two vessels.

3. Bourdon pressure gage. Although manometers are used to measure pressures, the most common pressure-measuring device is the mechanical Bourdon-tube pressure gage. A coiled hollow tube in the gage tends to straighten out when subjected to internal pressure, and the degree of straightening depends on the pressure difference between the inside and outside pressures. The tube is connected to a pointer on a calibrated dial.

4. Gravity separator for two immiscible liquids. In Fig. 2.2-6 a continuous gravity separator (decanter) is shown for the separation of two immiscible liquids A (heavy liquid) and B (light liquid). The feed mixture of the two liquids enters at one end of the separator vessel and the liquids flow slowly to the other end and separate into two distinct layers. Each liquid flows through a separate overflow line as shown. Assuming

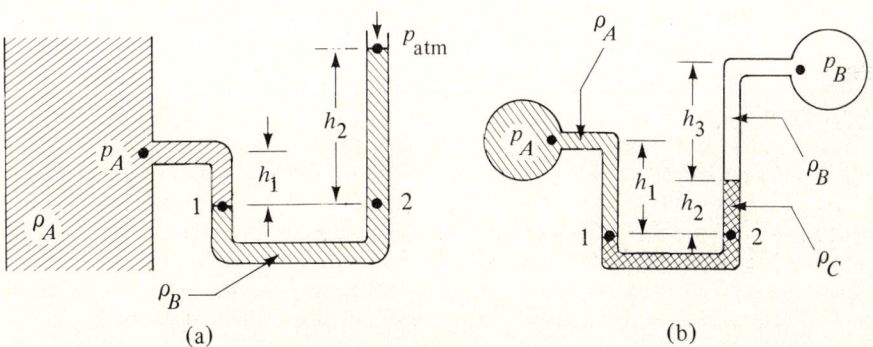

(a) (b)

FIGURE 2.2-5. *Measurements of pressure in vessels : (a) measurement of pressure in a vessel, (b) measurement of differential pressure.*

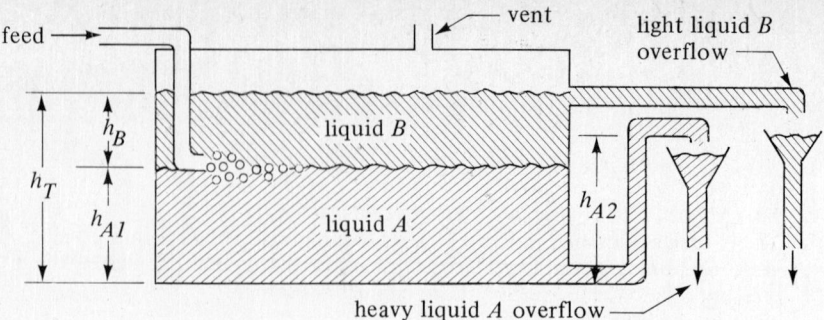

FIGURE 2.2-6. *Continuous atmospheric gravity separator for immiscible liquids.*

the frictional resistance to the flow of the liquids is essentially negligible, the principles of fluid statics can be used to analyze the performance.

In Fig. 2.2-6, the depth of the layer of heavy liquid A is h_{A1} m and that of B is h_B. The total depth $h_T = h_{A1} + h_B$ and is fixed by position of the overflow line for B. The heavy liquid A discharges through an overflow leg h_{A2} m above the vessel bottom. The vessel and the overflow lines are vented to the atmosphere. A hydrostatic balance gives

$$h_B \rho_B g + h_{A1} \rho_A g = h_{A2} \rho_A g \qquad (2.2\text{-}20)$$

Substituting $h_B = h_T - h_{A1}$ into Eq. (2.2-20) and solving for h_{A1},

$$h_{A1} = \frac{h_{A2} - h_T \rho_B/\rho_A}{1 - \rho_B/\rho_A} \qquad (2.2\text{-}21)$$

This shows that the position of the interface or height h_{A1} depends on the ratio of the densities of the two liquids and on the elevations h_{A2} and h_T of the two overflow lines. Usually, the height h_{A2} is movable so that the interface level can be adjusted.

2.3 GENERAL MOLECULAR TRANSPORT EQUATION FOR MOMENTUM, HEAT, AND MASS TRANSFER

2.3A General Molecular Transport Equation and General Property Balance

1. Introduction to transport processes. In molecular transport processes in general we are concerned with the transfer or movement of a given property or entity by molecular movement through a system or medium which can be a fluid (gas or liquid) or a solid. This property that is being transferred can be mass, thermal energy (heat), or momentum. Each molecule of a system has a given quantity of the property mass, thermal energy, or momentum associated with it. When a difference of concentration of the property exists for any of these properties from one region to an adjacent region, a net transport of this property occurs. In dilute fluids such as gases where the molecules are relatively far apart, the rate of transport of the property should be relatively fast since few molecules are present to block the transport or interact. In dense fluids such as liquids the molecules are close together and transport or diffusion procedes more slowly. The molecules in solids are even more close-packed than in liquids and molecular migration is even more restricted.

2. *General molecular transport equation.* All three of the molecular transport processes of momentum, heat or thermal energy, and mass are characterized in the elementary sense by the same general type of transport equation. First we start by noting the following:

$$\text{rate of a transfer process} = \frac{\text{driving force}}{\text{resistance}} \tag{2.3-1}$$

This states what is quite obvious—that we need a driving force to overcome a resistance in order to transport a property. This is similar to Ohm's law in electricity, where the rate of flow of electricity is proportional to the voltage drop (driving force) and inversely proportional to the resistance.

We can formalize Eq. (2.3-1) by writing an equation as follows for molecular transport or diffusion of a property:

$$\psi_z = -\delta \frac{d\Gamma}{dz} \tag{2.3-2}$$

where ψ_z is defined as the flux of the property as amount of property being transferred per unit time per unit cross-sectional area perpendicular to the z direction of flow in amount of property/s $\cdot$ m^2, δ is a proportionality constant called diffusivity in m^2/s, Γ is concentration of the property in amount of property/m^3, and z is the distance in the direction of flow in m.

If the process is at steady state, then the flux ψ_z is constant. Rearranging Eq. (2.3-2) and integrating,

$$\psi_z \int_{z_1}^{z_2} dz = -\delta \int_{\Gamma_1}^{\Gamma_2} d\Gamma \tag{2.3-3}$$

$$\psi_z = \frac{\delta(\Gamma_1 - \Gamma_2)}{z_2 - z_1} \tag{2.3-4}$$

A plot of the concentration Γ versus z is shown in Fig. 2.3-1a and is a straight line. Since the flux is in the direction 1 to 2 of decreasing concentration, the slope $d\Gamma/dz$ is negative and the negative sign in Eq. (2.3-2) gives a positive flux in the direction 1 to 2. In Section 2.3B the specialized equations for momentum, heat, and mass transfer will be shown to be the same as Eq. (2.3-4) for the general property transfer.

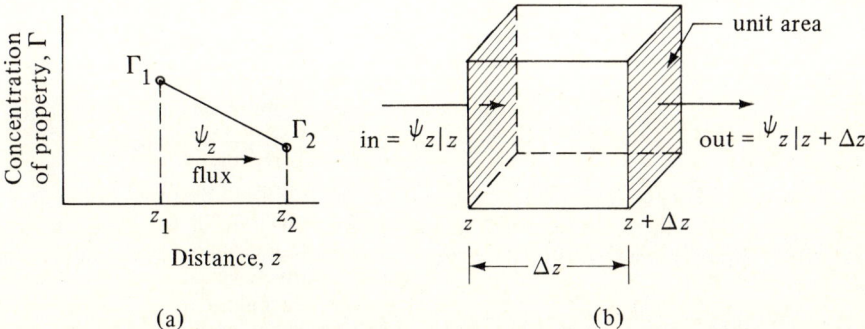

(a)   (b)

FIGURE 2.3-1. *Molecular transport of a property: (a) plot of concentration versus distance for steady state, (b) unsteady-state general property balance.*

Principles of Momentum Transfer and Overall Balances

EXAMPLE 2.3-1. *Molecular Transport of a Property at Steady State*
A property is being transported by diffusion through a fluid at steady state. At a given point 1 the concentration is 1.37×10^{-2} amount of property/m^3 and 0.72×10^{-2} at point 2 at a distance $z_2 = 0.40$ m. The diffusivity $\delta = 0.013$ m^2/s and the cross-sectional area is constant.

   (a) Calculate the flux.
   (b) Derive the equation for Γ as a function of distance.
   (c) Calculate Γ at the midpoint of the path.

Solution: For part (a) substituting into Eq. (2.3-4),

$$\psi_z = \frac{\delta(\Gamma_1 - \Gamma_2)}{z_2 - z_1} = \frac{(0.013)(1.37 \times 10^{-2} - 0.72 \times 10^{-2})}{0.40 - 0}$$

$$= 2.113 \times 10^{-4} \text{ amount of property/s} \cdot \text{m}^2$$

For part (b), integrating Eq. (2.3-2) between Γ_1 and Γ and z_1 and z and rearranging,

$$\psi_z \int_{z_1}^{z} dz = -\delta \int_{\Gamma_1}^{\Gamma} d\Gamma \qquad (2.3\text{-}5)$$

$$\Gamma = \Gamma_1 + \frac{\psi_z}{\delta} (z_1 - z) \qquad (2.3\text{-}6)$$

For part (c), using the midpoint $z = 0.20$ m and substituting into Eq. (2.3-6),

$$\Gamma = 1.37 \times 10^{-2} + \frac{2.113 \times 10^{-4}}{0.013} (0 - 0.2)$$

$$= 1.045 \times 10^{-2} \text{ amount of property/m}^3$$

3. General property balance for unsteady state. In calculating the rates of transport in a system using the molecular transport equation (2.3-2), it is necessary to account for the amount of this property being transported in the entire system. This is done by writing a general property balance or conservation equation for the property (momentum, thermal energy, or mass) at unsteady state. We start by writing an equation for the z direction only, which accounts for all the property entering by molecular transport, leaving, being generated, and accumulating in a system shown in Fig. 2.3-1b, which is an element of volume $\Delta z(1)$ m^3 fixed in space.

$$\begin{pmatrix} \text{rate of} \\ \text{property in} \end{pmatrix} + \begin{pmatrix} \text{rate of generation} \\ \text{of property} \end{pmatrix}$$

$$= \begin{pmatrix} \text{rate of} \\ \text{property out} \end{pmatrix} + \begin{pmatrix} \text{rate of accum-} \\ \text{ulation of property} \end{pmatrix} \qquad (2.3\text{-}7)$$

The rate of input is $(\psi_{z|z}) \cdot 1$ amount of property/s and the rate of output is $(\psi_{z|z+\Delta z}) \cdot 1$, where the cross-sectional area is 1.0 m^2. The rate of generation of the property is $R(\Delta z \cdot 1)$, where R is rate of generation of property/s $\cdot$ m^3. The accumulation term is

$$\text{rate of accumulation of property} = \frac{\partial \Gamma}{\partial t} (\Delta z \cdot 1) \qquad (2.3\text{-}8)$$

Substituting the various terms into Eq. (2.3-7),

$$(\psi_{z|z}) \cdot 1 + R(\Delta z \cdot 1) = (\psi_{z|z + \Delta z}) \cdot 1 + \frac{\partial \Gamma}{\partial t} (\Delta z \cdot 1) \qquad \textbf{(2.3-9)}$$

Dividing by Δz and letting Δz go to zero,

$$\frac{\partial \Gamma}{\partial t} + \frac{\partial \psi_z}{\partial z} = R \qquad \textbf{(2.3-10)}$$

Substituting Eq. (2.3-2) for ψ_z into (2.3-10) and assuming that δ is constant,

$$\frac{\partial \Gamma}{\partial t} - \delta \frac{\partial^2 \Gamma}{\partial z^2} = R \qquad \textbf{(2.3-11)}$$

For the case where no generation is present,

$$\frac{\partial \Gamma}{\partial t} = \delta \frac{\partial^2 \Gamma}{\partial z^2} \qquad \textbf{(2.3-12)}$$

This final equation relates the concentration of the property Γ to position z and time t.

Equations (2.3-11) and (2.3-12) are general equations for the conservation of momentum, thermal energy, or mass and will be used in many sections of this text. The equations consider here only molecular transport occurring and the equations do not consider other transport mechanisms such as convection, and so on, which will be considered when the specific conservation equations are derived in later sections of this text for momentum, energy, or mass.

2.3B Introduction to Molecular Transport

The kinetic theory of gases gives us a good physical interpretation of the motion of individual molecules in fluids. Because of their kinetic energy the molecules are in rapid random movement, often colliding with each other. Molecular transport or molecular diffusion of a property such as momentum, heat, or mass occurs in a fluid because of these random movements of individual molecules. Each individual molecule containing the property being transferred moves randomly in all directions and there are fluxes in all directions. Hence, if there is a concentration gradient of the property, there will be a net flux of the property from high to low concentration. This occurs because equal numbers of molecules diffuse in each direction between the high-concentration and low-concentration regions.

1. Momentum transport and Newton's law. When a fluid is flowing in the x direction parallel to a solid surface, a velocity gradient exists where the velocity v_x in the x direction decreases as we approach the surface in the z direction. The fluid has x-directed momentum and its concentration is $v_x \rho$ momentum/m³, where the momentum has units of kg·m/s. Hence, the units of $v_x \rho$ are (kg·m/s)/m³. By random diffusion of molecules there is an exchange of molecules in the z direction, an equal number moving in each direction ($+z$ and $-z$ directions) between the faster-moving layer of molecules and the slower adjacent layer. Hence, the x-directed momentum has been transferred in the z direction from the faster- to the slower-moving layer. The equation for this transport of momentum is similar to Eq. (2.3-2) and is Newton's law of viscosity written as follows for constant density ρ:

$$\tau_{zx} = -v \frac{d(v_x \rho)}{dz} \qquad (2.3\text{-}13)$$

where τ_{zx} is flux of x-directed momentum in the z direction, $(kg \cdot m/s)/s \cdot m^2$; v is μ/ρ, the momentum diffusivity in m^2/s; z is the direction of transport or diffusion in m; ρ is the density in kg/m^3; and μ is the viscosity in $kg/m \cdot s$.

2. Heat transport and Fourier's law. Fourier's law for molecular transport of heat or heat conduction in a fluid or solid can be written as follows for constant density ρ and heat capacity c_p.

$$\frac{q_z}{A} = -\alpha \frac{d(\rho c_p T)}{dz} \qquad (2.3\text{-}14)$$

where q_z/A is the heat flux in $J/s \cdot m^2$, α is the thermal diffusivity in m^2/s, and $\rho c_p T$ is the concentration of heat or thermal energy in J/m^3. When there is a temperature gradient in a fluid, equal numbers of molecules diffuse in each direction between the hot and the colder region. In this way energy is transferred in the z direction.

3. Mass transport and Fick's law. Fick's law for molecular transport of mass in a fluid or solid for constant total concentration in the fluid is

$$J_{Az}^* = -D_{AB} \frac{dc_A}{dz} \qquad (2.3\text{-}15)$$

where J_{Az}^* is the flux of A in kg mol $A/s \cdot m^2$, D_{AB} is the molecular diffusivity of the molecule A in B in m^2/s, and c_A is the concentration of A in kg mol A/m^3. In a manner similar to momentum and heat transport, when there is a concentration gradient in a fluid, equal numbers of molecules diffuse in each direction between the high- and the low-concentration region and a net flux of mass occurs.

Hence, Eqs. (2.3-13), (2.3-14), and (2.3-15) for momentum, heat, and mass transfer are all similar to each other and to the general molecular transport equation (2.3-2). All equations have a flux on the left-hand side of each equation, a diffusivity in m^2/s, and the derivative of the concentration with respect to distance. All three of the molecular transport equations are mathematically identical. Thus, we state we have an analogy or similarity among them. It should be emphasized, however, that even though there is a mathematical analogy, the actual physical mechanisms occurring can be totally different. For example, in mass transfer two components are often being transported by relative motion through one another. In heat transport in a solid, the molecules are relatively stationary and the transport is done mainly by the electrons. Transport of momentum can occur by several types of mechanisms. More detailed considerations of each of the transport processes of momentum, energy, and mass are presented in this and succeeding chapters.

2.4 VISCOSITY OF FLUIDS

2.4A Newton's Law and Viscosity

When a fluid is flowing through a closed channel such as a pipe or between two flat plates, either of two types of flow may occur, depending on the velocity of this fluid. At low velocities the fluid tends to flow without lateral mixing, and adjacent layers slide past one another like playing cards. There are no cross currents perpendicular to the direction of flow, nor eddies or swirls of fluid. This regime or type of flow is called *laminar flow*. At higher velocities eddies form, which leads to lateral mixing. This is called *turbulent flow*. The discussion in this section is limited to laminar flow.

A fluid can be distinguished from a solid in this discussion of viscosity by its behavior when subjected to a stress (force per unit area) or applied force. An elastic solid deforms by an amount proportional to the applied stress. However, a fluid when subjected to a similar applied stress will continue to deform, i.e., to flow at a velocity that increases with increasing stress. A fluid exhibits resistance to this stress. Viscosity is that property of a fluid which gives rise to forces that resist the relative movement of adjacent layers in the fluid. These *viscous forces* arise from forces existing between the molecules in the fluid and are of similar character as the *shear forces* in solids.

The ideas above can be clarified by a more quantitative discussion of viscosity. In Fig. 2.4-1 a fluid is contained between two infinite (very long and very wide) parallel plates. Suppose that the bottom plate is moving parallel to the top plate and at a constant velocity Δv_z m/s faster relative to the top plate because of a steady force F newtons being applied. This force is called the *viscous drag*, and it arises from the viscous forces in the fluid. The plates are Δy m apart. Each layer of liquid moves in the z direction. The layer immediately adjacent to the bottom plate is carried along at the velocity of this plate. The layer just above is at a slightly slower velocity, each layer moving at a slower velocity as we go up in the y direction. This velocity profile is linear, with y direction as shown in Fig. 2.4-1. An analogy to a fluid is a deck of playing cards, where, if the bottom card is moved, all the other cards above will slide to some extent.

It has been found experimentally for many fluids that the force F in newtons is directly proportional to the velocity Δv_z in m/s, to the area A in m^2 of the plate used, and inversely proportional to the distance Δy in m. Or, as given by Newton's law of viscosity when the flow is laminar,

$$\frac{F}{A} = -\mu \frac{\Delta v_z}{\Delta y} \qquad (2.4\text{-}1)$$

where μ is a proportionality constant called the *viscosity* of the fluid, in Pa·s or kg/m·s. If we let Δy approach zero, then, using the definition of the derivative,

$$\tau_{yz} = -\mu \frac{dv_z}{dy} \qquad \text{(SI units)} \qquad (2.4\text{-}2)$$

where $\tau_{yz} = F/A$ and is the shear stress or force per unit area in newtons/m^2 (N/m^2). In the cgs system, F is in dynes, μ in g/cm·s, v_z in cm/s, and y in cm. We can also write

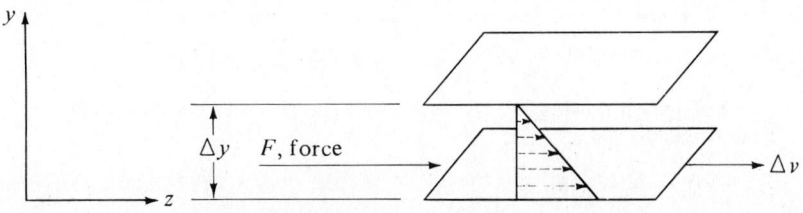

FIGURE 2.4-1. *Fluid shear between two parallel plates.*

Eq. (2.2-2) as

$$\tau_{yz} g_c = -\mu \frac{dv_z}{dy} \qquad \text{(English units)} \qquad \textbf{(2.4-3)}$$

where τ_{yz} is in units of lb_f/ft^2.

The units of viscosity in the cgs system are $\text{g/cm} \cdot \text{s}$, called *poise* or centipoise (cp). In the SI system, viscosity is given in $\text{Pa} \cdot \text{s} (\text{N} \cdot \text{s/m}^2 \text{ or kg/m} \cdot \text{s})$.

$$1 \text{ cp} = 1 \times 10^{-3} \text{ kg/m} \cdot \text{s} = 1 \times 10^{-3} \text{ Pa} \cdot \text{s} = 1 \times 10^{-3} \text{ N} \cdot \text{s/m}^2 \text{ (SI)}$$

$$1 \text{ cp} = 0.01 \text{ poise} = 0.01 \text{ g/cm} \cdot \text{s}$$

$$1 \text{ cp} = 6.7197 \times 10^{-4} \text{ lb}_m/\text{ft} \cdot \text{s}$$

Other conversion factors for viscosity are given in Appendix A.1. Sometimes the viscosity is given as μ/ρ, kinematic viscosity, in m^2/s or cm^2/s, where ρ is the density of the fluid.

EXAMPLE 2.4-1. Calculation of Shear Stress in a Liquid

Referring to Fig. 2.4-1, the distance between plates is $\Delta y = 0.5$ cm, $\Delta v_z = 10$ cm/s, and the fluid is ethyl alcohol at 273 K having a viscosity of 1.77 cp (0.0177 g/cm · s).

(a) Calculate the shear stress τ_{yz} and the velocity gradient or shear rate dv_z/dy using cgs units.
(b) Repeat, using lb force, s, and ft units (English units).
(c) Repeat, using SI units.

Solution: We can substitute directly into Eq. (2.4-1) or we can integrate Eq. (2.4-2). Using the latter method, rearranging Eq. (2.4-2), calling the bottom plate point 1, and integrating,

$$\tau_{yz} \int_{y_1=0}^{y_2=0.5} dy = -\mu \int_{v_1=10}^{v_2=0} dv_z \qquad \textbf{(2.4-4)}$$

$$\tau_{yz} = \mu \frac{v_1 - v_2}{y_2 - y_1} \qquad \textbf{(2.4-5)}$$

Substituting the known values,

$$\tau_{yz} = \mu \frac{v_1 - v_2}{y_2 - y_1} = \left(0.0177 \frac{\text{g}}{\text{cm} \cdot \text{s}} \right) \frac{(10-0) \text{ cm/s}}{(0.5-0) \text{ cm}}$$

$$= 0.354 \frac{\text{g} \cdot \text{cm/s}^2}{\text{cm}^2} = 0.354 \frac{\text{dyn}}{\text{cm}^2} \qquad \textbf{(2.4-6)}$$

To calculate the shear rate dv_z/dy, since the velocity change is linear with y,

$$\text{shear rate} = \frac{dv_z}{dy} = \frac{\Delta v_z}{\Delta y} = \frac{(10-0) \text{ cm/s}}{(0.5-0) \text{ cm}} = 20.0 \text{ s}^{-1} \qquad \textbf{(2.4-7)}$$

For part (b), using lb force units and the viscosity conversion factor from Appendix A.1,

$$\mu = 1.77 \text{ cp}(6.7197 \times 10^{-4} \text{ lb}_m/\text{ft} \cdot \text{s})/\text{cp}$$

$$= 1.77(6.7197 \times 10^{-4}) \text{ lb}_m/\text{ft} \cdot \text{s}$$

Integrating Eq. (2.4-3),

$$\tau_{yz} = \frac{\mu \; \text{lb}_\text{m}/\text{ft} \cdot \text{s}}{g_c \; \dfrac{\text{lb}_\text{m} \cdot \text{ft}}{\text{lb}_\text{f} \cdot \text{s}^2}} \frac{(v_1 - v_2)\text{ft/s}}{(y_2 - y_1) \; \text{ft}} \qquad\qquad \textbf{(2.4-8)}$$

Substituting known values into Eq. (2.4-8) and converting Δv_z to ft/s and Δy to ft, $\tau_{yz} = 7.39 \times 10^{-4} \; \text{lb}_\text{f}/\text{ft}^2$. Also, $dv_z/dy = 20 \; \text{s}^{-1}$.

For part (c), $\Delta y = 0.5/100 = 0.005$ m, $\Delta v_z = 10/100 = 0.1$ m/s, and $\mu = 1.77 \times 10^{-3}$ kg/m $\cdot$ s $= 1.77 \times 10^{-3}$ Pa $\cdot$ s. Substituting into Eq. (2.4-5),

$$\tau_{yz} = (1.77 \times 10^{-3})(0.10)/0.005 = 0.0354 \; \text{N/m}^2$$

The shear rate will be the same at $20.0 \, \text{s}^{-1}$.

2.4B Momentum Transfer in a Fluid

The shear stress τ_{yz} in Eqs. (2.4-1)-(2.4-3) can also be interpreted as a *flux of z-directed momentum in the y direction*, which is the rate of flow of momentum per unit area. The units of momentum are mass times velocity in kg $\cdot$ m/s. The shear stress can be written

$$\tau_{yz} = \frac{\text{kg} \cdot \text{m/s}}{\text{m}^2 \cdot \text{s}} = \frac{\text{momentum}}{\text{m}^2 \cdot \text{s}} \qquad\qquad \textbf{(2.4-9)}$$

This gives an amount of momentum transferred per second per unit area.

This can be shown by considering the interaction between two adjacent layers of a fluid in Fig. 2.4-1 which have different velocities, and hence different momentum, in the z direction. The random motions of the molecules in the faster-moving layer send some of the molecules into the slower-moving layer, where they collide with the slower-moving molecules and tend to speed them up or increase their momentum in the z direction. Also, in the same fashion, molecules in the slower layer tend to retard those in the faster layer. This exchange of molecules between layers produces a transfer or flux of z-directed momentum from high-velocity to low-velocity layers. The negative sign in Eq. (2.4-2) indicates that momentum is transferred down the gradient from high- to low-velocity regions. This is similar to the transfer of heat from high- to low-temperature regions.

2.4C Viscosities of Newtonian Fluids

Fluids that follow Newton's law of viscosity, Eqs. (2.4-1)–(2.4-3), are called *Newtonian fluids*. For a Newtonian fluid, there is a linear relation between the shear stress τ_{yz} and the velocity gradient dv_z/dy (rate of shear). This means that the viscosity μ is a constant and independent of the rate of shear. For non-Newtonian fluids, the relation between τ_{yz} and dv_z/dy is not linear; i.e., the viscosity μ does not remain constant but is a function of shear rate. Certain liquids do not obey this simple Newton's law. These are primarily pastes, slurries, high polymers, and emulsions. The science of the flow and deformation of fluids is often called *rheology*. A discussion of non-Newtonian fluids, will not be given here but will be included in Section 3.5.

The viscosity of gases, which are Newtonian fluids, increases with temperature and is approximately independent of pressure up to a pressure of about 1000 kPa. At higher pressures, the viscosity of gases increases with increase in pressure. For example, the viscosity of N_2 gas at 298 K approximately doubles in going from 100 kPa to about 5×10^4 kPa (R1). In liquids, the viscosity decreases with increasing temperature. Since liquids are essentially incompressible, the viscosity is not affected by pressure.

TABLE 2.4-1. *Viscosities of Some Gases and Liquids at 101.32 kPa Pressure*

	Gases				Liquids		
Substance	Temp., K	Viscosity $(Pa \cdot s)10^3$ or $(kg/m \cdot s)$ 10^3	Ref.	Substance	Temp., K	Viscosity $(Pa \cdot s)10^3$ or $(kg/m \cdot s)$ 10^3	Ref.
Air	293	0.01813	N1	Water	293	1.0019	S1
CO_2	273	0.01370	R1		373	0.2821	S1
	373	0.01828	R1	Benzene	278	0.826	R1
CH_4	293	0.01089	R1				
				Glycerol	293	1069	L1
SO_2	373	0.01630	R1	Hg	293	1.55	R2
				Olive oil	303	84	E1

In Table 2.4-1 some experimental viscosity data are given for some typical pure fluids at 101.32 kPa. The viscosities for gases are the lowest and do not differ markedly from gas to gas, being about 5×10^{-6} to 3×10^{-5} Pa·s. The viscosities for liquids are much greater. The value for water at 293 K is about 1×10^{-3} and for glycerol 1.069 Pa·s. Hence, there is a great difference between viscosities of liquids. More complete tables of viscosities are given for water in Appendix A.2, for inorganic and organic liquids and gases in Appendix A.3, and for biological and food liquids in Appendix A.4. Extensive data are available in other references (P1, R1, W1, L1). Methods of estimating viscosities of gases and liquids when experimental data are not available are summarized elsewhere (R1). These estimation methods for gases at pressures below 100 kPa are reasonably accurate, with an error within about $\pm 5\%$, but the methods for liquids are often quite inaccurate.

2.5 TYPES OF FLUID FLOW AND REYNOLDS NUMBER

2.5A Introduction and Types of Fluid Flow

The principles of the statics of fluids, treated in Section 2.2, are almost an exact science. On the other hand, the principles of the motions of fluids are quite complex. The basic relations describing the motions of a fluid are the equations for the overall balances of mass, energy, and momentum, which will be covered in the following sections.

These overall or macroscopic balances will be applied to a finite enclosure or control volume fixed in space. We use the term "overall" because we wish to describe these balances from outside the enclosure. The changes inside the enclosure are determined in terms of the properties of the streams entering and leaving and the exchanges of energy between the enclosure and its surroundings.

When making overall balances on mass, energy, and momentum we are not interested in the details of what occurs inside the enclosure. For example, in an overall balance average inlet and outlet velocities are considered. However, in a differential balance the velocity distribution inside an enclosure can be obtained with the use of Newton's law of viscosity.

In this section we first discuss the two types of fluid flow that can occur: laminar and turbulent flow. Also, the Reynolds number used to characterize the regimes of flow are considered. Then in Sections 2.6, 2.7, and 2.8 the overall mass balance, energy balance, and momentum balance are covered together with a number of applications. Finally, a

Transport Processes: Momentum, Heat, and Mass

discussion is given in Section 2.9 on the methods of making a shell balance on an element to obtain the velocity distribution in the element and pressure drop.

2.5B Laminar and Turbulent Flow

The type of flow occurring in a fluid in a channel is important in fluid dynamics problems. When fluids move through a closed channel of any cross section, either of two distinct types of flow can be observed according to the conditions present. These two types of flow can be commonly seen in a flowing open stream or river. When the velocity of flow is slow, the flow patterns are smooth. However, when the velocity is quite high, an unstable pattern is observed in which eddies or small packets of fluid particles are present moving in all directions and at all angles to the normal line of flow.

The first type of flow at low velocities where the layers of fluid seem to slide by one another without eddies or swirls being present is called *laminar flow* and Newton's law of viscosity holds, as discussed in Section 2.4A. The second type of flow at higher velocities where eddies are present giving the fluid a fluctuating nature is called *turbulent flow*.

The existence of laminar and turbulent flow is most easily visualized by the experiments of Reynolds. His experiments are shown in Fig. 2.5-1. Water was allowed to flow at steady state through a transparent pipe with the flow rate controlled by a valve at the end of the pipe. A fine steady stream of dye-colored water was introduced from a fine jet as shown and its flow pattern observed. At low rates of water flow, the dye pattern was regular and formed a single line or stream similar to a thread, as shown in Fig. 2.5-1a. There was no lateral mixing of the fluid, and it flowed in streamlines down the tube. By putting in additional jets at other points in the pipe cross section, it was shown that there was no mixing in any parts of the tube and the fluid flowed in straight parallel lines. This type of flow is called *laminar* or *viscous flow*.

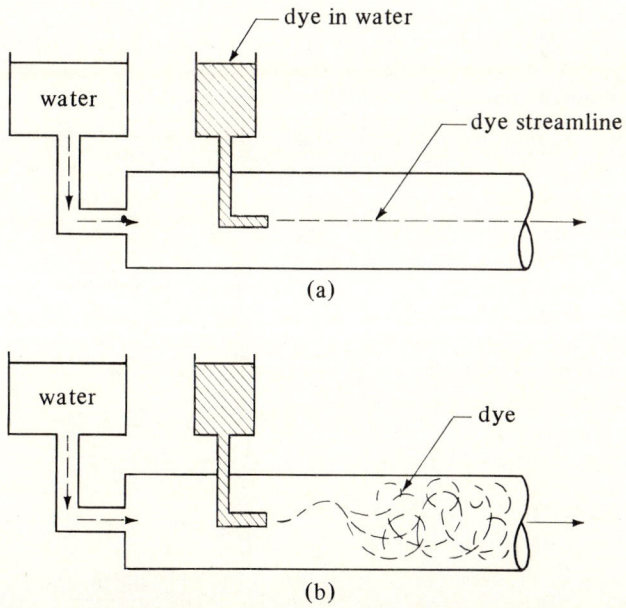

FIGURE 2.5-1.   *Reynolds' experiment for different types of flow: (a) laminar flow. (b) turbulent flow.*

As the velocity was increased, it was found that at a definite velocity the thread of dye became dispersed and the pattern was very erratic, as shown in Fig. 2.5-1b. This type of flow is known as turbulent flow. The velocity at which the flow changes is known as the *critical velocity*.

2.5C Reynolds Number

Studies have shown that the transition from laminar to turbulent flow in tubes is not only a function of velocity but also of density and viscosity of the fluid and the tube diameter. These variables are combined into the Reynolds number, which is dimensionless.

$$N_{Re} = \frac{Dv\rho}{\mu} \tag{2.5-1}$$

where N_{Re} is the Reynolds number, D the diameter in m, ρ the fluid density in kg/m^3, μ the fluid viscosity in Pa·s, and v the average velocity of the fluid in m/s (where average velocity is defined as the volumetric rate of flow divided by the cross-sectional area of the pipe). Units in the cgs system are D in cm, ρ in g/cm^3, μ in g/cm·s, and v in cm/s. In the English system D is in ft, ρ in lb_m/ft^3, μ in $lb_m/ft·s$, and v in ft/s.

The instability of the flow that leads to disturbed or turbulent flow is determined by the ratio of the kinetic or inertial forces to the viscous forces in the fluid stream. The inertial forces are proportional to ρv^2 and the viscous forces to $\mu v/D$, and the ratio $\rho v^2/(\mu v/D)$ is the Reynolds number $Dv\rho/\mu$. Further explanation and derivation of dimensionless numbers is given in Section 3.8.

For a straight circular pipe when the value of the Reynolds number is less than 2100, the flow is always laminar. When the value is over 4000, the flow will be turbulent, except in very special cases. In between, which is called the *transition region*, the flow can be viscous or turbulent, depending upon the apparatus details, which cannot be predicted.

EXAMPLE 2.5-1 Reynolds Number in a Pipe
Water at 303 K is flowing at the rate of 10 gal/min in a pipe having an inside diameter (ID) of 2.067 in. Calculate the Reynolds number using both English units and SI units.

Solution: From Appendix A.1, 7.481 gal = 1 ft^3. The flow rate is calculated as

$$\text{flow rate} = \left(10.0 \frac{gal}{min}\right)\left(\frac{1\ ft^3}{7.481\ gal}\right)\left(\frac{1\ min}{60\ s}\right) = 0.0223\ ft^3/s$$

$$\text{pipe diameter, } D = \frac{2.067}{12} = 0.172\ ft$$

$$\text{cross-sectional area of pipe} = \frac{\pi D^2}{4} = \frac{\pi(0.172)^2}{4} = 0.0233\ ft^2$$

$$\text{velocity in pipe, } v = \left(0.0223\ \frac{ft^3}{s}\right)\left(\frac{1}{0.0233\ ft^2}\right) = 0.957\ ft/s$$

From Appendix A.2 for water at 303 K (30°C),

$$\text{density, } \rho = 0.996(62.43)\ lb_m/ft^3$$

$$\text{viscosity, } \mu = (0.8007\ cp)\left(6.7197 \times 10^{-4} \frac{lb_m/ft·s}{cp}\right)$$

$$= 5.38 \times 10^{-4}\ lb_m/ft·s$$

Substituting into Eq. (2.5-1),

$$N_{Re} = \frac{Dv\rho}{\mu} = \frac{(0.172 \text{ ft})(0.957 \text{ ft/s})(0.996 \times 62.43 \text{ lb}_m/\text{ft}^3)}{5.38 \times 10^{-4} \text{ lb}_m/\text{ft} \cdot \text{s}}$$

$$= 1.905 \times 10^4$$

Hence, the flow is turbulent. Using SI units,

$$\rho = (0.996)(100 \text{ kg/m}^3) = 996 \text{ kg/m}^3$$

$$D = (2.067 \text{ in.})(1 \text{ ft/12 in.})(1 \text{ m/3.2808 ft}) = 0.0525 \text{ m}$$

$$v = \left(0.957 \frac{\text{ft}}{\text{s}}\right)(1 \text{ m/3.2808 ft}) = 0.2917 \text{ m/s}$$

$$\mu = (0.8007 \text{ cp})\left(1 \times 10^{-3} \frac{\text{kg/m} \cdot \text{s}}{\text{cp}}\right) = 8.007 \times 10^{-4} \frac{\text{kg}}{\text{m} \cdot \text{s}}$$

$$= 8.007 \times 10^{-4} \text{ Pa} \cdot \text{s}$$

$$N_{Re} = \frac{Dv\rho}{\mu} = \frac{(0.0525 \text{ m})(0.2917 \text{ m/s})(996 \text{ kg/m}^3)}{8.007 \times 10^{-4} \text{ kg/m} \cdot \text{s}} = 1.905 \times 10^4$$

2.6 OVERALL MASS BALANCE AND CONTINUITY EQUATION

2.6A Introduction and Simple Mass Balances

In fluid dynamics fluids are in motion. Generally, they are moved from place to place by means of mechanical devices such as pumps or blowers, by gravity head, or by pressure, and flow through systems of piping and/or process equipment. The first step in the solution of flow problems is generally to apply the principles of the conservation of mass to the whole system or to any part of the system. First, we will consider an elementary balance on a simple geometry, and later we shall derive the general mass-balance equation.

Simple mass or material balances were introduced in Section 1.5, where

$$\text{input} = \text{output} + \text{accumulation} \tag{1.5-1}$$

Since, in fluid flow, we are usually working with rates of flow and usually at steady state, the rate of accumulation is zero and we obtain

$$\text{rate of input} = \text{rate of output (steady state)} \tag{2.6-1}$$

In Fig. 2.6-1 a simple flow system is shown where fluid enters section 1 with an

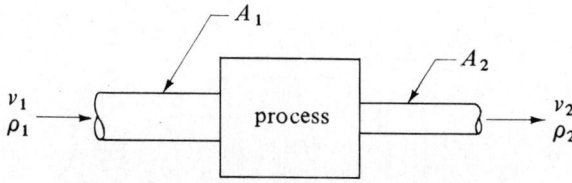

FIGURE 2.6-1. *Mass balance on flow system.*

average velocity v_1 m/s and density ρ_1 kg/m³. The cross-sectional area is A_1 m². The fluid leaves section 2 with average velocity v_2. The mass balance, Eq. (2.6-1), becomes

$$m = \rho_1 A_1 v_1 = \rho_2 A_2 v_2 \qquad (2.6\text{-}2)$$

where $m =$ kg/s. Often, $v\rho$ is expressed as $G = v\rho$, where G is mass velocity or mass flux in kg/s·m². In English units, v is in ft/s, ρ in lb_m/ft^3, A in ft², m in lb_m/s, and G in $lb_m/s \cdot ft^2$.

EXAMPLE 2.6-1. *Flow of Crude Oil and Mass Balance*

A petroleum crude oil having a density of 892 kg/m³ is flowing through the piping arrangement shown in Fig. 2.6-2 at a total rate of 1.388×10^{-3} m³/s entering pipe 1.

The flow divides equally in each of pipes 3. The steel pipes are schedule 40 pipe (see Appendix A.5 for actual dimensions). Calculate the following using SI units.

(a) The total mass flow rate m in pipe 1 and pipes 3.
(b) The average velocity v in 1 and 3.
(c) The mass velocity G in 1.

Solution: From Appendix A.5, the dimensions of the pipes are as follows: 2-in. pipe: D_1 (ID) = 2.067 in., cross-sectional area

$$A_1 = 0.02330 \text{ ft}^2 = 0.02330(0.0929) = 2.165 \times 10^{-3} \text{ m}^2$$

$1\frac{1}{2}$-in. pipe: D_3 (ID) = 1.610 in., cross-sectional area

$$A_3 = 0.01414 \text{ ft}^2 = 0.01414(0.0929) = 1.313 \times 10^{-3} \text{ m}^2$$

The total mass flow rate is the same through pipes 1 and 2 and is

$$m_1 = (1.388 \times 10^{-3} \text{ m}^3/\text{s})(892 \text{ kg/m}^3) = 1.238 \text{ kg/s}$$

Since the flow divides equally in each of pipes 3,

$$m_3 = \frac{m_1}{2} = \frac{1.238}{2} = 0.619 \text{ kg/s}$$

For part (b), using Eq. (2.6-2) and solving for v,

$$v_1 = \frac{m_1}{\rho_1 A_1} = \frac{1.238 \text{ kg/s}}{(892 \text{ kg/m}^3)(2.165 \times 10^{-3} \text{ m}^2)} = 0.641 \text{ m/s}$$

$$v_3 = \frac{m_3}{\rho_3 A_3} = \frac{0.619}{(892)(1.313 \times 10^{-3})} = 0.528 \text{ m/s}$$

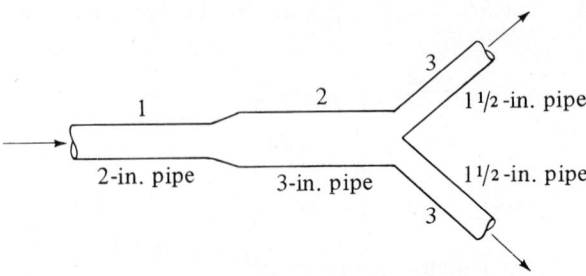

FIGURE 2.6-2. *Piping arrangement for Example 2.6-1.*

Transport Processes: Momentum, Heat, and Mass

For part (c),

$$G_1 = v_1 \rho_1 = \frac{m_1}{A_1} = \frac{1.238}{2.165 \times 10^{-3}} = 572 \frac{\text{kg}}{\text{s} \cdot \text{m}^2}$$

2.6B Control Volume for Balances

The laws for the conservation of mass, energy, and momentum are all stated in terms of a system, and these laws give the interaction of a system with its surroundings. A *system* is defined as a collection of fluid of fixed identity. However, in flow of fluids, individual particles are not easily identifiable. As a result, attention is focused on a given space through which the fluid flows rather than to a given mass of fluid. The method used, which is more convenient, is to select a control volume, which is a region fixed in space through which the fluid flows.

In Fig. 2.6-3 the case of a fluid flowing through a conduit is shown. The control surface shown as a dashed line is the surface surrounding the control volume. In most problems part of the control surface will coincide with some boundary, such as the wall of the conduit. The remaining part of the control surface is a hypothetical surface through which the fluid can flow, shown as point 1 and point 2 in Fig. 2.6-3. The control-volume representation is analogous to the open system of thermodynamics.

2.6C Overall Mass-Balance Equation

In deriving the general equation for the overall balance of the property mass, the law of conservation of mass may be stated as follows for a control volume where no mass is being generated.

$$\begin{pmatrix} \text{rate of mass output} \\ \text{from control volume} \end{pmatrix} - \begin{pmatrix} \text{rate of mass input} \\ \text{from control volume} \end{pmatrix}$$

$$+ \begin{pmatrix} \text{rate of mass accumulation} \\ \text{in control volume} \end{pmatrix} = 0 \text{ (rate of mass generation)}$$

(2.6-3)

We now consider the general control volume fixed in space and located in a fluid flow field, as shown in Fig. 2.6-4. For a small element of area dA m^2 on the control surface, the rate of mass efflux from this element $= (\rho v)(dA \cos \alpha)$, where $(dA \cos \alpha)$ is the area dA projected in a direction normal to the velocity vector $\mathbf{v}$, α is the angle between the velocity vector $\mathbf{v}$ and the outward-directed unit normal vector $\mathbf{n}$ to dA, and ρ is the

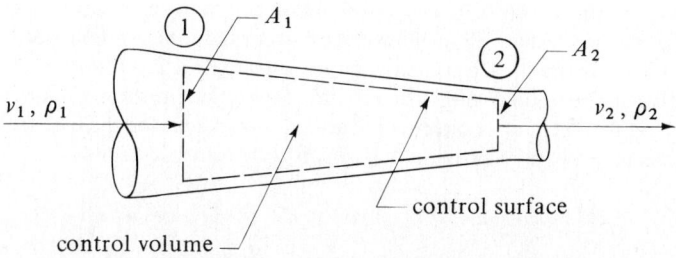

FIGURE 2.6-3. *Control volume for flow through a conduit.*

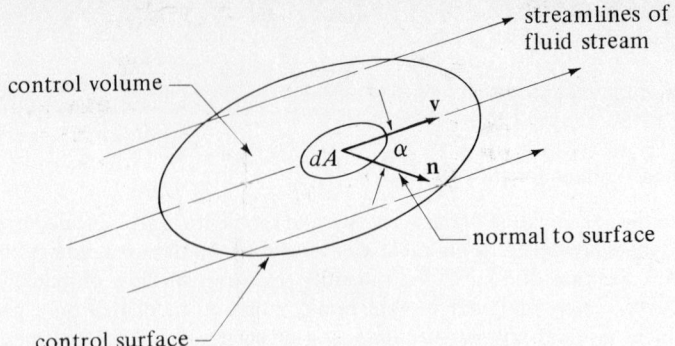

FIGURE 2.6-4. *Flow through a differential area dA on a control surface.*

density in kg/m³. The quantity ρv has units of kg/s·m² and is called *a flux or mass velocity G*.

From vector algebra we recognize that $(\rho v)(dA \cos \alpha)$ is the scalar or dot product $\rho(\mathbf{v} \cdot \mathbf{n}) \, dA$. If we now integrate this quantity over the entire control surface A we have the net outflow of mass across the control surface, or the net mass efflux in kg/s from the entire control volume V:

$$\binom{\text{net mass efflux}}{\text{from control volume}} = \iint_A v\rho \cos \alpha \, dA = \iint_A \rho(\mathbf{v} \cdot \mathbf{n}) \, dA \qquad \textbf{(2.6-4)}$$

We should note that if mass is entering the control volume, i.e., flowing inward across the control surface, the net efflux of mass in Eq. (2.6-4) is negative since $\alpha > 90°$ and cos α is negative. Hence, there is a net influx of mass. If $\alpha < 90°$, there is a net efflux of mass.

The rate of accumulation of mass within the control volume V can be expressed as follows.

$$\binom{\text{rate of mass accumulation}}{\text{in control volume}} = \frac{\partial}{\partial t} \iiint_V \rho \, dV = \frac{dM}{dt} \qquad \textbf{(2.6-5)}$$

where M is the mass of fluid in the volume in kg. Substituting Eqs. (2.6-4) and (2.6-5) into (2.6-3), we obtain the general form of the overall mass balance.

$$\iint_A \rho(\mathbf{v} \cdot \mathbf{n}) \, dA + \frac{\partial}{\partial t} \iiint_V \rho \, dV = 0 \qquad \textbf{(2.6-6)}$$

The use of Eq. (2.6-6) can be shown for a common situation for steady-state one-dimensional flow, where all the flow inward is normal to A_1 and outward normal to A_2, as shown in Fig. 2.6-3. When the velocity v_2 leaving (Fig. 2.6-3) is normal to A_2, the angle α_2 between the normal to the control surface and the direction of the velocity is 0° and cos $\alpha_2 = 1.0$. Where v_1 is directed inward, $\alpha_1 > \pi/2$, and for the case in Fig. 2.6-3, α_1 is 180° (cos $\alpha_1 = -1.0$). Since α_2 is 0° and α_1 is 180°, using Eq. (2.6-4),

$$\iint_A v\rho \cos \alpha \, dA = \iint_{A_2} v\rho \cos \alpha_2 \, dA + \iint_{A_1} v\rho \cos \alpha_1 \, dA$$

$$= v_2 \rho_2 A_2 - v_1 \rho_1 A_1 \qquad \textbf{(2.6-7)}$$

Transport Processes: Momentum, Heat, and Mass

For steady state, $dM/dt = 0$ in Eq. (2.6-5), and Eq. (2.6-6) becomes

$$m = \rho_1 v_1 A_1 = \rho_2 v_2 A_2 \qquad \text{(2.6-2)}$$

which is Eq. (2.6-2), derived earlier.

In Fig. 2.6-3 and Eqs. (2.6-3)–(2.6-7) we were not concerned with the composition of any of the streams. These equations can easily be extended to represent an overall mass balance for component i in a multicomponent system. For the case shown in Fig. 2.6-3 we combine Eqs. (2.6-5), (2.6-6), and (2.6-7), add a generation term, and obtain

$$m_{i2} - m_{i1} + \frac{dM_i}{dt} = R_i \qquad \text{(2.6-8)}$$

where m_{i2} is the mass flow rate of component i leaving the control volume and R_i is the rate of generation of component i in the control volume in kg per unit time. (Diffusion fluxes are neglected here or are assumed negligible.) In some cases, of course, $R_i = 0$ for no generation. Often it is more convenient to use Eq. (2.6-8) written in molar units.

EXAMPLE 2.6-2. Overall Mass Balance in Stirred Tank
Initially, a tank contains 500 kg of salt solution containing 10% salt. At point (1) in the control volume in Fig. 2.6-5, a stream enters at a constant flow rate of 10 kg/h containing 20% salt. A stream leaves at point (2) at a constant rate of 5 kg/h. The tank is well stirred. Derive an equation relating the weight fraction w_A of the salt in the tank at any time t in hours.

Solution: First we make a total mass balance using Eq. (2.6-7) for the net total mass efflux from the control volume.

$$\iint_A v\rho \cos \alpha \, dA = m_2 - m_1 = 5 - 10 = -5 \text{ kg solution/h} \quad \text{(2.6-9)}$$

From Eq. (2.6-5), where M is total kg of solution in control volume at time t,

$$\frac{\partial}{\partial t} \iiint_V \rho \, dV = \frac{dM}{dt} \qquad \text{(2.6-5)}$$

Substituting Eqs. (2.6-5) and (2.6-9) into (2.6-6), and then integrating,

$$-5 + \frac{dM}{dt} = 0$$

$$\int_{M=500}^{M} dM = 5 \int_{t=0}^{t} dt \qquad \text{(2.6-10)}$$

$$M = 5t + 500 \qquad \text{(2.6-11)}$$

Eq. (2.6-11) relates the total mass M in the tank at any time t.

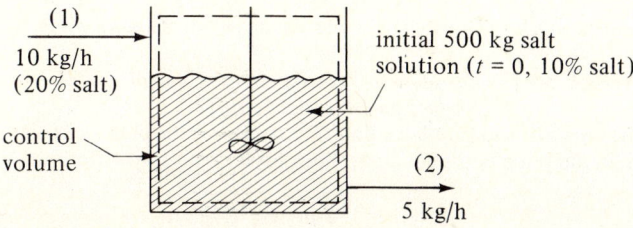

FIGURE 2.6-5. *Control volume for flow in a stirred tank for Example 2.6-2.*

Next, making a component A salt balance, let w_A = weight fraction of salt in tank at time t and also the concentration in the stream m_2 leaving at time t. Again using Eq. (2.6-7) but for a salt balance,

$$\iint_A v\rho \cos \alpha \, dA = (5)w_A - 10(0.20) = 5w_A - 2 \text{ kg salt/h} \quad (2.6\text{-}12)$$

Using Eq. (2.6-5) for a salt balance,

$$\frac{\partial}{\partial t} \iiint_V \rho \, dV = \frac{d}{dt}(Mw_A) = \frac{M \, dw_A}{dt} + w_A \frac{dM}{dt} \text{ kg salt/h} \quad (2.6\text{-}13)$$

Substituting Eqs. (2.6-12) and (2.6-13) into (2.6-6),

$$5w_A - 2 + M \frac{dw_A}{dt} + w_A \frac{dM}{dt} = 0 \quad (2.6\text{-}14)$$

Substituting the value for M from Eq. (2.6-11) into (2.6-14), separating variables, integrating, and solving for w_A,

$$5w_A - 2 + (500 + 5t) \frac{dw_A}{dt} + w_A \frac{d(500 + 5t)}{dt} = 0$$

$$5w_A - 2 + (500 + 5t) \frac{dw_A}{dt} + 5w_A = 0$$

$$\int_{w_A = 0.10}^{w_A} \frac{dw_A}{2 - 10w_A} = \int_{t=0}^{t} \frac{dt}{500 + 5t}$$

$$-\frac{1}{10} \ln \left(\frac{2 - 10w_A}{1} \right) = \frac{1}{5} \ln \left(\frac{500 + 5t}{500} \right) \quad (2.6\text{-}15)$$

$$w_A = -0.1 \left(\frac{100}{100 + t} \right)^2 + 0.20$$

$$(2.6\text{-}16)$$

Note that Eq. (2.6-8) for component i could have been used for the salt balance with $R_i = 0$ (no generation).

2.6D Average Velocity to Use in Overall Mass Balance

In solving the case in Eq. (2.6-7) we assumed a constant velocity v_1 at section 1 and constant v_2 at section 2. If the velocity is not constant but varies across the surface area, an average or bulk velocity is defined by

$$v_{av} = \frac{1}{A} \iint_A v \, dA \quad (2.6\text{-}17)$$

for a surface over which v is normal to A and the density ρ is assumed constant.

EXAMPLE 2.6-3. *Variation of Velocity Across Control Surface and Average Velocity*
For the case of imcompressible flow (ρ is constant) through a circular pipe of radius R, the velocity profile is parabolic for laminar flow as follows:

$$v = v_{max} \left[1 - \left(\frac{r}{R} \right)^2 \right] \quad (2.6\text{-}18)$$

where v_{max} is the maximum velocity at the center where $r = 0$ and v is the velocity at a radial distance r from the center. Derive an expression for the average or bulk velocity v_{av} to use in the overall mass-balance equation.

Solution: The average velocity is represented by Eq. (2.6-17). In Cartesian coordinates dA is $dx\,dy$. However, using polar coordinates which are more appropriate for a pipe, $dA = r\,dr\,d\theta$, where θ is the angle in polar coordinates. Substituting Eq. (2.6-18), $dA = r\,dr\,d\theta$, and $A = \pi R^2$ into Eq. (2.6-17) and integrating,

$$v_{av} = \frac{1}{\pi R^2} \int_0^{2\pi} \int_0^R v_{max} \left[1 - \left(\frac{r}{R} \right)^2 \right] r\,dr\,d\theta$$

$$= \frac{v_{max}}{\pi R^4} \int_0^{2\pi} \int_0^R (R^2 - r^2) r\,dr\,d\theta$$

$$= \frac{v_{max}}{\pi R^4} (2\pi - 0) \left(\frac{R^4}{2} - \frac{R^4}{4} \right) \tag{2.6-19}$$

$$v_{av} = \frac{v_{max}}{2} \tag{2.6-20}$$

In this discussion overall or macroscopic mass balances were made because we wish to describe these balances from outside the enclosure. In this section on overall mass balances, some of the equations presented may have seemed quite obvious. However, the purpose was to develop the methods which should be helpful in the next sections. Overall balances will also be made on energy and momentum in the next sections. These overall balances do not tell us the details of what happens inside. However, in Section 2.9 a shell momentum balance will be made to obtain these details, which will give us the velocity distribution and pressure drop. To further study these details of the processes occurring inside the enclosure, differential balances rather than shell balances can be written and these are discussed in selected topic Sections 3.6 and 3.7 on differential equations of continuity and momentum transfer, Sections 5.6 and 5.7 on differential equations of energy change and boundary-layer flow, and Section 7.5B on differential equations of continuity for a binary mixture.

2.7 OVERALL ENERGY BALANCE

2.7A Introduction

The second property to be considered in the overall balances on a control volume is energy. We shall apply the principle of the conservation of energy to a control volume fixed in space in much the same manner as the principle of conservation of mass was used to obtain the overall mass balance. The energy-conservation equation will then be combined with the first law of thermodynamics to obtain the final overall energy-balance equation.

We can write the first law of thermodynamics as

$$\Delta E = Q - W \tag{2.7-1}$$

where E is the total energy per unit mass of fluid, Q is the heat *absorbed* per unit mass of fluid, and W is the work of all kinds done per unit mass of fluid *upon* the surroundings.

In the calculations, each term in the equation must be expressed in the same units, such as J/kg (SI), btu/lb$_m$, or ft · lb$_f$/lb$_m$ (English).

Since mass carries with it associated energy because of its position, motion, or physical state, we will find that each of these types of energy will appear in the energy balance. In addition, we can also transport energy across the boundary of the system without transferring mass.

2.7B Derivation of Overall Energy-Balance Equation

The entity balance for a conserved quantity such as energy is similar to Eq. (2.6-3) and is as follows for a control volume.

$$\text{rate of entity output} - \text{rate of entity input}$$
$$+ \text{rate of entity accumulation} = 0 \qquad \textbf{(2.7-2)}$$

The energy E present within a system can be classified in three ways.

1. *Potential energy* zg of a unit mass of fluid is the energy present because of the position of the mass in a gravitational field g, where z is the relative height in meters from a reference plane. The units for zg for the SI system are m · m/s^2. Multiplying and dividing by kg mass, the units can be expressed as (kg · m/s^2) · (m/kg), or J/kg. In English units the potential energy is zg/g_c in ft · lb$_f$/lb$_m$.
2. *Kinetic energy* $v^2/2$ of a unit mass of fluid is the energy present because of translational or rotational motion of the mass, where v is the velocity in m/s relative to the boundary of the system at a given point. Again in the SI system the units of $v^2/2$ are J/kg. In the English system the kinetic energy is $v^2/2g_c$ in ft · lb$_f$/lb$_m$.
3. *Internal energy* U of a unit mass of a fluid is all of the other energy present, such as rotational and vibrational energy in chemical bonds. Again the units are in J/kg or ft · lb$_f$/lb$_m$.

The total energy of the fluid per unit mass is then

$$E = U + \frac{v^2}{2} + zg \qquad \text{(SI)}$$

$$\qquad\qquad\qquad\qquad\qquad\qquad\qquad\qquad \textbf{(2.7-3)}$$

$$E = U + \frac{v^2}{2g_c} + \frac{zg}{g_c} \qquad \text{(English)}$$

The rate of accumulation of energy within the control volume V in Fig. 2.6-4 is

$$\begin{pmatrix} \text{rate of energy accumulation} \\ \text{in control volume} \end{pmatrix} = \frac{\partial}{\partial t} \iiint_V \left(U + \frac{v^2}{2} + zg \right) \rho \, dV \qquad \textbf{(2.7-4)}$$

Next we consider the rate of energy input and output associated with mass in the control volume. The mass added or removed from the system carries internal, kinetic, and potential energy. In addition, energy is transferred when mass flows into and out of the control volume. Net work is done by the fluid as it flows into and out of the control volume. This pressure–volume work per unit mass fluid is pV. The contribution of shear work is usually neglected. The pV term and U term are combined using the definition of enthalpy, H.

$$H = U + pV \qquad \textbf{(2.7-5)}$$

Hence, the total energy carried with a unit mass is $(H + v^2/2 + zg)$.

For a small area dA on the control surface in Fig. 2.6-4, the rate of energy efflux is $(H + v^2/2 + zg)(\rho v)(dA \cos \alpha)$, where $(dA \cos \alpha)$ is the area dA projected in a direction normal to the velocity vector $\mathbf{v}$ and α is the angle between the velocity vector $\mathbf{v}$ and the outward-directed unit normal vector $\mathbf{n}$. We now integrate this quantity over the entire control surface to obtain

$$\binom{\text{net energy efflux}}{\text{from control volume}} = \iint\limits_{A} \left(H + \frac{v^2}{2} + zg \right)(\rho v) \cos \alpha \, dA \qquad (2.7\text{-}6)$$

Now we have accounted for all energy associated with mass in the system and moving across the boundary in the entity balance, Eq. (2.7-2). Next we take into account heat and work energy which transfers across the boundary and is not associated with mass. The term q is the heat transferred per unit time across the boundary to the fluid because of a temperature gradient. Heat absorbed by the system is positive by convention.

The work $\dot{W}$, which is energy per unit time, can be divided into $\dot{W}_S$, purely mechanical shaft work identified with a rotating shaft crossing the control surface, and the pressure–volume work, which has been included in the enthalpy term H in Eq. (2.7-6). By convention, work done by the fluid upon the surroundings, i.e., work out of the system, is positive.

To obtain the overall energy balance, we substitute Eqs. (2.7-4) and (2.7-6) into the entity balance Eq. (2.7-2) and equate the resulting equation to $q - \dot{W}_S$.

$$\iint\limits_{A} \left(H + \frac{v^2}{2} + zg \right)(\rho v) \cos \alpha \, dA + \frac{\partial}{\partial t} \iiint\limits_{V} \left(U + \frac{v^2}{2} + zg \right) \rho \, dV = q - \dot{W}_S \qquad (2.7\text{-}7)$$

2.7C Overall Energy Balance for Steady-State Flow System

A common special case of the overall or macroscopic energy balance is that of a steady-state system with one-dimensional flow across the boundaries, a single inlet, a single outlet, and negligible variation of height z, density ρ, and enthalpy H across either inlet or outlet area. This is shown in Fig. 2.7-1. Setting the accumulation term in Eq. (2.7-7) equal to zero and integrating.

$$H_2 m_2 - H_1 m_1 + \frac{m_2 (v_2^3)_{av}}{2 v_{2\,av}} - \frac{m_1 (v_1^3)_{av}}{2 v_{1\,av}} + g m_2 z_2 - g m_1 z_1 = q - \dot{W}_S \qquad (2.7\text{-}8)$$

For steady state, $m_1 = \rho_1 v_{1\,av} A_1 = m_2 = m$. Dividing through by m so that the equation is on a unit mass basis,

$$H_2 - H_1 + \frac{1}{2} \left[\frac{(v_2^3)_{av}}{v_{2\,av}} - \frac{(v_1^3)_{av}}{v_{1\,av}} \right] + g(z_2 - z_1) = Q - W_S \qquad \text{(SI)} \qquad (2.7\text{-}9)$$

The term $(v^3)_{av}/(2v_{av})$ can be replaced by $v_{av}^2/2\alpha$, where α is the kinetic-energy velocity correction factor and is equal to $v_{av}^3/(v^3)_{av}$. The term α has been evaluated for various flows in pipes and is $\frac{1}{2}$ for laminar flow and close to 1.0 for turbulent flow. (See Section 2.7D.) Hence, Eq. (2.7-9) becomes

$$H_2 - H_1 + \frac{1}{2\alpha} (v_{2\,av}^2 - v_{1\,av}^2) + g(z_2 - z_1) = Q - W_S \qquad \text{(SI)}$$

$$(2.7\text{-}10)$$

$$H_2 - H_1 + \frac{1}{2\alpha g_c} (v_{2\,av}^2 - v_{1\,av}^2) + \frac{g}{g_c} (z_2 - z_1) = Q - W_S \qquad \text{(English)}$$

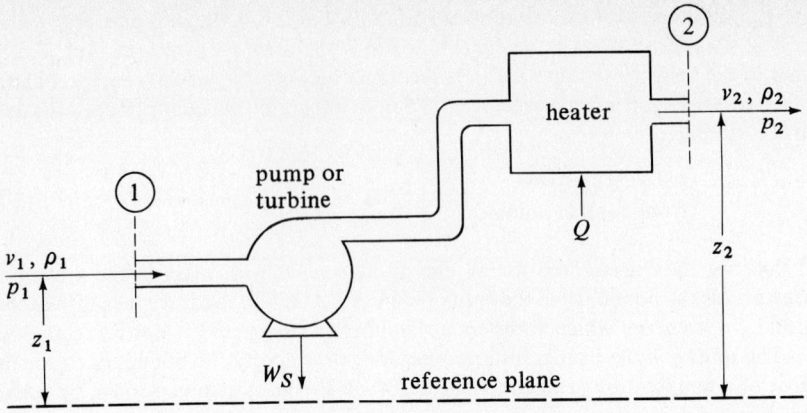

FIGURE 2.7-1.   *Steady-state flow system for a fluid.*

Some useful conversion factors to be used are as follows from Appendix A.1:

$$1 \text{ btu} = 778.17 \text{ ft} \cdot \text{lb}_f = 1055.06 \text{ J} = 1.05506 \text{ kJ}$$

$$1 \text{ hp} = 550 \text{ ft} \cdot \text{lb}_f/\text{s} = 0.7457 \text{ kW}$$

$$1 \text{ ft} \cdot \text{lb}_f/\text{lb}_m = 2.9890 \text{ J/kg}$$

$$1 \text{ J} = 1 \text{ N} \cdot \text{m} = 1 \text{ kg} \cdot \text{m}^2/\text{s}^2$$

2.7D   Kinetic-Energy Velocity Correction Factor α

1. Introduction.   In obtaining Eq. (2.7-8) it was necessary to integrate the kinetic-energy term,

$$\text{kinetic energy} = \iint_A \left(\frac{v^2}{2}\right)(\rho v) \cos \alpha \, dA \tag{2.7-11}$$

which appeared in Eq. (2.7-7). To do this we first take ρ as a constant and $\cos \alpha = 1.0$. Then multiplying the numerator and denominator by $v_{av} A$, where v_{av} is the bulk or average velocity and noting that $m = \rho v_{av} A$, Eq. (2.7-11) becomes

$$\frac{\rho}{2} \iint_A (v^3) \, dA = \frac{\rho v_{av} A}{2 v_{av} A} \iint_A (v^3) \, dA = \frac{m}{2 v_{av}} \frac{1}{A} \iint_A (v^3) \, dA \tag{2.7-12}$$

Dividing through by m so that Eq. (2.7-12) is on a unit mass basis,

$$\left(\frac{1}{2 v_{av}}\right) \frac{1}{A} \iint_A (v^3) \, dA = \frac{(v^3)_{av}}{2 v_{av}} = \frac{v^2_{av}}{2\alpha} \tag{2.7-13}$$

where α is defined as

$$\alpha = \frac{v^3_{av}}{(v^3)_{av}} \tag{2.7-14}$$

Transport Processes: Momentum, Heat, and Mass

and $(v^3)_{av}$ is defined as follows:

$$(v^3)_{av} = \frac{1}{A} \iint_A (v^3) \, dA \qquad (2.7\text{-}15)$$

The local velocity v varies across the cross-sectional area of a pipe. To evaluate $(v^3)_{av}$ and, hence, the value of α, we must have an equation relating v as a function of position in the cross-sectional area.

2. Laminar flow. In order to determine the value of α for laminar flow, we first combine Eqs. (2.6-18) and (2.6-20) for laminar flow to obtain v as a function of position r.

$$v = 2v_{av}\left[1 - \left(\frac{r}{R}\right)^2\right] \qquad (2.7\text{-}16)$$

Substituting Eq. (2.7-16) into (2.7-15) and noting that $A = \pi R^2$ and $dA = r \, dr \, d\theta$ (see Example 2.6-3), Eq. (2.7-15) becomes

$$(v^3)_{av} = \frac{1}{\pi R^2} \int_0^{2\pi} \int_0^R \left[2v_{av}\left(1 - \frac{r^2}{R^2}\right)\right]^3 r \, dr \, d\theta$$

$$= \frac{(2\pi)2^3 v_{av}^3}{\pi R^2} \int_0^R \frac{(R^2 - r^2)^3}{R^6} r \, dr = \frac{16v_{av}^3}{R^8} \int_0^R (R^2 - r^2)^3 r \, dr \qquad (2.7\text{-}17)$$

Integrating Eq. (2.7-17) and rearranging,

$$(v^3)_{av} = \frac{16v_{av}^3}{R^8} \int_0^R (R^6 - 3r^2 R^4 + 3r^4 R^2 - r^6) r \, dr$$

$$= \frac{16v_{av}^3}{R^8}\left(\frac{R^8}{2} - \frac{3}{4} R^8 + \frac{1}{2} R^8 - \frac{1}{8} R^8\right)$$

$$= 2v_{av}^3 \qquad (2.7\text{-}18)$$

Substituting Eq. (2.7-18) into (2.7-14),

$$\alpha = \frac{v_{av}^3}{(v^3)_{av}} = \frac{v_{av}^3}{2v_{av}^3} = 0.50 \qquad (2.7\text{-}19)$$

Hence, for laminar flow the value of α to use in the kinetic-energy term of Eq. (2.7-10) is 0.50.

3. Turbulent flow. For turbulent flow a relationship is needed between v and position. This can be approximated by the following expression:

$$v = v_{max}\left(\frac{R - r}{R}\right)^{1/7} \qquad (2.7\text{-}20)$$

where r is the radial distance from the center. This Eq. (2.7-20) is substituted into Eq. (2.7-15) and the resultant integrated to obtain the value of $(v^3)_{av}$. Next, Eq. (2.7-20) is substituted into Eq. (2.6-17) and this equation integrated to obtain v_{av} and $(v_{av})^3$. Combining the results for $(v^3)_{av}$ and $(v_{av})^3$ into Eq. (2.7-14), the value of α is 0.945. (See Problem 2.7-1 for solution.) The value of α for turbulent flow varies from about 0.90 to 0.99. In most cases (except for precise work) the value of α is taken to be 1.0.

2.7E Applications of Overall Energy-Balance Equation

The total energy balance, Eq. (2.7-10), in the form given is not often used when appreciable enthalpy changes occur or appreciable heat is added (or subtracted) since the kinetic- and potential-energy terms are usually small and can be neglected. As a result, when appreciable heat is added or subtracted or large enthalpy changes occur, the methods of doing heat balances described in Section 1.7 are generally used. Examples will be given to illustrate this and other cases.

EXAMPLE 2.7-1. Energy Balance on Steam Boiler
Water enters a boiler at 18.33°C and 137.9 kPa through a pipe at an average velocity of 1.52 m/s. Exit steam at a height of 15.2 m above the liquid inlet leaves at 137.9 kPa, 148.9°C, and 9.14 m/s in the outlet line. At steady state how much heat must be added per kg mass of steam? The flow in the two pipes is turbulent.

Solution: The process flow diagram is shown in Fig. 2.7-2. Rearranging Eq. (2.7-10) and sitting $\alpha = 1$ for turbulent flow and $W_S = 0$ (no external work),

$$Q = (z_2 - z_1)g + \frac{v_2^2 - v_1^2}{2} + (H_2 - H_1) \qquad (2.7\text{–}21)$$

To solve for the kinetic-energy terms,

$$\frac{v_1^2}{2} = \frac{(1.52)^2}{2} = 1.115 \text{ J/kg}$$

$$\frac{v_2^2}{2} = \frac{(9.14)^2}{2} = 41.77 \text{ J/kg}$$

Taking the datum height z_1 at point $1, z_2 = 15.2$ m. Then,

$$z_2 g = (15.2)(9.80665) = 149.1 \text{ J/kg}$$

From Appendix A.2, steam tables in SI units, H_1 at 18.33°C = 76.97 kJ/kg, H_2 of superheated steam at 148.9°C = 2771.4 kJ/kg, and

$$H_2 - H_1 = 2771.4 - 76.97 = 2694.4 \text{ kJ/kg} = 2.694 \times 10^6 \text{ J/kg}$$

Substituting these values into Eq. (2.7-21),

$$Q = (149.1 - 0) + (41.77 - 1.115) + 2.694 \times 10^6$$

$$Q = 189.75 + 2.694 \times 10^6 = 2.6942 \times 10^6 \text{ J/kg}$$

Hence, the kinetic-energy and potential-energy terms totaling 189.75 J/kg are negligible compared to the enthalpy change of 2.694×10^6 J/kg. This 189.75 J/kg would raise the temperature of liquid water about 0.0453°C, a negligible amount.

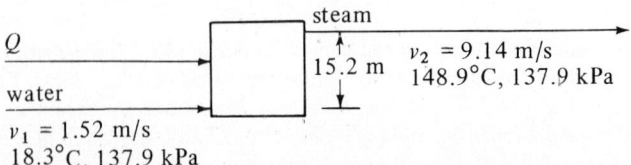

FIGURE 2.7-2. *Process flow diagram for Example 2.7-1.*

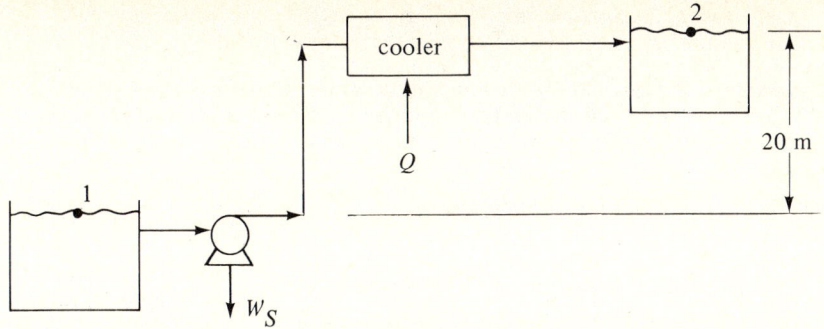

FIGURE 2.7-3. *Process flow diagram for energy balance for Example 2.7-2.*

EXAMPLE 2.7–2. Energy Balance on a Flow System with a Pump

Water at 85.0°C is being stored in a large, insulated tank at atmospheric pressure as shown in Fig. 2.7-3. It is being pumped at steady state from this tank at point 1 by a pump at the rate of 0.567 m³/min. The motor driving the pump supplies energy at the rate of 7.45 kW. The water passes through a heat exchanger, where it gives up 1408 kW of heat. The cooled water is then delivered to a second, large open tank at point 2, which is 20 m above the first tank. Calculate the final temperature of the water delivered to the second tank. Neglect any kinetic-energy changes since the initial and final velocities in the tanks are essentially zero.

Solution: From Appendix A.2, steam tables, H_1 (85°C) = 355.90 × 10³ J/kg, $\rho_1 = 1/0.0010325 = 968.5$ kg/m³. Then, for steady state,

$$m_1 = m_2 = (0.567)(968.5)(\tfrac{1}{60}) = 9.152 \text{ kg/s}$$

Also, $z_1 = 0$ and $z_2 = 20$ m. The work done by the fluid is W_S, but in this case work is done on the fluid and W_S is negative.

$$W_S = -(7.45 \times 10^3 \text{ J/s})(1/9.152 \text{ kg/s}) = -0.8140 \times 10^3 \text{ J/kg}$$

The heat added to the fluid is also negative since it gives up heat and is

$$Q = -(1408 \times 10^3 \text{ J/s})(1/9.152 \text{ kg/s}) = -153.8 \times 10^3 \text{ J/kg}$$

Setting $(v_1^2 - v_2^2)/2 = 0$ and substituting into Eq. (2.7-10),

$$H_2 - 355.90 \times 10^3 + 0 + 9.80665(20 - 0)$$
$$= (-153.8 \times 10^3) - (-0.814 \times 10^3)$$

Solving, $H_2 = 202.71 \times 10^3$ J/kg. From the steam tables this corresponds to $t_2 = 48.41$°C. Note that in this example, W_S and $g(z_2 - z_1)$ are very small compared to Q.

EXAMPLE 2.7–3. Energy Balance in Flow Calorimeter

A flow calorimeter is being used to measure the enthalpy of steam. The calorimeter, which is a horizontal insulated pipe, consists of an electric heater immersed in a fluid flowing at steady state. Liquid water at 0°C at a rate of 0.3964 kg/min enters the calorimeter at point 1. The liquid is vaporized completely by the heater, where 19.63 kW is added and steam leaves point 2 at 250°C and 150 kPa absolute. Calculate the exit enthalpy H_2 of the steam if the liquid enthalpy at 0°C is set arbitrarily as 0. The kinetic-energy changes are small and can be neglected. (It will be assumed that pressure has a negligible effect on the enthalpy of the liquid.)

Solution: For this case, $W_S = 0$ since there is no shaft work between points 1 and 2. Also, $(v_2^2/2\alpha - v_1^2/2\alpha) = 0$ and $g(z_2 - z_1) = 0$. For steady state, $m_1 = m_2 = 0.3964/60 = 6.607 \times 10^{-3}$ kg/s. Since heat is added to the system,

$$Q = +\frac{19.63 \text{ kJ/s}}{6.607 \times 10^{-3} \text{ kg/s}} = 2971 \text{ kJ/kg}$$

The value of $H_1 = 0$. Equation (2.7-10) becomes

$$H_2 - H_1 + 0 + 0 = Q - 0$$

The final equation for the calorimeter is

$$H_2 = Q + H_1 \tag{2.7-22}$$

Substituting $Q = 2971$ kJ/kg and $H_1 = 0$ into Eq. (2.7-22), $H_2 = 2971$ kJ/kg at 250°C and 150 kPa, which is close to the value from the steam table of 2972.7 kJ/kg.

2.7F Overall Mechanical-Energy Balance

A more useful type of energy balance for flowing fluids, especially liquids, is a modification of the total energy balance to deal with mechanical energy. Engineers are often concerned with this special type of energy, called *mechanical energy*, which includes the work term, kinetic energy, potential energy, and the flow work part of the enthalpy term. Mechanical energy is a form of energy that is either work or a form that can be directly converted into work. The other terms in the energy-balance equation (2.7-10), heat terms and internal energy, do not permit simple conversion into work because of the second law of thermodynamics and the efficiency of conversion, which depends on the temperatures. Mechanical-energy terms have no such limitation and can be converted almost completely into work. Energy converted to heat or internal energy is lost work or a loss in mechanical energy which is caused by frictional resistance to flow.

It is convenient to write an energy balance in terms of this loss, $\sum F$, which is the sum of all frictional losses per unit mass. For the case of steady-state flow, when a unit mass of fluid passes from inlet to outlet, the batch work done by the fluid, W', is expressed as

$$W' = \int_{V_1}^{V_2} p \, dV - \sum F \qquad (\sum F > 0) \tag{2.7-23}$$

This work W' differs from the W of Eq. (2.7-1), which also includes kinetic- and potential-energy effects. Writing the first law of thermodynamics for this case, where ΔE becomes ΔU,

$$\Delta U = Q - W' \tag{2.7–24}$$

The equation defining enthalpy, Eq. (2.7-5), can be written as

$$\Delta H = \Delta U + \Delta pV = \Delta U + \int_{V_1}^{V_2} p \, dV + \int_{p_1}^{p_2} V \, dp \tag{2.7-25}$$

Substituting Eq. (2.7-23) into (2.7-24) and then combining the resultant with Eq. (2.7-25), we obtain

$$\Delta H = Q + \sum F + \int_{p_1}^{p_2} V\, dp \qquad (2.7\text{-}26)$$

Finally, we substitute Eq. (2.7-26) into (2.7-10) and $1/\rho$ for V, to obtain the overall mechanical-energy-balance equation:

$$\frac{1}{2\alpha}\left[v_{2\,\mathrm{av}}^2 - v_{1\,\mathrm{av}}^2\right] + g(z_2 - z_1) + \int_{p_1}^{p_2}\frac{dp}{\rho} + \sum F + W_S = 0 \qquad (2.7\text{-}27)$$

For English units the kinetic- and potential-energy terms of Eq. (2.7-27) are divided by g_c.

The value of the integral in Eq. (2.7-27) depends on the equation of state of the fluid and the path of the process. If the fluid is an incompressible liquid, the integral becomes $(p_2 - p_1)/\rho$ and Eq. (2.7-27) becomes

$$\frac{1}{2\alpha}(v_{2\,\mathrm{av}}^2 - v_{1\,\mathrm{av}}^2) + g(z_2 - z_1) + \frac{p_2 - p_1}{\rho} + \sum F + W_S = 0 \qquad (2.7\text{-}28)$$

EXAMPLE 2.7-4. Mechanical-Energy Balance on Pumping System
Water with a density of 998 kg/m³ is flowing at a steady mass flow rate through a uniform-diameter pipe. The entrance pressure of the fluid is 68.9 kN/m² abs in the pipe, which connects to a pump which actually supplies 155.4 J/kg of fluid flowing in the pipe. The exit pipe from the pump is the same diameter as the inlet pipe. The exit section of the pipe is 3.05 m higher than the entrance, and the exit pressure is 137.8 kN/m² abs. The Reynolds number in the pipe is above 4000 in the system. Calculate the frictional loss $\sum F$ in the pipe system.

Solution: First a flow diagram is drawn of the system (Fig. 2.7-4), with 155.4 J/kg mechanical energy added to the fluid. Hence, $W_S = -155.4$, since the work done by the fluid is positive.

Setting the datum height $z_1 = 0$, $z_2 = 3.05$ m. Since the pipe is of

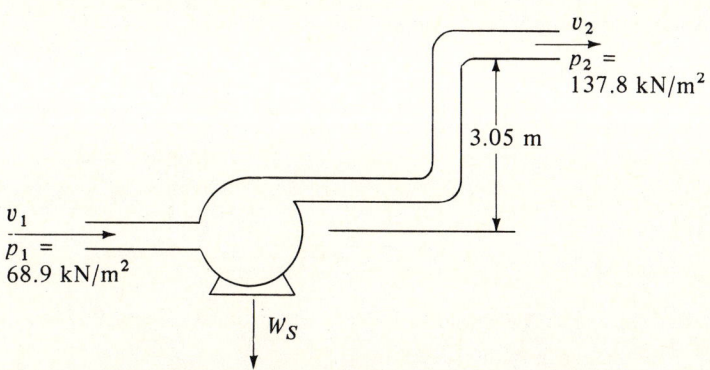

FIGURE 2.7-4. *Process flow diagram for Example 2.7-4.*

constant diameter, $v_1 = v_2$. Also, for turbulent flow $\alpha = 1.0$ and

$$\frac{1}{2(1)}(v_2^2 - v_1^2) = 0$$

$$z_2 g = (3.05 \text{ m})(9.806 \text{ m/s}^2) = 29.9 \text{ J/kg}$$

Since the liquid can be considered incompressible, Eq. (2.7-28) is used.

$$\frac{p_1}{\rho} = \frac{68.9 \times 1000}{998} = 69.0 \text{ J/kg}$$

$$\frac{p_2}{\rho} = \frac{137.8 \times 1000}{998} = 138.0 \text{ J/kg}$$

Using Eq. (2.7-28) and solving for $\sum F$, the frictional losses,

$$\sum F = -W_s + \frac{1}{2\alpha}(v_1^2 - v_2^2) + g(z_1 - z_2) + \frac{p_1 - p_2}{\rho} \qquad \textbf{(2.7-29)}$$

Substituting the known values, and solving for the frictional losses,

$$\sum F = -(-155.4) + 0 - 29.9 + 69.0 - 138.0$$

$$= 56.5 \text{ J/kg}\left(18.9 \ \frac{\text{ft} \cdot \text{lb}_f}{\text{lb}_m}\right)$$

EXAMPLE 2.7-5. Pump Horsepower in Flow System

A pump draws 69.1 gal/min of a liquid solution having a density of 114.8 lb_m/ft^3 from an open storage feed tank of large cross-sectional area through a 3.068-in.-ID suction line. The pump discharges its flow through a 2.067-in.-ID line to an open overhead tank. The end of the discharge line is 50 ft above the level of the liquid in the feed tank. The friction losses in the piping system are $\sum F = 10.0$ ft-lb force/lb mass. What pressure must the pump develop and what is the horsepower of the pump if its efficiency is 65% ($\eta = 0.65$)? The flow is turbulent.

Solution: First, a flow diagram of the system is drawn (Fig. 2.7-5). Equation (2.7-28) will be used. The term W_s in Eq. (2.7-28) becomes

$$W_S = -\eta W_p \qquad \textbf{(2.7-30)}$$

where $-W_S$ = mechanical energy actually delivered to the fluid by the

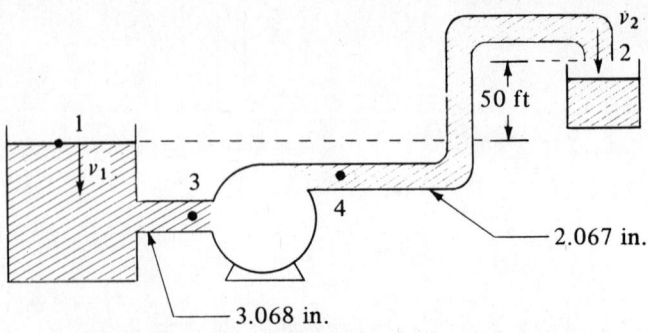

FIGURE 2.7-5. *Process flow diagram for Example 2.7-5.*

pump or net mechanical work, η = fractional efficiency, and W_p is the energy or shaft work delivered to the pump.

From Appendix A.5, the cross-sectional area of the 3.068-in. pipe is $0.05134\ \text{ft}^2$ and of the 2.067-in. pipe, $0.0233\ \text{ft}^2$. The flow rate is

$$\text{flow rate} = \left(69.1\ \frac{\text{gal}}{\text{min}}\right)\left(\frac{1\ \text{min}}{60\ \text{s}}\right)\left(\frac{1\ \text{ft}^3}{7.481\ \text{gal}}\right) = 0.1539\ \text{ft}^3/\text{s}$$

$$v_2 = \left(0.1539\ \frac{\text{ft}^3}{\text{s}}\right)\left(\frac{1}{0.0233\ \text{ft}^2}\right) = 6.61\ \text{ft/s}$$

$v_1 = 0$, since the tank is very large. Then $v_1^2/2g_c = 0$. The pressure $p_1 = 1$ atm and $p_2 = 1$ atm. Also, $\alpha = 1.0$ since the flow is turbulent. Hence,

$$\frac{p_1}{\rho} - \frac{p_2}{\rho} = 0$$

$$\frac{v_2^2}{2g_c} = \frac{(6.61)^2}{2(32.174)} = 0.678\ \frac{\text{ft} \cdot \text{lb}_f}{\text{lb}_m}$$

Using the datum of $z_1 = 0$,

$$z_2 \frac{g}{g_c} = (50.0)\frac{32.174}{32.174} = 50.0\ \frac{\text{ft} \cdot \text{lb}_f}{\text{lb}_m}$$

Using Eq. (2.4-28), solving for W_S, and substituting the known values,

$$W_S = z_1 \frac{g}{g_c} - z_2 \frac{g}{g_c} + \frac{v_1^2}{2g_c} - \frac{v_2^2}{2g_c} + \frac{p_1 - p_2}{\rho} - \sum F$$

$$= 0 - 50.0 + 0 - 0.678 + 0 - 10 = -60.678\ \frac{\text{ft} \cdot \text{lb}_f}{\text{lb}_m}$$

Using Eq. (2.7-30) and solving for W_p,

$$W_p = -\frac{W_S}{\eta} = \frac{60.678}{0.65}\ \frac{\text{ft} \cdot \text{lb}_f}{\text{lb}_m} = 93.3\ \frac{\text{ft} \cdot \text{lb}_f}{\text{lb}_m}$$

$$\text{mass flow rate} = \left(0.1539\ \frac{\text{ft}^3}{\text{s}}\right)\left(114.8\ \frac{\text{lb}_m}{\text{ft}^3}\right) = 17.65\ \frac{\text{lb}_m}{\text{s}}$$

$$\text{pump horsepower} = \left(17.65\ \frac{\text{lb}_m}{\text{s}}\right)\left(93.3\ \frac{\text{ft} \cdot \text{lb}_f}{\text{lb}_m}\right)\left(\frac{1\ \text{hp}}{550\ \text{ft} \cdot \text{lb}_f/\text{s}}\right)$$

$$= 3.00\ \text{hp}$$

To calculate the pressure the pump must develop, Eq. (2.7-28) must be written over the pump itself between points 3 and 4 as shown on the diagram.

$$v_3 = \left(0.1539\ \frac{\text{ft}^3}{\text{s}}\right)\left(\frac{1}{0.05134\ \text{ft}^2}\right) = 3.00\ \text{ft/s}$$

$$v_4 = v_2 = 6.61\ \text{ft/s}$$

Since the difference in level between z_3 and z_4 of the pump itself is negligible, it will be neglected. Rewriting Eq. (2.7-28) between points 3 and 4 and substituting known values $\left(\sum F = 0\ \text{since this is for the piping system}\right)$,

$$\frac{p_4 - p_3}{\rho} = z_3 \frac{g}{g_c} - z_4 \frac{g}{g_c} + \frac{v_3^2}{2g_c} - \frac{v_4^2}{2g_c} - W_s - \sum F \qquad (2.7\text{-}31)$$

$$= 0 - 0 + \frac{(3.00)^2}{2(32.174)} - \frac{(6.61)^2}{2(32.174)} + 60.678 - 0$$

$$= 0 - 0 + 0.140 - 0.678 + 60.678 = 60.14 \frac{\text{ft} \cdot \text{lb}_f}{\text{lb}_m}$$

$$p_4 - p_3 = \left(60.14 \frac{\text{ft} \cdot \text{lb}_f}{\text{lb}_m}\right)\left(114.8 \frac{\text{lb}_m}{\text{ft}^3}\right)\left(\frac{1}{144 \text{ in.}^2/\text{ft}^2}\right)$$

$$= 48.0 \text{ lb force/in.}^2 \text{ (psia pressure developed by pump)} \quad (331 \text{ kPa})$$

2.7G Bernoulli Equation for Mechanical-Energy Balance

In the special case where no mechanical energy is added ($W_s = 0$) and for no friction ($\sum F = 0$), then Eq. (2.7-28) becomes the Bernoulli equation, Eq. (2.7-32), for turbulent flow, which is of sufficient importance to deserve further discussion.

$$z_1 g + \frac{v_1^2}{2} + \frac{p_1}{\rho} = z_2 g + \frac{v_2^2}{2} + \frac{p_2}{\rho} \qquad (2.7\text{-}32)$$

This equation covers many situations of practical importance and is often used in conjunction with the mass-balance equation (2.6-2) for steady state.

$$m = \rho_1 A_1 v_1 = \rho_2 A_2 v_2 \qquad (2.6\text{-}2)$$

Several examples of its use will be given.

EXAMPLE 2.7-6. Rate of Flow from Pressure Measurements
A liquid with a constant density ρ kg/m^3 is flowing at an unknown velocity v_1 m/s through a horizontal pipe of cross-sectional area A_1 m^2 at a pressure p_1 N/m^2, and then it passes to a section of the pipe in which the area is reduced gradually to A_2 m^2 and the pressure is p_2. Assuming no friction losses, calculate the velocity v_1 and v_2 if the pressure difference ($p_1 - p_2$) is measured.

Solution: In Fig. 2.7-6, the flow diagram is shown with pressure taps to measure p_1 and p_2. From the mass-balance continuity equation (2.6-2), for constant ρ where $\rho_1 = \rho_2 = \rho$,

$$v_2 = \frac{v_1 A_1}{A_2} \qquad (2.7\text{-}33)$$

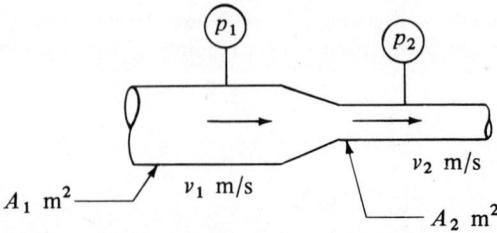

FIGURE 2.7-6. *Process flow diagram for Example 2.7-6.*

Transport Processes: Momentum, Heat, and Mass

For the items in the Bernoulli equation (2.7-32), for a horizontal pipe,

$$z_1 = z_2 = 0$$

Then Eq. (2.7-32) becomes, after substituting Eq. (2.7-33) for v_2,

$$0 + \frac{v_1^2}{2} + \frac{p_1}{\rho} = 0 + \frac{v_1^2 A_1^2 / A_2^2}{2} + \frac{p_2}{\rho} \qquad \text{(2.7-34)}$$

Rearranging,

$$p_1 - p_2 = \frac{\rho v_1^2 [(A_1/A_2)^2 - 1]}{2} \qquad \text{(2.7-35)}$$

$$v_1 = \sqrt{\frac{p_1 - p_2}{\rho} \frac{2}{[(A_1/A_2)^2 - 1]}} \qquad \text{(SI)}$$

$$\hspace{11cm}\text{(2.7-36)}$$

$$v_1 = \sqrt{\frac{p_1 - p_2}{\rho} \frac{2g_c}{[(A_1/A_2)^2 - 1]}} \qquad \text{(English)}$$

Performing the same derivation but in terms of v_2,

$$v_2 = \sqrt{\frac{p_1 - p_2}{\rho} \frac{2}{1 - (A_2/A_1)^2}} \qquad \text{(2.7-37)}$$

EXAMPLE 2.7-7. Rate of Flow from a Nozzle in a Tank
A nozzle of cross-sectional area A_2 is discharging to the atmosphere and is located in the side of a large tank, in which the open surface of the liquid in the tank is H m above the center line of the nozzle. Calculate the velocity v_2 in the nozzle and the volumetric rate of discharge if no friction losses are assumed.

Solution: The process flow is shown in Fig. 2.7-7, with point 1 taken in the liquid at the entrance to the nozzle and point 2 at the exit of the nozzle.
Since A_1 is very large compared to A_2, $v_1 \cong 0$. The pressure p_1 is greater than 1 atm (101.3 kN/m²) by the head of fluid of H m. The pressure p_2, which is at the nozzle exit, is at 1 atm. Using point 2 as a datum, $z_2 = 0$ and $z_1 = 0$ m. Rearranging Eq. (2.7-32),

$$z_1 g + \frac{v_1^2}{2} + \frac{p_1 - p_2}{\rho} = z_2 g + \frac{v_2^2}{2} \qquad \text{(2.7-38)}$$

Substituting the known values,

$$0 + 0 + \frac{p_1 - p_2}{\rho} = 0 + \frac{v_2^2}{2} \qquad \text{(2.7-39)}$$

Solving for v_2,

$$v_2 = \sqrt{\frac{2(p_1 - p_2)}{\rho}} \qquad \text{m/s} \qquad \text{(2.7-40)}$$

FIGURE 2.7-7. *Nozzle flow diagram for Example 2.7-7.*

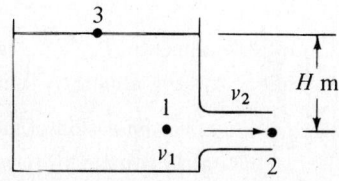

Since $p_1 - p_3 = H\rho g$ and $p_3 = p_2$ (both at 1 atm),

$$H = \frac{p_1 - p_2}{\rho g} \quad \text{m} \tag{2.7-41}$$

where H is the head of liquid with density ρ. Then Eq. (2.4-40) becomes

$$v_2 = \sqrt{2gH} \tag{2.7-42}$$

The volumetric flow rate is

$$\text{flow rate} = v_2 A_2 \quad \text{m}^3/\text{s} \tag{2.7-43}$$

To illustrate the fact that different points can be used in the balance, points 3 and 2 will be used. Writing Eq. (2.7-32),

$$z_2 g + \frac{v_2^2}{2} + \frac{p_2 - p_3}{\rho} = z_3 g + \frac{v_3^2}{2} \tag{2.7-44}$$

Since $p_2 = p_3 = 1$ atm, $v_3 = 0$, and $z_2 = 0$,

$$v_2 = \sqrt{2gz_3} = \sqrt{2gH} \tag{2.7-45}$$

2.8 OVERALL MOMENTUM BALANCE

2.8A Derivation of General Equation

A momentum balance can be written for the control volume shown in Fig. 2.6-3, which is somewhat similar to the overall mass-balance equation. Momentum, in contrast to mass and energy, is a vector quantity. The total linear momentum vector $\mathbf{P}$ of the total mass M of a moving fluid having a velocity of $\mathbf{v}$ is

$$\mathbf{P} = M\mathbf{v} \tag{2.8-1}$$

The term $M\mathbf{v}$ is the momentum of this moving mass M enclosed at a particular instant in the control volume shown in Fig. 2.6-4. The units of $M\mathbf{v}$ are kg $\cdot$ m/s in the SI system.

Starting with Newton's second law we will develop the integral momentum-balance equation for linear momentum. Angular momentum will not be considered here. *Newton's law* may be stated: The time rate of change of momentum of a system is equal to the summation of all forces acting on the system and takes place in the direction of the net force.

$$\sum \mathbf{F} = \frac{d\mathbf{P}}{dt} \tag{2.8-2}$$

where $\mathbf{F}$ is force. In the SI system $\mathbf{F}$ is in newtons (N) and $1 \text{ N} = 1 \text{ kg} \cdot \text{m/s}^2$. Note that in the SI system g_c is not needed, but it is needed in the English system.

The equation for the conservation of momentum with respect to a control volume can be written as follows:

$$\begin{pmatrix} \text{sum of forces acting} \\ \text{on control volume} \end{pmatrix} = \begin{pmatrix} \text{rate of momentum} \\ \text{out of control volume} \end{pmatrix} - \begin{pmatrix} \text{rate of momentum} \\ \text{into control volume} \end{pmatrix}$$

$$+ \begin{pmatrix} \text{rate of accumulation of momentum} \\ \text{in control volume} \end{pmatrix} \tag{2.8-3}$$

This is in the same form as the general mass-balance equation (2.6-3), with the sum of the forces as the generation rate term. Hence, momentum is not conserved, since it is generated by external forces on the system. If external forces are absent, momentum is conserved.

Using the general control volume shown in Fig. 2.6-4, we shall evaluate the various terms in Eq. (2.8-3), using methods very similar to the development of the general mass balance. For a small element of area dA on the control surface, we write

$$\text{rate of momentum efflux} = \mathbf{v}(\rho v)(dA \cos \alpha) \tag{2.8-4}$$

Note that the rate of mass efflux is $(\rho v)(dA \cos \alpha)$. Also, note that $(dA \cos \alpha)$ is the area dA projected in a direction normal to the velocity vector $\mathbf{v}$ and α is the angle between the velocity vector $\mathbf{v}$ and the outward-directed-normal vector $\mathbf{n}$. From vector algebra the product in Eq. (2.8-4) becomes

$$\mathbf{v}(\rho v)(dA \cos \alpha) = \rho \mathbf{v}(\mathbf{v} \cdot \mathbf{n}) \, dA \tag{2.8-5}$$

Integrating over the entire control surface A,

$$\binom{\text{net momentum efflux}}{\text{from control volume}} = \iint_A \mathbf{v}(\rho v)\cos \alpha \, dA = \iint_A \rho \mathbf{v}(\mathbf{v} \cdot \mathbf{n}) \, dA \tag{2.8-6}$$

The net efflux represents the first two terms on the right-hand side of Eq. (2.8-3).

Similarly to Eq. (2.6-5), the rate of accumulation of linear momentum within the control volume V is

$$\binom{\text{rate of accumulation of momentum}}{\text{in control volume}} = \frac{\partial}{\partial t} \iiint_V \rho \mathbf{v} \, dV \tag{2.8-7}$$

Substituting Equations (2.8-2), (2.8-6), and (2.8-7) into (2.8-3), the overall linear momentum balance for a control volume becomes

$$\sum \mathbf{F} = \iint_A \rho \mathbf{v}(\mathbf{v} \cdot \mathbf{n}) \, dA + \frac{\partial}{\partial t} \iiint_V \rho \mathbf{v} \, dV \tag{2.8-8}$$

We should note that $\sum \mathbf{F}$ in general may have a component in any direction, and the $\mathbf{F}$ is the force the surroundings exert on the control-volume fluid. Since Eq. (2.8-8) is a vector equation, we may write the component scalar equations for the x, y, and z directions.

$$\sum F_x = \iint_A v_x \rho v \cos \alpha \, dA + \frac{\partial}{\partial t} \iiint_V \rho v_x \, dV \quad \text{(SI)}$$
$$\tag{2.8-9}$$

$$\sum F_x = \iint_A v_x \frac{\rho}{g_c} v \cos \alpha \, dA + \frac{\partial}{\partial t} \iiint_V \frac{\rho}{g_c} v_x \, dV \quad \text{(English)}$$

$$\sum F_y = \iint_A v_y \rho v \cos \alpha \, dA + \frac{\partial}{\partial t} \iiint_V \rho v_y \, dV \tag{2.8-10}$$

$$\sum F_z = \iint_A v_z \rho v \cos \alpha \, dA + \frac{\partial}{\partial t} \iiint_V \rho v_z \, dV \tag{2.8-11}$$

Principles of Momentum Transfer and Overall Balances

The force term $\sum F_x$ in Eq. (2.8-9) is composed of the sum of several forces. These are given as follows.

1. *Body force.* The body force F_{xg} is the x-directed force caused by gravity acting on the total mass M in the control volume. This force, F_{xg}, is Mg_x. It is zero if the x direction is horizontal.
2. *Pressure force.* The force F_{xp} is the x-directed force caused by the pressure forces acting on the surface of the fluid system. When the control surface cuts through the fluid, the pressure is taken to be directed inward and perpendicular to the surface. In some cases part of the control surface may be a solid, and this wall is included inside the control surface. Then there is a contribution to F_{xp} from the pressure on the outside of this wall, which is typically atmospheric pressure. If gage pressure is used, the integral of the constant external pressure over the entire outer surface can be automatically ignored.
3. *Friction force.* When the fluid is flowing, an x-directed shear or friction force F_{xs} is present, which is exerted on the fluid by a solid wall when the control surface cuts between the fluid and the solid wall. In some or many cases this frictional force may be negligible compared to the other forces and is neglected.
4. *Solid surface force.* In cases where the control surface cuts through a solid, there is present force R_x, which is the x component of the resultant of the forces acting on the control volume at these points. This occurs in typical cases when the control volume includes a section of pipe and the fluid it contains. This is the force exerted by the solid surface on the fluid.

The force terms of Eq. (2.8-9) can then be represented as

$$\sum F_x = F_{xg} + F_{xp} + F_{xs} + R_x \tag{2.8-12}$$

Similar equations can be written for the y and z directions. Then Eq. (2.8-9) becomes for the x direction,

$$\sum F_x = F_{xg} + F_{xp} + F_{xs} + R_x$$
$$= \iint_A v_x \, \rho v \cos \alpha \, dA + \frac{\partial}{\partial t} \iiint_V \rho v_x \, dV \tag{2.8-13}$$

2.8B Overall Momentum Balance in Flow System in One Direction

A quite common application of the overall momentum-balance equation is the case of a section of a conduit with its axis in the x direction. The fluid will be assumed to be flowing at steady state in the control volume shown in Fig. 2.6-3 and also shown in Fig. 2.8-1. Equation (2.8-13) for the x direction becomes as follows since $v = v_x$.

$$\sum F_x = F_{xg} + F_{xp} + F_{xs} + R_x = \iint_A v_x \, \rho v_x \cos \alpha \, dA \tag{2.8-14}$$

Integrating with $\cos \alpha = \pm 1.0$ and $\rho A = m/v_{av}$,

$$F_{xg} + F_{xp} + F_{xs} + R_x = m \frac{(v_{x2}^2)_{av}}{v_{x2 \ av}} - m \frac{(v_{x1}^2)_{av}}{v_{x1 \ av}} \tag{2.8-15}$$

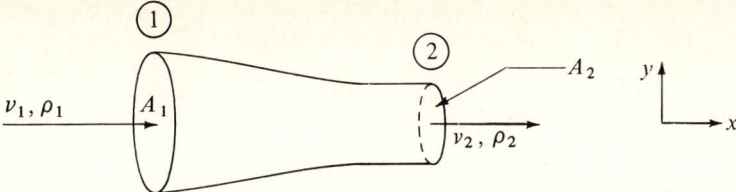

FIGURE 2.8-1. *Flow through a horizontal nozzle in the x direction only.*

where if the velocity is not constant and varies across the surface area,

$$(v_x^2)_{av} = \frac{1}{A} \iint\limits_A v_x^2 \, dA \qquad (2.8\text{-}16)$$

The ratio $(v_x^2)_{av}/v_{x\,av}$ is replaced by $v_{x\,av}/\beta$, where β, which is the momentum velocity correction factor, has a value of 0.95 to 0.99 for turbulent flow and $\frac{3}{4}$ for laminar flow. For most applications in turbulent flow, $(v_x)_{av}^2/v_{x\,av}$ is replaced by $v_{x\,av}$, the average bulk velocity. Note that the subscript x on v_x and F_x can be dropped since $v_x = v$ and $F_x = F$ for one directional flow.

The term F_{xp}, which is the force caused by the pressures acting on the surface of the control volume, is

$$F_{xp} = p_1 A_1 - p_2 A_2 \qquad (2.8\text{-}17)$$

The friction force will be neglected in Eq. (2.8-15), so $F_{xs} = 0$. The body force $F_{xg} = 0$ since gravity is acting only in the y direction. Substituting F_{xp} from Eq. (2.8-17) into (2.8-15), replacing $(v_x^2)_{av}/v_{x\,av}$ by v/β (where $v_{x\,av} = v$), setting $\beta = 1.0$, and solving for R_x in Eq. (2.8-15),

$$R_x = mv_2 - mv_1 + p_2 A_2 - p_1 A_1 \qquad (2.8\text{-}18)$$

where R_x is the force exerted by the solid on the fluid. The force of the fluid on the solid (reaction force) is the negative of this or $-R_x$.

EXAMPLE 2.8-1. Momentum Velocity Correction Factor β for Laminar Flow

The momentum velocity correction factor β is defined as follows for flow in one direction where the subscript x is dropped.

$$\frac{(v^2)_{av}}{v_{av}} = \frac{v_{av}}{\beta} \qquad (2.8\text{-}19)$$

$$\beta = \frac{(v_{av})^2}{(v^2)_{av}} \qquad (2.8\text{-}20)$$

Determine β for laminar flow in a tube.

Solution: Using Eq. (2.8-16),

$$(v^2)_{av} = \frac{1}{A} \iint\limits_A v^2 \, dA \qquad (2.8\text{-}21)$$

Substituting Eq. (2.7-16) for laminar flow into Eq. (2.8-21) and noting that $A = \pi R^2$ and $dA = r\,dr\,d\theta$, we obtain (see Example 2.6-3)

$$(v^2)_{av} = \frac{1}{\pi R^2} \int_0^{2\pi} \int_0^R \left[2v_{av}\left(1 - \frac{r^2}{R^2}\right) \right]^2 r\,dr\,d\theta$$

$$= \frac{(2\pi)2^2 v_{av}^2}{\pi R^2} \int_0^R \frac{(R^2 - r^2)^2}{R^4} r\,dr \qquad (2.8\text{-}22)$$

Integrating Eq. (2.8-22) and rearranging,

$$(v^2)_{av} = \frac{8v_{av}^2}{R^6}\left(\frac{R^6}{2} - \frac{R^6}{2} + \frac{R^6}{6}\right) = \tfrac{4}{3}v_{av}^2 \qquad (2.8\text{-}23)$$

Substituting Eq. (2.8-23) into (2.8-20), $\beta = \tfrac{3}{4}$.

EXAMPLE 2.8-2.   Momentum Balance for Horizontal Nozzle
Water is flowing at a rate of 0.03154 m³/s through a horizontal nozzle shown in Fig. 2.8-1 and discharges to the atmosphere at point 2. The nozzle is attached at the upstream end at point 1 and frictional forces are considered negligible. The upstream ID is 0.0635 m and the downstream 0.0286 m. Calculate the resultant force on the nozzle. The density of the water is 1000 kg/m³.

Solution: First, the mass flow and average or bulk velocities at points 1 and 2 are calculated. The area at point 1 is $A_1 = (\pi/4)(0.0635)^2 = 3.167 \times 10^{-3}$ m² and $A_2 = (\pi/4)(0.0286)^2 = 6.424 \times 10^{-4}$ m². Then,

$$m_1 = m_2 = m = (0.03154)(1000) = 31.54 \text{ kg/s}$$

The velocity at point 1 is $v_1 = 0.03154/(3.167 \times 10^{-3}) = 9.96$ m/s and $v_2 = 0.03154/(6.424 \times 10^{-4}) = 49.1$ m/s.

To evaluate the upstream pressure p_1 we use the mechanical-energy balance equation (2.7-28) assuming no frictional losses and turbulent flow. (This can be checked by calculating the Reynolds number.) This equation then becomes, for $\alpha = 1.0$,

$$\frac{v_1^2}{2} + \frac{p_1}{\rho} = \frac{v_2^2}{2} + \frac{p_2}{\rho} \qquad (2.8\text{-}24)$$

Setting $p_2 = 0$ gage pressure, $\rho = 1000$ kg/m³, $v_1 = 9.96$ m/s, $v_2 = 49.1$ m/s, and solving for p_1,

$$p_1 = \frac{(1000)(49.1^2 - 9.96^2)}{2} = 1.156 \times 10^6 \text{ N/m}^2 \qquad \text{(gage pressure)}$$

For the x direction, the momentum balance equation (2.8-18) is used. Substituting the known values and solving for R_x,

$$R_x = 31.54(49.10 - 9.96) + 0 - (1.156 \times 10^6)(3.167 \times 10^{-3})$$

$$= -2427 \text{ N}(-546 \text{ lb}_f)$$

Since the force is negative, it is acting in the negative x direction or to the left. This is the force of the nozzle on the fluid. The force of the fluid on the solid is $-R_x$ or $+2427$ N.

2.8C   Overall Momentum Balance in Two Directions

Another application of the overall momentum balance is shown in Fig. 2.8-2 for a flow system with fluid entering a conduit at point 1 inclined at an angle of α_1 relative to the

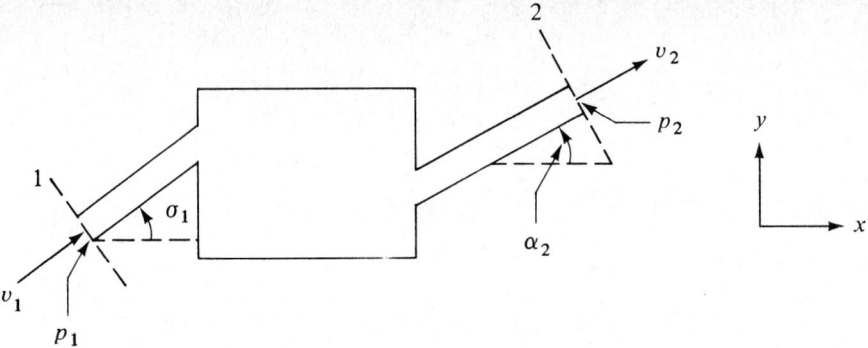

FIGURE 2.8-2. *Overall momentum balance for flow system with fluid entering at point 1 and leaving at 2.*

horizontal x direction and leaving a conduit at point 2 at an angle α_2. The fluid will be assumed to be flowing at steady state and the frictional force F_{xs} will be neglected. Then Eq. (2.8-13) for the x direction becomes as follows for no accumulation:

$$F_{xg} + F_{xp} + R_x = \iint_A v_x \rho v \cos \alpha \, dA \qquad (2.8\text{-}25)$$

Integrating the surface (area) integral,

$$F_{xg} + F_{xp} + R_x = m \frac{(v_2^2)_{av}}{v_{2\,av}} \cos \alpha_2 - m \frac{(v_1^2)_{av}}{v_{1\,av}} \cos \alpha_1 \qquad (2.8\text{-}26)$$

The term $(v^2)_{av}/v_{av}$ can again be replaced by v_{av}/β with β being set at 1.0. From Fig. 2.8-2, the term F_{xp} is

$$F_{xp} = p_1 A_1 \cos \alpha_1 - p_2 A_2 \cos \alpha_2 \qquad (2.8\text{-}27)$$

Then Eq. (2.8-26) becomes as follows after solving for R_x:

$$R_x = mv_2 \cos \alpha_2 - mv_1 \cos \alpha_1 + p_2 A_2 \cos \alpha_2 - p_1 A_1 \cos \alpha_1 \qquad (2.8\text{-}28)$$

The term $F_{xg} = 0$ in this case.

For R_y the body force F_{yg} is in the negative y direction and $F_{yg} = -m_t g$, where m_t is the total mass fluid in the control volume. Replacing $\cos \alpha$ by $\sin \alpha$, the equation for the y direction becomes

$$R_y = mv_2 \sin \alpha_2 - mv_1 \sin \alpha_1 + p_2 A_2 \sin \alpha_2 - p_1 A_1 \sin \alpha_1 + m_t g \qquad (2.8\text{-}29)$$

EXAMPLE 2.8-3. Momentum Balance in a Pipe Bend

Fluid is flowing at steady state through a reducing pipe bend, as shown in Fig. 2.8-3. Turbulent flow will be assumed with frictional forces negligible. The volumetric flow rate of the liquid and the pressure p_2 at point 2 are known as are the pipe diameters at both ends. Derive the equations to calculate the forces on the bend. Assume that the density ρ is constant.

Solution: The velocities v_1 and v_2 can be obtained from the volumetric flow rate and the areas. Also, $m = \rho_1 v_1 A_1 = \rho_2 v_2 A_2$. As in Example 2.8-2, the mechanical-energy balance equation (2.8-24) is used to obtain the upstream pressure, p_1. For the x direction Eq. (2.8-28) is used for the mo-

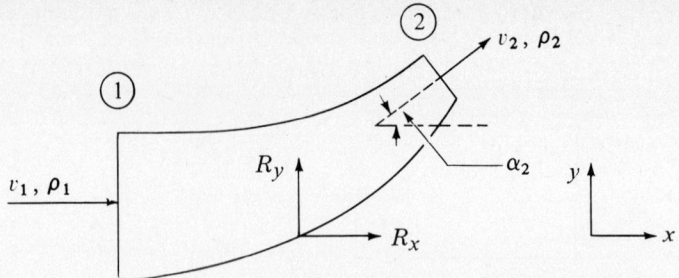

FIGURE 2.8-3. *Flow through a reducing bend in Example 2.8-3.*

mentum balance. Since $\alpha_1 = 0°$, $\cos \alpha_1 = 1.0$. Equation (2.8-28) becomes

$$R_x = mv_2 \cos \alpha_2 - mv_1 + p_2 A_2 \cos \alpha_2 - p_1 A_1 \qquad \text{(SI)}$$

$$R_x = \frac{m}{g_c} v_2 \cos \alpha_2 - \frac{m}{g_c} v_1 + p_2 A_2 \cos \alpha_2 - p_1 A_1 \qquad \text{(English)}$$

(2.8-30)

For the y direction the momentum balance Eq. (2.8-29) is used where $\sin \alpha_1 = 0$.

$$R_y = mv_2 \sin \alpha_2 + p_2 A_2 \sin \alpha_2 + m_t g \qquad \text{(SI)} \qquad \textbf{(2.8-31)}$$

where m_t is total mass fluid in the pipe bend. The pressures at points 1 and 2 are gage pressures since the atmospheric pressures acting on all surfaces cancel. The magnitude of the resultant force of the bend acting on the control volume fluid is

$$|\mathbf{R}| = \sqrt{R_x^2 + R_y^2} \qquad \textbf{(2.8-32)}$$

The angle this makes with the vertical is $\theta = \arctan (R_x/R_y)$. Often the gravity force F_{yg} is small compared to the other terms in Eq. (2.8-31) and is neglected.

EXAMPLE 2.8-4. *Friction Loss in a Sudden Enlargement*
A mechanical-energy loss occurs when a fluid flows from a small pipe to a large pipe through an abrupt expansion, as shown in Fig. 2.8-4. Use the momentum balance and mechanical-energy balance to obtain an expression for the loss for a liquid. (*Hint :* Assume that $p_0 = p_1$ and $v_0 = v_1$. Make a mechanical-energy balance between points 0 and 2 and a momentum balance between points 1 and 2. It will be assumed that p_1 and p_2 are uniform over the cross-sectional area.)

FIGURE 2.8-4. *Losses in expansion flow.*

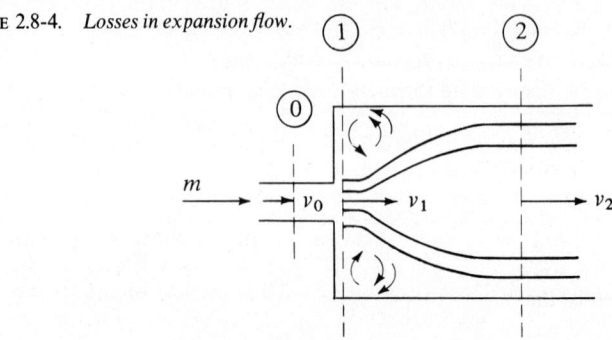

Transport Processes: Momentum, Heat, and Mass

Solution: The control volume is selected so that it does not include the pipe wall, so R_x drops out. The boundaries selected are points 1 and 2. The flow through plane 1 occurs only through an area of A_0. The frictional drag force will be neglected and all the loss is assumed to be from eddies in this volume. Making a momentum balance between points 1 and 2 using Eq. (2.8-18) and noting that $p_0 = p_1$, $v_1 = v_0$, and $A_1 = A_2$,

$$p_1 A_2 - p_2 A_2 = mv_2 - mv_1 \tag{2.8-33}$$

The mass-flow rate is $m = v_0 \rho A_0$ and $v_2 = (A_0/A_2)v_0$. Substituting these terms into Eq. (2.8-33) and rearranging,

$$v_0^2 \frac{A_0}{A_2} \left(1 - \frac{A_0}{A_2}\right) = \frac{p_2 - p_1}{\rho} \tag{2.8-34}$$

Applying the mechanical-energy-balance equation (2.7-28) to points 1 and 2,

$$\frac{v_0^2 - v_2^2}{2} - \sum F = \frac{p_2 - p_1}{\rho} \tag{2.8-35}$$

Finally combining Eqs. (2.8-34) and (2.8-35),

$$\sum F = \frac{v_0^2}{2}\left(1 - \frac{A_0}{A_2}\right)^2 \tag{2.8-36}$$

2.8D Overall Momentum Balance for Free Jet Striking a Fixed Vane

When a free jet impinges on a fixed vane as in Fig. 2.8-5 the overall momentum balance can be applied to determine the force on the smooth vane. Since there are no changes in elevation or pressure before and after impact, there is no loss in energy and application of the Bernoulli equation shows that the magnitude of the velocity is unchanged. Losses due to impact are neglected. The frictional resistance between the jet and the smooth vane is also neglected. The velocity is assumed to be uniform throughout the jet upstream and downstream. Since the jet is open to the atmosphere, the pressure is the same at all ends of the vane.

In making a momentum balance for the control volume shown for the curved vane in Fig. 2.8-5a, Eq. (2.8-28) is written as follows for steady state, where the pressure terms are zero, $v_1 = v_2$, $A_1 = A_2$, and $m = v_1 A_1 \rho_1 = v_2 A_2 \rho_2$:

$$R_x = mv_2 \cos \alpha_2 - mv_1 + 0 = mv_1 (\cos \alpha_2 - 1) \tag{2.8-37}$$

Using Eq. (2.8-29) for the y direction and neglecting the body force,

$$R_y = mv_2 \sin \alpha_2 - 0 = mv_1 \sin \alpha_2 \tag{2.8-38}$$

Hence, R_x and R_y are the force components of the vane on the control volume fluid. The force components on the vane are $-R_x$ and $-R_y$.

EXAMPLE 2.8-5. Force of Free Jet on a Curved, Fixed Vane

A jet of water having a velocity of 30.5 m/s and a diameter of 2.54×10^{-2} m is deflected by a smooth, curved vane as shown in Fig. 2.8-5a, where $\alpha_2 = 60°$. What is the force of the jet on the vane? Assume that $\rho = 1000$ kg/m^3.

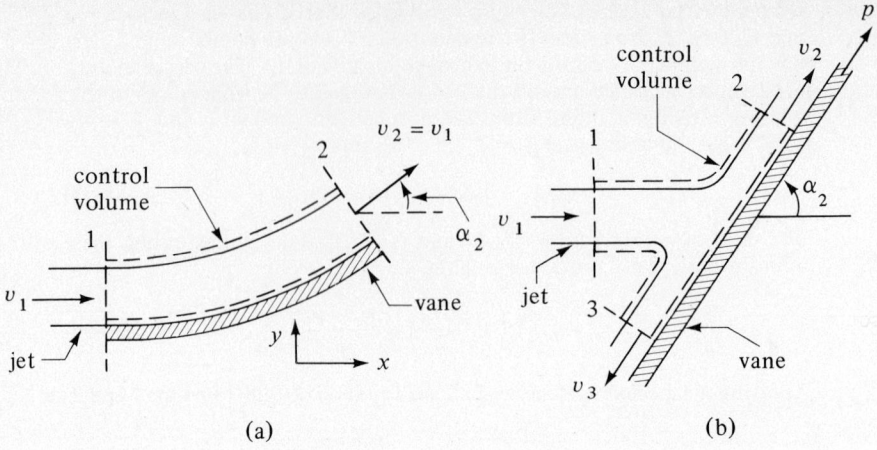

FIGURE 2.8-5. *Free jet impinging on a fixed vane: (a) smooth, curved vane, (b) smooth, flat vane.*

Solution: The cross-sectional area of the jet is $A_1 = \pi(2.54 \times 10^{-2})^2/4 = 5.067 \times 10^{-4}$ m^2. Then, $m = v_1 A_1 \rho_1 = 30.5 \times 5.067 \times 10^{-4} \times 1000 = 15.45$ kg/s. Substituting into Eqs. (2.8-37) and (2.8-38),

$$R_x = 15.45 \times 30.5 \,(\cos 60° - 1) = -235.6 \text{ N}(-52.97 \text{ lb}_f)$$

$$R_y = 15.45 \times 30.5 \sin 60° = 408.1 \text{ N}(91.74 \text{ lb}_f)$$

The force on the vane is $-R_x = +235.6$ N and $-R_y = -408.1$ N. The resultant force is calculated using Eq. (2.8-32).

In Fig. 2.8-5b a free jet at velocity v_1 strikes a smooth, inclined flat plate and the flow divides into two separate streams whose velocities are all equal $(v_1 = v_2 = v_3)$ since there is no loss in energy. It is convenient to make a momentum balance in the p direction parallel to the plate. No force is exerted on the fluid by the flat plate in this direction; i.e., there is no tangential force. Then, the initial momentum component in the p direction must equal the final momentum component in this direction. This means $\sum F_p = 0$. Writing an equation similar to Eq. (2.8-26), where m_1 is kg/s entering at 1 and m_2 leaves at 2 and m_3 at 3,

$$\sum F_p = 0 = m_2 v_2 - m_1 v_1 \cos \alpha_2 - m_3 v_3$$

$$0 = m_2 v_1 - m_1 v_1 \cos \alpha_2 - m_3 v_1 \qquad \text{(2.8-39)}$$

By the continuity equation,

$$m_1 = m_2 + m_3 \qquad \text{(2.8-40)}$$

Combining and solving,

$$m_2 = \frac{m_1}{2}(1 + \cos \alpha_2), \qquad m_3 = \frac{m_1}{2}(1 - \cos \alpha_2) \qquad \text{(2.8-41)}$$

The resultant force exerted by the plate on the fluid must be normal to it. This means the resultant force is simply $m_1 v_1 \sin \alpha_2$. Alternatively, the resultant force on the fluid can be calculated by determining R_x and R_y from Eqs. (2.8-28) and 2.8-29) and then using Eq. (2.8-32). The force on the bend is the opposite of this.

2.9 SHELL MOMENTUM BALANCE AND VELOCITY PROFILE IN LAMINAR FLOW

2.9A Introduction

In Section 2.8 we analyzed momentum balances using an overall, macroscopic control volume. From this we obtained the total or overall changes in momentum crossing the control surface. This overall momentum balance did not tell us the details of what happens inside the control volume. In the present section we analyze a small control volume and then shrink this control volume to differential size. In doing this we make a shell momentum balance using the momentum-balance concepts of the preceding section, and then, using the equation for the definition of viscosity, we obtain an expression for the velocity profile inside the enclosure and the pressure drop. The equations are derived for flow systems of simple geometry in laminar flow at steady state.

In many engineering problems a knowledge of the complete velocity profile is not needed, but a knowledge of the maximum velocity, average velocity, or the shear stress on a surface is needed. In this section we show how to obtain these quantities from the velocity profiles.

2.9B Shell Momentum Balance Inside a Pipe

Engineers often deal with the flow of fluids inside a circular conduit or pipe. In Fig. 2.9-1 we have a horizontal section of pipe in which an incompressible Newtonian fluid is flowing in one-dimensional, steady-state, laminar flow. The flow is fully developed; i.e., it is not influenced by entrance effects and the velocity profile does not vary along the axis of flow in the x direction.

The cylindrical control volume is a shell with an inside radius r, thickness Δr, and length Δx. At steady state the conservation of momentum, Eq. (2.8-3), becomes as follows: sum of forces acting on control volume = rate of momentum out − rate of momentum into volume. The pressure forces become, from Eq. (2.8-17),

$$\text{pressure forces} = pA|_x - pA|_{x+\Delta x} = p(2\pi r\ \Delta r)|_x - p(2\pi r\ \Delta r)|_{x+\Delta x} \qquad (2.9\text{-}1)$$

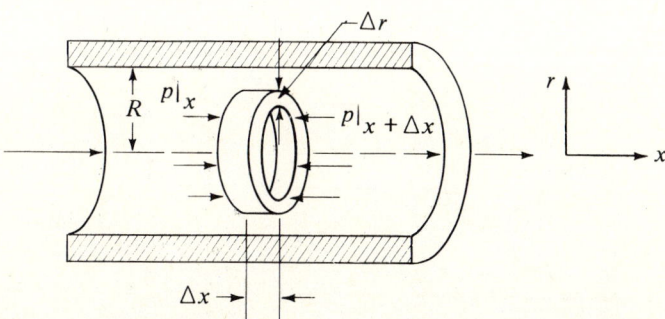

FIGURE 2.9-1. *Control volume for shell momentum balance on a fluid flowing in a circular tube.*

The shear force or drag force acting on the cylindrical surface at the radius r is the shear stress τ_{rx} times the area $2\pi r \, \Delta x$. However, this can also be considered as the rate of momentum flow into the cylindrical surface of the shell as described by Eq. (2.4-9). Hence, the net rate of momentum efflux is the rate of momentum out — rate of momentum in and is

$$\text{net efflux} = (\tau_{rx} 2\pi r \, \Delta x)|_{r+\Delta r} - (\tau_{rx} 2\pi r \, \Delta x)|_{r} \qquad (2.9\text{-}2)$$

The net convective momentum flux across the annular surface at x and $x + \Delta x$ is zero, since the flow is fully developed and the terms are independent of x. This is true since v_x at x is equal to v_x at $x + \Delta x$.

Equating Eq. (2.9-1) to (2.9-2) and rearranging,

$$\frac{(r\tau_{rx})|_{r+\Delta r} - (r\tau_{rx})|_{r}}{\Delta r} = \frac{r(p|_x - p|_{x+\Delta x})}{\Delta x} \qquad (2.9\text{-}3)$$

In fully developed flow, the pressure gradient $(\Delta p/\Delta x)$ is constant and becomes $(\Delta p/L)$, where Δp is the pressure drop for a pipe of length L. Letting Δr approach zero, we obtain

$$\frac{d(r\tau_{rx})}{dr} = \left(\frac{\Delta p}{L}\right) r \qquad (2.9\text{-}4)$$

Separating variables and integrating,

$$\tau_{rx} = \left(\frac{\Delta p}{L}\right) \frac{r}{2} + \frac{C_1}{r} \qquad (2.9\text{-}5)$$

The constant of integration C_1 must be zero if the momentum flux is not infinite at $r = 0$. Hence,

$$\tau_{rx} = \left(\frac{\Delta p}{2L}\right) r = \frac{p_0 - p_L}{2L} r \qquad (2.9\text{-}6)$$

This means that the momentum flux varies linearly with the radius, as shown in Fig. 2.9-2, and the maximum value occurs at $r = R$ at the wall.

Substituting Newton's law of viscosity,

$$\tau_{rx} = -\mu \frac{dv_x}{dr} \qquad (2.9\text{-}7)$$

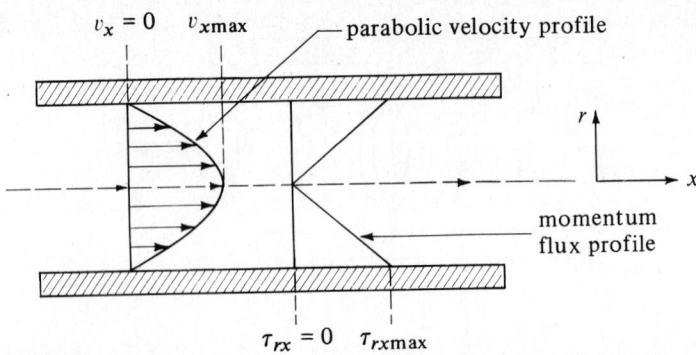

FIGURE 2.9-2. *Velocity and momentum flux profiles for laminar flow in a pipe.*

into Eq. (2.9-6), we obtain the following differential equation for the velocity:

$$\frac{dv_x}{dr} = -\frac{p_0 - p_L}{2\mu L} r \qquad (2.9\text{-}8)$$

Integrating using the boundary condition that at the wall, $v_x = 0$ at $r = R$, we obtain the equation for the velocity distribution.

$$v_x = \frac{p_0 - p_L}{4\mu L} R^2 \left[1 - \left(\frac{r}{R}\right)^2 \right] \qquad (2.9\text{-}9)$$

This result shows us that the velocity distribution is parabolic as shown in Fig. 2.9-2.

The average velocity $v_{x\,av}$ for a cross section is found by summing up all the velocities over the cross section and dividing by the cross-sectional area as in Eq. (2.6-17). Following the procedure given in Example 2.6-3, where $dA = r\ dr\ d\theta$ and $A = \pi R^2$,

$$v_{x\,av} = \frac{1}{A} \iint_A v_x\, dA = \frac{1}{\pi R^2} \int_0^{2\pi} \int_0^R v_x r\ dr\ d\theta = \frac{1}{\pi R^2} \int_0^R v_x\, 2\pi r\ dr \qquad (2.9\text{-}10)$$

Combining Eqs. (2.9-9) and (2.9-10) and integrating,

$$v_{x\,av} = \frac{(p_0 - p_L)R^2}{8\mu L} = \frac{(p_0 - p_L)D^2}{32\mu L} \qquad (2.9\text{-}11)$$

where diameter $D = 2R$. Hence, Eq. (2.9-11), which is the *Hagen–Poiseuille equation*, relates the pressure drop and average velocity for laminar flow in a horizontal pipe.

The maximum velocity for a pipe is found from Eq. (2.9-9) and occurs at $r = 0$.

$$v_{x\,max} = \frac{p_0 - p_L}{4\mu L} R^2 \qquad (2.9\text{-}12)$$

Combining Eqs. (2.9-11) and (2.9-12), we find that

$$v_{x\,av} = \frac{v_{x\,max}}{2} \qquad (2.9\text{-}13)$$

2.9C Shell Momentum Balance for Falling Film

We now use an approach similar to that used for laminar flow inside a pipe for the case of flow of a fluid as a film in laminar flow down a vertical surface. Falling films have been used to study various phenomena in mass transfer, coatings on surfaces, and so on. The control volume for the falling film is shown in Fig. 2.9-3a, where the shell of fluid considered is Δx thick and has a length of L in the vertical z direction. This region is sufficiently far from the entrance and exit regions so that the flow is not affected by these regions. This means the velocity $v_z(x)$ does not depend on position z.

To start we set up a momentum balance in the z direction over a system Δx thick, bounded in the z direction by the planes $z = 0$ and $z = L$, and extending a distance W in the y direction. First, we consider the momentum flux due to molecular transport. The rate of momentum out-rate of momentum in is the momentum flux at point $x + \Delta x$ minus that at x times the area LW.

$$\text{net efflux} = LW(\tau_{xz})|_{x+\Delta x} - LW(\tau_{xz})|_x \qquad (2.9\text{-}14)$$

The net convective momentum flux is the rate of momentum entering the area $\Delta x W$ at

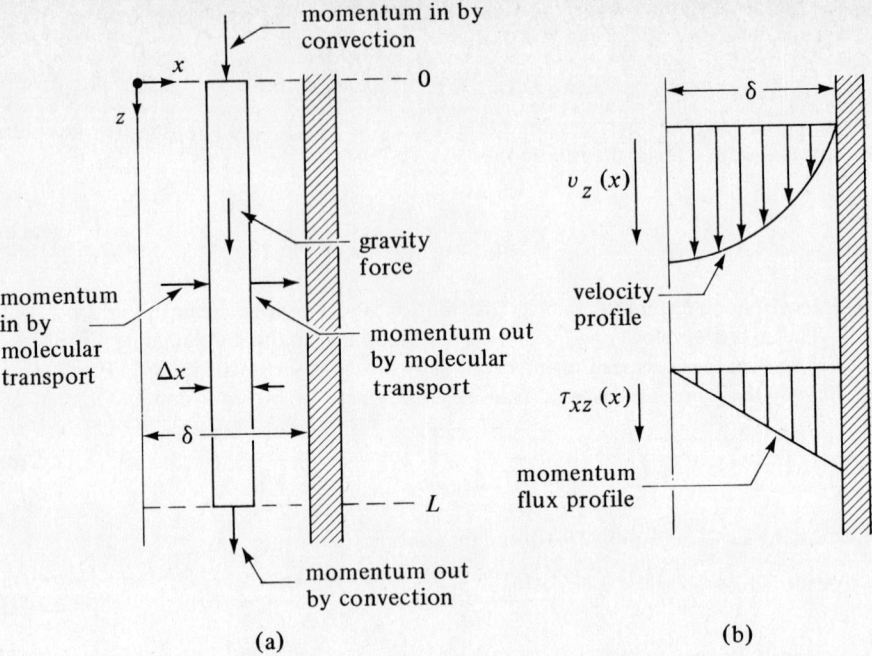

momentum in by convection

x

z

0

gravity force

momentum in by molecular transport

momentum out by molecular transport

Δx

δ

L

momentum out by convection

(a)

δ

$v_z(x)$

velocity profile

$\tau_{xz}(x)$

momentum flux profile

(b)

FIGURE 2.9-3. *Vertical laminar flow of a liquid film : (a) shell momentum balance for a control volume Δx thick, (b) velocity and momentum flux profiles.*

$z = L$ minus that leaving at $z = 0$. This net efflux is equal to 0 since v_z at $z = 0$ is equal to v_z at $z = L$ for each value of x.

$$\text{net efflux} = \Delta x W v_z (\rho v_z)|_{z=L} - \Delta x W v_z (\rho v_z)|_{z=0} = 0 \qquad (2.9\text{-}15)$$

The gravity force acting on the fluid is

$$\text{gravity force} = \Delta x W L (\rho g) \qquad (2.9\text{-}16)$$

Then using Eq. (2.8-3) for the conservation of momentum at steady state,

$$\Delta x W L(\rho g) = L W (\tau_{xz})|_{x+\Delta x} - L W (\tau_{xz})|_x + 0 \qquad (2.9\text{-}17)$$

Rearranging Eq. (2.9-17) and letting $\Delta x \to 0$,

$$\frac{\tau_{xz}|_{x+\Delta x} - \tau_{xz}|_x}{\Delta x} = \rho g \qquad (2.9\text{-}18)$$

$$\frac{d}{dx} \tau_{xz} = \rho g \qquad (2.9\text{-}19)$$

Integrating using the boundary conditions at $x = 0$, $\tau_{xz} = 0$ at the free liquid surface and at $x = x$, $\tau_{xz} = \tau_{xz}$,

$$\tau_{xz} = \rho g x \qquad (2.9\text{-}20)$$

This means the momentum-flux profile is linear as shown in Fig. 2.9-3b and the

Transport Processes: Momentum, Heat, and Mass

maximum value is at the wall. For a Newtonian fluid using Newton's law of viscosity,

$$\tau_{xz} = -\mu \frac{dv_z}{dx} \qquad (2.9\text{-}21)$$

Combining Eqs. (2.9-20) and (2.9-21) we obtain the following differential equation for the velocity:

$$\frac{dv_z}{dx} = -\left(\frac{\rho g}{\mu}\right)x \qquad (2.9\text{-}22)$$

Separating variables and integrating,

$$v_z = -\left(\frac{\rho g}{2\mu}\right)x^2 + C_1 \qquad (2.9\text{-}23)$$

Using the boundary condition that $v_z = 0$ at $x = \delta$, $C_1 = (\rho g/2\mu)\delta^2$. Hence, the velocity distribution equation becomes

$$v_z = \frac{\rho g \delta^2}{2\mu}\left[1 - \left(\frac{x}{\delta}\right)^2\right] \qquad (2.9\text{-}24)$$

This means the velocity profile is parabolic as shown in Fig. 2.9-3b. The maximum velocity occurs at $x = 0$ in Eq. (2.9-24) and is

$$v_{z\,max} = \frac{\rho g \delta^2}{2\mu} \qquad (2.9\text{-}25)$$

The average velocity can be found by using Eq. (2.6-17).

$$v_{z\,av} = \frac{1}{A}\iint_A v_z\,dA = \frac{1}{W\delta}\int_0^W\int_0^\delta v_z\,dx\,dy = \frac{W}{W\delta}\int_0^\delta v_z\,dx \qquad (2.9\text{-}26)$$

Substituting Eq. (2.9-24) into (2.9-26) and integrating,

$$v_{z\,av} = \frac{\rho g \delta^2}{3\mu} \qquad (2.9\text{-}27)$$

Combining Eqs. (2.9-25) and (2.9-27), we obtain $v_{z\,av} = (2/3)v_{z\,max}$. The volumetric flow rate q is obtained by multiplying the average velocity $v_{z\,av}$ times the cross-sectional area δW.

$$q = \frac{\rho g \delta^3 W}{3\mu} \text{ m}^3/\text{s} \qquad (2.9\text{-}28)$$

Often in falling films, the mass rate of flow per unit width of wall Γ in kg/s · m is defined as $\Gamma = \rho \delta v_{z\,av}$ and a Reynolds number is defined as

$$N_{Re} = \frac{4\Gamma}{\mu} = \frac{4\rho \delta v_{z\,av}}{\mu} \qquad (2.9\text{-}29)$$

Laminar flow occurs for $N_{Re} < 1200$. Laminar flow with rippling present occurs above a N_{Re} of 25.

EXAMPLE 2.9-1. Falling Film Velocity and Thickness
An oil is flowing down a vertical wall as a film 1.7 mm thick. The oil density is 820 kg/m^3 and the viscosity is 0.20 Pa · s. Calculate the mass flow rate per

unit width of wall, Γ, needed and the Reynolds number. Also calculate the average velocity.

Solution: The film thickness is $\delta = 0.0017$ m. Substituting Eq. (2.9-27) into the definition of Γ,

$$\Gamma = \rho \delta v_{z\,av} = \frac{(\rho\delta)\rho g \delta^2}{3\mu} = \frac{\rho^2 \delta^3 g}{3\mu}$$

(2.9-30)

$$= \frac{(820)^2(1.7 \times 10^{-3})^3(9.806)}{3 \times 0.20} = 0.05399 \text{ kg/s} \cdot \text{m}$$

Using Eq. (2.9-29),

$$N_{Re} = \frac{4\Gamma}{\mu} = \frac{4(0.05399)}{0.20} = 1.080$$

Hence, the film is in laminar flow. Using Eq. (2.9-27),

$$v_{z\,av} = \frac{\rho g \delta^2}{3\mu} = \frac{820(9.806)(1.7 \times 10^{-3})^2}{3(0.20)} = 0.03873 \text{ m/s}$$

2.10 DESIGN EQUATIONS FOR LAMINAR AND TURBULENT FLOW IN PIPES

2.10A Velocity Profiles in Pipes

One of the most important applications of fluid flow is flow inside circular conduits, pipes, and tubes. Appendix A.5 gives sizes of commercial standard steel pipe. Schedule 40 pipe in the different sizes is the standard usually used. Schedule 80 has a thicker wall and will withstand about twice the pressure of schedule 40 pipe. Both have the same outside diameter so that they will fit the same fittings. Pipes of other metals have the same outside diameters as steel pipe to permit interchanging parts of a piping system. Sizes of tubing are generally given by the outside diameter and wall thickness. Perry and Chilton (P1) give detailed tables of various types of tubing and pipes.

When fluid is flowing in a circular pipe and the velocities are measured at different distances from the pipe wall to the center of the pipe, it has been shown that in both laminar and turbulent flow, the fluid in the center of the pipe is moving faster than the fluid near the walls. These measurements are made at a reasonable distance from the entrance to the pipe. Figure 2.10-1 is a plot of the relative distance from the center of the pipe versus the fraction of maximum velocity v'/v_{max}, where v' is local velocity at the given position and v_{max} the maximum velocity at the center of the pipe. For viscous or laminar flow, the velocity profile is a true parabola, as derived in Eq. (2.9-9). The velocity at the wall is zero.

In many engineering applications the relation between the average velocity v_{av} in a pipe and the maximum velocity v_{max} is useful, since in some cases only the v_{max} at the center point of the tube is measured. Hence, from only one point measurement this relationship between v_{max} and v_{av} can be used to determine v_{av}. In Fig. 2.10-2 experimentally measured values of v_{av}/v_{max} are plotted as a function of the Reynolds numbers $Dv_{av}\rho/\mu$ and $Dv_{max}\rho/\mu$.

The average velocity over the whole cross section of the pipe is precisely 0.5 times the maximum velocity at the center as given by the shell momentum balance in Eq. (2.9-13) for laminar flow. On the other hand, for turbulent flow, the curve is somewhat flattened in the center (see Fig. 2.10-1) and the average velocity is about 0.8 times the

Transport Processes: Momentum, Heat, and Mass

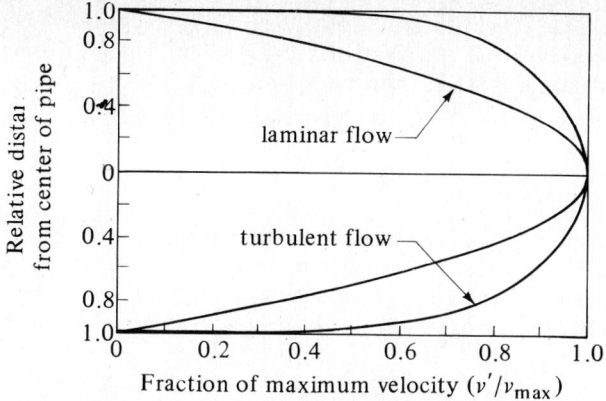

FIGURE 2.10-1. *Velocity distribution of a fluid across a pipe.*

maximum. This value of 0.8 varies slightly, depending upon the Reynolds number, as shown in the correlation in Fig. 2.10-2. (*Note :* See Problem 2.6-3, where a value of 0.817 is derived using the $\frac{1}{7}$-power law.)

2.10B Pressure Drop and Friction Loss in Laminar Flow

1. Pressure drop and loss due to friction. When the fluid is in steady-state laminar flow in a pipe, then for a Newtonian fluid the shear stress is given by Eq. (2.4-2), which is rewritten for change in radius dr rather than distance dy, as follows.

$$\tau_{rz} = -\mu \frac{dv_z}{dr} \qquad (2.10\text{-}1)$$

Using this relationship and making a shell momentum balance on the fluid over a

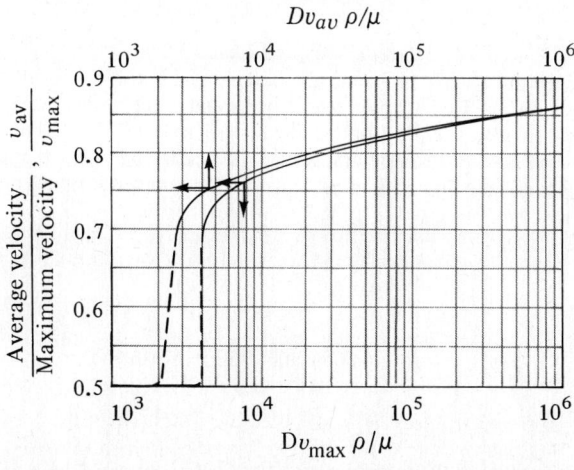

FIGURE 2.10-2. *Ratio v_{av}/v_{max} as a function of Reynolds number for pipes.*

Principles of Momentum Transfer and Overall Balances

cylindrical shell, the Hagen–Poiseuille equation (2.9-11) for laminar flow of a liquid in circular tubes was obtained. A derivation is also given in Section 3.6 using a differential momentum balance. This can be written as

$$\Delta p_f = (p_1 - p_2)_f = \frac{32\mu v(L_2 - L_1)}{D^2} \tag{2.10-2}$$

where p_1 is upstream pressure at point 1, N/m^2; p_2, pressure at point 2; v is average velocity in tube, m/s; D is inside diameter, m; and $(L_2 - L_1)$ or ΔL is length of straight tube, m. For English units, the right-hand side of Eq. (2.10-2) is divided by g_c.

The quantity $(p_1 - p_2)_f$ or Δp_f is the pressure loss due to skin friction. Then, for constant ρ, the friction loss F_f is

$$F_f = \frac{(p_1 - p_2)_f}{\rho} = \frac{N \cdot m}{kg} \quad \text{or} \quad \frac{J}{kg} \quad \text{(SI)}$$

$$\tag{2.10-3}$$

$$F_f = \frac{ft \cdot lb_f}{lb_m} \quad \text{(English)}$$

This is the mechanical-energy loss due to skin friction for the pipe in $N \cdot m/kg$ of fluid and is part of the $\sum F$ term for frictional losses in the mechanical-energy-balance equation (2.7-28). This term $(p_1 - p_2)_f$ for skin friction loss is different from the $(p_1 - p_2)$ term, owing to velocity head or potential head changes in Eq. (2.7-28). That part of $\sum F$ which arises from friction within the channel itself by laminar or turbulent flow is discussed in Sections 2.10B and in 2.10C. The part of friction loss due to fittings (valves, elbows, etc.), bends, and the like, which sometimes constitute a large part of the friction, is discussed in Section 2.10F. Note that if Eq. (2.7-28) is applied to steady flow in a straight, horizontal tube, we obtain $(p_1 - p_2)/\rho = \sum F$.

One of the uses of Eq. (2.10-2) is in the experimental measurement of the viscosity of a fluid by measuring the pressure drop and volumetric flow rate through a tube of known length and diameter. Slight corrections for kinetic energy and entrance effects are usually necessary in practice. Also, Eq. (2.10-2) is often used in the metering of small liquid flows.

EXAMPLE 2.10-1. Metering of Small Liquid Flows
A small capillary with an inside diameter of 2.22×10^{-3} m and a length 0.317 m is being used to continuously measure the flow rate of a liquid having a density of 875 kg/m^3 and $\mu = 1.13 \times 10^{-3}$ Pa·s. The pressure-drop reading across the capillary during flow is 0.0655 m water (density 996 kg/m^3). What is the flow rate in m^3/s if end-effect corrections are neglected?

Solution: Assuming that the flow is laminar, Eq. (2.10-2) will be used. First, to convert the height h of 0.0655 m water to a pressure drop using Eq. (2.2-4),

$$\Delta p_f = h\rho g = (0.0655 \text{ m})\left(996 \frac{kg}{m^3}\right)\left(9.80665 \frac{m}{s^2}\right)$$

$$= 640 \text{ kg} \cdot m/s^2 \cdot m^2 = 640 \text{ N}/m^2$$

Substituting the following values into Eq. (2.10-2) of $\mu = 1.13 \times 10^{-3}$ Pa·s,

$L_2 - L_1 = 0.317$ m, $D = 2.22 \times 10^{-3}$ m, and $\Delta p_f = 640$ N/m^2, and solving for v,

$$\Delta p_f = \frac{32\mu v(L_2 - L_1)}{D^2} \tag{2.10-2}$$

$$640 = \frac{32(1.13 \times 10^{-3})(v)(0.317)}{(2.22 \times 10^{-3})^2}$$

$$v = 0.275 \text{ m/s}$$

The volumetric rate is then

$$\text{volumetric flow rate} = v\pi \frac{D^2}{4} = \frac{0.275(\pi)(2.22 \times 10^{-3})^2}{4}$$

$$= 1.066 \times 10^{-6} \text{ m}^3/\text{s}$$

Since it was assumed that laminar flow is occurring, the Reynolds number will be calculated to check this.

$$N_{Re} = \frac{Dv\rho}{\mu} = \frac{(2.22 \times 10^{-3})(0.275)(875)}{1.13 \times 10^{-3}} = 473$$

Hence, the flow is laminar as assumed.

2. Use of friction factor for friction loss in laminar flow. A common parameter used in laminar and especially in turbulent flow is the *Fanning friction factor*, *f*, which is defined as the drag force per wetted surface unit area (shear stress τ_s at the surface) divided by the product of density times velocity head or $\frac{1}{2}\rho v^2$. The force is Δp_f times the cross-sectional area πR^2 and the wetted surface area is $2\pi R \, \Delta L$. Hence, the relation between the pressure drop due to friction and *f* is as follows for laminar and turbulent flow.

$$f = \frac{\tau_s}{\rho v^2/2} = \frac{\Delta p_f \, \pi R^2}{2\pi R \, \Delta L} \bigg/ \frac{\rho v^2}{2} \tag{2.10-4}$$

Rearranging, this becomes

$$\Delta p_f = 4f\rho \frac{\Delta L}{D} \frac{v^2}{2} \qquad \text{(SI)}$$

$$\Delta p_f = 4f\rho \frac{\Delta L}{D} \frac{v^2}{2g_c} \qquad \text{(English)} \tag{2.10-5}$$

$$F_f = \frac{\Delta p_f}{\rho} = 4f \frac{\Delta L}{D} \frac{v^2}{2} \qquad \text{(SI)}$$

$$F_f = 4f \frac{\Delta L}{D} \frac{v^2}{2g_c} \qquad \text{(English)} \tag{2.10-6}$$

For laminar flow only, combining Eqs. (2.10-2) and (2.10-5),

$$f = \frac{16}{N_{Re}} = \frac{16}{Dv\rho/\mu} \tag{2.10-7}$$

Equations (2.10-2), (2.10-5), (2.10-6), and (2.10-7) for laminar flow hold up to a Reynolds number of 2100. Beyond that at an N_{Re} above 2100, Eqs. (2.10-2) and (2.10-7) do not hold for turbulent flow. For turbulent flow Eqs. (2.10-5) and (2.10-6), however, are used extensively along with empirical methods of predicting the friction factor f, as discussed in the next section.

EXAMPLE 2.10-2. Use of Friction Factor in Laminar Flow
Assume the same known conditions as in Example 2.10-1 except that the velocity of 0.275 m/s is known and the pressure drop Δp_f is to be predicted. Use the Fanning friction factor method.

Solution: The Reynolds number is, as before,

$$N_{Re} = \frac{Dv\rho}{\mu} = \frac{(2.22 \times 10^{-3} \text{ m})(0.275 \text{ m/s})(875 \text{ kg/m}^3)}{1.13 \times 10^{-3} \text{ kg/m} \cdot \text{s}} = 473$$

From Eq. (2.10-7) the friction factor f is

$$f = \frac{16}{N_{Re}} = \frac{16}{473} = 0.0338 \quad \text{(dimensionless)}$$

Using Eq. (2.10-5) with $\Delta L = 0.317$ m, $v = 0.275$ m/s, $D = 2.22 \times 10^{-3}$ m, $\rho = 875$ kg/m^3,

$$\Delta p_f = 4f\rho\, \frac{\Delta L}{D} \frac{v^2}{2} = \frac{4(0.0338)(875)(0.317)(0.275)^2}{(2.22 \times 10^{-3})(2)} = 640 \text{ N/m}^2$$

This, of course, checks the value in Example 2.10-1.

When the fluid is a gas and not a liquid, the Hagen–Poiseuille equation (2.10-2) can be written as follows for laminar flow.

$$m = \frac{\pi D^4 M(p_1^2 - p_2^2)}{128(2RT)\mu(L_2 - L_1)} \quad \text{(SI)}$$

$$m = \frac{\pi D^4 g_c M(p_1^2 - p_2^2)}{128(2RT)\mu(L_2 - L_1)} \quad \text{(English)}$$

(2.10-8)

where m = kg/s, M = molecular weight in kg/kg mol, T = absolute temperature in K, and $R = 8314.3$ N $\cdot$ m/kg mol $\cdot$ K. In English units, $R = 1545.3$ ft $\cdot$ lb$_f$/lb mol $\cdot$ °R.

2.10C   Pressure Drop and Friction Factor in Turbulent Flow

In turbulent flow, as in laminar flow, the friction factor also depends on the Reynolds number. However, it is not possible to predict theoretically the Fanning friction factor f for turbulent flow as it was done for laminar flow. The friction factor must be determined empirically (experimentally) and it not only depends upon the Reynolds number but also on surface roughness of the pipe. In laminar flow the roughness has essentially no effect.

Dimensional analysis also shows the dependence of the friction factor on these factors. In Sections 3.8 and 4.14 (Selected Topic) methods of obtaining the dimensionless numbers and their importance are discussed.

A large number of experimental data on friction factors of smooth pipe and pipes of varying degrees of equivalent roughness have been obtained and the data correlated. For design purposes to predict the friction factor f and, hence, the frictional pressure drop of round pipe, the friction factor chart in Fig. 2.10-3 can be used. It is a log–log plot of f

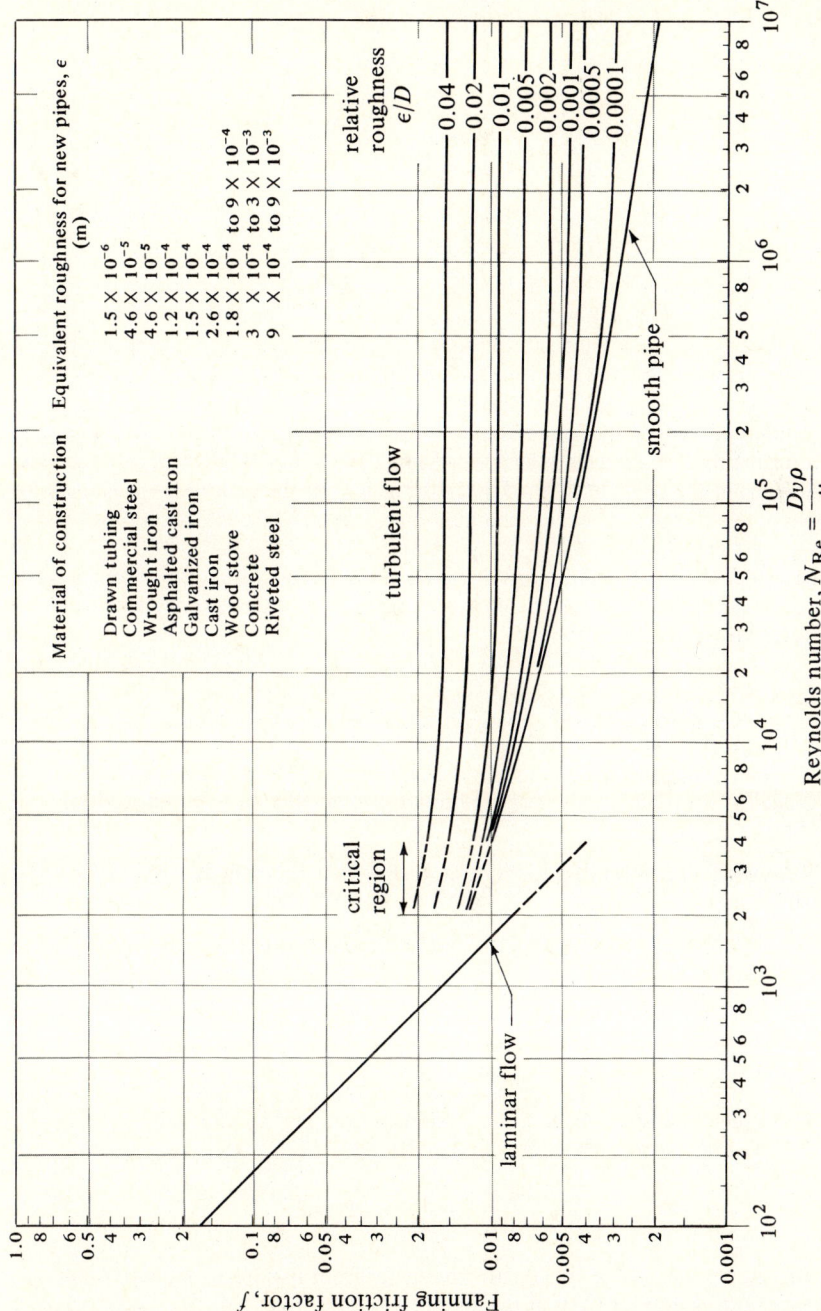

FIGURE 2.10-3. *Friction factors for fluids inside pipes.* [*Based on L. F. Moody, Trans. A.S.M.E.,* **66**, *671,* (*1944*); *Mech. Eng.* **69**, *1005* (*1947*). *With permission.*]

Material of construction	Equivalent roughness for new pipes, ϵ (m)
Drawn tubing	1.5×10^{-6}
Commercial steel	4.6×10^{-5}
Wrought iron	4.6×10^{-5}
Asphalted cast iron	1.2×10^{-4}
Galvanized iron	1.5×10^{-4}
Cast iron	2.6×10^{-4}
Wood stove	1.8×10^{-4} to 9×10^{-4}
Concrete	3×10^{-4} to 3×10^{-3}
Riveted steel	9×10^{-4} to 9×10^{-3}

relative
roughness
ϵ/D

0.04
0.02
0.01
0.005
0.002
0.001
0.0005
0.0001

turbulent flow

smooth pipe

critical
region

laminar flow

Reynolds number, $N_{\mathrm{Re}} = \dfrac{D\upsilon\rho}{\mu}$

Fanning friction factor, f

versus N_{Re}. This friction factor f is then used in Eqs. (2.10-5) and (2.10-6) to predict the friction loss Δp_f or F_f.

$$\Delta p_f = 4f\rho \frac{\Delta L}{D} \frac{v^2}{2} \quad \text{(SI)}$$

$$\Delta p_f = 4f\rho \frac{\Delta L}{D} \frac{v^2}{2g_c} \quad \text{(English)}$$

(2.10-5)

$$F_f = \frac{\Delta p_f}{\rho} = 4f \frac{\Delta L}{D} \frac{v^2}{2} \quad \text{(SI)}$$

$$F_f = 4f \frac{\Delta L}{D} \frac{v^2}{2g_c} \quad \text{(English)}$$

(2.10-6)

For the region with a Reynolds number below 2100, the line is the same as Eq. (2.10-7). For a Reynolds number above 4000 for turbulent flow, the lowest line in Fig. 2.10-3 represents the friction factor line for smooth pipes and tubes, such as glass tubes, and drawn copper and brass tubes. The other lines for higher friction factors represent lines for different relative roughness factors, ε/D, where D is the inside pipe diameter in m and ε is a roughness parameter, which represents the average height in m of roughness projections from the wall (M1). On Fig. 2.10-3 values of the equivalent roughness for new pipes are given (M1). The most common pipe, commercial steel, has a roughness of $\varepsilon = 4.6 \times 10^{-5}$ m $(1.5 \times 10^{-4}$ ft$)$.

The reader should be cautioned on using friction factors f from other sources. The Fanning friction factor of f in Eq. (2.10-6) is the one used here. Others use a friction factor that may be 4 times larger.

EXAMPLE 2.10-3. Use of Friction Factor in Turbulent Flow
A liquid is flowing through a horizontal straight commercial steel pipe at 4.57 m/s. The pipe used is commercial steel, schedule 40, 2-in. nominal diameter. The viscosity of the liquid is 4.46 cp and the density 801 kg/m³. Calculate the mechanical-energy friction loss F_f in J/kg for a 36.6-m section of pipe.

Solution: The following data are given: From Appendix A.5, $D = 0.0525$ m, $v = 4.57$ m/s, $\rho = 801$ kg/m³, $\Delta L = 36.6$ m, and

$$\mu = (4.46 \text{ cp})(1 \times 10^{-3}) = 4.46 \times 10^{-3} \text{ kg/m} \cdot \text{s}$$

The Reynolds number is calculated as

$$N_{Re} = \frac{Dv\rho}{\mu} = \frac{0.0525(4.57)(801)}{4.46 \times 10^{-3}} = 4.310 \times 10^4$$

Hence, the flow is turbulent. For commercial steel pipe from the table in Fig. 2.10-3, the equivalent roughness is 4.6×10^{-5} m.

$$\frac{\varepsilon}{D} = \frac{4.6 \times 10^{-5} \text{ m}}{0.0525 \text{ m}} = 0.00088$$

For an N_{Re} of 4.310×10^4, the friction factor from Fig. 2.10-3 is $f = 0.0060$. Substituting into Eq. (2.10-6), the friction loss is

$$F_f = 4f \frac{\Delta L}{D} \frac{v^2}{2} = \frac{4(0.0060)(36.6)(4.57)^2}{(0.0525)(2)} = 174.8 \frac{J}{kg} \left(58.5 \frac{ft \cdot lb_f}{lb_m}\right)$$

In problems involving the friction loss F_f in pipes, F_f is usually the unknown with the diameter D, velocity v, and pipe length ΔL known. Then a direct solution is possible as in Example 2.10-3. However, in some cases, the friction loss F_f is already set by the available head of liquid. Then if the volumetric flow rate and pipe length are set, the unknown to be calculated is the diameter. This solution is by trial and error since the velocity v appears in both N_{Re} and f, which are unknown. In another case, with the F_f being again already set, the diameter and pipe length are specified. This is also by trial and error, to calculate the velocity. Example 2.10-4 indicates the method to be used to calculate the pipe diameter with F_f set. Others (M2) give a convenient chart to aid in these types of calculations.

EXAMPLE 2.10-4.   *Trial-and-Error Solution to Calculate Pipe Diameter*
Water at 4.4°C is to flow through a horizontal commercial steel pipe having a length of 305 m at the rate of 150 gal/min. A head of water of 6.1 m is available to overcome the friction loss F_f. Calculate the pipe diameter.

Solution: From Appendix A.2 the density $\rho = 1000$ kg/m^3 and the viscosity μ is

$$\mu = (1.55 \text{ cp})(1 \times 10^{-3}) = 1.55 \times 10^{-3} \text{ kg/m} \cdot \text{s}$$

friction loss $F_f = (6.1 \text{ m}) g = (6.1)(9.80665) = 59.82$ J/kg

$$\text{flow rate} = \left(150 \frac{\text{gal}}{\text{min}} \right)\left(\frac{1 \text{ ft}^3}{7.481 \text{ gal}} \right)\left(\frac{1 \text{ min}}{60 \text{ s}} \right)(0.028317 \text{ m}^3/\text{ft}^3)$$

$$= 9.46 \times 10^{-3} \text{ m}^3/\text{s}$$

$$\text{area of pipe} = \frac{\pi D^2}{4} \text{ m}^2 \quad (D \text{ is unknown})$$

$$\text{velocity } v = (9.46 \times 10^{-3} \text{ m}^3/\text{s})\left(\frac{1}{\pi D^2/4 \text{ m}^2} \right) = \frac{0.01204}{D^2} \text{ m/s}$$

The solution is by trial and error since v appears in N_{Re} and f. Assume that $D = 0.089$ m for first trial.

$$N_{Re} = \frac{Dv\rho}{\mu} = (0.089) \frac{0.01204(1000)}{(0.089)^2(1.55 \times 10^{-3})} = 8.730 \times 10^4$$

For commercial steel pipe and using Fig. 2.10-3, $\varepsilon = 4.6 \times 10^{-5}$ m. Then,

$$\frac{\varepsilon}{D} = \frac{4.6 \times 10^{-5} \text{ m}}{0.089 \text{ m}} = 0.00052$$

From Fig. 2.10-3 for $N_{Re} = 8.730 \times 10^4$ and $\varepsilon/D = 0.00052$, $f = 0.0051$. Substituting into Eq. (2.10-6),

$$F_f = 59.82 = 4f \frac{\Delta L}{D} \frac{v^2}{2} = \frac{4(0.0051)(305)}{D(2)} \frac{(0.01204)^2}{D^4}$$

Solving for D, $D = 0.0945$ m. This does not check the assumed value of 0.089 m.

For the second trial D will be assumed as 0.0945 m.

$$N_{Re} = (0.0945) \frac{0.01204}{(0.0945)^2} \frac{1000}{1.55 \times 10^{-3}} = 8.220 \times 10^4$$

$$\frac{\varepsilon}{D} = \frac{4.6 \times 10^{-5}}{0.0945} = 0.00049$$

From Fig. 2.10-3, $f = 0.0052$. It can be seen that f does not change much with N_{Re} in the turbulent region.

$$F_f = 59.82 = \frac{4(0.0052)(305)}{D(2)} \frac{(0.01204)^2}{D^4}$$

Solving, $D = 0.0954$ m or 3.75 in. Hence, the solution checks the assumed value of D closely.

2.10D   Pressure Drop and Friction Factor in Flow of Gases

The equations and methods discussed in this section for turbulent flow in pipes hold for incompressible liquids. They also hold for a gas if the density (or the pressure) changes less than 10%. Then an average density, ρ_{av} in kg/m³, should be used and the errors involved will be less than the uncertainty limits in the friction factor f. For gases, Eq. (2.10-5) can be rewritten as follows for laminar and turbulent flow:

$$(p_1 - p_2)_f = \frac{4f\,\Delta L G^2}{D2\rho_{av}} \tag{2.10-9}$$

where ρ_{av} is the density at $p_{av} = (p_1 + p_2)/2$. Also, the N_{Re} used is DG/μ, where G is kg/m²·s and is a constant independent of the density and velocity variation for the gas. Equation (2.10-5) can also be written for gases as

$$p_1^2 - p_2^2 = \frac{4f\,\Delta L G^2 R T}{DM} \qquad \text{(SI)}$$

$$p_1^2 - p_2^2 = \frac{4f\,\Delta L G^2 R T}{g_c\,DM} \qquad \text{(English)} \tag{2.10-10}$$

where R is 8314.3 J/kg mol · K or 1545.3 ft · lb$_f$/lb mol · °R and M is molecular weight.

The derivation of Eqs. (2.10-9) and (2.10-10) applies only to cases with gases where the relative pressure change is small enough so that large changes in velocity do not occur. If the exit velocity becomes large, the kinetic-energy term, which has been omitted, becomes important. For pressure changes above about 10%, compressible flow is occurring and the reader should refer to Section 2.11 (Selected Topic). In adiabatic flow in a uniform pipe, the velocity in the pipe cannot exceed the velocity of sound.

EXAMPLE 2.10-5.   Flow of Gas in Line and Pressure Drop
Nitrogen gas at 25°C is flowing in a smooth tube having an inside diameter of 0.010 m at the rate of 9.0 kg/s · m². The tube is 200 m long and the flow can be assumed to be isothermal. The pressure at the entrance to the tube is 2.0265×10^5 Pa. Calculate the outlet pressure.

Solution: The viscosity of the gas from Appendix A.3 is $\mu = 1.77 \times 10^{-5}$ Pa·s at $T = 298.15$ K. Inlet gas pressure $p_1 = 2.0265 \times 10^5$ Pa, $G = 9.0$ kg/s·m², $D = 0.010$ m, $M = 28.02$ kg/kg mol, $\Delta L = 200$ m, and $R = 8314.3$ J/kg mol · K. Assuming that Eq. (2.10-10) holds for this case and that the pressure drop is less than 10%, the Reynolds number is

$$N_{Re} = \frac{DG}{\mu} = \frac{0.010(9.0)}{1.77 \times 10^{-5}} = 5085$$

Hence, the flow is turbulent. Using Fig. 2.10-3, $f = 0.0090$ for a smooth tube.

Substituting into Eq. (2.10-10),

$$p_1^2 - p_2^2 = \frac{4f \, \Delta L G^2 R T}{DM}$$

$$(2.0265 \times 10^5)^2 - p_2^2 = \frac{4(0.0090)(200)(9.0)^2(8314.3)(298.15)}{0.010(28.02)}$$

$$4.1067 \times 10^{10} - p_2^2 = 0.5160 \times 10^{10}$$

Solving, $p_2 = 1.895 \times 10^5$ Pa. Hence, Eq. (2.10-10) can be used since the pressure drop is less than 10%.

2.10E Effect of Heat Transfer on Friction Factor

The friction factor f given in Fig. 2.10-3 is given for isothermal flow, i.e., no heat transfer. When a fluid is being heated or cooled, the temperature gradient will cause a change in physical properties of the fluid, especially the viscosity. For engineering practice the following method of Sieder and Tate (P1, S3) can be used to predict the friction factor for nonisothermal flow for liquids and gases.

1. Calculate the mean bulk temperature t_a as the average of the inlet and outlet bulk fluid temperatures.
2. Calculate the N_{Re} using the viscosity μ_a at t_a and use Fig. 2.10-3 to obtain f.
3. Using the tube wall temperature t_w, determine μ_w at t_w.
4. Calculate ψ for the case occurring below.

$$\psi = \left(\frac{\mu_a}{\mu_w}\right)^{0.17} \qquad \text{(heating)} \ N_{Re} > 2100 \qquad \textbf{(2.10-11)}$$

$$\psi = \left(\frac{\mu_a}{\mu_w}\right)^{0.11} \qquad \text{(cooling)} \ N_{Re} > 2100 \qquad \textbf{(2.10-12)}$$

$$\psi = \left(\frac{\mu_a}{\mu_w}\right)^{0.38} \qquad \text{(heating)} \ N_{Re} < 2100 \qquad \textbf{(2.10-13)}$$

$$\psi = \left(\frac{\mu_a}{\mu_w}\right)^{0.23} \qquad \text{(cooling)} \ N_{Re} < 2100 \qquad \textbf{(2.10-14)}$$

5. The final friction factor is obtained by dividing the f from step 2 by the ψ from step 4.

Hence, when the liquid is being heated, ψ is greater than 1.0 and the final f decreases. The reverse occurs on cooling the liquid.

2.10F Friction Losses in Expansion, Contraction, and Pipe Fittings

Skin friction losses in flow through straight pipe are calculated by using the Fanning friction factor. However, if the velocity of the fluid is changed in direction or magnitude, additional friction losses occur. This results from additional turbulence which develops because of vortices and other factors. Methods to estimate these losses are discussed below.

1. Sudden enlargement losses. If the cross section of a pipe enlarges very gradually, very little or no extra losses are incurred. If the change is sudden, it results in additional losses due to eddies formed by the jet expanding in the enlarged section. This friction loss can be calculated by the following for turbulent flow in both sections. This Eq. (2.8-36) was derived in Example 2.8-4.

$$h_{ex} = \frac{(v_1 - v_2)^2}{2\alpha} = \left(1 - \frac{A_1}{A_2}\right)^2 \frac{v_1^2}{2\alpha} = K_{ex} \frac{v_1^2}{2\alpha} \frac{J}{kg} \tag{2.10-15}$$

where h_{ex} is the friction loss in J/kg, K_{ex} is the expansion-loss coefficient and $= (1 - A_1/A_2)^2$, v_1 is the upstream velocity in the smaller area in m/s, v_2 is the downstream velocity, and $\alpha = 1.0$. If the flow is laminar in both sections, the factor α in the equation becomes $\frac{1}{2}$. For English units the right-hand side of Eq. (2.10-15) is divided by g_c. Also, $h = \text{ft} \cdot \text{lb}_f/\text{lb}_m$.

2. *Sudden contraction losses.* When the cross section of the pipe is suddenly reduced, the stream cannot follow around the sharp corner, and additional frictional losses due to eddies occur. For turbulent flow, this is given by

$$h_c = 0.55 \left(1 - \frac{A_2}{A_1}\right) \frac{v_2^2}{2\alpha} = K_c \frac{v_2^2}{2\alpha} \frac{J}{kg} \tag{2.10-16}$$

where h_c is the friction loss, $\alpha = 1.0$ for turbulent flow, v_2 is the average velocity in the smaller or downstream section, and K_c is the contraction-loss coefficient (P1) and approximately equals $0.55\,(1 - A_2/A_1)$. For laminar flow, the same equation can be used with $\alpha = \frac{1}{2}$(S2). For English units the right side is divided by g_c.

TABLE 2.10-1. *Friction Loss for Turbulent Flow Through Valves and Fittings*

Type of Fitting or Valve	Frictional Loss, Number of Velocity Heads, K_f	Frictional Loss, Equivalent Length of Straight Pipe in Pipe Diameters, L_e/D
Elbow, 45°	0.35	17
Elbow, 90°	0.75	35
Tee	1	50
Return bend	1.5	75
Coupling	0.04	2
Union	0.04	2
Gate valve		
Wide open	0.17	9
Half open	4.5	225
Globe valve		
Wide open	6.0	300
Half open	9.5	475
Angle valve, wide open	2.0	100
Check valve		
Ball	70.0	3500
Swing	2.0	100
Water meter, disk	7.0	350

Source: R. H. Perry and C. H. Chilton, *Chemical Engineers' Handbook*, 5th ed. New York: McGraw-Hill Book Company, 1973. With permission.

Transport Processes: Momentum, Heat, and Mass

3. *Losses in fittings and valves.* Pipe fittings and valves also disturb the normal flow lines in a pipe and cause additional friction losses. In a short pipe with many fittings, the friction loss from these fittings could be greater than in the straight pipe. The friction loss for fittings and valves is given by the following equation:

$$h_f = K_f \frac{v_1^2}{2} \qquad (2.10\text{-}17)$$

where K_f is the loss factor for the fitting or valve and v_1 is the average velocity in the pipe leading to the fitting. Experimental values for K_f are given in Table 2.10-1 for turbulent flow (P1) and in Table 2.10-2 for laminar flow.

As an alternative method, some texts and references (B1) give data for losses in fittings as an equivalent pipe length in pipe diameters. These data, also given in Table 2.10-1, are presented as L_e/D, where L_e is the equivalent length of straight pipe in m having the same frictional loss as the fitting, and D is the inside pipe diameter in m. The K values in Eqs. (2.10-15) and (2.10-16) can be converted to L_e/D values by multiplying the K by 50 (P1). The L_e values for the fittings are simply added to the length of the straight pipe to get the total length of equivalent straight pipe to use in Eq. (2.10-6).

4. *Frictional losses in mechanical-energy-balance equation.* The frictional losses from the friction in the straight pipe (Fanning friction), enlargement losses, contraction losses, and losses in fittings and valves are all incorporated in the $\sum F$ term of Eq. (2.7-28) for the mechanical-energy balance, so that

$$\sum F = 4f \frac{\Delta L}{D} \frac{v^2}{2} + K_{ex} \frac{v_1^2}{2} + K_c \frac{v_2^2}{2} + K_f \frac{v_1^2}{2} \qquad (2.10\text{-}18)$$

If all the velocities, v, v_1, and v_2, are the same, then factoring, Eq. (2.10-18) becomes, for this special case,

$$\sum F = \left(4f \frac{\Delta L}{D} + K_{ex} + K_c + K_f \right) \frac{v^2}{2} \qquad (2.10\text{-}19)$$

The use of the mechanical-energy-balance equation (2.7-28) along with Eq. (2.10-18) will be shown in the following examples.

EXAMPLE 2.10-6. *Friction Losses and Mechanical-Energy Balance*
An elevated storage tank contains water at 82.2°C as shown in Fig. 2.10-4. It is desired to have a discharge rate at point 2 of 0.223 ft³/s. What must be the height H in ft of the surface of the water in the tank relative to the

TABLE 2.10-2. *Friction Loss for Laminar Flow Through Valves and Fittings (K1)*

Type of Fitting or Valve	Frictional Loss, Number of Velocity Heads, K_f Reynolds Number					
	50	100	200	400	1000	Turbulent
Elbow, 90°	17	7	2.5	1.2	0.85	0.75
Tee	9	4.8	3.0	2.0	1.4	1.0
Globe valve	28	22	17	14	10	6.0
Check valve, swing	55	17	9	5.8	3.2	2.0

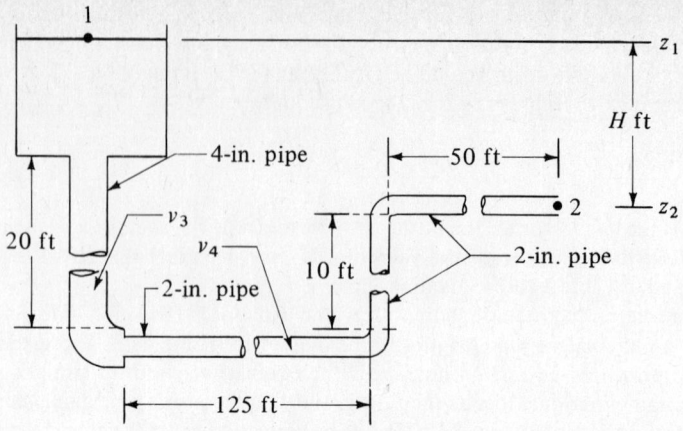

FIGURE 2.10-4. *Process flow diagram for Example 2.10-6.*

discharge point? The pipe used is commercial steel pipe, schedule 40, and the lengths of the straight portions of pipe are shown.

Solution: The mechanical-energy-balance equation (2.7-28) is written between points 1 and 2.

$$z_1 \frac{g}{g_c} + \frac{v_1^2}{2\alpha g_c} + \left(\frac{p_1}{\rho_1} - \frac{p_2}{\rho_2}\right) - W_S = z_2 \frac{g}{g_c} + \frac{v_2^2}{2\alpha g_c} + \sum F \quad \textbf{(2.10-20)}$$

From Appendix A.2, for water, $\rho = 0.970(62.43) = 60.52$ lb$_m$/ft^3 and $\mu = 0.347$ cp $= 0.347(6.7197 \times 10^{-4}) = 2.33 \times 10^{-4}$ lb$_m$/ft·s. The diameters of the pipes are

$$\text{For 4-in. pipe:} \quad D_3 = \frac{4.026}{12} = 0.3353 \text{ ft}; \qquad A_3 = 0.0884 \text{ ft}^2$$

$$\text{For 2-in. pipe:} \quad D_4 = \frac{2.067}{12} = 0.1722 \text{ ft}; \qquad A_4 = 0.02330 \text{ ft}^2$$

The velocities in the 4-in. and 2-in. pipe are

$$v_3 = \frac{0.223 \text{ ft}^3/\text{s}}{0.0884 \text{ ft}^2} = 2.523 \text{ ft/s} \quad \text{(4-in. pipe)}$$

$$v_4 = \frac{0.223}{0.02330} = 9.57 \text{ ft/s} \quad \text{(2-in. pipe)}$$

The $\sum F$ term for frictional losses in the system includes the following: (1) contraction loss at tank exit, (2) friction in the 4-in. straight pipe, (3) friction in 4-in. elbow, (4) contraction loss from 4-in. to 2-in. pipe, (5) friction in the 2-in. straight pipe, and (6) friction in the two 2-in. elbows. Calculations for the six items are as follows.

1. Contraction loss at tank exit. From Eq. (2.10-16) for contraction from

Transport Processes: Momentum, Heat, and Mass

A_1 to A_3 cross-sectional area since A_1 of the tank is very large compared to A_3,

$$K_c = 0.55\left(1 - \frac{A_3}{A_1}\right) = 0.55(1 - 0) = 0.55$$

$$h_c = K_c \frac{v_3^2}{2g_c} = 0.55 \frac{(2.523)^2}{2(32.174)} = 0.054 \text{ ft} \cdot \text{lb}_f/\text{lb}_m$$

2. Friction in the 4-in. pipe. The Reynolds number is

$$N_{Re} = \frac{D_3 v_3 \rho}{\mu} = \frac{0.3353(2.523)(60.52)}{2.33 \times 10^{-4}} = 2.193 \times 10^5$$

Hence, the flow is turbulent. From Fig. 2.10-3, $\varepsilon = 4.6 \times 10^{-5}$ m $(1.5 \times 10^{-4}$ ft$)$.

$$\frac{\varepsilon}{D_3} = \frac{0.00015}{0.3353} = 0.000448$$

Then, for $N_{Re} = 219\,300$, the Fanning friction factor $f = 0.0047$. Substituting to Eq. (2.10-6) for $\Delta L = 20.0$ ft of 4-in. pipe,

$$F_f = 4f \frac{\Delta L}{D} \frac{v^2}{2g_c} = 4(0.0047) \frac{20.0}{0.3353} \frac{(2.523)^2}{2(32.174)} = 0.111 \frac{\text{ft} \cdot \text{lb}_f}{\text{lb}_m}$$

3. Friction in 4-in. elbow. From Table 2.10-1, $K_f = 0.75$. Then, substituting into Eq. (2.10-17),

$$h_f = K_f \frac{v^2}{2g_c} = 0.75 \frac{(2.523)^2}{2(32.174)} = 0.074 \frac{\text{ft} \cdot \text{lb}_f}{\text{lb}_m}$$

4. Contraction loss from 4- to 2-in. pipe. Using Eq. (2.10-16) again for contraction from A_3 to A_4 cross-sectional area,

$$K_c = 0.55\left(1 - \frac{A_4}{A_3}\right) = 0.55\left(1 - \frac{0.02330}{0.0884}\right) = 0.405$$

$$h_c = K_c \frac{v_4^2}{2g_c} = 0.405 \frac{(9.57)^2}{2(32.174)} = 0.575 \frac{\text{ft} \cdot \text{lb}_f}{\text{lb}_m}$$

5. Friction in the 2-in. pipe. The Reynolds number is

$$N_{Re} = \frac{D_4 v_4 \rho}{\mu} = \frac{0.1722(9.57)(60.52)}{2.33 \times 10^{-4}} = 4.280 \times 10^5$$

$$\frac{\varepsilon}{D} = \frac{0.00015}{0.1722} = 0.00087$$

The Fanning friction factor from Fig. 2.10-3 is $f = 0.0048$. The total length $\Delta L = 125 + 10 + 50 = 185$ ft. Substituting into Eq. (2.10-6),

$$F_f = 4f \frac{\Delta L}{D} \frac{v^2}{2g_c} = 4(0.0048) \frac{185(9.57)^2}{(0.1722)(2)(32.174)} = 29.4 \frac{\text{ft} \cdot \text{lb}_f}{\text{lb}_m}$$

6. Friction in the two 2-in. elbows. For a $K_f = 0.75$ and two elbows,

$$h_f = 2K_f \frac{v^2}{2g_c} = \frac{2(0.75)(9.57)^2}{2(32.174)} = 2.136 \frac{\text{ft} \cdot \text{lb}_f}{\text{lb}_m}$$

The total frictional loss $\sum F$ is the sum of items (1) through (6).

$$\sum F = 0.054 + 0.111 + 0.074 + 0.575 + 29.4 + 2.136$$

$$= 32.35 \text{ ft} \cdot \text{lb}_f/\text{lb}_m$$

Using as a datum level z_2, $z_1 = H$ ft, $z_2 = 0$. Since turbulent flow exists, $\alpha = 1.0$. Also, $v_1 = 0$ and $v_2 = v_4 = 9.57$ ft/s. Since p_1 and p_2 are both at 1 atm abs pressure and $\rho_1 = \rho_2$,

$$\frac{p_1}{\rho} - \frac{p_2}{\rho} = 0$$

Also, since no pump is used, $W_S = 0$. Substituting these values into Eq. (2.10-20),

$$H \frac{g}{g_c} + 0 + 0 - 0 = 0 + \frac{1(9.57)^2}{2(32.174)} + 32.35$$

Solving, $H(g/g_c) = 33.77$ ft · lb$_f$/lb$_m$ (100.9 J/kg) and H is 33.77 ft (10.3 m) height of water level above the discharge outlet.

EXAMPLE 2.10-7. Friction Losses with Pump in Mechanical-Energy Balance

Water at 20°C is being pumped from a tank to an elevated tank at the rate of 5.0×10^{-3} m³/s. All of the piping in Fig. 2.10-5 is 4-in. schedule 40 pipe. The pump has an efficiency of 65%. Calculate the kW power needed for the pump.

Solution: The mechanical-energy-balance equation (2.7-28) is written between points 1 and 2, with point 1 being the reference plane.

$$\frac{1}{2\alpha} (v_{2\,av}^2 - v_{1\,av}^2) + g(z_2 - z_1) + \frac{p_2 - p_1}{\rho} + \sum F + W_S = 0 \quad \textbf{(2.7-28)}$$

From Appendix A.2 for water, $p = 998.2$ kg/m³, $\mu = 1.005 \times 10^{-3}$ Pa·s. For 4-in. pipe from Appendix A.5, $D = 0.1023$ m and $A = 8.219 \times 10^{-3}$ m². The velocity in the pipe is $v = 5.0 \times 10^{-3}/(8.219 \times 10^{-3}) = 0.6083$ m/s. The Reynolds number is

$$N_{Re} = \frac{Dv_\rho}{\mu} = \frac{0.1023(0.6083)(998.2)}{1.005 \times 10^{-3}} = 6.181 \times 10^4$$

Hence, the flow is turbulent.

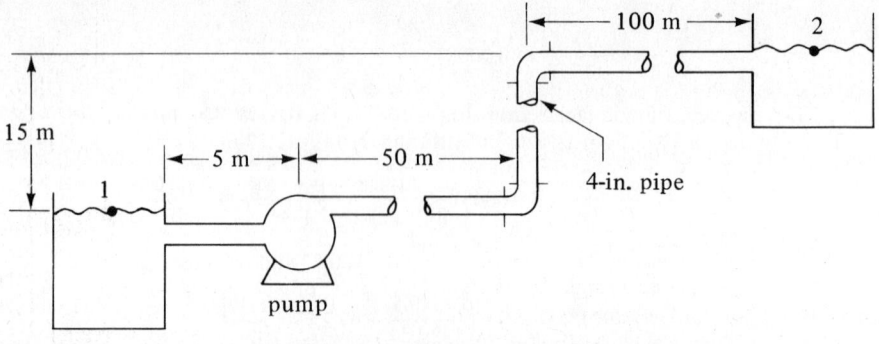

15 m

5 m · 50 m · 100 m · 2

1 · 4-in. pipe

pump

FIGURE 2.10-5. Process flow diagram for Example 2.10-7.

The $\sum F$ term for frictional losses includes the following: (1) contraction loss at tank exit, (2) friction in the straight pipe, (3) friction in the two elbows, and (4) expansion loss at the tank entrance.

1. *Contraction loss at tank exit.* From Eq. (2.10-16) for contraction from a large A_1 to a small A_2,

$$k_c = 0.55\left(1 - \frac{A_2}{A_1}\right) = 0.55(1 - 0) = 0.55$$

$$h_c = K_c \frac{v^2}{2\alpha} = (0.55)\frac{(0.6083)^2}{2(1.0)} = 0.102 \text{ J/kg}$$

2. *Friction in the straight pipe.* From Fig. 2.10-3, $\varepsilon = 4.6 \times 10^{-5}$ m and $\varepsilon/D = 4.6 \times 10^{-5}/0.1023 = 0.00045$. Then for $N_{Re} = 6.181 \times 10^4$, $f = 0.0051$. Substituting into Eq. (2.10-6) for $\Delta L = 5 + 50 + 15 + 100 = 170$ m,

$$F_f = 4f \frac{\Delta L}{D} \frac{v^2}{2} = 4(0.0051)\frac{170}{0.1023}\frac{(0.6083)^2}{2} = 6.272 \text{ J/kg}$$

3. *Friction in the two elbows.* From Table 2.10-1, $K_f = 0.75$. Then, substituting into Eq. (2.10-7) for two elbows,

$$h_f = 2K_f \frac{v^2}{2} = 2(0.75)\frac{(0.6083)^2}{2} = 0.278 \text{ J/kg}$$

4. *Expansion loss at the tank entrance.* Using Eq. (2.10-15),

$$K_{ex} = \left(1 - \frac{A_1}{A_2}\right)^2 = (1 - 0)^2 = 1.0$$

$$h_{ex} = K_{ex} \frac{v^2}{2} = 1.0 \frac{(0.6083)^2}{2} = 0.185 \text{ J/kg}$$

The total frictional loss is $\sum F$.

$$\sum F = 0.102 + 6.272 + 0.278 + 0.185 = 6.837 \text{ J/kg}$$

Substituting into Eq. (2.7-28), where $(v_1^2 - v_2^2) = 0$ and $(p_2 - p_1) = 0$,

$$0 + 9.806(15.0 - 0) + 0 + 6.837 + W_S = 0$$

Solving, $W_S = -153.93$ J/kg. The mass flow rate is $m = 5.0 \times 10^{-3}(998.2) = 4.991$ kg/s. Using Eq. (2.7-30),

$$W_S = -\eta W_p$$

$$-153.93 = -0.65W_p$$

Solving, $W_p = 236.8$ J/kg. The pump kW power is

$$\text{pump kW} = mW_p = \frac{4.991(236.8)}{1000} = 1.182 \text{ kW}$$

2.10G   Friction Loss in Noncircular Conduits

The friction loss in long straight channels or conduits of noncircular cross section can be estimated by using the same equations employed for circular pipes if the diameter in the Reynolds number and in the friction factor equation (2.10-6) is taken as the equivalent diameter. The equivalent diameter D is defined as four times the hydraulic radius r_H. The hydraulic radius is defined as the ratio of the cross-sectional area of the channel to the wetted perimeter of the channel for turbulent flow only. Hence,

$$D = 4r_H = 4 \frac{\text{cross-sectional area of channel}}{\text{wetted perimeter of channel}} \qquad (2.10\text{-}21)$$

For example, for a circular tube,

$$D = \frac{4(\pi D^2/4)}{\pi D} = D$$

For an annular space with outside diameter D_1 and inside D_2,

$$D = \frac{4(\pi D_1^2/4 - \pi D_2^2/4)}{\pi D_1 + \pi D_2} = D_1 - D_2 \qquad (2.10\text{-}22)$$

For a rectangular duct of sides a and b ft,

$$D = \frac{4(ab)}{2a + 2b} = \frac{2ab}{a + b} \qquad (2.10\text{-}23)$$

For open channels and partly filled ducts in turbulent flow, the equivalent diameter and Eq. (2.10-6) are also used (P1). For a rectangle with depth of liquid y and width b,

$$D = \frac{4(by)}{b + 2y} \qquad (2.10\text{-}24)$$

For a wide, shallow stream of depth y,

$$D = 4y \qquad (2.10\text{-}25)$$

For laminar flow in ducts running full and in open channels with various cross-sectional shapes other than circular, equations are given elsewhere (P1).

2.10H Entrance Section of a Pipe

If the velocity profile at the entrance region of a tube is flat, a certain length of the tube is necessary for the velocity profile to be fully established. This length for the establishment of fully developed flow is called the transition length or entry length. This is shown in Fig. 2.10-6 for laminar flow. At the entrance the velocity profile is flat; i.e., the velocity is the same at all positions. As the fluid progresses down the tube, the boundary-layer thickness increases until finally they meet at the center of the pipe and the parabolic velocity profile is fully established.

The approximate entry length L_e of a pipe having a diameter of D for a fully

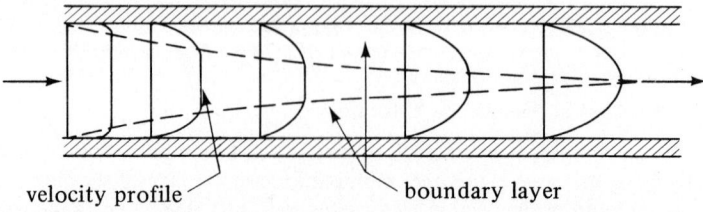

velocity profile   boundary layer

FIGURE 2.10-6.   *Velocity profiles near a pipe entrance for laminar flow.*

developed velocity profile to be formed in laminar flow is (L2)

$$\frac{L_e}{D} = 0.0575 N_{Re} \qquad\qquad (2.10\text{-}26)$$

For turbulent flow, no relation is available to predict the entry length for a fully developed turbulent velocity profile to form. As an approximation, the entry length is nearly independent of the Reynolds number and is fully developed after 50 diameters downstream.

> **EXAMPLE 2.10-8. Entry Length for a Fluid in a Pipe**
> Water at 20°C is flowing through a tube having a diameter of 0.010 m at a velocity of 0.10 m/s.
>   (a) Calculate the entry length.
>   (b) Calculate the entry length for turbulent flow.
>
> **Solution:** For part (a), from Appendix A.2, $\rho = 998.2$ kg/m^3, $\mu = 1.005 \times 10^{-3}$ Pa·s. The Reynolds number is
>
> $$N_{Re} = \frac{Dv\rho}{\mu} = \frac{0.010(0.10)(998.2)}{1.005 \times 10^{-3}} = 993.2$$
>
> Using Eq. (2.10-26) for laminar flow,
>
> $$\frac{L_e}{D} = \frac{L_e}{0.01} = 0.0575(993.2) = 57.1$$
>
> Hence, $L_e = 0.571$ m.
>   For turbulent flow in part (b), $L_e = 50(0.01) = 0.50$ m.

The pressure drop or friction factor in the entry length is greater than in fully developed flow. For laminar flow the friction factor is highest at the entrance (L2) and then decreases smoothly to the fully developed flow value. For turbulent flow there will be some portion of the entrance over which the boundary layer is laminar and the friction factor profile is difficult to express. As an approximation the friction factor for the entry length can be taken as two to three times the value of the friction factor in fully developed flow.

2.10I Selection of Pipe Sizes

In large or complex process piping systems, the optimum size of pipe to use for a specific situation depends upon the relative costs of capital investment, power, maintenance, and so on. Charts are available for determining these optimum sizes (P1). However, for small installations approximations are usually sufficiently accurate. A table of representative values of ranges of velocity in pipes are shown in Table 2.10-3.

TABLE 2.10-3. *Representative Ranges of Velocities in Steel Pipes*

		Velocity	
Type of Fluid	*Type of Flow*	*ft/s*	*m/s*
Nonviscous liquid	Inlet to pump	2–3	0.6–0.9
	Process line or pump discharge	5–8	1.5–2.5
Viscous liquid	Inlet to pump	0.2–0.8	0.06–0.25
	Process line or pump discharge	0.5–2	0.15–0.6
Gas		30–120	9–36
Steam		30–75	9–23

2.11 (SELECTED TOPIC) COMPRESSIBLE FLOW OF GASES

2.11A Introduction and Basic Equation for Flow in Pipes

When pressure changes in gases occur which are greater than about 10%, the friction-loss equations (2.10-9) and (2.10-10) may be in error since compressible flow is occurring. Then the solution of the energy balance is more complicated because of the variation of the density or specific volume with changes in pressure. The field of compressible flow is very large and covers a very wide range of variations in geometry, pressure, velocity, and temperature. In this section we restrict our discussion to isothermal and adiabatic flow in uniform, straight pipes and do not cover flow in nozzles, which is discussed in some detail in other references (M2, P1).

The general mechanical energy-balance equation (2.7-27) can be used as a starting point. Assuming turbulent flow, so that $\alpha = 1.0$; no shaft work, so that $W_S = 0$; and writing the equation for a differential length dL, Eq. (2.7-27) becomes

$$v \, dv + g \, dz + \frac{dp}{\rho} + dF = 0 \tag{2.11-1}$$

For a horizontal duct, $dz = 0$. Using only the wall shear frictional term for dF and writing Eq. (2.10-6) in differential form,

$$v \, dv + V \, dp + \frac{4 f v^2 \, dL}{2D} = 0 \tag{2.11-2}$$

where $V = 1/\rho$. Assuming steady-state flow and a uniform pipe diameter, G is constant and

$$G = v\rho = \frac{v}{V} \tag{2.11-3}$$

$$dv = G \, dV \tag{2.11-4}$$

Substituting Eqs. (2.11-3) and (2.11-4) into (2.11-2) and rearranging,

$$G^2 \frac{dV}{V} + \frac{dp}{V} + \frac{2fG^2}{D} \, dL = 0 \tag{2.11-5}$$

This is the basic differential equation that is to be integrated. To do this the relation between V and p must be known so that the integral of dp/V can be evaluated. This integral depends upon the nature of the flow and two important conditions used are isothermal and adiabatic flow in pipes.

2.11B Isothermal Compressible Flow

To integrate Eq. (2.11-5) for isothermal flow, an ideal gas will be assumed where

$$pV = \frac{1}{M} RT \tag{2.11-6}$$

Solving for V in Eq. (2.11-6) and substituting it into Eq. (2.11-5), and integrating assuming f is constant,

$$G^2 \int_1^2 \frac{dV}{V} + \frac{M}{RT} \int_1^2 p \, dp + 2f \frac{G^2}{D} \int_1^2 dL = 0 \tag{2.11-7}$$

$$G^2 \ln \frac{V_2}{V_1} + \frac{M}{2RT} (p_2^2 - p_1^2) + 2f \frac{G^2}{D} \Delta L = 0 \tag{2.11-8}$$

Substituting p_1/p_2 for V_2/V_1 and rearranging,

$$p_1^2 - p_2^2 = \frac{4f\Delta L G^2 RT}{DM} + \frac{2G^2 RT}{M} \ln \frac{p_1}{p_2} \qquad \text{(2.11-9)}$$

where M = molecular weight in kg mass/kg mol, $R = 8314.34$ N·m/kg mol·K, and T = temperature K. The quantity $RT/M = p_{av}/\rho_{av}$, where $p_{av} = (p_1 + p_2)/2$ and ρ_{av} is the average density at T and p_{av}. In English units, $R = 1545.3$ ft·lb$_f$/lb mol·°R and the right-hand terms are divided by g_c. Equation (2.11-9) then becomes

$$(p_1 - p_2)_f = \frac{4f\Delta L G^2}{2D\rho_{av}} + \frac{G^2}{\rho_{av}} \ln \frac{p_1}{p_2} \qquad \text{(2.11-10)}$$

The first term on the right of Eqs. (2.11-9) and (2.11-10) represents the frictional loss as given by Eqs. (2.10-9) and (2.10-10). The last term in both equations is generally negligible in ducts of appreciable lengths unless the pressure drop is very large.

EXAMPLE 2.11-1. Compressible Flow of a Gas in a Pipe Line

Natural gas, which is essentially methane, is being pumped through a 1.016-m-ID pipeline for a distance of 1.609×10^5 m (D1) at a rate of 2.077 kg mol/s. It can be assumed that the line is isothermal at 288.8 K. The pressure p_2 at the discharge end of the line is 170.3×10^3 Pa absolute. Calculate the pressure p_1 at the inlet of the line. The viscosity of methane at 288.8 K is 1.04×10^{-5} Pa·s.

Solution: $D = 1.016$ m, $A = \pi D^2/4 = \pi(1.016)^2/4 = 0.8107$ m^2. Then,

$$G = \left(2.077 \frac{\text{kg mol}}{\text{s}}\right)\left(16.0 \frac{\text{kg}}{\text{kg mol}}\right)\left(\frac{1}{0.8107 \text{ m}^2}\right) = 41.00 \frac{\text{kg}}{\text{s·m}^2}$$

$$N_{Re} = \frac{DG}{\mu} = \frac{1.016(41.00)}{1.04 \times 10^{-5}} = 4.005 \times 10^6$$

From Fig. 2.10-3, $\varepsilon = 4.6 \times 10^5$ m.

$$\frac{\varepsilon}{D} = \frac{4.6 \times 10^{-5}}{1.016} = 0.0000453$$

The friction factor $f = 0.0027$.

In order to solve for p_1 in Eq. (2.11-9), trial and error must be used. Estimating p_1 at 620.5×10^3 Pa, $R = 8314.34$ N·m/kg mol·K, and $\Delta L = 1.609 \times 10^5$ m. Substituting into Eq. (2.11-9),

$$p_1^2 - p_2^2 = \frac{4(0.0027)(1.609 \times 10^5)(41.00)^2(8314.34)(288.8)}{1.016(16.0)}$$

$$+ \frac{2(41.00)^2(8314.34)(288.8)}{(16.0)} \ln \frac{620.5 \times 10^3}{170.3 \times 10^3}$$

$$= 4.375 \times 10^{11} + 0.00652 \times 10^{11} = 4.382 \times 10^{11} \text{ (Pa)}^2$$

Now, $P_2 = 170.3 \times 10^3$ Pa. Substituting this into the above and solving for p_1, $p_1 = 683.5 \times 10^3$ Pa. Substituting this new value of p_1 into Eq. (2.11-9) again and solving for p_1, the final result is $p_1 = 683.5 \times 10^3$ Pa. Note that the last term in Eq. (2.11-9) in this case is almost negligible.

When the upstream pressure p_1 remains constant, the mass flow rate G changes as the downstream pressure p_2 is varied. From Eq. (2.11-9), when $p_1 = p_2$, $G = 0$ and when $p_2 = 0$, $G = 0$. This indicates that at some intermediate value of p_2, the flow G must be a

Principles of Momentum Transfer and Overall Balances

maximum. This means that the flow is a maximum when $dG/dp_2 = 0$. Performing this differentiation on Eq. (2.11-9) for constant p_1 and f and solving for G,

$$G_{max} = \sqrt{\frac{Mp_2^2}{RT}} \tag{2.11-11}$$

Using Eqs. (2.11-3) and (2.11-6),

$$v_{max} = \sqrt{\frac{RT}{M}} = \sqrt{p_2 V_2} \tag{2.11-12}$$

This is the equation for the velocity of sound in the fluid at the conditions for isothermal flow. Thus, for isothermal compressible flow there is a maximum flow for a given upstream p_1 and further reduction of p_2 will not give any further increase in flow. Further details as to the length of pipe and the pressure at the maximum flow conditions are discussed elsewhere (D1, M2, P1).

EXAMPLE 2.11-2.   Maximum Flow for Compressible Flow of a Gas
For the conditions of Example 2.11-1, calculate the maximum velocity that can be obtained and the velocity of sound at these conditions. Compare with Example 2.11-1.

Solution: Using Eq. (2.11-12) and the conditions in Example 2.11-1,

$$v_{max} = \sqrt{\frac{RT}{M}} = \sqrt{\frac{8314(288.8)}{16.0}} = 387.4 \text{ m/s}$$

This is the maximum velocity obtainable if p_2 is decreased. This is also the velocity of sound in the fluid at the conditions for isothermal flow. To compare with Example 2.11-1, the actual velocity at the exit pressure p_2 is obtained by combining Eqs. (2.11-3) and (2.11-6) to give

$$v_2 = \frac{RTG}{p_2 M} \tag{2.11-13}$$

$$= \frac{8314.34(288.8)(41.00)}{(170.3 \times 10^3)16.0} = 36.13 \text{ m/s}$$

2.11   Adiabatic Compressible Flow

When heat transfer through the wall of the pipe is negligible, the flow of gas in compressible flow in a straight pipe of constant cross section is adiabatic. Equation (2.11-5) has been integrated for adiabatic flow and details are given elsewhere (D1, M1, P1). Convenient charts to solve this case are also available (P1). The results for adiabatic flow often deviate very little from isothermal flow, especially in long lines. For very short pipes and relatively large pressure drops, the adiabatic flow rate is greater than the isothermal, but the maximum possible difference is about 20% (D1). For pipes of length of about 1000 diameters or longer, the difference is generally less than 5%. Equation (2.11-8) can also be used when the temperature change over the conduit is small by using an arithmetic average temperature.

Using the same procedures for finding a maximum flow that were used in the isothermal case, the maximum flow occurs when the velocity at the downstream end of the pipe is the sonic velocity for adiabatic flow. This is

$$v_{max} = \sqrt{\gamma p_2 V_2} = \sqrt{\frac{\gamma RT}{M}} \tag{2.11-14}$$

where, $\gamma = c_p/c_v$, the ratio of heat capacities. For air, $\gamma = 1.4$. Hence, the maximum velocity for adiabatic flow is about 20% greater than for isothermal flow. The rate of flow may not be limited by the flow conditions in the pipe, in practice, but by the development of sonic velocity in a fitting or valve in the pipe. Hence, care should be used in selection of fittings in such pipes for compressible flow. Further details as to the length of pipe and pressure at the maximum flow conditions are given elsewhere (D1, M2, P1).

A convenient parameter often used in compressible flow equations is the Mach number, N_{Ma}, which is defined as the ratio of v, the speed of the fluid in the conduit, to v_{max}, the speed of sound in the fluid at the actual flow conditions.

$$N_{Ma} = \frac{v}{v_{max}} \tag{2.11-15}$$

At a Mach number of 1.0, the flow is sonic. At a value less than 1.0, the flow is subsonic, and supersonic at a number above 1.0.

PROBLEMS

2.2-1. *Pressure in a Spherical Tank.* Calculate the pressure in psia and kN/m² in a spherical tank at the bottom of the tank filled with oil having a diameter of 8.0 ft. The top of the tank is vented to the atmosphere having a pressure of 14.72 psia. The density of the oil is 0.922 g/cm³.

      Ans.   17.92 lb$_f$/in.² (psia), 123.5 kN/m²

2.2-2. *Pressure with Two Liquids, Hg and Water.* An open test tube at 293 K is filled at the bottom with 12.1 cm of Hg and 5.6 cm of water is placed above the Hg. Calculate the pressure at the bottom of the test tube if the atmospheric pressure is 756 mm Hg. Use a density of 13.55 g/cm³ for Hg and 0.998 g/cm³ for water. Give the answer in terms of dyn/cm², psia, and kN/m². See Appendix A.1 for conversion factors.

      Ans. 1.175×10^6 dyn/cm², 17.0 psia, 2.3 psig, 117.5 kN/m²

2.2-3. *Head of a Fluid of Jet Fuel and Pressure.* The pressure at the top of a tank of jet fuel is 180.6 kN/m². The depth of liquid in the tank is 6.4 m. The density of the fuel is 825 kg/m³. Calculate the head of the liquid in m which corresponds to the absolute pressure at the bottom of the tank.

2.2-4 *Measurement of Pressure.* An open U-tube manometer similar to Fig. 2.2-4a is being used to measure the absolute pressure p_a in a vessel containing air. The pressure p_b is atmospheric pressure, which is 754 mm Hg. The liquid in the manometer is water having a density of 1000 kg/m³. Assume that the density ρ_B is 1.30 kg/m³ and that the distance Z is very small. The reading R is 0.415 m. Calculate p_a in psia and kPa.

      Ans. $p_a = 15.17$ psia, 104.6 kPa

2.2-5. *Measurement of Small Pressure Differences.* The two-fluid U-tube manometer is being used to measure the difference in pressure at two points in a line containing air at 1 atm abs pressure. The value of $R_0 = 0$ for equal pressures. The lighter fluid is a hydrocarbon with a density of 812 kg/m³ and the heavier water has a density of 998 kg/m³. The inside diameters of the U tube and reservoir are 3.2 mm and 54.2 mm, respectively. The reading R of the manometer is 117.2 mm. Calculate the pressure difference in mm Hg and pascal.

2.2-6. *Pressure in a Sea Lab.* A sea lab 5.0 m high is to be designed to withstand submersion to 150 m, measured from the sea level to the top of the sea lab. Calculate the pressure on top of the sea lab and also the pressure variation on the side of the container measured as the distance x in m from the top of the sea lab downward. The density of seawater is 1020 kg/m³.

      Ans. $p = 10.00(150 + x)$ kN/m²

2.2-7. Measurement of Pressure Difference in Vessels. In Fig. 2.2-5b the differential manometer is used to measure the pressure difference between two vessels. Derive the equation for the pressure difference $p_A - p_B$ in terms of the liquid heights and densities.

2.2-8. Design of Settler and Separator for Immiscible Liquids. A vertical cylindrical settler-separator is to be designed for separating a mixture flowing at $20.0 \, \text{m}^3/\text{h}$ and containing equal volumes of a light petroleum liquid ($\rho_B = 875 \, \text{kg/m}^3$) and a dilute solution of wash water ($\rho_A = 1050 \, \text{kg/m}^3$). Laboratory experiments indicate a settling time of 15 min is needed to adequately separate the two phases. For design purposes use a 25-min settling time and calculate the size of the vessel needed, the liquid levels of the light and heavy liquids in the vessel, and the height h_{A2} of the heavy liquid overflow. Assume that the ends of the vessel are approximately flat, that the vessel diameter equals its height, and that one-third of the volume is vapor space vented to the atmosphere. Use the nomenclature given in Fig. 2.2-6.

Ans. $h_{A2} = 1.537 \, \text{m}$

2.3-1. Molecular Transport of a Property with a Variable Diffusivity. A property is being transported through a fluid at steady state through a constant cross-sectional area. At point 1 the concentration Γ_1 is 2.78×10^{-2} amount of property/m^3 and 1.50×10^{-2} at point 2 at a distance of 2.0 m away. The diffusivity depends on concentration Γ as follows.

$$\delta = A + B\Gamma = 0.150 + 1.65\Gamma.$$

(a) Derive the integrated equation for the flux in terms of Γ_1 and Γ_2. Then, calculate the flux.
(b) Calculate Γ at $z = 1.0$ m and plot Γ versus z for the three points.

Ans. (a) $\psi_z = [A(\Gamma_1 - \Gamma_2) + (B/2)(\Gamma_1^2 - \Gamma_2^2)]/(z_2 - z_1)$

2.3-2. Integration of General Property Equation for Steady State. Integrate the general property equation (2.3-11) for steady state and no generation between the points Γ_1 at z_1 and Γ_2 at z_2. The final equation should relate Γ to z.

Ans. $\Gamma = (\Gamma_2 - \Gamma_1)(z - z_1)/(z_2 - z_1) + \Gamma_1$

2.4-1. Shear Stress in Soybean Oil. Using Fig. 2.4-1, the distance between the two parallel plates is 0.00914 m and the lower plate is being pulled at a relative velocity of 0.366 m/s greater than the top plate. The fluid used is soybean oil with viscosity of 4×10^{-2} Pa·s at 303 K (Appendix A.4).

(a) Calculate the shear stress τ and the shear rate using lb force, ft, and s units.
(b) Repeat, using SI units.
(c) If glycerol at 293 K having a viscosity of 1.069 kg/m·s is used instead of soybean oil, what relative velocity in m/s is needed using the same distance between plates so that the same shear stress is obtained as in part (a)? Also, what is the new shear rate?

Ans. (a) Shear stress = $3.34 \times 10^{-2} \, \text{lb}_f/\text{ft}^2$, shear rate = $40.0 \, \text{s}^{-1}$;
(b) $1.60 \, \text{N/m}^2$; (c) relative velocity = 0.01369 m/s, shear rate = $1.50 \, \text{s}^{-1}$

2.4-2. Shear Stress and Shear Rate in Fluids. Using Fig. 2.4-1, the lower plate is being pulled at a relative velocity of 0.40 m/s greater than the top plate. The fluid used is water at 24°C.

(a) How far apart should the two plates be placed so that the shear stress τ is $0.30 \, \text{N/m}^2$? Also, calculate the shear rate.
(b) If oil with a viscosity of 2.0×10^{-2} Pa·s is used instead at the same plate spacing and velocity as in part (a), what is the shear stress and the shear rate?

2.5-1. Reynolds Number for Milk Flow. Whole milk at 293 K having a density of $1030 \, \text{kg/m}^3$ and viscosity of 2.12 cp is flowing at the rate of 0.605 kg/s in a glass pipe having a diameter of 63.5 mm.

Transport Processes: Momentum, Heat, and Mass

(a) Calculate the Reynolds number. Is this turbulent flow?

(b) Calculate the flow rate needed in m^3/s for a Reynolds number of 2100 and the velocity in m/s.

$\qquad$**Ans.** (a) $N_{Re} = 5723$, turbulent flow

2.5-2. *Pipe Diameter and Reynolds Number.* An oil is being pumped inside a 10.0-mm-diameter pipe at a Reynolds number of 2100. The oil density is $855 \, kg/m^3$ and the viscosity is $2.1 \times 10^{-2} \, Pa \cdot s$.

(a) What is the velocity in the pipe?

(b) It is desired to maintain the same Reynolds number of 2100 and the same velocity as in part (a) using a second fluid with a density of $925 \, kg/m^3$ and a viscosity of $1.5 \times 10^{-2} \, Pa \cdot s$. What pipe diameter should be used?

2.6-1. *Average Velocity for Mass Balance in Flow Down Vertical Plate.* For a layer of liquid flowing in laminar flow in the z direction down a vertical plate or surface, the velocity profile is

$$v_z = \frac{\rho g \delta^2}{2\mu}\left[1 - \left(\frac{x}{\delta}\right)^2\right]$$

where δ is the thickness of the layer, x is the distance from the free surface of the liquid toward the plate, and v_z is the velocity at a distance x from the free surface.

(a) What is the maximum velocity $v_{z\,max}$?

(b) Derive the expression for the average velocity $v_{z\,av}$ and also relate it to $v_{z\,max}$.

$\qquad$**Ans.** (a) $v_{z\,max} = \rho g \delta^2/2\mu$, (b) $v_{z\,av} = \frac{2}{3} v_{z\,max}$

2.6-2. *Flow of Liquid in a Pipe and Mass Balance.* A hydrocarbon liquid enters a simple flow system shown in Fig. 2.6-1 at an average velocity of 1.282 m/s, where $A_1 = 4.33 \times 10^{-3} \, m^2$ and $\rho_1 = 902 \, kg/m^3$. The liquid is heated in the process and the exit density is $875 \, kg/m^3$. The cross-sectional area at point 2 is $5.26 \times 10^{-3} \, m^2$. The process is steady state.

(a) Calculate the mass flow rate m at the entrance and exit.

(b) Calculate the average velocity v in 2 and the mass velocity G in 1.

$\qquad$**Ans.** (a) $m_1 = m_2 = 5.007$ kg/s, (b) $G_1 = 1156$ kg/s $\cdot m^2$

2.6-3. *Average Velocity for Mass Balance in Turbulent Flow.* For turbulent flow in a smooth circular tube with a radius of R, the velocity profile varies according to the following expression at a Reynolds number of about 10^5:

$$v = v_{max}\left(\frac{R-r}{R}\right)^{1/7}$$

where r is the radial distance from the center and v_{max} the maximum velocity at the center. Derive the equation relating the average velocity (bulk velocity) v_{av} to v_{max} for an incompressible fluid. (*Hint*: The integration can be simplified by substituting z for $R - r$.)

$\qquad$**Ans.** $v_{av} = \left(\dfrac{49}{60}\right) v_{max} = 0.817 v_{max}$

2.6-4. *Bulk Velocity for Flow Between Parallel Plates.* A fluid flowing in laminar flow in the x direction between two parallel plates has a velocity profile given by the following.

$$v_x = v_{x\,max}\left[1 - \left(\frac{y}{y_0}\right)^2\right]$$

where $2y_0$ is the distance between the plates, y is the distance from the center line, and v_x is the velocity in the x direction at position y. Derive an equation relating $v_{x\,\text{av}}$ (bulk or average velocity) with $v_{x\,\text{max}}$.

2.6-5. Overall Mass Balance for Dilution Process. A well-stirred storage vessel contains 10 000 kg of solution of a dilute methanol solution ($w_A = 0.05$ mass fraction alcohol). A constant flow of 500 kg/min of pure water is suddenly introduced into the tank and a constant rate of withdrawal of 500 kg/min of solution is started. These two flows are continued and remain constant. Assuming that the densities of the solutions are the same and that the total contents of the tank remain constant at 10 000 kg of solution, calculate the time for the alcohol content to drop to 1.0 wt %.

Ans. 32.2 min

2.6-6. Overall Mass Balance for Unsteady-State Process. A storage vessel is well stirred and contains 500 kg of total solution with a concentration of 5.0 % salt. A constant flow rate of 900 kg/h of salt solution containing 16.67 % salt is suddenly introduced into the tank and a constant withdrawal rate of 600 kg/h is also started. These two flows remain constant thereafter. Derive an equation relating the outlet withdrawal concentration as a function of time. Also, calculate the concentration after 2.0 h.

2.6-7 Mass Balance for Flow of Sucrose Solution. A 20 wt % sucrose (sugar) solution having a density of 1074 kg/m^3 is flowing through the same piping system as Example 2.6-1 (Fig. 2.6-2). The flow rate entering pipe 1 is 1.892 m^3/h. The flow divides equally in each of pipes 3. Calculate the following:
(a) The velocity in m/s in pipes 2 and 3.
(b) The mass velocity G kg/m$^2 \cdot$ s in pipes 2 and 3.

2.7-1. Kinetic-Energy Velocity Correction Factor for Turbulent Flow. Derive the equation to determine the value of α, the kinetic-energy velocity correction factor, for turbulent flow. Use Eq. (2.7-20) to approximate the velocity profile and substitute this into Eq. (2.7-15) to obtain $(v^3)_{\text{av}}$. Then use Eqs. (2.7-20), (2.6-17), and (2.7-14) to obtain α.

Ans. $\alpha = 0.9448$

2.7-2. Flow Between Parallel Plates and Kinetic-Energy Correction Factor. The equation for the velocity profile for a fluid flowing in laminar flow between two parallel plates is given in Problem 2.6-4. Derive the equation to determine the value of the kinetic-energy velocity correction factor α. [*Hint*: First derive an equation relating v to v_{av}. Then derive the equation for $(v^3)_{\text{av}}$ and, finally, relate these results to α.]

2.7-3. Temperature Drop in Throttling Valve and Energy Balance. Steam is flowing through an adiabatic throttling valve (no heat loss or external work). Steam enters point 1 upstream of the valve at 689 kPa abs and 171.1°C and leaves the valve (point 2) at 359 kPa. Calculate the temperature t_2 at the outlet. [*Hint*: Use Eq. (2.7-21) for the energy balance and neglect the kinetic-energy and potential-energy terms as shown in Example 2.7-1. Obtain the enthalpy H_1 from Appendix A.2, steam tables. For H_2, linear interpolation of the values in the table will have to be done to obtain t_2.] Use SI units.

Ans. $t_2 = 160.6$°C

2.7-4. Energy Balance on a Heat Exchanger and a Pump. Water at 93.3°C is being pumped from a large storage tank at 1 atm abs at a rate of 0.189 m^3/min by a pump. The motor that drives the pump supplies energy to the pump at the rate of 1.49 kW. The water is pumped through a heat exchanger, where it gives up 704 kW of heat and is then delivered to a large open storage tank at an elevation of 15.24 m above the first tank. What is the final temperature of the water to the second tank? Also, what is the gain in enthalpy of the water due to the work input? (*Hint :* Be sure and use the steam tables for the enthalpy of the water. Neglect any kinetic-energy changes, but not potential-energy changes.)

Ans. $t_2 = 38.2°C$, work input gain $= 0.491$ kJ/kg

2.7-5. *Steam Boiler and Overall Energy Balance.* Liquid water under pressure at 150 kPa enters a boiler at 24°C through a pipe at an average velocity of 3.5 m/s in turbulent flow. The exit steam leaves at a height of 25 m above the liquid inlet at 150°C and 150 kPa absolute and the velocity in the outlet line is 12.5 m/s in turbulent flow. The process is steady state. How much heat must be added per kg of steam?

2.7-6. *Energy Balance on a Flow System with a Pump and Heat Exchanger.* Water stored in a large, well-insulated storage tank at 21.0°C and atmospheric pressure is being pumped at steady state from this tank by a pump at the rate of 40 m^3/h. The motor driving the pump supplies energy at the rate of 8.5 kW. The water is used as a cooling medium and passes through a heat exchanger where 255 kW of heat is added to the water. The heated water then flows to a second, large vented tank, which is 25 m above the first tank. Determine the final temperature of the water delivered to the second tank.

2.7-7. *Mechanical-Energy Balance in Pumping Soybean Oil.* Soybean oil is being pumped through a uniform-diameter pipe at a steady mass-flow rate. A pump supplies 209.2 J/kg mass of fluid flowing. The entrance abs pressure in the inlet pipe to the pump is 103.4 kN/m^2. The exit section of the pipe downstream from the pump is 3.35 m above the entrance and the exit pressure is 172.4 kN/m^2. Exit and entrance pipes are the same diameter. The fluid is in turbulent flow. Calculate the friction loss in the system. See Appendix A.4 for the physical properties of soybean oil. The temperature is 303 K.

Ans. $\sum F = 101.3$ J/kg

2.7-8. *Pump Horsepower in Brine System.* A pump pumps 0.200 ft^3/s of brine solution having a density of 1.15 g/cm^3 from an open feed tank having a large cross-sectional area. The suction line has an inside diameter of 3.548 in. and the discharge line from the pump a diameter of 2.067 in. The discharge flow goes to an open overhead tank and the open end of this line is 75 ft above the liquid level in the feed tank. If the friction losses in the piping system are 18.0 ft · lb_f/lb_m, what pressure must the pump develop and what is the horsepower of the pump if the efficiency is 70%? The flow is turbulent.

2.7-9. *Pressure Measurements from Flows.* Water having a density of 998 kg/m^3 is flowing at the rate of 1.676 m/s in a 3.068-in.-diameter horizontal pipe at a pressure p_1 of 68.9 kPa abs. It then passes to a pipe having an inside diameter of 2.067 in.

(a) Calculate the new pressure p_2 in the 2.067-in. pipe. Assume no friction losses.

(b) If the piping is vertical and the flow is upward, calculate the new pressure p_2. The pressure tap for p_2 is 0.457 m above the tap for p_1.

Ans.   (a) $p_2 = 63.5$ kPa; (b) $p_2 = 59.1$ kPa

2.7-10. *Draining Cotton Seed Oil from a Tank.* A cylindrical tank 1.52 m in diameter and 7.62 m high contains cotton seed oil having a density of 917 kg/m^3. The tank is open to the atmosphere. A discharge nozzle of inside diameter 15.8 mm and cross-sectional area A_2 is located near the bottom of the tank. The surface of the liquid is located at $H = 6.1$ m above the center line of the nozzle. The discharge nozzle is opened, draining the liquid level from $H = 6.1$ m to $H = 4.57$ m. Calculate the time in seconds to do this. [*Hint :* The velocity on the surface of the reservoir is small and can be neglected. The velocity v_2 m/s in the nozzle can be calculated for a given H by Eq. (2.7-36). However, H, and hence v_2, are varying. Set up an unsteady-state mass balance as follows. The volumetric flow rate in the tank is $(A_t \, dH)/dt$, where A_t is the tank cross section in m^2

and $A_t \, dH$ ts the m^3 liquid flowing in dt s. This rate must equal the negative of the volumetric rate in the nozzle, or $-A_2 v_2 \, m^3/s$. The negative sign is present since dH is the negative of v_2. Rearrange this equation and integrate between $H = 6.1$ m at $t = 0$ and $H = 4.57$ m at $t = t_F$.]

Ans. $t_F = 1380\,s$

2.7-11. *Friction Loss in Turbine Water Power System.* Water is stored in an elevated reservoir. To generate power, water flows from this reservoir down through a large conduit to a turbine and then through a similar-sized conduit. At a point in the conduit 89.5 m above the turbine, the pressure is 172.4 kPa and at a level 5 m below the turbine, the pressure is 89.6 kPa. The water flow rate is $0.800 \, m^3/s$. The output of the shaft of the turbine is 658 kW. The water density is $1000 \, kg/m^3$. If the efficiency of the turbine in converting the mechanical energy given up by the fluid to the turbine shaft is 89% ($\eta_t = 0.89$), calculate the friction loss in the turbine in J/kg. Note that in the mechanical-energy-balance equation, the W_s is equal to the output of the shaft of the turbine over η_t.

Ans. $\sum F = 85.3$ J/kg

2.7-12. *Pipeline Pumping of Oil.* A pipeline laid cross country carries oil at the rate of $795 \, m^3/d$. The pressure of the oil is 1793 kPg gage leaving pumping station 1. The pressure is 862 kPa gage at the inlet to the next pumping station, 2. The second station is 17.4 m higher than the first station. Calculate the lost work ($\sum F$ friction loss) in J/kg mass oil. The oil density is $769 \, kg/m^3$.

2.7-13. *Test of Centrifugal Pump and Mechanical-Energy Balance.* A centrifugal pump is being tested for performance and during the test the pressure reading in the 0.305-m-diameter suction line just adjacent to the pump casing is -20.7 kPa (vacuum below atmospheric pressure). In the discharge line with a diameter of 0.254 m at a point 2.53 m above the suction line, the pressure is 289.6 kPa gage. The flow of water from the pump is measured as $0.1133 m^3/s$. (The density can be assumed as $1000 \, kg/m^3$.) Calculate the kW input of the pump.

Ans. 38.11 kW

2.7-14. *Friction Loss in Pump and Flow System.* Water at 20°C is pumped from the bottom of a large storage tank where the pressure is 310.3 kPa gage to a nozzle which is 15.25 m above the tank bottom and discharges to the atmosphere with a velocity in the nozzle of 19.81 m/s. The water flow rate is 45.4 kg/s. The efficiency of the pump is 80% and 7.5 kW are furnished to the pump shaft. Calculate the following.
(a) The friction loss in the pump.
(b) The friction loss in the rest of the process.

2.7-15. *Power for Pumping in Flow System.* Water is being pumped from an open water reservoir at the rate of 2.0 kg/s at 10°C to an open storage tank 1500 m away. The pipe used is schedule 40 $3\frac{1}{2}$-in. pipe and the frictional losses in the system are 625 J/kg. The surface of the water reservoir is 20 m above the level of the storage tank. The pump has an efficiency of 75%.
(a) What is the kW power required for the pump?
(b) If the pump is not present in the system, will there be a flow?

Ans. (a) 1.143 kW

2.8-1. *Momentum Balance in a Reducing Bend.* Water is flowing at steady state through the reducing bend in Fig. 2.8-3. The angle $\alpha_2 = 90°$ (a right-angle bend). The pressure at point 2 is 1.0 atm abs. The flow rate is $0.020 \, m^3/s$ and the diameters at points 1 and 2 are 0.050 m and 0.030 m, respectively. Neglect frictional and gravitational forces. Calculate the resultant forces on the bend in newtons and lb force. Use $\rho = 1000 \, kg/m^3$.

Ans. $-R_x = +450.0$ N, $-R_y = -565.8$ N.

2.8-2. *Forces on Reducing Bend.* Water is flowing at steady state and 363 K at a rate of $0.0566 \, m^3/s$ through a 60° reducing bend ($\alpha_2 = 60°$) in Fig. 2.8-3. The inlet pipe diameter is 0.1016 m and the outlet 0.0762 m. The friction loss in the pipe bend can be estimated as $v_2^2/5$. Neglect gravity forces. The exit pressure $p_2 = 111.5 \text{ kN/m}^2$ gage. Calculate the forces on the bend in newtons.

Ans. $-R_x = +1344 \text{ N}, -R_y = -1026 \text{ N}$

2.8-3. Force of Water Stream on a Wall. Water at 298 K discharges from a nozzle and travels horizontally hitting a flat vertical wall. The nozzle has a diameter of 12 mm and the water leaves the nozzle with a flat velocity profile at a velocity of 6.0 m/s. Neglecting frictional resistance of the air on the jet, calculate the force in newtons on the wall.

Ans. $-R_x = 4.059 \text{ N}$

2.8-4. Flow Through an Expanding Bend. Water at a steady-state rate of $0.050 \text{ m}^3/\text{s}$ is flowing through an expanding bend that changes direction by 120°. The upstream diameter is 0.0762 m and downstream is 0.2112 m. The upstream pressure is 68.94 kPa gage. Neglect energy losses within the elbow and calculate the downstream pressure at 298 K. Also calculate R_x and R_y.

2.8-5. Force of Stream on a Wall. Repeat Problem 2.8-3 for the same conditions except that the wall is inclined 45° with the vertical. The flow is frictionless. Assume no loss in energy. The amount of fluid splitting in each direction along the plate can be determined by using the continuity equation and a momentum balance. Calculate this flow division and the force on the wall.

Ans. $m_2 = 0.5774 \text{ kg/s}, m_3 = 0.09907 \text{ kg/s}, -R_x = 2.030 \text{ N}, -R_y = -2.030 \text{ N}$ (force on wall).

2.8-6. Momentum Balance for Free Jet on a Curved, Fixed Vane. A free jet having a velocity of 30.5 m/s and a diameter of 5.08×10^{-2} m is deflected by a curved, fixed vane as in Fig. 2.8-5a. However, the vane is curved downward at an angle of 60° instead of upward. Calculate the force of the jet on the vane. The density is 1000 kg/m³.

Ans. $-R_x = 942.8 \text{ N}, -R_y = 1633 \text{ N}$

2.8-7. Momentum Balance for Free Jet on a U-Type, Fixed Vane. A free jet having a velocity of 30.5 m/s and a diameter of 1.0×10^{-2} m is deflected by a smooth, fixed vane as in Fig. 2.8-5a. However, the vane is in the form of a U so that the exit jet travels in a direction exactly opposite to the entering jet. Calculate the force of the jet on the vane. Use $\rho = 1000 \text{ kg/m}^3$.

Ans. $-R_x = 146.1 \text{ N}, -R_v = 0$

2.8-8. Momentum Balance on Reducing Elbow and Friction Losses. Water at 20°C is flowing through a reducing bend, where α_2 (see Fig. 2.8-3) is 120°. The inlet pipe diameter is 1.829 m, the outlet is 1.219 m, and the flow rate is 8.50 m³/s. The exit point z_2 is 3.05 m above the inlet and the inlet pressure is 276 kPa gage. Friction losses are estimated as $0.5v_2^2/2$ and the mass of water in the elbow is 8500 kg. Calculate the forces R_x and R_y and the resultant force on the control volume fluid.

2.8-9. Momentum Velocity Correction Factor β for Turbulent Flow. Determine the momentum velocity correction factor β for turbulent flow in a tube. Use Eq. (2.7-20) for the relationship between v and position.

2.9-1. Film of Water on Wetted-Wall Tower. Pure water at 20°C is flowing down a vertical wetted-wall column at a rate of 0.124 kg/s·m. Calculate the film thickness and the average velocity.

Ans. $\delta = 3.370 \times 10^{-4} \text{ m}, v_{z \text{ av}} = 0.3687 \text{ m/s}$

2.9-2. Shell Momentum Balance for Flow Between Parallel Plates. A fluid of constant density is flowing in laminar flow at steady state in the horizontal x direction between two flat and parallel plates. The distance between the two plates in the vertical y direction is $2y_0$. Using a shell momentum balance, derive the equation for the velocity profile within this fluid and the maximum velocity for a distance

L m in the x direction. [*Hint :* See the method used in Section 2.9B to derive Eq. (2.9-9). One boundary condition used is $dv_x/dy = 0$ at $y = 0$.]

$$\textbf{Ans.} \quad v_x = \frac{p_0 - p_L}{2\mu L} y_0^2 \left[1 - \left(\frac{y}{y_0} \right)^2 \right]$$

2.9-3. *Velocity Profile for Non-Newtonian Fluid.* The stress rate of shear for a non-Newtonian fluid is given by

$$\tau_{rx} = K \left(-\frac{dx_x}{dr} \right)^n$$

where K and n are constants. Find the relation between velocity and radial position r for this incompressible fluid at steady state. [*Hint :* Combine the equation given here with Eq. (2.9-6). Then raise both sides of the resulting equation to the $1/n$ power and integrate.]

$$\textbf{Ans.} \quad v_x = \frac{n}{n+1} \left(\frac{p_0 - p_L}{2KL} \right)^{1/n} (R_0)^{(n+1)/n} \left[1 - \left(\frac{r}{R_0} \right)^{(n+1)/n} \right]$$

2.9-4. *Shell Momentum Balance for Flow Down an Inclined Plane.* Consider the case of a Newtonian fluid in steady-state laminar flow down an inclined plane surface that makes an angle θ with the horizontal. Using a shell momentum balance, find the equation for the velocity profile within the liquid layer having a thickness L and the maximum velocity of the free surface. (*Hint :* The convective momentum terms cancel for fully developed flow and the pressure-force terms also cancel, because of the presence of a free surface. Note that there is a gravity force on the fluid.)

$$\textbf{Ans.} \quad v_{x\,max} = \rho g L^2 \sin \theta / 2\mu$$

2.10-1. *Viscosity Measurement of a Liquid.* One use of the Hagen–Poiseuille equation (2.10-2) is in determining the viscosity of a liquid by measuring the pressure drop and velocity of the liquid in a capillary of known dimensions. The liquid used has a density of 912 kg/m^3 and the capillary has a diameter of 2.222 mm and a length of 0.1585 m. The measured flow rate was 5.33×10^{-7} m^3/s of liquid and the pressure drop 131 mm of water (density 996 kg/m^3). Neglecting end effects, calculate the viscosity of the liquid in Pa·s.

$$\textbf{Ans.} \quad \mu = 9.06 \times 10^{-3} \text{ Pa·s}$$

2.10-2. *Frictional Pressure Drop in Flow of Olive Oil.* Calculate the frictional pressure drop in pascal for olive oil at 293 K flowing through a commercial pipe having an inside diameter of 0.0525 m and a length of 76.2 m. The velocity of the fluid is 1.22 m/s. Use the friction factor method. Is the flow laminar or turbulent? Use physical data from Appendix A.4.

2.10-3. *Frictional Loss in Straight Pipe and Effect of Type of Pipe.* A liquid having a density of 801 kg/m^3 and a viscosity of 1.49×10^{-3} Pa·s is flowing through a horizontal straight pipe at a velocity of 4.57 m/s. The commercial steel pipe is $1\frac{1}{2}$-in. nominal pipe size, schedule 40. For a length of pipe of 61 m, do as follows.
(a) Calculate the friction loss F_f.
(b) For a smooth tube of the same inside diameter, calculate the friction loss. What is the percent reduction of the F_f for the smooth tube?

$$\textbf{Ans.} \quad \text{(a) 348.9 J/kg; (b) 274.2 J/kg (91.7 ft·lb}_f\text{/lb}_m\text{), 21.4\% reduction}$$

2.10-4. *Trial-and-Error Solution for Hydraulic Drainage.* In a hydraulic project a cast iron pipe having an inside diameter of 0.156 m and a 305-m length is used to drain wastewater at 293 K. The available head is 4.57 m of water. Neglecting any losses in fittings and joints in the pipe, calculate the flow rate in m^3/s.

(*Hint:* Assume the physical properties of pure water. The solution is trial and error since the velocity appears in N_{Re}, which is needed to determine the friction factor. As a first trial, assume that $v = 1.7$ m/s.)

2.10-5. *Mechanical-Energy Balance and Friction Losses.* Hot water is being discharged from a storage tank at the rate of 0.223 ft^3/s. The process flow diagram and conditions are the same as given in Example 2.10-6, except for different nominal pipe sizes of schedule 40 steel pipe as follows. The 20-ft-long outlet pipe from the storage tank is $1\frac{1}{2}$-in. pipe instead of 4-in. pipe. The other piping, which was 2-in. pipe, is now 2.5-in. pipe. Note that now a sudden expansion occurs after the elbow in the $1\frac{1}{2}$-in. pipe to a $2\frac{1}{2}$-in pipe.

2.10-6. *Friction Losses and Pump Horsepower.* Hot water in an open storage tank at 82.2°C is being pumped at the rate of 0.379 m^3/min from this storage tank. The line from the storage tank to the pump suction is 6.1 m of 2-in. schedule 40 steel pipe and it contains three elbows. The discharge line after the pump is 61 m of 2-in. pipe and contains two elbows. The water discharges to the atmosphere at a height of 6.1 m above the water level in the storage tank.
(a) Calculate all frictional losses $\sum F$.
(b) Make a mechanical-energy balance and calculate W_S of the pump in J/kg.
(c) What is the kW power of the pump if its efficiency is 75%?
$$\textbf{Ans.} \quad \text{(a) } \sum F = 122.8 \text{ J/kg.}$$
$$\text{(b) } W_S = -186.9 \text{ J/kg, (c) } 1.527 \text{ kW}$$

2.10-7. *Pressure Drop of a Flowing Gas.* Nitrogen gas is flowing through a 4-in. schedule 40 commerical steel pipe at 298 K. The total flow rate is 7.40×10^{-2} kg/s and the flow can be assumed as isothermal. The pipe is 3000 m long and the inlet pressure is 200 kPa. Calculate the outlet pressure.
$$\textbf{Ans.} \quad p_2 = 188.5 \text{ kPa}$$

2.10-8. *Entry Length for Flow in a Pipe.* Air at 10°C and 1.0 atm abs pressure is flowing at a velocity of 2.0 m/s inside a tube having a diameter of 0.012 m.
(a) Calculate the entry length.
(b) Calculate the entry length for water at 10°C and the same velocity.

2.10-9. *Friction Loss in Pumping Oil to Pressurized Tank.* An oil having a density of 833 kg/m^3 and a viscosity of 3.3×10^{-3} Pa·s is pumped from an open tank to a pressurized tank held at 345 kPa gage. The oil is pumped from an inlet at the side of the open tank through a line of commercial steel pipe having an inside diameter of 0.07792 m at the rate of 3.494×10^{-3} m^3/s. The length of straight pipe is 122 m and the pipe contains two elbows (90°) and a globe valve half open. The level of the liquid in the open tank is 20 m above the liquid level in the pressurized tank. The pump efficiency is 65%. Calculate the kW power of the pump.

2.10-10. *Flow in an Annulus and Pressure Drop.* Water flows in the annulus of a horizontal, concentric-pipe heat exchanger and is being heated from 40°C to 50°C in the exchanger which has a length of 30 m of equivalent straight pipe. The flow rate of the water is 2.90×10^{-3} m^3/s. The inner pipe is 1-in. schedule 40 and the outer is 2-in. schedule 40. What is the pressure drop? Use an average temperature of 45°C for bulk physical properties. Assume that the wall temperature is an average of 4°C higher than the average bulk temperature so that a correction can be made for the effect of heat transfer on the friction factor.

2.11-1. *(Selected Topic) Derivation of Maximum Velocity for Isothermal Compressible Flow.* Starting with Eq. (2.11-9), derive Eqs. (2.11-11) and (2.11-12) for the maximum velocity in isothermal compressible flow.

2.11-2. *(Selected Topic) Pressure Drop in Compressible Flow.* Methane gas is being pumped through a 305-m length of 52.5-mm-ID steel pipe at the rate of 41.0 kg/m^2·s. The inlet pressure is $p_1 = 345$ kPa abs. Assume isothermal flow at 288.8 K.

(a) Calculate the pressure p_2 at the end of the pipe. The viscosity is 1.04×10^{-5} Pa·s.

(b) Calculate the maximum velocity that can be attained at these conditions and compare with the velocity in part (a).

Ans. (a) $p_2 = 298.4$ kPa, (b) $v_{max} = 387.4$ m/s, $v_2 = 20.62$ m/s

2.11-3. *(Selected Topic) Pressure Drop in Isothermal Compressible Flow.* Air at 288 K and 275 kPa abs enters a pipe and is flowing in isothermal compressible flow in a commercial pipe having an ID of 0.080 m. The length of the pipe is 60 m. The mass velocity at the entrance to the pipe is 165.5 kg/m²·s. Assume 29 for the molecular weight of air. Calculate the pressure at the exit. Also, calculate the maximum allowable velocity that can be attained and compare with the actual.

REFERENCES

(D1) DODGE, B. F. *Chemical Engineering Thermodynamics.* New York: McGraw-Hill Book Company, 1944.

(E1) EARLE, R. L. *Unit Operations in Food Processing.* Oxford: Pergamon Press, Inc., 1966.

(K1) KITTRIDGE, C. P., and ROWLEY, D. S. *Trans. A.S.M.E.,* **79,** 1759 (1957).

(L1) LANGE, N. A. *Handbook of Chemistry,* 10th ed. New York: McGraw-Hill Book Company, 1967.

(L2) LANGHAAR, H. L. *Trans. A.S.M.E,* **64,** A-55 (1942).

(M1) MOODY, L. F. *Trans. A.S.M.E.,* **66,** 671 (1944); *Mech Eng.,* **69,** 1005 (1947).

(M2) MCCABE, W. L., and SMITH, J. C. *Unit Operations of Chemical Engineering,* 3rd ed. New York: McGraw-Hill Book Company, 1976.

(P1) PERRY, R. H., and CHILTON, C. H. *Chemical Engineers' Handbook,* 5th ed. New York: McGraw-Hill Book Company, 1973.

(R1) REID, R. C., and SHERWOOD, T. K. *The Properties of Gases and Liquids,* 2nd ed. New York: McGraw-Hill Book Company, 1966.

(R2) *Reactor Handbook,* vol. 2, AECD-3646. Washington D.C.: Atomic Energy Commission, May 1955.

(S1) SWINDELLS, J. F., COE, J. R. Jr., and GODFREY, T. B. *J. Res. Nat. Bur. Standards,* **48,** 1 (1952).

(S2) SKELLAND, A. H. P. *Non-Newtonian Flow and Heat Transfer.* New York: John Wiley & Sons, Inc., 1967.

(S3) SIEDER, E. N., and TATE, G. E. *Ind. Eng. Chem.,* **28,** 1429 (1936).

(W1) WEAST, R. C. *Handbook of Chemistry and Physics,* 48th ed. Boca Raton, Fla.: Chemical Rubber Co., Inc., 1967–1968.

CHAPTER 3

Principles of
Momentum Transfer
and Applications

3.1 FLOW PAST IMMERSED OBJECTS AND PACKED AND FLUIDIZED BEDS

3.1A Definition of Drag Coefficient for Flow Past Immersed Objects

1. Introduction and types of drag.   In Chapter 2 we were concerned primarily with the momentum transfer and the frictional losses for flow of fluids inside of conduits or pipes. In this section we consider in some detail the flow of fluids around solid, immersed objects.

The flow of fluids outside immersed bodies appears in many chemical engineering applications and other processing applications. These occur, for example, in flow past spheres in settling, flow through packed beds in drying and filtration, flow past tubes in heat exchangers, and so on. It is useful to be able to predict the frictional losses and/or the force on the submerged objects in these various applications.

In the examples of fluid friction inside conduits that we considered in Chapter 2, the transfer of momentum perpendicular to the surface resulted in a tangential shear stress or drag on the smooth surface parallel to the direction of flow. This force exerted by the fluid on the solid in the direction of flow is called *skin* or *wall drag*. For any surface in contact with a flowing fluid, skin friction will exist. In addition to skin friction, if the fluid is not flowing parallel to the surface but must change directions to pass around a solid body such as a sphere, significant additional frictional losses will occur and this is called *form drag*.

In Fig. 3.1-1a the flow of fluid is parallel to the smooth surface of the flat, solid plate, and the force F in newtons on an element of area dA m^2 of the plate is the wall shear stress τ_w times the area dA or $\tau_w\, dA$. The total force is the sum of the integrals of these quantities evaluated over the entire area of the plate. Here the transfer of momentum to the surface results in a tangential stress or skin drag on the surface.

In many cases, however, the immersed body is a blunt-shaped solid which presents various angles to the direction of the fluid flow. As shown in Fig. 3.1-1b, the free-stream velocity is v_0 and is uniform on approaching the blunt-shaped body suspended in a very

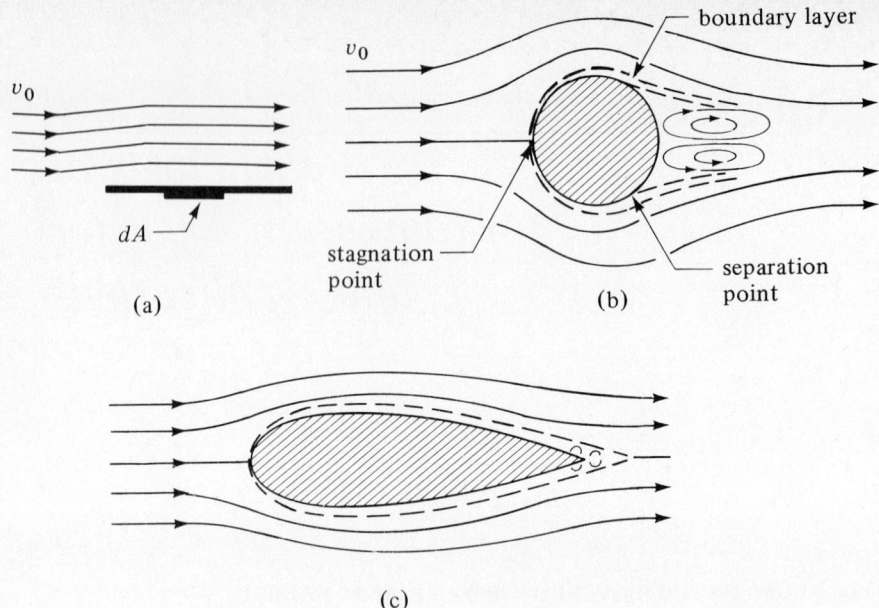

FIGURE 3.1-1. *Flow past immersed objects: (a) flat plate, (b) sphere, (c) streamlined object.*

large duct. Lines called *streamlines* represent the path of fluid elements around the suspended body. The thin boundary layer adjacent to the solid surface is shown as a dashed line and at the edge of this layer the velocity is essentially the same as the bulk fluid velocity adjacent to it. At the front center of the body, called the *stagnation point*, the fluid velocity will be zero and boundary-layer growth begins at this point and continues over the surface until it separates. The tangential stress on the body because of the velocity gradient in the boundary layer is the skin friction. Outside the boundary layer the fluid changes direction to pass around the solid and also accelerates near the front and then decelerates. Because of these effects, an additional force is exerted by the fluid on the body. This phenomenon, called *form drag*, is in addition to the skin drag in the boundary layer.

In Fig. 3.1-1b, as shown, separation of the boundary layer occurs and a wake, covering the entire rear of the object, occurs where large eddies are present and contribute to the form drag. The point of separation depends on the shape of the particle, Reynolds number, and so on, and is discussed in detail elsewhere (S3).

Form drag for bluff bodies can be minimized by streamlining the body (Fig. 3.1-1c), which forces the separation point toward the rear of the body, which greatly reduces the size of the wake. Additional discussion of turbulence and boundary layers is given in Section 3.7.

2. Drag coefficient. From the previous discussions it is evident that the geometry of the immersed solid is a main factor in determining the amount of total drag force exerted on the body. Correlations of the geometry and flow characteristics for solid objects suspended or held in a free stream (immersed objects) are similar in concept and form to the friction factor–Reynolds number correlation given for flow inside conduits. In flow

Transport Processes: Momentum, Heat, and Mass

through conduits, the friction factor was defined as the ratio of the drag force per unit area (shear stress) to the product of density times velocity head as given in Eq. (2.10-4).

In a similar manner for flow past immersed objects, the drag coefficient C_D is defined as the ratio of the total drag force per unit area to $\rho v_0^2/2$.

$$C_D = \frac{F_D/A_p}{\rho v_0^2/2} \quad \text{(SI)}$$

$$C_D = \frac{F_D/A_p}{\rho v_0^2/2g_c} \quad \text{(English)}$$

(3.1-1)

where F_D is the total drag force in N, A_p is an area in m^2, C_D is dimensionless, v_0 is free-stream velocity in m/s, and ρ is density of fluid in kg/m^3. In English units, F_D is in lb_f, v_0 is in ft/s, ρ is in lb_m/ft^3, and A_p in ft^2. The area A_p used is the area obtained by projecting the body on a plane perpendicular to the line of flow. For a sphere, $A_p = \pi D_p^2/4$, where D_p is sphere diameter; for a cylinder whose axis is perpendicular to the flow direction, $A_p = LD_p$, where L = cylinder length. Solving Eq. (3.1-1) for the total drag force,

$$F_D = C_D \frac{v_0^2}{2} \rho A_p$$

(3.1-2)

The Reynolds number for a given solid immersed in a flowing liquid is

$$N_{Re} = \frac{D_p v_0 \rho}{\mu} = \frac{D_p G_0}{\mu}$$

(3.1-3)

where $G_0 = v_0 \rho$.

3.1B Flow Past Sphere, Long Cylinder, and Disk

For each particular shape of object and orientation of the object with the direction of flow, a different relation of C_D versus N_{Re} exists. Correlations of drag coefficient versus Reynolds number are shown in Fig. 3.1-2 for spheres, long cylinders, and disks. The face of the disk and the axis of the cylinder are perpendicular to the direction of flow. These curves have been determined experimentally. However, in the laminar region for low Reynolds numbers less than about 1.0, the experimental drag force for a sphere is the same as the theoretical Stokes' law equation as follows.

$$F_D = 3\pi\mu D_p v_0$$

(3.1-4)

Combining Eqs. (3.1-2) and (3.1-4) and solving for C_D, the drag coefficient predicted by Stokes' law is

$$C_D = \frac{24}{D_p v_0 \rho/\mu} = \frac{24}{N_{Re}}$$

(3.1-5)

The variation of C_D with N_{Re} (Fig. 3.1-2) is quite complicated because of the interaction of the factors that control skin drag and form drag. For a sphere as the Reynolds number is increased beyond the Stokes' law range, separation occurs and a wake is formed. Further increases in N_{Re} cause shifts in the separation point. At about $N_{Re} = 3 \times 10^5$ the sudden drop in C_D is the result of the boundary layer becoming completely turbulent and the point of separation moving downstream. In the region of N_{Re} of about 1×10^3 to 2×10^5 the drag coefficient is approximately constant for each shape and $C_D = 0.44$ for a sphere. Above a N_{Re} of about 5×10^5 the drag coefficients are again approximately constant with C_D for a sphere being 0.20, 0.33 for a cylinder, and

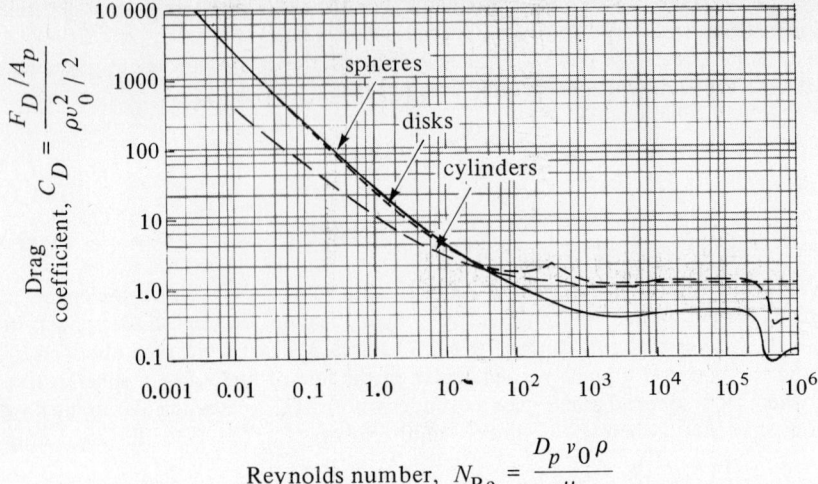

$$N_{Re} = \frac{D_p v_0 \rho}{\mu}$$

FIGURE 3.1-2. *Drag coefficients for flow past immersed spheres, long cylinders, and disks. (Reprinted with permission from C. E. Lapple and C. B. Shepherd, Ind. Eng. Chem., 32, 606 (1940). Copyright by the American Chemical Society.)*

1.12 for a disk. Additional discussions and theory on flow past spheres are given in Section 3.6G.

For derivations of theory and detailed discussions of the drag force for flow parallel to a flat plate, Section 3.7 on boundary-layer flow and turbulence should be consulted. The flow of fluids normal to banks of cylinders or tubes occurs in heat exchangers and other processing applications. The banks of tubes can be arranged in a number of different geometries. Because of the many possible geometric tube configurations and spacings, it is not possible to have one correlation of the data on pressure drop and friction factors. Details of the many correlations available are given elsewhere (P1).

EXAMPLE 3.1-1. Force on Submerged Sphere

Air at 37.8°C and 101.3 kPa absolute pressure flows past a sphere having a diameter of 42 mm at a velocity of 23 m/s. What is the drag coefficient C_D and the force on the sphere?

Solution: From Appendix A.3 for air at 37.8°C, $\rho = 1.137 \text{ kg/m}^3$, $\mu = 1.90 \times 10^{-5} \text{ Pa} \cdot \text{s}$. Also, $D_p = 0.042$ m and $v_0 = 23.0$ m/s. Using Eq. (3.1-3),

$$N_{Re} = \frac{D_p v_0 \rho}{\mu} = \frac{0.042(23.0)(1.137)}{1.90 \times 10^{-5}} = 5.781 \times 10^4$$

From Fig. 3.1-2 for a sphere, $C_D = 0.47$. Substituting into Eq. (3.1-2), where $A_p = \pi D_p^2/4$ for a sphere,

$$F_D = C_D \frac{v_0^2}{2} \rho A_p = (0.47) \frac{(23.0)^2}{2} (1.137)(\pi) \frac{(0.042)^2}{4} = 0.1958 \text{ N}$$

EXAMPLE 3.1-2. Force on a Cylinder in a Tunnel

Water at 24°C is flowing past a long cylinder at a velocity of 1.0 m/s in a large tunnel. The axis of the cylinder is perpendicular to the direction of

Transport Processes: Momentum, Heat, and Mass

flow. The diameter of the cylinder is 0.090 m. What is the force per meter length on the cylinder?

Solution: From Appendix A.2 for water at 24°C, $\rho = 997.2$ kg/m^3, $\mu = 0.9142 \times 10^{-3}$ Pa·s. Also, $D_p = 0.090$ m, $L = 1.0$ m, and $v_0 = 1.0$ m/s. Using Eq. (3.1-3),

$$N_{Re} = \frac{D_p v_0 \rho}{\mu} = \frac{0.090(1.0)(997.2)}{0.9142 \times 10^{-3}} = 9.817 \times 10^4$$

From Fig. 3.1-2 for a long cylinder, $C_D = 1.4$. Substituting into Eq. (3.1-2), where $A_p = LD_p = 1.0(0.090) = 0.090$ m^2,

$$F_D = C_D \frac{v_0^2}{2} \rho A_p = (1.4) \frac{(1.0)^2}{2} (997.2)(0.09) = 62.82 \text{ N}$$

3.1C Flow in Packed Beds

1. Introduction. A system of considerable importance in chemical and other process engineering fields is the packed bed or packed column which is used for a fixed-bed catalytic reactor, adsorption of a solute, absorption, filter bed, and so on. The packing material in the bed may be spheres, irregular particles, cylinders, or various kinds of commercial packings. In the discussion to follow it is assumed that the packing is everywhere uniform and that little or no channeling occurs. The ratio of diameter of the tower to the packing diameter should be a minimum of 8 : 1 to 10 : 1 for wall effects to be small. In the theoretical approach used, the packed column is regarded as a bundle of crooked tubes of varying cross-sectional area. The theory developed in Chapter 2 for single straight tubes is used to develop the results for the bundle of crooked tubes.

2. Laminar flow in packed beds. Certain geometric relations for particles in packed beds are used in the derivations for flow. The void fraction ε in a packed bed is defined as

$$\varepsilon = \frac{\text{volume of voids in bed}}{\text{total volume of bed (voids plus solids)}} \tag{3.1-6}$$

The specific surface of a particle a_v in m^{-1} is defined as

$$a_v = \frac{S_p}{v_p} \tag{3.1-7}$$

where S_p is the surface area of a particle in m^2 and v_p the volume of a particle in m^3. For a spherical particle,

$$a_v = \frac{6}{D_p} \tag{3.1-8}$$

where D_p is diameter in m. For a packed bed of nonspherical particles, the effective particle diameter D_p is defined as

$$D_p = \frac{6}{a_v} \tag{3.1-9}$$

Since $(1 - \varepsilon)$ is the volume fraction of particles in the bed,

$$a = a_v(1 - \varepsilon) = \frac{6}{D_p}(1 - \varepsilon) \tag{3.1-10}$$

where a is the ratio of total surface area in the bed to total volume of bed (void volume plus particle volume) in m^{-1}.

EXAMPLE 3.1-3. *Surface Area in Packed Bed of Cylinders*
A packed bed is composed of cylinders having a diameter $D = 0.02$ m and a length $h = D$. The bulk density of the overall packed bed is 962 kg/m^3 and the density of the solid cylinders is 1600 kg/m^3.

    (a) Calculate the void fraction ε.
    (b) Calculate the effective diameter D_p of the particles.
    (c) Calculate the value of a in Eq. (3.1-10).

Solution: For part (a), taking 1.00 m^3 of packed bed as a basis, the total mass of the bed is (962 kg/m^3) (1.00 m^3) = 962 kg. This mass of 962 kg is also the mass of the solid cylinders. Hence, volume of cylinders = 962 kg/ (1600 kg/m^3) = 0.601 m^3. Using Eq. (3.1-6),

$$\varepsilon = \frac{\text{volume of voids in bed}}{\text{total volume of bed}} = \frac{1.000 - 0.601}{1.000} = 0.399$$

For the effective particle diameter D_p in part (b), for a cylinder where $h = D$, the surface area of a particle is

$$S_p = (2)\frac{\pi D^2}{4}\ (\text{ends}) + \pi D(D)\ (\text{sides}) = \tfrac{3}{2}\,\pi D^2$$

The volume v_p of a particle is

$$v_p = \frac{\pi}{4}\,D^2(D) = \frac{\pi D^3}{4}$$

Substituting into Eq. (3.1-7),

$$a_v = \frac{S_p}{v_p} = \frac{\tfrac{3}{2}\pi D^2}{\tfrac{1}{4}\pi D^3} = \frac{6}{D}$$

Finally substituting into Eq. (3.1-9),

$$D_p = \frac{6}{a_v} = \frac{6}{6/D} = D = 0.02 \text{ m}$$

Hence, the effective diameter to use is $D_p = D = 0.02$ m. For part (c), using Eq. (3.1-10),

$$a = \frac{6}{D_p}\,(1 - \varepsilon) = \frac{6}{0.02}\,(1 - 0.399) = 180.3 \text{ m}^{-1}$$

The average interstitial velocity in the bed is v m/s and it is related to the superficial velocity v' based on the cross section of the empty container by

$$v' = \varepsilon v \tag{3.1-11}$$

The hydraulic radius r_H for flow defined in Eq. (2.10-21) is modified as follows (B2).

$$r_H = \frac{\left(\begin{array}{c}\text{cross-sectional area}\\ \text{available for flow}\end{array}\right)}{(\text{wetted perimeter})}$$

$$= \frac{\text{void volume available for flow}}{\text{total wetted surface of solids}}$$

$$= \frac{\text{volume of voids/volume of bed}}{\text{wetted surface/volume of bed}} = \frac{\varepsilon}{a} \tag{3.1-12}$$

Combining Eqs. (3.1-10) and (3.1-12),

$$r_H = \frac{\varepsilon}{6(1 - \varepsilon)} D_p \qquad (3.1\text{-}13)$$

Since the equivalent diameter D for a channel is $D = 4r_H$, the Reynolds number for a packed bed is as follows using Eq. (3.1-13) and $v' = \varepsilon v$.

$$N_{Re} = \frac{(4r_H)v\rho}{\mu} = \frac{4\varepsilon}{6(1 - \varepsilon)} D_p \frac{v'}{\varepsilon} \frac{\rho}{\mu} = \frac{4}{6(1 - \varepsilon)} \frac{D_p v'\rho}{\mu} \qquad (3.1\text{-}14)$$

For packed beds Ergun (E1) defined the Reynolds number as above but without the 4/6 term.

$$N_{Re, \, p} = \frac{D_p v'\rho}{(1 - \varepsilon)\mu} = \frac{D_p G'}{(1 - \varepsilon)\mu} \qquad (3.1\text{-}15)$$

where $G' = v'\rho$.

For laminar flow, the Hagen–Poiseuille equation (2.10-2) can be combined with Eq. (3.1-13) for r_H and Eq. (3.1-11) to give

$$\Delta p = \frac{32\mu v \, \Delta L}{D^2} = \frac{32\mu(v'/\varepsilon) \, \Delta L}{(4r_H)^2} = \frac{(72)\mu v' \, \Delta L(1 - \varepsilon)^2}{\varepsilon^3 D_p^2} \qquad (3.1\text{-}16)$$

The true ΔL is larger because of the tortuous path and use of the hydraulic radius predicts too large a v'. Experimental data show that the constant should be 150, which gives the Blake–Kozeny equation for laminar flow, void fractions less than 0.5, effective particle diameter D_p, and $N_{Re, \, p} < 10$.

$$\Delta p = \frac{150\mu v' \, \Delta L}{D_p^2} \frac{(1 - \varepsilon)^2}{\varepsilon^3} \qquad (3.1\text{-}17)$$

3. Turbulent flow in packed beds. For turbulent flow we use the same procedure by starting with Eq. (2.10-5) and substituting Eqs. (3.1-11) and (3.1-13) into this equation to obtain

$$\Delta p = \frac{3 f\rho(v')^2 \, \Delta L}{D_p} \frac{1 - \varepsilon}{\varepsilon^3} \qquad (3.1\text{-}18)$$

For highly turbulent flow the friction factor should approach a constant value. Also, it is assumed that all packed beds should have the same relative roughness. Experimental data indicated that $3f = 1.75$. Hence, the final equation for turbulent flow for $N_{Re, \, p} > 1000$, which is called the Burke–Plummer equation, becomes

$$\Delta p = \frac{1.75\rho(v')^2 \, \Delta L}{D_p} \frac{1 - \varepsilon}{\varepsilon^3} \qquad (3.1\text{-}19)$$

Adding Eq. (3.1-17) for laminar flow and Eq. (3.1-19) for turbulent flow, Ergun (E1) proposed the following general equation for low, intermediate, and high Reynolds numbers which has been tested experimentally.

$$\Delta p = \frac{150\mu v' \, \Delta L}{D_p^2} \frac{(1 - \varepsilon)^2}{\varepsilon^3} + \frac{1.75\rho(v')^2 \, \Delta L}{D_p} \frac{1 - \varepsilon}{\varepsilon^3} \qquad (3.1\text{-}20)$$

Rewriting Eq. (3.1-20) in terms of dimensionless groups,

$$\frac{\Delta p\rho}{(G')^2} \frac{D_p}{\Delta L} \frac{\varepsilon^3}{1 - \varepsilon} = \frac{150}{N_{Re, \, p}} + 1.75 \qquad (3.1\text{-}21)$$

Principles of Momentum Transfer and Applications   131

The Ergun equation (3.1-21) can be used for gases by using the density ρ of the gas as the arithmetic average of the inlet and outlet pressures. The velocity v' changes throughout the bed for a compressible fluid, but G' is a constant. At high values of $N_{Re, p}$, Eqs. (3.1-20) and (3.1-21) reduce to Eq. (3.1-19) and to Eq. (3.1-17) for low values. For large pressure drops with gases, Eq. (3.1-20) can be written in differential form (P1).

EXAMPLE 3.1-4. *Pressure Drop and Flow of Gases in Packed Bed*
Air at 311 K is flowing through a packed bed of spheres having a diameter of 12.7 mm. The void fraction ε of the bed is 0.38 and the bed has a diameter of 0.61 m and a height of 2.44 m. The air enters the bed at 1.10 atm abs at the rate of 0.358 kg/s. Calculate the pressure drop of the air in the packed bed. The average molecular weight of air is 28.97.

Solution: From Appendix A.3 for air at 311 K, $\mu = 1.90 \times 10^{-5}$ Pa·s. The cross-sectional area of the bed is $A = (\pi/4)D^2 = (\pi/4)(0.61)^2 = 0.2922$ m^2. Hence, $G' = 0.358/0.2922 = 1.225$ kg/m^2·s (based on empty cross section of container or bed). $D_p = 0.0127$ m, $\Delta L = 2.44$ m, inlet pressure $p_1 = 1.1(1.01325 \times 10^5) = 1.115 \times 10^5$ Pa.
From Eq. (3.1-15),

$$N_{Re, p} = \frac{D_p G'}{(1 - \varepsilon)\mu} = \frac{0.0127(1.225)}{(1 - 0.38)(1.90 \times 10^{-5})} = 1321$$

To use Eq. (3.1-21) for gases, the density ρ to use is the average at the inlet p_1 and outlet p_2 pressures or at $(p_1 + p_2)/2$. This is trial and error since p_2 is unknown. Assuming that $\Delta p = 0.05 \times 10^5$ Pa, $p_2 = 1.115 \times 10^5 - 0.05 \times 10^5 = 1.065 \times 10^5$ Pa. The average pressure is $p_{av} = (1.115 \times 10^5 + 1.065 \times 10^5)/2 = 1.090 \times 10^5$ Pa. The average density to use is

$$\rho_{av} = \frac{M}{RT} p_{av} \tag{3.1-22}$$

$$= \frac{28.97(1.090 \times 10^5)}{8314.34(311)} = 1.221 \text{ kg/m}^3$$

Substituting into Eq. (3.1-21) and solving for Δp,

$$\frac{\Delta p(1.221)}{(1.225)^2} \frac{0.0127}{2.44} \frac{(0.38)^3}{1 - 0.38} = \frac{150}{1321} + 1.75$$

Solving, $\Delta p = 0.0497 \times 10^5$ Pa. This is close enough to the assumed value, so a second trial is not needed.

4. Shape factors and mixtures of particles. Many particles in packed beds are often irregular in shape. The *shape factor* or *sphericity* ϕ_S of a particle is the ratio of surface area of a sphere having the same volume as the particle to the actual surface area of the particle. For a sphere, the surface area is $S_p = \pi D_p^2$ and the volume is $v_p = \pi D_p^3/6$. Hence, for any particle, $\phi_S = \pi D_p^2/S_p$, where S_p is the surface area of the particle. Then

$$\frac{S_p}{v_p} = \frac{\pi D_p^2/\phi_S}{\pi D_p^3/6} = \frac{6}{\phi_S D_p} \tag{3.1-23}$$

From Eq. (3.1-7),

$$a_v = \frac{S_p}{v_p} = \frac{6}{\phi_S D_p} \tag{3.1-24}$$

Then Eq. (3.1-10) becomes

$$a = \frac{6}{\phi_S D_p} (1 - \varepsilon)$$ (3.1-25)

For a cylinder where the diameter = length, $\phi_S = 1.0$ (See Example 3.1-3). Some typical values of ϕ_S for irregular particles are given in Table 3.1-1. Typical values for ϕ_S for many crushed materials are between 0.6 to 0.7.

For a mixture of particles of various sizes we can define a mean specific surface a_{vm} as

$$a_{vm} = \sum x_i a_{vi}$$ (3.1-26)

where x_i is volume fraction. Combining Eqs. (3.1-24) and (3.1-26),

$$D_{pm} = \frac{6}{a_{vm}} = \frac{6}{\sum x_i (6/\phi_S D_{pi})} = \frac{1}{\sum x_i/(\phi_S D_{pi})}$$ (3.1-27)

where D_{pm} is the effective mean diameter for the mixture.

EXAMPLE 3.1-5.   Mean Diameter for a Particle Mixture

A mixture contains three sizes of particles: 25% by volume of 25 mm size, 40% of 50 mm, and 35% of 75 mm. The sphericity is 0.68. Calculate the effective mean diameter.

Solution: The following data are given: $x_1 = 0.25$, $D_{p1} = 25$ mm; $x_2 = 0.40$, $D_{p2} = 50$; $x_3 = 0.35$, $D_{p3} = 75$; $\phi_S = 0.68$. Substituting into Eq. (3.1-27),

$$D_{pm} = \frac{1}{0.25/(0.68 \times 25) + 0.40/(0.68 \times 50) + 0.35/(0.68 \times 75)}$$

$$= 30.0 \text{ mm}$$

5. *Darcy's empirical law for laminar flow.*   Equation (3.1-17) for laminar flow in packed beds shows that the flow rate is proportional to Δp and inversely proportional to the viscosity μ and length ΔL. This is the basis for Darcy's law as follows for purely viscous flow in consolidated porous media.

$$v' = \frac{q'}{A} = -\frac{k}{\mu} \frac{\Delta p}{\Delta L}$$ (3.1-28)

TABLE 3.1-1.   *Shape Factors (Sphericity) of Some Materials*

Material	Shape Factor, ϕ_S	Reference
Spheres	1.0	
Cubes	1.0	
Cylinders, $D_p = h$ (length)	1.0	
Berl saddles	0.3	(B4)
Raschig rings	0.3	(C2)
Coal dust, pulverized	0.73	(C2)
Sand, average	0.75	(C2)
Crushed glass	0.65	(C2)

where v' is superficial velocity based on the empty cross section in cm/s, q' is flow rate cm^3/s, A is empty cross section in cm^2, μ is viscosity in cp, Δp is pressure drop in atm, ΔL is length in cm, and k is permeability in (cm^3 flow/s)·(cp)·(cm length)/(cm^2 area)·(atm pressure drop). The units used for k of cm^2·cp/s·atm are often given in darcy or in millidarcy (1/1000 darcy). Hence, if a porous medium has a permeability of 1 darcy, a fluid of 1-cp viscosity will flow at 1 cm^3/s per 1 cm^2 cross section with a Δp of 1 atm per cm length. This equation is often used in measuring permeabilities of underground oil reservoirs.

3.1D Flow in Fluidized Beds

1. Minimum velocity and porosity for fluidization.   When a fluid flows upward through a packed bed of particles at low velocities, the particles remain stationary. As the fluid velocity is increased, the pressure drop increases according to the Ergun equation (3.1-20). Upon further increases in velocity, conditions finally occur where the force of the pressure drop times the cross-sectional area just equals the gravitational force on the mass of particles. Then the particles just begin to move, and this is the onset of fluidization or minimum fluidization. The gas velocity at which fluidization begins is the minimum fluidization velocity v'_{mf} in m/s based on the empty cross section of the tower (superficial velocity).

The porosity of the bed when true fluidization occurs is the minimum porosity for fluidization and is ε_{mf}. Some typical values of ε_{mf} for various materials are given in Table 3.1-2. The bed expands to this voidage or porosity before particle motion appears. This minimum voidage can be experimentally determined by subjecting the bed to a rising gas stream and measuring the height of the bed L_{mf} in m. Generally, it appears best to use gas as the fluid rather than a liquid since liquids give somewhat higher values of ε_{mf}.

As stated earlier, the pressure drop increases as the gas velocity is increased until the onset of minimum fluidization. Then as the velocity is further increased, the pressure drop decreases very slightly and then remains practically unchanged as the bed continues to expand or increase in porosity with increases in velocity. The bed resembles a boiling liquid. As the bed expands with increase in velocity, the bed continues to retain its top horizontal surface. Eventually, as the velocity is increased much further, entrainment of particles from the actual fluidized bed becomes appreciable.

The relation between bed height L and porosity ε is as follows for a bed having a uniform cross-sectional area A. Since the volume $LA(1 - \varepsilon)$ is equal to the total volume

TABLE 3.1-2.   *Void Fraction, ε_{mf}, at Minimum Fluidization Conditions (L2)*

Type of Particles	Particle Size, D_p (mm)			
	0.06	0.10	0.20	0.40
	Void fraction, ε_{mf}			
Sharp sand ($\phi_S = 0.67$)	0.60	0.58	0.53	0.49
Round sand ($\phi_S = 0.86$)	0.53	0.48	0.43	(0.42)
Anthracite coal ($\phi_S = 0.63$)	0.61	0.60	0.56	0.52

of solids if they existed as one piece,

$$L_1 A(1 - \varepsilon_1) = L_2 A(1 - \varepsilon_2) \tag{3.1-29}$$

$$\frac{L_1}{L_2} = \frac{1 - \varepsilon_2}{1 - \varepsilon_1} \tag{3.1-30}$$

where L_1 is height of bed with porosity ε_1 and L_2 is height with porosity ε_2.

2. Pressure drop and minimum fluidizing velocity. As a first approximation, the pressure drop at the start of fluidization can be determined as follows. The force obtained from the pressure drop times the cross-sectional area must equal the gravitational force exerted by the mass of the particles minus the buoyant force of the displaced fluid.

$$\Delta p A = L_{mf} A(1 - \varepsilon_{mf})(\rho_p - \rho)g \tag{3.1-31}$$

Hence,

$$\frac{\Delta p}{L_{mf}} = (1 - \varepsilon_{mf})(\rho_p - \rho)g \quad \text{(SI)}$$

$$\tag{3.1-32}$$

$$\frac{\Delta p}{L_{mf}} = (1 - \varepsilon_{mf})(\rho_p - \rho)\frac{g}{g_c} \quad \text{(English)}$$

Often we have irregular-shaped particles in the bed, and it is more convenient to use the particle size and shape factor in the equations. First we substitute for the effective mean diameter D_p the term $\phi_S D_P$ where D_P now represents the particle size of a sphere having the same volume as the particle and ϕ_S the shape factor. Then Eq. (3.1-20) for pressure drop in a packed bed becomes

$$\frac{\Delta p}{L} = \frac{150\mu v'}{\phi_S^2 D_P^2} \frac{(1 - \varepsilon)^2}{\varepsilon^3} + \frac{1.75\rho(v')^2}{\phi_S D_P} \frac{1 - \varepsilon}{\varepsilon^3} \tag{3.1-33}$$

where $\Delta L = L$, bed length in m.

Equation (3.1-33) can now be used by a small extrapolation for packed beds to calculate the minimum gas velocity v'_{mf} at which fluidization begins by substituting v'_{mf} for v', ε_{mf} for ε, and L_{mf} for L and combining the result with Eq. (3.1-32) to give

$$\frac{1.75 D_P^2 (v'_{mf})^2 \rho^2}{\phi_S \varepsilon_{mf}^3 \mu^2} + \frac{150(1 - \varepsilon_{mf}) D_P v'_{mf} \rho}{\phi_S^2 \varepsilon_{mf}^3 \mu} - \frac{D_P^3 \rho(\rho_p - \rho)g}{\mu^2} = 0 \tag{3.1-34}$$

Defining a Reynolds number as

$$N_{\text{Re}, mf} = \frac{D_P v'_{mf} \rho}{\mu} \tag{3.1-35}$$

Eq. (3.1-34) becomes

$$\frac{1.75(N_{\text{Re}, mf})^2}{\phi_S \varepsilon_{mf}^3} + \frac{150(1 - \varepsilon_{mf})(N_{\text{Re}, mf})}{\phi_S^2 \varepsilon_{mf}^3} - \frac{D_P^3 \rho(\rho_p - \rho)g}{\mu^2} = 0 \tag{3.1-36}$$

When $N_{\text{Re}, mf} < 20$ (small particles), the first term of Eq. (3.1-36) can be dropped and when $N_{\text{Re}, mf} > 1000$ (large particles), the second term drops out.

If the terms ε_{mf} and/or ϕ_S are not known, Wen and Yu (W4) found for a variety of systems that

$$\phi_S \varepsilon_{mf}^3 \cong \frac{1}{14}, \qquad \frac{1 - \varepsilon_{mf}}{\phi_S^2 \varepsilon_{mf}^3} \cong 11 \tag{3.1-37}$$

Substituting into Eq. (3.1-36), the following simplified equation is obtained.

$$N_{\text{Re}, mf} = \left[(33.7)^2 + 0.0408 \frac{D_P^3 \rho (\rho_p - \rho) g}{\mu^2} \right]^{1/2} - 33.7 \qquad \text{(3.1-38)}$$

This equation holds for a Reynolds number range of 0.001 to 4000 with an average deviation of $\pm 25\%$. Alternative equations are available in the literature (K1, W4).

EXAMPLE 3.1-6. Minimum Velocity for Fluidization

Solid particles having a size of 0.12 mm, a shape factor ϕ_S of 0.88, and a density of 1000 kg/m^3 are to be fluidized using air at 2.0 atm abs and 25°C. The voidage at minimum fluidizing conditions is 0.42.

  (a) If the cross section of the empty bed is 0.30 m^2 and the bed contains 300 kg of solid, calculate the minimum height of the fluidized bed.
  (b) Calculate the pressure drop at minimum fluidizing conditions.
  (c) Calculate the minimum velocity for fluidization.
  (d) Use Eq. (3.1-38) to calculate v'_{mf} assuming that data for ϕ_S and ε_{mf} are not available.

Solution: For part (a), the volume of solids = 300 kg/(1000 kg/m^3) = 0.300 m^3. The height the solids would occupy in the bed if $\varepsilon_1 = 0$ is $L_1 = 0.300$ m^3/(0.30 m^2 cross section) = 1.00 m. Using Eq. (3.1-30) and calling $L_{mf} = L_2$ and $\varepsilon_{mf} = \varepsilon_2$,

$$\frac{L_1}{L_{mf}} = \frac{1 - \varepsilon_{mf}}{1 - \varepsilon_1}$$

$$\frac{1.00}{L_{mf}} = \frac{1 - 0.42}{1 - 0}$$

Solving, $L_{mf} = 1.724$ m.

The physical properties of air at 2.0 atm and 25°C (Appendix A.3) are $\mu = 1.845 \times 10^{-5}$ Pa·s, $\rho = 1.187 \times 2 = 2.374$ kg/m^3, $p = 2.0265 \times 10^5$ Pa. For the particle, $D_P = 0.00012$ m, $\rho_P = 1000$ kg/m^3, $\phi_S = 0.88$, $\varepsilon_{mf} = 0.42$.

For part (b) using Eq. (3.1-32) to calculate Δp,

$$\Delta p = L_{mf}(1 - \varepsilon_{mf})(\rho_p - \rho)g$$

$$= 1.724(1 - 0.42)(1000 - 2.374)(9.80665) = 0.0978 \times 10^5 \text{ Pa}$$

To calculate v'_{mf} for part (c), Eq. (3.1-36) is used.

$$\frac{1.75(N_{\text{Re}, mf})^2}{(0.88)(0.42)^3} + \frac{150(1 - 0.42)(N_{\text{Re}, mf})}{(0.88)^2(0.42)^3}$$

$$- (0.00012)^3 \frac{2.374(1000 - 2.374)(9.80665)}{(1.845 \times 10^{-5})^2}$$

Solving,

$$N_{\text{Re}, mf} = 0.07764 = \frac{D_P v'_{mf} \rho}{\mu} = \frac{0.00012(v'_{mf})(2.374)}{1.845 \times 10^{-5}}$$

$$v'_{mf} = 0.005029 \text{ m/s}$$

Using the simplified Eq. (3.1-38) for part (d),

$$N_{\text{Re}, mf} = \left[(33.7)^2 + \frac{0.0408(0.00012)^3(2.374)(1000 - 2.374)(9.80665)}{(1.845 \times 10^{-5})^2} \right]^{1/2} - 33.7$$

$$= 0.07129$$

Solving, $v'_{mf} = 0.004618$ m/s.

3. *Expansion of fluidized beds.* For the case of small particles and where $N_{Re, f} = D_P v' \rho / \mu < 20$, we can estimate the variation of porosity or bed height L as follows. We assume that Eq. (3.1-36) applies over the whole range of fluid velocities with the first term being neglected. Then, solving for v',

$$v' = \frac{D_P^2(\rho_p - \rho)g\phi_s^2}{150\mu} \frac{\varepsilon^3}{1 - \varepsilon} = K_1 \frac{\varepsilon^3}{1 - \varepsilon} \tag{3.1-39}$$

We find that all terms except ε are constant for the particular system and ε depends upon v'. This equation can be used with liquids to estimate ε with $\varepsilon < 0.80$. However, because of clumping and other factors, errors can occur when used for gases.

The flow rate in a fluidized bed is limited on one hand by the minimum v'_{mf} and on the other by entrainment of solids from the bed proper. This maximum allowable velocity is approximated as the terminal settling velocity v'_t of the particles. (See Section 13.3 for methods to calculate this settling velocity.) Approximate equations to calculate the operating range are as follows (P2). For fine solids and $N_{Re, f} < 0.4$,

$$\frac{v'_t}{v'_{mf}} \cong \frac{90}{1} \tag{3.1-40}$$

For large solids and $N_{Re, f} > 1000$,

$$\frac{v'_t}{v'_{mf}} \cong \frac{9}{1} \tag{3.1-41}$$

EXAMPLE 3.1-7.   *Expansion of Fluidized Bed*
Using the data from Example 3.1-6, estimate the maximum allowable velocity v'_t. Using an operating velocity of 3.0 times the minimum, estimate the voidage of the bed.

Solution: From Example 3.1-6, $N_{Re, mf} = 0.07764$, $v'_{mf} = 0.005029$ m/s, $\varepsilon_{mf} = 0.42$. Using Eq. (3.1-40), the maximum allowable velocity is

$$v'_t \cong 90(v'_{mf}) = 90(0.005029) = 0.4526 \text{ m/s}$$

Using an operating velocity v' of 3.0 times the minimum,

$$v' = 3.0(v'_{mf}) = 3.0(0.005029) = 0.01509 \text{ m/s}$$

To determine the voidage at this new velocity, we substitute into Eq. (3.1-39) using the known values at minimum fluidizing conditions to determine K_1.

$$0.005029 = K_1 \frac{(0.42)^3}{1 - 0.42}$$

Solving, $K_1 = 0.03938$. Then using the operating velocity in Eq. (3.1-39),

$$0.01509 = (0.03938) \frac{\varepsilon^3}{1 - \varepsilon}$$

Solving, the voidage of the bed $\varepsilon = 0.555$ at the operating velocity.

3.2   MEASUREMENT OF FLOW OF FLUIDS

It is important to be able to measure and control the amount of material entering and leaving a chemical and other processing plants. Since many of the materials are in the form of fluids, they are flowing in pipes or conduits. Many different types of devices are used to measure the flow of fluids. The most simple are those that measure directly the volume of the fluids, such as ordinary gas and water meters and positive-displacement pumps. Current meters make use of an element, such as a propeller or cups on a rotating arm, which rotates at a speed determined by the velocity of the fluid passing through it. Very widely used for fluid metering are the pitot tube, venturi meter, orifice meter, and open-channel weirs.

3.2A   Pitot Tube

The pitot tube is used to measure the local velocity at a given point in the flow stream and not the average velocity in the pipe or conduit. In Fig. 3.2-1a a sketch of this simple device is shown. One tube, the impact tube, has its opening normal to the direction of flow and the static tube has its opening parallel to the direction of flow.

The fluid flows into the opening at point 2, pressure builds up, and then remains stationary at this point, called the *stagnation point*. The difference in the stagnation pressure at this point 2 and the static pressure measured by the static tube represents the pressure rise associated with the deceleration of the fluid. The manometer measures this small pressure rise. If the fluid is incompressible, we can write the Bernoulli equation (2.7-32) between point 1, where the velocity v_1 is undisturbed before the fluid decelerates, and point 2, where the velocity v_2 is zero.

$$\frac{v_1^2}{2} - \frac{v_2^2}{2} + \frac{p_1 - p_2}{\rho} = 0 \qquad (3.2\text{-}1)$$

Setting $v_2 = 0$ and solving for v_1,

$$v = C_p \sqrt{\frac{2(p_2 - p_1)}{\rho}} \qquad (3.2\text{-}2)$$

where v is the velocity v_1 in the tube at point 1 in m/s, p_2 is the stagnation pressure, ρ is the density of the flowing fluid at the static pressure p_1, and C_p is a dimensionless coefficient to take into account deviations from Eq. (3.2-1) and generally varies between about 0.98 to 1.0. For accurate use, the coefficient should be determined by calibration of the pitot tube. This equation applies to incompressible fluids but can be used to

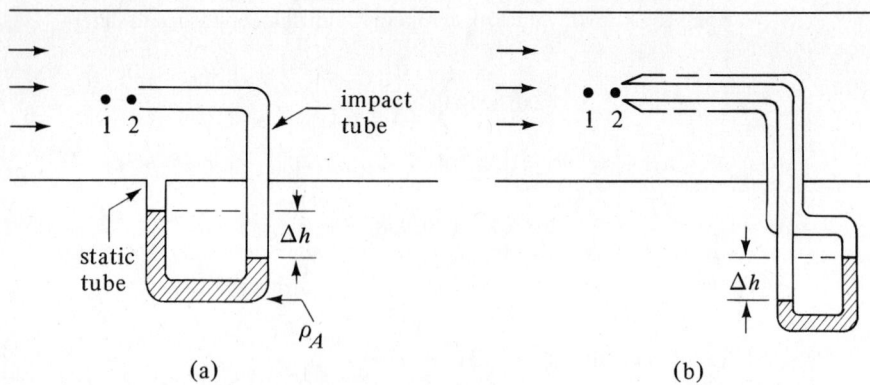

(a)   (b)

FIGURE 3.2-1.   *Diagram of pitot tube: (a) simple tube, (b) tube with static pressure holes.*

approximate the flow of gases at moderate velocities and pressure changes of about 10% or less of the total pressure. For gases the pressure change is often quite low and, hence, accurate measurement of velocities is difficult.

The value of the pressure drop $p_2 - p_1$ or Δp in Pa is related to Δh, the reading on the manometer, by Eq. (2.2-14) as follows:

$$\Delta p = \Delta h(\rho_A - \rho)g \qquad (3.2-3)$$

where ρ_A is the density of the fluid in the manometer in kg/m^3 and Δh is the manometer reading in m. In Fig. 3.2-1b, a more compact design is shown with concentric tubes. In the outer tube, static pressure holes are parallel to the direction of flow. Further details are given elsewhere (P1).

Since the pitot tube measures velocity at one point only in the flow, several methods can be used to obtain the average velocity in the pipe. In the first method the velocity is measured at the exact center of the tube to obtain v_{max}. Then by using Fig. 2.10-2 the v_{av} can be obtained. Care should be taken to have the pitot tube at least 100 diameters downstream from any pipe obstruction. In the second method, readings are taken at several known positions in the pipe cross section and then using Eq. (2.6-17), a graphical or numerical integration is performed to obtain v_{av}.

EXAMPLE 3.2-1.   Flow Measurement Using a Pitot Tube

A pitot tube similar to Fig. 3.2-1a is used to measure the airflow in a circular duct 600 mm in diameter. The flowing air temperature is 65.6°C. The pitot tube is placed at the center of the duct and the reading Δh on the manometer is 10.7 mm of water. A static pressure measurement obtained at the pitot tube position is 205 mm of water above atmospheric. The pitot tube coefficient $C_p = 0.98$.

   (a) Calculate the velocity at the center and the average velocity.
   (b) Calculate the volumetric flow rate of the flowing air in the duct.

Solution: For part (a), the properties of air at 65.6°C from Appendix A.3 are $\mu = 2.03 \times 10^{-5}$ Pa·s, $\rho = 1.043$ kg/m^3 (at 101.325 kPa). To calculate the absolute static pressure, the manometer reading $\Delta h = 0.205$ m of water indicates the pressure above 1 atm abs. Using Eq. (2.2-14), the water density as 1000 kg/m^3, and assuming 1.043 kg/m^3 as the air density,

$$\Delta p = 0.205(1000 - 1.043)9.80665 = 2008 \text{ Pa}$$

Then the absolute static pressure $p_1 = 1.01325 \times 10^5 + 0.02008 \times 10^5 = 1.0333 \times 10^5$ Pa. The correct air density in the flowing air is $(1.0333 \times 10^5/1.01325 \times 10^5)(1.043) = 1.063$ kg/m^3. This correct value when used instead of 1.043 would have a negligible effect on the recalculation of p_1.

To calculate the Δp for the pitot tube, Eq. (3.2-3) is used.

$$\Delta p = \Delta h(\rho_A - \rho)g = \frac{10.7}{1000}(1000 - 1.063)(9.80665) = 104.8 \text{ Pa}$$

Using Eq. (3.2-2), the maximum velocity at the center is

$$v = 0.98\sqrt{\frac{2(104.8)}{1.063}} = 13.76 \text{ m/s}$$

The Reynolds number using the maximum velocity is

$$N_{Re} = \frac{D v_{max} \rho}{\mu} = \frac{0.600(13.76)(1.063)}{2.03 \times 10^{-5}} = 4.323 \times 10^5$$

From Fig. 2.10-2, $v_{av}/v_{max} = 0.85$. Then, $v_{av} = 0.85(13.76) = 11.70$ m/s.

To calculate the flow rate for part (b), the cross-sectional area of the duct, $A = (\pi/4)(0.600)^2 = 0.2827$ m^2. The flow rate $= 0.2827(11.70) = 3.308$ m^3/s.

3.2B Venturi Meter

A *venturi meter* is shown in Fig. 3.2-2 and is usually inserted directly into a pipeline. A manometer or other device is connected to the two pressure taps shown and measures the pressure difference $p_1 - p_2$ between points 1 and 2. The average velocity at point 1 where the diameter is D_1 m is v_1 m/s, and at point 2 or the throat the velocity is v_2 and diameter D_2. Since the narrowing down from D_1 to D_2 and the expansion from D_2 back to D_1 is gradual, little frictional loss due to contraction and expansion is incurred.

To derive the equation for the venturi meter, friction is neglected and the pipe is assumed horizontal. Assuming turbulent flow and writing the mechanical-energy-balance equation (2.7-28) between points 1 and 2 for an incompressible fluid,

$$\frac{v_1^2}{2} + \frac{p_1}{\rho} = \frac{v_2^2}{2} + \frac{p_2}{\rho} \tag{3.2-4}$$

The continuity equation for constant p is

$$v_1 \frac{\pi D_1^2}{4} = v_2 \frac{\pi D_2^2}{4} \tag{3.2-5}$$

Combining Eqs. (3.2-4) and (3.2-5) and eliminating v_1,

$$v_2 = \frac{1}{\sqrt{1 - (D_2/D_1)^4}} \sqrt{\frac{2(p_1 - p_2)}{\rho}} \tag{3.2-6}$$

To account for the small friction loss an experimental coefficient C_v is introduced to give

$$v_2 = \frac{C_v}{\sqrt{1 - (D_2/D_1)^4}} \sqrt{\frac{2(p_1 - p_2)}{\rho}} \qquad \text{(SI)}$$

$$v_2 = \frac{C_v}{\sqrt{1 - (D_2/D_1)^4}} \sqrt{\frac{2g_c(p_1 - p_2)}{\rho}} \qquad \text{(English)} \tag{3.2-7}$$

For many meters and a Reynolds number $> 10^4$ at point 1, C_v is about 0.98 for pipe diameters below 0.2 m and 0.99 for larger sizes. However, these coefficients can vary and individual calibration is recommended if the manufacturer's calibration is not available.

To calculate the volumetric flow rate, the velocity v_2 is multiplied by the area A_2.

$$\text{flow rate} = v_2 \frac{\pi D_2^2}{4} \qquad \text{m}^3/\text{s} \tag{3.2-8}$$

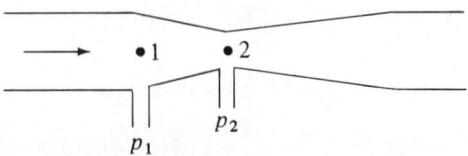

FIGURE 3.2-2. *Venturi flow meter.*

For the measurement of compressible flow of gases, the adiabatic expansion from p_1 to p_2 pressure must be allowed for in Eq. (3.2-7). A similar equation and the same coefficient C_v are used along with the dimensionless expansion correction factor Y (shown in Fig. 3.2-3 for air) as follows:

$$m = \frac{C_v A_2 Y}{\sqrt{1 - (D_2/D_1)^4}} \sqrt{2(p_1 - p_2)\rho_1} \tag{3.2-9}$$

where m is flow rate in kg/s, ρ_1 is density of the fluid upstream at point 1 in kg/m³, and A_2 is cross-sectional area at point 2 in m².

The pressure difference $p_1 - p_2$ occurs because the velocity is increased from v_1 to v_2. However, farther down the tube the velocity returns to its original value of v_1 for liquids. Because of some frictional losses, some of the difference $p_1 - p_2$ is not fully recovered. In a properly designed venturi meter, the permanent loss is about 10% of the differential $p_1 - p_2$, and this represents power loss. A venturi meter is often used to measure flows in large lines, such as city water systems.

3.2C Orifice Meter

For ordinary installations in a process plant the venturi meter has several disadvantages. It occupies considerable space and is expensive. Also, the throat diameter is fixed so that if the flow-rate range is changed considerably, inaccurate pressure differences may result. The *orifice meter* overcomes these objections but at the price of a much larger permanent head or power loss.

A typical sharp-edged orifice is shown in Fig. 3.2-4. A machined and drilled plate having a hole of diameter D_0 is mounted between two flanges in a pipe of diameter D_1. Pressure taps at point 1 upstream and 2 downstream measure $p_1 - p_2$. The exact positions of the two taps are somewhat arbitrary, and in one type of meter the taps are installed about 1 pipe diameter upstream and 0.3 to 0.8 pipe diameter downstream. The fluid stream, once past the orifice plate, forms a vena contracta or free-flowing jet.

The equation for the orifice is similar to Eq. (3.2-7).

$$v_0 = \frac{C_0}{\sqrt{1 - (D_0/D_1)^4}} \sqrt{\frac{2(p_1 - p_2)}{\rho}} \tag{3.2-10}$$

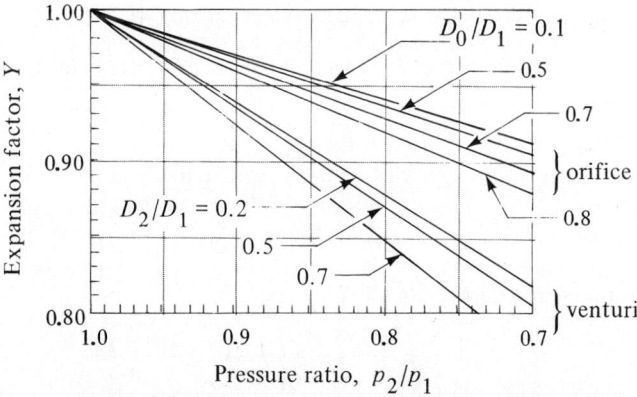

FIGURE 3.2-3. *Expansion factor for air in venturi and orifice. [Calculated from equations and data in references (A2, M2, S3).]*

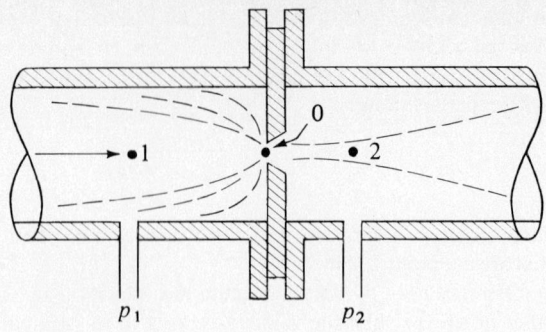

FIGURE 3.2-4. *Orifice flow meter.*

where v_0 is the velocity in the orifice in m/s, D_0 is the orifice diameter in m, and C_0 is the dimensionless orifice coefficient. The orifice coefficient C_0 is always determined experimentally. If the N_{Re} at the orifice is above 20 000 and D_0/D_1 is less than about 0.5, the value of C_0 is approximately constant and it has the value of 0.61, which is adequate for design for liquids (M2, P1). Below 20 000 the coefficient rises sharply and then drops and a correlation for C_0 is given elsewhere (P1).

As in the case of the venturi, for the measurement of compressible flow of gases in an orifice, a correction factor Y given in Fig. 3.2-3 for air is used as follows.

$$m = \frac{C_0 A_0 Y}{\sqrt{1 - (D_0/D_1)^4}} \sqrt{2(p_1 - p_2)\rho_1} \qquad (3.2\text{-}11)$$

where m is flow rate in kg/s, ρ_1 is upstream density in kg/m^3, and A_0 is the cross-sectional area of the orifice.

The permanent pressure loss is much higher than for a venturi because of the eddies formed when the jet expands below the vena contracta. This loss depends on D_0/D_1 and is 73% of $p_1 - p_2$ for $D_0/D_1 = 0.5$, 56% for $D_0/D_1 = 0.65$, and 38% for $D_0/D_1 = 0.8$ (P1).

EXAMPLE 3.2-2. *Metering Oil Flow by an Orifice*
A sharp-edged orifice having a diameter of 0.0566 m is installed in a 0.1541-m pipe through which oil having a density of 878 kg/m^3 and a viscosity of 4.1 cp is flowing. The measured pressure difference across the orifice is 93.2 kN/m^2. Calculate the volumetric flow rate in m^3/s. Assume that $C_0 = 0.61$.

Solution:

$$p_1 - p_2 = 93.2 \text{ kN/m}^2 = 9.32 \times 10^4 \text{ N/m}^2$$

$$D_1 = 0.1541 \text{ m} \qquad D_0 = 0.0566 \text{ m} \qquad \frac{D_0}{D_1} = \frac{0.0566}{0.1541} = 0.368$$

Substituting into Eq. (3.2-10),

$$v_0 = \frac{C_0}{\sqrt{1 - (D_0/D_1)^4}} \sqrt{\frac{2(p_1 - p_2)}{\rho}}$$

$$= \frac{0.61}{\sqrt{1 - (0.368)^4}} \sqrt{\frac{2(9.32 \times 10^4)}{878}}$$

Transport Processes: Momentum, Heat, and Mass

$$v_0 = 8.97 \text{ m/s}$$

$$\text{volumetric flow rate} = v_0 \frac{\pi D_0^2}{4} = (8.97)\frac{(\pi)(0.0566)^2}{4}$$

$$= 0.02257 \text{ m}^3/\text{s} \ (0.797 \text{ ft}^3/\text{s})$$

The N_{Re} is calculated to see if it is greater than 2×10^4 for $C_0 = 0.61$.

$$\mu = 4.1 \times 1 \times 10^{-3} = 4.1 \times 10^{-3} \text{ kg/m} \cdot \text{s} = 4.1 \times 10^{-3} \text{ Pa} \cdot \text{s}$$

$$N_{Re} = \frac{D_0 v_0 \rho}{\mu} = \frac{0.0566(8.97)(878)}{4.1 \times 10^{-3}} = 1.087 \times 10^5$$

Hence, the Reynolds number is above 2×10^4.

Other measuring devices for flow in closed conduits, such as rotameters, flow nozzles, and so on, are discussed elsewhere (P1).

3.2D Flow in Open Channels and Weirs

In many instances in process engineering and in agriculture, liquids are flowing in open channels rather than closed conduits. To measure the flow rates, weir devices are often used. A *weir* is a dam over which the liquid flows. The two main types of weirs are the rectangular weir and the triangular weir shown in Fig. 3.2-5. The liquid flows over the weir and the height h_0 (weir head) in m is measured above the flat base or the notch as shown. This head should be measured at a distance of about $3h_0$ m upstream of the weir by a level or float gage.

The equation for the volumetric flow rate q in m^3/s for a rectangular weir is given by

$$q = 0.415(L - 0.2h_0)h_0^{1.5}\sqrt{2g} \tag{3.2-12}$$

where L = crest length in m, $g = 9.80665$ m/s^2, and h_0 = weir head in m. This is called the *modified Francis weir formula* and it agrees with experimental values within 3% if $L > 2h_0$, velocity upstream is <0.6 m/s, $h_0 > 0.09$ m, and the height of the crest above the bottom of the channel is $>3h_0$. In English units L and h are in ft, q in ft^3/s, and $g = 32.174$ ft/s^2.

For the triangular notch weir,

$$q = \frac{0.31h_0^{2.5}}{\tan \phi}\sqrt{2g} \tag{3.2-13}$$

Both Eqs. (3.2-12) and (3.2-13) apply only to water. For other liquids, see data given elsewhere (P1).

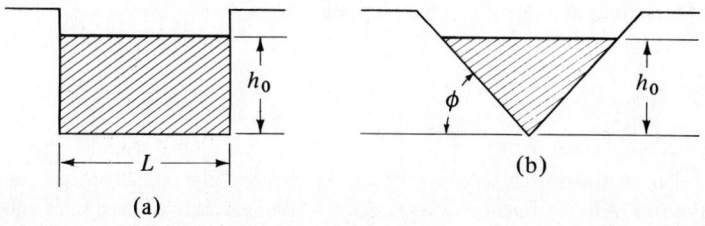

$$(a)$$

$$(b)$$

FIGURE 3.2-5. *Types of weirs: (a) rectangular, (b) triangular.*

3.3 PUMPS AND GAS-MOVING EQUIPMENT

3.3A Introduction

In order to make a fluid flow from one point to another in a closed conduit or pipe, it is necessary to have a driving force. Sometimes this force is supplied by gravity, where differences in elevation occur. Usually, the energy or driving force is supplied by a mechanical device such as a pump or blower, which increases the mechanical energy of the fluid. This energy may be used to increase the velocity (move the fluid), the pressure, or the elevation of the fluid, as seen in the mechanical-energy-balance equation (2.7-28), which relates v, p, ρ, and work. The most common methods of adding energy are by positive displacement or centrifugal action.

Generally, the word "pump" designates a machine or device for moving an incompressible liquid. Fans, blowers, and compressors are devices for moving gas (usually air). Fans discharge large volumes of gas at low pressures of the order of several hundred mm of water. Blowers and compressors discharge gases at higher pressures. In pumps and fans the density of the fluid does not change appreciably, and incompressible flow can be assumed. Compressible flow theory is used for blowers and compressors.

3.3B Pumps

1. Power and work required. Using the total mechanical-energy-balance equation (2.7-28) on a pump and piping system, the actual or theoretical mechanical energy W_S J/kg added to the fluid by the pump can be calculated. Example 2.7-5 shows such a case. If η is the fractional efficiency and W_p the shaft work delivered to the pump, Eq. (2.7-30) gives

$$W_p = -\frac{W_S}{\eta} \tag{3.3-1}$$

The actual or brake power of a pump is as follows.

$$\text{brake kW} = \frac{W_p m}{1000} = -\frac{W_S m}{\eta \times 1000} \quad \text{(SI)}$$

$$\text{brake hp} = \frac{W_p m}{550} = \frac{W_S m}{\eta \times 550} \quad \text{(English)} \tag{3.3-2}$$

where W_p is J/kg, m is the flow rate in kg/s, and 1000 is the conversion factor W/kW. In English units, W_S is in ft · lb$_f$/lb$_m$ and m in lb$_m$/s. The theoretical or fluid power is

$$\text{theoretical power} = (\text{brake kW})(\eta) \tag{3.3-3}$$

The mechanical energy W_S in J/kg added to the fluid is often expressed as the developed head H of the pump in m of fluid being pumped, where

$$-W_S = Hg \quad \text{(SI)}$$

$$-W_S = H\frac{g}{g_c} \quad \text{(English)} \tag{3.3-4}$$

To calculate the power of a fan where the pressure difference is of the order of a few hundred mm of water, a linear average density of the gas between the inlet and outlet of the fan is used to calculate W_S and brake kW or horsepower.

Since most pumps are driven by electric motors, the efficiency of the electric motor must be taken into account to determine the total electric power input to the motor. Typical efficiencies η_e of electric motors are as follows: 75% for $\frac{1}{2}$-kW motors, 80% for 2 kW, 84% for 5 kW, 87% for 15 kW, and about 93% for over 150-kW motors. Hence, the total electric power input equals the brake power divided by the electric motor drive efficiency η_e.

$$\text{electric power input (kW)} = \frac{\text{brake kW}}{\eta_e} = \frac{-W_S m}{\eta \eta_e \cdot 1000} \qquad (3.3\text{-}5)$$

2. Suction lift. The power calculated by Eq. (2.7-3) depends on the differences in pressures and not on the actual pressures being above or below atmospheric pressure. However, the lower limit of the absolute pressure in the suction (inlet) line to the pump is fixed by the vapor pressure of the liquid at the temperature of the liquid in the suction line. If the pressure on the liquid in the suction line drops to the vapor pressure, some of the liquid flashes into vapor (cavitation). Then no liquid can be drawn into the pump.

For the special case where the liquid is nonvolatile, the friction in the suction line to the pump is negligible, and the liquid is being pumped from an open reservoir, the maximum possible vertical suction lift which the pump can perform occurs. For cold water this would be about 10.4 m of water. Practically, however, because of friction, vapor pressure, dissolved gases, and the entrance loss, the actual value is much less. For details, see references elsewhere (P1, M2).

3. Centrifugal pumps. Process industries commonly use centrifugal pumps. They are available in sizes of about 0.004 to 380 m³/min (1 to 100 000 gal/min) and for discharge pressures from a few m of head to 5000 kPa or so. A centrifugal pump in its simplest form consists of an impeller rotating inside a casing. Figure 3.3-1 shows a schematic diagram of a simple centrifugal pump.

The liquid enters the pump axially at point 1 in the suction line and then enters the rotating eye of the impeller, where it spreads out radially. On spreading radially it enters the channels between the vanes at point 2 and flows through these channels to point 3 at the periphery of the impeller. From here it is collected in the volute chamber 4 and flows out the pump discharge at 5. The rotation of the impeller imparts a high-velocity head to the fluid, which is changed to a pressure head as the liquid passes into the volute chamber and out the discharge. Some pumps are also made as two-stage or even multistage pumps.

Many complicating factors determine the actual efficiency and performance charac-

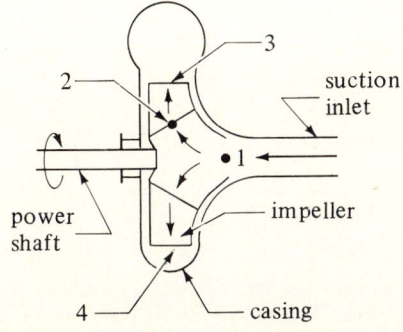
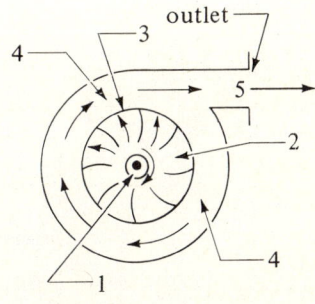

FIGURE 3.3-1.   *Simple centrifugal pump.*

teristics of a pump. Hence, the actual experimental performance of the pump is usually employed. The performance is usually expressed by the pump manufacturer by means of curves called *characteristic curves* and are usually for water. The head H in m produced will be the same for any liquid of the same viscosity. The pressure produced, which is $p = H\rho g$, will be in proportion to the density. Viscosities of less than 0.05 Pa·s (50 cp) have little effect on the head produced. The brake kW varies directly as the density.

As rough approximations, the following can be used for a given pump. The *capacity* q_1 in m^3/s is directly proportional to the rpm N_1, or

$$\frac{q_1}{q_2} = \frac{N_1}{N_2} \tag{3.3-6}$$

The *head* H_1 is proportional to q_1^2, or

$$\frac{H_1}{H_2} = \frac{q_1^2}{q_2^2} = \frac{N_1^2}{N_2^2} \tag{3.3-7}$$

The *power consumed* W_1 is proportional to the product of $H_1 q_1$, or

$$\frac{W_1}{W_2} = \frac{H_1 q_1}{H_2 q_2} = \frac{N_1^3}{N_2^3} \tag{3.3-8}$$

In most pumps, the speed is generally not varied. Characteristic curves for a typical single-stage centrifugal pump operating at a constant speed are given in Fig. 3.3-2. Most pumps are usually rated on the basis of head and capacity at the point of peak efficiency. The efficiency reaches a peak at about 50 gal/min flow rate. As the discharge rate in gal/min increases, the developed head drops. The brake hp increases, as expected, with flow rate.

EXAMPLE 3.3-1. Calculation of Brake Horsepower of a Pump

In order to see how the brake-hp curve is determined, calculate the brake hp at 40 gal/min flow rate for the pump in Fig. 3.3-2.

Solution: At 40 gal/min, the efficiency η from the curve is about 60% and the head H is 38.5 ft. A flow rate of 40 gal/min of water with a density of

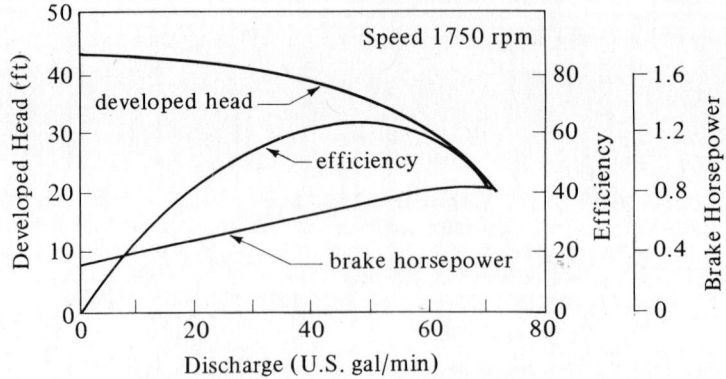

FIGURE 3.3-2.   *Characteristic curves for a single-stage centrifugal pump with water. (From W. L. Badger and J. T. Banchero, Introduction to Chemical Engineering. New York: McGraw-Hill Book Company, 1955. With permission.)*

Transport Processes: Momentum, Heat, and Mass

62.4 lb mass/ft^3 is

$$m = \left(\frac{40 \text{ gal/min}}{60 \text{ s/min}}\right)\left(\frac{1 \text{ ft}^3}{7.481 \text{ gal}}\right)\left(62.4 \frac{\text{lb}_m}{\text{ft}^3}\right) = 5.56 \frac{\text{lb}_m}{\text{s}}$$

The work W_S is as follows, from Eq. (3.3-4):

$$W_S = -H\frac{g}{g_c} = -38.5 \frac{\text{ft} \cdot \text{lb}_f}{\text{lb}_m}$$

The brake hp from Eq. (3.3-2) is

$$\text{brake hp} = \frac{-W_S m}{\eta \cdot 550} = \frac{38.5(5.56)}{0.60(550)} = 0.65 \text{ hp (0.48 kW)}$$

This value checks the value on the curve in Fig. 3.3-2.

4. Positive-displacement pumps. In this class of pumps a definite volume of liquid is drawn into a chamber and then forced out of the chamber at a higher pressure. There are two main types of positive-displacement pumps. In the *reciprocating pump* the chamber is a stationary cylinder and liquid is drawn into the cylinder by withdrawal of a piston in the cylinder. Then the liquid is forced out by the piston on the return stroke. In the *rotary pump* the chamber moves from inlet to discharge and back again. In a gear rotary pump two intermeshing gears rotate and liquid is trapped in the spaces between the teeth and forced out the discharge.

Reciprocating and rotary pumps can be used to very high pressures, whereas centrifugal pumps are limited in their head and are used for lower pressures. Centrifugal pumps deliver liquid at uniform pressure without shocks or pulsations and can handle liquids with large amounts of suspended solids. In general, in chemical and biological processing plants, centrifugal pumps are primarily used.

Equations (3.3-1) through (3.3-5) hold for calculation of the power of positive displacement pumps. At a constant speed, the flow capacity will remain constant with different liquids. In general, the discharge rate will be directly dependent upon the speed. The power increases directly as the head, and the discharge rate remains nearly constant as the head increases.

3.3C Gas-Moving Machinery

Gas-moving machinery comprises mechanical devices used for compressing and moving gases. They are often classified or considered from the standpoint of the pressure heads produced and are fans for low pressures, blowers for intermediate pressures, and compressors for high pressures.

1. Fans. The commonest method for moving small volumes of gas at low pressures is by means of a *fan*. Large fans are usually centrifugal and their operating principle is similar to that of centrifugal pumps. The discharge heads are low, from about 0.1 m to 1.5 m H$_2$O. However, in some cases much of the added energy of the fan is converted to velocity energy and a small amount to pressure head.

In a centrifugal fan, the centrifugal force produced by the rotor produces a compression of the gas, called the *static pressure head*. Also, since the velocity of the gas is increased, a velocity head is produced. Both the static-pressure-head increase and velocity-head increase must be included in estimating efficiency and power. Operating efficiencies are in the range 40 to 70%. The operating pressure of a fan is generally given

Principles of Momentum Transfer and Applications

as inches of water gage and is the sum of the velocity head and static pressure of the gas leaving the fan. Incompressible flow theory can be used to calculate the power of fans.

EXAMPLE 3.3-2. Brake-kW Power of a Centrifugal Fan
It is desired to use 28.32 m^3/min of air (metered at a pressure of 101.3 kPa and 294.1 K) in a process. This amount of air, which is at rest, enters the fan suction at a pressure of 741.7 mm Hg and a temperature of 366.3 K and is discharged at a pressure of 769.6 mm Hg and a velocity of 45.7 m/s. A centrifugal fan having a fan efficiency of 60% is to be used. Calculate the brake-kW power needed.

Solution: Incompressible flow can be assumed, since the pressure drop is only (27.9/741.7)100, or 3.8% of the upstream pressure. The average density of the flowing gas can be used in the mechanical-energy-balance equation.
The density at the suction, point 1, is

$$\rho_1 = \left(28.97 \ \frac{\text{kg air}}{\text{kg mol}}\right)\left(\frac{1 \ \text{kg mol}}{22.414 \ \text{m}^3}\right)\left(\frac{273.2}{366.3}\right)\left(\frac{741.7}{760}\right)$$

$$= 0.940 \ \text{kg/m}^3$$

(The molecular weight of 28.97 for air, the volume of 22.414 m^3/kg mol at 101.3 kPa, and 273.2 K were obtained from Appendix A.1.) The density at the discharge, point 2, is

$$\rho_2 = (0.940)\frac{769.6}{741.7} = 0.975 \ \text{kg/m}^3$$

The average density of the gas is

$$\rho_{\text{av}} = \frac{\rho_1 + \rho_2}{2} = \frac{0.940 + 0.975}{2} = 0.958 \ \text{kg/m}^3$$

The mass flow rate of the gas is

$$m = \left(28.32 \ \frac{\text{m}^3}{\text{min}}\right)\left(\frac{1 \ \text{min}}{60 \ \text{s}}\right)\left(\frac{1 \ \text{kg mol}}{22.414 \ \text{m}^3}\right)\left(\frac{273.2}{294.1}\right)\left(28.97 \ \frac{\text{kg}}{\text{kg mol}}\right)$$

$$= 0.5663 \ \text{kg/s}$$

The developed pressure head is

$$\frac{p_2 - p_1}{\rho_{\text{av}}} = \frac{(769.6 - 741.7) \ \text{mm Hg}}{760 \ \text{mm/atm}}\left(1.01325 \times 10^5 \ \frac{\text{N/m}^2}{\text{atm}}\right)\left(\frac{1}{0.958 \ \text{kg/m}^3}\right)$$

$$= 3883 \ \text{J/kg}$$

The developed velocity head for $v_1 = 0$ is

$$\frac{v_2^2}{2} = \frac{(45.7)^2}{2} = 1044 \ \text{J/kg}$$

Writing the mechanical-energy-balance equation (2.7-28),

$$z_1 g + \frac{v_1^2}{2} + \frac{p_1}{\rho} - W_S = z_2 g + \frac{v_2^2}{2} + \frac{p_2}{\rho} + \sum F$$

Setting $z_1 = 0, z_2 = 0, v_1 = 0$, and $\sum F = 0$, and solving for W_S,

$$-W_S = \frac{p_2 - p_1}{\rho_{\text{av}}} + \frac{v_2^2}{2} = 3883 + 1044 = 4927 \ \text{J/kg}$$

Substituting into Eq. (3.3-2),

$$\text{brake kW} = \frac{-W_S m}{\eta \cdot 1000} = \frac{4927(0.5663)}{0.60(1000)} = 4.65 \text{ kW (6.23 hp)}$$

2. *Blowers and compressors.*   For handling gas volumes at higher pressure rises than fans, several distinct types of equipment are used. *Turboblowers* or *centrifugal compressors* are widely used to move large volumes of gas for pressure rises from about 5 kPa to several thousand kPa. The principles of operation of a turboblower are the same as those of a centrifugal pump. The turboblower resembles the centrifugal pump in appearance, the main difference being that the gas in the blower is compressible. The head of the turboblower, as in a centrifugal pump, is independent of the fluid handled. Multistage turboblowers are often used to go to the higher pressures.

Rotary blowers and compressors are machines of the positive-displacement type and are essentially constant-volume flow-rate machines with variable discharge pressure. Changing the speed will change the volume flow rate. Details of construction of the various types (P1) vary considerably and pressures up to about 1000 kPa can be obtained, depending on the type.

Reciprocating compressers which are of the positive displacement type using pistons are available for higher pressures. Multistage machines are also available for pressures up to 10 000 kPa or more.

3.3D   Equations for Compression of Gases

In blowers and compressors pressure changes are large and compressible flow occurs. Since the density changes markedly, the mechanical-energy-balance equation must be written in differential form and then integrated to obtain the work of compression. In compression of gases the static-head terms, the velocity-head terms, and the friction terms are dropped and only the work term dW and the dp/ρ term remain in the differential form of the mechanical-energy equation; or,

$$dW = \frac{dp}{\rho} \tag{3.3-9}$$

Integration between the suction pressure p_1 and discharge pressure p_2 gives the work of compression.

$$W = \int_{p_1}^{p_2} \frac{dp}{\rho} \tag{3.3-10}$$

To integrate Eq. (3.3-10) for a perfect gas, either adiabatic or isothermal compression is assumed. For isothermal compression, where the gas is cooled on compression, p/ρ is a constant equal to RT/M, where $R = 8314.3$ J/kg mol · K in SI units and 1545.3 ft · lb$_f$/lb mol · °R in English units. Then,

$$\frac{p_1}{\rho_1} = \frac{p}{\rho} \tag{3.3-11}$$

Solving for ρ in Eq. (3.3-11) and substituting it in Eq. (3.3-10), the work for isothermal compression is

$$-W_S = \frac{p_1}{\rho_1} \int_{p_1}^{p_2} \frac{dp}{p} = \frac{p_1}{\rho_1} \ln \frac{p_2}{p_1} = \frac{2.3026 RT_1}{M} \log \frac{p_2}{p_1} \tag{3.3-12}$$

Also, $T_1 = T_2$, since the process is isothermal.

For adiabatic compression, the fluid follows an isentropic path and

$$\frac{p_1}{p} = \left(\frac{\rho_1}{\rho}\right)^\gamma \tag{3.3-13}$$

where $\gamma = c_p/c_v$, the ratio of heat capacities. By combining Eqs. (3.3-10) and (3.3-13) and integrating,

$$-W_S = \frac{\gamma}{\gamma - 1} \frac{RT_1}{M} \left[\left(\frac{p_2}{p_1}\right)^{(\gamma - 1)/\gamma} - 1\right] \tag{3.3-14}$$

The adiabatic temperatures are related by

$$\frac{T_2}{T_1} = \left(\frac{p_2}{p_1}\right)^{(\gamma - 1)/\gamma} \tag{3.3-15}$$

To calculate the brake power when the efficiency is η,

$$\text{brake kW} = \frac{-W_S m}{(\eta)(1000)} \tag{3.3-16}$$

where $m =$ kg gas/s and $W_S =$ J/kg.

The values of γ are approximately 1.40 for air, 1.31 for methane, 1.29 for SO_2, 1.20 for ethane, and 1.40 for N_2 (P1). For a given compression ratio, the work in isothermal compression in Eq. (3.3-12) is less than the work for adiabatic compression in Eq. (3.3-14). Hence, cooling is sometimes used in compressors.

EXAMPLE 3.3-3. Compression of Methane
A single-stage compressor is to compress 7.56×10^{-3} kg mol/s of methane gas at 26.7°C and 137.9 kPa abs to 551.6 kPa abs.
   (a) Calculate the power required if the mechanical efficiency is 80% and the compression is adiabatic.
   (b) Repeat, but for isothermal compression.

Solution: For part (a), $p_1 = 137.9$ kPa, $p_2 = 551.6$ kPa, $M = 16.0$ kg mass/kg mol, and $T_1 = 273.2 + 26.7 = 299.9$ K. The mass flow rate per sec is

$$m = (7.56 \times 10^{-3} \text{ kg mol/s})(16.0 \text{ kg/kg mol}) = 0.121 \frac{\text{kg}}{\text{s}}$$

Substituting into Eq. (3.3-14) for $\gamma = 1.31$ for methane and $p_2/p_1 = 551.6/137.9 = 4.0/1$,

$$-W_S = \frac{\gamma}{\gamma - 1} \frac{RT_1}{M} \left[\left(\frac{p_2}{p_1}\right)^{(\gamma - 1)/\gamma} - 1\right]$$

$$= \left(\frac{1.31}{1.31 - 1}\right) \frac{8314.3(299.9)}{16.0} \left[\left(\frac{4}{1}\right)^{(1.31 - 1)/1.31} - 1\right]$$

$$= 256\,300 \text{ J/kg}$$

Using Eq. (3.3-16),

$$\text{brake kW} = \frac{-W_S m}{\eta \cdot 1000} = \frac{(256\,300)0.121}{0.80(1000)} = 38.74 \text{ kW } (52.0 \text{ hp})$$

For part (b) using Eq. (3.3-12) for isothermal compression,

$$-W_S = \frac{2.3026RT_1}{M} \log \frac{p_2}{p_1} = \frac{2.3026(8314.3)(299.9)}{16.0} \log \frac{4}{1}$$

$$= 216\,000 \text{ J/kg}$$

$$\text{brake kW} = \frac{-W_S m}{\eta \cdot 1000} = \frac{(216\,000)0.121}{0.80(1000)} = 32.67 \text{ kW (43.8 hp)}$$

Hence, isothermal compression uses 15.8% less power.

3.4 AGITATION AND MIXING OF FLUIDS AND POWER REQUIREMENTS

3.4A Purposes of Agitation

In the chemical and other processing industries, many operations are dependent to a great extent on effective agitation and mixing of fluids. Generally, *agitation* refers to forcing a fluid by mechanical means to flow in a circulatory or other pattern inside a vessel. *Mixing* usually implies the taking of two or more separate phases, such as a fluid and a powdered solid, or two fluids, and causing them to be randomly distributed through one another.

There are a number of purposes for agitating fluids and some of these are briefly summarized.

1. Blending of two miscible liquids, such as ethyl alcohol and water.
2. Dissolving solids in liquids, such as salt in water.
3. Dispersing a gas in a liquid as fine bubbles, such as oxygen from air in a suspension of microorganisms for fermentation or for the activated sludge process in waste treatment.
4. Suspending of fine solid particles in a liquid, such as in the catalytic hydrogenation of a liquid where solid catalyst particles and hydrogen bubbles are dispersed in the liquid.
5. Agitation of the fluid to increase heat transfer between the fluid and a coil or jacket in the vessel wall.

3.4B Equipment for Agitation

Generally, liquids are agitated in a cylindrical vessel which can be closed or open to the air. The height of liquid is approximately equal to the tank diameter. An impeller mounted on a shaft is driven by an electric motor. A typical agitator assembly is shown in Fig. 3.4-1.

1. Three-blade propeller agitator.   There are several types of agitators commonly used. A common type, shown in Fig. 3.4-1, is a three-bladed marine-type propeller similar to the propeller blade used in driving boats. The propeller can be a side-entering type in a tank or be clamped on the side of an open vessel in an off-center position. These propellers turn at high speeds of 400 to 1750 rpm (revolutions per minute) and are used for liquids of low viscosity. The flow pattern in a baffled tank with a propeller positioned on the center of the tank is shown in Fig. 3.4-1. This type of flow pattern is called *axial flow* since the fluid flows axially down the center axis or propeller shaft and up on the sides of the tank as shown.

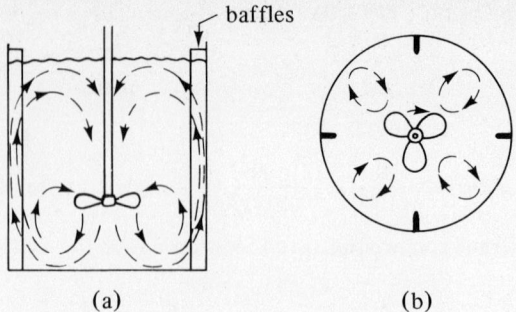

baffles

(a)   (b)

FIGURE 3.4-1. *Baffled tank and three-blade propeller agitator with axial-flow pattern:*
(a) side view, (b) bottom view.

2. Paddle agitators.   Various types of paddle agitators are often used at low speeds between about 20 and 200 rpm. Two-bladed and four-bladed flat paddles are often used, as shown in Fig. 3.4-2a. The total length of the paddle impeller is usually 60 to 80% of the tank diameter and the width of the blade $\frac{1}{6}$ to $\frac{1}{10}$ of its length. At low speeds mild agitation is obtained in an unbaffled vessel. At higher speeds baffles are used, since, without baffles, the liquid is simply swirled around with little actual mixing. The paddle agitator is ineffective for suspending solids since good radial flow is present but little vertical or axial flow. An anchor or gate paddle, shown in Fig. 3.4-2b, is often used. It sweeps or scrapes the tank walls and sometimes the tank bottom. It is used with viscous liquids where deposits on walls can occur and to improve heat transfer to the walls. However, it is a poor mixer. These are often used to process starch pastes, paints, adhesives, and cosmetics.

3. Turbine agitators.   Turbines that resemble multibladed paddle agitators with shorter blades are used at high speeds for liquids with a very wide range of viscosities. The

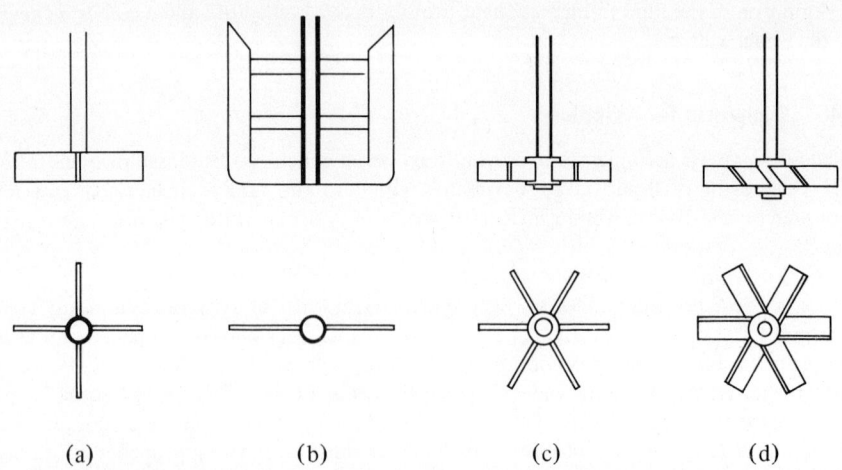

(a)   (b)   (c)   (d)

FIGURE 3.4-2. *Various types of agitators: (a) four-blade paddle, (b) gate or anchor*
paddle, (c) six-blade open turbine, (d) pitched-blade (45°) turbine.

diameter of a turbine is normally between 30 and 50% of the tank diameter. Normally, the turbines have four or six blades. Figure 3.4-3 shows a flat, six-blade turbine agitator with disk. In Fig. 3.4-2c a flat, six-blade open turbine is shown. The turbines with flat blades give radial flow, as shown in Fig. 3.4-3. They are also useful for good gas dispersion where the gas is introduced just below the impeller at its axis and is drawn up to the blades and chopped into fine bubbles. In the pitched-blade turbine shown in Fig. 3.4-2d with the blades at 45°, some axial flow is imparted so that a combination of axial and radial flow is present. This type is useful in suspending solids since the currents flow downward and then sweep up the solids.

4. Agitator selection and viscosity ranges.   The viscosity of the fluid is one of several factors affecting the selection of the type of agitator. Indications of the viscosity ranges of these agitators are as follows. Propellers are used for viscosities of the fluid below about 3 Pa·s (3000 cp); turbines can be used below about 100 Pa·s (100 000 cp); modified paddles such as anchor agitators can be used above 50 Pa·s to about 500 Pa·s (500 000 cp); helical and ribbon-type agitators are often used above this range to about 1000 Pa·s and have been used up to 25 000 Pa·s. For viscosities greater than about 2.5 to 5 Pa·s (5000 cp) and above, baffles are not needed since little swirling is present above these viscosities.

3.4C   Flow Patterns in Agitation

The flow patterns in an agitated tank depend upon the fluid properties, the geometry of the tank, types of baffles in the tank, and the agitator itself. If a propeller or other agitator is mounted vertically in the center of a tank with no baffles, a swirling flow pattern usually develops. Generally, this is undesirable, because of excessive air entrainment, development of a large vortex, surging, and the like, especially at high speeds. To prevent this, an angular off-center position can be used with propellors with small

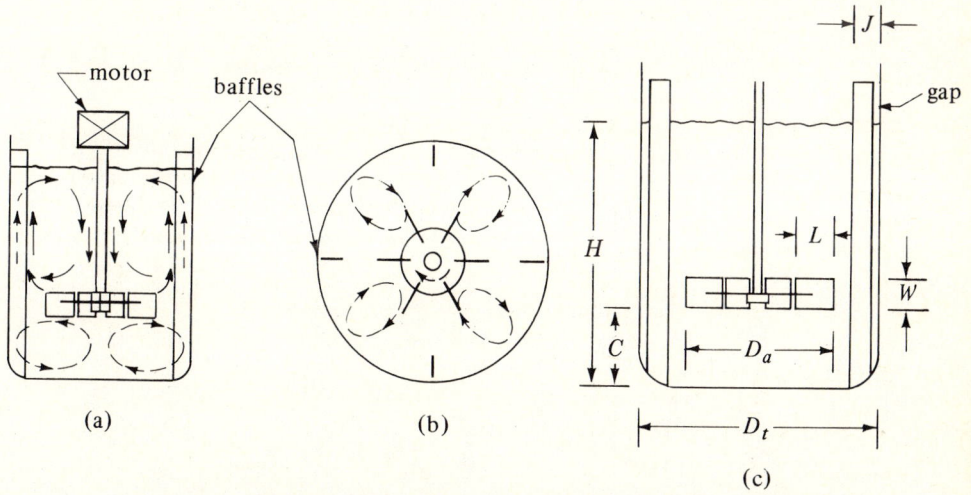

FIGURE 3.4-3.   *Baffled tank with six-blade turbine agitator with disk showing flow patterns: (a) side view, (b) bottom view, (c) dimensions of turbine and tank.*

Principles of Momentum Transfer and Applications   **153**

horsepower. However, for vigorous agitation at higher power, unbalanced forces can become severe and limit the use of higher power.

For vigorous agitation with vertical agitators, baffles are generally used to reduce swirling and still promote good mixing. Baffles installed vertically on the walls of the tank are shown in Fig. 3.4-3. Usually four baffles are sufficient, with their width being about $\frac{1}{12}$ of the tank diameter for turbines and for propellers. The turbine impeller drives the liquid radially against the wall, where it divides, with one portion flowing upward near the surface and back to the impeller from above and the other flowing downward. Sometimes, in tanks with large liquid depths much greater than the tank diameter, two or three impellers are mounted on the same shaft, each acting as a separate mixer. The bottom impeller is about 1.0 impeller diameter above the tank bottom.

In an agitation system, the volume flow rate of fluid moved by the impeller, or circulation rate, is important to sweep out the whole volume of the mixer in a reasonable time. Also, turbulence in the moving stream is important for mixing, since it entrains the material from the bulk liquid in the tank into the flowing stream. Some agitation systems require high turbulence with low circulation rates, and others low turbulence and high circulation rates. This often depends on the types of fluids being mixed and on the amount of mixing needed.

3.4D Typical "Standard" Design of Turbine

The turbine agitator shown in Fig. 3.4-3 is the most commonly used agitator in the process industries. For design of an ordinary agitation system, this type of agitator is often used in the initial design. The geometric proportions of the agitation system which are considered as a typical "standard" design are given in Table 3.4-1. These relative proportions are the basis of the major correlations of agitator performance in numerous publications. (See Fig. 3.4-3c for nomenclature.)

In some cases $W/D_a = \frac{1}{8}$ for agitator correlations. The number of baffles is 4 in most uses. The clearance between the baffles and the wall is usually 0.10 to 0.15 J to ensure that liquid does not form stagnant pockets next to the baffle and wall. In a few correlations the ratio of baffle to tank diameter is $J/D_t = \frac{1}{10}$ instead of $\frac{1}{12}$.

3.4E Power Used in Agitated Vessels

In the design of an agitated vessel, an important factor is the power required to drive the impeller. Since the power required for a given system cannot be predicted theoretically, empirical correlations have been developed to predict the power required. The presence or absence of turbulence can be correlated with the impeller Reynolds number N'_{Re},

TABLE 3.4-1. *Geometric Proportions for a "Standard" Agitation System*

$\dfrac{D_a}{D_t} = 0.3 \text{ to } 0.5$	$\dfrac{H}{D_t} = 1$	$\dfrac{C}{D_t} = \dfrac{1}{3}$
$\dfrac{W}{D_a} = \dfrac{1}{5}$	$\dfrac{L}{D_a} = \dfrac{1}{4}$	$\dfrac{J}{D_t} = \dfrac{1}{12}$

defined as

$$N'_{Re} = \frac{D_a^2 N \rho}{\mu} \tag{3.4-1}$$

where D_a is the impeller (agitator) diameter in m, N is rotational speed in rev/s, ρ is fluid density in kg/m^3, and μ is viscosity in kg/m·s. The flow is laminar in the tank for $N'_{Re} < 10$, turbulent for $N'_{Re} > 10^4$, and for a range between 10 and 10^4, the flow is transitional, being turbulent at the impeller and laminar in remote parts of the vessel.

Power consumption is related to fluid density ρ, fluid viscosity μ, rotational speed N, and impeller diameter D_a by plots of power number N_p versus N'_{Re}. The power number is

$$N_p = \frac{P}{\rho N^3 D_a^5} \quad \text{(SI)}$$

$$N_p = \frac{P g_c}{\rho N^3 D_a^5} \quad \text{(English)} \tag{3.4-2}$$

where P = power in J/s or W. In English units, $P = \text{ft} \cdot \text{lb}_f/\text{s}$.

Figure 3.4-4 is a correlation (B3, R1) for frequently used impellers with Newtonian

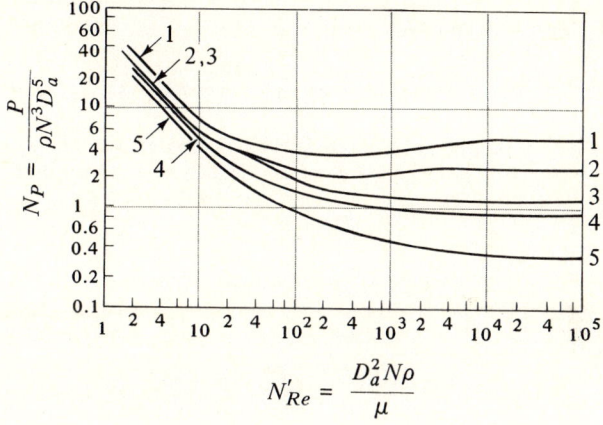

FIGURE 3.4-4. *Power correlations for various impellers and baffles (see Fig. 3.4-3c for dimensions D_a, D_t, J, and W).*

Curve 1. Flat six-blade turbine with disk (like Fig. 3.4-3 but six blades); $D_a/W = 5$; four baffles each $D_t/J = 12$.

Curve 2. Flat six-blade open turbine (like Fig. 3.4-2c); $D_a/W = 8$; four baffles each $D_t/J = 12$.

Curve 3. Six-blade open turbine but blades at 45° (like Fig. 3.4-2d); $D_a/W = 8$; four baffles each $D_t/J = 12$.

Curve 4. Propeller (like Fig. 3.4-1); pitch = $2D_a$; four baffles each $D_t/J = 10$; also holds for same propeller in angular off-center position with no baffles.

Curve 5. Propeller; pitch = D_a; four baffles each $D_t/J = 10$; also holds for same propeller in angular off-center position with no baffles.

*[Curves 1, 2, and 3 reprinted with permission from R. L. Bates, P. L. Fondy, and R. R. Corpstein, Ind. Eng. Chem. Proc. Des. Dev., **2**, 310 (1963). Copyright by the American Chemical Society. Curves 4 and 5 from J. H. Rushton, E. W. Costich, and H. J. Everett, Chem. Eng. Progr., **46**, 395, 467 (1950). With permission.]*

liquids contained in baffled, cylindrical vessels. Dimensional measurements of baffle, tank, and impeller sizes are given in Fig. 3.4-3c. These curves may also be used for the same impellers in unbaffled tanks when N'_{Re} is 300 or less (B3, R1). When N'_{Re} is above 300, the power consumption for an unbaffled vessel is considerably less than for a baffled vessel. Curves for other impellers are also available (B3, R1).

EXAMPLE 3.4-1. Power Consumption in an Agitator

A flat-blade turbine agitator with disk having six blades is installed in a tank similar to Fig. 3.4-3. The tank diameter D_t is 1.83 m, the turbine diameter D_a is 0.61 m, $D_t = H$, and the width W is 0.122 m. The tank contains four baffles, each having a width J of 0.15 m. The turbine is operated at 90 rpm and the liquid in the tank has a viscosity of 10 cp and a density of 929 kg/m³.
   (a) Calculate the required kW of the mixer.
   (b) For the same conditions, except for the solution having a viscosity of 100 000 cp, calculate the required kW.

Solution: For part (a) the following data are given: $D_a = 0.61$ m, $W = 0.122$ m, $D_t = 1.83$ m, $J = 0.15$ m, $N = 90/60 = 1.50$ rev/s, $\rho = 929$ kg/m³, and

$$\mu = (10.0 \text{ cp})(1 \times 10^{-3}) = 0.01 \frac{\text{kg}}{\text{m} \cdot \text{s}} = 0.01 \text{ Pa} \cdot \text{s}$$

Using Eq. (3.4-1), the Reynolds number is

$$N'_{Re} = \frac{D_a^2 N \rho}{\mu} = \frac{(0.61)^2 (1.50) 929}{0.01} = 5.185 \times 10^4$$

Using curve 1 in Fig. 3.4-4 since $D_a/W = 5$ and $D_t/J = 12$, $N_P = 5$ for $N'_{Re} = 5.185 \times 10^4$. Solving for P in Eq. (3.4-2) and substituting known values,

$$P = N_P \rho N^3 D_a^5 = 5(929)(1.50)^3(0.61)^5$$

$$= 1324 \text{ J/s} = 1.324 \text{ kW} (1.77 \text{ hp})$$

For part (b),

$$\mu = 100\,000(1 \times 10^{-3}) = 100 \frac{\text{kg}}{\text{m} \cdot \text{s}}$$

$$N'_{Re} = \frac{(0.61)^2 (1.50) 929}{100} = 5.185.$$

This is in the laminar flow region. From Fig. 3.4-4, $N_P = 14$.

$$P = 14(929)(1.50)^3(0.61)^5 = 3707 \text{ J/s} = 3.71 \text{ kW} (4.98 \text{ hp})$$

Hence, a 10 000-fold increase is viscosity only increases the power from 1.324 to 3.71 kW.

Variations of various geometric ratios from the "standard" design can have different effects on the power number N_P in the turbulent region of the various turbine agitators as follows (B3).

1. For the flat six-blade open turbine, $N_P \propto (W/D_a)^{1.0}$.
2. For the flat, six-blade open turbine, varying D_a/D_t from 0.25 to 0.50 has practically no effect on N_P.
3. With two six-blade open turbines installed on the same shaft and the spacing between the two impellors (vertical distance between the bottom edges of the two turbines)

being at least equal to D_a, the total power is 1.9 times a single flat-blade impeller. For two six-blade pitched-blade (45°) turbines, the power is also about 1.9 times that of a single pitched-blade impeller.

4. A baffled vertical square tank or a horizontal cylindrical tank has the same power number as a vertical cylindrical tank. However, marked changes in the flow patterns occur.

The power number for a plain anchor-type agitator similar to Fig. 3.4-2b but without the two hroizontal cross bars is as follows for $N'_{Re} < 100$ (H2):

$$N_P = 215(N'_{Re})^{-0.955} \tag{3.4-3}$$

where $D_a/D_t = 0.90$, $W/D_t = 0.10$, and $C/D_t = 0.05$. Equations for the power number for helical-ribbon and helical-screw agitators for very viscous liquids are also available (H2, P3).

3.4F Agitator Scale-Up

1. Introduction. In the process industries experimental data are often available on a laboratory-size or pilot-unit-size agitation system and it is desired to scale-up the results to design a full-scale unit. Since there is much diversity in processes to be scaled up, no single method can handle all types of scale-up problems, and many approaches to scale-up exist. Geometric similarity is, of course, important and simplest to achieve. Kinematic similarity can be defined in terms of ratios of velocities or of times (R2). Dynamic similarity requires fixed ratios of viscous, inertial, or gravitational forces. Even if geometric similarity is achieved, dynamic and kinematic similarity cannot often be obtained at the same time. Hence, it is often up to the designer to rely on judgment and experience in the scale-up.

In many cases, the main objectives usually present in an agitation process are as follows: *equal liquid motion*, such as in liquid blending, where the liquid motion or corresponding velocities are approximately the same in both cases; *equal suspension of solids*, where the levels of suspension are the same; and *equal rates of mass transfer*, where mass transfer is occurring between a liquid and a solid phase, liquid–liquid phases, and so on, and the rates are the same.

2. Scale-up procedure. A suggested step-by-step procedure to follow in the scale-up is detailed as follows for scaling up from the initial conditions where the geometric sizes given in Table 3.4-1 are D_{a1}, D_{T1}, H_1, W_1, and so on, to the final conditions of D_{a2}, D_{T2}, and so on.

1. Calculate the scale-up ratio R. Assuming that the original vessel is a standard cylinder with $D_{T1} = H_1$, the volume V_1 is

$$V_1 = \left(\frac{\pi D_{T1}^2}{4}\right)(H_1) = \left(\frac{\pi D_{T1}^3}{4}\right) \tag{3.4-4}$$

Then the ratio of the volumes is

$$\frac{V_2}{V_1} = \frac{\pi D_{T2}^3/4}{\pi D_{T1}^3/4} = \frac{D_{T2}^3}{D_{T1}^3} \tag{3.4-5}$$

The scale-up ratio is then

$$R = \left(\frac{V_2}{V_1}\right)^{1/3} = \frac{D_{T2}}{D_{T1}} \tag{3.4-6}$$

2. Using this value of R, apply it to all of the dimensions in Table 3.4-1 to calculate the new dimensions. For example,

$$D_{a2} = RD_{a1}, \qquad J_2 = RJ_1, \cdots \tag{3.4-7}$$

3. Then a scale-up rule must be selected and applied to determine the agitator speed N_2 to use to duplicate the small-scale results using N_1. This equation is as follows (R2):

$$N_2 = N_1 \left(\frac{1}{R}\right)^n = N_1 \left(\frac{D_{T1}}{D_{T2}}\right)^n \tag{3.4-8}$$

where $n = 1$ for equal liquid motion, $n = \frac{3}{4}$ for equal suspension of solids, and $n = \frac{2}{3}$ for equal rates of mass transfer (which is equivalent to equal power per unit volume). This value of n is based on empirical and theoretical considerations.

4. Knowing N_2, the power required can be determined using Eq. (3.4-2) and Fig. 3.4-4.

EXAMPLE 3.4-2.   Derivation of Scale-Up Rule Exponent
For the scale-up rule exponent n in Eq. (3.4-8), show the following for turbulent agitation.

(a) That when $n = \frac{2}{3}$, the power per unit volume is constant in the scale-up.

(b) That when $n = 1.0$, the tip speed is constant in the scale-up.

Solution: For part (a), from Fig. 3.4-4, N_P is constant for the turbulent region. Then, from Eq. (3.4-2),

$$P_1 = N_P \rho N_1^3 D_{a1}^5 \tag{3.4-9}$$

Then for equal power per unit volume, $P_1/V_1 = P_2/V_2$, or using Eq. (3.4-4),

$$\frac{P_1}{V_1} = \frac{P_1}{\pi D_{T1}^3/4} = \frac{P_2}{V_2} = \frac{P_2}{\pi D_{T2}^3/4} \tag{3.4-10}$$

Substituting P_1 from Eq. (3.4-9) and also a similar equation for P_2 into Eq. (3.4-10) and combining with Eq. (3.4-6),

$$N_2 = N_1 \left(\frac{1}{R}\right)^{2/3} \tag{3.4-11}$$

For part (b), using Eq. (3.4-8) with $n = 1.0$, rearranging, and multiplying by π,

$$N_2 = N_1 \left(\frac{D_{T1}}{D_{T2}}\right)^{1.0} \tag{3.4-12}$$

$$\pi D_{T2} N_2 = \pi D_{T1} N_1 \tag{3.4-13}$$

where $\pi D_{T2} N_2$ is the tip speed in m/s.

To aid the designer of new agitation systems and to serve as a guide for evaluating existing systems, some approximate guidelines are given as follows for liquids of normal viscosities (M2): for mild agitation and blending, 0.1 to 0.2 kW/m^3 of fluid (0.0005 to 0.001 hp/gal); for vigorous agitation, 0.4 to 0.6 kW/m^3 (0.002 to 0.003 hp/gal); for intense agitation or where mass transfer is important, 0.8 to 2.0 kW/m^3 (0.004 to 0.010 hp/gal). This power in kW is the actual power delivered to the fluid as given in Fig. 3.4-4 and Eq. (3.4-2). This does not include the power used in the gear boxes and bearings. Typical efficiencies of electric motors are given in Section 3.3B. As an approximation the power lost in the gear boxes, bearings, and in inefficiency of the electric motor is about 30 to 40% of P, the actual power input to the fluid.

EXAMPLE 3.4-3. Scale-Up of Turbine Agitation System

An existing agitation system is the same as given in Example 3.4-1a for a flat-blade turbine with a disk and six blades. The given conditions and sizes are $D_{T1} = 1.83$ m, $D_{a1} = 0.61$ m, $W_1 = 0.122$ m, $J_1 = 0.15$ m, $N_1 = 90/60 = 1.50$ rev/s, $\rho = 929$ kg/m^3, and $\mu = 0.01$ Pa·s. It is desired to scale up these results for a vessel whose volume is 3.0 times as large. Do this for the following two process objectives.

    (a) Where equal rate of mass transfer is desired.

    (b) Where equal liquid motion is needed.

Solution: Since $H_1 = D_{T1} = 1.83$ m, the original tank volume $V_1 = (\pi D_{T1}^2/4)(H_1) = \pi(1.83)^3/4 = 4.813$ m^3. Volume $V_2 = 3.0(4.813) = 14.44$ m^3. Following the steps in the scale-up procedure, and using Eq. (3.4-6),

$$R = \left(\frac{V_2}{V_1}\right)^{1/3} = \left(\frac{14.44}{4.813}\right)^{1/3} = 1.442$$

The dimensions of the larger agitation system are as follows: $D_{T2} = RD_{T1} = 1.442(1.83) = 2.64$ m, $D_{a2} = 1.442(0.61) = 0.880$ m, $W_2 = 1.442(0.122) = 0.176$ m, and $J_2 = 1.442(0.15) = 0.216$ m.

For part (a) for equal mass transfer, $n = \frac{2}{3}$ in Eq. (3.4-8).

$$N_2 = N_1\left(\frac{1}{R}\right)^{2/3} = (1.50)\left(\frac{1}{1.442}\right)^{2/3} = 1.175 \text{ rev/s (70.5 rpm)}$$

Using Eq. (3.4-1),

$$N'_{Re} = \frac{D_a^2 N \rho}{\mu} = \frac{(0.880)^2(1.175)929}{0.01} = 8.453 \times 10^4$$

Using $N_P = 5.0$ in Eq. (3.4-2),

$$P_2 = N_P \rho N_2^3 D_{a2}^5 = 5.0(929)(1.175)^3(0.880)^5 = 3977 \text{ J/s} = 3.977 \text{ kW}$$

The power per unit volume is

$$\frac{P_1}{V_1} = \frac{1.324}{4.813} = 0.2752 \text{ kW/m}^3$$

$$\frac{P_2}{V_2} = \frac{3.977}{14.44} = 0.2752 \text{ kW/m}^3$$

The value of 0.2752 kW/m^3 is somewhat lower than the approximate guidelines of 0.8 to 2.0 for mass transfer.

For part (b) for equal liquid motion, $n = 1.0$.

$$N_2 = (1.50)\left(\frac{1}{1.442}\right)^{1.0} = 1.040 \text{ rev/s}$$

$$P_2 = 5.0(929)(1.040)^3(0.880)^5 = 2757 \text{ J/s} = 2.757 \text{ kW}$$

$$\frac{P_2}{V_2} = \frac{2.757}{14.44} = 0.1909 \text{ kW/m}^3$$

3.4G Special Agitation Systems

1. Suspension of solids. In some agitation systems a solid is suspended in the agitated liquid. Examples are where a finely dispersed solid is to be dissolved in the liquid, microorganisms are suspended in fermentation, a homogeneous liquid–solid mixture is to be produced for feed to a process, and a suspended solid is used as a catalyst to speed up a reaction. The suspension of solids is somewhat similar to a fluidized bed. In the agitated system circulation currents of the liquid keep the particles in suspension. The amount and type of agitation needed depends mainly on the terminal settling velocity of the particles, which can be calculated using the equations in Section 13.3. Empirical equations to predict the power required to suspend particles are given in references (P1, W1).

2. Blending time of miscible liquids.   In order to study the blending time of two miscible liquids, an amount of HCl acid is added to an equivalent amount of NaOH and the time required for the indicator to change color is noted. Empirical equations to predict this blending time t_T in seconds have been obtained for turbines (M5, N1) and anchor and helical-type mixers (H2, M5).

3. Dispersion of gases and liquids in liquids.   In gas–liquid dispersion processes, the gas is introduced below the impeller, which chops the gas into very fine bubbles. The type and degree of agitation affects the size of the bubbles and the total interfacial area. Typical of such processes are aeration in sewage treatment plants, hydrogenation of liquids by hydrogen gas in the presence of a catalyst, absorption of a solute from the gas by the liquid, and fermentation. Correlations are available to predict the bubble size, holdup, and kW power needed (C3, L1, P1, Z1). For liquids dispersed in inmiscible liquids, see references (P1, T1).

3.4H   Mixing of Powders, Viscous Material, and Pastes

1. Powders.   In mixing of solid particles or powders it is necessary to displace parts of the powder mixture with respect to other parts. The simplest class of devices suitable for gentle blending is the tumbler. However, it is not usually used for breaking up agglomerates. A common type of tumbler used is the *double-cone blender*, in which two cones are mounted with their open ends fastened together and rotated as shown in Fig. 3.4-5a. Baffles can also be used internally. If an internal rotating device is also used in the double cone, agglomerates can also be broken up. Other geometries used are a cylindrical drum with internal baffles or twin-shell V type. Tumblers used specifically for breaking up agglomerates are a rotating cylindrical or conical shell charged with metal or porcelain steel balls or rods.

Another class of devices for solids blending is the *stationary shell device*, in which the container is stationary and the material displacement is accomplished by single or multiple rotating inner devices. In the ribbon mixer in Fig. 3.4-5b, a shaft with two open helical screws numbers 1 and 2 attached to it rotates. One screw is left-handed and one right-handed. As the shaft rotates, sections of powder move in opposite directions and mixing occurs. Other types of internal rotating devices are available for special situations (P1). Also, in some devices both the shell and the internal device rotate.

2. Dough, pastes, and viscous materials.   In the mixing of dough, pastes, and viscous materials, large amounts of power are required so that the material is divided, folded, or recombined, and also different parts of the material should be displaced relative to each other so that fresh surfaces recombine as often as possible. Some machines may require jacketed cooling to remove the heat generated.

The first class of device is somewhat similar to those for agitating fluids, with an impeller slowly rotating in a tank. The impeller can be a close-fitting anchor agitator as in Fig. 3.4-5b, where the outer sweep assembly may have scraper blades. A gate impeller can also be used which has horizontal and vertical bars that cut the paste at various levels and at the wall, which may have stationary bars. A modified gate mixer is the shear-bar mixer, which contains vertical rotating bars or paddles passing between vertical stationary fingers. Other modifications of these types are those where the can or container will rotate as well as the bars and scrapers. These are called *change-can mixers*.

The most commonly used mixer for heavy pastes and dough is the *double-arm kneader mixer*. The mixing action is bulk movement, smearing, stretching, dividing, folding, and recombining. The most widely used design employs two contrarotating arms of sigmoid shape which may rotate at different speeds, as shown in Fig. 3.4-5c.

3.5 NON-NEWTONIAN FLUIDS

3.5A Types of Non-Newtonian Fluids

As discussed in Section 2.4, Newtonian fluids are those which follow Newton's law, Eq. (3.5-1).

$$\tau = -\mu \frac{dv}{dr} \qquad \text{(SI)}$$

$$\tau = -\frac{\mu}{g_c} \frac{dv}{dr} \qquad \text{(English)}$$

(3.5-1)

where μ is the viscosity and is a constant independent of shear rate. In Fig. 3.5-1 a plot is shown of shear stress τ versus shear rate $-dv/dr$. The line for a Newtonian fluid is straight, the slope being μ.

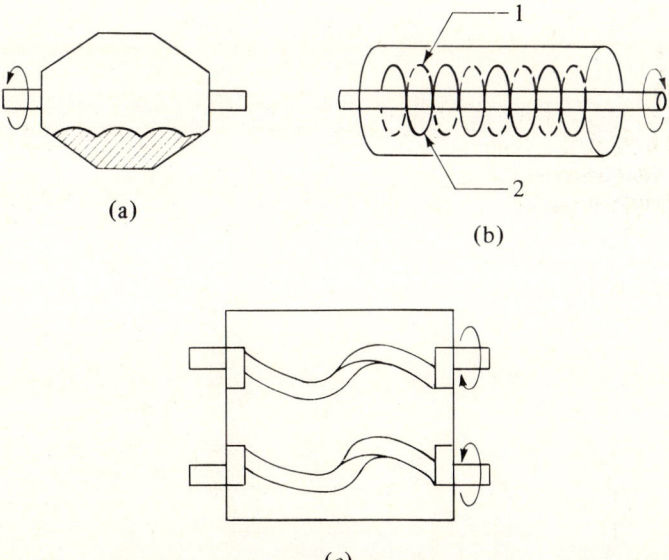

(a)

(b)

(c)

FIGURE 3.4-5. *Mixers for powders and pastes: (a) double-cone powder mixer, (b) ribbon powder mixer with two ribbons, (c) kneader mixer for pastes.*

If a fluid does not follow Eq. (3.5-1), it is a non-Newtonian fluid. Then a plot of τ versus $-dv/dr$ is not linear through the origin for these fluids. Non-Newtonian fluids can be divided into two broad categories on the basis of their shear stress/shear rate behavior: those whose shear stress is independent of time or duration of shear (time-independent) and those whose shear stress is dependent on time or duration of shear (time-dependent). In addition to unusual shear-stress behavior, some non-Newtonian fluids also exhibit elastic (rubberlike) behavior which is a function of time and results in their being called *viscoelastic fluids*. These fluids exhibit normal stresses perpendicular to the direction of flow in addition to the usual tangential stresses. Most of the emphasis will be put on the time-independent class, which includes the majority of non-Newtonian fluids.

3.5B Time-Independent Fluids

1. Bingham plastic fluids. These are the simplest because, as shown in Fig. 3.5-1, they differ from Newtonian only in that the linear relationship does not go through the origin. A finite shear stress τ_y (called *yield stress*) in N/m^2 is needed to initiate flow. Some fluids have a finite yield (shear) stress τ_y, but the plot of τ versus $-dv/dr$ is curved upward or downward. However, this departure from exact Bingham plasticity is often small. Examples of fluids with a yield stress are drilling muds, peat slurries, margarine, chocolate mixtures, greases, soap, grain-water suspensions, toothpaste, paper pulp, and sewage sludge.

2. Pseudoplastic fluids. The majority of non-Newtonian fluids are in this category and include polymer solutions or melts, greases, starch suspensions, mayonnaise, biological fluids, detergent slurries, dispersion media in certain pharmaceuticals, and paints. The shape of the flow curve is shown in Fig. 3.5-1, and it generally can be represented by a power-law equation (sometimes called *Ostwald-deWaele equation*).

$$\tau = K\left(-\frac{dv}{dr} \right)^n \qquad (n < 1) \qquad (3.5\text{-}2)$$

where K is the consistency index in $N \cdot s^n/m^2$ or $lb_f \cdot s^n/ft^2$, and n is the flow behavior index, dimensionless. The apparent viscosity decreases with increasing shear rate.

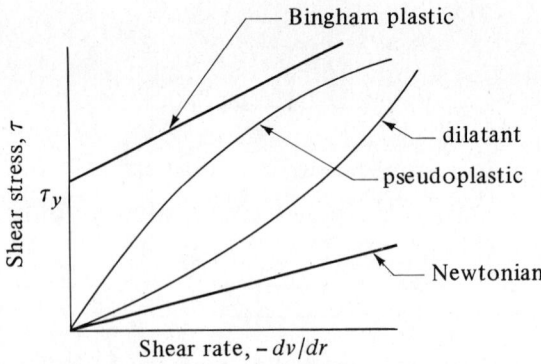

FIGURE 3.5-1. *Shear diagram for Newtonian and time-independent non-Newtonian fluids.*

Transport Processes: Momentum, Heat, and Mass

3. *Dilatant fluids.* These fluids are far less common than pseudoplastic fluids and their flow behavior in Fig. 3.5-1 shows an increase in apparent viscosity with increasing shear rate. The power law equation (3.5-2) is often applicable, but with $n > 1$.

$$\tau = K\left(-\frac{dv}{dr}\right)^n \qquad (n > 1) \tag{3.5-3}$$

For a Newtonian fluid, $n = 1$. Solutions showing dilatancy are some corn flour-sugar solutions, wet beach sand, starch in water, potassium silicate in water, and some solutions containing high concentrations of powder in water.

3.5C Time-Dependent Fluids

1. *Thixotropic fluids.* These fluids exhibit a reversible decrease in shear stress with time at a constant rate of shear. This shear stress approaches a limiting value that depends on the shear rate. Examples include some polymer solutions, shortening, some food materials, and paints. The theory for time-dependent fluids at present is still not completely developed.

2. *Rheopectic fluids.* These fluids are quite rare in occurrence and exhibit a reversible increase in shear stress with time at a constant rate of shear. Examples are bentonite clay suspensions, certain sols, and gypsum suspensions. In design procedures for thixotropic and rheopectic fluids for steady flow in pipes, the limiting flow-property values at a constant rate of shear are sometimes used (S2, W3).

3.5D Viscoelastic Fluids

Viscoelastic fluids exhibit elastic recovery from the deformations that occur during flow. They show both viscous and elastic properties. Part of the deformation is recovered upon removal of the stress. Examples are flour dough, napalm, polymer melts, and bitumens.

3.5E Laminar Flow of Time-Independent Non-Newtonian Fluids

1. *Flow properties of a fluid.* In determining the flow properties of a time-independent non-Newtonian fluid, a capillary-tube viscometer is often used. The pressure drop ΔP N/m^2 for a given flow rate q m^3/s is measured in a straight tube of length L m and diameter D m. This is repeated for different flow rates or average velocities V m/s. If the fluid is time-independent, these flow data can be used to predict the flow in any other pipe size.

A plot of $D\,\Delta p/4L$, which is τ_w, the shear stress at the wall in N/m^2, versus $8V/D$, which is proportional to the shear rate at the wall, is shown in Fig. 3.5-2 for a power-law fluid following Eq. (3.5-4).

$$\frac{D\,\Delta p}{4L} = K'\left(\frac{8V}{D}\right)^{n'} \tag{3.5-4}$$

where n' is the slope of the line when the data are plotted on logarithmic coordinates and K' has units of $N \cdot s^{n'}/m^2$. For $n' = 1$, the fluid is Newtonian; for $n' < 1$, pseudoplastic, or Bingham plastic if the curve does not go through the origin; and for $n' > 1$, dilatant. The

FIGURE 3.5-2. *General flow curve for a power-law fluid in laminar flow in a tube.*

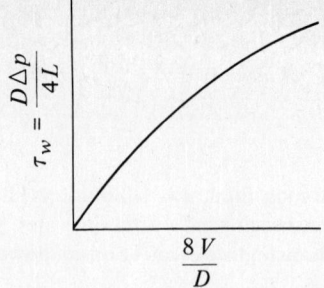

term K', the consistency index in Eq. (3.5-4), is the value of $D\,\Delta p/4L$ for $8V/D = 1$. Also, $K' = \mu$ for Newtonian fluids.

Equation (3.5-4) is simply another statement of the power-law model of Eq. (3.5-2) applied to flow in round tubes (D2), and is more convenient to use for pipe-flow situations (D2). Hence, Eq. (3.5-4) defines the flow characteristics just as completely as Eq. (3.5-2). It has been found experimentally (M3) that for most fluids K' and n' are constant over wide ranges of $8V/D$ or $D\,\Delta p/4L$. For some fluids this is not the case, and K' and n' vary. Then the particular values of K' and n' used must be valid for the actual $8V/D$ or $D\,\Delta P/4L$ with which one is dealing in a design problem.

In many cases the flow properties of a fluid are determined using a rotational viscometer. The flow properties K and n in Eq. (3.5-2) are determined in this manner. When the flow properties are constant over a range of shear stresses which occurs for many fluids, the following equations hold (M3):

$$n' = n \tag{3.5-5}$$

$$K' = K\left(\frac{3n' + 1}{4n'}\right)^{n'} \tag{3.5-6}$$

Often a generalized viscosity coefficient γ is defined as

$$\gamma = K'8^{n'-1} \quad \text{(SI)}$$

$$\gamma = g_c\,K'8^{n'-1} \quad \text{(English)} \tag{3.5-7}$$

where γ has units of $N \cdot s^{n'}/m^2$ or $lb_m/ft \cdot s^{2-n'}$.

Typical flow-property constants (rheological constants) for some fluids are given in Table 3.5-1. Some data give γ values instead of K' values, but Eq. (3.5-7) can be used to convert these values if necessary. In some cases in the literature, K or K' values are given as $dyn \cdot s^{n'}/cm^2$ or $lb_f \cdot s^{n'}/ft^2$. From Appendix A.1, the conversion factors are

$$1\ lb_f \cdot s^{n'}/ft^2 = 47.880\ N \cdot s^{n'}/m^2$$

$$1\ dyn \cdot s^{n'}/cm^2 = 2.0886 \times 10^{-3}\ lb_f \cdot s^{n'}/ft^2 \tag{3.5-8}$$

2. Equations for flow in a tube. In order to predict the frictional pressure drop Δp in laminar flow in a tube, Eq. (3.5-4) is solved for Δp.

$$\Delta p = \frac{K'4L}{D}\left(\frac{8V}{D}\right)^{n'} \tag{3.5-9}$$

TABLE 3.5-1. *Flow-Property Constants for Non-Newtonian Fluids*

Fluid	n'	Flow-Property Constants $\gamma\left(\dfrac{N \cdot s^{n'}}{m^2}\right)$	$K\left(\dfrac{N \cdot s^{n'}}{m^2}\right)$	Ref.
1.5% carboxymethylcellulose in water	0.554	1.369		(S1)
3.0% CMC in water	0.566	4.17		(S1)
4% paper pulp in water	0.575	9.12		(A1)
14.3% clay in water	0.350	0.0512		(W2)
10% napalm in kerosene	0.520	1.756		(S1)
25% clay in water	0.185	0.3036		(W2)
Applesauce, brand A (297 K), density = 1.10 g/cm³	0.645		0.500	(C1)
Banana purée, brand A (297 K), density = 0.977 g/cm³	0.458		6.51	(C1)
Honey (297 K)	1.00		5.61	(C1)
Cream, 30% fat (276 K)	1.0		0.01379	(M4)
Tomato concentrate, 5.8% total solids (305 K)	0.59		0.2226	(H1)

If the average velocity is desired, Eq. (3.5-4) can be rearranged to give

$$V = \frac{D}{8}\left(\frac{\Delta p D}{K' 4L}\right)^{1/n'} \tag{3.5-10}$$

If the equations are desired in terms of K instead of K', Eqs. (3.5-5) and (3.5-6) can be substituted into (3.5-9) and (3.5-10). The flow must be laminar and the generalized Reynolds number has been defined as

$$N_{\text{Re, gen}} = \frac{D^{n'} V^{2-n'} \rho}{\gamma} = \frac{D^{n'} V^{2-n'} \rho}{K' 8^{n'-1}} = \frac{D^{n} V^{2-n} \rho}{K 8^{n-1}\left(\dfrac{3n+1}{4n}\right)^n} \quad \text{(SI)} \tag{3.5-11}$$

3. Friction factor method. Alternatively, using the Fanning friction factor method given in Eqs. (2.10-5) to (2.10-7) for Newtonian fluids, but using the generalized Reynolds numbers,

$$f = \frac{16}{N_{\text{Re, gen}}} \tag{3.5-12}$$

$$\Delta p = 4fp\,\frac{L}{D}\frac{V^2}{2} \quad \text{(SI)}$$

$$\Delta p = 4fp\,\frac{L}{D}\frac{V^2}{2g_c} \quad \text{(English)} \tag{3.5-13}$$

EXAMPLE 3.5-1. *Pressure Drop of Power-Law Fluid in Laminar Flow*
A power-law fluid having a density of 1041 kg/m³ is flowing through 14.9 m of a tubing having an inside diameter of 0.0524 m at an average velocity of

0.0728 m/s. The rheological or flow properties of the fluid are $K' = 15.23$ $N \cdot s^{n'}/m^2$ (0.318 $lb_f \cdot s^{n'}/ft^2$) and $n' = 0.40$.

    (a) Calculate the pressure drop and friction loss using Eq. (3.5-9) for laminar flow. Check the generalized Reynolds number to make sure that the flow is laminar.

    (b) Repeat part (a) but use the friction factor method.

Solution: The known data are as follows: $K' = 15.23$, $n' = 0.40$, $D = 0.0524$ m, $V = 0.0728$ m/s, $L = 14.9$ m, and $\rho = 1041$ kg/m^3. For part (a), using Eq. (3.5-9),

$$\Delta p = \frac{K'4L}{D} \left(\frac{8V}{D}\right)^{n'} = \frac{15.23(4)(14.9)}{0.0524} \left(\frac{8 \times 0.0728}{0.0524}\right)^{0.4} = 45\,390 \text{ N/m}^2$$

Also, to calculate the friction loss,

$$F_f = \frac{\Delta p}{\rho} = \frac{45\,390}{1041} = 43.60 \text{ J/kg}$$

Using Eq. (3.5-11),

$$N_{\text{Re, gen}} = \frac{D^{n'}V^{2-n'}\rho}{K'8^{n'-1}} = \frac{(0.0524)^{0.40}(0.0728)^{1.60}(1041)}{15.23(8)^{-0.6}} = 1.106$$

Hence, the flow is laminar.

    For part (b), using Eq. (3.5-12),

$$f = \frac{16}{N_{\text{Re, gen}}} = \frac{16}{1.106} = 14.44$$

Substituting into Eq. (3.5-13),

$$\Delta p = 4fp \frac{L}{D} \frac{V^2}{2} = 4(14.44)(1041) \frac{14.9}{0.0524} \frac{(0.0728)^2}{2}$$

$$= 45.39 \frac{\text{kN}}{\text{m}^2} \left(946 \frac{\text{lb}_f}{\text{ft}^2}\right)$$

    To calculate the pressure drop for a Bingham plastic fluid with a yield stress, methods are available for laminar flow and are discussed in detail elsewhere (C1, P1, S2).

3.5F   Friction Losses in Contractions, Expansions, and Fittings in Laminar Flow

Since non-Newtonian power-law fluids flowing in conduits are often in laminar flow because of their usually high effective viscosity, losses in sudden changes of velocity and fittings are important in laminar flow.

1. Kinetic energy in laminar flow.   In application of the total mechanical-energy balance in Eq. (2.7-28), the average kinetic energy per unit mass of fluid is needed. For fluids, this is (S2)

$$\text{average kinetic energy/kg} = \frac{V^2}{2\alpha} \qquad (3.5\text{-}14)$$

For Newtonian fluids, $\alpha = \frac{1}{2}$ for laminar flow. For power-law non-Newtonian fluids,

$$\alpha = \frac{(2n + 1)(5n + 3)}{3(3n + 1)^2} \qquad (3.5\text{-}15)$$

For example, if $n = 0.50$, $\alpha = 0.585$. If $n = 1.00$, $\alpha = \frac{1}{2}$. For turbulent flow for Newtonian and non-Newtonian flow, $\alpha = 1.0$ (D1).

2. Losses in contractions and fittings. Skelland (S2) and Dodge and Metzner (D2) state that when a fluid leaves a tank and flows through a sudden contraction to a pipe of diameter D_2 or flows from a pipe of diameter D_1 through a sudden contraction to a pipe of D_2, a vena contracta is usually formed downstream from the contraction. General indications are that the frictional pressure losses for pseudoplastic and Bingham plastic fluids are very similar to those for Newtonian fluids at the same generalized Reynolds numbers in laminar and turbulent flow for contractions and also for fittings and valves.

For contraction losses, Eq. (2.10-16) can be used where $\alpha = 1.0$ for turbulent flow and for laminar flow Eq. (3.5-15) can be used to determine α, since n is not 1.00.

For fittings and valves, frictional losses should be determined using Eq. (2.10-17) and values from Table 2.10-1.

3. Losses in sudden expansion. For the frictional loss for a non-Newtonian fluid in laminar flow through a sudden expansion from D_1 to D_2 diameter, Skelland (S2) gives

$$ h_{ex} = \frac{3n + 1}{2n + 1} V_1^2 \left[\frac{n + 3}{2(5n + 3)} \left(\frac{D_1}{D_2} \right)^4 - \left(\frac{D_1}{D_2} \right)^2 + \frac{3(3n + 1)}{2(5n + 3)} \right] \qquad (3.5\text{-}16) $$

where h_{ex} is the frictional loss in J/kg. In English units Eq. (3.5-16) is divided by g_c and h_{ex} is in ft · lb_f/lb_m.

Equation (2.10-15) for laminar flow with $\alpha = \frac{1}{2}$ for a Newtonian fluid gives values reasonably close to those of Eq. (3.5-16) for $n = 1$ (Newtonian fluid). For turbulent flow the frictional loss can be approximated by Eq. (2.10-15), with $\alpha = 1.0$ for non-Newtonian fluids (S2).

3.5G Turbulent Flow and Generalized Friction Factors

In turbulent flow of time-independent fluids the Reynolds number at which turbulent flow occurs varies with the flow properties of the non-Newtonian fluid. Dodge and Metzner (D2) in a comprehensive study derived a theoretical equation for turbulent flow of non-Newtonian fluids through smooth round tubes. The final equation is plotted in Fig. 3.5-3, where the Fanning friction factor is plotted versus the generalized Reynolds number, $N_{Re, gen}$, given in Eq. (3.5-11). Power-law fluids with flow-behavior indexes n' between 0.36 and 1.0 were experimentally studied at Reynolds numbers up to 3.5×10^4 and confirmed the derivation.

The curves for different n' values break off from the laminar line at different Reynolds numbers to enter the transition region. For $n' = 1.0$ (Newtonian), the transition region starts at $N_{Re, gen} = 2100$. Since many non-Newtonian power-law fluids have high effective viscosities, they are often in laminar flow. The correlation for a smooth tube also holds for a rough pipe in laminar flow.

For rough commercial pipes with various values of roughness ε/D, Fig. 3.5-3 cannot be used for turbulent flow, since it is derived for smooth tubes. The functional dependence of the roughness values ε/D on n' requires experimental data which are not yet available. Metzner and Reed (M3, S3) recommend use of the existing relationship, Fig. 2.10-3, for Newtonian fluids in rough tubes using the generalized Reynolds number $N_{Re, gen}$. This is somewhat conservative since preliminary data indicate that friction factors for pseudoplastic fluids may be slightly smaller than for Newtonian fluids. This is also consistent with Fig. 3.5-3 for smooth tubes that indicate lower f values for fluids with n' below 1.0 (S2).

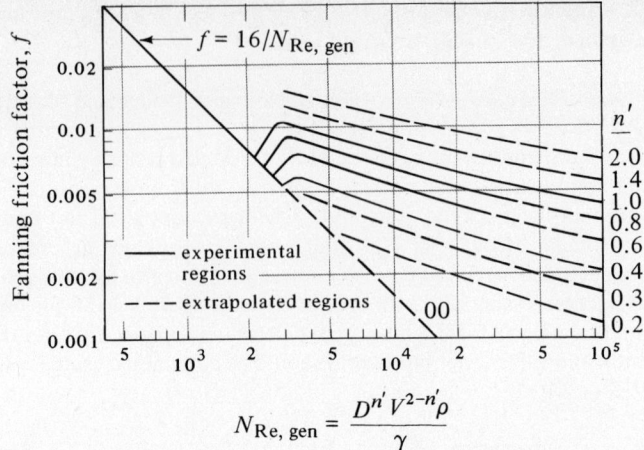

$$N_{Re,\,gen} = \frac{D^{n'} V^{2-n'} \rho}{\gamma}$$

FIGURE 3.5-3. *Fanning friction factor versus generalized Reynolds number for time-independent non-Newtonian and Newtonian fluids flowing in smooth tubes. [From D. W. Dodge and A. B. Metzner, A.I.Ch.E. J., 5, 189 (1959). With permission.]*

EXAMPLE 3.5-2. Turbulent Flow of Power-Law Fluid

A pseudoplastic fluid that follows the power law, having a density of 961 kg/m³, is flowing through a smooth circular tube having an inside diameter of 0.0508 m at an average velocity of 6.10 m/s. The flow properties of the fluid are $n' = 0.30$ and $K' = 2.744 \text{ N} \cdot \text{s}^{n'}/\text{m}^2$. Calculate the frictional pressure drop for a tubing 30.5 m long.

Solution: The data are as follows: $K' = 2.744$, $n' = 0.30$, $D = 0.0508$ m, $V = 6.10$ m/s, $\rho = 961 \text{ kg/m}^3$, and $L = 30.5$ m. Using the general Reynolds-number equation (3.5-11),

$$N_{Re,\,gen} = \frac{D^{n'} V^{2-n'} \rho}{K' 8^{n'-1}} = \frac{(0.0508)^{0.3}(6.10)^{1.7}(961)}{2.744(8^{-0.7})}$$

$$= 1.328 \times 10^4$$

Hence, the flow is turbulent. Using Fig. 3.5-3 for $N_{Re,\,gen} = 1.328 \times 10^4$ and $n' = 0.30$, $f = 0.0032$.

Substituting into Eq. (3.5-13),

$$\Delta p = 4fp \frac{L}{D} \frac{V^2}{2} = 4(0.0032)(961)\left(\frac{30.5}{0.0508}\right)\frac{(6.10)^2}{2}$$

$$= 137.4 \text{ kN/m}^2 \ (2870 \text{ lb}_f/\text{ft}^2)$$

3.5H Velocity Profiles for Non-Newtonian Fluids

Starting with Eq. (3.5-2) written as

$$\tau_{rx} = K\left(-\frac{dv_x}{dr}\right) \tag{3.5-17}$$

the following equation can be derived relating the velocity v_x with the radial position r, which is the distance from the center. (See Problem 2.9-3 for this derivation.)

$$v_x = \frac{n}{n+1}\left(\frac{p_0 - p_L}{2KL}\right)^{1/n} (R_0)^{(n+1)/n}\left[1 - \left(\frac{r}{R_0}\right)^{(n+1)/n}\right] \qquad (3.5\text{-}18)$$

At $r = 0$, $v_x = v_{x\,max}$ and Eq. (3.5-18) becomes

$$v_x = v_{x\,max}\left[1 - \left(\frac{r}{R_0}\right)^{(n+1)/n}\right] \qquad (3.5\text{-}19)$$

The velocity profile can be calculated for laminar flow of a non-Newtonian fluid to show that the velocity profile for a Newtonian fluid given in Eq. (2.9-9) can differ greatly from that of a non-Newtonian fluid. For pseudoplastic fluids ($n < 1$), a relatively flat velocity profile is obtained compared to the parabolic profile for a Newtonian fluid. For $n = 0$, rodlike flow is obtained. For dilitant fluids ($n > 1$), a much sharper profile is obtained and for $n = \infty$, the velocity is a linear function of the radius.

3.5I Power Requirements in Agitation and Mixing of Non-Newtonian Fluids

For correlating the power requirements in agitation and mixing of non-Newtonian fluids the power number N_P is defined by Eq. (3.4-2), which is also the same equation used for Newtonian fluids. However, the definition of the Reynolds number is much more complicated than for Newtonian fluids since the apparent viscosity is not constant for non-Newtonian fluids and varies with the shear rates or velocity gradients in the vessel. Several investigators (G1, M1) have used an average apparent viscosity μ_a, which is used in the Reynolds number as follows:

$$N'_{Re,\,n} = \frac{D_a^2 N \rho}{\mu_a} \qquad (3.5\text{-}20)$$

The average apparent viscosity can be related to the average shear rate or average velocity gradient by the following method. For a power-law fluid,

$$\tau = K\left(-\frac{dv}{dy}\right)^n_{av} \qquad (3.5\text{-}21)$$

For a Newtonian fluid,

$$\tau = \mu_a\left(-\frac{dv}{dy}\right)_{av} \qquad (3.5\text{-}22)$$

Combining Eqs. (3.5-21) and (3.5-22),

$$\mu_a = K\left(\frac{dv}{dy}\right)^{n-1}_{av} \qquad (3.5\text{-}23)$$

Metzner and others (G1, M1) found experimentally that the average shear rate $(dv/dy)_{av}$ for pseudoplastic liquids ($n < 1$) varies approximately as follows with the rotational speed:

$$\left(\frac{dv}{dy}\right)_{av} = 11N \qquad (3.5\text{-}24)$$

Hence, combining Eqs. (3.5-23) and (3.5-24),

$$\mu_a = (11N)^{n-1}K \qquad (3.5\text{-}25)$$

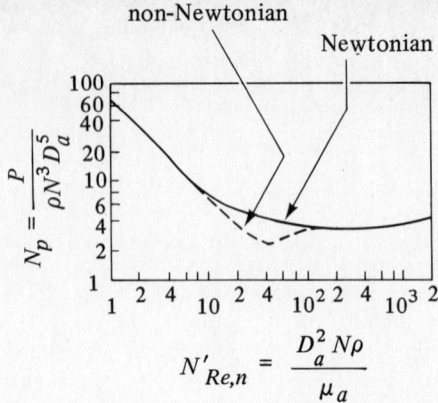

$$N'_{Re,n} = \frac{D_a^2 N \rho}{\mu_a}$$

FIGURE 3.5-4. *Power correlation in agitation for a flat, six-blade turbine with disk in pseudoplastic non-Newtonian and Newtonian fluids (G1, M1, R1):* $D_a/W = 5, L/W = 5/4, D_t/J = 10.$

Substituting into Eq. (3.5-20),

$$N'_{Re,n} = \frac{D_a^2 N^{2-n} \rho}{11^{n-1} K} \tag{3.5-26}$$

Equation (3.5-26) has been used to correlate data for a flat six-blade turbine with disk in pseudoplastic liquids, and the dashed curve in Fig. 3.5-4 shows the correlation (M1). The solid curve applies to Newtonian fluids (R1). Both sets of data were obtained for four baffles with $D_t/J = 10$, $D_a/W = 5$, and $L/W = 5/4$. However. since it has been shown that the difference in results for $D_t/J = 10$ and $D_t/J = 12$ is very slight (R1), this Newtonian line can be considered the same as curve 1, Fig. 3.4-4. The curves in Fig. 3.5-4 show that the results are identical for the Reynolds number range 1 to 2000 except that they differ only in the Reynolds number range 10 to 100, where the pseudoplastic fluids use less power than the Newtonian fluids. The flow patterns for the pseudoplastic fluids show much greater velocity gradient changes than do the Newtonian fluids in the agitator. The fluid far from the impeller may be moving in slow laminar flow with a high apparent viscosity. Data for fan turbines and propellers are also available (M1).

3.6 (SELECTED TOPIC) DIFFERENTIAL EQUATIONS OF CONTINUITY AND MOMENTUM TRANSFER

3.6A Introduction

In Sections 2.6, 2.7, and 2.8 overall mass, energy, and momentum balances allowed us to solve many elementary problems on fluid flow. These balances were done on an arbitrary finite volume sometimes called a *control volume*. In these total energy, mechanical energy, and momentum balances, we only needed to know the state of the inlet and outlet streams and the exchanges with the surroundings.

These overall balances were powerful tools in solving various flow problems because they did not require knowledge of what goes on inside the finite control volume. Also, in the simple shell momentum balances made in Section 2.9, expressions were obtained for the velocity distribution and pressure drop. However, to advance in our study of these flow systems, we must investigate in greater detail what goes on inside this finite control volume. To do this, we now use a differential element for a control volume. The differential balances will be somewhat similar to the overall and shell balances, but now we shall make the balance in a single phase and integrate to the phase boundary using the boundary conditions. In these balances done earlier, a balance was made for each new system studied. It is not necessary to formulate new balances for each new flow problem. It is often easier to start with the differential equations of the conservation of mass (equation of continuity) and the conservation of momentum in general form. Then these equations are simplified by discarding unneeded terms for each particular problem.

For nonisothermal systems a general differential equation of conservation of energy will be considered in Chapter 5 (Selected Topic). Also in Chapter 7 (Selected Topic) a general differential equation of continuity for a binary mixture will be derived. The differential-momentum-balance equation to be derived is based on Newton's second law and allows us to determine the way velocity varies with position and time and the pressure drop in laminar flow. The equation of momentum balance can be used for turbulent flow with certain modifications.

Often these conservation equations are called *equations of change*, since they describe the variations in the properties of the fluid with respect to position and time. Before we derive these equations, a brief review of the different types of derivatives with respect to time which occur in these equations and a brief description of vector notation will be given.

3.6B   Types of Time Derivatives and Vector Notation

1. Partial time derivative.   Various types of time derivatives are used in the derivations to follow. The most common type of derivative is the partial time derivative. For example, suppose that we are interested in the mass concentration or density ρ in kg/m^3 in a flowing stream as a function of position x, y, z and time t. The partial time derivative of ρ is $\partial\rho/\partial t$. This is the local change of density with time at a fixed point x, y, and z.

2. Total time derivative.   Suppose that we want to measure the density in the stream while we are moving about in the stream with velocities in the x, y, and z directions of dx/dt, dy/dt and dz/dt, respectively.   The total derivative $d\rho/dt$ is

$$\frac{d\rho}{dt} = \frac{\partial\rho}{\partial t} + \frac{\partial\rho}{\partial x}\frac{dx}{dt} + \frac{\partial\rho}{\partial y}\frac{dy}{dt} + \frac{\partial\rho}{\partial z}\frac{dz}{dt} \tag{3.6-1}$$

This means that the density is a function of t and of the velocity components dx/dt, dy/dt, and dz/dt at which the observer is moving.

3. Substantial time derivative.   Another useful type of time derivative is obtained if the observer floats along with the velocity $\mathbf{v}$ of the flowing stream and notes the change in density with respect to time. This is called the derivative that follows the motion, or the *substantial time derivative*, $D\rho/Dt$.

$$\frac{D\rho}{Dt} = \frac{\partial\rho}{\partial t} + v_x\frac{\partial\rho}{\partial x} + v_y\frac{\partial\rho}{\partial y} + v_z\frac{\partial\rho}{\partial z} = \frac{\partial\rho}{\partial t} + (\mathbf{v}\cdot\nabla\rho) \tag{3.6-2}$$

where v_x, v_y, and v_z are the velocity components of the stream velocity $\mathbf{v}$, which is a vector. This substantial derivative is applied to both scalar and vector variables. The term $(\mathbf{v}\cdot\nabla\rho)$ will be discussed in part 6 of Section 3.6B.

4. Scalars. The physical properties encountered in momentum, heat, and mass transfer can be placed in several categories: scalars, vectors, and tensors. Scalars are quantities such as concentration, temperature, length, volume, time, and energy. They have magnitude but no direction and are considered to be zero-order tensors. The common mathematical algebraic laws hold for the algebra of scalars. For example, $bc = cd$, $b(cd) = (bc)d$, and so on.

5. Vectors. Velocity, force, momentum, and acceleration are considered vectors since they have magnitude and direction. They are regarded as first-order tensors and are written in boldface letters in this text, such as $\mathbf{v}$ for velocity. The addition of the two vectors $\mathbf{B} + \mathbf{C}$ by parallelogram construction and the subtraction of two vectors $\mathbf{B} - \mathbf{C}$ is shown in Fig. 3.6-1. The vector $\mathbf{B}$ is represented by its three projections B_x, B_y, and B_z on the x, y, and z axes and

$$\mathbf{B} = \mathbf{i}B_x + \mathbf{j}B_y + \mathbf{k}B_z \qquad (3.6\text{-}3)$$

where $\mathbf{i}, \mathbf{j}$, and $\mathbf{k}$ are unit vectors along the axes x, y, and z, respectively.

In multiplying a scalar quantity r or s by a vector $\mathbf{B}$, the following hold.

$$r\mathbf{B} = \mathbf{B}r \qquad (3.6\text{-}4)$$

$$(rs)\mathbf{B} = r(s\mathbf{B}) \qquad (3.6\text{-}5)$$

$$r\mathbf{B} + s\mathbf{B} = (r + s)\mathbf{B} \qquad (3.6\text{-}6)$$

The following also hold:

$$(\mathbf{B} \cdot \mathbf{C}) = (\mathbf{C} \cdot \mathbf{B}) \qquad (3.6\text{-}7)$$

$$\mathbf{B} \cdot (\mathbf{C} + \mathbf{D}) = (\mathbf{B} \cdot \mathbf{C}) + (\mathbf{B} \cdot \mathbf{D}) \qquad (3.6\text{-}8)$$

$$(\mathbf{B} \cdot \mathbf{C})\mathbf{D} \neq \mathbf{B}(\mathbf{C} \cdot \mathbf{D}) \qquad (3.6\text{-}9)$$

$$(\mathbf{B} \cdot \mathbf{C}) = BC \cos \phi_{BC} \qquad (3.6\text{-}10)$$

where ϕ_{BC} is the angle between two vectors and is $< 180°$.

Second-order tensors τ arise primarily in momentum transfer and have nine components. They are discussed elsewhere (B2).

6. Differential operations with scalars and vectors. The gradient or "grad" of a scalar field is

$$\nabla\rho = \mathbf{i}\frac{\partial\rho}{\partial x} + \mathbf{j}\frac{\partial\rho}{\partial y} + \mathbf{k}\frac{\partial\rho}{\partial z} \qquad (3.6\text{-}11)$$

where ρ is a scalar such as density.

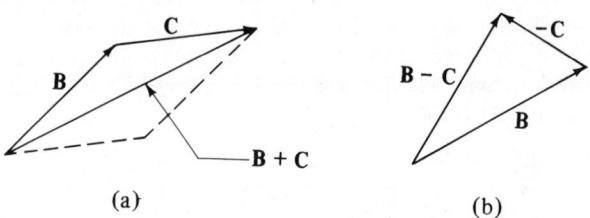

(a)   (b)

FIGURE 3.6-1.   *Addition and subtraction of vectors: (a) addition of vectors, $\mathbf{B} + \mathbf{C}$;*
(b) subtraction of vectors, $\mathbf{B} - \mathbf{C}$.

The divergence or "div" of a vector **v** is

$$(\nabla \cdot \mathbf{v}) = \frac{\partial v_x}{\partial x} + \frac{\partial v_y}{\partial y} + \frac{\partial v_z}{\partial z} \qquad (3.6\text{-}12)$$

where **v** is a function of v_x, v_y, and v_z.

The Laplacian of a scalar field is

$$\nabla^2 \rho = \frac{\partial^2 \rho}{\partial x^2} + \frac{\partial^2 \rho}{\partial y^2} + \frac{\partial^2 \rho}{\partial z^2} \qquad (3.6\text{-}13)$$

Other operations that may be useful are

$$\nabla rs = r\nabla s + s\nabla r \qquad (3.6\text{-}14)$$

$$(\nabla \cdot s\mathbf{v}) = (\nabla s \cdot \mathbf{v}) + s(\nabla \cdot \mathbf{v}) \qquad (3.6\text{-}15)$$

$$(\mathbf{v} \cdot \nabla s) = v_x \frac{\partial s}{\partial x} + v_y \frac{\partial s}{\partial y} + v_z \frac{\partial s}{\partial z} \qquad (3.6\text{-}16)$$

3.6C   Differential Equation of Continuity

1. Derivation of equation of continuity.   A mass balance will be made for a pure fluid flowing through a stationary volume element $\Delta x\, \Delta y\, \Delta z$ which is fixed in space as in Fig. 3.6-2. The mass balance for the fluid with a concentration of ρ kg/m^3 is

(rate of mass in) − (rate of mass out) = (rate of mass accumulation)   (3.6-17)

In the x direction the rate of mass entering the face at x having an area of $\Delta y\, \Delta z$ m^2 is $(\rho v_x)_x\, \Delta y\, \Delta z$ kg/s and that leaving at $x + \Delta x$ is $(\rho v_x)_{x+\Delta x}\, \Delta y\, \Delta z$. The term (ρv_x) is a mass flux in kg/s · m^2. Mass entering and that leaving in the y and the z directions are also shown in Fig. 3.6-2.

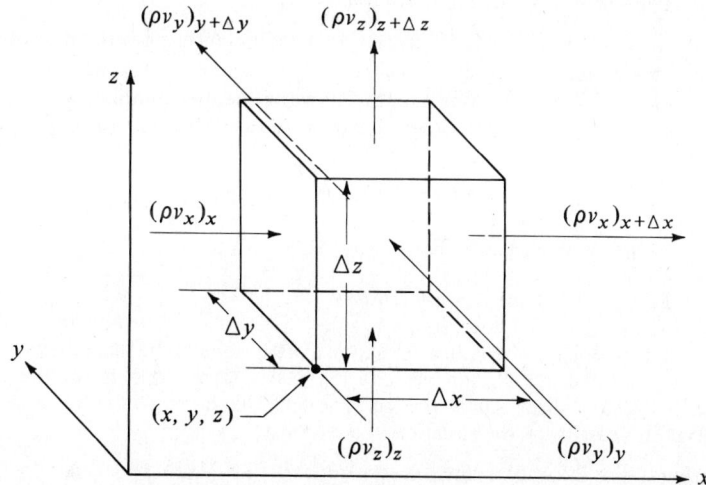

FIGURE 3.6-2.   *Mass balance for a pure fluid flowing through a fixed volume $\Delta x\, \Delta y\, \Delta z$ in space.*

The rate of mass accumulation in the volume $\Delta x\ \Delta y\ \Delta z$ is

$$\text{rate of mass accumulation} = \Delta x\ \Delta y\ \Delta z\ \frac{\partial \rho}{\partial t} \qquad (3.6\text{-}18)$$

Substituting all these expressions into Eq. (3.6-17) and dividing both sides by $\Delta x\ \Delta y\ \Delta z$,

$$\frac{[(\rho v_x)_x - (\rho v_x)_{x+\Delta x}]}{\Delta x} + \frac{[(\rho v_y)_y - (\rho v_y)_{y+\Delta y}]}{\Delta y} + \frac{[(\rho v_z)_z - (\rho v_z)_{z+\Delta z}]}{\Delta z} = \frac{\partial \rho}{\partial t} \quad (3.6\text{-}19)$$

Taking the limit as Δx, Δy, and Δz approach zero, we obtain the equation of continuity or conservation of mass for a pure fluid.

$$\frac{\partial \rho}{\partial t} = -\left[\frac{\partial(\rho v_x)}{\partial x} + \frac{\partial(\rho v_y)}{\partial y} + \frac{\partial(\rho v_z)}{\partial z}\right] = -(\nabla \cdot \rho \mathbf{v}) \qquad (3.6\text{-}20)$$

The vector notation on the right side of Eq. (3.6-20) comes from the fact that $\mathbf{v}$ is a vector. Equation (3.6-20) tells us how density ρ changes with time at a fixed point resulting from the changes in the mass velocity vector $\rho \mathbf{v}$.

We can convert Eq. (3.6-20) into another form by carrying out the actual partial differentiation.

$$\frac{\partial \rho}{\partial t} = -\rho\left(\frac{\partial v_x}{\partial x} + \frac{\partial v_y}{\partial y} + \frac{\partial v_z}{\partial z}\right) - \left(v_x \frac{\partial \rho}{\partial x} + v_y \frac{\partial \rho}{\partial y} + v_z \frac{\partial \rho}{\partial z}\right) \qquad (3.6\text{-}21)$$

Rearranging Eq. (3.6-21),

$$\frac{\partial \rho}{\partial t} + v_x \frac{\partial \rho}{\partial x} + v_y \frac{\partial \rho}{\partial y} + v_z \frac{\partial \rho}{\partial z} = -\rho\left(\frac{\partial v_x}{\partial x} + \frac{\partial v_y}{\partial y} + \frac{\partial v_z}{\partial z}\right) \qquad (3.6\text{-}22)$$

The left-hand side of Eq. (3.6-22) is the same as the substantial derivative in Eq. (3.6-2). Hence, Eq. (3.6-22) becomes

$$\frac{D\rho}{Dt} = -\rho\left(\frac{\partial v_x}{\partial x} + \frac{\partial v_y}{\partial y} + \frac{\partial v_z}{\partial z}\right) = -\rho(\nabla \cdot \mathbf{v}) \qquad (3.6\text{-}23)$$

2. Equation of continuity for constant density. Often in engineering with liquids that are relatively incompressible, the density ρ is essentially constant. Then ρ remains constant for a fluid element as it moves along a path following the fluid motion, or $D\rho/Dt = 0$. Hence, Eq. (3.6-23) becomes for a fluid of constant density at steady or unsteady state,

$$(\nabla \cdot \mathbf{v}) = \frac{\partial v_x}{\partial x} + \frac{\partial v_y}{\partial y} + \frac{\partial v_z}{\partial z} = 0 \qquad (3.6\text{-}24)$$

At steady state, $\partial \rho/\partial t = 0$ in Eq. (3.6-22).

EXAMPLE 3.6-1. Flow over a Flat Plate
An incompressible fluid flows past one side of a flat plate. The flow in the x direction is parallel to the flat plate. At the leading edge of the plate the flow is uniform at the free stream velocity v_{x0}. There is no velocity in the z direction. The y direction is the perpendicular distance from the plate. Analyze this case using the equation of continuity.

Solution: For this case where ρ is constant, Eq. (3.6-24) holds.

$$\frac{\partial v_x}{\partial x} + \frac{\partial v_y}{\partial y} + \frac{\partial v_z}{\partial z} = 0 \qquad (3.6\text{-}24)$$

Transport Processes: Momentum, Heat, and Mass

Since there is no velocity in the z direction, we obtain

$$\frac{\partial v_x}{\partial x} = -\frac{\partial v_y}{\partial y} \qquad (3.6\text{-}25)$$

At a given small value of y close to the plate, the value of v_x must decrease from its free stream velocity v_{x0} as it passes the leading edge in the x direction because of fluid friction. Hence, $\partial v_x/\partial x$ is negative. Then from Eq. (3.6-25), $\partial v_y/\partial y$ is positive and there is a component of velocity away from the plate.

3. Continuity equation in cylindrical and spherical coordinates. It is often convenient to use cylindrical coordinates to solve the equation of continuity if fluid is flowing in a cylinder. The coordinate system as related to rectangular coordinates is shown in Fig. 3.6-3a. The relations between rectangular x, y, z and cylindrical r, θ, z coordinates are

$$x = r \cos \theta \qquad y = r \sin \theta \qquad z = z$$

$$\qquad\qquad\qquad\qquad\qquad\qquad\qquad\qquad (3.6\text{-}26)$$

$$r = +\sqrt{x^2 + y^2} \qquad \theta = \tan^{-1} \frac{y}{x}$$

Using the relations from Eq. (3.6-26) with Eq. (3.6-20), the equation of continuity in cylindrical coordinates is

$$\frac{\partial \rho}{\partial t} + \frac{1}{r}\frac{\partial(\rho r v_r)}{\partial r} + \frac{1}{r}\frac{\partial(\rho v_\theta)}{\partial \theta} + \frac{\partial(\rho v_z)}{\partial z} = 0 \qquad (3.6\text{-}27)$$

For spherical coordinates the variables r, θ, and ϕ are related to x, y, z by the following as shown in Fig. 3.6-3b.

$$x = r \sin \theta \cos \phi \qquad y = r \sin \theta \sin \phi \qquad z = r \cos \theta$$

$$\qquad\qquad\qquad\qquad\qquad\qquad\qquad\qquad (3.6\text{-}28)$$

$$r = +\sqrt{x^2 + y^2 + z^2} \qquad \theta = \tan^{-1} \frac{\sqrt{x^2 + y^2}}{z} \qquad \phi = \tan^{-1} \frac{y}{x}$$

The equation of continuity in spherical coordinates becomes

$$\frac{\partial \rho}{\partial t} + \frac{1}{r^2}\frac{\partial(\rho r^2 v_r)}{\partial r} + \frac{1}{r \sin \theta}\frac{\partial(\rho v_\theta \sin \theta)}{\partial \theta} + \frac{1}{r \sin \theta}\frac{\partial(\rho v_\phi)}{\partial \phi} = 0 \qquad (3.6\text{-}29)$$

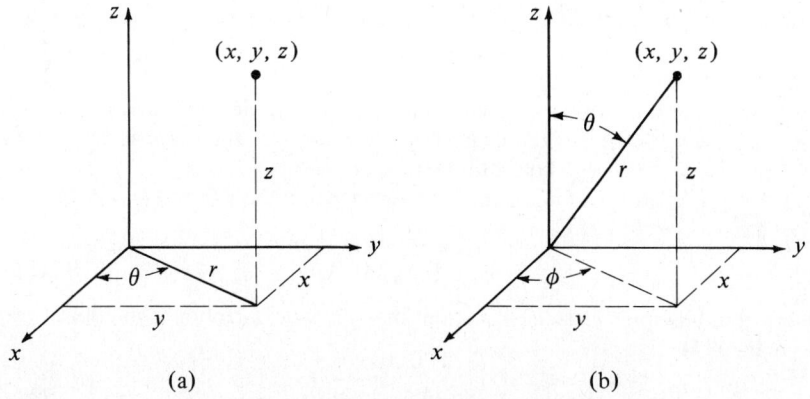

(a)   (b)

FIGURE 3.6-3. *Curvilinear coordinate systems : (a) cylindrical coordinates, (b) spherical coordinates.*

3.6D Differential Equation of Momentum Transfer or Motion

1. Derivation of equation of momentum transfer. The equation of motion is really the equation for the conservation-of-momentum equation (2.8-3), which we can write as

$$
\begin{pmatrix} \text{rate of} \\ \text{momentum in} \end{pmatrix} - \begin{pmatrix} \text{rate of} \\ \text{momentum out} \end{pmatrix}
$$

$$
+ \begin{pmatrix} \text{sum of forces} \\ \text{acting on system} \end{pmatrix} = \begin{pmatrix} \text{rate of momentum} \\ \text{accumulation} \end{pmatrix} \quad \text{(3.6-30)}
$$

We will make a balance on an element as in Fig. 3.6-2. First we shall consider only the x component of each term in Eq. (3.6-30). The y and z components can be described in an analogous manner.

The rate at which the x component of momentum enters the face at x in the x direction by convection is $(\rho v_x v_x)_x \, \Delta y \, \Delta z$, and the rate at which it leaves at $x + \Delta x$ is $(\rho v_x v_x)_{x+\Delta x} \, \Delta y \, \Delta z$. The quantity (ρv_x) is the concentration in momentum/m^3 or (kg · m/s)/m^3, and it is multiplied by v_x to give the momentum flux as momentum/s · m^2.

The x component of momentum entering the face at y is $(\rho v_y v_x)_y \, \Delta x \, \Delta z$, and leaving at $y + \Delta y$ is $(\rho v_y v_x)_{y+\Delta y} \, \Delta x \, \Delta z$. For the face at z we have $(\rho v_z v_x)_z \, \Delta x \, \Delta y$ entering, and at $z + \Delta z$ we have $(\rho v_z v_x)_{z+\Delta z} \, \Delta x \, \Delta y$ leaving. Hence, the net convective x momentum flow into the volume element $\Delta x \, \Delta y \, \Delta z$ is

$$
[(\rho v_x v_x)_x - (\rho v_x v_x)_{x+\Delta x}]\Delta y \, \Delta z + [(\rho v_y v_x)_y - (\rho v_y v_x)_{y+\Delta y}]\Delta x \, \Delta z
$$

$$
+ [(\rho v_z v_x)_z - (\rho v_z v_x)_{z+\Delta z}]\Delta x \, \Delta y \quad \text{(3.6-31)}
$$

Momentum flows in and out of the volume element by the mechanisms of convection or bulk flow as given in Eq. (3.6-31) and also by molecular transfer (by virtue of the velocity gradients in laminar flow). The rate at which the x component of momentum enters the face at x by molecular transfer is $(\tau_{xx})_x \, \Delta y \, \Delta z$, and the rate at which it leaves the surface at $x + \Delta x$ is $(\tau_{xx})_{x+\Delta x} \, \Delta y \, \Delta z$. The rate at which it enters the face at y is $(\tau_{yx})_y \, \Delta x \, \Delta z$, and it leaves at $y + \Delta y$ at a rate of $(\tau_{yx})_{y+\Delta y} \, \Delta x \, \Delta z$. Note that τ_{yx} is the flux of x momentum through the face perpendicular to the y axis. Writing a similar equation for the remaining faces the net x component of momentum by molecular transfer is

$$
[(\tau_{xx})_x - (\tau_{xx})_{x+\Delta x}]\Delta y \, \Delta z + [(\tau_{yx})_y - (\tau_{yx})_{y+\Delta y}]\Delta x \, \Delta z + [(\tau_{zx})_z - (\tau_{zx})_{z+\Delta z}]\Delta x \, \Delta y
$$

$$
\text{(3.6-32)}
$$

These molecular fluxes of momentum may be considered as shear stresses and normal stresses. Hence, τ_{yx} is the x direction shear stress on the y face and τ_{zx} the shear stress on the z face. Also, τ_{xx} is the normal stress on the x face.

The net fluid pressure force acting on the element in the x direction is the difference between the force acting at x and $x + \Delta x$.

$$
(p_x - p_{x+\Delta x})\Delta y \, \Delta z \quad \text{(3.6-33)}
$$

The gravitational force g_x acting on a unit mass in the x direction is multiplied by the mass of the element to give

$$
\rho g_x \, \Delta x \, \Delta y \, \Delta z \quad \text{(3.6-34)}
$$

where g_x is the x component of the gravitational vector **g**.

Transport Processes: Momentum, Heat, and Mass

The rate of accumulation of x momentum in the element is

$$\Delta x\,\Delta y\,\Delta z\,\frac{\partial(\rho v_x)}{\partial t} \qquad \text{(3.6-35)}$$

Substituting Eqs. (3.6-31)–(3.6-35) into (3.6-30), dividing by $\Delta x\,\Delta y\,\Delta z$, and taking the limit as Δx, Δy, and Δz approach zero,

$$\frac{\partial(\rho v_x)}{\partial t} = -\left[\frac{\partial(\rho v_x v_x)}{\partial x} + \frac{\partial(\rho v_y v_x)}{\partial y} + \frac{\partial(\rho v_z v_x)}{\partial z}\right] - \left(\frac{\partial \tau_{xx}}{\partial x} + \frac{\partial \tau_{yx}}{\partial y} + \frac{\partial \tau_{zx}}{\partial z}\right) - \frac{\partial p}{\partial x} + \rho g_x$$

$$\text{(3.6-36)}$$

This is the x component of the differential equation of motion. The y and z components of the differential equation of motion are, respectively,

$$\frac{\partial(\rho v_y)}{\partial t} = -\left[\frac{\partial(\rho v_x v_y)}{\partial x} + \frac{\partial(\rho v_y v_y)}{\partial y} + \frac{\partial(\rho v_z v_y)}{\partial z}\right] - \left(\frac{\partial \tau_{xy}}{\partial x} + \frac{\partial \tau_{yy}}{\partial y} + \frac{\partial \tau_{zy}}{\partial z}\right) - \frac{\partial p}{\partial y} + \rho g_y$$

$$\text{(3.6-37)}$$

$$\frac{\partial(\rho v_z)}{\partial t} = -\left[\frac{\partial(\rho v_x v_z)}{\partial x} + \frac{\partial(\rho v_y v_z)}{\partial y} + \frac{\partial(\rho v_z v_z)}{\partial z}\right] - \left(\frac{\partial \tau_{xz}}{\partial x} + \frac{\partial \tau_{yz}}{\partial y} + \frac{\partial \tau_{zz}}{\partial z}\right) - \frac{\partial p}{\partial z} + \rho g_z$$

$$\text{(3.6-38)}$$

We can use Eq. (3.6-20), which is the continuity equation, and Eq. (3.6-36) and obtain an equation of motion for the x component as follows.

$$\rho\,\frac{Dv_x}{Dt} = -\left[\frac{\partial \tau_{xx}}{\partial x} + \frac{\partial \tau_{yx}}{\partial y} + \frac{\partial \tau_{zx}}{\partial z}\right] + \rho g_x - \frac{\partial p}{\partial x} \qquad \text{(3.6-39)}$$

Similar equations can be written for the y and z components and adding we obtain an equation of motion for a pure fluid.

$$\rho\,\frac{D\mathbf{v}}{Dt} = -[\nabla \cdot \tau] - \nabla p + \rho \mathbf{g} \qquad \text{(3.6-40)}$$

2. Equation of motion for Newtonian fluids. In order to use Eqs. (3.6-36) to (3.6-40) to determine velocity distributions, expressions must be used for the various stresses in terms of velocity gradients and fluid properties. For Newtonian fluids the expressions for the stresses τ_{xx}, τ_{yx}, τ_{zx}, and so on have been related to the velocity gradients and the fluid viscosity μ (B1, B2, D1). After these expressions are substituted into Eq. (3.6-39), for the x component of momentum we obtain

$$\rho\,\frac{Dv_x}{Dt} = \frac{\partial}{\partial x}\left[2\mu\,\frac{\partial v_x}{\partial x} - \frac{2}{3}\mu(\nabla \cdot \mathbf{v})\right] + \frac{\partial}{\partial y}\left[\mu\left(\frac{\partial v_x}{\partial y} + \frac{\partial v_y}{\partial x}\right)\right]$$

$$+ \frac{\partial}{\partial z}\left[\mu\left(\frac{\partial v_z}{\partial x} + \frac{\partial v_x}{\partial z}\right)\right] - \frac{\partial p}{\partial x} + \rho g_x \qquad \text{(3.6-41)}$$

Similar equations are obtained for the y and z components of momentum.

3. Navier–Stokes equations. The equations above are seldom used in their complete forms. In many cases the density ρ and viscosity μ are constant, and since $(\nabla \cdot \mathbf{v}) = 0$, we obtain the Navier–Stokes equations for the x component, y component, and z component, respectively.

$$\rho\left(\frac{\partial v_x}{\partial t} + v_x\frac{\partial v_x}{\partial x} + v_y\frac{\partial v_x}{\partial y} + v_z\frac{\partial v_x}{\partial z}\right) = \mu\left(\frac{\partial^2 v_x}{\partial x^2} + \frac{\partial^2 v_x}{\partial y^2} + \frac{\partial^2 v_x}{\partial z^2}\right) - \frac{\partial p}{\partial x} + \rho g_x \qquad \textbf{(3.6-42)}$$

$$\rho\left(\frac{\partial v_y}{\partial t} + v_x\frac{\partial v_y}{\partial x} + v_y\frac{\partial v_y}{\partial y} + v_z\frac{\partial v_y}{\partial z}\right) = \mu\left(\frac{\partial^2 v_y}{\partial x^2} + \frac{\partial^2 v_y}{\partial y^2} + \frac{\partial^2 v_y}{\partial z^2}\right) - \frac{\partial p}{\partial y} + \rho g_y \qquad \textbf{(3.6-43)}$$

$$\rho\left(\frac{\partial v_z}{\partial t} + v_x\frac{\partial v_z}{\partial x} + v_y\frac{\partial v_z}{\partial y} + v_z\frac{\partial v_z}{\partial z}\right) = \mu\left(\frac{\partial^2 v_z}{\partial x^2} + \frac{\partial^2 v_z}{\partial y^2} + \frac{\partial^2 v_z}{\partial z^2}\right) - \frac{\partial p}{\partial z} + \rho g_z \qquad \textbf{(3.6-44)}$$

Combining the three equations for the three components, we obtain

$$\rho\frac{D\mathbf{v}}{Dt} = -\nabla p + \rho\mathbf{g} + \mu\nabla^2\mathbf{v} \qquad \textbf{(3.6-45)}$$

3.6E Use of Equations of Motion and Continuity

The purpose and uses of the equations of motion and continuity, as mentioned previously, are to apply these equations to any viscous-flow problem. For a given specific problem, the terms that are zero or near zero are simply discarded and the remaining equations used in the solution to solve for the velocity, density, and pressure distributions. Of course, it is necessary to know the initial conditions and the boundary conditions to solve the equations. Several examples will be given to illustrate the general methods used.

EXAMPLE 3.6-2. Laminar Flow Between Parallel Plates
Derive the equation giving the velocity distribution at steady state for laminar flow of a constant-density fluid with constant viscosity which is flowing between two flat and parallel plates. The velocity profile desired is at a point far from the inlet or outlet of the channel. The two plates will be considered to be fixed and of infinite width, with the flow driven by the pressure gradient in the x direction.

Solution: Assuming that the channel is horizontal, Fig. 3.6-4 shows the axes selected with flow in the x direction and the width in the z direction. The velocities v_y and v_z are then zero. The plates are a distance $2y_0$ apart.
    The continuity equation (3.6-24) for constant density is

$$\frac{\partial v_x}{\partial x} + \frac{\partial v_y}{\partial y} + \frac{\partial v_z}{\partial z} = 0 \qquad \textbf{(3.6-24)}$$

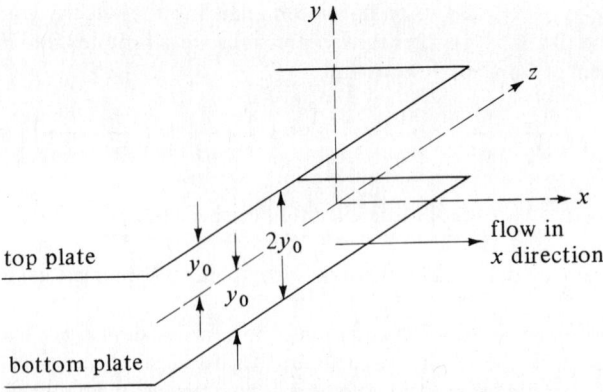

FIGURE 3.6-4. *Flow between two parallel plates in Example 3.6-2.*

Since v_y and v_z are zero, Eq. (3.6-24) becomes

$$\frac{\partial v_x}{\partial x} = 0 \qquad \text{(3.6-46)}$$

The Navier–Stokes equation for the x component is

$$\rho\left(\frac{\partial v_x}{\partial t} + v_x \frac{\partial v_x}{\partial x} + v_y \frac{\partial v_x}{\partial y} + v_z \frac{\partial v_x}{\partial z}\right) = \mu\left(\frac{\partial^2 v_x}{\partial x^2} + \frac{\partial^2 v_x}{\partial y^2} + \frac{\partial^2 v_x}{\partial z^2}\right) - \frac{\partial p}{\partial x} + \rho g_x$$

$$\text{(3.6-42)}$$

Also, $\partial v_x/\partial t = 0$ for steady state, $v_y = 0$, $v_z = 0$, $\partial v_x/\partial x = 0$, $\partial^2 v_x/\partial x^2 = 0$. We can see that $\partial v_x/\partial z = 0$, since there is no change of v_x with z. Then $\partial^2 v_x/\partial z^2 = 0$. Making these substitutions into Eq. (3.6-42), we obtain

$$\frac{\partial p}{\partial x} - \rho g_x = \mu \frac{\partial^2 v_x}{\partial y^2} \qquad \text{(3.6-47)}$$

In fluid-flow problems we will be concerned with gravitational force only in the vertical direction for g_x, which is g, the gravitational force, in m/s^2. We shall combine the static pressure p and the gravitational force and call them simply p, as follows (note that $g_x = 0$ for the present case of a horizontal pipe but is not zero for the general case of a nonhorizontal pipe):

$$p = p + \rho g h \qquad \text{(3.6-48)}$$

where h is the distance upward from any chosen reference plane (h is in the direction opposed to gravity). Then Eq. (3.6-47) becomes

$$\frac{\partial p}{\partial x} = \mu \frac{\partial^2 v_x}{\partial y^2} \qquad \text{(3.6-49)}$$

We can see that p is not a function of z. Also, assuming that $2y_0$ is small, p is not a function of y. (Some references avoid this problem and simply use p as a dynamic pressure, which is rigorously correct since dynamic pressure gradients cause flow. In a fluid at rest the total pressure gradient is the hydrostatic pressure gradient and the dynamic pressure gradient is zero.) Also, $\partial p/\partial x$ is a constant in this problem since v_x is not a function of x. Then Eq. (3.6-49) becomes an ordinary differential equation.

$$\frac{d^2 v_x}{dy^2} = \frac{1}{\mu}\frac{dp}{dx} = \text{const} \qquad \text{(3.6-50)}$$

Integrating Eq. (3.6-50) once using the condition $dv_x/dy = 0$ at $y = 0$ for symmetry,

$$\frac{dv_x}{dy} = \left(\frac{1}{\mu}\frac{dp}{dx}\right) y \qquad \text{(3.6-51)}$$

Integrating again using $v_x = 0$ at $y = y_0$,

$$v_x = \frac{1}{2\mu}\frac{dp}{dx}(y^2 - y_0^2) \qquad \text{(3.6-52)}$$

The maximum velocity in Eq. (3.6-52) occurs when $y = 0$, giving

$$v_{x\,\text{max}} = \frac{1}{2\mu}\frac{dp}{dx}(-y_0^2) \qquad \text{(3.6-53)}$$

Combining Eqs. (3.6-52) and (3.6-53),

$$v_x = v_{x\,max}\left[1 - \left(\frac{y}{y_0}\right)^2\right]$$ (3.6-54)

Hence, a parabolic velocity profile is obtained. This result was also obtained in Eq. (2.9-9) when using a shell momentum balance.

The results obtained in Example 3.6-2 could also have been obtained by making a force balance on a differential element of fluid and using the symmetry of the system to omit certain terms.

EXAMPLE 3.6-3. *Laminar Flow in a Circular Tube*
Derive the equation for steady-state viscous flow in a horizontal tube of radius r_0, where the fluid is far from the tube inlet. The fluid is incompressible and μ is a constant. The flow is driven in one direction by a constant-pressure gradient.

Solution: The fluid will be assumed to flow in the z direction in the tube, as shown in Fig. 3.6-5. The y direction is vertical and the x direction horizontal. Since v_x and v_y are zero, the continuity equation becomes $\partial v_z/\partial z = 0$. For steady state $\partial v_z/\partial t = 0$. Then substituting into Eq. (3.6-44) for the z component, we obtain

$$\frac{dp}{dz} = \mu\left(\frac{\partial^2 v_z}{\partial x^2} + \frac{\partial^2 v_z}{\partial y^2}\right)$$ (3.6-55)

To solve Eq. (3.6-55) we can use cylindrical coordinates from Eq. (3.6-26), giving

$$z = z \qquad x = r\cos\theta \qquad y = r\sin\theta$$

$$r = +\sqrt{x^2 + y^2} \qquad \theta = \tan^{-1}\frac{y}{x}$$ (3.6-26)

Substituting these into Eq. (3.6-55),

$$\frac{1}{\mu}\frac{dp}{dz} = \frac{\partial^2 v_z}{\partial r^2} + \frac{1}{r}\frac{\partial v_z}{\partial r} + \frac{1}{r^2}\frac{\partial^2 v_z}{\partial \theta^2}$$ (3.6-56)

The flow is symmetrical about the z axis so $\partial^2 v_z/\partial \theta^2$ is zero in Eq. (3.6-56). As before, dp/dz is a constant, so Eq. (3.6-56) becomes

$$\frac{1}{\mu}\frac{dp}{dz} = \text{const} = \frac{d^2 v_z}{dr^2} + \frac{1}{r}\frac{dv_z}{dr} = \frac{1}{r}\frac{d}{dr}\left(r\frac{dv_z}{dr}\right)$$ (3.6-57)

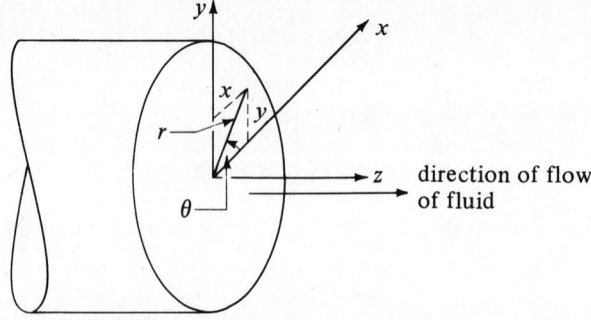

FIGURE 3.6-5. *Horizontal flow in a tube in Example 3.6-3.*

The boundary conditions for the first integration are $dv_z/dr = 0$ at $r = 0$. For the second integration, $v_z = 0$ at $r = r_0$ (tube radius). The result is

$$v_z = \frac{1}{4\mu}\frac{dp}{dz}(r^2 - r_0^2) \qquad \text{(3.6-58)}$$

Converting to the maximum velocity as before,

$$v_z = v_{z\,max}\left[1 - \left(\frac{r}{r_0}\right)^2\right] \qquad \text{(3.6-59)}$$

If Eq. (3.6-58) is integrated over the pipe cross section using Eq. (2.9-10) to give the average velocity $v_{z\,av}$,

$$v_{z\,av} = -\frac{r_0^2}{8\mu}\frac{dp}{dz} \qquad \text{(3.6-60)}$$

Integrating to obtain the pressure drop from $z = 0$ for $p = p_1$ to $z = L$ for $p = p_2$, we obtain

$$p_1 - p_2 = \frac{8\mu v_{z\,av}L}{r_0^2} = \frac{32\mu v_{z\,av}L}{D^2} \qquad \text{(3.6-61)}$$

where $D = 2r_0$. This is the Hagen-Poiseuille equation derived previously as Eq. (2.9-11).

EXAMPLE 3.6-4. *Laminar Flow in a Cylindrical Annulus*
Derive the equation for steady-state laminar flow inside the annulus between two concentric horizontal pipes. This type of flow occurs often in concentric pipe heat exchangers.

Solution: In this case Eq. (3.6-57) also still holds. However, the velocity in the annulus will reach a maximum at some radius $r = r_{max}$ which is between r_1 and r_2, as shown in Fig. 3.6-6. For the first integration of Eq. (3.6-57), the boundary conditions are $dv_z/dr = 0$ at $r = r_{max}$, which gives

$$\frac{r\,dv_z}{dr} = \left(\frac{1}{\mu}\frac{dp}{dz}\right)\left(\frac{r^2}{2} - \frac{r_{max}^2}{2}\right) \qquad \text{(3.6-62)}$$

Also, for the second integration of Eq. (3.6-62), $v_z = 0$ at the inner wall where $r = r_1$, giving

$$v_z = \left(\frac{1}{2\mu}\frac{dp}{dz}\right)\left(\frac{r^2}{2} - \frac{r_1^2}{2} - r_{max}^2\ln\frac{r}{r_1}\right) \qquad \text{(3.6-63)}$$

Repeating the second integration but for $v_z = 0$ at the outer wall where

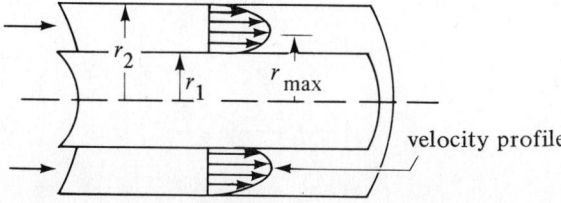

FIGURE 3.6-6. *Flow through a cylindrical annulus.*

$r = r_2$, we obtain

$$v_z = \left(\frac{1}{2\mu}\frac{dp}{dz}\right)\left(\frac{r^2}{2} - \frac{r_2^2}{2} - r^2_{max} \ln \frac{r}{r_2}\right) \qquad (3.6\text{-}64)$$

Combining Eqs. (3.6-63) and (3.6-64) and solving for r_{max},

$$r_{max} = \sqrt{\frac{1}{\ln(r_2/r_1)}(r_2^2 - r_1^2)/2} \qquad (3.6\text{-}65)$$

In Fig. 3.6-6 the velocity profile predicted by Eq. (3.6-64) is plotted. For the case where $r_1 = 0$, r_{max} in Eq. (3.6-65) becomes zero and Eq. (3.6-64) reduces to Eq. (3.6-58) for a single circular pipe.

3.6F Euler Equations of Motion for Ideal Fluids

Special equations for *ideal fluids* can be obtained for a fluid having a constant density and zero viscosity. These are called the *Euler equations*. Equations (3.6-42)–(3.6-44) for the x, y, and z components of momentum become

$$\rho\left(\frac{\partial v_x}{\partial t} + v_x\frac{\partial v_x}{\partial x} + v_y\frac{\partial v_x}{\partial y} + v_z\frac{\partial v_x}{\partial z}\right) = -\frac{\partial p}{\partial x} + \rho g_x \qquad (3.6\text{-}66)$$

$$\rho\left(\frac{\partial v_y}{\partial t} + v_x\frac{\partial v_y}{\partial x} + v_y\frac{\partial v_y}{\partial y} + v_z\frac{\partial v_y}{\partial y}\right) = -\frac{\partial p}{\partial y} + \rho g_y \qquad (3.6\text{-}67)$$

$$\rho\left(\frac{\partial v_z}{\partial t} + v_x\frac{\partial v_z}{\partial x} + v_y\frac{\partial v_z}{\partial y} + v_z\frac{\partial v_z}{\partial z}\right) = -\frac{\partial p}{\partial z} + \rho g_z \qquad (3.6\text{-}68)$$

These equations are useful in calculating pressure distribution at the outer edge of the thin boundary layer in flow past immersed bodies. This theory is particularly useful in aerodynamics in the study of airfoils.

3.6G Equations for Creeping Flow and Stokes' Law

At very low Reynolds numbers below about 1, the term *creeping flow* is used to describe flow at very low velocities. This type of flow applies for the fall or settling of small particles through a fluid. Stokes' law is derived using this type of flow in problems of settling and sedimentation.

In flow around a sphere, for example, the fluid changes velocity and direction in a complex manner. If the inertia effects in this case were important, it would be necessary to keep all the terms in the three Navier–Stokes equations. Experiments show that at a Reynolds number below about 1, the inertia effects are small and can be omitted. Hence, the equations of motion, Eqs. (3.6-42)–(3.6-44) for creeping flow of an incompressible fluid, become

$$\frac{\partial p}{\partial x} = \mu\left(\frac{\partial^2 v_x}{\partial x^2} + \frac{\partial^2 v_x}{\partial y^2} + \frac{\partial^2 v_x}{\partial z^2}\right) \qquad (3.6\text{-}69)$$

$$\frac{\partial p}{\partial y} = \mu\left(\frac{\partial^2 v_y}{\partial x^2} + \frac{\partial^2 v_y}{\partial y^2} + \frac{\partial^2 v_y}{\partial z^2}\right) \qquad (3.6\text{-}70)$$

$$\frac{\partial p}{\partial z} = \mu\left(\frac{\partial^2 v_z}{\partial x^2} + \frac{\partial^2 v_z}{\partial y^2} + \frac{\partial^2 v_z}{\partial z^2}\right) \qquad (3.6\text{-}71)$$

For flow past a sphere, these equations have been used to describe the velocity distribution and the pressure distribution over the sphere. Then by integration over the whole sphere, the form drag, caused by the pressure distribution, and the skin friction or viscous drag, caused by the shear stress at the surface, can be summed to give the total drag.

$$F_D = 3\pi\mu D_p v \qquad \text{(SI)}$$

$$F_D = \frac{3\pi\mu D_p v}{g_c} \qquad \text{(English)}$$

(3.6-72)

where F_D is total drag force in N, D_p is particle diameter in m, v is free stream velocity of fluid approaching the sphere in m/s, and μ is viscosity in kg/m · s. This is Stokes' equation for the drag force on a sphere.

Often Eq. (3.6-72) is rewritten as follows:

$$F_D = C_D \frac{v^2}{2} \rho A \qquad \text{(SI)}$$

$$F_D = C_D \frac{v^2}{2g_c} \rho A \qquad \text{(English)}$$

(3.6-73)

where C_D is a drag coefficient, which is equal to $24/N_{Re}$ for Stokes' law, and A is the projected area of the sphere, which is $\pi D_p^2/4$. This is discussed in more detail in Section 3.1 for flow past spheres.

3.7 (SELECTED TOPIC) BOUNDARY-LAYER FLOW AND TURBULENCE

3.7A   Boundary-Layer Flow

In Section 3.6 the Navier–Stokes equations were used to find relations that described laminar flow between flat plates and inside circular tubes, flow of ideal fluids, and creeping flow. In this section the flow of fluids around objects will be considered in more detail, with particular attention being given to the region close to the solid surface, called the *boundary layer*.

In the boundary-layer region near the solid, the fluid motion is greatly affected by this solid surface. In the bulk of the fluid away from the boundary layer the flow can often be adequately described by the theory of ideal fluids with zero viscosity. However, in the thin boundary layer, viscosity is important. Since the region is thin, simplified solutions can be obtained for the boundary-layer region. Prandtl originally suggested this division of the problem into two parts, which has been used extensively in fluid dynamics.

In order to help explain boundary layers, an example of boundary-layer formation in the steady-state flow of a fluid past a flat plate is given in Fig. 3.7-1. The velocity of the fluid upstream of the leading edge at $x = 0$ of the plate is uniform across the entire fluid stream and has the value v_∞. The velocity of the fluid at the interface is zero and the velocity v_x in the x direction increases as one goes farther from the plate. The velocity v_x approaches asymptotically the velocity v_∞ of the bulk of the stream.

The dashed line L is drawn so that the velocity at that point is 99% of the bulk velocity v_∞. The layer or zone between the plate and the dashed line constitutes the boundary layer. When the flow is laminar, the thickness δ of the boundary layer increases

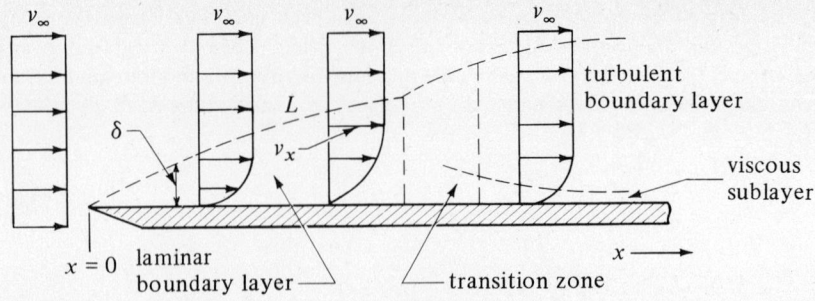

FIGURE 3.7-1. *Boundary layer for flow past a flat plate.*

with the $\sqrt{x}$ as we move in the x direction. The Reynolds number is defined as $N_{Re, x} = x v_\infty \rho / \mu$, where x is the distance downstream from the leading edge. When the Reynolds number is less than 2×10^5 the flow is laminar, as shown in Fig. 3.7-1.

The transition from laminar to turbulent flow on a smooth plate occurs in the Reynolds number range 2×10^5 to 3×10^6, as shown in Fig. 3.7-1. When the boundary layer is turbulent, a thin viscous sublayer persists next to the plate. The drag caused by the viscous shear in the boundary layers is called *skin friction* and it is the only drag present for flow past a flat plate.

The type of drag occurring when fluid flows by a bluff or blunt shape such as a sphere or cylinder, which is mostly caused by a pressure difference, is termed *form drag*. This drag predominates in flow past such objects at all except low values of the Reynolds numbers, and often a wake is present. Skin friction and form drag both occur in flow past a bluff shape, and the total drag is the sum of the skin friction and the form drag. (See also Section 3.1).

3.7B Boundary-Layer Separation and Formation of Wakes

We discussed the growth of the boundary layer at the leading edge of a plate as shown in Fig. 3.7-2. However, some important phenomena also occur at the trailing edge of this plate and other objects. At the trailing edge or rear edge of the flat plate, the boundary layers are present at the top and bottom sides of the plate. On leaving the plate, the boundary layers gradually intermingle and disappear.

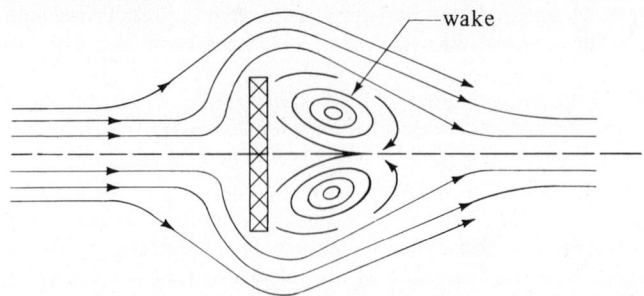

FIGURE 3.7-2. *Flow perpendicular to a flat plate and boundary-layer separation.*

Transport Processes: Momentum, Heat, and Mass

If the direction of flow is at right angles to the plate as shown in Fig. 3.7-2, a boundary layer forms as before in the fluid that is flowing over the upstream face. Once at the edge of the plate, however, the momentum in the fluid prevents it from making the abrupt turn around the edge of the plate, and it separates from the plate. A zone of decelerated fluid is present behind the plate and large eddies (vortices), called the *wake*, are formed in this area. The eddies consume large amounts of mechanical energy. This separation of boundary layers occurs when the change in velocity of the fluid flowing by an object is too large in direction or magnitude for the fluid to adhere to the surface.

Since formation of a wake causes large losses in mechanical energy, it is often necessary to minimize or prevent boundary-layer separation by streamlining the objects or by other means. This is also discussed in Section 3.1A for flow past immersed objects.

3.7C   Laminar Flow and Boundary-Layer Theory

1. Boundary-layer equations.   When laminar flow is occurring in a boundary layer, certain terms in the Navier-Stokes equations become negligible and can be neglected. The thickness of the boundary layer δ is arbitrarily taken as the distance away from the surface where the velocity reaches 99% of the free stream velocity. The concept of a relatively thin boundary layer leads to some important simplifications of the Navier-Stokes equations.

For two-dimensional laminar flow in the x and y directions of a fluid having a constant density, Eqs. (3.6-42) and (3.6-43) become as follows for flow at steady state as shown in Fig. 3.7-1 when we neglect the body forces g_x and g_y.

$$v_x \frac{\partial v_x}{\partial x} + v_y \frac{\partial v_x}{\partial y} = -\frac{1}{\rho}\frac{\partial p}{\partial x} + \frac{\mu}{\rho}\left(\frac{\partial^2 v_x}{\partial x^2} + \frac{\partial^2 v_x}{\partial y^2}\right) \tag{3.7-1}$$

$$v_x \frac{\partial v_y}{\partial x} + v_y \frac{\partial v_y}{\partial y} = -\frac{1}{\rho}\frac{\partial p}{\partial y} + \frac{\mu}{\rho}\left(\frac{\partial^2 v_y}{\partial x^2} + \frac{\partial^2 v_y}{\partial y^2}\right) \tag{3.7-2}$$

The continuity equation for two-dimensional flow becomes

$$\frac{\partial v_x}{\partial x} + \frac{\partial v_y}{\partial y} = 0 \tag{3.7-3}$$

In Eq. (3.7-1), the term $\mu/\rho(\partial^2 v_x/\partial x^2)$ is negligible in comparison with the other terms in the equation. Also, it can be shown that all the terms containing v_y and its derivatives are small. Hence, the final two boundary-layer equations to be solved are Eqs. (3.7-3) and (3.7-4).

$$v_x \frac{\partial v_x}{\partial x} + v_y \frac{\partial v_x}{\partial y} = -\frac{1}{\rho}\frac{dp}{dx} + \frac{\mu}{\rho}\frac{\partial^2 v_x}{\partial y^2} \tag{3.7-4}$$

2. Solution for laminar boundary layer on a flat plate.   An important case in which an analytical solution has been obtained for the boundary-layer equations is for the laminar boundary layer on a flat plate in steady flow, as shown in Fig. 3.7-1. A further simplification can be made in Eq. (3.7-4) in that dp/dx is zero since v_∞ is constant.

The final boundary-layer equations reduce to the equation of motion for the x direction and the continuity equation as follows:

$$v_x \frac{\partial v_x}{\partial x} + v_y \frac{\partial v_x}{\partial y} = \frac{\mu}{\rho}\frac{\partial^2 v_x}{\partial y^2} \tag{3.7-5}$$

$$\frac{\partial v_x}{\partial x} + \frac{\partial v_y}{\partial y} = 0 \tag{3.7-3}$$

The boundary conditions are $v_x = v_y = 0$ at $y = 0$ (y is distance from plate), and $v_x = v_\infty$ at $y = \infty$.

The solution of this problem for laminar flow over a flat plate giving v_x and v_y as a function of x and y was first obtained by Blasius and later elaborated by Howarth (B1, B2, S3). The mathematical details of the solution are quite tedious and complex and will not be given here. The general procedure will be outlined. Blasius reduced the two equations to a single ordinary differential equation which is nonlinear. The equation could not be solved to give a closed form but a series solution was obtained.

The results of the work by Blasius are given as follows. The boundary-layer thickness δ, where $v_x \cong 0.99v_\infty$, is given approximately by

$$\delta = \frac{5.0x}{\sqrt{N_{Re,x}}} = 5.0\sqrt{\frac{\mu x}{\rho v_\infty}} \qquad (3.7\text{-}6)$$

where $N_{Re,x} = xv_\infty\rho/\mu$. Hence, the thickness δ varies as $\sqrt{x}$.

The drag in flow past a flat plate consists only of skin friction and is calculated from the shear stress at the surface at $y = 0$ for any x as follows.

$$\tau_0 = \mu\left(\frac{\partial v_x}{\partial y}\right)_{y=0} \qquad (3.7\text{-}7)$$

From the relation of v_x as a function of x and y obtained from the series solution, Eq. (3.7-7) becomes

$$\tau_0 = 0.332\mu v_\infty\sqrt{\frac{\rho v_\infty}{\mu x}} \qquad (3.7\text{-}8)$$

The total drag is given by the following for a plate of length L and width b:

$$F_D = b\int_0^L \tau_0\,dx \qquad (3.7\text{-}9)$$

Substituting Eq. (3.7-8) into (3.7-9) and integrating,

$$F_D = 0.664b\sqrt{\mu\rho v_\infty^3\, L} \qquad (3.7\text{-}10)$$

The drag coefficient C_D related to the total drag on one side of the plate having an area $A = bL$ is defined as

$$F_D = C_D\frac{v_\infty^2}{2}\rho A \qquad (3.7\text{-}11)$$

Substituting the value for A and Eq. (3.7-10) into (3.7-11),

$$C_D = 1.328\sqrt{\frac{\mu}{Lv_\infty\rho}} = \frac{1.328}{N_{Re,L}^{1/2}} \qquad (3.7\text{-}12)$$

where $N_{Re,L} = Lv_\infty\rho/\mu$. A form of Eq. (3.7-11) is used in Section 13.3 for particle movement through a fluid. The definition of C_D in Eq. (3.7-12) is similar to the Fanning friction factor f for pipes.

The equation derived for C_D applies only to the laminar boundary layer for $N_{Re,L}$ less than about 5×10^5. Also, the results are valid only for positions where x is sufficiently far from the leading edge so that x or L is much greater than δ. Experimental results on the drag coefficient to a flat plate confirm the validity of Eq. (3.7-12). Boundary-layer flow past many other shapes has been successfully analyzed using similar methods.

3.7D Nature and Intensity of Turbulence

1. Nature of turbulence. Since turbulent flow is important in many areas of engineering, the nature of turbulence has been extensively investigated. Measurements of the velocity fluctuations of the eddies in turbulent flow have helped explain turbulence.

For turbulent flow there are no exact solutions of flow problems as there are in laminar flow, since the approximate equations used depend on many assumptions. However, useful relations have been obtained by using a combination of experimental data and theory. Some of these relations will be discussed.

Turbulence can be generated by contact of two layers of fluid moving at different velocities or by a flowing stream in contact with a solid boundary, such as a wall or sphere. When a jet of fluid from an orifice flows into a mass of fluid, turbulence can arise. In turbulent flow at a given place and time large eddies are continually being formed which break down into smaller eddies and which finally disappear. Eddies are as small as about 0.1 or 1 mm or so and as large as the smallest dimension of the turbulent stream. Flow inside an eddy is laminar because of its large size.

In turbulent flow the velocity is fluctuating in all directions. In Fig. 3.7-3 a typical plot of the variation of the instantaneous velocity v_x in the x direction at a given point in turbulent flow is shown. The velocity v'_x is the deviation of the velocity from the mean velocity $\bar{v}_x$ in the x-direction of flow of the stream. Similar relations also hold for the y and z directions.

$$v_x = \bar{v}_x + v'_x, \qquad v_y = \bar{v}_y + v'_y, \; v_z = \bar{v}_z + v'_z \qquad (3.7\text{-}13)$$

$$\bar{v}_x = \frac{1}{t}\int_0^t v_x \, dt \qquad (3.7\text{-}14)$$

where the mean velocity $\bar{v}_x$ is the time-averaged velocity for time t, v_x the instantaneous total velocity in the x direction, and v'_x the instantaneous deviating or fluctuating velocity in the x direction. These fluctuations can also occur in the y and z directions. The value of v'_x fluctuates about zero as an average and, hence, the time-averaged values $\bar{v}'_x = 0$, $\bar{v}'_y = 0$, $\bar{v}'_z = 0$. However, the values of v'^2_x, v'^2_y, and v'^2_z will not be zero. Similar expressions can also be written for pressure, which also fluctuates.

2. Intensity of turbulence. The time average of the fluctuating components vanishes over a time period of a few seconds. However, the time average of the mean square of the fluctuating components is a positive value. Since the fluctuations are random, the data have been analyzed by statistical methods. The level or intensity of turbulence can be

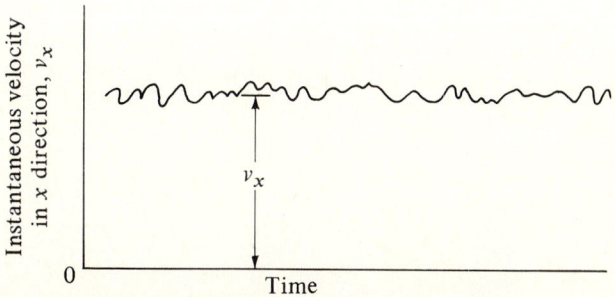

FIGURE 3.7-3. *Velocity fluctuations in turbulent flow.*

related to the square root of the sum of the mean squares of the fluctuating components. This intensity of turbulence is an important parameter in testing of models and theory of boundary layers.

The intensity of turbulence I can be defined mathematically as

$$I = \frac{\sqrt{\frac{1}{3}(\overline{v_x'^2} + \overline{v_y'^2} + \overline{v_z'^2})}}{\bar{v}_x} \tag{3.7-15}$$

This parameter I is quite important. Such factors as boundary-layer transition, separation, and heat- and mass-transfer coefficients depend upon the intensity of turbulence. Simulation of turbulent flows in testing of models requires that the Reynolds number and the intensity of turbulence be the same. One method used to measure intensity of turbulence is to utilize a hot-wire anemometer.

3.7E Turbulent Shear or Reynolds Stresses

In a fluid flowing in turbulent flow shear forces occur wherever there is a velocity gradient across a shear plane and these are much larger than those occurring in laminar flow. The velocity fluctuations in Eq. (3.7-13) give rise to turbulent shear stresses. The equations of motion and the continuity equation are still valid for turbulent flow. For an incompressible fluid having a constant density ρ and viscosity μ, the continuity equation (3.6-24) holds.

$$\frac{\partial v_x}{\partial x} + \frac{\partial v_y}{\partial y} + \frac{\partial v_z}{\partial z} = 0 \tag{3.6-24}$$

Also, the x component of the equation of motion, Eq. (3.6-42), can be written as follows if Eq. (3.6-24) holds:

$$\frac{\partial(\rho v_x)}{\partial t} + \frac{\partial(\rho v_x v_x)}{\partial_x} + \frac{\partial(\rho v_x v_y)}{\partial y} \frac{\partial(\rho v_x v_z)}{\partial z} = \mu\left(\frac{\partial^2 v_x}{\partial x^2} + \frac{\partial^2 v_x}{\partial y^2} + \frac{\partial^2 v_x}{\partial z^2}\right) - \frac{\partial p}{\partial x} + \rho g_x \tag{3.7-16}$$

We can rewrite the continuity equation (3.6-24) and Eq. (3.7-16) by replacing v_x by $\bar{v}_x + v_x'$, v_y by $\bar{v}_y + v_y'$, v_z by $\bar{v}_z + v_z'$, and p by $\bar{p} + p'$.

$$\frac{\partial(\bar{v}_x + v_x')}{\partial x} + \frac{\partial(\bar{v}_y + v_y')}{\partial y} + \frac{\partial(\bar{v}_z + v_z')}{\partial z} = 0 \tag{3.7-17}$$

$$\frac{\partial[\rho(\bar{v}_x + v_x')]}{\partial t} + \frac{\partial[\rho(\bar{v}_x + v_x')(\bar{v}_x + v_x')]}{\partial x} + \frac{\partial[\rho(\bar{v}_x + v_x')(\bar{v}_y + v_y')]}{\partial y}$$

$$+ \frac{\partial[\rho(\bar{v}_x + v_x')(\bar{v}_z + v_z')]}{\partial z} = \mu\nabla^2(\bar{v}_x + v_x') - \frac{\partial(\bar{p} + p')}{\partial x} + \rho g_x \tag{3.7-18}$$

Now we use the fact that the time-averaged value of the fluctuating velocities is zero ($\bar{v}_x'$, $\bar{v}_y'$, $\bar{v}_z'$ are zero), and that the time-averaged product $\overline{v_x' v_y'}$ is not zero. Then Eqs. (3.7-17) and (3.7-18) become

$$\frac{\partial \bar{v}_x}{\partial x} + \frac{\partial \bar{v}_y}{\partial y} + \frac{\partial \bar{v}_z}{\partial z} = 0 \tag{3.7-19}$$

$$\frac{\partial(\rho \bar{v}_x)}{\partial t} + \frac{\partial(\rho \bar{v}_x \bar{v}_x)}{\partial x} + \frac{\partial(\rho \bar{v}_x \bar{v}_y)}{\partial y} + \frac{\partial(\rho \bar{v}_x \bar{v}_z)}{\partial z}$$

$$+ \left[\frac{\partial(\rho \overline{v_x' v_x'})}{\partial x} + \frac{\partial(\rho \overline{v_x' v_y'})}{\partial y} + \frac{\partial(\rho \overline{v_x' v_z'})}{\partial z}\right] = \mu\nabla^2 \bar{v}_x - \frac{\partial \bar{p}}{\partial x} + \rho g_x \tag{3.7-20}$$

Transport Processes: Momentum, Heat, and Mass

By comparing these two time-smoothed equations with Eqs. (3.6-24) and (3.7-16) we see that the time-smoothed values everywhere replace the instantaneous values. However, in Eq. (3.7-20) new terms arise in the set of brackets which are related to turbulent velocity fluctuations. For convenience we use the notation

$$\bar{\tau}_{xx}^t = \rho \overline{v_x' v_x'}, \quad \bar{\tau}_{yx}^t = \rho \overline{v_x' v_y'}, \quad \bar{\tau}_{zx}^t = \rho \overline{v_x' v_z'} \tag{3.7-21}$$

These are the components of the turbulent momentum flux and are called *Reynolds stresses*.

3.7F Prandtl Mixing Length

The equations derived for turbulent flow must be solved to obtain velocity profiles. To do this, more simplifications must be made before the expressions for the Reynolds stresses can be evaluated. A number of semiempirical equations have been used and the eddy diffusivity model of Boussinesq is one early attempt to evaluate these stresses. By analogy to the equation for shear stress in laminar flow, $\tau_{yx} = -\mu(dv_x/dy)$, the turbulent shear stress can be written as

$$\bar{\tau}_{yx}^t = -\eta_t \frac{d\bar{v}_x}{dy} \tag{3.7-22}$$

where η_t is a turbulent or eddy viscosity, which is a strong function of position and flow. This equation can also be written as follows:

$$\bar{\tau}_{yx}^t = -\rho \varepsilon_t \frac{d\bar{v}_x}{dy} \tag{3.7-23}$$

where $\varepsilon_t = \eta_t/\rho$ and ε_t is eddy diffusivity of momentum in m²/s by analogy to the momentum diffusivity μ/ρ for laminar flow.

Prandtl in his mixing-length model developed an expression to evaluate these stresses by assuming that eddies move in a fluid in a manner similar to the movement of molecules in a gas. The eddies move a distance called the mixing length L before they lose their identity.

Actually, the moving eddy or "lump" of fluid will gradually lose its identity. However, in the definition of the Prandtl mixing-length L, this small packet of fluid is assumed to retain its identity while traveling the entire length L and then to lose its identity or be absorbed in the host region.

Prandtl assumed that the velocity fluctuation v_x' is due to a "lump" of fluid moving a distance L in the y direction and retaining its mean velocity. At point L, the lump of fluid will differ in mean velocity from the adjacent fluid by $\bar{v}_{x|y+L} - \bar{v}_{x|y}$. Then, the value of $v_{x|y}'$ is

$$v_{x|y}' = \bar{v}_{x|y+L} - \bar{v}_{x|y} \tag{3.7-24}$$

The length L is small enough so that the velocity difference can be written as

$$v_{x|y}' = \bar{v}_{x|y+L} - \bar{v}_{x|y} = L \frac{d\bar{v}_x}{dy} \tag{3.7-25}$$

Hence,

$$v_x' = L \frac{d\bar{v}_x}{dy} \tag{3.7-26}$$

Prandtl also assumed $v'_x \simeq v'_y$. Then the time average, $\overline{v'_x x'_y}$, is

$$\overline{v'_x v'_y} = -L^2 \left| \frac{d\bar{v}_x}{dy} \right| \frac{d\bar{v}_x}{dy} \tag{3.7-27}$$

The minus sign and the absolute value were used to make the quantity $\overline{v'_x v'_y}$ agree with experimental data. Substituting Eq. (3.7-27) into (3.7-21),

$$\bar{\tau}^t_{yx} = -\rho L^2 \left| \frac{d\bar{v}_x}{dy} \right| \frac{d\bar{v}_x}{dy} \tag{3.7-28}$$

Comparing with Eq. (3.7-23),

$$\varepsilon_t = L^2 \left| \frac{d\bar{v}_x}{dy} \right| \tag{3.7-29}$$

3.7G Universal Velocity Distribution in Turbulent Flow

To determine the velocity distribution for turbulent flow at steady state inside a circular tube, we divide the fluid inside the pipe into two regions: a central core where the Reynolds stress approximately equals the shear stress; and a thin, viscous sublayer adjacent to the wall where the shear stress is due only to viscous shear and the turbulence effects are assumed negligible. Later we include a third region, the buffer zone, where both stresses are important.

Dropping the subscripts and superscripts on the shear stresses and velocity, and considering the thin, viscous sublayer, we can write

$$\tau_0 = -\mu \frac{dv}{dy} \tag{3.7-30}$$

where τ_0 is assumed constant in this region. On integration,

$$\tau_0 y = \mu v \tag{3.7-31}$$

Defining a friction velocity as follows and substituting into Eq. (3.7-31),

$$v^* = \sqrt{\frac{\tau_0}{\rho}} \tag{3.7-32}$$

$$\frac{v}{v^*} = \frac{yv^*}{\mu/\rho} \tag{3.7-33}$$

The dimensionless velocity ratio on the left can be written as

$$v^+ = v\sqrt{\frac{\rho}{\tau_0}} \quad \text{(SI)}$$
$$\tag{3.7-34}$$
$$v^+ = v\sqrt{\frac{\rho}{\tau_0 g_c}} \quad \text{(English)}$$

The dimensionless number on the right can be written as

$$y^+ = \frac{\sqrt{\tau_0 \rho}}{\mu} y \quad \text{(SI)}$$
$$\tag{3.7-35}$$
$$y^+ = \frac{\sqrt{\tau_0 g_c \rho}}{\mu} y \quad \text{(English)}$$

where y is the distance from the wall of the tube. For a tube of radius r_0, $y = r_0 - r$, where r is the distance from the center. Hence, for the viscous sublayer, the velocity distribution is

$$v^+ = y^+ \tag{3.7-36}$$

Next, considering the turbulent core where any viscous stresses are neglected, Eq. (3.7-28) becomes

$$\tau = \rho L^2 \left(\frac{dv}{dy}\right)^2 \tag{3.7-37}$$

where dv/dy is always positive and the absolute value sign is dropped. Prandtl assumed that the mixing length is proportional to the distance from the wall, or

$$L = Ky \tag{3.7-38}$$

and that $\tau = \tau_0 = $ constant. Equation (3.7-37) now becomes

$$\tau_0 = \rho K^2 y^2 \left(\frac{dv}{dy}\right)^2 \tag{3.7-39}$$

Hence,

$$v^* = Ky \frac{dv}{dy} \tag{3.7-40}$$

Upon integration,

$$v^* \ln y = Kv + K_1 \tag{3.7-41}$$

where K_1 is a constant. The constant K_1 can be found by assuming that v is zero at a small value of y, say y_0.

$$\frac{v}{v^*} = v^+ = \frac{1}{K} \ln \frac{y}{y_0} \tag{3.7-42}$$

Introducing the variable y^+ by multiplying the numerator and the denominator of the term y/y_0 by v^*/v, where $v = \mu/\rho$, we obtain

$$v^+ = \frac{1}{K} \left(\ln \frac{yv^*}{v} - \ln \frac{y_0 v^*}{v} \right) \tag{3.7-43}$$

$$v^+ = \frac{1}{K} \ln y^+ + C_1 \tag{3.7-44}$$

A large amount of velocity distribution data by Nikuradse and others for a range of Reynolds numbers of 4000 to 3.2×10^6 have been obtained and the data fit Eq. (3.7-36) in the region up to y^+ of 5 and also fit Eq. (3.7-44) above y^+ of 30 with K and C_1 being universal constants. For the region of y^+ from 5 to 30, which is defined as the buffer region, an empirical equation of the form of Eq. (3.7-44) fits the data. In Fig. 3.7-4 the following relations which are valid are plotted to give a universal velocity profile for fluids flowing in smooth circular tubes.

$$v^+ = y^+ \qquad (0 < y^+ < 5) \tag{3.7-45}$$

$$v^+ = 5.0 \ln y^+ - 3.05 \qquad (5 < y^+ < 30) \tag{3.7-46}$$

$$v^+ = 2.5 \ln y^+ + 5.5 \qquad (30 < y^+) \tag{3.7-47}$$

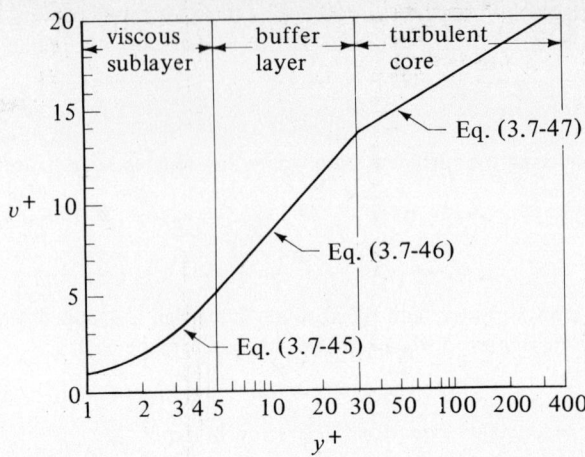

FIGURE 3.7-4. *Universal velocity profile for turbulent flow in smooth circular tubes.*

Three distinct regions are apparent in Fig. 3.7-4. The first region next to the wall is the *viscous sublayer* (historically called "laminar" sublayer), given by Eq. (3.7-45), where the velocity is proportional to the distance from the wall. The second region, called the *buffer layer*, is given by Eq. (3.7-46), which is a region of transition between the viscous sublayer with practically no eddy activity and the violent eddy activity in the *turbulent core region* given by Eq. (3.7-47). These equations can then be used and related to the Fanning friction factor discussed earlier in the chapter. They can also be used in solving turbulent boundary-layer problems.

3.7H   Integral Momentum Balance for Boundary-Layer Analysis

1. Introduction and derivation of integral expression.   In the solution for the laminar boundary layer on a flat plate, the Blasius solution is quite restrictive, since it is for laminar flow over a flat plate. Other more complex systems cannot be solved by this method. An approximate method developed by von Kármán can be used when the configuration is more complicated or the flow is turbulent. This is an approximate momentum integral analysis of the boundary layer using an empirical or assumed velocity distribution.

In order to derive the basic equation for a laminar or turbulent boundary layer, a small control volume in the boundary layer on a flat plate is used as shown in Fig. 3.7-5. The depth in the z direction is b. Flow is only through the surfaces A_1 and A_2 and also from the top curved surface at δ. An overall integral momentum balance using Eq. (2.8-8) and overall integral mass balance using Eq. (2.6-6) are applied to the control volume inside the boundary layer at steady state and the final integral expression by von Kármán is (B2, S3)

$$\frac{\tau_0}{\rho} = \frac{d}{dx} \int_0^\delta v_x(v_\infty - v_x)\, dy \tag{3.7-48}$$

where τ_0 is the shear stress at the surface $y = 0$ at point x along the plate. Also, δ and τ_0 are functions of x.

Equation (3.7-48) is an expression whose solution requires knowledge of the velocity

v_x as a function of the distance from the surface, y. The accuracy of the results will, of course, depend on how closely the assumed velocity profile approaches the actual profile.

2. *Integral momentum balance for laminar boundary layer.* Before we use Eq. (3.7-48) for the turbulent boundary layer, this equation will be applied to the laminar boundary layer over a flat plate so that the results can be compared with the exact Blasius solution in Eqs. (3.7-6)–(3.7-12).

In this analysis certain boundary conditions must be satisfied in the boundary layer.

$$v_x = 0 \qquad \text{at } y = 0$$

$$v_x \cong v_\infty \qquad \text{at } y = \delta \tag{3.7-49}$$

$$\frac{dv_x}{dy} \cong 0 \qquad \text{at } y = \delta$$

The conditions above are fulfilled in the following simple, assumed velocity profile.

$$\frac{v_x}{v_\infty} = \frac{3}{2}\frac{y}{\delta} - \frac{1}{2}\left(\frac{y}{\delta}\right)^3 \tag{3.7-50}$$

The shear stress τ_0 at a given x can be obtained from

$$\tau_0 = \mu\left(\frac{dv_x}{dy}\right)_{y=0} \tag{3.7-51}$$

Differentiating Eq. (3.7-50) with respect to y and setting $y = 0$,

$$\left(\frac{dv_x}{dy}\right)_{y=0} = \frac{3v_\infty}{2\delta} \tag{3.7-52}$$

Substituting Eq. (3.7-52) into (3.7-51),

$$\tau_0 = \frac{3\mu v_\infty}{2\delta} \tag{3.7-53}$$

Substituting Eq. (3.7-50) into Eq. (3.7-48) and integrating between $y = 0$ and $y = \delta$, we obtain

$$\frac{d\delta}{dx} = \frac{280}{39}\frac{\tau_0}{v_\infty^2 \rho} \tag{3.7-54}$$

Combining Eqs. (3.7-53) and (3.7-54) and integrating between $\delta = 0$ and $\delta = \delta$, and $x = 0$

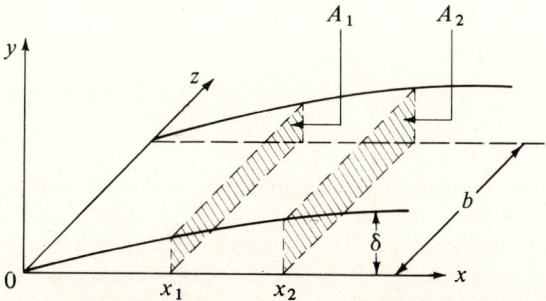

FIGURE 3.7-5. *Control volume for integral analysis of the boundary-layer flow.*

and $x = L$,

$$\delta = 4.64 \sqrt{\frac{\mu L}{\rho v_\infty}} \qquad (3.7\text{-}55)$$

where the length of plate is $x = L$. Proceeding in a manner similar to Eqs. (3.7-6)–(3.7-12), the drag coefficient is

$$C_D = 1.292 \sqrt{\frac{\mu}{L v_\infty \rho}} = \frac{1.292}{N_{\text{Re}, L}^{1/2}} \qquad (3.7\text{-}56)$$

A comparison of Eq. (3.7-6) with (3.7-55) and (3.7-12) with (3.7-56) shows the success of this method. Only the numerical constants differ slightly. This method can be used with reasonable accuracy for cases where an exact analysis is not feasible.

3. Integral momentum analysis for turbulent boundary layer. The procedures used for the integral momentum analysis for a laminar boundary layer can be applied to the turbulent boundary layer on a flat plate. A simple empirical velocity distribution for pipe flow which is valid up to a Reynolds number of 10^5 can be adapted for the boundary layer on a flat plate, to become

$$\frac{v_x}{v_\infty} = \left(\frac{y}{\delta}\right)^{1/7} \qquad (3.7\text{-}57)$$

This is the Blasius $\frac{1}{7}$-power law often used.

Equation (3.7-57) is substituted into the integral relation equation (3.7-48).

$$\frac{d}{dx} \int_0^\delta v_\infty^2 \left[\left(\frac{y}{\delta}\right)^{1/7} - \left(\frac{y}{\delta}\right)^{2/7} \right] dy = \frac{\tau_0}{\rho} \qquad (3.7\text{-}58)$$

The power-law equation does not hold, as y goes to zero at the wall. Another useful relation is the Blasius correlation for shear stress for pipe flow, which is consistent at the wall for the wall shear stress τ_0. For boundary-layer flow over a flat plate, it becomes

$$\frac{\tau_0}{\rho v_\infty^2} = 0.023 \left(\frac{\delta v_\infty \rho}{\mu}\right)^{-1/4} \qquad (3.7\text{-}59)$$

Integrating Eq. (3.7-58), combining the result with Eq. (3.7-59), and integrating between $\delta = 0$ and $\delta = \delta$, and $x = 0$ and $x = L$,

$$\delta = 0.376 \left(\frac{L v_\infty \rho}{\mu}\right)^{-1/5} L = \frac{0.376 L}{N_{\text{Re}, L}^{1/5}} \qquad (3.7\text{-}60)$$

Integration of the drag force as before gives

$$C_D = \frac{0.072}{N_{\text{Re}, L}^{1/5}} \qquad (3.7\text{-}61)$$

In this development the turbulent boundary layer was assumed to extend to $x = 0$. Actually, a certain length at the front has a laminar boundary layer. Experimental data check Eq. (3.7-61) reasonably well from a Reynolds number of 5×10^5 to 10^7. More accurate results at higher Reynolds numbers can be obtained by using a logarithmic velocity distribution, Eqs. (3.7-45)–(3.7-47).

3.8 (SELECTED TOPIC) DIMENSIONAL ANALYSIS IN MOMENTUM TRANSFER

3.8A Dimensional Analysis of Differential Equations

In this chapter we have derived several differential equations describing various flow situations. Dimensional homogeneity requires that each term in a given equation have the same units. Then, the ratio of one term in the equation to another term is dimensionless. Knowing the physical meaning of each term in the equation, we are then able to give a physical interpretation to each of the dimensionless parameters or numbers formed. These dimensionless numbers, such as the Reynolds number and others, are useful in correlating and predicting transport phenomena in laminar and turbulent flow.

Often it is not possible to integrate the differential equation describing a flow situation. However, we can use the equation to find out which dimensionless numbers can be used in correlating experimental data for this physical situation.

An important example of this involves the use of the Navier–Stokes equation, which often cannot be integrated for a given physical situation. To start, we use Eq. (3.6-42) for the x component of the Navier–Stokes equation. At steady state this becomes

$$v_x \frac{\partial v_x}{\partial x} + v_y \frac{\partial v_x}{\partial y} + v_z \frac{\partial v_x}{\partial z} = g_x - \frac{1}{\rho} \frac{\partial p}{\partial x} + \frac{\mu}{\rho} \left(\frac{\partial^2 v_x}{\partial x^2} + \frac{\partial^2 v_x}{\partial y^2} + \frac{\partial^2 v_x}{\partial z^2} \right) \qquad (3.8\text{-}1)$$

Each term in this equation has the units length/time^2 or (L/t^2).

In this equation each term has a physical significance. First we use a single characteristic velocity v and a single characteristic length L for all terms. Then the expression of each term in Eq. (3.8-1) is as follows. The left-hand side can be expressed as v^2/L and the right-hand terms, respectively, as g, $p/\rho L$, and $\mu v/\rho L^2$. We then write

$$\left[\frac{v^2}{L} \right] = [g] - \left[\frac{p}{\rho L} \right] + \left[\frac{\mu v}{\rho L^2} \right] \qquad (3.8\text{-}2)$$

This expresses a dimensional equality and not a numerical equality. Each term has dimensions L/t^2.

The left-hand term in Eq. (3.8-2) represents the inertia force and the terms on the right-hand side represent, respectively, the gravity force, pressure force, and viscous force. Dividing each of the terms in Eq. (3.8-2) by the inertia force $[v^2/L]$, the following dimensionless groups or their reciprocals are obtained.

$$\frac{[v^2/L]}{[g]} = \frac{\text{inertia force}}{\text{gravity force}} = \frac{v^2}{gL} = N_{Fr} \qquad \text{(Froude number)} \qquad (3.8\text{-}3)$$

$$\frac{[p/\rho L]}{[v^2/L]} = \frac{\text{pressure force}}{\text{inertia force}} = \frac{p}{\rho v^2} = N_{Eu} \qquad \text{(Euler number)} \qquad (3.8\text{-}4)$$

$$\frac{[v^2/L]}{[\mu v/\rho L^2]} = \frac{\text{inertia force}}{\text{viscous force}} = \frac{L v \rho}{\mu} = N_{Re} \qquad \text{(Reynolds number)} \qquad (3.8\text{-}5)$$

Note that this method not only gives the various dimensionless groups for a differential equation but also gives physical meaning to these dimensionless groups. The length, velocity, etc., to be used in a given case will be that value which is most significant. For example, the length may be the diameter of a sphere, the length of a flat plate, and so on.

Systems that are geometrically similar are said to be *dynamically similar* if the parameters representing ratios of forces pertinent to the situation are equal. This means that the Reynolds, Euler, or Froude numbers must be equal between the two systems.

This dynamic similarity is an important requirement in obtaining experimental data on a small model and extending these data to scale up to the large prototype. Since experiments with full-scale prototypes would often be difficult and/or expensive, it is customary to study small models. This is done in the scaleup of chemical process equipment and in the design of ships and airplanes.

3.8B Dimensional Analysis Using Buckingham Method

The method of obtaining the important dimensionless numbers from the basic differential equations is generally the preferred method. In many cases, however, we are not able to formulate a differential equation which clearly applies. Then a more general procedure is required, which is known as the *Buckingham method*. In this method the listing of the important variables in the particular physical problem is done first. Then we determine the number of dimensionless parameters into which the variables may be combined by using the Buckingham pi theorem.

The *Buckingham theorem* states that the functional relationship among q quantities or variables whose units may be given in terms of u fundamental units or dimensions may be written as $(q - u)$ independent dimensionless groups, often called π's. [This quantity u is actually the maximum number of these variables which will not form a dimensionless group. However, only in a few cases is this u not equal to the number of fundamental units (B1).]

Let us consider the following example, to illustrate the use of this method. An incompressible fluid is flowing inside a circular tube of inside diameter D. The significant variables are pressure drop Δp, velocity v, diameter D, tube length L, viscosity μ, and density ρ. The total number of variables is $q = 6$.

The fundamental units or dimensions are $u = 3$ and are mass M, length L, and time t. The units of the variables are as follows: Δp in M/Lt^2, v in L/t, D in L, L in L, μ in M/Lt, and ρ in M/L^3. The number of dimensionless groups or π's is $q - u$, or $6 - 3 = 3$. Thus,

$$\pi_1 = f(\pi_2, \pi_3) \tag{3.8-6}$$

Next, we must select a core group of u (or 3) variables which will appear in each π group and among them contain all the fundamental dimensions. Also, no two of the variables selected for the core can have the same dimensions. In choosing the core, the variable whose effect one desires to isolate is often excluded (for example, Δp). This leaves us with the variables v, D, μ, and ρ to be used. (L and D have the same dimensions.)

We will select D, v, and ρ to be the core variables common to all three groups. Then the three dimensionless groups are

$$\pi_1 = D^a v^b \rho^c \, \Delta p^1 \tag{3.8-7}$$

$$\pi_2 = D^d v^e \rho^f L^1 \tag{3.8-8}$$

$$\pi_3 = D^g v^h \rho^i \mu^1 \tag{3.8-9}$$

To be dimensionless, the variables must be raised to certain exponents a, b, c, etc.

First we consider the π_1 group.

$$\pi_1 = D^a v^b \rho^c \, \Delta p^1 \tag{3.8-7}$$

To evaluate these exponents, we write Eq. (3.8-7) dimensionally by substituting the dimensions for each variable.

$$M^0 L^0 t^0 = 1 = L^a \left(\frac{L}{t}\right)^b \left(\frac{M}{L^3}\right)^c \frac{M}{Lt^2} \tag{3.8-10}$$

Next we equate the exponents of L on both sides of this equation, of M, and finally of t.

$$(L) \quad 0 = a + b - 3c - 1$$

$$(M) \quad 0 = c + 1 \tag{3.8-11}$$

$$(t) \quad 0 = -b - 2$$

Solving these equations, $a = 0$, $b = -2$, and $c = -1$.
Substituting these values into Eq. (3.8-7),

$$\pi_1 = \frac{\Delta p}{v^2 \rho} = N_{Eu} \tag{3.8-12}$$

Repeating this procedure for π_2 and π_3,

$$\pi_2 = \frac{L}{D} \tag{3.8-13}$$

$$\pi_3 = \frac{Dv\rho}{\mu} = N_{Re} \tag{3.8-14}$$

Finally, substituting π_1, π_2, and π_3 into Eq. (3.8-6),

$$\frac{\Delta p}{v^2 \rho} = f\left(\frac{L}{D}, \frac{Dv\rho}{\mu}\right) \tag{3.8-15}$$

Combining Eq. (2.10-5) with the left-hand side of Eq. (3.8-15), the result obtained shows that the friction factor is a function of the Reynolds number (as was shown before in the empirical correlation of friction factor and Reynolds number) and of length/diameter ratio. In pipes with $L/D \gg 1$ or pipes with fully developed flow, the friction factor is found to be independent of L/D.

This type of analysis is useful in empirical correlations of data. However, it does not tell us the importance of each dimensionless group, which must be determined by experimentation, nor does it select the variables to be used.

PROBLEMS

3.1-1. Force on a Cylinder in a Wind Tunnel. Air at 101.3 kPa absolute and 25°C is flowing at a velocity of 10 m/s in a wind tunnel. A long cylinder having a diameter of 90 mm is placed in the tunnel and the axis of the cylinder is held perpendicular to the air flow. What is the force on the cylinder per meter length?
Ans. $C_D = 1.3$, $F_D = 6.94$ N

3.1-2. Wind Force on a Steam Boiler Stack. A cylindrical steam boiler stack has a diameter of 1.0 m and is 30.0 m high. It is exposed to a wind at 25°C having a velocity of 50 miles/h. Calculate the force exerted on the boiler stack.
Ans. $C_D = 0.33$, $F_D = 2935$ N

3.1-3. Effect of Velocity on Force on a Sphere and Stokes' Law. A sphere is held in a small wind tunnel where air at 37.8°C and 1 atm abs and various velocities is forced by the sphere having a diameter of 0.042 m.

(a) Determine the drag coefficient and force on the sphere for a velocity of 2.30×10^{-4} m/s. Use Stokes' law here if it is applicable.

(b) Also determine the force for velocities of 2.30×10^{-3}, 2.30×10^{-2}, 2.30×10^{-1}, and 2.30 m/s. Make a plot of F_D versus velocity.

3.1-4. Drag Force on Bridge Pier in River. A cylindrical bridge pier 1.0 m in diameter is submerged to a depth of 10 m. Water in the river at 20°C is flowing past at a velocity of 1.2 m/s. Calculate the force on the pier.

3.1-5. Surface Area in a Packed Bed. A packed bed is composed of cubes 0.020 m on a side and the bulk density of the packed bed is 980 kg/m³. The density of the solid cubes is 1500 kg/m³.

(a) Calculate ε, effective diameter D_p, and a.

(b) Repeat for the same conditions but for cylinders having a diameter of $D = 0.02$ m and a length $h = 1.5D$.

Ans. (a) $\varepsilon = 0.3467$, $D_p = 0.020$ m, $a = 196.0\,\mathrm{m}^{-1}$

3.1-6. Derivation for Number of Particles in a Bed of Cylinders. For a packed bed containing cylinders where the diameter D of the cylinders is equal to the length h, do as follows for a bed having a void fraction ε.

(a) Calculate the effective diameter.

(b) Calculate the number, n, of cylinders in 1 m³ of the bed.

Ans. (a) $D_p = D$

3.1-7. Derivation of Dimensionless Equation for Packed Bed. Starting with Eq. (3.1-20), derive the dimensionless equation (3.1-21). Show all steps in the derivation.

3.1-8. Flow and Pressure Drop of Gases in Packed Bed. Air at 394.3 K flows through a packed bed of cylinders having a diameter of 0.0127 m and length the same as the diameter. The bed void fraction is 0.40 and the length of the packed bed is 3.66 m. The air enters the bed at 2.20 atm abs at the rate of 2.45 kg/m² · s based on the empty cross section of the bed. Calculate the pressure drop of air in the bed.

Ans. $\Delta p = 0.1547 \times 10^5$ Pa

3.1-9. Flow of Water in a Filter Bed. Water at 24°C is flowing by gravity through a filter bed of small particles having an equivalent diameter of 0.0060 m. The void fraction of the bed is measured as 0.42. The packed bed has a depth of 1.50 m. The liquid level of water above the bed is held constant at 0.40 m. What is the water velocity v' based on the empty cross section of the bed?

3.1-10. Mean Diameter of Particles in Packed Bed. A mixture of particles in a packed bed contains the following volume percent of particles and sizes: 15%, 10 mm; 25%, 20 mm; 40%, 40 mm; 20%, 70 mm. Calculate the effective mean diameter, D_{pm}, if the shape factor is 0.74.

Ans. $D_{pm} = 18.34$ mm

3.1-11. Permeability and Darcy's Law. A sample core of a porous rock obtained from an oil reservoir is 8 cm long and has a diameter of 2.0 cm. It is placed in a core holder. With a pressure drop of 1.0 atm, the water flow at 20.2°C through the core was measured as 2.60 cm³/s. What is the permeability in darcy?

3.1-12. Minimum Fluidization and Expansion of Fluid Bed. Particles having a size of 0.10 mm, a shape factor of 0.86, and a density of 1200 kg/m³ are to be fluidized using air at 25°C and 202.65 kPa abs pressure. The void fraction at minimum fluidizing conditions is 0.43. The bed diameter is 0.60 m and the bed contains 350 kg of solids.

(a) Calculate the minimum height of the fluidized bed.

(b) Calculate the pressure drop at minimum fluidizing conditions.

(c) Calculate the minimum velocity for fluidization.

(d) Using 4.0 times the minimum velocity, estimate the porosity of the bed.

Ans. (a) $L_{mf} = 1.810$ m, (b) $\Delta p = 0.1212 \times 10^5$ Pa,
(c) $v'_{mf} = 0.004374$ m/s, (d) $\varepsilon = 0.604$

3.1-13. Minimum Fluidization Velocity Using a Liquid. A tower having a diameter of 0.1524 m is being fluidized with water at 20.2°C. The uniform spherical beads in the tower bed have a diameter of 4.42 mm and a density of 1603 kg/m^3. Estimate the minimum fluidizing velocity and compare with the experimental value of 0.02307 m/s of Wilhelm and Kwauk (W5).

3.1-14. Fluidization of a Sand Bed Filter. To clean a sand bed filter it is fluidized at minimum conditions using water at 24°C. The round sand particles have a density of 2550 kg/m^3 and an average size of 0.40 mm. The sand has the properties given in Table 3.1-2.
(a) The bed diameter is 0.40 m and the desired height of the bed at these minimum fluidizing conditions is 1.75 m. Calculate the amount of solids needed.
(b) Calculate the pressure drop at these conditions and the minimum velocity for fluidization.
(c) Using 4.0 times the minimum velocity, estimate the porosity and height of the expanded bed.

3.2-1. Flow Measurement Using a Pitot Tube. A pitot tube is used to measure the flow rate of water at 20°C in the center of a pipe having an inside diameter of 102.3 mm. The manometer reading is 78 mm of carbon tetrachloride at 20°C. The pitot tube coefficient is 0.98.
(a) Calculate the velocity at the center and the average velocity.
(b) Calculate the volumetric flow rate of the water.
Ans. (a) $v_{max} = 0.9372$ m/s, $v_{av} = 0.773$ m/s, (b) 6.35×10^{-3} m^3/s

3.2-2. Gas Flow Rate Using a Pitot Tube. The flow rate of air at 37.8°C is being measured at the center of a duct having a diameter of 800 mm by a pitot tube. The pressure difference reading on the manometer is 12.4 mm of water. At the pitot tube position, the static pressure reading is 275 mm of water above one atmosphere absolute. The pitot tube coefficient is 0.97. Calculate the velocity at the center and the volumetric flow rate of the air.

3.2-3. Pitot-Tube Traverse for Flow Rate Measurement. In a pitot tube traverse of a pipe having an inside diameter of 155.4 mm in which water at 20°C is flowing, the following data were obtained.

Distance from Wall (mm)	Reading in Manometer (mm of Carbon Tetrachloride)
26.9	122
52.3	142
77.7	157
103.1	137
128.5	112

The pitot tube coefficient is 0.98.
(a) Calculate the maximum velocity at the center.
(b) Calculate the average velocity. [*Hint:* Use Eq. (2.6-17) and do a graphical integration.]

3.2-4. Metering Flow by a Venturi. A venturi meter having a throat diameter of 38.9 mm is installed in a line having an inside diameter of 102.3 mm. It meters water having a density of 999 kg/m^3. The measured pressure drop across the venturi is 156.9 kPa. The venturi coefficient C_v is 0.98. Calculate the gal/min and m^3/s flow rate.

Ans.   330 gal/min, 0.0208 m^3/s

3.2-5. *Use of a Venturi to Meter Water Flow.* Water at 20°C is flowing in a 2-in. schedule 40 steel pipe. Its flow rate is measured by a venturi meter having a throat diameter of 20 mm. The manometer reading is 214 mm of mercury. The venturi coefficient is 0.98. Calculate the flow rate.

3.2-6. *Metering of Oil Flow by an Orifice.* A heavy oil at 20°C having a density of 900 kg/m^3 and a viscosity of 6 cp is flowing in a 4-in. schedule 40 steel pipe. When the flow rate is 0.0174 m^3/s it is desired to have a pressure drop reading across the manometer equivalent to 0.93×10^5 Pa. What size orifice should be used if the orifice coefficient is assumed as 0.61? What is the permanent pressure loss?

3.2-7. *Water Flow Rate in an Irrigation Ditch.* Water is flowing in an open channel in an irrigation ditch. A rectangular weir having a crest length $L = 1.75$ ft is used. The weir head is measured as $h_0 = 0.47$ ft. Calculate the flow rate in ft^3/s and m^3/s.

Ans.   Flow rate = 1.776 ft^3/s, 0.0503 m^3/s

3.3-1. *Brake Horsepower of Centrifugal Pump.* Using Fig. 3.3-2 and a flow rate of 60 gal/min, do as follows.
(a) Calculate the brake hp of the pump using water with a density of 62.4 lb_m/ft^3. Compare with the value from the curve.
(b) Do the same for a nonviscous liquid having a density of 0.85 g/cm^3.

Ans.   (b) 0.69 brake hp (0.51 kW)

3.3-2. *kW-Power of a Fan.* A centrifugal fan is to be used to take a flue gas at rest (zero velocity) and at a temperature of 352.6 K and a pressure of 749.3 mm Hg and to discharge this gas at a pressure of 800.1 mm Hg and a velocity of 38.1 m/s. The volume flow rate of gas is 56.6 std m^3/min of gas (at 294.3 K and 760 mm Hg). Calculate the brake kW of the fan if its efficiency is 65% and the gas has a molecular weight of 30.7. Assume incompressible flow.

3.3-3. *Adiabatic Compression of Air.* A compressor operating adiabatically is to compress 2.83 m^3/min of air at 29.4°C and 102.7 kN/m^2 to 311.6 kN/m^2. Calculate the power required if the efficiency of the compressor is 75%. Also, calculate the outlet temperature.

3.4-1. *Power for Liquid Agitation.* It is desired to agitate a liquid having a viscosity of 1.5×10^{-3} Pa·s and a density of 969 kg/m^3 in a tank having a diameter of 0.91 m. The agitator will be a six-blade open turbine having a diameter of 0.305 m operating at 180 rpm. The tank has four vertical baffles each with a width J of 0.076 m. Also, $W = 0.0381$ m. Calculate the required kW. Use curve 2, Fig. 3.4-4.

Ans. $N_P = 2.5$, power = 0.172 kW (0.231 hp)

3.4-2. *Power for Agitation and Scale-Up.* A turbine agitator having six flat blades and a disk has a diameter of 0.203 m and is used in a tank having a diameter of 0.61 m and height of 0.61 m. The width $W = 0.0405$ m. Four baffles are used having a width of 0.051 m. The turbine operates at 275 rpm in a liquid having a density of 909 kg/m^3 and viscosity of 0.020 Pa·s.
(a) Calculate the kW power of the turbine and kW/m^3 of volume.
(b) Scale up this system to a vessel having a volume of 100 times the original for the case of equal mass transfer rates.

Ans.   (a) $P = 0.1508$ kW, $P/V = 0.845$ kW/m^3,
(b) $P_2 = 15.06$ kW, $P_2/V_2 = 0.845$ kW/m^3

3.4-3. *Scale-Down of Process Agitation System.* An existing agitation process operates using the same agitation system and fluid as described in Example 3.4-1a. It is desired to design a small pilot unit with a vessel volume of 2.0 liters so that effects of various process variables on the system can be studied in the laboratory. The rates of mass transfer appear to be important in this system, so the scale-down should be on this basis. Design the new system specifying sizes, rpm, and kW power.

3.4-4. *Anchor Agitation System.* An anchor-type agitator similar to that described for Eq. (3.4-3) is to be used to agitate a fluid having a viscosity of 100 Pa·s and a density of 980 kg/m³. The vessel size is $D_t = 0.90$ m and $H = 0.90$ m. The rpm is 50. Calculate the power required.

3.4-5. *Design of Agitation System.* An agitation system is to be designed for a fluid having a density of 950 kg/m³ and viscosity of 0.005 Pa·s. The vessel volume is 1.50 m³ and a standard six-blade open turbine with blades at 45° (curve 3, Fig. 3.4-4) is to be used with $D_a/W = 8$ and $D_a/D_t = 0.35$. For the preliminary design a power of 0.5 kW/m³ volume is to be used. Calculate the dimensions of the agitation system, rpm, and kW power.

3.5-1. *Pressure Drop of Power-Law Fluid, Banana Purée.* A power-law biological fluid, banana purée, is flowing at 23.9°C, with a velocity of 1.018 m/s, through a smooth tube 6.10 m long having an inside diameter of 0.01267 m. The flow properties of the fluid are $K = 6.00$ N·s$^{0.454}$/m² and $n = 0.454$. The density of the fluid is 976 kg/m³.
(a) Calculate the generalized Reynolds number and also the pressure drop using Eq. (3.5-9). Be sure to convert K to K' first.
(b) Repeat part (a), but use the friction factor method.
    Ans. (a) $N_{Re, gen} = 63.6$, $\Delta p = 245.2$ kN/m² (5120 lb$_f$/ft²)

3.5-2. *Pressure Drop of Pseudoplastic Fluid.* A pseudoplastic power-law fluid having a density of 63.2 lb$_m$/ft³ is flowing through 100 ft of a pipe having an inside diameter of 2.067 in. at an average velocity of 0.500 ft/s. The flow properties of the fluid are $K = 0.280$ lb$_f$·s^n/ft² and $n = 0.50$. Calculate the generalized Reynolds number and also the pressure drop, using the friction factor method.

3.5-3. *Turbulent Flow of Non-Newtonian Fluid, Applesauce.* Applesauce having the flow properties given in Table 3.5-1 is flowing in a smooth tube having an inside diameter of 50.8 mm and a length of 3.05 m at a velocity of 4.57 m/s.
(a) Calculate the friction factor and the pressure drop in the smooth tube.
(b) Repeat, but for a commercial pipe having the same inside diameter with a roughness of $\varepsilon = 4.6 \times 10^{-5}$ m.
    Ans. (a) $N_{Re, gen} = 4855$, $f = 0.0073$, (b) $f = 0.0100$

3.5-4. *Agitation of a Non-Newtonian Liquid.* A pseudoplastic liquid having the following properties of $n = 0.53$, $K = 26.49$ N·s^n/m², and $\rho = 975$ kg/m³ is being agitated in a system such as in Fig. 3.5-4 where $D_t = 0.304$ m, $D_a = 0.151$ m, and $N = 5$ rev/s. Calculate μ_a, $N'_{Re, n}$, and the kW power for this system.
    Ans. $\mu_a = 4.028$ Pa·s, $N'_{Re, n} = 27.60$, $N_P = 3.1$, $P = 0.02966$ kW

3.6-1. (*Selected Topic*) *Equation of Continuity in a Cylinder.* Fluid having a constant density ρ is flowing in the z direction through a circular pipe with axial symmetry. The radial direction is designated by r.
(a) Using a cylindrical shell balance with dimensions dr and dz, derive the equation of continuity for this system.
(b) Use the equation of continuity in cylindrical coordinates to derive the equation.

3.6-2. (*Selected Topic*) *Change of Coordinates for Continuity Equation.* Using the general equation of continuity given in rectangular coordinates, convert it to Eq. (3.6-27), which is the equation of continuity in cylindrical coordinates. Use the relationships in Eq. (3.6-26) to do this.

3.6-3. (*Selected Topic*) *Combining Equations of Continuity and Motion.* Using the continuity equation and the equation of motion for the x component, derive Eq. (3.6-39). [*Hint:* Use Eq. (3.6-2) but write it for the substantial time derivative using v_x instead of ρ. Then perform the actual partial differentiation on Eq. (3.6-36). Combine these equations with Eq. (3.6-20), the continuity equation.]

3.6-4. (*Selected Topic*) *Euler's Equation of Motion for an Ideal Fluid.* Using the Euler equations (3.6-62)–(3.6-64) for ideal fluids with constant density and zero viscosity, obtain the following equation:

$$\rho \frac{D\mathbf{v}}{Dt} = -\nabla p + \rho \mathbf{g}$$

3.6-5. (*Selected Topic*) *Average Velocity in a Circular Tube.* Using Eq. (3.6-58) for the velocity in a circular tube as a function of radius r,

$$v_z = \frac{1}{4\mu} \frac{dp}{dz} (r^2 - r_0^2) \qquad \text{(3.6-58)}$$

derive Eq. (3.6-60) for the average velocity.

$$v_{z\,av} = -\frac{r_0^2}{8\mu} \frac{dp}{dz} \qquad \text{(3.6-60)}$$

3.6-6. (*Selected Topic*) *Laminar Flow in a Cylindrical Annulus.* Derive all the equations given in Example 3.6-4 showing all the steps. Also, derive the equation for the average velocity $v_{z\,av}$. Finally, integrate to obtain the pressure drop from $z = 0$ for $p = p_0$ to $z = L$ for $p = p_L$.

$$\textbf{Ans.} \quad v_{z\,av} = -\frac{1}{8\mu} \frac{dp}{dz} \left[r_2^2 + r_1^2 - \frac{r_2^2 - r_1^2}{\ln(r_2/r_1)} \right],$$

$$v_{z\,av} = \frac{p_0 - p_L}{8\mu L} \left[r_2^2 + r_1^2 - \frac{r_2^2 - r_1^2}{\ln(r_2/r_1)} \right]$$

3.6-7. (*Selected Topic*) *Velocity Profile in Wetted-Wall Tower.* In a vertical wetted-wall tower, the fluid flows down the inside as a thin film δ m thick in laminar flow in the vertical z direction. Derive the equation for the velocity profile v_z as a function of x, the distance from the liquid surface toward the wall. The fluid is at a large distance from the entrance. Also, derive expressions for $v_{z\,av}$ and $v_{z\,max}$. (*Hint :* At $x = \delta$, which is at the wall, $v_z = 0$. At $x = 0$, the surface of the flowing liquid, $v_z = v_{z\,max}$.) Show all steps.
$$\textbf{Ans.} \quad v_z = (\rho g \delta^2 / 2\mu)[1 - (x/\delta)^2], \, v_{z\,av} = \rho g \delta^2 / 3\mu, \, v_{z\,max} = \rho g \delta^2 / 2\mu$$

3.7-1. (*Selected Topic*) *Laminar Boundary Layer on Flat Plate.* Water at 20°C is flowing past a flat plate at 0.914 m/s. The plate is 0.305 m wide.
(a) Calculate the Reynolds number 0.305 m from the leading edge to determine if the flow is laminar.
(b) Calculate the boundary-layer thickness at $x = 0.152$ and $x = 0.305$ m from the leading edge.
(c) Calculate the total drag on the 0.305-m-long plate.
$$\textbf{Ans.} \quad \text{(a) } N_{Re,\,L} = 2.77 \times 10^5, \text{(b) } \delta = 0.0029 \text{ m at } x = 0.305 \text{ m}$$

3.7-2. (*Selected Topic*) *Air Flow Past a Plate.* Air at 294.3 K and 101.3 kPa is flowing past a flat plate at 6.1 m/s. Calculate the thickness of the boundary layer at a distance of 0.3 m from the leading edge and the total drag for a 0.3-m-wide plate.

3.7-3. (*Selected Topic*) *Boundary-Layer Flow Past a Plate.* Water at 293 K is flowing past a flat plate at 0.5 m/s. Do as follows.
(a) Calculate the boundary-layer thickness in m at a point 0.1 m from the leading edge.
(b) At the same point, calculate the point shear stress τ_0. Also calculate the total drag coefficient.

Transport Processes: Momentum, Heat, and Mass

3.7-4. (*Selected Topic*) *Transition Point to Turbulent Boundary Layer.* Air at 101.3 kPa and 293 K is flowing past a smooth flat plate at 100 ft/s. The turbulence in the air stream is such that the transition from a laminar to a turbulent boundary layer occurs at $N_{Re, L} = 5 \times 10^5$.
  (a) Calculate the distance from the leading edge where the transition occurs.
  (b) Calculate the boundary-layer thickness δ at a distance of 0.5 ft and 3.0 ft from the leading edge. Also calculate the drag coefficient for both distances $L = 0.5$ and 3.0 ft.

3.8-1. (*Selected Topic*) *Dimensional Analysis for Flow Past a Body.* A fluid is flowing external to a solid body. The force F exerted on the body is a function of the fluid velocity v, fluid density ρ, fluid viscosity μ, and a dimension of the body L. By dimensional analysis, obtain the dimensionless groups formed from the variables given. (*Note :* Use the M, L, t system of units. The units of F are ML/t^2. Select v, ρ, and L as the core variables.)

$$\text{Ans.} \quad \pi_1 = (F/L^2)/\rho v^2, \pi_2 = Lv\rho/\mu$$

3.8-2. (*Selected Topic*) *Dimensional Analysis for Bubble Formation.* Dimensional analysis is to be used to correlate data on bubble size with the properties of the liquid when gas bubbles are formed by a gas issuing from a small orifice below the liquid surface. Assume that the significant variables are bubble diameter D, orifice diameter d, liquid density ρ, surface tension σ in N/m, liquid viscosity μ, and g. Select d, ρ, and g as the core variables.

$$\text{Ans.} \quad \pi_1 = D/d, \pi_2 = \sigma/\rho d^2 g, \pi_3 = \mu^2/\rho^2 d^3 g$$

REFERENCES

(A1) Allis Chalmers Mfg. Co. *Bull. 1659.*

(A2) American Gas Association, "Orifice Metering of Natural Gas," *Gas Measurement Rept. 3*, New York, 1955.

(B1) BENNETT, C. O., and MYERS, J. E. *Momentum, Heat, and Mass Transfer*, 2nd ed. New York: McGraw-Hill Book Company, 1974.

(B2) BIRD, R. B., STEWART, W. E., and LIGHTFOOT, E. N. *Transport Phenomena.* New York: John Wiley & Sons, Inc., 1960.

(B3) BATES, R. L., FONDY, P. L., and CORPSTEIN, R. R. *I.E.C. Proc. Des. Dev.*, **2**, 310 (1963).

(B4) BROWN, G. G., et al. *Unit Operations.* New York: John Wiley & Sons, Inc., 1950.

(C1) CHARM, S. E. *The Fundamentals of Food Engineering.* 2nd ed. Westport, Conn.: Avi Publishing Co., Inc., 1971.

(C2) CARMAN, P. C. *Trans. Inst. Chem. Eng. (London)*, **15**, 150 (1937).

(C3) CALDERBANK, P. H. In *Mixing : Theory and Practice*, Vol. 2, V. W. Uhl and J. B. Gray, eds. New York: Academic Press, Inc., 1967.

(D1) DREW, T. B., and HOOPES, J. W., Jr. *Advances in Chemical Engineering.* New York: Academic Press, Inc., 1956.

(D2) DODGE, D. W., and METZNER, A. B. *A.I.Ch.E. J.*, **1**, 434 (1955).

(E1) ERGUN, S. *Chem. Eng. Progr.*, **48**, 89 (1952).

(G1) GODLESKI, E. S., and SMITH, J. C. *A.I.Ch.E. J.*, **8**, 617 (1962).

(H1) HARPER, J. C., and EL SAHRIGI. *J. Food Sci.*, **30**, 470 (1965).

(H2) HO, F. C., and KWONG, A. *Chem. Eng.*, July 23, 94 (1973).

(K1) KUNII, D., and LEVENSPIEL, O. *Fluidization Engineering.* New York: John Wiley & Sons, Inc., 1969.

(L1) LEAMY, G. H. *Chem. Eng.*, Oct. 15, 115 (1973).

(L2)   LEVA, M., WEINTRAUB, M., GRUMMER, M., POLLCHIK, M., and STORCH, H. H. *U.S. Bur. Mines Bull.*, 504 (1951).

(M1)   METZNER, A. B., FEEHS, R. H., RAMOS, H. L., OTTO, R. E., and TUTHILL, J. D. *A.I.Ch.E. J.*, **7**, 3 (1961).

(M2)   McCABE, W. L., and SMITH, J. C. *Unit Operations of Chemical Engineering*, 3rd ed. New York: McGraw-Hill Book Company, 1976.

(M3)   METZNER, A. B., and REED, J. C. *A.I.Ch.E. J.*, **1**, 434 (1955).

(M4)   MOHSENIN, N. N. *Physical Properties of Plant and Animal Materials*, Vol. 1, Part II. New York: Gordon & Breach, Inc., 1970.

(M5)   MOO-YUNG, TICHAR, M. K., and DULLIEN, F. A. L. *A.I.Ch.E. J.*, **18**, 178 (1972).

(N1)   NORWOOD, K. W., and METZNER, A. B. *A.I.Ch.E. J.*, **6**, 432 (1960).

(P1)   PERRY, R. H., and CHILTON, C. H. *Chemical Engineers' Handbook*, 5th ed. New York: McGraw-Hill Book Company, 1973.

(P2)   PINCHBECK, P. H., and POPPER, F. *Chem. Eng. Sci.*, **6**, 57 (1956).

(P3)   PATTERSON, W. I., CARREAU, P. J., and YAP, C. Y. *A.I.Ch.E. J.*, **25**, 208 (1979).

(R1)   RUSHTON, J. H., COSTICH, D. W., and EVERETT, H. J. *Chem. Eng. Progr.*, **46**, 395, 467 (1950).

(R2)   RAUTZEN, R. R., CORPSTEIN, R. R., and DICKEY, D. S. *Chem. Eng.*, Oct. 25, 119 (1976).

(S1)   STEVENS, W. E. Ph.D. thesis, University of Utah, 1953.

(S2)   SKELLAND, A. H. P. *Non-Newtonian Flow and Heat Transfer*. New York: John Wiley & Sons, Inc., 1967.

(S3)   STREETER, V. L. *Handbook of Fluid Dynamics*. New York: McGraw-Hill Book Company, 1961.

(T1)   TREYBAL, R. E. *Liquid Extraction*. 2nd ed. New York: McGraw-Hill Book Company, 1953.

(W1)   WEISMAN, J., and EFFERDING, L. E. *A.I.Ch.E. J.*, **6**, 419 (1960).

(W2)   WINNING, M. D. M.Sc. thesis, University of Alberta, 1948.

(W3)   WALTERS, K. *Rheometry*. London: Chapman & Hall Ltd., 1975.

(W4)   WEN, C. Y., and YU, Y. H. *A.I.Ch.E. J.*, **12**, 610 (1966).

(W5)   WILHELM, R. H., and KWAUK, M. *Chem. Eng. Progr.*, **44**, 201 (1948).

(Z1)   ZLOKARNIK, M., and JUDAT, H. *Chem. Eng. Tech.*, **39**, 1163 (1967).

CHAPTER 4

Principles of
Steady-State
Heat Transfer

4.1 INTRODUCTION AND MECHANISMS OF HEAT TRANSFER

4.1A Introduction to Steady-State Heat Transfer

The transfer of energy in the form of heat occurs in many chemical and other types of processes. Heat transfer often occurs in combination with other unit operations, such as drying of lumber or foods, alcohol distillation, burning of fuel, and evaporation. The heat transfer occurs because of a temperature difference driving force and heat flows from the high- to the low-temperature region.

In Section 2.3 we derived an equation for a general property balance of momentum, thermal energy, or mass at unsteady state by writing Eq. (2.3-7). Writing a similar equation but specifically for heat transfer,

$$\begin{pmatrix} \text{rate of} \\ \text{heat in} \end{pmatrix} + \begin{pmatrix} \text{rate of gener-} \\ \text{ation of heat} \end{pmatrix} = \begin{pmatrix} \text{rate of} \\ \text{heat out} \end{pmatrix} + \begin{pmatrix} \text{rate of accumu-} \\ \text{lation of heat} \end{pmatrix} \tag{4.1-1}$$

Assuming the rate of transfer of heat occurs only by conduction, we can rewrite Eq. (2.3-14), which is *Fourier's law*, as

$$\frac{q_x}{A} = -k\frac{dT}{dx} \tag{4.1-2}$$

Making an unsteady-state heat balance for the x direction only on the element of volume or control volume in Fig. 4.1-1 by using Eqs. (4.1-1) and (4.1-2) with the cross-sectional area being A m^2,

$$q_{x|x} + \dot{q}(\Delta x \cdot A) = q_{x|x+\Delta x} + \rho c_p \frac{\partial T}{\partial t}(\Delta x \cdot A) \tag{4.1-3}$$

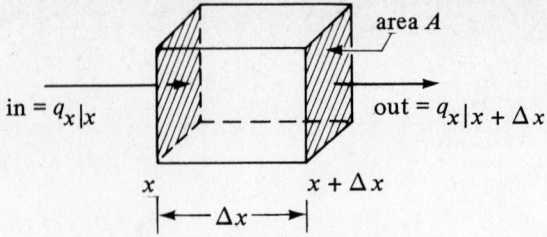

FIGURE 4.1-1. *Unsteady-state balance for heat transfer in control volume.*

where $\dot{q}$ is rate of heat generated per unit volume. Assuming no heat generation and also assuming steady-state heat transfer where the rate of accumulation is zero, Eq. (4.1-3) becomes

$$q_{x|x} = q_{x|x + \Delta x} \qquad (4.1\text{-}4)$$

This means the rate of heat input by conduction = the rate of heat output by conduction; or q_x is a constant with time for steady-state heat transfer.

In this chapter we are concerned with a control volume where the rate of accumulation of heat is zero and we have steady-state heat transfer. The rate of heat transfer is then constant with time, and the temperatures at various points in the system do not change with time. To solve problems in steady-state heat transfer, various mechanistic expressions in the form of differential equations for the different modes of heat transfer such as Fourier's law are integrated. Expressions for the temperature profile and heat flux are then obtained in this chapter.

In Chapter 5 the conservation-of-energy equations (2.7-2) and (4.1-3) will be used again when the rate of accumulation is not zero and unsteady-state heat transfer occurs. The mechanistic expression for Fourier's law in the form of a partial differential equation will be used where temperature at various points and the rate of heat transfer change with time. In Section 5.6 (Selected Topic) a general differential equation of energy change will be derived and integrated for various specific cases to determine the temperature profile and heat flux.

4.1B Basic Mechanisms of Heat Transfer

Heat transfer may occur by any one or more of the three basic mechanisms of heat transfer: conduction, convection, or radiation.

1. Conduction. In conduction, heat can be conduced through solids, liquids, and gases. The heat is conducted by the transfer of the energy of motion between adjacent molecules. In a gas the "hotter" molecules, which have greater energy and motions, impart energy to the adjacent molecules at lower energy levels. This type of transfer is present to some extent in all solids, gases, or liquids in which a temperature gradient exists. In conduction, energy can also be transferred by "free" electrons, which is quite important in metallic solids. Examples of heat transfer mainly by conduction are heat transfer through walls of exchangers or a refrigerator, heat treatment of steel forgings, freezing of the ground during the winter, and so on.

2. Convection. The transfer of heat by convection implies the transfer of heat by bulk transport and mixing of macroscopic elements of warmer portions with cooler portions of a gas or a liquid. It also often involves the energy exchange between a solid surface and a fluid. A distinction must be made between forced-convection heat transfer, where a fluid is forced to flow past a solid surface by a pump, fan, or other mechanical means, and natural or free convection, where warmer or cooler fluid next to the solid surface causes a circulation because of a density difference resulting from the temperature differences in the fluid. Examples of heat transfer by convection are loss of heat from a car radiator where the air is being circulated by a fan, cooking of foods in a vessel being stirred, cooling of a hot cup of coffee by blowing over the surface, and so on.

3. *Radiation.*   Radiation differs from heat transfer by conduction and convection in that no physical medium is needed for its propagation. Radiation is the transfer of energy through space by means of electromagnetic waves in much the same way as electromagnetic light waves transfer light. The same laws which govern the transfer of light govern the radiant transfer of heat. Solids and liquids tend to absorb the radiation being transferred through it, so that radiation is important primarily in transfer through space or gases. The most important example of radiation is the transport of heat to the earth from the sun. Other examples are cooking of food when passed below red-hot electric heaters, heating of fluids in coils of tubing inside a combustion furnace, and so on.

4.1C   Fourier's Law of Heat Conduction

As discussed in Section 2.3 for the general molecular transport equation, all three main types of rate-transfer processes—momentum transfer, heat transfer, and mass transfer—are characterized by the same general type of equation. The transfer of electric current can also be included in this category. This basic equation is as follows:

$$\text{rate of a transfer process} = \frac{\text{driving force}}{\text{resistance}} \qquad \textbf{(2.3-1)}$$

This equation states what we know intuitively: that in order to transfer a property such as heat or mass, we need a driving force to overcome a resistance.

The transfer of heat by conduction also follows this basic equation and is written as Fourier's law for heat conduction in fluids or solids.

$$\frac{q_x}{A} = -k \frac{dT}{dx} \qquad \textbf{(4.1-2)}$$

where q_x is the heat-transfer rate in the x direction in watts (W), A is the cross-sectional area normal to the direction of flow of heat in m^2, T is temperature in K, x is distance in m, and k is the thermal conductivity in $W/m \cdot K$ in the SI system. The quantity q_x/A is called the heat flux in W/m^2. The quantity dT/dx is the temperature gradient in the x direction. The minus sign in Eq. (4.1-2) is required because if the heat flow is positive in a given direction, the temperature decreases in this direction (becomes negative).

The units in Eq. (4.1-2) may also be expressed in the cgs system with q_x in cal/s, A in cm^2, k in $cal/s \cdot °C \cdot cm$, T in °C, and x in cm. In the English system, q_x is in btu/h, A in ft^2, T in °F, x in ft, k in $btu/h \cdot °F \cdot ft$, and q_x/A in $btu/h \cdot ft^2$. From Appendix A.1, the conversion factors are, for thermal conductivity,

$$1 \text{ btu/h} \cdot \text{ft} \cdot °F = 4.1365 \times 10^{-3} \text{ cal/s} \cdot \text{cm} \cdot °C \qquad \textbf{(4.1-5)}$$

$$1 \text{ btu/h} \cdot \text{ft} \cdot °F = 1.73073 \text{ W/m} \cdot \text{K} \qquad \textbf{(4.1-6)}$$

For heat flux and power,

$$1 \text{ btu/h} \cdot \text{ft}^2 = 3.1546 \text{ W/m}^2 \qquad \text{(4.1-7)}$$

$$1 \text{ btu/h} = 0.29307 \text{ W} \qquad \text{(4.1-8)}$$

Fourier's law, Eq. (4.1-2), can be integrated for the case of steady-state heat transfer through a flat wall of constant cross-sectional area A, where the inside temperature at point 1 is T_1 and T_2 at point 2 a distance of $x_2 - x_1$ m away. Rearranging Eq. (4.1-2),

$$\frac{q_x}{A} \int_{x_1}^{x_2} dx = -k \int_{T_1}^{T_2} dT \qquad \text{(4.1-9)}$$

Integrating, assuming that k is constant and does not vary with temperature and dropping the subscript x on q_x for convenience,

$$\frac{q}{A} = \frac{k}{x_2 - x_1} (T_1 - T_2) \qquad \text{(4.1-10)}$$

EXAMPLE 4.1-1. *Heat Loss Through an Insulating Wall*
Calculate the heat loss per m^2 of surface area for an insulating wall composed of 25.4-mm-thick fiber insulating board, where the inside temperature is 352.7 K and the outside temperature is 297.1 K.

Solution: From Appendix A.3, the thermal conductivity of fiber insulating board is 0.048 W/m·K. The thickness $x_2 - x_1 = 0.0254$ m. Substituting into Eq. (4.1-10),

$$\frac{q}{A} = \frac{k}{x_2 - x_1} (T_1 - T_2) = \frac{0.048}{0.0254} (352.7 - 297.1)$$

$$= 105.1 \text{ W/m}^2$$

$$= (105.1 \text{ W/m}^2) \frac{1}{(3.1546 \text{ W/m}^2)/(\text{btu/h} \cdot \text{ft}^2)} = 33.30 \text{ btu/h} \cdot \text{ft}^2$$

4.1D Thermal Conductivity

The defining equation for thermal conductivity is given as Eq. (4.1-2), and with this definition, experimental measurements have been made to determine the thermal conductivity of different materials. In Table 4.1-1 thermal conductivities are given for a few materials for the purpose of comparison. More detailed data are given in Appendix A.3 for inorganic and organic materials and A.4 for food and biological materials. As seen in Table 4.1-1, gases have quite low values of thermal conductivity, liquids intermediate values, and solid metals very high values.

1. Gases. In gases the mechanism of thermal conduction is relatively simple. The molecules are in continuous random motion, colliding with one another and exchanging energy and momentum. If a molecule moves from a high-temperature region to a region of lower temperature, it transports kinetic energy to this region and gives up this energy through collisions with lower-energy molecules. Since smaller molecules move faster, gases such as hydrogen should have higher thermal conductivities, as shown in Table 4.1-1.

Theories to predict thermal conductivities of gases are reasonably accurate and are given elsewhere (R1). The thermal conductivity increases approximately as the square root of the absolute temperature and is independent of pressure up to a few atmospheres. At very low pressures (vacuum), however, the thermal conductivity approaches zero.

2. *Liquids.* The physical mechanism of conduction of energy in liquids is somewhat similar to that of gases, where higher-energy molecules collide with lower-energy molecules. However, the molecules are packed so closely together that molecular force fields exert a strong effect on the energy exchange. Since an adequate molecular theory of liquids is not available, most correlations to predict the thermal conductivities are empirical. Reid et al. (R1) discuss these in detail. The thermal conductivity of liquids varies moderately with temperature and often can be expressed as a linear variation,

$$k = a + bT \qquad\qquad\qquad (4.1\text{-}11)$$

where a and b are empirical constants. Thermal conductivities of liquids are essentially independent of pressure.

Water has a high thermal conductivity compared to organic-type liquids such as benzene. As shown in Table 4.1-1, the thermal conductivities of most unfrozen foodstuffs, such as skim milk and applesauce, which contain large amounts of water have thermal conductivities near that of pure water.

3. *Solids.* The thermal conductivity of homogeneous solids varies quite widely, as may be seen for some typical values in Table 4.1-1. The metallic solids of copper and aluminum have very high thermal conductivities, and some insulating nonmetallic materials such as rock wool and corkboard have very low conductivities.

Heat or energy is conducted through solids by two mechanisms. In the first, which applies primarily to metallic solids, heat, like electricity, is conducted by free electrons which move through the metal lattice. In the second mechanism, present in all solids, heat is conducted by the transmission of energy of vibration between adjacent atoms.

TABLE 4.1-1. *Thermal Conductivities of Some Materials at 101.325 kPa (1 Atm) Pressure (k in W/m·K)*

Substance	Temp. (K)	k	Ref.	Substance	Temp. (K)	k	Ref.
Gases				Solids			
Air	273	0.0242	(K2)	Ice	273	2.25	(C1)
	373	0.0316		Fire claybrick	473	1.00	(P1)
H_2	273	0.167	(K2)	Paper	—	0.130	(M1)
n-Butane	273	0.0135	(P2)	Hard rubber	273	0.151	(M1)
Liquids				Cork board	303	0.043	(M1)
Water	273	0.569	(P1)	Asbestos	311	0.168	(M1)
	366	0.680		Rock wool	266	0.029	(K1)
Benzene	303	0.159	(P1)	Steel	291	45.3	(P1)
	333	0.151			373	45	
Biological materials				Copper	273	388	(P1)
and foods					373	377	
Olive oil	293	0.168	(P1)	Aluminum	273	202	(P1)
	373	0.164					
Lean beef	263	1.35	(C1)				
Skim milk	275	0.538	(C1)				
Applesauce	296	0.692	(C1)				
Salmon	277	0.502	(C1)				
	248	1.30					

Thermal conductivities of insulating materials such as rock wool approach that of air since the insulating materials contain large amounts of air trapped in void spaces. Superinsulations to insulate cryogenic materials such as liquid hydrogen are composed of multiple layers of highly reflective materials separated by evacuated insulating spacers. Values of thermal conductivity are considerably lower than for air alone.

Ice has a thermal conductivity much greater than water. Hence, the thermal conductivities of frozen foods such as lean beef and salmon given in Table 4.1-1 are much higher than for unfrozen foods.

4.1E   Convective-Heat-Transfer Coefficient

It is well known that a hot piece of material will cool faster when air is blown or forced by the object. When the fluid outside the solid surface is in forced or natural convective motion, we express the rate of heat transfer from the solid to the fluid, or vice versa, by the following equation:

$$q = hA(T_w - T_f) \qquad (4.1\text{-}12)$$

where q is the heat-transfer rate in W, A is the area in m^2, T_w is the temperature of the solid surface in K, T_f is the average or bulk temperature of the fluid flowing by in K, and h is the convective heat-transfer coefficient in $W/m^2 \cdot K$. In English units, h is in $btu/h \cdot ft^2 \cdot °F$.

The coefficient h is a function of the system geometry, fluid properties, flow velocity, and temperature difference. In many cases, empirical correlations are available to predict this coefficient, since it often cannot be predicted theoretically. Since we know that when a fluid flows by a surface there is a thin, almost stationary layer or film of fluid adjacent to the wall which presents most of the resistance to heat transfer, we often call the coefficient h a *film coefficient*.

In Table 4.1-2 some order-of-magnitude values of h for different convective mechanisms of free or natural convection, forced convection, boiling, and condensation are given. Water gives the highest values of the heat-transfer coefficients.

To convert the heat-transfer coefficient h from English to SI units,

$$1 \text{ btu/h} \cdot ft^2 \cdot °F = 5.6783 \text{ W/m}^2 \cdot K$$

TABLE 4.1-2.   *Approximate Magnitude of Some Heat-Transfer Coefficients*

	Range of Values of h	
Mechanism	*btu/h·ft²·°F*	*W/m²·K*
Condensing steam	1000–5000	5700–28 000
Condensing organics	200–500	1100–2800
Boiling liquids	300–5000	1700–28 000
Moving water	50–3000	280–17 000
Moving hydrocarbons	10–300	55–1700
Still air	0.5–4	2.8–23
Moving air	2–10	11.3–55

4.2 CONDUCTION HEAT TRANSFER

4.2A Conduction Through a Flat Slab or Wall

In this section Fourier's equation (4.1-2) will be used to obtain equations for one-dimensional steady-state conduction of heat through some simple geometries. For a flat slab or wall where the cross-sectional area A and k in Eq. (4.1-2) are constant, we obtained Eq. (4.1-10), which we rewrite as

$$\frac{q}{A} = \frac{k}{x_2 - x_1}(T_1 - T_2) = \frac{k}{\Delta x}(T_1 - T_2) \tag{4.2-1}$$

This is shown in Fig. 4.2-1, where $\Delta x = x_2 - x_1$. Equation (4.2-1) indicates that if T is substituted for T_2 and x for x_2, the temperature varies linearly with distance as shown in Fig. 4.2-1b.

   If the thermal conductivity is not constant but varies linearly with temperature, then substituting Eq. (4.1-11) into Eq. (4.1-2) and integrating,

$$\frac{q}{A} = \frac{a + b\dfrac{T_1 + T_2}{2}}{\Delta x}(T_1 - T_2) = \frac{k_m}{\Delta x}(T_1 - T_2) \tag{4.2-2}$$

where

$$k_m = a + b\frac{T_1 + T_2}{2} \tag{4.2-3}$$

This means that the mean value of k (i.e., k_m) to use in Eq. (4.2-2) is the value of k evaluated at the linear average of T_1 and T_2.

   As stated in the introduction in Eq. (2.3-1), the rate of a transfer process equals the driving force over the resistance. Equation (4.2-1) can be rewritten in that form.

$$q = \frac{T_1 - T_2}{\Delta x / kA} = \frac{T_1 - T_2}{R} = \frac{\text{driving force}}{\text{resistance}} \tag{4.2-4}$$

where $R = \Delta x / kA$ and is the resistance in K/W or h · °F/btu.

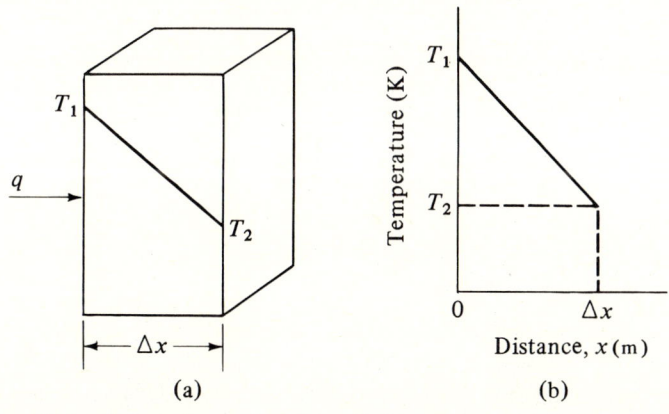

FIGURE 4.2-1.   *Heat conduction in a flat wall : (a) geometry of wall, (b) temperature plot.*

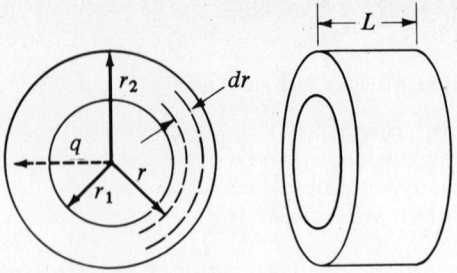

FIGURE 4.2-2.   *Heat conduction in a cylinder.*

4.2B Conduction Through a Hollow Cylinder

In many instances in the process industries, heat is being transferred through the walls of a thick-walled cylinder as in a pipe that may or may not be insulated. Consider the hollow cylinder in Fig. 4.2-2 with an inside radius of r_1, where the temperature is T_1, an outside radius of r_2 having a temperature of T_2, and a length of L m. Heat is flowing radially from the inside surface to the outside. Rewriting Fourier's law, Eq. (4.1-2), with distance dr instead of dx,

$$\frac{q}{A} = -k\frac{dT}{dr} \tag{4.2-5}$$

The cross-sectional area normal to the heat flow is

$$A = 2\pi r L \tag{4.2-6}$$

Substituting Eq. (4.2-6) into (4.2-5), rearranging, and integrating,

$$\frac{q}{2\pi L}\int_{r_1}^{r_2}\frac{dr}{r} = -k\int_{T_1}^{T_2} dT \tag{4.2-7}$$

$$q = k\frac{2\pi L}{\ln(r_2/r_1)}(T_1 - T_2) \tag{4.2-8}$$

Multiplying numerator and denominator by $(r_2 - r_1)$,

$$q = kA_{lm}\frac{T_1 - T_2}{r_2 - r_1} = \frac{T_1 - T_2}{(r_2 - r_1)/(kA_{lm})} = \frac{T_1 - T_2}{R} \tag{4.2-9}$$

where

$$A_{lm} = \frac{(2\pi L r_2) - (2\pi L r_1)}{\ln(2\pi L r_2/2\pi L r_1)} = \frac{A_2 - A_1}{\ln(A_2/A_1)} \tag{4.2-10}$$

$$R = \frac{r_2 - r_1}{kA_{lm}} = \frac{\ln(r_2/r_1)}{2\pi kL} \tag{4.2-11}$$

The log mean area is A_{lm}. In engineering practice, if $A_2/A_1 < 1.5/1$, the linear mean area of $(A_1 + A_2)/2$ is within 1.5% of the log mean area. From Eq. (4.2-8), if r is substituted for r_2 and T for T_2, the temperature is seen to be a linear function of $\ln r$ instead of r as in the case of a flat wall. If the thermal conductivity varies with temperature as in Eq. (4.1-10), it can be shown that the mean value to use in a cylinder is still k_m of Eq. (4.2-3).

EXAMPLE 4.2-1.   Length of Tubing for Cooling Coil

A thick-walled cylindrical tubing of hard rubber having an inside radius of 5 mm and an outside radius of 20 mm is being used as a temporary cooling coil in a bath. Ice water is flowing rapidly inside and the inside wall temperature is 274.9 K. The outside surface temperature is 297.1 K. A total of 14.65 W must be removed from the bath by the cooling coil. How many m of tubing are needed?

Solution: From Appendix A.3, the thermal conductivity at 0°C (273 K) is $k = 0.151$ W/m $\cdot$ K. Since data at other temperatures are not available, this value will be used for the range of 274.9 to 297.1 K.

$$r_1 = \frac{5}{1000} = 0.005 \text{ m} \qquad r_2 = \frac{20}{1000} = 0.02 \text{ m}$$

The calculation will be done first for a length of 1.0 m of tubing. Solving for the areas A_1, A_2, and A_{1m} in Eq. (4.2-10),

$$A_1 = 2\pi L r_1 = 2\pi(1.0)(0.005) = 0.0314 \text{ m}^2 \qquad A_2 = 0.1257 \text{ m}^2$$

$$A_{1m} = \frac{A_2 - A_1}{\ln (A_2/A_1)} = \frac{0.1257 - 0.0314}{2.303 \log (0.1257/0.0314)} = 0.0680 \text{ m}^2$$

Substituting into Eq. (4.2-9) and solving,

$$q = kA_{1m} \frac{T_1 - T_2}{r_2 - r_1} = 0.151(0.0682) \left(\frac{274.9 - 297.1}{0.02 - 0.005} \right)$$

$$= -15.2 \text{ W (51.9 btu/h)}$$

The negative sign indicates that the heat flow is from r_2 on the outside to r_1 on the inside. Since 15.2 W is removed for a 1-m length, the needed length is

$$\text{length} = \frac{14.65 \text{ W}}{15.2 \text{ W/m}} = 0.964 \text{ m}$$

Note that the thermal conductivity of rubber is quite small. Generally, metal cooling coils are used, since the thermal conductivity of metals is quite high. The liquid film resistances in this case are quite small and are neglected.

4.2C   Conduction Through a Hollow Sphere

Heat conduction through a hollow sphere is another case of one-dimensional conduction. Using Fourier's law for constant thermal conductivity with distance dr, where r is the radius of the sphere,

$$\frac{q}{A} = -k \frac{dT}{dr} \tag{4.2-5}$$

The cross-sectional area normal to the heat flow is

$$A = 4\pi r^2 \tag{4.2-12}$$

Substituting Eq. (4.2-12) into (4.2-5), rearranging, and integrating,

$$\frac{q}{4\pi} \int_{r_1}^{r_2} \frac{dr}{r^2} = -k \int_{T_1}^{T_2} dt \tag{4.2-13}$$

$$q = \frac{4\pi k(T_1 - T_2)}{1/r_1 - 1/r_2} = \frac{T_1 - T_2}{(1/r_1 - 1/r_2)/4\pi k} \tag{4.2-14}$$

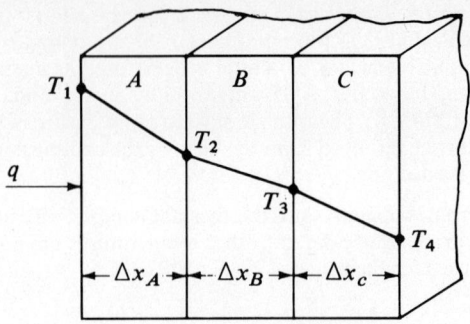

FIGURE 4.3-1. *Heat flow through a multilayer wall.*

It can be easily shown that the temperature varies hyperbolically with the radius. (See Problem 4.2-5.)

4.3 CONDUCTION THROUGH SOLIDS IN SERIES

4.3A Plane Walls in Series

In the case where there is a multilayer wall of more than one material present as shown in Fig. 4.3-1, we proceed as follows. The temperature profiles in the three materials A, B, and C are shown. Since the heat flow q must be the same in each layer, we can write Fourier's equation for each layer as

$$q = \frac{k_A A}{\Delta x_A}(T_1 - T_2) = \frac{k_B A}{\Delta x_B}(T_2 - T_3) = \frac{k_C A}{\Delta x_C}(T_3 - T_4) \tag{4.3-1}$$

Solving each equation for ΔT,

$$T_1 - T_2 = q\frac{\Delta x_A}{k_A A} \qquad T_2 - T_3 = q\frac{\Delta x_B}{k_B A} \qquad T_3 - T_4 = q\frac{\Delta x_C}{k_C A} \tag{4.3-2}$$

Adding the equations for $T_1 - T_2$, $T_2 - T_3$, and $T_3 - T_4$, the internal temperatures T_2 and T_3 drop out and the final rearranged equation is

$$q = \frac{T_1 - T_4}{\Delta x_A/(k_A A) + \Delta x_B/(k_B A) + \Delta x_C/(k_C A)} = \frac{T_1 - T_4}{R_A + R_B + R_C} \tag{4.3-3}$$

where the resistance $R_A = \Delta x_A/k_A A$, and so on.

Hence, the final equation is in terms of the overall temperature drop $T_1 - T_4$ and the total resistance, $R_A + R_B + R_C$.

EXAMPLE 4.3-1. Heat Flow Through an Insulated Wall of a Cold Room
A cold-storage room is constructed of an inner layer of 12.7 mm of pine, a middle layer of 101.6 mm of cork board, and an outer layer of 76.2 mm of concrete. The wall surface temperature is 255.4 K inside the cold room and 297.1 K at the outside surface of the concrete. Use conductivities from Appendix A.3 for pine, 0.151; for cork board, 0.0433; and for concrete, 0.762 W/m·K. Calculate the heat loss in W for 1 m² and the temperature at the interface between the wood and cork board.

Solution: Calling $T_1 = 255.4$, $T_4 = 297.1$ K, pine as material A, cork as B, and concrete as C, a tabulation of the properties and dimensions is as follows:

$$k_A = 0.151 \qquad k_B = 0.0433 \qquad k_C = 0.762$$

$$\Delta x_A = 0.0127 \text{ m}$$

$$\Delta x_B = 0.1016 \text{ m}$$

$$\Delta x_C = 0.0762 \text{ m}$$

The resistances for each material are, from Eq. (4.3-3), for an area of 1 m²,

$$R_A = \frac{\Delta x_A}{k_A A} = \frac{0.0127}{0.151(1)} = 0.0841 \text{ K/W}$$

$$R_B = \frac{\Delta x_B}{k_B A} = \frac{0.1016}{0.0433(1)} = 2.346$$

$$R_C = \frac{\Delta x_C}{k_C A} = \frac{0.0762}{0.762(1)} = 0.100$$

Substituting into Eq. (4.3-3),

$$q = \frac{T_1 - T_4}{R_A + R_B + R_C} = \frac{255.4 - 297.1}{0.0841 + 2.346 + 0.100}$$

$$= \frac{-41.7}{2.530} = -16.48 \text{ W} \ (-56.23 \text{ btu/h})$$

Since the answer is negative, heat flows in from the outside.

To calculate the temperature T_2 at the interface between the pine wood and cork,

$$q = \frac{T_1 - T_2}{R_A}$$

Substituting the known values and solving,

$$-16.48 = \frac{255.4 - T_2}{0.0841} \quad \text{and} \quad T_2 = 256.79 \text{ K at the interface}$$

An alternative procedure to use to calculate T_2 is to use the fact that the temperature drop is proportional to the resistance.

$$T_1 - T_2 = \frac{R_A}{R_A + R_B + R_C}(T_1 - T_4) \qquad (4.3\text{-}4)$$

Substituting,

$$255.4 - T_2 = \frac{0.0841(255.4 - 297.1)}{2.530} = -1.39 \text{ K}$$

Hence, $T_2 = 256.79$ K, as calculated before.

4.3B   Multilayer Cylinders

In the process industries, heat transfer often occurs through multilayers of cylinders, as for example when heat is being transferred through the walls of an insulated pipe. Figure

FIGURE 4.3-2. *Radial heat flow through multiple cylinders in series.*

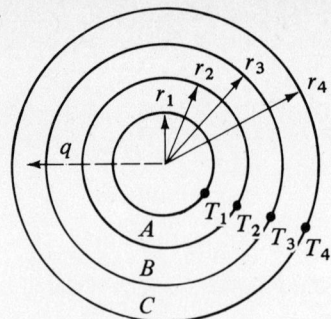

4.3-2 shows a pipe with two layers of insulation around it, i.e., a total of three concentric hollow cylinders. The temperature drop is $T_1 - T_2$ across material A, $T_2 - T_3$ across B, and $T_3 - T_4$ across C.

The heat-transfer rate q will, of course, be the same for each layer, since we are at steady state. Writing an equation similar to Eq. (4.2-9) for each concentric cylinder,

$$q = \frac{T_1 - T_2}{(r_2 - r_1)/(k_A A_{A\,\mathrm{lm}})} = \frac{T_2 - T_3}{(r_3 - r_2)/(k_B A_{B\,\mathrm{lm}})} = \frac{T_3 - T_4}{(r_4 - r_3)/(k_C A_{C\,\mathrm{lm}})} \qquad \text{(4.3-5)}$$

where

$$A_{A\,\mathrm{lm}} = \frac{A_2 - A_1}{\ln (A_2/A_1)} \quad A_{B\,\mathrm{lm}} = \frac{A_3 - A_2}{\ln (A_3/A_2)} \quad A_{C\,\mathrm{lm}} = \frac{A_4 - A_3}{\ln (A_4/A_3)} \qquad \text{(4.3-6)}$$

Using the same method to combine the equations to eliminate T_2 and T_3 as was done for the flat walls in series, the final equations are

$$q = \frac{T_1 - T_4}{(r_2 - r_1)/(k_A A_{A\,\mathrm{lm}}) + (r_3 - r_2)/(k_B A_{B\,\mathrm{lm}}) + (r_4 - r_3)/(k_C A_{C\,\mathrm{lm}})} \qquad \text{(4.3-7)}$$

$$q = \frac{T_1 - T_4}{R_A + R_B + R_C} = \frac{T_1 - T_4}{\sum R} \qquad \text{(4.3-8)}$$

Hence, the overall resistance is again the sum of the individual resistances in series.

EXAMPLE 4.3-2. *Heat Loss from an Insulated Pipe*
A thick-walled tube of stainless steel (A) having a $k = 21.63$ W/m · K with dimensions of 0.0254 m ID and 0.0508 m OD is covered with a 0.0254-m layer of asbestos (B) insulation, $k = 0.2423$ W/m · K. The inside wall temperature of the pipe is 811 K and the outside surface of the insulation is at 310.8 K. For a 0.305-m length of pipe, calculate the heat loss and also the temperature at the interface between the metal and the insulation.

Solution: Calling $T_1 = 811$ K, T_2 the interface, and $T_3 = 310.8$ K, the dimensions are

$$r_1 = \frac{0.0254}{2} = 0.0127 \text{ m} \qquad r_2 = \frac{0.0508}{2} = 0.0254 \text{ m} \qquad r_3 = 0.0508 \text{ m}$$

The areas are as follows for $L = 0.305$ m.

$$A_1 = 2\pi L r_1 = 2\pi(0.305)(0.0127) = 0.0243 \text{ m}^2$$

$$A_2 = 2\pi L r_2 = 2\pi(0.305)(0.0254) = 0.0487 \text{ m}^2$$

216   *Transport Processes: Momentum, Heat, and Mass*

$$A_3 = 2\pi L r_3 = 2\pi(0.305)(0.0508) = 0.0974 \text{ m}^2$$

From Eq. (4.3-6), the log mean areas for the stainless steel (A) and asbestos (B) are

$$A_{A \text{ lm}} = \frac{A_2 - A_1}{\ln (A_2/A_1)} = \frac{0.0487 - 0.0243}{\ln (0.0487/0.0243)} = 0.0351 \text{ m}^2$$

$$A_{B \text{ lm}} = \frac{A_3 - A_2}{\ln (A_3/A_2)} = \frac{0.0974 - 0.0487}{\ln (0.0974/0.0487)} = 0.0703 \text{ m}^2$$

From Eq. (4.3-7) the resistances are

$$R_A = \frac{r_2 - r_1}{k_A A_{A \text{ lm}}} = \frac{0.0127}{21.63(0.0351)} = 0.01673 \text{ K/W}$$

$$R_B = \frac{r_3 - r_2}{k_B A_{B \text{ lm}}} = \frac{0.0254}{0.2423(0.0703)} = 1.491 \text{ K/W}$$

Hence, the heat-transfer rate is

$$q = \frac{T_1 - T_3}{R_A + R_B} = \frac{811 - 310.8}{0.01673 + 1.491} = 331.7 \text{ W (1132 btu/h)}$$

To calculate the temperature T_2,

$$q = \frac{T_1 - T_2}{R_A} \quad \text{or} \quad 331.7 = \frac{811 - T_2}{0.01673}$$

Solving, $811 - T_2 = 5.5$ K and $T_2 = 805.5$ K. Only a small temperature drop occurs across the metal wall because of its high thermal conductivity.

4.3C Conduction Through Materials in Parallel

Suppose that two plane solids A and B are placed side by side in parallel, and the direction of heat flow is perpendicular to the plane of the exposed surface of each solid. Then the total heat flow is the sum of the heat flow through solid A plus that through B. Writing Fourier's equation for each solid and summing,

$$q_T = q_A + q_B = \frac{k_A A_A}{\Delta x_A}(T_1 - T_2) + \frac{k_B A_B}{\Delta x_B}(T_3 - T_4) \qquad \text{(4.3-9)}$$

where q_T is total heat flow, T_1 and T_2 are the front and rear surface temperatures of solid A; T_3 and T_4, for solid B.

If we assume that $T_1 = T_3$ (front temperatures the same for A and B) and $T_2 = T_4$ (equal rear temperatures),

$$q_T = \frac{T_1 - T_2}{\Delta x_A/k_A A_A} + \frac{T_1 - T_2}{\Delta x_B/k_B A_B} = \left(\frac{1}{R_A} + \frac{1}{R_B}\right)(T_1 - T_2) \qquad \text{(4.3-10)}$$

An example would be an insulated wall (A) of a brick oven where steel reinforcing members (B) are in parallel and penetrate the wall. Even though the area A_B of the steel would be small compared to the insulated brick area A_A, the higher conductivity of the metal (which could be several hundred times larger than that of the brick) could allow a large portion of the heat lost to be conducted by the steel.

Another example is a method of increasing heat conduction to accelerate the freeze drying of meat. Spikes of metal in the frozen meat conduct heat more rapidly into the insides of the meat.

It should be mentioned that in some cases some two-dimensional heat flow can occur if the thermal conductivities of the materials in parallel differ markedly. Then the results using Eq. (4.3-10) would be affected somewhat.

4.3D   Combined Convection and Conduction and Overall Coefficients

In many practical situations the surface temperatures (or boundary conditions at the surface) are not known, but there is a fluid on both sides of the solid surfaces. Consider the plane wall in Fig. 4.3-3a with a hot fluid at temperature T_1 on the inside surface and a cold fluid at T_4 on the outside surface. The outside convective coefficient is h_o W/m$^2 \cdot$ K and h_i on the inside. (Methods to predict the convective h will be given later in Section 4.4 of this chapter.)

The heat-transfer rate using Eqs. (4.1-12) and (4.3-1) is given as

$$q = h_i A(T_1 - T_2) = \frac{k_A A}{\Delta x_A}(T_2 - T_3) = h_o A(T_3 - T_4) \qquad \textbf{(4.3-11)}$$

Expressing $1/h_i A$, $\Delta x_A/k_A A$, and $1/h_o A$ as resistances and combining the equations as before,

$$q = \frac{T_1 - T_4}{1/h_i A + \Delta x_A/k_A A + 1/h_o A} = \frac{T_1 - T_4}{\sum R} \qquad \textbf{(4.3-12)}$$

The overall heat transfer by combined conduction and convection is often expressed in terms of an overall heat-transfer coefficient U defined by

$$q = UA\Delta T_{\text{overall}} \qquad \textbf{(4.3-13)}$$

where $\Delta T_{\text{overall}} = T_1 - T_4$ and U is

$$U = \frac{1}{1/h_i + \Delta x_A/k_A + 1/h_o} \frac{\text{W}}{\text{m}^2 \cdot \text{K}} \left(\frac{\text{btu}}{\text{h} \cdot \text{ft}^2 \cdot {}^\circ\text{F}} \right) \qquad \textbf{(4.3-14)}$$

A more important application is heat transfer from a fluid outside a cylinder, through a metal wall, and to a fluid inside the tube, as often occurs in heat exchangers. In Fig. 4.3-3b, such a case is shown.

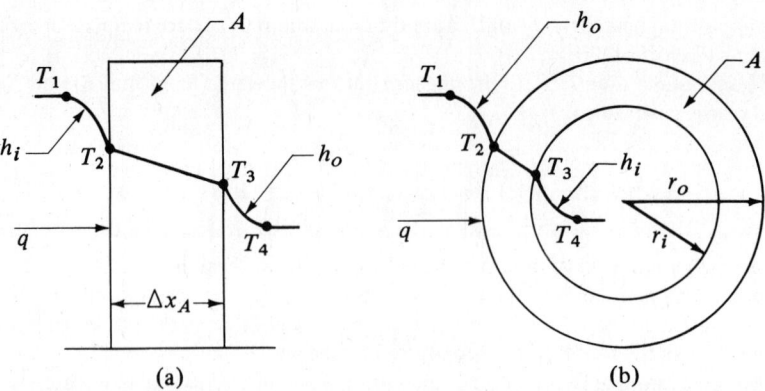

(a)   (b)

FIGURE 4.3-3.   *Heat flow with convective boundaries: (a) plane wall, (b) cylindrical wall.*

Using the same procedure as before, the overall heat-transfer rate through the cylinder is

$$q = \frac{T_1 - T_4}{1/h_i A_i + (r_o - r_i)/k_A A_{A\,\text{lm}} + 1/h_o A_o} = \frac{T_1 - T_4}{\sum R} \qquad (4.3\text{-}15)$$

where A_i represents $2\pi L r_i$, the inside area of the metal tube; $A_{A\,\text{lm}}$ the log mean area of the metal tube; and A_o the outside area.

The overall heat-transfer coefficient U for the cylinder may be based on the inside area A_i or the outside area A_o of the tube. Hence,

$$q = U_i A_i (T_1 - T_4) = U_o A_o (T_1 - T_4) = \frac{T_1 - T_4}{\sum R} \qquad (4.3\text{-}16)$$

$$U_i = \frac{1}{1/h_i + (r_o - r_i)A_i/k_A A_{A\,\text{lm}} + A_i/A_o h_o} \qquad (4.3\text{-}17)$$

$$U_o = \frac{1}{A_o/A_i h_i + (r_o - r_i)A_o/k_A A_{A\,\text{lm}} + 1/h_o} \qquad (4.3\text{-}18)$$

EXAMPLE 4.3-3. *Heat Loss by Convection and Conduction and Overall U*
Saturated steam at 267°F is flowing inside a $\frac{3}{4}$-in. steel pipe having an ID of 0.824 in. and an OD of 1.050 in. The pipe is insulated with 1.5 in. of insulation on the outside. The convective coefficient for the inside steam surface of the pipe is estimated as $h_i = 1000$ btu/h · ft² · °F, and the convective coefficient on the outside of the lagging is estimated as $h_0 = 2$ btu/h · ft² · °F. The mean thermal conductivity of the metal is 45 W/m · K or 26 btu/h · ft · °F and 0.064 W/m · K or 0.037 btu/h · ft · °F for the insulation.
  (a) Calculate the heat loss for 1 ft of pipe using resistances if the surrounding air is at 80°F.
  (b) Repeat using the overall U_i based on the inside area A_i.

Solution: Calling r_i the inside radius of the steel pipe, r_1 the outside radius of the pipe, and r_o the outside radius of the lagging, then

$$r_i = \frac{0.412}{12}\ \text{ft} \qquad r_i = \frac{0.525}{12}\ \text{ft} \qquad r_o = \frac{2.025}{12}\ \text{ft}$$

For 1 ft of pipe, the areas are as follows.

$$A_i = 2\pi L r_i = 2\pi(1)\left(\frac{0.412}{12}\right) = 0.2157\ \text{ft}^2$$

$$A_1 = 2\pi L r_1 = 2\pi(1)\left(\frac{0.525}{12}\right) = 0.2750\ \text{ft}^2$$

$$A_o = 2\pi L r_o = 2\pi(1)\left(\frac{2.025}{12}\right) = 1.060\ \text{ft}^2$$

From Eq. (4.3-6) the log mean areas for the steel (A) pipe and lagging (B) are

$$A_{A\,\text{lm}} = \frac{A_1 - A_i}{\ln\,(A_1/A_i)} = \frac{0.2750 - 0.2157}{\ln\,(0.2750/0.2157)} = 0.245$$

$$A_{B\,\text{lm}} = \frac{A_o - A_1}{\ln\,(A_o/A_1)} = \frac{1.060 - 0.2750}{\ln\,(1.060/0.2750)} = 0.583$$

From Eq. (4.3-15) the various resistances are

$$R_i = \frac{1}{h_i A_i} = \frac{1}{1000(0.2157)} = 0.00464$$

$$R_A = \frac{r_1 - r_i}{k_A A_{A\,\text{lm}}} = \frac{(0.525 - 0.412)/12}{26(0.245)} = 0.00148$$

$$R_B = \frac{r_o - r_1}{k_B A_{B\,\text{lm}}} = \frac{(2.025 - 0.525)/12}{0.037(0.583)} = 5.80$$

$$R_o = \frac{1}{h_o A_o} = \frac{1}{2(1.060)} = 0.472$$

Using an equation similar to Eq. (4.3-15),

$$q = \frac{T_i - T_o}{R_i + R_A + R_B + R_o} = \frac{267 - 80}{0.00464 + 0.00148 + 5.80 + 0.472} \qquad \textbf{(4.3-19)}$$

$$= \frac{267 - 80}{6.278} = 29.8 \text{ btu/h}$$

For part (b), the equation relating U_i to q is Eq. (4.3-16), which can be equated to Eq. (4.3-19).

$$q = U_i A_i (T_i - T_o) = \frac{T_i - T_o}{\sum R} \qquad \textbf{(4.3-20)}$$

Solving for U_i,

$$U_i = \frac{1}{A_i \sum R} \qquad \textbf{(4.3-21)}$$

Substituting known values,

$$U_i = \frac{1}{0.2157(6.278)} = 0.738 \ \frac{\text{btu}}{\text{h} \cdot \text{ft}^2 \cdot {}^\circ\text{F}}$$

Then to calculate q,

$$q = U_i A_i (T_i - T_o) = 0.738(0.2157)(267 - 80) = 29.8 \text{ btu/h} \ (8.73 \text{ W})$$

4.3E Conduction with Internal Heat Generation

In certain systems heat is generated inside the conducting medium; i.e., a uniformly distributed heat source is present. Examples of this are electric resistance heaters and nuclear fuel rods. Also, if a chemical reaction is occurring uniformly in a medium, a heat of reaction is given off. In the agricultural and sanitation fields, compost heaps and trash heaps in which biological activity is occurring will have heat given off.

Other important examples are in food processing, where the heat of respiration of fresh fruits and vegetables is present. These heats of generation can be as high as 0.3 to 0.6 W/kg or 0.5 to 1 btu/h $\cdot$ lb$_{\text{m}}$.

1. Heat generation in plane wall. In Fig. 4.3-4 a plane wall is shown with internal heat generation. Heat is conducted only in the one x direction. The other walls are assumed to be insulated. The temperature T_w in K at $x = L$ and $x = -L$ is held constant. The volumetric rate of heat generation is $\dot{q}$ W/m^3 and the thermal conductivity of the medium is k W/m $\cdot$ K.

To derive the equation for this case of heat generation at steady state, we start with Eq. (4.1-3) but drop the accumulation term.

$$q_{x|x} + \dot{q}(\Delta x \cdot A) = q_{x|x+\Delta x} + 0 \qquad (4.3\text{-}22)$$

where A is the cross-sectional area of the plate. Rearranging, dividing by Δx, and letting Δx approach zero,

$$\frac{-dq_x}{dx} + \dot{q} \cdot A = 0 \qquad (4.3\text{-}23)$$

Substituting Eq. (4.1-2) for q_x,

$$\frac{d^2 T}{dx^2} + \frac{\dot{q}}{k} = 0 \qquad (4.3\text{-}24)$$

Integration gives the following for $\dot{q}$ constant:

$$T = -\frac{\dot{q}}{2k} x^2 + C_1 x + C_2 \qquad (4.3\text{-}25)$$

where C_1 and C_2 are integration constants.

$x = L$ or $-L$, $T = T_w$, and at $x = 0$, $T = T_0$ (center temperature). Then, the temperature profile is

$$T = -\frac{\dot{q}}{2k} x^2 + T_0 \qquad (4.3\text{-}26)$$

The center temperature is

$$T_0 = \frac{\dot{q}L^2}{2k} + T_w \qquad (4.3\text{-}27)$$

The total heat lost from the two faces at steady state is equal to the total heat generated, $\dot{q}_T$, in W.

$$\dot{q}_T = \dot{q}(2LA) \qquad (4.3\text{-}28)$$

where A is the cross-sectional area (surface area at T_w) of the plate.

2. *Heat generation in cylinder.* In a similar manner an equation can be derived for a cylinder of radius R with uniformly distributed heat sources and constant thermal

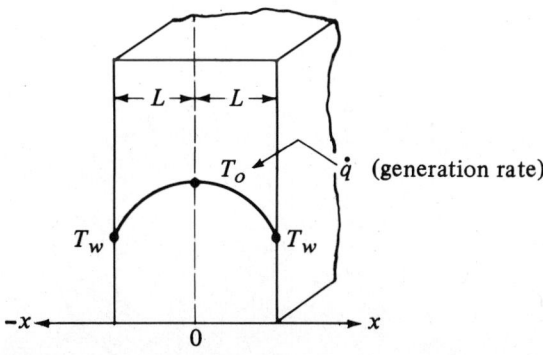

FIGURE 4.3-4. *Plane wall with internal heat generation at steady state.*

conductivity. The heat is assumed to flow only radially; i.e., the ends are neglected or insulated. The final equation for the temperature profile is

$$T = \frac{\dot{q}}{4k}(R^2 - r^2) + T_w \qquad (4.3\text{-}29)$$

where r is distance from the center. The center temperature T_0 is

$$T_0 = \frac{\dot{q}R^2}{4k} + T_w \qquad (4.3\text{-}30)$$

EXAMPLE 4.3-4.   *Heat Generation in a Cylinder*
An electric current of 200 A is passed through a stainless steel wire having a radius R of 0.001268 m. The wire is $L = 0.91$ m long and has a resistance R of 0.126 Ω. The outer surface temperature T_w is held at 422.1 K. The average thermal conductivity is $k = 22.5$ W/m · K. Calculate the center temperature.

Solution: First the value of $\dot{q}$ must be calculated. Since power $= I^2R$, where I is current in amps and R is resistance in ohms,

$$I^2R = \text{watts} = \dot{q}\pi R^2 L \qquad (4.3\text{-}31)$$

Substituting known values and solving,

$$(200)^2(0.126) = \dot{q}\pi(0.001268)^2(0.91)$$

$$\dot{q} = 1.096 \times 10^9 \text{ W/m}^3$$

Substituting into Eq. (4.3-30) and solving, $T_0 = 441.7$ K.

4.3F   Critical Thickness of Insulation for a Cylinder

In Fig. 4.3-5 a layer of insulation is installed around the outside of a cylinder whose radius r_1 is fixed with a length L. The cylinder has a high thermal conductivity and the inner temperature T_1 at point r_1 outside the cylinder is fixed. An example is the case where the cylinder is a metal pipe with saturated steam inside. The outer surface of the insulation at T_2 is exposed to an environment at T_0 where convective heat transfer occurs. It is not obvious if adding more insulation with a thermal conductivity of k will decrease the heat transfer rate.

At steady state the heat-transfer rate q through the cylinder and the insulation equals the rate of convection from the surface.

$$q = h_o A(T_2 - T_0) \qquad (4.3\text{-}32)$$

As insulation is added, the outside area, which is $A = 2\pi r_2 L$, increases but T_2 decreases.

FIGURE 4.3-5.   *Critical radius for insulation of cylinder or pipe.*

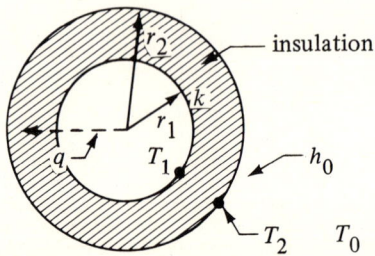

Transport Processes: Momentum, Heat, and Mass

However, it is not apparent whether q increases or decreases. To determine this, an equation similar to Eq. (4.3-15) with the resistance of the insulation represented by Eq. (4.2-11) is written using the two resistances.

$$q = \frac{2\pi L(T_1 - T_0)}{\dfrac{\ln (r_2/r_1)}{k} + \dfrac{1}{r_2 h_o}} \qquad (4.3\text{-}33)$$

To determine the effect of the thickness of insulation on q, we take the derivative of q with respect to r_2, equate this result to zero, and obtain the following for maximum heat flow.

$$\frac{dq}{dr_2} = \frac{-2\pi L(T_1 - T_0)(1/r_2 k - 1/r_2^2 h_o) = 0}{\left[\dfrac{\ln (r_2/r_1)}{k} + \dfrac{1}{r_2 h_o} \right]^2} \qquad (4.3\text{-}34)$$

Solving,

$$(r_2)_{\text{cr}} = \frac{k}{h_o} \qquad (4.3\text{-}35)$$

where $(r_2)_{\text{cr}}$ is the value of the critical radius when the heat-transfer rate is a maximum. Hence, if the outer radius r_2 is less than the critical value, adding more insulation will actually increase the heat-transfer rate q. Also, if the outer radius is greater than the critical, adding more insulation will decrease the heat-transfer rate. Using typical values of k and h_o often encountered, the critical radius is only a few mm. As a result, adding insulation on small electrical wires could increase the heat loss. Adding insulation to large pipes decreases the heat-transfer rate.

EXAMPLE 4.3-5.   Insulating an Electrical Wire and Critical Radius
An electric wire having a diameter of 1.5 mm and covered with a plastic insulation (thickness = 2.5 mm) is exposed to air at 300 K and $h_o = 20 \ W/m^2 \cdot K$. The insulation has a k of 0.4 $W/m \cdot K$. It is assumed that the wire surface temperature is constant at 400 K and is not affected by the covering.
(a) Calculate the value of the critical radius.
(b) Calculate the heat loss per m of wire length with no insulation.
(c) Repeat (b) for the insulation present.

Solution: For part (a) using Eq. (4.3-35),

$$(r_2)_{\text{cr}} = \frac{k}{h_o} = \frac{0.4}{20} = 0.020 \text{ m} = 20 \text{ mm}$$

For part (b), $L = 1.0$ m, $r_2 = 1.5/(2 \times 1000) = 0.75 \times 10^{-3}$ m, $A = 2\pi r_2 L$. Substituting into Eq. (4.3-32),

$$q = h_o A(t_2 - T_0) = (20)(2\pi \times 0.75 \times 10^{-3} \times 1)(400 - 300) = 9.42 \text{ W}$$

For part (c) with insulation, $r_1 = 1.5/(2 \times 1000) = 0.75 \times 10^{-3}$ m, $r_2 = (2.5 + 1.5/2)/1000 = 3.25 \times 10^{-3}$ m. Substituting into Eq. (4.3-33),

$$q = \frac{2\pi(1.0)(400 - 300)}{\dfrac{\ln (3.25 \times 10^{-3}/0.75 \times 10^{-3})}{0.4} + \dfrac{1}{(3.25 \times 10^{-3})(20)}} = 32.98 \text{ W}$$

Hence, adding insulation greatly increases the heat loss.

4.3G Contact Resistance at an Interface

In the equations derived in this section for conduction through solids in series (see Fig. 4.3-1) it has been assumed that the adjacent touching surfaces are at the same temperature; i.e., completely perfect contact is made between the surfaces. For many engineering designs in industry, this assumption is reasonably accurate. However, in cases such as in nuclear power plants where very high heat fluxes are present, a significant drop in temperature may be present at the interface. This interface resistance, called *contact resistance*, occurs when the two solids do not fit tightly together and a thin layer of stagnant fluid is trapped between the two surfaces. At some points the solids touch at peaks in the surfaces and at other points the fluid occupies the open space.

This interface resistance is a complex function of the roughness of the two surfaces, the pressure applied to hold the surfaces in contact, the interface temperature, and the interface fluid. Heat transfer takes place by conduction, radiation, and convection across the trapped fluid and also by conduction through the points of contact of the solids. No completely reliable empirical correlations or theories are available to predict contact resistances for all types of materials. See references (C7, R2) for detailed discussions.

The equation for the contact resistance is often given as follows:

$$q = h_c \, A \, \Delta T = \frac{\Delta T}{1/h_c \, A} = \frac{\Delta T}{R_c} \qquad (4.3\text{-}36)$$

where h_c is the contact resistance coefficient in $W/m^2 \cdot K$, ΔT the temperature drop across the contact resistance in K, and R_c the contact resistance. The contact resistance R_c can be added with the other resistances in Eq. (4.3-3) to include this effect for solids in series. For contact between two ground metal surfaces h_c values of the order of magnitude of about 0.2×10^4 to $1 \times 10^4 \, W/m^2 \cdot K$ have been obtained.

An approximation of the maximum contact resistance can be obtained if the maximum gap Δx between the surfaces can be estimated. Then, assuming that the heat transfer across the gap is by conduction only through the stagnant fluid, h_c is estimated as

$$h_c = \frac{k}{\Delta x} \qquad (4.3\text{-}37)$$

If any actual convection, radiation, or point-to-point contact is present, this will reduce this assumed resistance.

4.4 STEADY-STATE CONDUCTION AND SHAPE FACTORS

4.4A Introduction and Graphical Method for Two-Dimensional Conduction

In previous sections of this chapter we discussed steady-state heat conduction in one direction. In many cases, however, steady-state heat conduction is occurring in two directions; i.e., two-dimensional conduction is occurring. The two-dimensional solutions are more involved and in most cases analytical solutions are not available. One important approximate method to solve such problems is to use a numerical method discussed in detail in Section 4.15 (Selected Topic). Another important approximate method is the graphical method, which is a simple method that can provide reasonably accurate answers for the heat-transfer rate. This method is particularly applicable to systems having isothermal boundaries.

In the graphical method we first note that for one-dimensional heat conduction through a flat slab (see Fig. 4.2-1) the direction of the heat flux or flux lines is always perpendicular to the isotherms. The graphical method for two-dimensional conduction is also based on the requirement that the heat flux lines and the isotherm lines intersect each other at right angles while forming a network of curvilinear squares. This means, as shown in Fig. 4.4-1, that we can sketch the isotherms and also the flux lines until they intersect at right angles (are perpendicular to each other). With care and experience we can obtain reasonably accurate results. General steps to use in this graphical method are as follows.

1. Draw a model to scale of the two-dimensional solid. Lable the isothermal boundaries. In Fig. 4.4-1, T_1 and T_2 are isothermal boundaries.
2. Select a number N that is the number of equal temperature subdivisions between the isothermal boundaries. In Fig. 4.4-1, $N = 4$ subdivisions between T_1 and T_2. Sketch in the isotherm lines and the heat flow or flux lines so that they are perpendicular to each other at the intersections. Note that isotherms are perpendicular to adiabatic (insulated) boundaries and also lines of symmetry.
3. Keep adjusting the isotherm and flux lines until for each curvilinear square the condition $\Delta x = \Delta y$ is satisfied.

In order to calculate the heat flux using the results of the graphical plot, we first assume unit depth of the material. The heat flow q' through the curvilinear section shown in Fig. 4.4-1 is given by Fourier's law.

$$q' = -kA \frac{dT}{dy} = k(\Delta x \cdot 1) \frac{\Delta T}{\Delta y} \qquad (4.4\text{-}1)$$

This heat flow q' will be the same through each curvilinear square within this heat-flow lane. Since $\Delta x = \Delta y$, each temperature subdivision ΔT is equal. This temperature subdivision can be expressed in terms of the overall temperature difference $T_1 - T_2$ and N, which is the number of equal subdivisions.

$$\Delta T = \frac{T_1 - T_2}{N} \qquad (4.4\text{-}2)$$

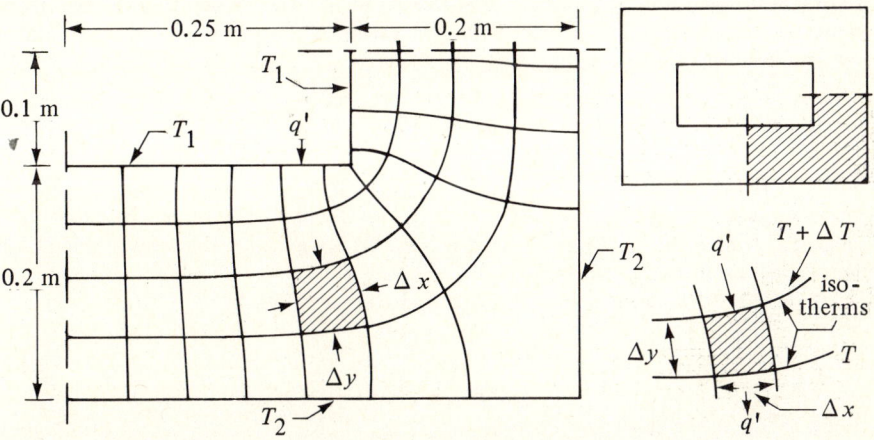

FIGURE 4.4-1. *Graphical curvilinear-square method for two-dimensional heat conduction in a rectangular flue.*

Also, the heat flow q' through each lane is the same since $\Delta x = \Delta y$ in the construction and in Eq. (4.4-1). Hence, the total heat transfer q through all of the lanes is

$$q = Mq' = Mk\,\Delta T \qquad (4.4\text{-}3)$$

where M is the total number of heat-flow lanes as determined by the graphical procedure. Substituting Eq. (4.4-2) into (4.4-3),

$$q = \frac{M}{N}\,k(T_1 - T_2) \qquad (4.4\text{-}4)$$

EXAMPLE 4.4-1.   Two-Dimensional Conduction by Graphical Procedure
Determine the total heat transfer through the walls of the flue shown in Fig. 4.4-1 if $T_1 = 600$ K, $T_2 = 400$ K, $k = 0.90$ W/m · K, and L (length of flue) = 5 m.

Solution: In Fig. 4.4-1, $N = 4$ temperature subdivisions and $M = 9.25$. The total heat-transfer rate through the four identical sections with a depth or length L of 5 m is obtained by using Eq. (4.4-4).

$$q = 4\left[\frac{M}{N}\,kL(T_1 - T_2)\right] = 4\left[\frac{9.25}{4}\,(0.9)(5.0)(600\text{–}400)\right]$$
$$= 8325\text{ W}$$

4.4B   Shape Factors in Conduction

In Eq. (4.4-4) the factor M/N is called the conduction shape factor S, where

$$S = \frac{M}{N} \qquad (4.4\text{-}5)$$

$$q = kS(T_1 - T_2) \qquad (4.4\text{-}6)$$

This shape factor S has units of m and is used in two-dimensional heat conduction where only two temperatures are involved. The shape factors for a number of geometries have been obtained and some are given in Table 4.4-1.

For a three-dimensional geometry such as a furnace, separate shape factors are used to obtain the heat flow through the edge and corner sections. When each of the interior dimensions is greater than one-fifth of the wall thickness, the shape factors are as follows for a uniform wall thickness T_w:

$$S_{\text{wall}} = \frac{A}{T_w} \qquad S_{\text{edge}} = 0.54L \qquad S_{\text{corner}} = 0.15T_w \qquad (4.4\text{-}7)$$

where A is the inside area of wall and L the length of inside edge. For a completely enclosed geometry, there are 6 wall sections, 12 edges, and 8 corners. Note that for a single flat wall, $q = kS_{\text{wall}}(T_1 - T_2) = k(A/T_w)(T_1 - T_2)$, which is the same as Eq. (4.2-1) for conduction through a single flat slab.

For a long hollow cylinder of length L such as that in Fig. 4.2-2,

$$S = \frac{2\pi L}{\ln\,(r_2/r_1)} \qquad (4.4\text{-}8)$$

For a hollow sphere from Eq. (4.2-14),

$$S = \frac{4\pi r_2 r_1}{r_2 - r_1} \qquad (4.4\text{-}9)$$

TABLE 4.4-1. *Conduction Shape Factors for* $q = kS(T_1 - T_2)$*

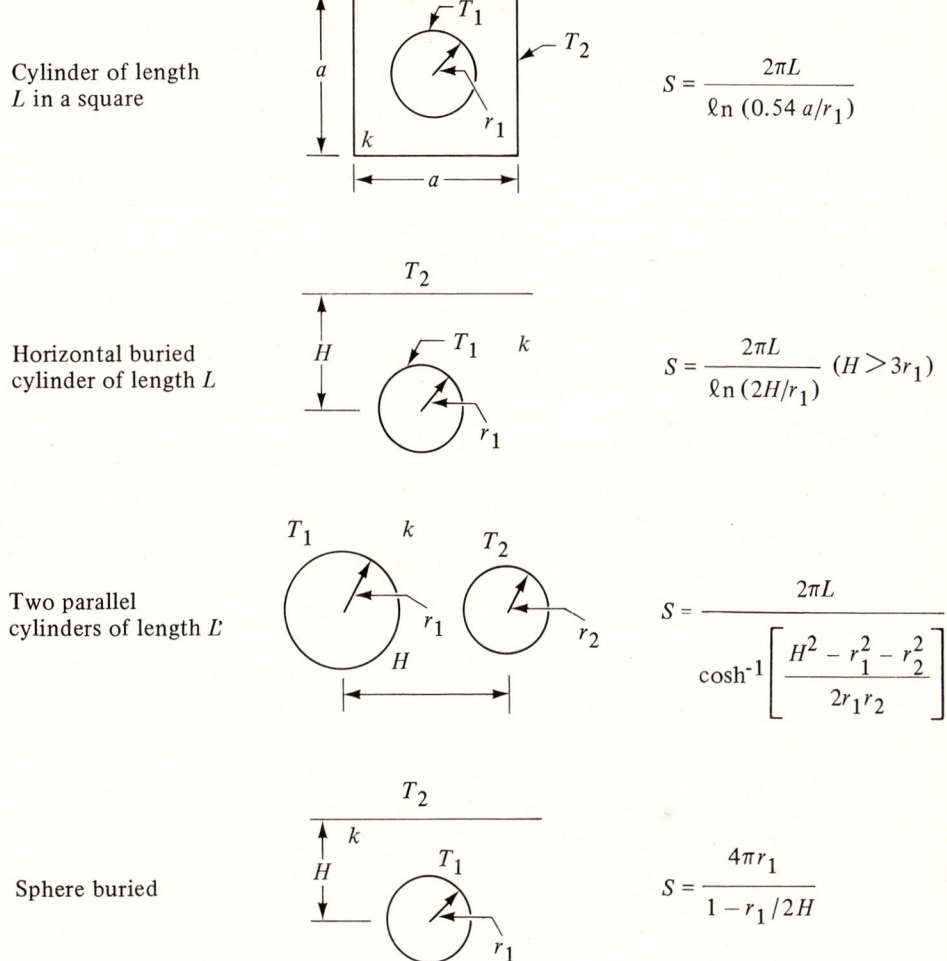

Cylinder of length L in a square

$$S = \frac{2\pi L}{\ln (0.54\, a/r_1)}$$

Horizontal buried cylinder of length L

$$S = \frac{2\pi L}{\ln (2H/r_1)} \quad (H > 3r_1)$$

Two parallel cylinders of length L

$$S = \frac{2\pi L}{\cosh^{-1} \left[\dfrac{H^2 - r_1^2 - r_2^2}{2r_1 r_2} \right]}$$

Sphere buried

$$S = \frac{4\pi r_1}{1 - r_1/2H}$$

* The thermal conductivity of the medium is k.

4.5 FORCED CONVECTION HEAT TRANSFER INSIDE PIPES

4.5A Introduction and Dimensionless Numbers

In most situations involving a liquid or a gas in heat transfer, convective heat transfer usually occurs as well as conduction. In most industrial processes where heat transfer is occurring, heat is being transformed from one fluid through a solid wall to a second fluid. In Fig. 4.5-1 heat is being transferred from the hot flowing fluid to the cold flowing fluid. The temperature profile is shown.

The velocity gradient, when the fluid is in turbulent flow, is very steep next to the wall in the thin viscous sublayer where turbulence is absent. Here the heat transfer is mainly by conduction with a large temperature difference of $T_2 - T_3$ in the warm fluid.

Principles of Steady-State Heat Transfer

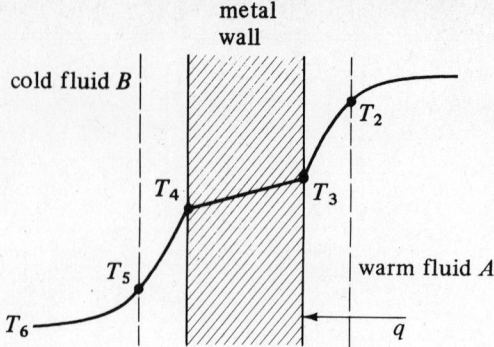

FIGURE 4.5-1. *Temperature profile for heat transfer by convection from one fluid to another.*

As we move farther away from the wall, we approach the turbulent region, where rapidly moving eddies tend to equalize the temperature. Hence, the temperature gradient is less and the difference $T_1 - T_2$ is small. The average temperature of fluid A is slightly less than the peak value T_1. A similar explanation can be given for the temperature profile in the cold fluid.

The convective coefficient for heat transfer through a fluid is given by

$$q = hA(T - T_w) \tag{4.5-1}$$

where h is the convective coefficient in $W/m^2 \cdot K$, A is the area in m^2, T is the bulk or average temperature of the fluid in K, T_w is the temperature of the wall in contact with the fluid in K, and q is the heat-transfer rate in W. In English units, q is in btu/h, h in btu/h $\cdot$ ft$^2 \cdot$ °F, A in ft^2, and T and T_w in °F.

The type of fluid flow, whether laminar or turbulent, of the individual fluid has a great effect on the heat-transfer coefficient h, which is often called a film coefficient, since most of the resistance to heat transfer is in a thin film close to the wall. The more turbulent the flow, the greater the heat-transfer coefficient.

There are two main classifications of convective heat transfer. The first is free or *natural convection*, where the motion of the fluid results from the density changes in heat transfer. The buoyant effect produces a natural circulation of the fluid, so it moves past the solid surface. In the second type, *forced convection*, the fluid is forced to flow by pressure differences, a pump, a fan, and so on.

Most of the correlations for predicting film coefficients h are semiempirical in nature and are affected by the physical properties of the fluid, the type and velocity of flow, the temperature difference, and by the geometry of the specific physical system. Some approximate values of convective coefficients were presented in Table 4.1-2. In the following correlations, SI or English units can be used since the equations are dimensionless.

To correlate these data for heat-transfer coefficients, dimensionless numbers such as the Reynolds and Prandtl numbers are used. The Prandtl number is the ratio of the shear component of diffusivity for momentum μ/ρ to the diffusivity for heat $k/\rho c_p$ and physically relates the relative thickness of the hydrodynamic layer and thermal boundary layer.

$$N_{Pr} = \frac{\mu/\rho}{k/\rho c_p} = \frac{c_p \mu}{k} \tag{4.5-2}$$

Values of the N_{Pr} for gases are given in Appendix A.3 and range from about 0.5 to 1.0. Values for liquids range from about 2 to well over 10^4. The dimensionless Nusselt number, N_{Nu}, is used to relate data for the heat-transfer coefficient h to the thermal conductivity k of the fluid and a characteristic dimension D.

$$N_{Nu} = \frac{hD}{k} \tag{4.5-3}$$

For example, for flow inside a pipe, D is the diameter.

4.5B Heat-Transfer Coefficient for Laminar Flow Inside a Pipe

Certainly, the most important convective heat-transfer process industrially is that of cooling or heating a fluid flowing inside a closed circular conduit or pipe. Different types of correlations for the convective coefficient are needed for laminar flow (N_{Re} below 2100), for fully turbulent flow (N_{Re} above 10^4), and for the transition region (N_{Re} between 2100 and 10^4).

For laminar flow of fluids inside horizontal tubes or pipes, the following equation of Sieder and Tate (S1) can be used for $N_{Re} < 2100$:

$$(N_{Nu})_a = \frac{h_a D}{k} = 1.86 \left(N_{Re} N_{Pr} \frac{D}{L} \right)^{1/3} \left(\frac{\mu_b}{\mu_w} \right)^{0.14} \tag{4.5-4}$$

where D = pipe diameter in m, L = pipe length before mixing occurs in the pipe in m, μ_b = fluid viscosity at bulk average temperature in Pa·s, μ_w = viscosity at the wall temperature, c_p = heat capacity in J/kg·K, k = thermal conductivity in W/m·K, h_a = average heat-transfer coefficient in W/m²·K, and N_{Nu} = dimensionless Nusselt number. All the physical properties are evaluated at the bulk fluid temperature except μ_w. The Reynolds number is

$$N_{Re} = \frac{D v \rho}{\mu} \tag{4.5-5}$$

and the Prandtl number,

$$N_{Pr} = \frac{c_p \mu}{k} \tag{4.5-6}$$

This equation holds for $(N_{Re} N_{Pr} D/L) > 100$. If used down to a $(N_{Re} N_{Pr} D/L) > 10$, it still holds to $\pm 20\%$ (B1). For $(N_{Re} N_{Pr} D/L) < 100$, another expression is available (P1).

In laminar flow the average coefficient h_a depends strongly on heated length. The average (arithmetic mean) temperature drop ΔT_a is used in the equation to calculate the heat-transfer rate q.

$$q = h_a A \, \Delta T_a = h_a A \frac{(T_w - T_{bi}) + (T_w - T_{bo})}{2} \tag{4.5-7}$$

where T_w is the wall temperature in K, T_{bi} the inlet bulk fluid temperature, and T_{bo} the outlet bulk fluid temperature.

For large pipe diameters and large temperature differences ΔT between pipe wall and bulk fluid, natural convection effects can increase h(P1). Equations are also available for laminar flow in vertical tubes.

4.5C Heat-Transfer Coefficient for Turbulent Flow Inside a Pipe

When the Reynolds number is above 2100, the flow is turbulent. Since the rate of heat transfer is greater in the turbulent region, many industrial heat-transfer processes are in the turbulent region.

The following equation has been found to hold for a $N_{Re} > 10^4$, a N_{Pr} between 0.7 and 700, and $L/D > 60$.

$$N_{Nu} = \frac{h_L D}{k} = 0.023 N_{Re}^{0.8} N_{Pr}^{1/3} \left(\frac{\mu_b}{\mu_w}\right)^{0.14} \tag{4.5-8}$$

where h_L is the heat-transfer coefficient based on the log mean driving force ΔT_{lm} (see Section 4.5H). The fluid properties except for μ_w are evaluated at the mean bulk temperature. If the bulk fluid temperature varies from the inlet to the outlet of the pipe, the mean of the inlet and outlet temperatures is used. To correct for an $L/D < 60$, where the entry is sharp-edged, an approximate correction is to multiply the right-hand side of Eq. (4.5-8) by $[1 + (D/L)^{0.7}]$.

The use of Eq. (4.5-8) may be trial and error, since the value of h_L must be known to evaluate T_w, and hence μ_w, at the wall temperature. Also, if the mean bulk temperature increases or decreases in the tube length L because of heat transfer, the bulk temperature at length L must be estimated in order to have a mean bulk temperature of the entrance and exit to use.

The heat-transfer coefficient for turbulent flow is somewhat greater for a rough tube than for a smooth one. This effect is much less than in fluid friction, and it is usually neglected in calculations. Also, for liquid metals that have Prandtl numbers $\ll 1$, other correlations must be used to predict the heat-transfer coefficient. (See Section 4.5G.) For shapes of tubes other than circular, the equivalent diameter can be used as discussed in Section 4.5E.

For air at 1 atm total pressure, the following simplified equation holds for turbulent flow in pipes.

$$h_L = \frac{3.52 v^{0.8}}{D^{0.2}} \quad \text{(SI)}$$

$$h_L = \frac{0.5 v_s^{0.8}}{(D')^{0.2}} \quad \text{(English)} \tag{4.5-9}$$

where D is in m, v in m/s, and h_L in W/m² · K for SI units; and D' is in in., v_s in ft/s, and h_L in btu/h · ft² · °F in English units.

Water is often used in heat-transfer equipment. A simplified equation to use for a temperature range of $T = 4$ to $105°C$ $(40 - 220°F)$ is

$$h_L = 1429(1 + 0.0146T°C) \frac{v^{0.8}}{D^{0.2}} \quad \text{(SI)}$$

$$h_L = 150(1 + 0.011T°F) \frac{v_s^{0.8}}{(D')^{0.2}} \quad \text{(English)} \tag{4.5-10}$$

A very simplified equation for organic liquids to use for approximations is as follows (P1):

$$h_L = 423 \frac{v^{0.8}}{D^{0.2}} \quad \text{(SI)}$$

$$h_L = 60 \frac{v_s^{0.8}}{(D')^{0.2}} \quad \text{(English)} \tag{4.5-11}$$

For flow inside helical coils and N_{Re} above 10^4, the predicted film coefficient for straight pipes should be increased by the factor $(1 + 3.5D/D_{coil})$.

EXAMPLE 4.5-1. Heating of Air in Turbulent Flow
Air at 206.8 kPa and an average of 477.6 K is being heated as it flows through a tube of 25.4 mm inside diameter at a velocity of 7.62 m/s. The heating medium is 488.7 K steam condensing on the outside of the tube. Since the heat-transfer coefficient of condensing steam is several thousand $W/m^2 \cdot K$ and the resistance of the metal wall is very small, it will be assumed that the surface wall temperature of the metal in contact with the air is 488.7 K. Calculate the heat-transfer coefficient for an $L/D > 60$ and also the heat-transfer flux q/A.

Solution: From Appendix A.3 for physical properties of air at 477.6 K (204.4°C), $\mu_b = 2.60 \times 10^{-5}$ Pa·s, $k = 0.03894$ W/m, $N_{Pr} = 0.686$. At 488.7 K (215.5°C), $\mu_w = 2.64 \times 10^{-5}$ Pa·s.

$$\mu_b = 2.60 \times 10^{-5} \text{ Pa·s} = 2.60 \times 10^{-5} \text{ kg/m·s}$$

$$\rho = (28.97)\left(\frac{1}{22.414}\right)\left(\frac{206.8}{101.33}\right)\left(\frac{273.2}{477.6}\right) = 1.509 \text{ kg/m}^3$$

The Reynolds number calculated at the bulk fluid temperature of 477.6 K is

$$N_{Re} = \frac{Dv\rho}{\mu} = \frac{0.0254(7.62)(1.509)}{2.6 \times 10^{-5}} = 1.122 \times 10^4$$

Hence, the flow is turbulent and Eq. (4.5-8) will be used. Substituting into Eq. (4.5-8),

$$N_{Nu} = \frac{h_L D}{k} = 0.023 N_{Re}^{0.8} N_{Pr}^{1/3} \left(\frac{\mu_B}{\mu_W}\right)^{0.14}$$

$$\frac{h_L(0.0254)}{0.03894} = 0.023(1.122 \times 10^4)^{0.8}(0.686)^{1/3}\left(\frac{0.0260}{0.0264}\right)^{0.14}$$

Solving, $h_L = 53.8$ W/m²·K (9.48 btu/h·ft²·°F).
To solve for the flux q/A,

$$\frac{q}{A} = h_L(T_w - T) = 53.8(488.7 - 477.6)$$

$$= 597.2 \text{ W/m}^2 \text{ (189.3 btu/h·ft}^2\text{)}$$

4.5D Heat-Transfer Coefficient for Transition Flow Inside a Pipe

In the transition region for a N_{Re} between 2100 and 10^4, the empirical equations are not well defined just as in the case of fluid friction factors. No simple equation exists for accomplishing a smooth transition from heat transfer in laminar flow to turbulent flow, i.e., a transition from Eq. (4.5-4) at a $N_{Re} = 2100$ to Eq. (4.5-8) at a $N_{Re} = 10^4$.

The plot in Fig. 4.5-2 represents an approximate relationship to use between the various heat-transfer parameters and the Reynolds number between 2100 and 10^4. For below a N_{Re} of 2100, the curves represent Eq. (4.5-4) and above 10^4, Eq. (4.5-8). The mean ΔT_a of Eq. (4.5-7) should be used with the h_a in Fig. 4.5-2.

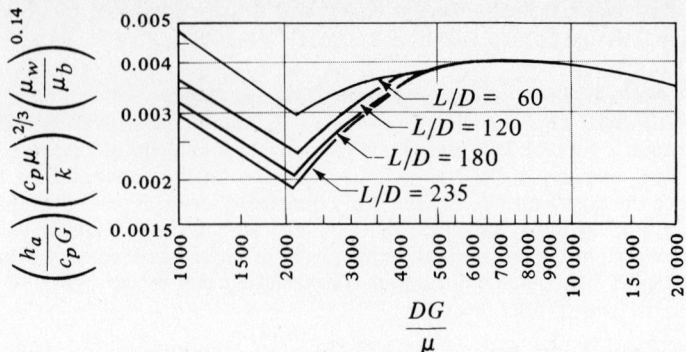

FIGURE 4.5-2. *Correlation of heat-transfer parameters for transition region for Reynolds numbers between 2100 and 10^4. (From R. H. Perry and C. H. Chilton, Chemical Engineers' Handbook, 5th ed. New York: McGraw-Hill Book Company, 1973. With permission.)*

4.5E Heat-Transfer Coefficient for Noncircular Conduits

A heat-transfer system often used is that in which fluids flow at different temperatures in concentric pipes. The heat-transfer coefficient of the fluid in the annular space can be predicted using the same equations as for circular pipes. However, the equivalent diameter defined in Section 2.10G must be used. For an annular space, D_{eq} is the ID of the outer pipe D_1 minus the OD of the inner pipe D_2. For other geometries, an equivalent diameter can also be used.

> *EXAMPLE 4.5-2. Water Heated by Steam and Trial-and-Error Solution*
> Water is flowing in a horizontal 1-in. schedule 40 steel pipe at an average temperature of 65.6°C and a velocity of 2.44 m/s. It is being heated by condensing steam at 107.8°C on the outside of the pipe wall. The steam side coefficient has been estimated as $h_o = 10\,500$ W/m$^2 \cdot$ K.
>   (a) Calculate the convective coefficient h_i for water inside the pipe.
>   (b) Calculate the overall coefficient U_i based on the inside surface area.
>   (c) Calculate the heat-transfer rate q for 0.305 m of pipe with the water at an average temperature of 65.6°C.
>
> *Solution:* From Appendix A.5 the various dimensions are $D_i = 0.0266$ m and $D_o = 0.0334$ m. For water at a bulk average temperature of 65.6°C from Appendix A.2, $N_{Pr} = 2.72$, $\rho = 0.980(1000) = 980$ kg/m^3, $k = 0.633$ W/m$\cdot$K, and $\mu = 4.32 \times 10^{-4}$ Pa$\cdot$s $= 4.32 \times 10^{-4}$ kg/m$\cdot$s.
>   The temperature of the inside metal wall is needed and will be assumed as about one third the way between 65.6 and 107.8 or 80°C $= T_w$ for the first trial. Hence, μ_w at 80°C $= 3.56 \times 10^{-4}$ Pa$\cdot$s.
>   First, the Reynolds number of the water is calculated at the bulk average temperature.
>
> $$N_{Re} = \frac{D_i v \rho}{\mu} = \frac{0.0266(2.44)(980)}{4.32 \times 10^{-4}} = 1.473 \times 10^5$$

Hence, the flow is turbulent. Using Eq. (4.5-8) and substituting known values,

$$\frac{h_L D}{k} = 0.023 N_{Re}^{0.8} N_{Pr}^{1/3} \left(\frac{\mu_b}{\mu_w} \right)^{0.14}$$

$$\frac{h_L (0.0266)}{0.663} = 0.023(1.473 \times 10^5)^{0.8}(2.72)^{1/3} \left(\frac{4.32 \times 10^{-4}}{3.56 \times 10^{-4}}\right)^{0.14}$$

Solving, $h_L = h_i = 11\,350$ W/m$^2 \cdot$ K.

For part (b), the various areas are as follows for 0.305-m pipe.

$$A_i = \pi D_i L = \pi(0.0266)(0.305) = 0.0255 \text{ m}^2$$

$$A_{1m} = \pi \frac{(0.0266) + 0.0334)0.305}{2} = 0.0287 \text{ m}^2$$

$$A_o = \pi(0.0334)(0.305) = 0.0320 \text{ m}^2$$

The k for steel is 45.0 W/m $\cdot$ K. The resistances are

$$R_i = \frac{1}{h_i A_i} = \frac{1}{(11\,350)0.0255} = 0.003455$$

$$R_m = \frac{r_o - r_i}{k A_{1m}} = \frac{0.0334 - 0.0266}{2} \frac{1}{45.0(0.0287)} = 0.002633$$

$$R_o = \frac{1}{h_o A_o} = \frac{1}{(10\,500)(0.0320)} = 0.002976$$

$$\sum R = 0.003455 + 0.002633 + 0.002976 = 0.009064$$

The overall temperature difference is $(107.8 - 65.6)°C = 42.2°C = 42.2$ K. The temperature drop across the water film is

$$\text{temperature drop} = \frac{R_i}{\sum R} (42.2) = \left(\frac{0.003455}{0.009064}\right)(42.2) = 16.1 \text{ K} = 16.1°C$$

Hence, $T_w = 65.6 + 16.1 = 81.7°C$. This is quite close to the original estimate of 80°C. The only physical property changing in the second estimate would be μ_w, which changes from 3.56×10^{-4} at 80°C to 3.52×10^{-4} at 81.7°C. This will have a negligible effect on h_i and a second trial is not necessary.

For part (b), the overall coefficient is, by Eq. (4.3-16),

$$q = U_i A_i (T_o - T_i) = \frac{T_o - T_i}{\sum R}$$

$$U_i = \frac{1}{A_i \sum R} = \frac{1}{0.0255(0.009064)} = 4327 \text{ W/m}^2 \cdot \text{K}$$

For part (c), with the water at an average temperature of 65.6°C,

$$T_o - T_i = 107.8 - 65.6 = 42.2°C = 42.2 \text{ K}$$

$$q = U_i A_i (T_o - T_i) = 4327(0.0255)(42.2) = 4656 \text{ W}$$

4.5F Entrance-Region Effect on Heat-Transfer Coefficient

Near the entrance of a pipe where the fluid is being heated, the temperature profile is not fully developed and the local coefficient h is greater than the fully developed heat-transfer coefficient h_L for turbulent flow. At the entrance itself where no temperature gradient has been established, the value of h is infinite. The value of h drops rapidly and is approximately the same as h_L at $L/D \cong 60$, where L is the entrance length. These relations for turbulent flow inside a pipe are as follows.

$$\frac{h}{h_L} = 1 + \left(\frac{D}{L}\right)^{0.7} \qquad 2 < \frac{L}{D} < 20 \qquad\qquad \text{(4.5-12)}$$

$$\frac{h}{h_L} = 1 + 6 \left(\frac{D}{L}\right) \qquad 20 < \frac{L}{D} < 60 \qquad\qquad \text{(4.5-13)}$$

4.5G   Liquid-Metals Heat-Transfer Coefficient

Liquid metals are sometimes used as a heat-transfer fluid in cases where a fluid is needed over a wide temperature range at relatively low pressures. Liquid metals are often used in nuclear reactors and have high heat-transfer coefficients as well as a high heat capacity per unit volume. The high heat-transfer coefficients are due to the very high thermal conductivities and, hence, low Prandtl numbers. In liquid metals in pipes, the heat transfer by conduction is very important in the entire turbulent core because of the high thermal conductivity and is often more important than the convection effects.

For fully developed turbulent flow in tubes with uniform heat flux the following equation can be used (L1):

$$N_{Nu} = \frac{h_L D}{k} = 0.625 N_{Pe}^{0.4} \qquad\qquad \text{(4.5-14)}$$

where the Peclet number $N_{Pe} = N_{Re} N_{Pr}$. This holds for $L/D > 60$ and N_{Pe} between 100 and 10^4. For constant wall temperatures,

$$N_{Nu} = \frac{h_L D}{k} = 5.0 + 0.025 N_{Pe}^{0.8} \qquad\qquad \text{(4.5-15)}$$

for $L/D > 60$ and $N_{Pe} > 100$. All physical properties are evaluated at the average bulk temperature.

EXAMPLE 4.5-3.   Liquid-Metal Heat Transfer Inside a Tube
A liquid metal flows at a rate of 4.00 kg/s through a tube having an inside diameter of 0.05 m. The liquid enters at 500 K and is heated to 505 K in the tube. The tube wall is maintained at a temperature of 30 K above the fluid bulk temperature and constant heat flux is maintained. Calculate the required tube length. The average physical properties are as follows: $\mu = 7.1 \times 10^{-4}$ Pa·s, $\rho = 7400$ kg/m^3, $c_p = 120$ J/kg·K, $k = 13$ W/m·K.

Solution: The area is $A = \pi D^2/4 = \pi(0.05)^2/4 = 1.963 \times 10^{-3}$ m^2. Then $G = 4.0/1.963 \times 10^{-3} = 2.038 \times 10^3$ kg/m^2·s. The Reynolds number is

$$N_{Re} = \frac{DG}{\mu} = \frac{0.05(2.038 \times 10^3)}{7.1 \times 10^{-4}} = 1.435 \times 10^5$$

$$N_{Pr} = \frac{c_p \mu}{k} = \frac{120(7.1 \times 10^{-4})}{13} = 0.00655$$

Using Eq. (4.5-14),

$$h_L = \frac{k}{D}(0.625)N_{Pe}^{0.4} = \frac{13}{0.05}(0.625)(1.435 \times 10^5 \times 0.00655)^{0.4}$$

$$= 2512 \text{ W/m}^2 \cdot \text{K}$$

Using a heat balance,

$$q = mc_p \, \Delta T$$

$$= 4.00(120)(505 - 500) = 2400 \text{ W} \qquad\qquad \text{(4.5-16)}$$

Substituting into Eq. (4.5-1),

$$\frac{q}{A} = \frac{2400}{A} = h_L(T_w - T) = 2512(30) = 75\,360 \text{ W/m}^2$$

Hence, $A = 2400/75\,360 = 3.185 \times 10^{-2}$ m^2. Then,

$$A = 3.185 \times 10^{-2} = \pi D L = \pi(0.05)(L)$$

Solving, $L = 0.203$ m.

4.5H Log Mean Temperature Difference and Varying Temperature Drop

Equations (4.5-1) and (4.3-12) as written apply only when the temperature drop $(T_i - T_o)$ is constant for all parts of the heating surface. Hence, the equation

$$q = U_i A_i (T_i - T_o) = U_o A_o (T_i - T_o) = UA(\Delta T) \qquad (4.5\text{-}17)$$

only holds at one point in the apparatus when the fluids are being heated or cooled. However, as the fluids travel through the heat exchanger, they become heated or cooled and both T_i and T_o or either T_i and T_o vary. Then $(T_i - T_o)$ or ΔT varies with position, and some mean ΔT_m must be used over the whole apparatus.

In a typical heat exchanger a hot fluid inside a pipe is cooled from T_1' to T_2' by a cold fluid which is flowing on the outside in a double pipe countercurrently (in the reverse direction) and is heated from T_2 to T_1 as shown in Fig. 4.5-3a. The ΔT shown is varying with distance. Hence, ΔT in Eq. (4.5-17) varies as the area A goes from 0 at the inlet to A at the outlet of the exchanger.

For countercurrent flow of the two fluids as in Fig. 4.5-3a, the heat-transfer rate is

$$q = UA\,\Delta T_m \qquad (4.5\text{-}18)$$

where ΔT_m is a suitable mean temperature difference to be determined. For a dA area, a heat balance on the hot and the cold fluids gives

$$dq = -m'c_p'\,dT' = mc_p\,dT \qquad (4.5\text{-}19)$$

where m is flow rate in kg/s. The values of m, m', c_p, c_p', and U are assumed constant. Also,

$$dq = U(T' - T)\,dA \qquad (4.5\text{-}20)$$

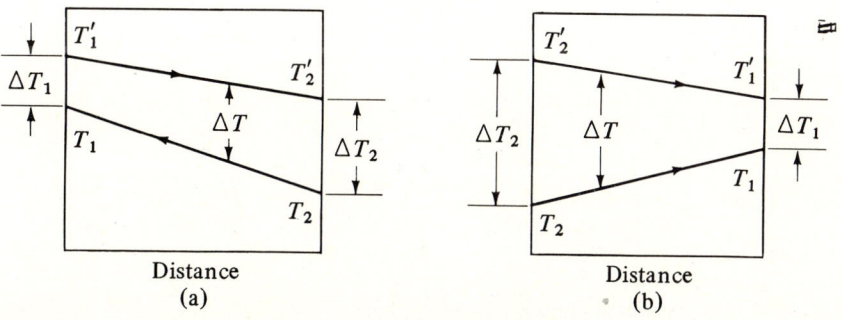

FIGURE 4.5-3. *Temperature profiles for one-pass double-pipe heat exchangers: (a) countercurrent flow, (b) cocurrent or parallel flow.*

From Eq. (4.5-19), $dT' = -dq/m'c'_p$ and $dT = dq/mc_p$. Then,

$$dT' - dT = d(T' - T) = -dq \left(\frac{1}{m'c'_p} + \frac{1}{mc_p} \right) \tag{4.5-21}$$

Substituting Eq. (4.5-20) into (4.5-21),

$$\frac{d(T' - T)}{T' - T} = -U \left(\frac{1}{m'c'_p} + \frac{1}{mc_p} \right) dA \tag{4.5-22}$$

Integrating between points 1 and 2,

$$\ln \left(\frac{T'_2 - T_2}{T'_1 - T_1} \right) = -UA \left(\frac{1}{m'c'_p} + \frac{1}{mc_p} \right) \tag{4.5-23}$$

Making a heat balance between the inlet and outlet,

$$q = m'c'_p(T'_1 - T'_2) = mc_p(T_2 - T_1) \tag{4.5-24}$$

Solving for $m'c'_p$ and mc_p in Eq. (4.5-24) and substituting into Eq. (4.5-23),

$$q = \frac{UA[(T'_2 - T_2) - (T'_1 - T_1)]}{\ln [(T'_2 - T_2)/(T'_1 - T_1)]} \tag{4.5-25}$$

Comparing Eqs. (4.5-18) and (4.5-25), we see that ΔT_m is the log mean temperature difference ΔT_{lm}. Hence, in the case where the overall heat-transfer coefficient U is constant throughout the equipment and the heat capacity of each fluid is constant, the proper temperature driving force to use over the entire apparatus is the log mean driving force,

$$q = UA\Delta T_{lm} \tag{4.5-26}$$

where

$$\Delta T_{lm} = \frac{\Delta T_2 - \Delta T_1}{\ln (\Delta T_2/\Delta T_1)} \tag{4.5-27}$$

It can be also shown that for parallel flow as pictured in Fig. 4.5-3b, the log mean temperature difference should be used. In some cases where steam is condensing, T'_1 and T'_2 may be the same. The equations still hold for this case. When U varies with distance or other complicating factors occur, other references should be consulted (B2, P1, W1).

EXAMPLE 4.5-4. Heat-Transfer Area and Log Mean Temperature Difference

A heavy hydrocarbon oil which has a $c_{pm} = 2.30$ kJ/kg · K is being cooled in a heat exchanger from 371.9 K to 349.7 K and flows inside the tube at a rate of 3630 kg/h. A flow of 1450 kg water/h enters at 288.6 K for cooling and flows outside the tube.

(a) Calculate the water outlet temperature and heat-transfer area if the overall $U_i = 340$ W/m² · K and the streams are countercurrent.
(b) Repeat for parallel flow.

Solution: Assume a $c_{pm} = 4.187$ kJ/kg · K for water. The water inlet $T_2 = 288.6$ K, outlet $= T_1$; oil inlet $T'_1 = 371.9$, outlet $T'_2 = 349.7$ K. Calculating the heat lost by the oil,

$$q = \left(3630 \, \frac{\text{kg}}{\text{h}} \right) \left(2.30 \, \frac{\text{kJ}}{\text{kg} \cdot \text{K}} \right)(371.9 - 349.7)\text{K}$$

$$= 185\,400 \text{ kJ/h} \quad \text{or} \quad 51\,490 \text{ W (175\,700 btu/h)}$$

By a heat balance, the q must also equal the heat gained by the water.

$$q = 185\,400 \text{ kJ/h} = \left(1450\,\frac{\text{kg}}{\text{h}}\right)\left(4.187\,\frac{\text{kJ}}{\text{kg}\cdot\text{K}}\right)(T_1 - 288.6) \text{ K}$$

Solving, $T_1 = 319.1$ K.

To solve for the log mean temperature difference, $\Delta T_2 = T_2' - T_2 = 349.7 - 288.6 = 61.1$ K, $\Delta T_1 = T_1' - T_1 = 371.9 - 319.1 = 52.8$ K. Substituting into Eq. (4.5-27),

$$\Delta T_{\text{lm}} = \frac{\Delta T_2 - \Delta T_1}{\ln(\Delta T_2/\Delta T_1)} = \frac{61.1 - 52.8}{\ln(61.1/52.8)} = 56.9 \text{ K}$$

Using Eq. (4.5-26),

$$q = U_i A_i \Delta T_{\text{lm}}$$

$$51\,490 = 340(A_i)(56.9)$$

Solving, $A_i = 2.66$ m².

For part (b), the water outlet is still $T_1 = 319.1$ K. Referring to Fig. 4.5-3b, $\Delta T_2 = 371.9 - 288.6 = 83.3$ K and $\Delta T_1 = 349.7 - 319.1 = 30.6$ K. Again, using Eq. (4.5-27) and solving, $\Delta T_{\text{lm}} = 52.7$ K. Substituting into Eq. (4.5-26), $A_i = 2.87$ m². This is a larger area than for counterflow. This occurs because counterflow gives larger temperature driving forces and is usually preferred over parallel flow for this reason.

EXAMPLE 4.5-5. Laminar Heat Transfer and Trial and Error

A hydrocarbon oil at 150°F enters inside a pipe with an inside diameter of 0.0303 ft and a length of 15 ft with a flow rate of 80 lb$_m$/h. The inside pipe surface is assumed constant at 350°F since steam is condensing outside the pipe wall and has a very large heat-transfer coefficient. The properties of the oil are $c_{pm} = 0.50$ btu/lb$_m \cdot$ °F and $k_m = 0.083$ btu/h·ft·°F. The viscosity of the oil varies with temperature as follows: 150°F, 6.50 cp; 200°F, 5.05 cp; 250°F, 3.80 cp; 300°F, 2.82 cp; 350°F, 1.95 cp. Predict the heat-transfer coefficient and the oil outlet temperature, T_{bo}.

Solution: This is a trial-and-error solution since the outlet temperature of the oil T_{bo} is unknown. The value of $T_{bo} = 250$°F will be assumed and checked later. The bulk mean temperature of the oil to use for the physical properties is $(150 + 250)/2$ or 200°F. The viscosity at 200°F is

$$\mu_b = 5.05(2.4191) = 12.23\,\frac{\text{lb}_m}{\text{ft}\cdot\text{h}}$$

At the wall temperature of 350°F,

$$\mu_w = 1.95(2.4191) = 4.72\,\frac{\text{lb}_m}{\text{ft}\cdot\text{h}}$$

The cross-section area of the pipe A is

$$A = \frac{\pi D_i^2}{4} = \frac{\pi(0.0303)^2}{4} = 0.000722 \text{ ft}^2$$

$$G = \frac{m}{A} = \frac{80\text{ lb}_m/\text{h}}{0.000722\text{ ft}^2} = 111\,000\,\frac{\text{lb}_m}{\text{ft}^2\cdot\text{h}}$$

The Reynolds number at the bulk mean temperature is

$$N_{\text{Re}} = \frac{D_i v \rho}{\mu} = \frac{D_i G}{\mu} = \frac{0.0303(111\,000)}{12.23} = 275.5$$

The Prandtl number is

$$N_{Pr} = \frac{c_p \mu}{k} = \frac{0.50(12.23)}{0.083} = 73.7$$

Since the N_{Re} is below 2100, the flow is in the laminar region and Eq. (4.5-4) will be used. Even at the outlet temperature of 250°F, the flow is still laminar. Substituting,

$$(N_{Nu})_a = \frac{h_a D}{k} = 1.86\left(N_{Re}\, N_{Pr}\, \frac{D}{L} \right)^{1/3} \left(\frac{\mu_b}{\mu_w} \right)^{0.14}$$

$$\frac{h_a(0.0303)}{0.083} = 1.86\left[275.5(73.7)\left(\frac{0.0303}{15.0}\right) \right]^{1/3} \left(\frac{12.23}{4.72}\right)^{0.14}$$

Solving, $h_a = 20.1$ btu/h $\cdot$ ft^2 $\cdot$ °F (114 W/m^2 $\cdot$ K).

Next, making a heat balance on the oil,

$$q = mc_{pm}(T_{bo} - T_{bi}) = 80.0(0.50)(T_{bo} - 150) \qquad \text{(4.5-28)}$$

Using Eq. (4.5-7),

$$q = h_a A \Delta T_a \qquad \text{(4.5-7)}$$

For ΔT_a,

$$\Delta T_a = \frac{(T_w - T_{bi}) + (T_w - T_{bo})}{2}$$

$$= \frac{(350 - 150) + (350 - T_{bo})}{2}$$

$$= 275 - 0.5 T_{bo}$$

Equating Eq. (4.5-28) to (4.5-7) and substituting,

$$80.0(0.50)(T_{bo} - 150) = hA\Delta T_a$$

$$= 20.1[\pi(0.0303)(15)](275 - 0.5T_{bo})$$

Solving, $T_{bo} = 255$°F.

This is higher than the assumed value of 250°F. For the second trial the mean bulk temperature of the oil would be $(150 + 255)/2$ or 202.5°F. The new viscosity is 5.0 cp compared with 5.05 for the first estimate. This only affects the $(\mu_b/\mu_w)^{0.14}$ factor in Eq. (4.5-4), since the viscosity effect in the $(N_{Re})(N_{Pr})$ factor cancels out. The heat-transfer coefficient will change by less than 0.2%, which is negligible. Hence, the outlet temperature of $T_1 = 255$°F (123.9°C) is correct.

4.6 HEAT TRANSFER OUTSIDE VARIOUS GEOMETRIES IN FORCED CONVECTION

4.6A Introduction

In many cases a fluid is flowing over completely immersed bodies such as spheres, tubes, plates, and so on, and heat transfer is occurring between the fluid and the solid only. Many of these shapes are of practical interest in process engineering. The sphere, cylinder, and flat plate are perhaps of greatest importance with heat transfer between these surfaces and a moving fluid frequently encountered.

When heat transfer occurs during immersed flow, the flux is dependent upon the geometry of the body, the position on the body (front, side, back, etc.), the proximity of other bodies, the flow rate, and the fluid properties. The heat-transfer coefficient varies over the body. The average heat-transfer coefficient is given in the empirical relationships to be discussed in the following sections.

In general, the average heat-transfer coefficient on immersed bodies is given by

$$N_{Nu} = C N_{Re}^m N_{Pr}^{1/3} \tag{4.6-1}$$

where C and m are constants that depend on the various configurations. The fluid properties are evaluated at the film temperature $T_f = (T_w + T_b)/2$, where T_w is the surface or wall temperature and T_b the average bulk fluid temperature. The velocity in the N_{Re} is the undisturbed free stream velocity v of the fluid approaching the object.

4.6B   Flow Parallel to Flat Plate

When the fluid is flowing parallel to a flat plate and heat transfer is occurring between the whole plate of length L m and the fluid, the N_{Nu} is as follows for a $N_{Re, L}$ below 3×10^5 in the laminar region and a $N_{Pr} > 0.7$.

$$N_{Nu} = 0.664 N_{Re, L}^{0.5} N_{Pr}^{1/3} \tag{4.6-2}$$

where $N_{Re, L} = Lv\rho/\mu$.

For the completely turbulent region at a $N_{Re, L}$ above 3×10^5 (K1, K3) and $N_{Pr} > 0.7$,

$$N_{Nu} = 0.0366 N_{Re, L}^{0.8} N_{Pr}^{1/3} \tag{4.6-3}$$

However, turbulence can start at a $N_{Re, L}$ below 3×10^5 if the plate is rough (K3) and then Eq. (4.6-3) will hold and give a N_{Nu} greater than by Eq. (4.6-2). Below about a $N_{Re, L}$ of 2×10^4, Eq. (4.6-2) gives the larger value of N_{Nu}.

EXAMPLE 4.6-1.   Cooling a Copper Fin
A smooth, flat, thin fin of copper extending out from a tube is 51 mm by 51 mm square. Its temperature is approximately uniform at 82.2°C. Cooling air at 15.6°C and 1 atm abs flows parallel to the fin at a velocity of 12.2 m/s.
   (a) For laminar flow, calculate the heat-transfer coefficient, h.
   (b) If the leading edge of the fin is rough so that all of the boundary layer or film next to the fin is completely turbulent, calculate h.

Solution: The fluid properties will be evaluated at the film temperature $T_f = (T_w + T_b)/2$.

$$T_f = \frac{T_w + T_b}{2} = \frac{82.2 + 15.6}{2} = 48.9°C \text{ (322.1 K)}$$

The physical properties of air at 48.9°C from Appendix A.3 are $k = 0.0280$ W/m·K, $\rho = 1.097$ kg/m³, $\mu = 1.95 \times 10^{-5}$ Pa·s, $N_{Pr} = 0.704$. The Reynolds number is, for $L = 0.051$ m,

$$N_{Re, L} = \frac{Lv\rho}{\mu} = \frac{(0.051)(12.2)(1.097)}{1.95 \times 10^{-5}} = 3.49 \times 10^4$$

Substituting into Eq. (4.6-2),

$$N_{Nu} = \frac{hL}{k} = 0.664 N_{Re, L}^{0.5} N_{Pr}^{1/3}$$

$$\frac{h(0.051)}{(0.0280)} = 0.664(3.49 \times 10^4)^{0.5}(0.704)^{1/3}$$

Solving, $h = 60.7$ W/m²·K (10.7 btu/h·ft²·°F).

For part (b), substituting into Eq. (4.6-3) and solving, $h = 77.2 \ \text{W/m}^2 \cdot \text{K} \ (13.6 \ \text{btu/h} \cdot \text{ft}^2 \cdot {}^\circ\text{F})$.

4.6C Cylinder with Axis Perpendicular to Flow

Often a cylinder containing a fluid inside is being heated or cooled by a fluid flowing perpendicular to its axis. The equation for predicting the average heat-transfer coefficient of the outside of the cylinder for gases and liquids is (K3, P1) Eq. (4.6-1) with C and m as given in Table 4.6-1. The $N_{\text{Re}} = Dv\rho/\mu$, where D is the outside tube diameter and all physical properties are evaluated at the film temperature T_f. The velocity is the undisturbed free stream velocity approaching the cylinder.

4.6D Flow Past Single Sphere

When a single sphere is being heated or cooled by a fluid flowing past it, the following equation can be used to predict the average heat-transfer coefficient for a $N_{\text{Re}} = Dv\rho/\mu$ of 1 to 70 000 and a N_{Pr} of 0.6 to 400.

$$N_{\text{Nu}} = 2.0 + 0.60 N_{\text{Re}}^{0.5} N_{\text{Pr}}^{1/3} \qquad (4.6\text{-}4)$$

The fluid properties are evaluated at the film temperature T_f. A somewhat more accurate correlation is available for a N_{Re} range 1–17 000 by others (S2), which takes into account the effects of natural convection at these lower Reynolds numbers.

EXAMPLE 4.6-2. Cooling of a Sphere

Using the same conditions as Example 4.6-1, where air at 1 atm abs pressure and 15.6°C is flowing at a velocity of 12.2 m/s, predict the average heat-transfer coefficient for air flowing by a sphere having a diameter of 51 mm and an average surface temperature of 82.2°C. Compare this with the value of $h = 77.2 \ \text{W/m}^2 \cdot \text{K}$ for the flat plate in turbulent flow.

Solution: The physical properties at the average film temperature of 48.9°C are the same as for Example 4.6-1. The N_{Re} is

$$N_{\text{Re}} = \frac{Dv\rho}{\mu} = \frac{(0.051)(12.2)(1.097)}{1.95 \times 10^{-5}} = 3.49 \times 10^4$$

TABLE 4.6-1. *Constants for Use in Eq. (4.6-1) for Heat Transfer to Cylinders with Axis Perpendicular to Flow ($N_{\text{Pr}} > 0.6$)*

N_{Re}	m	C
1–4	0.330	0.989
4–40	0.385	0.911
40–4×10^3	0.466	0.683
4×10^3–4×10^4	0.618	0.193
4×10^4–2.5×10^5	0.805	0.0266

Transport Processes: Momentum, Heat, and Mass

Substituting into Eq. (4.6-4) for a sphere,

$$N_{Nu} = \frac{hD}{k} = \frac{h(0.051)}{0.0280} = 2.0 + 0.60 N_{Re}^{0.5} N_{Pr}^{1/3}$$

$$= 2.0 + (0.60)(3.49 \times 10^4)^{0.5}(0.704)^{1/3}$$

Solving, $h = 56.1$ W/m$^2 \cdot$ K (9.88 btu/h $\cdot$ ft$^2 \cdot$ °F). This value is somewhat smaller than the value of $h = 77.2$ W/m$^2 \cdot$ K (13.6 btu/h $\cdot$ ft$^2 \cdot$ °F) for a flat plate.

4.6E   Flow Past Banks of Tubes or Cylinders

Many types of commercial heat exchangers are constructed with multiple rows of tubes, where the fluid flows at right angles to the bank of tubes. An example is a gas heater in which a hot fluid inside the tubes heats a gas passing over the outside of the tubes. Another example is a cold liquid stream inside the tubes being heated by a hot fluid on the outside.

Figure 4.6-1 shows the arrangement for banks of tubes in-line and banks of tubes staggered where D is tube OD in m (ft), S_n is distance m (ft) between the centers of the tubes normal to the flow, and S_p parallel to the flow. The open area to flow for in-line tubes is $(S_n - D)$ and $(S_p - D)$, and for staggered tubes it is $(S_n - D)$ and $(S_p' - D)$. Values of C and m to be used in Eq. (4.6-1) for a Reynolds number range of 2000 to 40 000 for heat transfer to banks of tubes containing more than 10 transverse rows in the direction of flow are given in Table 4.6-2. For less than 10 rows, Table 4.6-3 gives correction factors.

For cases where S_n/D and S_p/D are not equal to each other, the reader should consult Grimison (G1) for more data. In baffled exchangers where there is normal leakage where all the fluid does not flow normal to the tubes, the average values of h obtained should be multiplied by about 0.6 (P1). The Reynolds number is calculated using the minimum area open to flow for the velocity. All physical properties are evaluated at T_f.

EXAMPLE 4.6-3.   Heating Air by a Bank of Tubes
Air at 15.6°C and 1 atm abs flows across a bank of tubes containing four transverse rows in the direction of flow and 10 rows normal to the flow at a

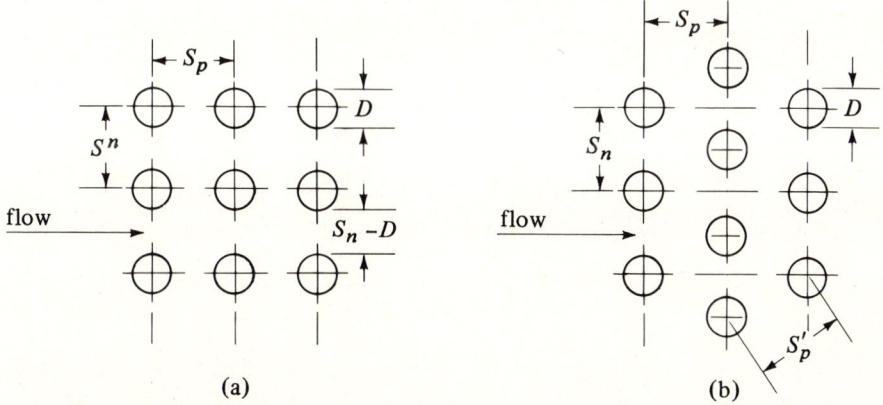

(a)   (b)

FIGURE 4.6-1.   *Nomenclature for banks of tubes in Table 4.6-2: (a) in-line tube rows, (b) staggered tube rows.*

TABLE 4.6-2.   *Values of C and m To Be Used in Eq. (4.6-1) for Heat Transfer to Banks of Tubes Containing More Than 10 Transverse Rows*

Arrangement	$\dfrac{S_n}{D} = \dfrac{S_p}{D} = 1.25$		$\dfrac{S_n}{D} = \dfrac{S_p}{D} = 1.50$		$\dfrac{S_n}{D} = \dfrac{S_p}{D} = 2.0$	
	C	m	C	m	C	m
In-line	0.386	0.592	0.278	0.620	0.254	0.632
Staggered	0.575	0.556	0.511	0.562	0.535	0.556

Source : E. D. Grimison, *Trans. ASME,* **59,** 583 (1937).

velocity of 7.62 m/s as the air approaches the bank of tubes. The tube surfaces are maintained at 57.2°C. The outside diameter of the tubes is 25.4 mm and the tubes are in-line to the flow. The spacing S_n of the tubes normal to the flow is 38.1 mm and also S_p is 38.1 mm parallel to the flow. For a 0.305 m length of the tube bank, calculate the heat-transfer rate.

Solution: Referring to Fig. 4.6-1a,

$$\frac{S_n}{D} = \frac{38.1}{25.4} = \frac{1.5}{1} \qquad \frac{S_p}{D} = \frac{38.1}{25.4} = \frac{1.5}{1}$$

Since the air is heated in passing through the four transverse rows, an outlet bulk temperature of 21.1°C will be assumed. The average bulk temperature is then

$$T_b = \frac{15.6 + 21.1}{2} = 18.3°C$$

The average film temperature is

$$T_f = \frac{T_w + T_b}{2} = \frac{57.2 + 18.3}{2} = 37.7°C$$

From Appendix A.3 for air at 37.7°C,

$$k = 0.02700 \text{ W/m} \cdot \text{K} \qquad N_{Pr} = 0.705$$

$$c_p = 1.0048 \text{ kJ/kg} \cdot \text{K} \qquad \rho = 1.137 \text{ kg/m}^3$$

$$\mu = 1.90 \times 10^{-5} \text{ Pa} \cdot \text{s}$$

TABLE 4.6-3.   *Ratio of h for N Transverse Rows Deep to h for 10 Transverse Rows Deep (for Use with Table 4.6-2)*

N	1	2	3	4	5	6	7	8	9	10
Ratio for staggered tubes	0.68	0.75	0.83	0.89	0.92	0.95	0.97	0.98	0.99	1.00
Ratio for in-line tubes	0.64	0.80	0.87	0.90	0.92	0.94	0.96	0.98	0.99	1.00

Source : W. M. Kays and R. K. Lo, *Stanford Univ. Tech. Rept. 15,* Navy Contract N6-ONR-251 T.O.6, 1952.

The ratio of the minimum-flow area to the total frontal area is $(S_n - D)/S_n$. The maximum velocity in the tube banks is then

$$v_{max} = \frac{vS_n}{S_n - D} = \frac{7.62(0.0381)}{(0.0381 - 0.0254)} = 22.86 \text{ m/s}$$

$$N_{Re} = \frac{D v_{max} \rho}{\mu} = \frac{0.0254(22.86)(1.137)}{1.90 \times 10^{-5}} = 3.47 \times 10^4$$

For $S_n/D = S_p/D = 1.5/1$, the values of C and m from Table 4.6-2 are 0.278 and 0.620, respectively. Substituting into Eq. (4.6-1) and solving for h,

$$h = \frac{k}{D} C N_{Re}^m N_{Pr}^{1/3} = \left(\frac{0.02700}{0.0254}\right)(0.278)(3.47 \times 10^4)^{0.62}(0.705)^{1/3}$$

$$= 171.8 \text{ W/m}^2 \cdot \text{K}$$

This h is for 10 rows. For only four rows in the transverse direction, the h must be multiplied by 0.90, as given in Table 4.6-3.

Since there are 10×4 or 40 tubes, the total heat-transfer area per 0.305 m length is

$$A = 40\pi DL = 40\pi(0.0254)(0.305) = 0.973 \text{ m}^2$$

The total heat-transfer rate q using an arithmetic average temperature difference between the wall and the bulk fluid is

$$q = hA(T_w - T_b) = (0.90 \times 171.8)(0.973)(57.2 - 18.3) = 5852 \text{ W}$$

Next, a heat balance on the air is made to calculate its temperature rise ΔT using the calculated q. First the mass flow rate of air m must be calculated. The total frontal area of the tube bank assembly of 10 rows of tubes each 0.305 m long is

$$A_t = 10S_n(1.0) = 10(0.0381)(0.305) = 0.1162 \text{ m}^2$$

The density of the entering air at 15.6°C is $\rho = 1.224 \text{ kg/m}^3$. The mass flow rate m is

$$m = v\rho A_t(3600) = 7.62(1.224)(0.1162) = 1.084 \text{ kg/s}$$

For the heat balance the mean c_p of air at 18.3°C is 1.0048 kJ/kg·K and then

$$q = 5852 = mc_p \Delta T = 1.084(1.0048 \times 10^3) \Delta T$$

Solving, $\Delta T = 5.37°C$.

Hence, the calculated outlet bulk gas temperature is $15.6 + 5.37 = 20.97°C$, which is close to the assumed value of 21.1°C. If a second trial were to be made, the new average T_b to use would be $(15.6 + 20.97)/2$ or 18.28°C.

4.6F Heat Transfer for Flow in Packed Beds

Correlations for heat-transfer coefficients for packed beds are useful in designing fixed-bed systems such as catalytic reactors, dryers for solids, and pebble-bed heat exchangers. In Section 3.1C the pressure drop in packed beds was considered and discussions of the geometry factors in these beds were given. For determining the rate of heat transfer in packed beds for a differential length dz in m,

$$dq = h(a \ S \ dz)(T_1 - T_2) \tag{4.6-5}$$

where a is the solid particle surface area per unit volume of bed in m^{-1}, S the empty cross-sectional area of bed in m^2, T_1 the bulk gas temperature in K, and T_2 the solid surface temperature.

For the heat transfer of gases in beds of spheres (G2, G3) and a Reynolds number range of 10–10 000,

$$\varepsilon J_H = \varepsilon \frac{h}{c_p v' \rho}\left(\frac{c_p \mu}{k}\right)^{2/3}_f = \frac{2.876}{N_{Re}} + \frac{0.3023}{N_{Re}^{0.35}} \tag{4.6-6}$$

where v' is the superficial velocity based on the cross section of the empty container in m/s [see Eq. (3.1-11)], ε is the void fraction, $N_{Re} = D_P G'/\mu_f$, and $G' = v' \rho$ is the superficial mass velocity in $kg/m^2 \cdot s$. The subscript f indicates properties evaluated at the film temperature with others at the bulk temperature. This correlation can also be used for a fluidized bed. An alternate equation to use in place of Eq. (4.6-6) for fixed and fluidized beds is Eq. (7.3-36) for a Reynolds number range of 10–4000. The term J_H is called the Colburn J factor and is defined as in Eq. (4.6-6) in terms of h.

Equations for heat transfer to noncircular cylinders such as a hexagon, etc., are given elsewhere (H1, J1, P1).

4.7 NATURAL CONVECTION HEAT TRANSFER

4.7A Introduction

Natural convection heat transfer occurs when a solid surface is in contact with a gas or liquid which is at a different temperature from the surface. Density differences in the fluid arising from the heating process provide the buoyancy force required to move the fluid. Free or natural convection is observed as a result of the motion of the fluid. An example of heat transfer by natural convection is a hot radiator used for heating a room. Cold air encountering the radiator is heated and rises in natural convection because of buoyancy forces. The theoretical derivation of equations for natural convection heat-transfer coefficients requires the solution of motion and energy equations.

An important heat-transfer system occurring in process engineering is that in which heat is being transferred from a hot vertical plate to a gas or liquid adjacent to it by natural convection. The fluid is not moving by forced convection but only by natural or free convection. In Fig. 4.7-1 the vertical flat plate is heated and the free-convection boundary layer is formed. The velocity profile differs from that in a forced-convection system in that the velocity at the wall is zero and also is zero at the other edge of the boundary layer since the free-stream velocity is zero for natural convection. The boundary layer initially is laminar as shown, but at some distance from the leading edge it starts to become turbulent. The wall temperature is T_w K and the bulk temperature T_b.

The differential momentum balance equation is written for the x and y directions for the control volume $(dx\,dy \cdot 1)$. The driving force is the buoyancy force in the gravitational field and is due to the density difference of the fluid. The momentum balance becomes

$$\rho\left(v_x \frac{\partial v_x}{\partial x} + v_y \frac{\partial v_x}{\partial y}\right) = g(\rho_b - \rho) + \mu \frac{\partial^2 v_x}{\partial y^2} \tag{4.7-1}$$

where ρ_b is the density at the bulk temperature T_b and ρ at T. The density difference can be expressed in terms of the volumetric coefficient of expansion β and substituted back into Eq. (4.7-1).

$$\beta = \frac{\rho_b - \rho}{\rho(T - T_b)} \tag{4.7-2}$$

For gases, $\beta = 1/T$. The energy-balance equation can be expressed as follows:

$$\rho c_p \left(v_x \frac{\partial T}{\partial x} + v_y \frac{\partial T}{\partial y} \right) = k \frac{\partial^2 T}{\partial y^2} \qquad (4.7\text{-}3)$$

The solutions of these equations have been obtained by using integral methods of analysis discussed in Section 3.7. Results for a vertical plate have been obtained, which is the most simple case and serves to introduce the dimensionless Grashof number discussed below. However, in other physical geometries the relations are too complex and empirical correlations have been obtained. These are discussed in the following sections.

4.7B Natural Convection from Various Geometries

1. Natural convection from vertical planes and cylinders. For an isothermal vertical surface or plate with height L less than 1 m (P1), the average natural convection heat-transfer coefficient can be expressed by the following general equation:

$$N_{\text{Nu}} = \frac{hL}{k} = a \left(\frac{L^3 \rho^2 g \beta \, \Delta T}{\mu^2} \frac{c_p \mu}{k} \right)^m = a(N_{\text{Gr}} N_{\text{Pr}})^m \qquad (4.7\text{-}4)$$

where a and m are constants from Table 4.7-1, N_{Gr} the Grashof number, ρ density in kg/m^3, μ viscosity in kg/m·s, ΔT the positive temperature difference between the wall and bulk fluid or vice versa in K, k the thermal conductivity in W/m·K, c_p the heat capacity in J/kg·K, β the volumetric coefficient of expansion of the fluid in 1/K [for gases β is $1/(T_f \text{ K})$], and g is 9.80665 m/s^2. All the physical properties are evaluated at the film temperature $T_f = (T_w + T_b)/2$. In general, for a vertical cylinder with length L m, the same equations can be used as for a vertical plate. In English units β is $1/(T_f\,°\text{F} + 460)$ in $1/°$R and g is $32.174 \times (3600)^2$ ft/h^2.

The Grashof number can be interpreted physically as a dimensionless number that represents the ratio of the buoyancy forces to the viscous forces in free convection and plays a role similar to that of the Reynolds number in forced convection.

> **EXAMPLE 4.7-1. Natural Convection from Vertical Wall of an Oven**
> A heated vertical wall 1.0 ft (0.305 m) high of an oven for baking food with the surface at 450°F (505.4 K) is in contact with air at 100°F (311 K). Calculate the heat-transfer coefficient and the heat transfer/ft (0.305 m) width of wall. Note that heat transfer for radiation will not be considered. Use English and SI units.

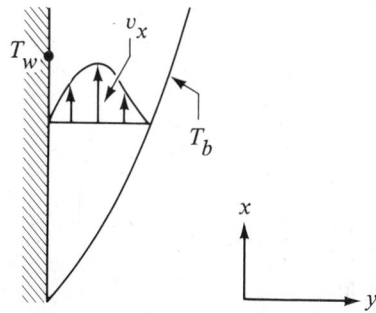

FIGURE 4.7-1. *Boundary-layer velocity profile for natural convection heat transfer from a heated, vertical plate.*

TABLE 4.7-1. *Constants for Use with Eq. (4.7-4) for Natural Convection*

Physical Geometry	$N_{Gr} N_{Pr}$	a	m	Ref.
Vertical planes and cylinders [vertical height $L < 1$ m (3 ft)]				
	$< 10^4$	1.36	$\frac{1}{5}$	(P1)
	10^4–10^9	0.59	$\frac{1}{4}$	(M1)
	$> 10^9$	0.13	$\frac{1}{3}$	(M1)
Horizontal cylinders [diameter D used for L and $D < 0.20$ m (0.66 ft)]				
	$< 10^{-5}$	0.49	0	(P1)
	10^{-5}–10^{-3}	0.71	$\frac{1}{25}$	(P1)
	10^{-3}–1	1.09	$\frac{1}{10}$	(P1)
	1–10^4	1.09	$\frac{1}{5}$	(P1)
	10^4–10^9	0.53	$\frac{1}{4}$	(M1)
	$> 10^9$	0.13	$\frac{1}{3}$	(P1)
Horizontal plates				
Upper surface of heated plates or lower surface of cooled plates	10^5–2×10^7	0.54	$\frac{1}{4}$	(M1)
	2×10^7–3×10^{10}	0.14	$\frac{1}{3}$	(M1)
Lower surface of heated plates or upper surface of cooled plates	10^5–10^{11}	0.58	$\frac{1}{5}$	(F1)

Solution: The film temperature is

$$T_f = \frac{T_w + T_b}{2} = \frac{450 + 100}{2} = 275°F = \frac{505.4 + 311}{2} = 408.2 \text{ K}$$

The physical properties of air at 275°F are $k = 0.0198$ btu/h · ft · °F, 0.0343 W/m · K; $\rho = 0.0541$ lb$_m$/ft^3, 0.867 kg/m^3; $N_{Pr} = 0.690$; $\mu = (0.0232 \text{ cp}) \times (2.4191) = 0.0562$ lb$_m$/ft · h $= 2.32 \times 10^{-5}$ Pa · s; $\beta = 1/408.2 = 2.45 \times 10^{-3}$ K^{-1}, $\beta = 1/(460 + 275) = 1.36 \times 10^{-3}$ °R^{-1}; $\Delta T = T_w - T_b = 450 - 100 = 350°F$ (194.4 K). The Grashof number is, in English units,

$$N_{Gr} = \frac{L^3 \rho^2 g \beta \Delta T}{\mu^2} = \frac{(1.0)^3 (0.0541)^2 (32.174)(3600)^2 (1.36 \times 10^{-3})(350)}{(0.0562)^2}$$

$$= 1.84 \times 10^8$$

In SI units,

$$N_{Gr} = \frac{(0.305)^3 (0.867)^2 (9.806)(2.45 \times 10^{-3})(194.4)}{(2.32 \times 10^{-5})^2} = 1.84 \times 10^8$$

The Grashof numbers calculated using English and SI units must, of course, be the same as shown.

$$N_{Gr} N_{Pr} = (1.84 \times 10^8)(0.690) = 1.270 \times 10^8$$

Hence, from Table 4.7-1, $a = 0.59$ and $m = \frac{1}{4}$ for use in Eq. (4.7-4).

Solving for h in Eq. (4.7-4) and substituting known values,

$$h = \frac{k}{L} a(N_{Gr} N_{Pr})^m = \left(\frac{0.0198}{1.0}\right)(0.59)(1.270 \times 10^8)^{1/4} = 1.24 \text{ btu/h} \cdot \text{ft}^2 \cdot °F$$

$$h = \left(\frac{0.0343}{0.305}\right)(0.59)(1.27 \times 10^8)^{1/4} = 7.03 \text{ W/m}^2 \cdot K$$

For a 1-ft width of wall, $A = 1 \times 1 = 1.0 \text{ ft}^2$ (0.305 × 0.305 m²). Then

$$q = hA(T_w - T_b) = (1.24)(1.0)(450 - 100) = 433 \text{ btu/h}$$

$$q = 7.03(0.305 \times 0.305)(194.4) = 127.1 \text{ W}$$

A considerable amount of heat will also be lost by radiation. This will be considered in Section 4.9.

Simplified equations for the natural convection heat transfer from air to vertical planes and cylinders at 1 atm abs pressure are given in Table 4.7-2. In SI units the equation for the range of $N_{Gr} N_{Pr}$ of 10^4 to 10^9 is the one usually encountered and this holds for ($L^3 \, \Delta T$) values below about 4.7 m³ · K and film temperatures between 255 and 533 K. To correct the value of h to pressures other than 1 atm, the values of h in Table

TABLE 4.7-2. *Simplified Equations for Natural Convection from Various Surfaces*

Physical Geometry	$N_{Gr} N_{Pr}$	Equation $h = btu/h \cdot ft^2 \cdot °F$ $L = ft, \Delta T = °F$ $D = ft$	$h = W/m^2 \cdot K$ $L = m, \Delta T = K$ $D = m$	Ref.
Air at 101.32 kPa (1 atm) abs pressure				
Vertical planes and cylinders	10^4–10^9	$h = 0.28(\Delta T/L)^{1/4}$	$h = 1.37(\Delta T/L)^{1/4}$	(P1)
	$> 10^9$	$h = 0.18(\Delta T)^{1/3}$	$h = 1.24 \, \Delta T^{1/3}$	(P1)
Horizontal cylinders	10^3–10^9	$h = 0.27(\Delta T/D)^{1/4}$	$h = 1.32(\Delta T/D)^{1/4}$	(M1)
	$> 10^9$	$h = 0.18(\Delta T)^{1/3}$	$h = 1.24 \, \Delta T^{1/3}$	(M1)
Horizontal plates				
Heated plate facing upward or cooled plate facing downward	10^5–2×10^7 2×10^7–3×10^{10}	$h = 0.27(\Delta T/L)^{1/4}$ $h = 0.22(\Delta T)^{1/3}$	$h = 1.32(\Delta T/L)^{1/4}$ $h = 1.52 \, \Delta T^{1/3}$	(M1) (M1)
Heated plate facing downward or cooled plate facing upward	3×10^5–3×10^{10}	$h = 0.12(\Delta T/L)^{1/4}$	$h = 0.59(\Delta T/L)^{1/4}$	(M1)
Water at 70°F (294 K)				
Vertical planes and cylinders	10^4–10^9	$h = 26(\Delta T/L)^{1/4}$	$h = 127(\Delta T/L)^{1/4}$	(P1)
Organic liquids at 70°F (294 K)				
Vertical planes and cylinders	10^4–10^9	$h = 12(\Delta T/L)^{1/4}$	$h = 59(\Delta T/L)^{1/4}$	(P1)

4.7-2 can be multiplied by $(p/101.32)^{1/2}$ for $N_{Gr} N_{Pr}$ 10^4 to 10^9 and by $(p/101.32)^{2/3}$ for $N_{Gr} N_{Pr} > 10^9$, where $p =$ pressure in kN/m^2. In English units the range of $N_{Gr} N_{Pr}$ of 10^4 to 10^9 is encountered when $(L^3 \Delta T)$ is less than about $300 \, ft^3 \cdot °F$. The value of h can be corrected to pressures other than 1.0 atm abs by multiplying the h at 1 atm by $p^{1/2}$ for $N_{Gr} N_{Pr}$ of 10^4 to 10^9 and by $p^{2/3}$ for $N_{Gr} N_{Pr}$ above 10^9, where $p =$ atm abs pressure. Simplified equations are also given for water and organic liquids.

EXAMPLE 4.7-2.   Natural Convections and Simplified Equation
Repeat Example 4.7-1 but use the simplified equation.

Solution: The film temperature of 408.2 K is in the range 255–533 K. Also,

$$L^3 \, \Delta T = (0.305)^3 (194.4) = 5.5$$

This is slightly greater than the value of 4.7 given as the approximate maximum for use of the simplified equation. However, in Example 4.7-1 the value of $N_{Gr} N_{Pr}$ is below 10^9, so the simplified equation from Table 4.7-2 will be used.

$$h = 1.37 \left(\frac{\Delta T}{L} \right)^{1/4} = 1.37 \left(\frac{194.4}{0.305} \right)^{1/4}$$

$$= 6.88 \, W/m^2 \cdot K \ (1.21 \, btu/h \cdot ft^2 \cdot °F)$$

The heat-transfer rate q is

$$q = hA(T_w - T_b) = 6.88(0.305 \times 0.305)(194.4) = 124.4 \, W \ (424 \, btu/h)$$

This value is reasonably close to the value of 127.1 W for Example 4.7-1.

2. Natural convection from horizontal cylinders.   For a horizontal cylinder with an outside diameter of D m, Eq. (4.7-4) is used with the constants being given in Table 4.7-1. The diameter D is used for L in the equation. Simplified equations are given in Table 4.7-2. The usual case for pipes is for a $N_{Gr} N_{Pr}$ range of 10^4 to 10^9 (M1).

3. Natural convection from horizontal plates.   For horizontal flat plates Eq. (4.7-4) is also used with the constants given in Table 4.7-1 and simplified equations in Table 4.7-2. The dimension L to be used is the length of a side of a square plate, the linear mean of the two dimensions for a rectangle, and 0.9 times the diameter of a circular disk.

4. Natural convection in enclosed spaces.   Free convection in enclosed spaces occurs in a number of processing applications. One example is in an enclosed double window in which two layers of glass are separated by a layer of air for energy conservation. The flow phenomena inside these enclosed spaces are complex since a number of different types of flow patterns can occur. At low Grashof numbers the heat transfer is mainly by conduction across the fluid layer. As the Grashof number is increased, different flow regimes are encountered.

The system for two vertical plates of height L m containing the fluid with a gap of δ m is shown in Fig. 4.7-2, where the plate surfaces are at T_1 and T_2 temperatures. The Grashof number is defined as

$$N_{Gr, \, \delta} = \frac{\delta^3 \rho^2 g \beta (T_1 - T_2)}{\mu^2} \tag{4.7-5}$$

The Nusselt number is defined as

$$N_{Nu, \, \delta} = \frac{h\delta}{k} \tag{4.7-6}$$

The heat flux is calculated from

$$\frac{q}{A} = h(T_1 - T_2) \tag{4.7-7}$$

The physical properties are all evaluated at the mean temperature between the two plates.

For gases enclosed between vertical plates and $L/\delta > 3$ (H1, J1, K1, P1),

$$N_{\text{Nu}, \delta} = \frac{h\delta}{k} = 1.0 \qquad (N_{\text{Gr}, \delta} N_{\text{Pr}} < 2 \times 10^3) \tag{4.7-8}$$

$$N_{\text{Nu}, \delta} = 0.20 \frac{(N_{\text{Gr}, \delta} N_{\text{Pr}})^{1/4}}{(L/\delta)^{1/9}} \qquad (6 \times 10^3 < N_{\text{Gr}, \delta} N_{\text{Pr}} < 2 \times 10^5) \tag{4.7-9}$$

$$N_{\text{Nu}, \delta} = 0.073 \frac{(N_{\text{Gr}, \delta} N_{\text{Pr}})^{1/3}}{(L/\delta)^{1/9}} \qquad (2 \times 10^5 < N_{\text{Gr}, \delta} N_{\text{Pr}} < 2 \times 10^7) \tag{4.7-10}$$

For liquids in vertical plates,

$$N_{\text{Nu}, \delta} = \frac{h\delta}{k} = 1.0 \qquad (N_{\text{Gr}, \delta} N_{\text{Pr}} < 1 \times 10^3) \tag{4.7-11}$$

$$N_{\text{Nu}, \delta} = 0.28 \frac{(N_{\text{Gr}, \delta} N_{\text{Pr}})^{1/4}}{(L/\delta)^{1/4}} \qquad (1 \times 10^3 < N_{\text{Gr}, \delta} N_{\text{Pr}} < 1 \times 10^7) \tag{4.7-12}$$

For gases or liquids in a vertical annulus, the same equations hold as for vertical plates.

For gases in horizontal plates with the lower plate hotter than the upper,

$$N_{\text{Nu}, \delta} = 0.21(N_{\text{Gr}, \delta} N_{\text{Pr}})^{1/4} \qquad (7 \times 10^3 < N_{\text{Gr}, \delta} N_{\text{Pr}} < 3 \times 10^5) \tag{4.7-13}$$

$$N_{\text{Nu}, \delta} = 0.061(N_{\text{Gr}, \delta} N_{\text{Pr}})^{1/3} \qquad (N_{\text{Gr}, \delta} N_{\text{Pr}} > 3 \times 10^5) \tag{4.7-14}$$

For liquids in horizontal plates with the lower plate hotter than the upper (G5),

$$N_{\text{Nu}, \delta} = 0.069(N_{\text{Gr}, \delta} N_{\text{Pr}})^{1/3} N_{\text{Pr}}^{0.074} \qquad (1.5 \times 10^5 < N_{\text{Gr}, \delta} N_{\text{Pr}} < 1 \times 10^9) \tag{4.7-15}$$

EXAMPLE 4.7-3. Natural Convection in Enclosed Vertical Space
Air at 1 atm abs pressure is enclosed between two vertical plates where $L = 0.6$ m and $\delta = 30$ mm. The plates are 0.4 m wide. The plate temperatures are $T_1 = 394.3$ K and $T_2 = 366.5$ K. Calculate the heat-transfer rate across the air gap.

FIGURE 4.7-2. *Natural convection in enclosed vertical space.*

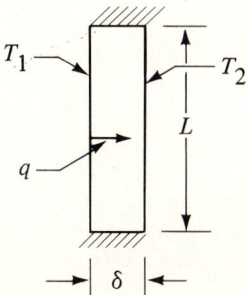

Solution: The mean temperature between the plates is used to evaluate the physical properties. $T_f = (T_1 + T_2)/2 = (394.3 + 366.5)/2 = 380.4$ K. Also, $\delta = 30/1000 = 0.030$ m. From Appendix A.3, $\rho = 0.9295$ kg/m^3, $\mu = 2.21 \times 10^{-5}$ Pa·s, $k = 0.03219$ W/m·K, $N_{Pr} = 0.693$, $\beta = 1/T_f = 1/380.4 = 2.629 \times 10^{-3}$K^{-1}.

$$N_{Gr,\delta} = \frac{(0.030)^3(0.9295)^2(9.806)(2.629 \times 10^{-3})(394.3 - 366.5)}{(2.21 \times 10^{-5})^2}$$

$$= 3.423 \times 10^4$$

Also, $N_{Gr,\delta}N_{Pr} = (3.423 \times 10^4)0.693 = 2.372 \times 10^4$. Using Eq. (4.7-9),

$$h = \frac{k}{\delta}\frac{(0.20)(N_{Gr,\delta}N_{Pr})^{1/4}}{(L/\delta)^{1/9}} = \frac{0.03219(0.20)(2.352 \times 10^4)^{1/4}}{0.030(0.6/0.030)^{1/9}}$$

$$= 1.909 \text{ W/m}^2 \cdot \text{K}$$

The area $A = (0.6 \times 0.4) = 0.24$ m^2. Substituting into Eq. (4.7-7),

$$q = hA(T_1 = T_2) = 1.909(0.24)(394.3 - 366.5) = 12.74 \text{ W}$$

5. Natural convection from other shapes. For spheres, blocks, and other types of enclosed air spaces, references elsewhere (H1, K1, M1, P1) should be consulted. In some cases when a fluid is forced over a heated surface at low velocity in the laminar region, combined forced-convection plus natural-convection heat transfer occurs. For further discussion of this, see (H1, K1, M1).

4.8 BOILING AND CONDENSATION

4.8A Boiling

1. Mechanisms of boiling. Heat transfer to a boiling liquid is very important in evaporation and distillation and also in other kinds of chemical and biological processing, such as petroleum processing, control of the temperature of chemical reactions, evaporation of liquid foods, and so on. The boiling liquid is usually contained in a vessel with a heating surface of tubes or vertical or horizontal plates which supply the heat for boiling. The heating surfaces can be heated electrically or by a hot or condensing fluid on the other side of the heated surface.

In boiling the temperature of the liquid is the boiling point of this liquid at the pressure in the equipment. The heated surface is, of course, at a temperature above the boiling point. Bubbles of vapor are generated at the heated surface and rise through the mass of liquid. The vapor accumulates in a vapor space above the liquid level and is withdrawn.

Boiling is a complex phenomenon. Suppose we consider a small heated horizontal tube or wire immersed in a vessel containing water boiling at 373.2 K (100°C). The heat flux is q/A W/m^2, $\Delta T = T_w - 373.2$ K, where T_w is the tube or wire wall temperature and h is the heat-transfer coefficient in W/m^2·K. Starting with a low ΔT, the q/A and h values are measured. This is repeated at higher values of ΔT and the data obtained are shown in Fig. 4.8-1 plotted as q/A versus ΔT.

In the first region A of the plot in Fig. 4.8-1, at low temperature drops, the mechanism of boiling is essentially that of heat transfer to a liquid in natural convection.

The variation of h with $\Delta T^{0.25}$ is approximately the same as that for natural convection to horizontal plates or cylinders. The very few bubbles formed are released from the surface of the metal and rise and do not disturb appreciably the normal natural convection.

In the region B of nucleate boiling for a ΔT of about $5 - 25$ K ($9 - 45°$F), the rate of bubble production increases so that the velocity of circulation of the liquid increases. The heat-transfer coefficient h increases rapidly and is proportional to ΔT^2 to ΔT^3 in this region.

In the region C of transition boiling, many bubbles are formed so quickly that they tend to coalesce and form a layer of insulating vapor. Increasing the ΔT increases the thickness of this layer and the heat flux and h drop as ΔT is increased. In region D or film boiling, bubbles detach themselves regularly and rise upward. At higher ΔT values radiation through the vapor layer next to the surface helps increase the q/A and h.

The curve of h versus ΔT has approximately the same shape as Fig. 4.8-1. The values of h are quite large. At the beginning of region B in Fig. 4.8-1 for nucleate boiling, h has a value of about 5700–11 400 W/m$^2 \cdot$ K, or 1000–2000 btu/h $\cdot$ ft$^2 \cdot$ °F, and at the end of this region h has a peak value of almost 57 000 W/m$^2 \cdot$ K, or 10 000 btu/hr $\cdot$ ft$^2 \cdot$ °F. These values are quite high, and in most cases the percent resistance of the boiling film is only a few percent of the overall resistance to heat transfer.

The regions of commercial interest are the nucleate and film-boiling regions (P1). Nucleate boiling occurs in kettle-type and natural-circulation reboilers.

2. *Nucleate boiling.* In the nucleate boiling region the heat flux is affected by ΔT, pressure, nature and geometry of the surface and system, and physical properties of the vapor and liquid. Equations have been derived by Rohsenow et al. (P1). They apply to single tubes or flat surfaces and are quite complex.

Simplified empirical equations to estimate the boiling heat-transfer coefficients for water boiling on the outside of submerged surfaces at 1.0 atm abs pressure have been developed (J2).

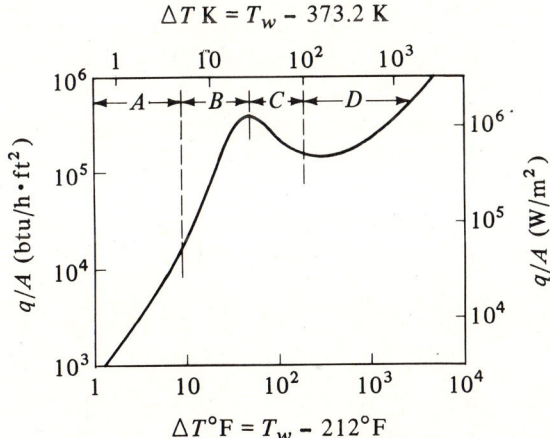

FIGURE 4.8-1. *Boiling mechanisms for water at atmospheric pressure, heat flux vs. temperature drop: (A) natural convection, (B) nucleate boiling, (C) transition boiling, (D) film boiling.*

For a horizontal surface (SI and English units),

$$h, \text{btu/h} \cdot \text{ft}^2 \cdot {}^\circ\text{F} = 151(\Delta T {}^\circ\text{F})^{1/3} \qquad q/A, \text{btu/h} \cdot \text{ft}^2, < 5000$$

$$h, \text{W/m}^2 \cdot \text{K} = 1043(\Delta T \text{K})^{1/3} \qquad q/A, \text{kW/m}^2, < 16$$

(4.8-1)

$$h, \text{btu/h} \cdot \text{ft}^2 \cdot {}^\circ\text{F} = 0.168(\Delta T {}^\circ\text{F})^3 \qquad 5000 < q/A, \text{btu/h} \cdot \text{ft}^2, < 75\,000$$

$$h, \text{W/m}^2 \cdot \text{K} = 5.56(\Delta T \text{K})^3 \qquad 16 < q/A, \text{kW/m}^2, < 240$$

(4.8-2)

For a vertical surface,

$$h, \text{btu/h} \cdot \text{ft}^2 \cdot {}^\circ\text{F} = 87(\Delta T {}^\circ\text{F})^{1/7} \qquad q/A, \text{btu/h} \cdot \text{ft}^2, < 1000$$

$$h, \text{W/m}^2 \cdot \text{K} = 537(\Delta T \text{K})^{1/7} \qquad q/A, \text{kW/m}^2, < 3$$

(4.8-3)

$$h, \text{btu/h} \cdot \text{ft}^2 \cdot {}^\circ\text{F} = 0.240(\Delta T {}^\circ\text{F})^3 \qquad 1000 < q/A, \text{btu/h} \cdot \text{ft}^2, < 20\,000$$

$$h, \text{W/m}^2 \cdot \text{K} = 7.95(\Delta T \text{K})^3 \qquad 3 < q/A, \text{kW/m}^2, < 63$$

(4.8-4)

where $\Delta T = T_w - T_{\text{sat}}$ K or ${}^\circ$F.

If the pressure is p atm abs, the values of h at 1 atm given above are multiplied by $(p/1)^{0.4}$. Equations (4.8-1) and (4.8-3) are in the natural convection region.

For forced convection boiling inside tubes, the following simplified relation can be used (J3).

$$h = 2.55(\Delta T \text{K})^3 e^{p/1551} \text{ W/m}^2 \cdot \text{K} \qquad \text{(SI)}$$

$$h = 0.077(\Delta T {}^\circ\text{F})^3 e^{p/225} \text{ btu/h} \cdot \text{ft}^2 \cdot {}^\circ\text{F} \qquad \text{(English)}$$

(4.8-5)

where p in this case is in kPa (SI units) and psia (English units).

3. *Film boiling.* In the film-boiling region the heat-transfer rate is low in view of the large temperature drop used, which is not utilized effectively. Film boiling has been subjected to considerable theoretical analysis. Bromley (B3) gives the following equation to predict the heat-transfer coefficient in the film-boiling region on a horizontal tube.

$$h = 0.62 \left[\frac{k_v^3 \rho_v(\rho_l - \rho_v)g(h_{fg} + 0.4c_{pv}\,\Delta T)}{D\mu_v\,\Delta T} \right]^{1/4}$$

(4.8-6)

where k_v is the thermal conductivity of the vapor in W/m $\cdot$ K, ρ_v the density of the vapor in kg/m^3, ρ_l the density of the liquid in kg/m^3, h_{fg} the latent heat of vaporization in J/kg, $\Delta T = T_w - T_{\text{sat}}$, T_{sat} the temperature of saturated vapor in K, D the outside tube diameter in m, μ_v the viscosity of the vapor in Pa $\cdot$ s, and g the acceleration of gravity in m/s^2. The physical properties of the vapor are evaluated at the film temperature of $T_f = (T_w + T_{\text{sat}})/2$ and h_{fg} at the saturation temperature. If the temperature difference is quite high, some additional heat transfer occurs by radiation (H1).

EXAMPLE 4.8-1. Rate of Heat Transfer in a Jacketed Kettle
Water is being boiled at 1 atm abs pressure in a jacketed kettle with steam condensing in the jacket at 115.6°C. The inside diameter of the kettle is 0.656 m and the height is 0.984 m. The bottom is slightly curved but it will be assumed to be flat. Both the bottom and the sides up to a height of 0.656 m are jacketed. The kettle surface for heat transfer is 3.2-mm stainless steel with a k of 16.27 W/m $\cdot$ K. The condensing steam coefficient h_i inside the jacket has been estimated as 10 200 W/m^2 $\cdot$ K. Predict the boiling heat-transfer coefficient h_0 for the bottom surface of the kettle.

Solution: A diagram of the kettle is shown in Fig. 4.8-2. The simplified equations will be used for the boiling coefficient h_0. The solution is trial and error, since the inside metal surface temperature T_w is unknown. Assuming that $T_w = 110°C$,

$$\Delta T = T_w - T_{sat} = 110 - 100 = 10°C = 10 \text{ K}$$

Substituting into Eq. (4.8-2),

$$h_o = 5.56(\Delta T)^3 = 5.56(10)^3 = 5560 \text{ W/m}^2 \cdot \text{K}$$

$$\frac{q}{A} = h\Delta T = 5560(10) = 55\,600 \text{ W/m}^2$$

To check the assumed T_w, the resistances R_i of the condensing steam, R_w of the metal wall, and R_o of the boiling liquid must be calculated. Assuming equal areas of the resistances for $A = 1 \text{ m}^2$, then by Eq. (4.3-12),

$$R_i = \frac{1}{h_i A} = \frac{1}{10\,200(1)} = 9.80 \times 10^{-5}$$

$$R_w = \frac{\Delta x}{kA} = \frac{3.2/1000}{16.27(1.0)} = 19.66 \times 10^{-5}$$

$$R_o = \frac{1}{h_o A} = \frac{1}{5560(1)} = 17.98 \times 10^{-5}$$

$$\sum R = 9.80 \times 10^{-5} + 19.66 \times 10^{-5} + 17.98 \times 10^{-5} = 47.44 \times 10^{-5}$$

The temperature drop across the boiling film is then

$$\Delta T = \frac{R_o}{\sum R}(115.6 - 100) = \frac{17.98 \times 10^{-5}}{47.44 \times 10^{-5}}(15.6) = 5.9°C$$

Hence, $T_w = 100 + 5.9 = 105.9°C$. This is lower than the assumed value of 110°C.

For the second trial $T_w = 108.3°C$ will be used. Then, $\Delta T = 108.3 - 100 = 8.3°C$ and, from Eq. (4.8-2), the new $h_0 = 3180$. Calculating the new $R_0 = 31.44 \times 10^{-5}$, and

$$\Delta T = \left(\frac{31.44 \times 10^{-5}}{60.90 \times 10^{-5}}\right)(115.6 - 100) = 8.1°C$$

and

$$T_w = 100 + 8.1 = 108.1°C$$

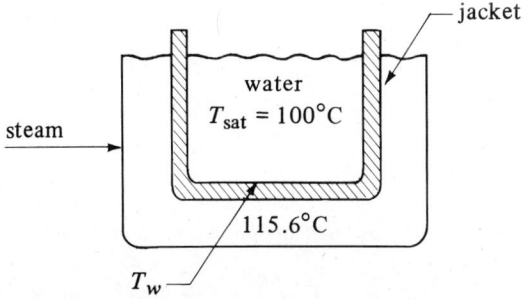

FIGURE 4.8-2.   *Steam-jacketed kettle and boiling water for Example 4.8-1.*

This value is reasonably close to the assumed value of 108.3°C, so no further trials will be made.

4.8B Condensation

1. Mechanisms of condensation. Condensation of a vapor to a liquid and vaporization of a liquid to a vapor both involve a change of phase of a fluid with large heat-transfer coefficients. Condensation occurs when a saturated vapor such as steam comes in contact with a solid whose surface temperature is below the saturation temperature, to form a liquid such as water.

Normally, when a vapor condenses on a surface such as a vertical or horizontal tube or other surfaces, a film of condensate is formed on the surface and flows over the surface by the action of gravity. It is this film of liquid between the surface and the vapor that forms the main resistance to heat transfer. This is called *film-type condensation.*

Another type of condensation, *dropwise condensation,* can occur where small drops are formed on the surface. These drops grow, coalesce, and the liquid flows from the surface. During this condensation, large areas of tube are devoid of any liquid and are exposed directly to the vapor. Very high rates of heat transfer occur on these bare areas. The average coefficient can be as high as $110\,000\ \text{W/m}^2 \cdot \text{K}$ ($20\,000\ \text{btu/h} \cdot \text{ft}^2 \cdot \text{°F}$), which is 5 to 10 times larger than film-type coefficients. Condensing film coefficients are normally much greater than those in forced convection and are of the order of magnitude of several thousand $\text{W/m}^2 \cdot \text{K}$ or more.

Dropwise condensation occurs on contaminated surfaces and when impurities are present. Film-type condensation is more dependable and more common. Hence, for normal design purposes, film-type condensation is assumed.

2. Film-condensation coefficients for vertical surfaces. Film-type condensation on a vertical wall or tube can be analyzed analytically by assuming laminar flow of the condensate film down the wall. The film thickness is zero at the top of the wall or tube and increases in thickness as it flows downward because of condensation. Nusselt (H1, W1) assumed that the heat transfer from the condensing vapor at T_{sat} K, through this liquid film, and to the wall at T_w K was by conduction. Equating this heat transfer by conduction to that from condensation of the vapor, a final expression can be obtained for the average heat-transfer coefficient over the whole surface.

In Fig. 4.8-3a vapor at T_{sat} is condensing on a wall whose temperature is T_w K. The condensate is flowing downward in laminar flow. Assuming unit thickness, the mass of the element with liquid density ρ_l in Fig. 4.8-3b is $(\delta - y)(dx \cdot 1)\rho_l$. The downward force on this element is the gravitational force minus the buoyancy force or $(\delta - y)(dx) \times (\rho_l - \rho_v)g$ where ρ_v is the density of the saturated vapor. This force is balanced by the viscous-shear force at the plane y of $\mu_l(dv/dy)(dx \cdot 1)$. Equating these forces,

$$(\delta - y)(dx)(\rho_l - \rho_v)g = \mu_l \left(\frac{dv}{dy}\right)(dx) \tag{4.8-7}$$

Integrating and using the boundary condition that $v = 0$ at $y = 0$,

$$v = \frac{g(\rho_l - \rho_v)}{\mu_l}(\delta y - y^2/2) \tag{4.8-8}$$

The mass flow rate of film condensate at any point x for unit depth is

$$m = \int_0^\delta \rho_l v\, dy = \int_0^\delta \rho_l \frac{g(\rho_l - \rho_v)}{\mu_l}(\delta y - y^2/2)\, dy \tag{4.8-9}$$

Transport Processes: Momentum, Heat, and Mass

Integrating,

$$m = \frac{\rho_l g(\rho_l - \rho_v)\delta^3}{3\mu_l}$$ (4.8-10)

At the wall for area $(dx \cdot 1)\, m^2$, the rate of heat transfer is as follows if a linear temperature distribution is assumed in the liquid between the wall and the vapor:

$$q_x = -k_l(dx \cdot 1) \frac{dT}{dy}\Big)_{y=0} = k_l\, dx \frac{T_{sat} - T_w}{\delta}$$ (4.8-11)

In a dx distance, the rate of heat transfer is q_x. Also, in this dx distance, the increase in mass from condensation is dm. Using Eq. (4.8-10),

$$dm = d\left[\frac{\rho_l g(\rho_l - \rho_v)\delta^3}{3\mu_l}\right] = \frac{\rho_l g(\rho_l - \rho_v)\delta^2\, d\delta}{\mu_l}$$ (4.8-12)

Making a heat balance for dx distance, the mass flow rate dm times the latent heat h_{fg} must equal the q_x from Eq. (4.8-11).

$$h_{fg} \frac{\rho_l g(\rho_l - \rho_v)\delta^2\, d\delta}{\mu_l} = k_l\, dx \frac{T_{sat} - T_w}{\delta}$$ (4.8-13)

Integrating with $\delta = 0$ at $x = 0$ and $\delta = \delta$ at $x = x$,

$$\delta = \left[\frac{4\mu_l k_l x(T_{sat} - T_w)}{gh_{fg}\rho_l(\rho_l - \rho_v)}\right]^{1/4}$$ (4.8-14)

Using the local heat-transfer coefficient h_x at x, a heat balance gives

$$h_x(dx \cdot 1)(T_{sat} - T_w) = k_l(dx \cdot 1) \frac{T_{sat} - T_w}{\delta}$$ (4.8-15)

This gives

$$h_x = \frac{k_l}{\delta}$$ (4.8-16)

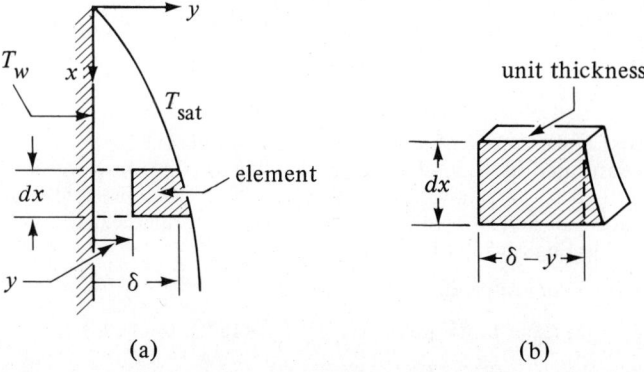

(a)   (b)

FIGURE 4.8-3.   Film condensation on a vertical plate : (a) increase in film thickness with position, (b) balance on element of condensate.

Combining Eqs. (4.8-14) and (4.8-16),

$$h_x = \left[\frac{\rho_l (\rho_l - \rho_v)gh_{fg} \, k_l^3}{4\mu_l \, x(T_{sat} - T_w)} \right]^{1/4} \tag{4.8-17}$$

By integrating over the total length L, the average value of h is obtained as follows.

$$h = \frac{1}{L} \int_0^L h_x \, dx = \tfrac{4}{3} h_{x=L} \tag{4.8-18}$$

$$h = 0.943 \left[\frac{\rho_l (\rho_l - \rho_v)gh_{fg} \, k_l^3}{\mu_l \, L(T_{sat} - T_w)} \right]^{1/4} \tag{4.8-19}$$

However, for laminar flow, experimental data show that the data are about 20% above Eq. (4.8-19).

Hence, the final recommended expression for vertical surfaces in laminar flow is (M1)

$$N_{Nu} = \frac{hL}{k_l} = 1.13 \left(\frac{\rho_l (\rho_l - \rho_v)gh_{fg} L^3}{\mu_l k_l \Delta T} \right)^{1/4} \tag{4.8-20}$$

where ρ_l is the density of liquid in kg/m^3 and ρ_v that of the vapor, g is 9.8066 m/s^2, L is the vertical height of the surface or tube in m, μ_l the viscosity of liquid in Pa·s, k_l the liquid thermal conductivity in W/m·K, $\Delta T = T_{sat} - T_w$ in K, and h_{fg} latent heat of condensation in J/kg at T_{sat}. All physical properties of the liquid except h_{fg} are evaluated at the film temperature $T_f = (T_{sat} + T_w)/2$. For long vertical surfaces the flow at the bottom can be turbulent. The Reynolds number is defined as

$$N_{Re} = \frac{4m}{\pi D \mu_l} = \frac{4\Gamma}{\mu_l} \qquad \text{(vertical tube, diameter } D) \tag{4.8-21}$$

$$N_{Re} = \frac{4m}{W \mu_l} = \frac{4\Gamma}{\mu_l} \qquad \text{(vertical plate, width } W) \tag{4.8-22}$$

where m is total kg mass/s of condensate at tube or plate bottom and $\Gamma = m/\pi D$ or m/W. The N_{Re} should be below about 1800 for Eq. (4.8-20) to hold. The reader should note that some references define N_{Re} as Γ/μ. Then this N_{Re} should be below 450.

For turbulent flow for $N_{Re} > 1800$ (M1),

$$N_{Nu} = \frac{hL}{k_l} = 0.0077 \left(\frac{g\rho_l^2 L^3}{\mu_l^2} \right)^{1/3} (N_{Re})^{0.4} \tag{4.8-23}$$

Solution of this equation is by trial and error since a value of N_{Re} must first be assumed to calculate h.

EXAMPLE 4.8-2. Condensation on a Vertical Tube
Steam saturated at 68.9 kPa (10 psia) is condensing on a vertical tube 0.305 m (1.0 ft) long having an OD of 0.0254 m (1.0 in.) and a surface temperature of 86.11°C (187°F). Calculate the average heat-transfer coefficient using English and SI units.

Solution: From Appendix A.2,

$$T_{sat} = 193°F \ (89.44°C) \qquad T_w = 187°F \ (86.11°C)$$

$$T_f = \frac{T_w + T_{sat}}{2} = \frac{187 + 193}{2} = 190°F \ (87.8°C)$$

latent heat $h_{fg} = 1143.3 - 161.0 = 982.3$ btu/lb$_m$

$$= 2657.8 - 374.6 = 2283.2 \text{ kJ/kg} = 2.283 \times 10^6 \text{ J/kg}$$

$$\rho_l = \frac{1}{0.01657} = 60.3 \text{ lb}_m/\text{ft}^3 = 60.3(16.018) = 966.7 \text{ kg/m}^3$$

$$\rho_v = \frac{1}{40.95} = 0.0244 \text{ lb}_m/\text{ft}^3 = 0.391 \text{ kg/m}^3$$

$$\mu_l = (0.324 \text{ cp})(2.4191) = 0.784 \text{ lb}_m/\text{ft} \cdot \text{h} = 3.24 \times 10^{-4} \text{ Pa} \cdot \text{s}$$

$$k_l = 0.390 \text{ btu/ft} \cdot \text{h} \cdot °\text{F} = (0.390)(1.7307) = 0.675 \text{ W/m} \cdot \text{K}$$

$$L = 1 \text{ ft} = 0.305 \text{ m} \qquad \Delta T = T_{\text{sat}} - T_w = 193 - 187 = 6°\text{F (3.33 K)}$$

Assuming a laminar film, using Eq. (4.8-20) in English and also SI units, and neglecting ρ_v as compared to ρ_l,

$$N_{\text{Nu}} = 1.13 \left(\frac{\rho_l^2 g h_{fg} L^3}{\mu_l k_l \Delta T} \right)^{1/4}$$

$$= 1.13 \left[\frac{(60.3)^2 (32.174)(3600)^2 (982.3)(1.0)^3}{(0.784)(0.390)(6)} \right]^{1/4} = 6040$$

$$N_{\text{Nu}} = 1.13 \left[\frac{(966.7)^2 (9.806)(2.283 \times 10^6)(0.305)^3}{(3.24 \times 10^{-4})(0.675)(3.33)} \right]^{1/4} = 6040$$

$$N_{\text{Nu}} = \frac{hL}{k_l} = \frac{h(1.0)}{0.390} = 6040 \qquad \text{SI units:} \frac{h(0.305)}{0.675} = 6040$$

Solving, $h = 2350$ btu/h $\cdot$ ft$^2 \cdot °$F $= 13\,350$ W/m$^2 \cdot$ K.

Next, the N_{Re} will be calculated to see if laminar flow occurs as assumed. To calculate the total heat transferred for a tube of area

$$A = \pi DL = \pi(1/12)(1.0) = \pi/12 \text{ ft}^2, \qquad A = \pi(0.0254)(0.305) \text{ m}^2$$

$$q = hA \, \Delta T \tag{4.8-24}$$

However, this q must also equal that obtained by condensation of m lb$_m$/h or kg/s. Hence,

$$q = hA \, \Delta T = h_{fg} m \tag{4.8-25}$$

Substituting the values given and solving for m,

$$2350(\pi/12)(193 - 187) = 982.3(m) \qquad m = 3.77 \text{ lb}_m/\text{h}$$

$$13.350(\pi)(0.0254)(0.305)(3.33) = 2.284 \times 10^6(m) \qquad m = 4.74 \times 10^{-4} \text{ kg/s}$$

Substituting into Eq. (4.8-21),

$$N_{\text{Re}} = \frac{4m}{\pi D \mu_l} = \frac{4(3.77)}{\pi(1/12)(0.784)} = 73.5 \qquad N_{\text{Re}} = \frac{4(4.74 \times 10^{-4})}{\pi(0.0254)(3.24 \times 10^{-4})} = 73.5$$

Hence, the flow is laminar as assumed.

3. Film-condensation coefficients outside horizontal cylinders. The analysis of Nusselt can also be extended to the practical case of condensation outside a horizontal tube. For a single tube the film starts out with zero thickness at the top of the tube and increases in thickness as it flows around to the bottom and then drips off. If there is a bank of horizontal tubes, the condensate from the top tube drips onto the one below; and so on.

For a vertical tier of N horizontal tubes placed one below the other with outside tube diameter D (M1),

$$N_{Nu} = \frac{hD}{k_l} = 0.725\left(\frac{\rho_l(\rho_l - \rho_v)gh_{fg}D^3}{N\mu_l k_l \Delta T}\right)^{1/4} \tag{4.8-26}$$

In most practical applications, the flow is in the laminar region and Eq. (4.8-26) holds (C3, M1).

4.9 INTRODUCTION TO RADIATION HEAT TRANSFER

4.9A Introduction and Basic Equation for Radiation

1. Nature of radiant heat transfer. In the preceding sections of this chapter we have studied conduction and convection heat transfer. In conduction heat is transferred from one part of a body to another, and the intervening material is heated. In convection the heat is transferred by the actual mixing of materials and by conduction. In radiant heat transfer the medium through which the heat is transferred usually is not heated. Radiation heat transfer is the transfer of heat by electromagnetic radiation.

Thermal radiation is a form of electromagnetic radiation similar to x rays, light waves, gamma rays, and so on, differing only in wavelength. It obeys the same laws as light: travels in straight lines, can be transmitted through space and vacuum, and so on. It is an important mode of heat transfer and is especially important where large temperature differences occur, as, for example, in a furnace with boiler tubes, in radiant dryers, and in an oven baking food. Radiation often occurs in combination with conduction and convection. An elementary discussion of radiant heat transfer will be given here, with a more advanced and comprehensive discussion being given in Section 4.11 (Selected Topic).

In an elementary sense the mechanism of radiant heat transfer is composed of three distinct steps or phases:

1. The thermal energy of a hot source, such as the wall of a furnace at T_1, is converted into the energy of electromagnetic radiation waves.
2. These waves travel through the intervening space in straight lines and strike a cold object at T_2 such as a furnace tube containing water to be heated.
3. The electromagnetic waves that strike the body are absorbed by the body and converted back to thermal energy or heat.

2. Absorptivity and black bodies. When thermal radiation (like light waves) falls upon a body, part is absorbed by the body in the form of heat, part is reflected back into space, and part may be actually transmitted through the body. For most cases in process engineering, bodies are opaque to transmission, so this will be neglected. Hence, for opaque bodies,

$$\alpha + \rho = 1.0 \tag{4.9-1}$$

where α is absorptivity or fraction absorbed and ρ is reflectivity or fraction reflected.

A *black body* is defined as one that absorbs all radiant energy and reflects none. Hence, ρ is 0 and $\alpha = 1.0$ for a black body. Actually, in practice there are no perfect black bodies, but a close approximation to this is a small hole in a hollow body, as shown in

FIGURE 4.9-1. *Concept of a perfect black body.*

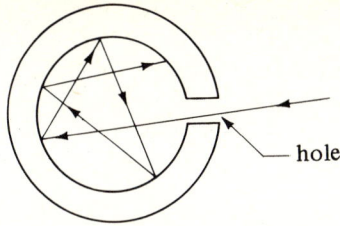

hole

Fig. 4.9-1. The inside surface of the hollow body is blackened by charcoal. The radiation enters the hole and impinges on the rear wall; part is absorbed there and part is reflected in all directions. The reflected rays impinge again, part is absorbed, and the process continues. Hence, essentially all of the energy entering is absorbed and the area of the hole acts as a perfect black body. The surface of the inside walls are "rough" and rays are scattered in all directions, unlike a mirror, where they are reflected at a definite angle.

As stated previously, a black body absorbs all radiant energy falling on it and reflects none. Such a black body also emits radiation, depending on its temperature, and does not reflect any. The ratio of the emissive power of a surface to that of a black body is called *emissivity* ε and it is 1.0 for a black body. Kirchhoff's law states that at the same temperature T_1, α_1 and ε_1 of a given surface are the same or

$$\alpha_1 = \varepsilon_1 \tag{4.9-2}$$

Equation (4.9-2) holds for any black or nonblack solid surface.

3. Radiation from a body and emissivity. The basic equation for heat transfer by radiation from a perfect black body with an emissivity $\varepsilon = 1.0$ is

$$q = A\sigma T^4 \tag{4.9-3}$$

where q is heat flow in W, A is m^2 surface area of body, σ a constant 5.676×10^{-8} W/m$^2 \cdot$ K^4 (0.1714×10^{-8} btu/h $\cdot$ ft$^2 \cdot$ °R^4), and T is temperature of the black body in K (°R).

For a body that is not a black body and has an emissivity $\varepsilon < 1.0$, the emissive power is reduced by ε, or

$$q = A\varepsilon\sigma T^4 \tag{4.9-4}$$

Substances that have emissivities of less than 1.0 are called *gray bodies* when the emissivity is independent of the wavelength. All real materials have an emissivity $\varepsilon < 1$.

Since the emissivity ε and absorptivity α of a body are equal at the same temperature, the emissivity, like absorptivity, is low for polished metal surfaces and high for oxidized metal surfaces. Typical values are given in Table 4.9-1 but do vary some with temperature. Most nonmetallic substances have high values. Additional data are tabulated in Appendix A.3.

4.9B Radiation to a Small Object from Surroundings

When we have the case of a small gray object of area A_1 m^2 at temperature T_1 in a large enclosure at a higher temperature T_2, there is a net radiation to the small object. The small body emits an amount of radiation to the enclosure given by Eq. (4.9-4) of $A_1\varepsilon_1\sigma T_1^4$. The emissivity ε_1 of this body is taken at T_1. The small body also absorbs energy from the surroundings at T_2 given by $A_1\alpha_{12}\sigma T_2^4$. The α_{12} is the absorptivity of

TABLE 4.9-1. *Total Emissivity, ε, of Various Surfaces*

Surface	$T(K)$	$T(°F)$	Emissivity, ε
Polished aluminum	500	440	0.039
	850	1070	0.057
Polished iron	450	350	0.052
Oxidized iron	373	212	0.74
Polished copper	353	176	0.018
Asbestos board	296	74	0.96
Oil paints, all colors	373	212	0.92–0.96
Water	273	32	0.95

body 1 for radiation from the enclosure at T_2. The value of α_{12} is approximately the same as the emissivity of this body at T_2. The net heat of absorption is then, by the Stefan-Boltzmann equation,

$$q = A_1\varepsilon_1\sigma T_1^4 - A_1\alpha_{12}\sigma T_2^4 = A_1\sigma(\varepsilon_1 T_1^4 - \alpha_{12} T_2^4) \qquad (4.9\text{-}5)$$

A further simplification of Eq. (4.9-5) is usually made for engineering purposes by using one emissivity of the small body at temperature T_2. Thus,

$$q = A_1\varepsilon\sigma(T_1^4 - T_2^4) \qquad (4.9\text{-}6)$$

EXAMPLE 4.9-1.   Radiation to a Metal Tube
A small oxidized horizontal metal tube with an OD of 0.0254 m (1 in.) and being 0.61 m (2 ft) long with a surface temperature at 588 K (600°F) is in a very large furnace enclosure with fire-brick walls and the surrounding air at 1088 K (1500°F). The emissivity of the metal tube is 0.60 at 1088 K and 0.46 at 588 K. Calculate the heat transfer to the tube by radiation using SI and English units.

Solution: Since the large furnace surroundings are very large compared with the small enclosed tube, the surroundings, even if gray, when viewed from the position of the small body appear black and Eq. (4.9-6) is applicable. Substituting the given values into Eq. (4.9-6) with an ε of 0.6 at 1088 K and

$$A_1 = \pi DL = \pi(0.0254)(0.61) \text{ m}^2 = \pi(1/12)(2.0) \text{ ft}^2$$

$$q = A_1\varepsilon\sigma(T_1^4 - T_2^4) = [\pi(0.0254)(0.61)](0.6)(5.676 \times 10^{-8})[(588)^4 - (1088)^4]$$

$$= -2130 \text{ W}$$

$$= [\pi(1/12)(2)](0.6)(0.1714 \times 10^{-8})[(1060)^4 - (1960)^4] = -7270 \text{ btu/h}$$

Other examples of small objects in large enclosures occurring in the process industries are a loaf of bread in an oven receiving radiation from the walls around it, a package of meat or food radiating heat to the walls of a freezing enclosure, a hot ingot of solid iron cooling and radiating heat in a large room, and a thermometer measuring the temperature in a large duct.

4.9C   Combined Radiation and Convection Heat Transfer

When radiation heat transfer occurs from a surface it is usually accompanied by convective heat transfer unless the surface is in a vacuum. When the radiating surface is

Transport Processes: Momentum, Heat, and Mass

at a uniform temperature, we can calculate the heat transfer for natural or forced convection using the methods described in the previous sections of this chapter. The radiation heat transfer is calculated by the Stefan-Boltzmann equation (4.9-6). Then the total rate of heat transfer is the sum of convection plus radiation.

As discussed before, the heat-transfer rate by convection and the convective coefficient are given by

$$q_{conv} = h_c A_1 (T_1 - T_2) \tag{4.9-7}$$

where q_{conv} is the heat-transfer rate by convection in W, h_c the natural or forced convection coefficient in $W/m^2 \cdot K$, T_1 the temperature of the surface, and T_2 the temperature of the air and the enclosure. A radiation heat-transfer coefficient h_r in $W/m^2 \cdot K$ can be defined as

$$q_{rad} = h_r A_1 (T_1 - T_2) \tag{4.9-8}$$

where q_{rad} is the heat-transfer rate by radiation in W. The total heat transfer is the sum of Eqs. (4.9-7) and (4.9-8),

$$q = q_{conv} + q_{rad} = (h_c + h_r) A_1 (T_1 - T_2) \tag{4.9-9}$$

To obtain an expression for h_r, we equate Eq. (4.9-6) to (4.9-8) and solve for h_r.

$$h_r = \frac{\varepsilon \sigma (T_1^4 - T_2^4)}{T_1 - T_2} = \varepsilon(5.676) \frac{(T_1/100)^4 - (T_2/100)^4}{T_1 - T_2} \qquad \text{(SI)}$$

$$h_r = \varepsilon(0.1714) \frac{(T_1/100)^4 - (T_2/100)^4}{T_1 - T_2} \qquad \text{(English)} \tag{4.9-10}$$

A convenient chart giving values of h_r in English units calculated from Eq. (4.9-10) with $\varepsilon = 1.0$ is given in Fig. 4.9-2. To use values from this figure, the value obtained from

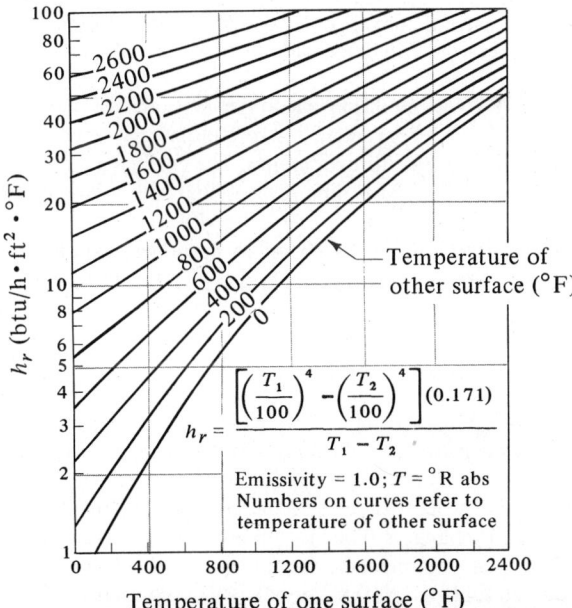

FIGURE 4.9-2. Radiation heat-transfer coefficient as a function of temperature. (From R. H. Perry and C. H. Chilton, Chemical Engineers' Handbook, 5th ed. New York: McGraw-Hill Book Company, 1973. With permission.)

this figure should be multiplied by ε to give the value of h_r to use in Eq. (4.9-9). If the air temperature is not the same as T_2 of the enclosure, Eqs. (4.9-7) and (4.9-8) must be used separately and not combined together as in (4.9-9).

EXAMPLE 4.9-2. Combined Convection Plus Radiation from a Tube
Recalculate Example 4.9-1 for combined radiation plus natural convection to the horizontal 0.0254-m tube.

Solution: The area A of the tube $= \pi(0.0254)(0.61) = 0.0487$ m^2. For the natural convection coefficient to the 0.0254-m horizontal tube, the simplified equation from Table 4.7-2 will be used as an approximation even though the film temperature is quite high.

$$h_c = 1.32\left(\frac{\Delta T}{D}\right)^{1/4}$$

Substituting the known values,

$$h_c = 1.32\left(\frac{1088 - 588}{0.0254}\right)^{1/4} = 15.64 \text{ W/m}^2 \cdot \text{K}$$

Using Eq. (4.9-10) and $\varepsilon = 0.6$,

$$h_r = (0.60)(5.676)\frac{(1088/100)^4 - (588/100)^4}{1088 - 588} = 87.3 \text{ W/m}^2 \cdot \text{K}$$

Substituting into Eq. (4.9-9),

$$q = (h_c + h_r)A_1(T_1 - T_2) = (15.64 + 87.3)(0.0487)(588 - 1088)$$

$$= -2507 \text{ W}$$

Hence, the heat loss of -2130 W for radiation is increased to only -2507 W when natural convection is also considered. In this case, because of the large temperature difference, radiation is the most important factor.

Perry and Chilton (P1, p. 10–11) give a convenient table of natural convection plus radiation coefficients $(h_c + h_r)$ from single horizontal oxidized steel pipes as a function of the outside diameter and temperature difference. The coefficients for insulated pipes are about the same as those for a bare pipe (except that lower surface temperatures are involved for the insulated pipes), since the emissivity of cloth insulation wrapping is about that of oxidized steel, approximately 0.8. A more detailed discussion of radiation will be given in Section 4.11 (Selected Topic).

4.10   HEAT EXCHANGERS

4.10A   Types of Exchangers

1. Introduction.   In the process industries the transfer of heat between two fluids is generally done in heat exchangers. The most common type is one in which the hot and the cold fluid do not come into direct contact with each other but are separated by a tube wall or a flat or curved surface. The transfer of heat is accomplished from the hot fluid to the wall or tube surface by convection, through the tube wall or plate by conduction, and then by convection to the cold fluid. In the preceding sections of this chapter we have discussed the calculation procedures for these various steps. Now we will discuss some of

Transport Processes: Momentum, Heat, and Mass

the types of equipment used and overall thermal analyses of exchangers. Complete detailed design methods have been highly developed and will not be considered here.

2. Double-pipe heat exchanger. The simplest exchanger is the double-pipe or concentric-pipe exchanger. This is shown in Fig. 4.10-1, where the one fluid flows inside one pipe and the other fluid in the annular space between the two pipes. The fluids can be in cocurrent or countercurrent flow. The exchanger can be made from a pair of single lengths of pipe with fittings at the ends or from a number of pairs interconnected in series. This type of exchanger is useful mainly for small flow rates.

3. Shell and tube exchanger. If larger flows are involved, a shell and tube exchanger is used, which is the most important type of exchanger in use in the process industries. In these exchangers the flows are continuous. Many tubes in parallel are used where one fluid flows inside these tubes. The tubes, arranged in a bundle, are enclosed in a single shell and the other fluid flows outside the tubes in the shell side. The simplest shell and tube exchanger is shown in Fig. 4.10-2a for 1 shell pass and 1 tube pass, or a 1–1 counterflow exchanger. The cold fluid enters and flows inside through all the tubes in parallel in one pass. The hot fluid enters at the other end and flows counterflow across the outside of the tubes. Cross baffles are used so that the fluid is forced to flow perpendicular across the tube bank rather than parallel with it. This added turbulence generated by this cross flow increases the shell-side heat-transfer coefficient.

In Fig. 4.10-2b a 1–2 parallel–counterflow exchanger is shown. The liquid on the tube side flows in two passes as shown and the shell-side liquid flows in one pass. In the first pass of the tube side the cold fluid is flowing counterflow to the hot shell-side fluid, and in the second pass of the tube side the cold fluid flows in parallel (cocurrent) with the hot fluid. Another type of exchanger has 2 shell-side passes and 4 tube passes. Other combinations of number of passes are also used sometimes, with the 1–2 and 2–4 types being the most common.

4. Cross-flow exchanger. When a gas such as air is being heated or cooled, a common device used is the cross-flow heat exchanger shown in Fig. 4.10-3a. One of the fluids, which is a liquid, flows inside through the tubes and the exterior gas flows across the tube bundle by forced or sometimes natural convection. The fluid inside the tubes is considered to be unmixed since it is confined and cannot mix with any other stream. The gas flow outside the tubes is mixed since it can move about freely between the tubes and there will be a tendency for the gas temperature to equalize in the direction normal to the flow. For the unmixed fluid inside the tubes there will be a temperature gradient both parallel and normal to the direction of flow.

A second type of cross-flow heat exchanger shown in Fig. 4.10-3b is used typically in

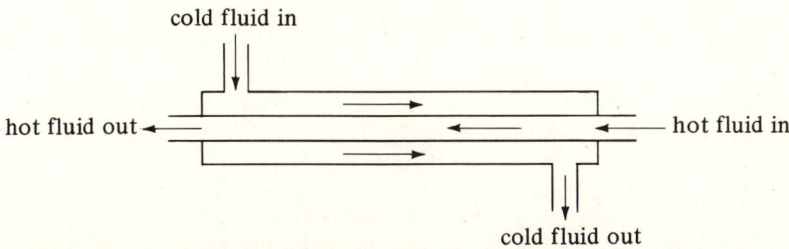

cold fluid in

hot fluid out ← — — — — — — — — — — ← — ← — hot fluid in

cold fluid out

FIGURE 4.10-1. *Flow in a double-pipe heat exchanger.*

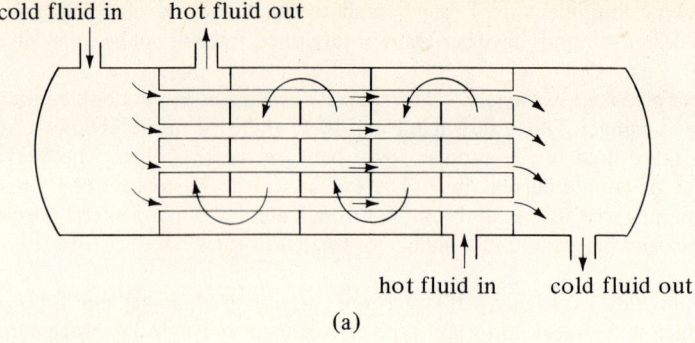

cold fluid in hot fluid out

hot fluid in cold fluid out

(a)

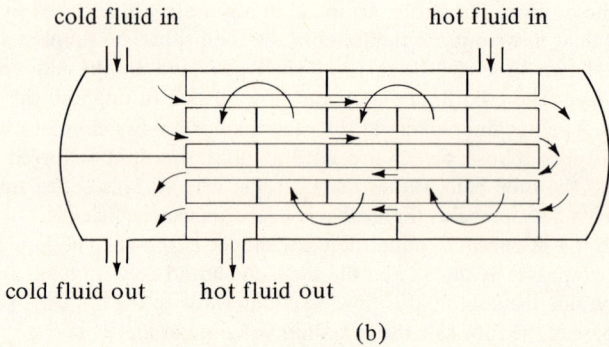

cold fluid in hot fluid in

cold fluid out hot fluid out

(b)

FIGURE 4.10-2. *Shell and tube heat exchangers : (a) 1 shell pass and 1 tube pass (1–1 exchanger), (b) 1 shell pass and 2 tube passes (1–2 exchanger).*

air-conditioning and space-heating applications. In this type the gas flows across a finned-tube bundle and is unmixed since it is confined in separate flow channels between the fins as it passes over the tubes. The fluid in the tubes is unmixed.

Discussions of other types of specialized heat-transfer equipment is deferred to

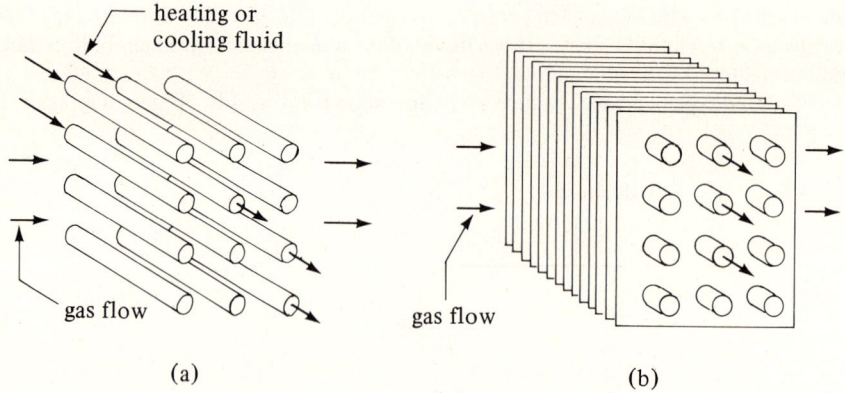

heating or cooling fluid

gas flow

(a)

gas flow

(b)

FIGURE 4.10-3. *Flow patterns of cross-flow heat exchangers : (a) one fluid mixed (gas) and one fluid unmixed, (b) both fluids unmixed.*

Transport Processes: Momentum, Heat, and Mass

Section 4.13. The remainder of this section deals primarily with a shell-and-tube and cross-flow heat exchangers.

4.10B Log Mean Temperature Difference Correction Factors

In Section 4.5H it was shown that when the hot and cold fluids in a heat exchanger are in true countercurrent flow or in cocurrent (parallel) flow, the log mean temperature difference should be used.

$$\Delta T_{1m} = \frac{\Delta T_2 - \Delta T_1}{\ln (\Delta T_2 / \Delta T_1)} \qquad (4.10\text{-}1)$$

where ΔT_2 is the temperature difference at one end of the exchanger and ΔT_1 at the other end. This ΔT_{1m} holds for a double-pipe heat exchanger and a 1–1 exchanger with 1 shell pass and 1 tube pass in parallel or counterflow.

In the cases where a multiple-pass heat exchanger is involved, it is necessary to obtain a different expression for the mean temperature difference to use, depending on the arrangement of the shell and tube passes. Considering first the one-shell-pass, two-tube-pass exchanger in Fig. 4.10-2b, the cold fluid in the first tube pass is in counterflow with the hot fluid. In the second tube pass the cold fluid is in parallel flow with the hot fluid. Hence, the log mean temperature difference, which applies to either parallel or counterflow but not to a mixture of both types, as in a 1–2 exchanger, cannot be used to calculate the true mean temperature drop without a correction.

The mathematical derivation of the equation for the proper mean temperature to use is quite complex. The usual procedure is to use a correction factor F_T which is so defined that when it is multiplied by the ΔT_{1m}, the product is the correct mean temperature drop ΔT_m to use. In using the correction factors F_T it is immaterial whether the warmer fluid flows through the tubes or shell (K1). The factor F_T has been calculated (B4) for a 1–2 exchanger and is shown in Fig. 4.10-4a. Two dimensionless ratios are used as follows.

$$Z = \frac{T_{hi} - T_{ho}}{T_{co} - T_{ci}} \qquad (4.10\text{-}2)$$

$$Y = \frac{T_{co} - T_{ci}}{T_{hi} - T_{ci}} \qquad (4.10\text{-}3)$$

where T_{hi} = inlet temperature of hot fluid in K (°F), T_{ho} = outlet of hot fluid, T_{ci} inlet of cold fluid, and T_{co} = outlet of cold fluid.

In Fig. 4.10-4b the factor F_T (B4) for a 2–4 exchanger is shown. In general, it is not recommended to use a heat exchanger for conditions under which $F_T < 0.75$. Another shell and tube arrangement should be used. Correction factors for two types of cross-flow exchangers are given in Fig. 4.10-5. Other types are available elsewhere (B4, P1).

Using the nomenclature of Eqs. (4.10-2) and (4.10-3), the ΔT_{1m} of Eq. (4.10-1) can be written as

$$\Delta T_{1m} = \frac{(T_{hi} - T_{co}) - (T_{ho} - T_{ci})}{\ln [(T_{hi} - T_{co})/(T_{ho} - T_{ci})]} \qquad (4.10\text{-}4)$$

Then the equation for an exchanger is

$$q = U_i A_i \Delta T_m = U_o A_o \Delta T_m \qquad (4.10\text{-}5)$$

where

$$\Delta T_m = F_T \Delta T_{1m} \qquad (4.10\text{-}6)$$

Principles of Steady-State Heat Transfer   **265**

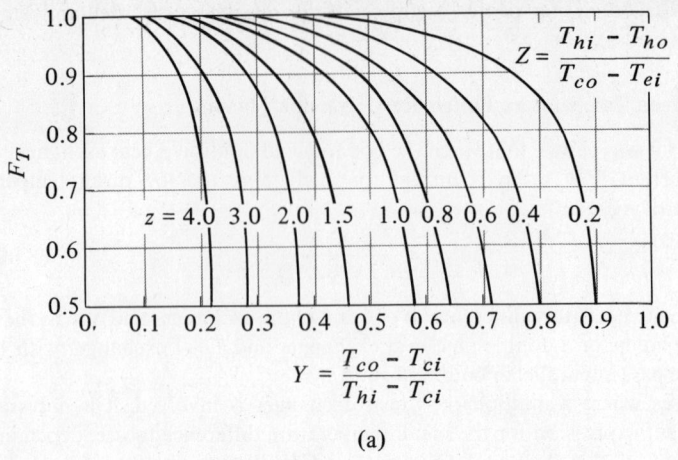

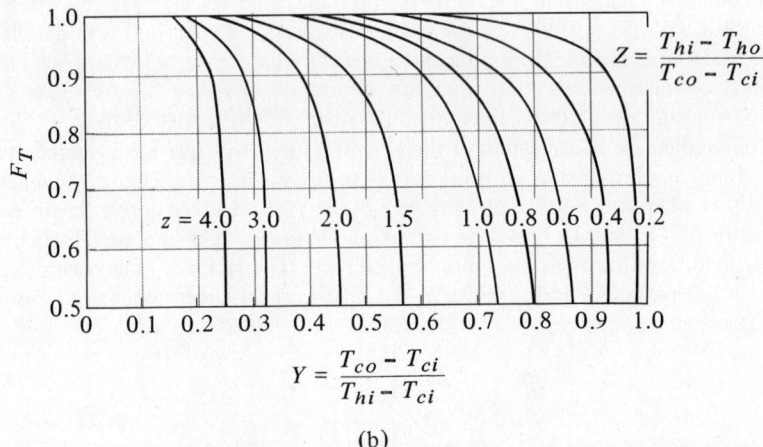

FIGURE 4.10-4. *Correction factor F_T to log mean temperature difference: (a) 1–2 exchangers, (b) 2–4 exchangers. [From R. A. Bowman, A. C. Mueller, and W. M. Nagle, Trans. A.S.M.E., **62**, 284, 285 (1940). With permission.]*

EXAMPLE 4.10-1. Temperature Correction Factor for a Heat Exchanger
A 1–2 heat exchanger containing one shell pass and two tube passes heats 2.52 kg/s of water from 21.1 to 54.4°C by using hot water under pressure entering at 115.6 and leaving at 48.9°C. The outside surface area of the tubes in the exchanger is $A_o = 9.30 \text{ m}^2$.

(a) Calculate the mean temperature difference ΔT_m in the exchanger and the overall heat-transfer coefficient U_o.

(b) For the same temperatures but using a 2–4 exchanger, what would be the ΔT_m?

Solution: The temperatures are as follows.

$$T_{hi} = 115.6°C \qquad T_{ho} = 48.9°C \qquad T_{ci} = 21.1°C \qquad T_{co} = 54.4°C$$

First making a heat balance on the cold water assuming a c_{pm} of water

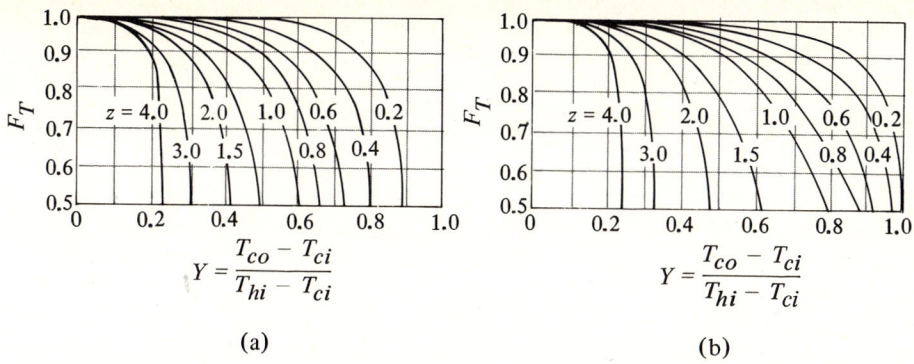

$$Y = \frac{T_{co} - T_{ci}}{T_{hi} - T_{ci}}$$

(a)

$$Y = \frac{T_{co} - T_{ci}}{T_{hi} - T_{ci}}$$

(b)

FIGURE 4.10-5. *Correction factor F_T to log mean temperature difference for cross-flow exchangers $[Z = (T_{hi} - T_{ho})/(T_{co} - T_{ci})]$: (a) single pass, shell fluid mixed, other fluid unmixed, (b) single pass, both fluids unmixed. [From R. A. Bowman, A. C. Mueller, and W. M. Nagle, Trans. A.S.M.E., 62, 288, 289 (1940). With permission.]*

of 4187 J/kg · K and $T_{co} - T_{ci} = (54.4 - 21.1)°C = 33.3°C = 33.3$ K,

$$q = mc_{pm}(T_{co} - T_{ci}) = (2.52)(4187)(54.4 - 21.1) = 348\,200 \text{ W}$$

The log mean temperature difference using Eq. (4.10-4) is

$$\Delta T_{lm} = \frac{(115.6 - 54.4) - (48.9 - 21.1)}{\ln\,[(115.6 - 54.4)/(48.9 - 21.1)]} = 42.3°C = 42.3 \text{ K}$$

Next substituting into Eqs. (4.10-2) and (4.10-3),

$$Z = \frac{T_{hi} - T_{ho}}{T_{co} - T_{ci}} = \frac{115.6 - 48.9}{54.4 - 21.1} = 2.00 \qquad \textbf{(4.10-2)}$$

$$Y = \frac{T_{co} - T_{ci}}{T_{hi} - T_{ci}} = \frac{54.4 - 21.1}{115.6 - 21.1} = 0.352 \qquad \textbf{(4.10-3)}$$

From Fig. 4.10-4a, $F_T = 0.74$. Then, by Eq. (4.10-6),

$$\Delta T_m = F_T \,\Delta T_{lm} = 0.74(42.3) = 31.3°C = 31.3 \text{ K} \qquad \textbf{(4.10-6)}$$

Rearranging Eq. (4.10-5) to solve for U_o and substituting the known values,

$$U_o = \frac{q}{A_o \,\Delta T_m} = \frac{348\,200}{(9.30)(31.3)} = 1196 \text{ W/m}^2 \cdot \text{K } (211 \text{ btu/h} \cdot \text{ft}^2 \cdot °\text{F})$$

For part (b), using a 2–4 exchanger and Fig. 4.10-4b, $F_T = 0.94$. Then,

$$\Delta T_m = F_T \,\Delta T_{lm} = 0.94(42.3) = 39.8°C = 39.8 \text{ K}$$

Hence, in this case the 2–4 exchanger utilizes more of the available temperature driving force.

4.10C   Heat-Exchanger Effectiveness

1. Introduction. In the preceding section the log mean temperature difference was used in the equation $q = UA \,\Delta T_{lm}$ in the design of heat exchangers. This form is convenient when the inlet and outlet temperatures of the two fluids are known or can be determined by a heat balance. Then the surface area can be determined if U is known. However, when the temperatures of the fluids leaving the exchanger are not known and a given exchanger is to be used, a tedious trial-and-error procedure is necessary. To solve these cases, a method called the heat-exchanger effectiveness ε is used which does not involve any of the outlet temperatures.

The heat-exchanger effectiveness is defined as the ratio of the actual rate of heat transfer in a given exchanger to the maximum possible amount of heat transfer if an infinite heat-transfer area were available. The temperature profile for a counterflow heat exchanger is shown in Fig. 4.10-6.

2. Derivation of effectiveness equation. The heat balance for the cold (C) and the hot (H) fluids is

$$q = (mc_p)_H(T_{Hi} - T_{Ho}) = (mc_p)_C(T_{Co} - T_{Ci}) \tag{4.10-7}$$

Calling $(mc_p)_H = C_H$ and $(mc_p)_C = C_C$, then in Fig. 4.10-6, $C_H > C_C$ and the cold fluid undergoes a greater temperature change than the hot fluid. Hence, we designate C_C as C_{min} or minimum heat capacity. Then, if there is an infinite area available for heat transfer, $T_{Co} = T_{Hi}$. Then the effectiveness ε is

$$\varepsilon = \frac{C_H(T_{Hi} - T_{Ho})}{C_C(T_{Hi} - T_{Ci})} = \frac{C_{max}(T_{Hi} - T_{Ho})}{C_{min}(T_{Hi} - T_{Ci})} \tag{4.10-8}$$

If the hot fluid is the minimum fluid, $T_{Ho} = T_{Ci}$, and

$$\varepsilon = \frac{C_C(T_{Co} - T_{Ci})}{C_H(T_{Hi} - T_{Ci})} = \frac{C_{max}(T_{Co} - T_{Ci})}{C_{min}(T_{Hi} - T_{Ci})} \tag{4.10-9}$$

In both equations the denominators are the same and the numerator gives the actual heat transfer.

$$q = \varepsilon C_{min}(T_{Hi} - T_{Ci}) \tag{4.10-10}$$

Note that Eq. (4.10-10) uses only inlet temperatures, which is an advantage when inlet temperatures are known and it is desired to predict the outlet temperatures for a given existing exchanger.

For the case of a single-pass, counterflow exchanger, combining Eqs. (4.10-8) and (4.10-9),

$$\varepsilon = \frac{C_H(T_{Hi} - T_{Ho})}{C_{min}(T_{Hi} - T_{Ci})} = \frac{C_C(T_{Co} - T_{Ci})}{C_{min}(T_{Hi} - T_{Ci})} \tag{4.10-11}$$

We consider first the case of the cold fluid to be the minimum fluid. Rewriting Eq.

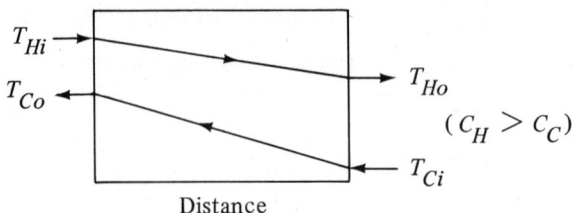

FIGURE 4.10-6.   *Temperature profile for countercurrent heat exchanger.*

Transport Processes: Momentum, Heat, and Mass

(4.5-25) using the present nomenclature,

$$q = C_c(T_{Co} - T_{Ci}) = UA \frac{(T_{Ho} - T_{Ci}) - (T_{Hi} - T_{Co})}{\ln\left[(T_{Ho} - T_{Ci})/(T_{Hi} - T_{Co})\right]} \qquad (4.10\text{-}12)$$

Combining Eq. (4.10-7) with the left side of Eq. (4.10-11) and solving for T_{Hi},

$$T_{Hi} = T_{Ci} + \frac{1}{\varepsilon}(T_{Co} - T_{Ci}) \qquad (4.10\text{-}13)$$

Subtracting T_{Co} from both sides,

$$T_{Hi} - T_{Co} = T_{Ci} - T_{Co} + \frac{1}{\varepsilon}(T_{Co} - T_{Ci}) = \left(\frac{1}{\varepsilon} - 1\right)(T_{Co} - T_{Ci}) \qquad (4.10\text{-}14)$$

From Eq. (4.10-7) for $C_{\min} = C_C$ and $C_{\max} = C_H$,

$$T_{Ho} = T_{Hi} - \frac{C_{\min}}{C_{\max}}(T_{Co} - T_{Ci}) \qquad (4.10\text{-}15)$$

This can be rearranged to give the following:

$$T_{Ho} - T_{Ci} = T_{Hi} - T_{Ci} - \frac{C_{\min}}{C_{\max}}(T_{Co} - T_{Ci}) \qquad (4.10\text{-}16)$$

Substituting Eq. (4.10-13) into (4.10-16),

$$T_{Ho} - T_{Ci} = \frac{1}{\varepsilon}(T_{Co} - T_{Ci}) - \frac{C_{\min}}{C_{\max}}(T_{Co} - T_{Ci}) \qquad (4.10\text{-}17)$$

Finally substituting Eqs. (4.10-14) and (4.10-17) into (4.10-12), rearranging, taking the antilog of both sides, and solving for ε,

$$\varepsilon = \frac{1 - \exp\left[-\dfrac{UA}{C_{\min}}\left(1 - \dfrac{C_{\min}}{C_{\max}}\right)\right]}{1 - \dfrac{C_{\min}}{C_{\max}}\exp\left[\dfrac{UA}{C_{\min}}\left(1 - \dfrac{C_{\min}}{C_{\max}}\right)\right]} \qquad (4.10\text{-}18)$$

We define NTU as the number of transfer units as follows:

$$\mathrm{NTU} = \frac{UA}{C_{\min}} \qquad (4.10\text{-}19)$$

The same result would have been obtained if $C_H = C_{\min}$.

For parallel flow we obtain

$$\varepsilon = \frac{1 - \exp\left[\dfrac{-UA}{C_{\min}}\left(1 + \dfrac{C_{\min}}{C_{\max}}\right)\right]}{1 + \dfrac{C_{\min}}{C_{\max}}} \qquad (4.10\text{-}20)$$

In Fig. 4.10-7, Eqs. (4.10-18) and (4.10-20) have been plotted in convenient graphical form. Additional charts are available for different shell-and-tube and cross-flow arrangements (K1, P1).

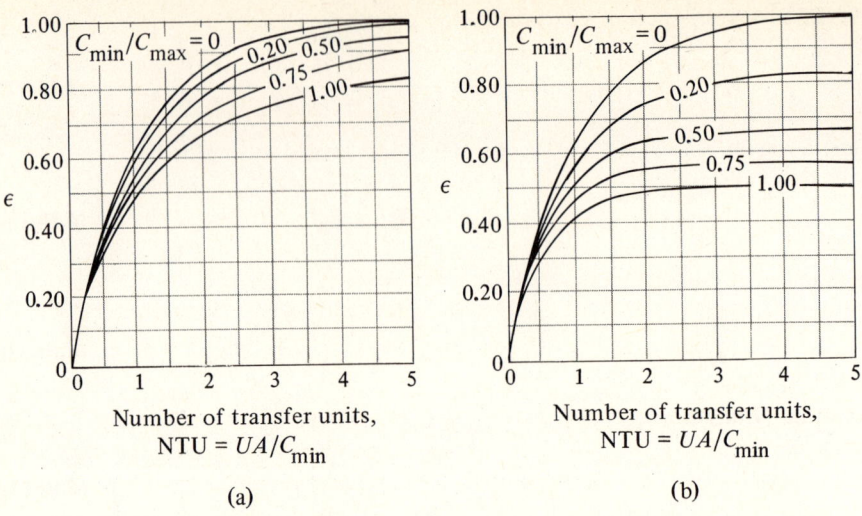

$C_{min}/C_{max} = 0$ 0.20   0.50
0.75
1.00

(a)

$C_{min}/C_{max} = 0$
0.20
0.50
0.75
1.00

(b)

Number of transfer units,
$NTU = UA/C_{min}$

Number of transfer units,
$NTU = UA/C_{min}$

FIGURE 4.10-7. *Heat-exchanger effectiveness ε : (a) counterflow exchanger, (b) parallel flow exchanger.*

EXAMPLE 4.10-2. Effectiveness of Heat Exchanger

Water flowing at a rate of 0.667 kg/s enters a countercurrent heat exchanger at 308 K and is heated by an oil stream entering at 383 K at a rate of 2.85 kg/s ($c_p = 1.89$ kJ/kg·K). The overall $U = 300$ W/m²·K and the area $A = 15.0$ m². Calculate the heat-transfer rate and the exit water temperature.

Solution: Assuming that the exit water temperature is about 370 K, the c_p for water at an average temperature of $(308 + 370)/2 = 339$ K is 4.192 kJ/kg·K (Appendix A.2). Then, $(mc_p)_H = C_H = 2.85(1.89 \times 10^3) = 5387$ W/K and $(mc_p)_C = C_C = 0.667(4.192 \times 10^3) = 2796$ W/K $= C_{min}$. Since C_C is the minimum, $C_{min}/C_{max} = 2796/5387 = 0.519$.

Using Eq. (4.10-19), NTU $= UA/C_{min} = 300(15.0)/2796 = 1.607$. Using Fig. (4.10-7a) for a counterflow exchanger, $\varepsilon = 0.71$. Substituting into Eq. (4.10-10),

$$q = \varepsilon C_{min}(T_{Hi} - T_{Ci}) = 0.71(2796)(383 - 308) = 148\,900 \text{ W}$$

Using Eq. (4.10-7),

$$q = 148\,900 = 2796(T_{Co} - 308)$$

Solving, $T_{Co} = 361.3$ K.

4.10D Fouling Factors and Typical Overall U Values

In actual practice, heat-transfer surfaces do not remain clean. Dirt, soot, scale, and other deposits form on one or both sides of the tubes of an exchanger and on other heat-transfer surfaces. These deposits form additional resistances to the flow of heat and reduce the overall heat-transfer coefficient U. In petroleum processes coke and other substances can deposit. Silting and deposits of mud and other materials can occur. Corrosion products may form on the surfaces which could form a serious resistance to heat transfer. Biological growth such as algae can occur with cooling water and in the biological industries.

Transport Processes: Momentum, Heat, and Mass

TABLE 4.10-1. *Typical Fouling Coefficients (P1, N1)*

	h_d $(W/m^2 \cdot K)$	h_d $(btu/h \cdot ft^2 \cdot °F)$
Distilled and seawater	11 350	2000
City water	5680	1000
Muddy water	1990–2840	350–500
Gases	2840	500
Vaporizing liquids	2840	500
Vegetable and gas oils	1990	350

To avoid or lessen these fouling problems chemical inhibitors are often added to minimize corrosion, salt deposition, and algae growth. Water velocities above 1 m/s are generally used to help reduce fouling. Large temperature differences may cause excessive deposition of solids on surfaces and should be avoided if possible.

The effect of such deposits and fouling is usually taken care of in design by adding a term for the resistance of the fouling on the inside and the outside of the tube in Eq. (4.3-17) as follows.

$$U_i = \frac{1}{1/h_i + 1/h_{di} + (r_o - r_i)A_i/k_A A_{A \, lm} + A_i/A_o h_o + A_i/A_o h_{do}} \qquad (4.10\text{-}21)$$

where h_{di} is the fouling coefficient for the inside and h_{do} the fouling coefficient for the outside of the tube in $W/m^2 \cdot K$. A similar expression can be written for U_o using Eq. (4.3-18).

Fouling coefficients recommended for use in designing heat-transfer equipment are available in many references (P1,N1). A short tabulation of some typical fouling coefficients is given in Table 4.10-1.

In order to do preliminary estimating of sizes of shell-and-tube heat exchangers, typical values of overall heat-transfer coefficients are given in Table 4.10-2. These values should be useful as a check on the results of the design methods described in this chapter.

TABLE 4.10-2. *Typical Values of Overall Heat-Transfer Coefficients in Shell-and-Tube Exchangers (H1, P1, W1)*

	U $(W/m^2 \cdot K)$	U $(btu/h \cdot ft^2 \cdot °F)$
Water to water	1140–1700	200–300
Water to brine	570–1140	100–200
Water to organic liquids	570–1140	100–200
Water to condensing steam	1420–2270	250–400
Water to gasoline	340–570	60–100
Water to gas oil	140–340	25–60
Water to vegetable oil	110–285	20–50
Gas oil to gas oil	110–285	20–50
Steam to boiling water	1420–2270	250–400
Water to air (finned tube)	110–230	20–40
Light organics to light organics	230–425	40–75
Heavy organics to heavy organics	55–230	10–40

4.11 (SELECTED TOPIC) RADIATION HEAT-TRANSFER PRINCIPLES

4.11A Introduction and Radiation Spectrum

1. Introduction. This section will cover some basic principles and also some advanced topics on radiation that were not covered in Section 4.9. The exchange of radiation between two surfaces depends upon the size, shape, and relative orientation of these two surfaces and also upon their emissivities and absorptivities. In the cases to be considered the surfaces are separated by nonabsorbing media such as air. When gases such as CO_2 and H_2O vapor are present, some absorption by the gases occurs, which is not considered until later in this section.

2. Radiation spectrum and thermal radiation. Energy can be transported in the form of electromagnetic waves and these waves travel at the speed of light. Bodies may emit many forms of radiant energy, such as gamma rays, thermal energy, radio waves, and so on. In fact, there is a continuous spectrum of electromagnetic radiation. This electromagnetic spectrum is divided into a number of wavelength ranges such as cosmic rays $(\lambda < 10^{-13}$ m), gamma rays $(\lambda, 10^{-13}$ to 10^{-10} m), thermal radiation $(\lambda, 10^{-7}$ to 10^{-4} m), and so on. The electromagnetic radiation produced solely because of the temperature of the emitter is called thermal radiation and exists between the wavelengths of 10^{-7} and 10^{-4} m. This portion of the electromagnetic spectrum is of importance in radiant thermal heat transfer. Electromagnetic waves having wavelengths between 3.8×10^{-7} and 7.6×10^{-7} m, called *visible radiation*, can be detected by the human eye. This visible radiation lies within the thermal radiation range.

When different surfaces are heated to the same temperature, they do not all emit or absorb the same amount of thermal radiant energy. A body that absorbs and emits the maximum amount of energy at a given temperature is called a *black body*. A black body is a standard to which other bodies can be compared.

3. Planck's law and emissive power. When a black body is heated to a temperature T, photons are emitted from the surface which have a definite distribution of energy. Planck's equation relates the monochromatic emissive power $E_{B\lambda}$ in W/m^3 at a temperature T in K and a wavelength λ in m.

$$E_{B\lambda} = \frac{3.7418 \times 10^{-16}}{\lambda^5 [e^{1.4388 \times 10^{-2}/\lambda T} - 1]} \tag{4.11-1}$$

A plot of Eq. (4.11-1) is given in Fig. 4.11-1 and shows that the energy given off increases with T. Also, for a given T, the emissive power reaches a maximum value at a wavelength that decreases as the temperature T increases. At a given temperature the radiation emitted extends over a spectrum of wavelengths. The visible light spectrum occurs in the low λ region. The sun has a temperature of about 5800 K and the solar spectrum straddles this visible range.

For a given temperature, the wavelength at which the black-body emissive power is a maximum can be determined by differentiating Eq. (4.11-1) with respect to λ at constant T and setting the result equal to zero. The result is as follows and is known as *Wien's displacement law*:

$$\lambda_{max} T = 2.898 \times 10^{-3} \text{ m} \cdot \text{K} \tag{4.11-2}$$

The locus of the maximum values is shown in Fig. 4.11-1.

4. *Stefan–Boltzmann law.* The total emissive power is the total amount of radiation energy per unit area leaving a surface with temperature T over all wavelengths. For a black body, the total emissive power is given by the integral of Eq. (4.11-1) at a given T over all wavelengths or the area under the curve in Fig. 4.11-1.

$$E_B = \int_0^\infty E_{B\lambda} \, d\lambda \qquad (4.11\text{-}3)$$

This gives

$$E_B = \sigma T^4 \qquad (4.11\text{-}4)$$

The result is the Stefan–Boltzmann law with $\sigma = 5.676 \times 10^{-8}$ W/m$^2 \cdot$ K^4. The units of E_B are W/m^2.

5. *Emissivity and Kirchhoff's law.* An important property in radiation is the emissivity of a surface. The *emissivity* ε of a surface is defined as the total emitted energy of the surface divided by the total emitted energy of a black body at the same temperature.

$$\varepsilon = \frac{E}{E_B} = \frac{E}{\sigma T^4} \qquad (4.11\text{-}5)$$

Since a black body emits the maximum amount of radiation, ε is always < 1.0.

We can derive a relationship between the absorptivity α_1 and emissitivy ε_1 of a material by placing this material in an isothermal enclosure and allowing the body and enclosure to reach the same temperature at thermal equilibrium. If G is the irradiation on the body, the energy absorbed must equal the energy emitted.

$$\alpha_1 G = E_1 \qquad (4.11\text{-}6)$$

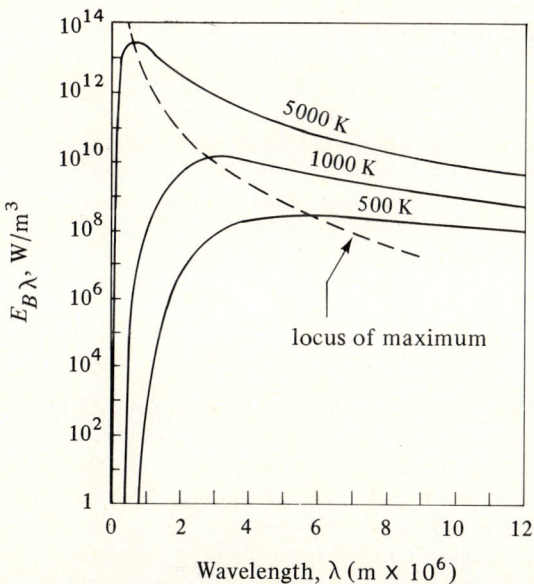

FIGURE 4.11-1.   *Spectral distribution of total energy emitted by a black body at various temperatures of the black body.*

If this body is removed and replaced by a black body of equal size, then at equilibrium,

$$\alpha_2 G = E_B \qquad\qquad \text{(4.11-7)}$$

Dividing Eq. (4.11-6) by (4.11-7),

$$\frac{\alpha_1}{\alpha_2} = \frac{E_1}{E_B} \qquad\qquad \text{(4.11-8)}$$

But $\alpha_2 = 1.0$ for a black body. Hence, since $E_1/E_B = \varepsilon_1$,

$$\alpha_1 = \frac{E_1}{E_B} = \varepsilon_1 \qquad\qquad \text{(4.11-9)}$$

This is Kirchhoff's law, which states that at thermal equilibrium $\alpha = \varepsilon$ of a body. When a body is not at equilibrium with its surroundings, the result is not valid.

6. *Concept of gray body.* A gray body is defined as a surface for which the monochromatic properties are constant over all wavelengths. For a gray surface,

$$\varepsilon_\lambda = \text{const.}, \qquad \alpha_\lambda = \text{const.} \qquad\qquad \text{(4.11-10)}$$

Hence, the total absorptivity α and the monochromatic absorptivity α_λ of a gray surface are equal, as are ε and ε_λ.

$$\alpha = \alpha_\lambda, \qquad \varepsilon = \varepsilon_\lambda \qquad\qquad \text{(4.11-11)}$$

Applying Kirchhoff's law to a gray body $\alpha_\lambda = \varepsilon_\lambda$ and

$$\alpha = \varepsilon \qquad\qquad \text{(4.11-12)}$$

As a result, the total absorptivity and emissivity are equal for a gray body even if the body is not in thermal equilibrium with its surroundings.

Gray bodies do not exist in practice and the concept of a gray body is an idealized one. The absorptivity of a surface actually varies with the wavelength of the incident radiation. Engineering calculations can often be based on the assumption of a gray body with reasonable accuracy. The α is assumed constant even with a variation in λ of the incident radiation. Also, in actual systems, various surfaces may be at different temperatures. In these cases, α for a surface is evaluated by determining the emissivity not at the actual surface temperature but at the temperature of the source of the other radiating surface or emitter since this is the temperature the absorbing surface would reach if the absorber and emitter were at thermal equilibrium. The temperature of the absorber has only a slight effect on the absorptivity.

4.11B Derivation of View Factors in Radiation for Various Geometries

1. Introduction. The concepts and definitions presented in Section 4.11A form a sufficient foundation so that the net radiant exchange between surfaces can be determined. If two surfaces are arranged so that radiant energy can be exchanged, a net flow of energy will occur from the hotter surface to the colder surface. The size, shape, and orientation of two radiating surfaces or a system of surfaces are factors in determining the net heat-flow rate between them. To simplify the discussion we assume that the surfaces are separated by a nonabsorbing medium such as air. This assumption is adequate for many engineering applications. However, in cases such as a furnace, the presence of CO_2 and H_2O vapor make such a simplification impossible because of their high absorptivities.

Transport Processes: Momentum, Heat, and Mass

The simplest geometrical configuration will be considered first, that of radiation exchange between parallel, infinite planes. This assumption implies that there are no edge effects in the case of finite surfaces. First, the simplest case will be treated in which the surfaces are black bodies and then more complicated geometries and gray bodies will be treated.

2. View factor for infinite parallel black planes. If two parallel and infinite black planes at T_1 and T_2 are radiating toward each other, plane 1 emits σT_1^4 radiation to plane 2, which is all absorbed. Also, plane 2 emits σT_2^4 radiation to plane 1, which is all absorbed. Then for plane 1, the net radiation is from plane 1 to 2,

$$q_{12} = A_1 \sigma (T_1^4 - T_2^4) \tag{4.11-13}$$

In this case all the radiation from 1 to 2 is intercepted by 2; i.e., the fraction of radiation leaving 1 that is intercepted by 2 is F_{12}, which is 1.0. The factor F_{12} is called the geometric view factor or view factor. Hence,

$$q_{12} = F_{12} A_1 \sigma (T_1^4 - T_2^4) \tag{4.11-14}$$

where F_{12} is fraction of radiation leaving surface 1 in all directions which is intercepted by surface 2. Also,

$$q_{21} = F_{21} A_2 \sigma (T_1^4 - T_2^4) \tag{4.11-15}$$

In the case for parallel plates $F_{12} = F_{21} = 1.0$ and the geometric factor is simply omitted.

3. View factor for infinite parallel gray planes. If both of the parallel plates A_1 and A_2 are gray with emissivities and absorptivities of $\varepsilon_1 = \alpha_1$ and $\varepsilon_2 = \alpha_2$, respectively, we can proceed as follows. Since each surface has an unobstructed view of each other, the view factor is 1.0. In unit time surface A_1 emits $\varepsilon_1 A_1 \sigma T_1^4$ radiation to A_2. Of this, the fraction ε_2 (where $\alpha_2 = \varepsilon_2$) is absorbed:

$$\text{absorbed by } A_2 = \varepsilon_2 (\varepsilon_1 A_1 \sigma T_1^4) \tag{4.11-16}$$

Also, the fraction $(1 - \varepsilon_2)$ or the amount $(1 - \varepsilon_2)(\varepsilon_1 A_1 \sigma T_1^4)$ is reflected back to A_1. Of this amount A_1 reflects back to A_2 a fraction $(1 - \varepsilon_1)$ or an amount $(1 - \varepsilon_1)(1 - \varepsilon_2)(\varepsilon_1 A_1 \sigma T_1^4)$. The surface A_2 absorbs the fraction ε_2, or

$$\text{absorbed by } A_2 = \varepsilon_2 (1 - \varepsilon_1)(1 - \varepsilon_2)(\varepsilon_1 A_1 \sigma T_1^4) \tag{4.11-17}$$

The amount reflected back to A_1 from A_2 is $(1 - \varepsilon_2)(1 - \varepsilon_1)(1 - \varepsilon_2)(\varepsilon_1 A_1 \sigma T_1^4)$. Then A_1 absorbs ε_1 of this and reflects back to A_2 an amount $(1 - \varepsilon_1)(1 - \varepsilon_2)(1 - \varepsilon_1)(1 - \varepsilon_2) \times (\varepsilon_1 A_1 \sigma T_1^4)$. The surface A_2 then absorbs

$$\text{absorbed by } A_2 = \varepsilon_2 (1 - \varepsilon_1)(1 - \varepsilon_2)(1 - \varepsilon_1)(1 - \varepsilon_2)(\varepsilon_1 A_1 \sigma T_1^4) \tag{4.11-18}$$

This continues and the total amount absorbed at A_2 is the sum of Eqs. (4.11-16), (4.11-17), (4.11-18), and so on.

$$q_{1 \to 2} = A_1 \sigma T_1^4 [\varepsilon_1 \varepsilon_2 + \varepsilon_1 \varepsilon_2 (1 - \varepsilon_1)(1 - \varepsilon_2) + \varepsilon_1 \varepsilon_2 (1 - \varepsilon_1)^2 (1 - \varepsilon_2)^2 + \cdots] \tag{4.11-19}$$

The result is a geometric series (M1).

$$q_{1 \to 2} = A_1 \sigma T_1^4 \frac{\varepsilon_1 \varepsilon_2}{1 - (1 - \varepsilon_1)(1 - \varepsilon_2)} = A_1 \sigma T_1^4 \frac{1}{1/\varepsilon_1 + 1/\varepsilon_2 - 1} \tag{4.11-20}$$

Repeating the above for the amount absorbed at A_1, which comes from A_2,

$$q_{2 \to 1} = A_1 \sigma T_2^4 \frac{1}{1/\varepsilon_1 + 1/\varepsilon_2 - 1} \tag{4.11-21}$$

The net radiation is the difference of Eqs. (4.11-20) and (4.11-21).

$$q_{12} = A_1 \sigma (T_1^4 - T_2^4) \frac{1}{1/\varepsilon_1 + 1/\varepsilon_2 - 1} \tag{4.11-22}$$

If $\varepsilon_1 = \varepsilon_2 = 1.0$ for black bodies, Eq. (4.11-22) becomes Eq. (4.11-13).

EXAMPLE 4.11-1. Radiation Between Parallel Planes
Two parallel gray planes which are very large have emissivities of $\varepsilon_1 = 0.8$ and $\varepsilon_2 = 0.7$ and surface 1 is at $1100°F$ (866.5 K) and surface 2 at $600°F$ (588.8 K). Use English and SI units for the following.
 (a) What is the net radiation from 1 to 2?
 (b) If the surfaces are both black, what is the net radiation?

Solution: For part (a), using Eq. (4.11-22) and substituting the known values,

$$\frac{q_{12}}{A_1} = \frac{\sigma(T_1^4 - T_2^4)}{1/\varepsilon_1 + 1/\varepsilon_2 - 1} = (0.1714 \times 10^{-8}) \frac{(1100 + 460)^4 - (600 + 460)^4}{1/0.8 + 1/0.7 - 1}$$

$$= 4750 \text{ btu/h} \cdot \text{ft}^2$$

$$\frac{q_{12}}{A_1} = (5.676 \times 10^{-8}) \frac{(866.5)^4 - (588.8)^4}{1/0.8 + 1/0.7 - 1} = 15\,010 \text{ W/m}^2$$

For black surfaces in part (b), using Eq. (4.11-13),

$$q = 7960 \text{ btu/h} \cdot \text{ft}^2 \quad \text{or} \quad 25\,110 \text{ W/m}^2$$

Note the large reduction in radiation when surfaces with emissivities less than 1.0 are used.

Example 4.11-1 shows the large influence that emissivities less than 1.0 have on radiation. This fact is used to reduce radiation loss or gain from a surface by using planes as a radiation shield. For example, for two parallel surfaces of emissivity ε at T_1 and T_2, the interchange is, by Eq. (4.11-22),

$$\frac{(q_{12})_0}{A} = \frac{\sigma(T_1^4 - T_2^4)}{2/\varepsilon - 1} \tag{4.11-23}$$

The subscript 0 indicates that there are no planes in between the two surfaces. Suppose that we now insert one or more radiation planes between the original surfaces. Then it can be shown that

$$\frac{(q_{12})_N}{A} = \frac{1}{N + 1} \frac{\sigma(T_1^4 - T_2^4)}{2/\varepsilon - 1} \tag{4.11-24}$$

where N is the number of radiation planes or shields between the original surfaces. Hence, a great reduction in radiation heat loss is obtained by using these shields.

4. Derivation of general equation for view factor between black bodies. Suppose that we consider radiation between two parallel black planes of finite size as in Fig. 4.11-2a. Since the planes are not infinite in size, some of the radiation from surface 1 does not strike surface 2, and vice versa. Hence, the net radiation interchange is less since some is

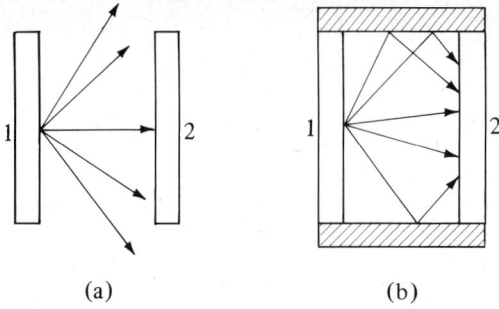

(a)   (b)

FIGURE 4.11-2. *Radiation between two black surfaces : (a) two planes alone, (b) two planes connected by refractory reradiating walls.*

lost to the surroundings. The fraction of radiation leaving surface 1 in all directions which is intercepted by surface 2 is called F_{12} and must be determined for each geometry by taking differential surface elements and integrating over the entire surfaces.

Before we can derive a general relationship for the view factor between two finite bodies we must consider and discuss two quantities, a solid angle and the intensity of radiation. A solid angle ω is a dimensionless quantity which is a measure of an angle in solid geometry. In Fig. 4.11-3a the differential solid angle $d\omega_1$ is equal to the normal projection of dA_2 divided by the square of the distance between the point P and area dA_2.

$$d\omega_1 = \frac{dA_2 \cos \theta_2}{r^2} \tag{4.11-25}$$

The units of a solid angle are steradian or sr. For a hemisphere the number of sr subtended by this surface is 2π.

The intensity of radiation for a black body, I_B, is the rate of radiation emitted per unit area projected in a direction normal to the surface and per unit solid angle in a

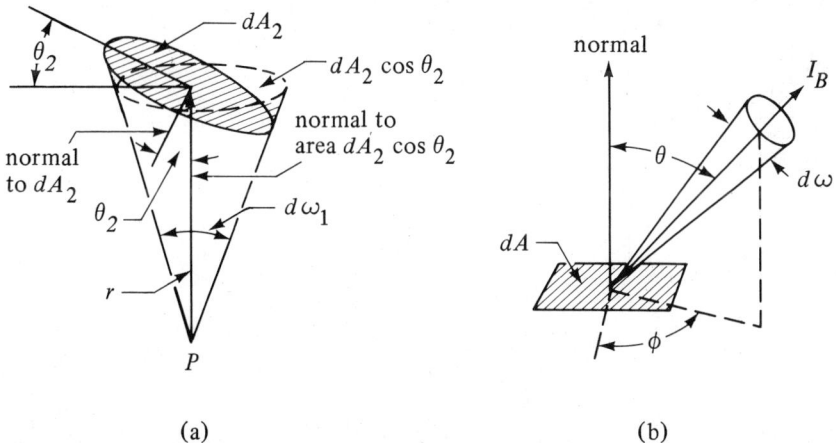

(a)   (b)

FIGURE 4.11-3. *Geometry for a solid angle and intensity of radiation : (a) solid-angle geometry, (b) intensity of radiation from emitting area dA.*

specified direction as shown in Fig. 4.11-3b. The projection of dA on the line between centers is $dA \cos \theta$.

$$I_B = \frac{dq}{dA \cos d\omega} \tag{4.11-26}$$

where q is in W and I_B is in W/m² · sr. We assume that the black body is a diffuse surface which emits with equal intensity in all directions, i.e., I = constant. The emissive power E_B which leaves a black-body plane surface is determined by integrating Eq. (4.11-26) over all solid angles subtended by a hemisphere covering the surface. The final result is as follows. [See references (C3, H1, K1) for details.]

$$E_B = \pi I_B \tag{4.11-27}$$

where E_B is in W/m².

In order to determine the radiation heat-transfer rates between two black surfaces we must determine the general case for the fraction of the total radiant heat that leaves a surface and arrives on a second surface. Using only black surfaces, we consider the case shown in Fig. 4.11-4, in which radiant energy is exchanged between area elements dA_1 and dA_2. The line r is the distance between the areas and the angles between this line and the normals to the two surfaces are θ_1 and θ_2. The rate of radiant energy that leaves dA_1 in the direction given by the angle θ_1 is $I_{B1} \, dA \cos \theta_1$. The rate that leaves dA_1 and arrives on dA_2 is given by Eq. (4.11-28).

$$dq_{1 \rightarrow 2} = I_{B1} \, dA \cos \theta_1 \, d\omega_1 \tag{4.11-28}$$

where $d\omega_1$ is the solid angle subtended by the area dA_2 as seen from dA_1. Combining Eqs. (4.11-25) and (4.11-28),

$$dq_{1 \rightarrow 2} = \frac{I_{B1} \, dA_1 \cos \theta_1 \cos \theta_2 \, dA_2}{r^2} \tag{4.11-29}$$

From Eq. (4.11-27), $I_{B1} = E_{B1}/\pi$. Substituting E_{B1}/π for I_{B1} into Eq. (4.11-29),

$$dq_{1 \rightarrow 2} = \frac{E_{B1} \cos \theta_1 \cos \theta_2 \, dA_1 \, dA_2}{\pi r^2} \tag{4.11-30}$$

The energy leaving dA_2 and arriving at dA_1 is

$$dq_{2 \rightarrow 1} = \frac{E_{B2} \cos \theta_2 \cos \theta_1 \, dA_2 \, dA_1}{\pi r^2} \tag{4.11-31}$$

Substituting σT_1^4 for E_{B1} and σT_2^4 for E_{B2} from Eq. (4.11-4) and taking the difference of

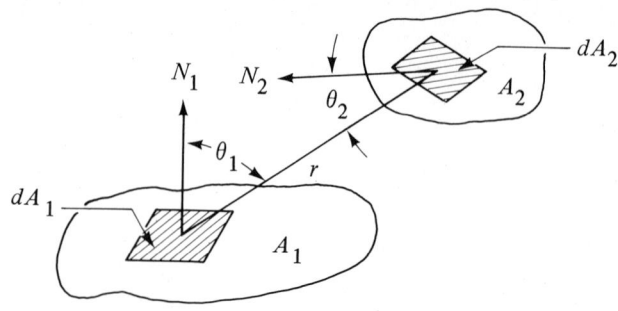

FIGURE 4.11-4.   *Area elements for radiation shape factor.*

Eqs. (4.11-30) and (4.11-31) for the net heat flow,

$$dq_{12} = \sigma(T_1^4 - T_2^4) \frac{\cos\theta_1 \cos\theta_2\, dA_1\, dA_2}{\pi r^2} \tag{4.11-32}$$

Performing the double integrations over surfaces A_1 and A_2 will yield the total net heat flow between the finite areas.

$$q_{12} = \sigma(T_1^4 - T_2^4) \int_{A_2} \int_{A_1} \frac{\cos\theta_1 \cos\theta_2\, dA_1\, dA_2}{\pi r^2} \tag{4.11-33}$$

Equation (4.11-33) can also be written as

$$q_{12} = A_1 F_{12}\, \sigma(T_1^4 - T_2^4) = A_2 F_{21}\sigma(T_1^4 - T_2^4) \tag{4.11-34}$$

where F_{12} is a geometric shape factor or view factor and designates the fraction of the total radiation leaving A_1 which strikes A_2 and F_{21} represents the fraction leaving A_2 which strikes A_1. Also, the following relation exists.

$$A_1 F_{12} = A_2 F_{21} \tag{4.11-35}$$

which is valid for black surfaces and also nonblack surfaces. The view factor F_{12} is then

$$F_{12} = \frac{1}{A_1} \int_{A_2} \int_{A_1} \frac{\cos\theta_1 \cos\theta_2\, dA_1\, dA_2}{\pi r^2} \tag{4.11-36}$$

Values of the view factor can be calculated for a number of geometrical arrangements.

5. *View factors between black bodies for various geometries.* A number of basic relationships between view factors are given below.

The reciprocity relationship given by Eq. (4.11-35) is

$$A_1 F_{12} = A_2 F_{21} \tag{4.11-35}$$

This relationship can be applied to any two surfaces i and j.

$$A_i F_{ij} = A_j F_{ji} \tag{4.11-37}$$

If surface A_1 can only see surface A_2, then $F_{12} = 1.0$.

If surface A_1 sees a number of surfaces $A_2, A_3, \ldots$, and all the surfaces form an enclosure then the enclosure relationship is

$$F_{11} + F_{12} + F_{13} + \cdots = 1.0 \tag{4.11-38}$$

If the surface A_1 cannot see itself (surface is flat or convex), $F_{11} = 0$.

EXAMPLE 4.11-2. View Factor from a Plane to a Hemisphere
Determine the view factors between a plane A_1 covered by a hemisphere A_2 as shown in Fig. 4.11-5.

Solution: Since surface A_1 sees only A_2, the view factor $F_{12} = 1.0$. Using Eq. (4.11-35),

$$A_1 F_{12} = A_2 F_{21} \tag{4.11-35}$$

The area $A_1 = \pi R^2$, $A_2 = 2\pi R^2$. Substituting into Eq. (4.11-35) and solving for F_{21},

$$F_{21} = F_{12}\frac{A_1}{A_2} = (1.0)\frac{\pi R^2}{2\pi R^2} = \frac{1}{2}$$

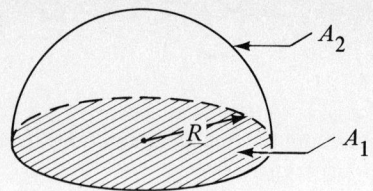

FIGURE 4.11-5. *Radiant exchange between a flat surface and a hemisphere for Example 4.11-2.*

Using Eq. (4.11-38) for surface A_1, $F_{11} = 1.0 - F_{12} = 1.0 - 1.0 = 0$. Also, writing Eq. (4.11-38) for surface A_2,

$$F_{22} + F_{21} = 1.0 \qquad\qquad \textbf{(4.11-39)}$$

Solving for F_{22}, $F_{22} = 1.0 - F_{21} = 1.0 - \frac{1}{2} = \frac{1}{2}$.

EXAMPLE 4.11-3. Radiation Between Parallel Disks

In Fig. 4.11-6 a small disk of area A_1 is parallel to a large disk of area A_2 and A_1 is centered directly below A_2. The distance between the centers of the disks is R and the radius of A_2 is a. Determine the view factor for radiant heat transfer from A_1 to A_2.

Solution: The differential area for A_2 is taken as the circular ring of radius x so that $dA_2 = 2\pi x\, dx$. The angle $\theta_1 = \theta_2$. Using Eq. (4.11-36),

$$F_{12} = \frac{1}{A_1} \int_{A_2} \int_{A_1} \frac{\cos\theta_1 \cos\theta_1\, dA_1 (2\pi x\, dx)}{\pi r^2}$$

In this case the area A_1 is very small compared to A_2, so dA_1 can be integrated to A_1 and the other terms inside the integral can be assumed constant. From the geometry shown, $r = (R^2 + x^2)^{1/2}$, $\cos\theta_1 = R/(R^2 + x^2)^{1/2}$. Making these substitutions into the equation for F_{12},

$$F_{12} = \int_0^a \frac{2R^2 x\, dx}{(R^2 + x^2)^2}$$

Integrating,

$$F_{12} = \frac{a^2}{R^2 + a^2}$$

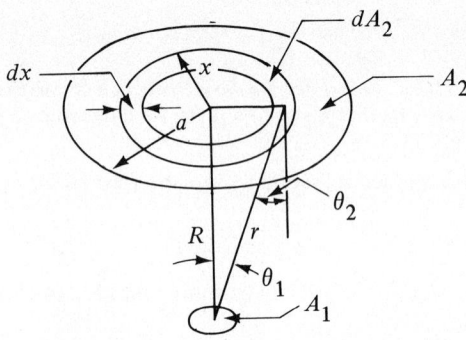

FIGURE 4.11-6. *View factor for radiation from a small element to a parallel disk for Example 4.11-3.*

The integration of Eq. (4.11-36) has been done for numerous geometrical configurations and values of F_{12} tabulated. Then,

$$q_{12} = F_{12} A_1 \sigma(T_1^4 - T_2^4) = F_{21} A_2 \sigma(T_1^4 - T_2^4) \qquad \textbf{(4.11-34)}$$

where F_{12} is the fraction of the radiation leaving A_1 which is intercepted by A_2 and F_{21} the fraction reaching A_1 from A_2. Since the flux from 1 to 2 must equal that from 2 to 1, Eq. (4.11-34) becomes Eq. (4.11-35) as given previously.

$$A_1 F_{12} = A_2 F_{21} \qquad \textbf{(4.11-35)}$$

Hence, one selects the surface whose view factor can be determined most easily. For example, the view factor F_{12} for a small surface A_1 completely enclosed by a large surface A_2 is 1.0, since all the radiation leaving A_1 is intercepted by A_2. In Fig. 4.11-7 the view factors F_{12} between parallel planes are given, and in Fig. 4.11-8 the view factors for adjacent perpendicular rectangles. View factors for other geometries are given elsewhere (H1, K1, P1, W1).

4.11C View Factors When Surfaces Are Connected by Reradiating Walls

If the two black-body surfaces A_1 and A_2 are connected by nonconducting (refractory) but reradiating walls as in Fig. 4.11-2b, a larger fraction of the radiation from surface 1 is intercepted by 2. This view factor is called $\bar{F}_{12}$. The case of two surfaces connected by the walls of an enclosure such as a furnace is a common example of this. The general equation for this case assuming a uniform refractory temperature has been derived (M1, C3) for two radiant sources A_1 and A_2, which are not concave, so they do not see themselves.

$$\bar{F}_{12} = \frac{A_2 - A_1 F_{12}^2}{A_1 + A_2 - 2A_1 F_{12}} = \frac{1 - (A_1/A_2)F_{12}^2}{A_1/A_2 + 1 - 2(A_1/A_2)F_{12}} \qquad \textbf{(4.11-40)}$$

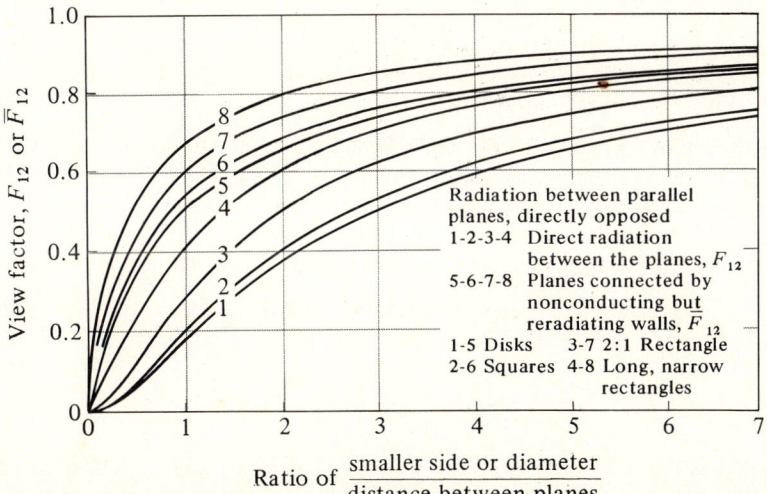

FIGURE 4.11-7. *View factor between parallel planes directly opposed. (From W. H. McAdams, Heat Transmission, 3rd ed. New York: McGraw-Hill Book Company, 1954. With permission.)*

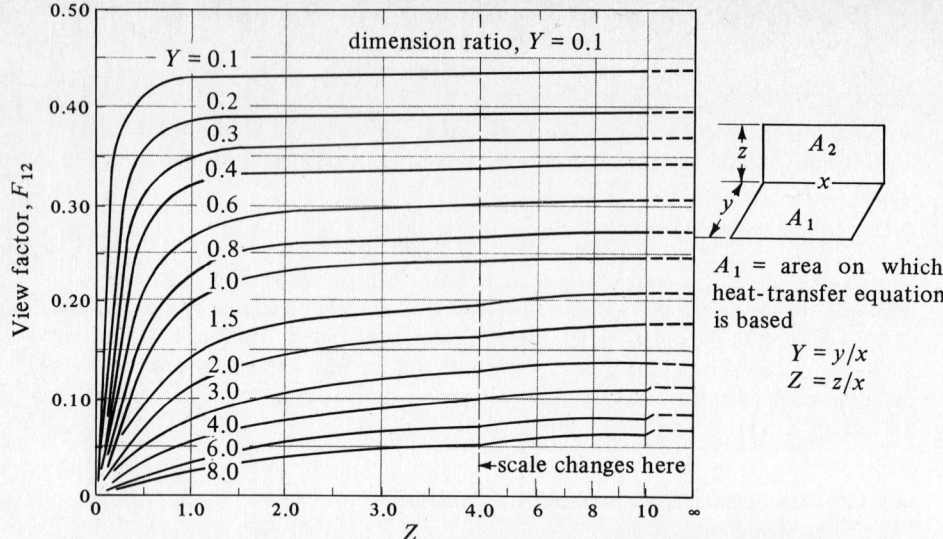

FIGURE 4.11-8. *View factor for adjacent perpendicular rectangles.* [*From H. C. Hottel, Mech. Eng.,* **52**, *699 (1930). With permission.*]

Also, as before,

$$A_1 \bar{F}_{12} = A_2 \bar{F}_{21} \qquad \text{(4.11-41)}$$

$$q_{12} = \bar{F}_{12} A_1 \sigma(T_1^4 - T_2^4) \qquad \text{(4.11-42)}$$

The factor $\bar{F}_{12}$ for parallel planes is given in Fig. 4.11-7 and for other geometries can be calculated from Eq. (4.11-36). For view factors F_{12} and $\bar{F}_{12}$ for parallel tubes adjacent to a wall as in a furnace and also for variation in refractory wall temperature, see elsewhere (M1, P1). If there are no reradiating walls,

$$F_{12} = \bar{F}_{12} \qquad \text{(4.11-43)}$$

4.11D View Factors and Gray Bodies

A general and more practical case, which is the same as for Eq. (4.11-40) but with the surfaces A_1 and A_2 being gray with emissivities ε_1 and ε_2, will be considered. Non-conducting reradiating walls are present as before. Since the two surfaces are now gray, there will be some reflection of radiation which will decrease the net radiant exchange between the surfaces below that for black surfaces. The final equations for this case are

$$q_{12} = \mathscr{F}_{12} A_1 \sigma(T_1^4 - T_2^4) \qquad \text{(4.11-44)}$$

$$\mathscr{F}_{12} = \cfrac{1}{\cfrac{1}{\bar{F}_{12}} + \cfrac{A_1}{A_2}\left(\cfrac{1}{\varepsilon_2} - 1\right) + \left(\cfrac{1}{\varepsilon_1} - 1\right)} \qquad \text{(4.11-45)}$$

where $\mathscr{F}_{12}$ is the new view factor for two gray surfaces A_1 and A_2 which cannot see themselves and are connected by reradiating walls. If no refractory walls are present, F_{12} is used in place of $\bar{F}_{12}$ in Eq. (4.11-41). Again,

$$A_1 \mathscr{F}_{12} = A_2 \mathscr{F}_{21} \qquad \text{(4.11-46)}$$

EXAMPLE 4.11-4. Radiation Between Infinite Parallel Gray Planes
Derive Eq. (4.11-22) by starting with the general equation for radiation between two gray bodies A_1 and A_2 which are infinite parallel planes having emissivities ε_1 and ε_2, respectively.

Solution: Since there are no reradiating walls, by Eq. (4.11-43), $\bar{F}_{12}$ becomes F_{12}. Also, since all the radiation from surface 1 is intercepted by surface 2, $F_{12} = 1.0$. Substituting into Eq. (4.11-45), noting that $A_1/A_2 = 1.0$,

$$\mathcal{F}_{12} = \cfrac{1}{\cfrac{1}{\bar{F}_{12}} + \cfrac{A_1}{A_2}\left(\cfrac{1}{\varepsilon_2} - 1\right) + \left(\cfrac{1}{\varepsilon_1} - 1\right)} = \cfrac{1}{\cfrac{1}{1} + 1\left(\cfrac{1}{\varepsilon_2} - 1\right) + \left(\cfrac{1}{\varepsilon_1} - 1\right)}$$

$$= \cfrac{1}{\cfrac{1}{\varepsilon_1} + \cfrac{1}{\varepsilon_2} - 1}$$

Then using Eq. (4.11-44),

$$q_{12} = \mathcal{F}_{12} A_1 \sigma(T_1^4 - T_2^4) = A_1 \sigma(T_1^4 - T_2^4) \cfrac{1}{\cfrac{1}{\varepsilon_1} + \cfrac{1}{\varepsilon_2} - 1}$$

This is identical to Eq. (4.11-22).

EXAMPLE 4.11-5. Complex View Factor for Perpendicular Rectangles
Find the view factor F_{12} for the configuration shown in Fig. 4.11-9 of the rectangle with area A_2 displaced from the common edge of rectangle A_1 and perpendicular to A_1. The temperature of A_1 is T_1 and that of A_2 and A_3 is T_2.

Solution: The area A_3 is a fictitious area between areas A_2 and A_1. Call the area A_2 plus A_3 as $A_{(23)}$. The view factor $F_{1(23)}$ for areas A_1 and $A_{(23)}$ can be obtained from Fig. 4.11-8 for adjacent perpendicular rectangles. Also, F_{13} can be obtained from Fig. 4.11-8. The radiation interchange between A_1 and $A_{(23)}$ is equal to that intercepted by A_2 and by A_3.

$$A_1 F_{1(23)} \sigma(T_1^4 - T_2^4) = A_1 F_{12}\, \sigma(T_1^4 - T_2^4) + A_1 F_{13}(T_1^4 - T_2^4) \quad \textbf{(4.11-47)}$$

Hence,

$$A_1 F_{1(23)} = A_1 F_{12} + A_1 F_{13} \quad \textbf{(4.11-48)}$$

Solving for F_{12},

$$F_{12} = F_{1(23)} - F_{13} \quad \textbf{(4.11-49)}$$

Methods similar to those used in this example can be employed to find the shape

FIGURE 4.11-9. *Configuration for Example 4.11-5.*

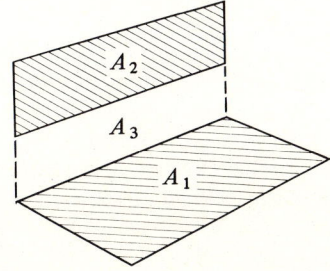

factors for a general orientation of two rectangles in perpendicular planes or parallel rectangles (C3, H1, K1).

EXAMPLE 4.11-6. Radiation to a Small Package

A small cold package having an area A_1 and emissivity ε_1 is at temperature T_1. It is placed in a warm room with the walls at T_2 and an emissivity ε_2. Derive the view factor for this using Eq. (4.11-45), and the equation for the radiation heat transfer.

Solution: For the small surface A_1 completely enclosed by the enclosure A_2, $\bar{F}_{12} = F_{12}$ by Eq. (4.11-43), since there are no reradiating (refractory) walls. Also, $F_{12} = 1.0$, since all the radiation from A_1 is intercepted by the enclosure A_2 because A_1 does not have any concave surfaces and cannot "see" itself. Since A_2 is very large compared to A_1, $A_1/A_2 = 0$. Substituting into Eq. (4.11-45),

$$\mathcal{F}_{12} = \cfrac{1}{\cfrac{1}{\bar{F}_{12}} + \cfrac{A_1}{A_2}\left(\cfrac{1}{\varepsilon_2} - 1\right) + \left(\cfrac{1}{\varepsilon_1} - 1\right)} = \cfrac{1}{1 + 0\left(\cfrac{1}{\varepsilon_2} - 1\right) + \cfrac{1}{\varepsilon_1} - 1} = \varepsilon_1$$

Substituting into Eq. (4.11-44),

$$q_{12} = \mathcal{F}_{12} A_1 \sigma(T_1^4 - T_2^4) = \varepsilon_1 A_1 \sigma(T_1^4 - T_2^4)$$

This is the same as Eq. (4.9-6) derived previously.

For methods to solve complicated radiation problems involving more than four or five heat-transfer surfaces, matrix methods to solve these problems have been developed and are discussed in detail elsewhere (H1, K1).

4.11E Radiation in Absorbing Gases

1. Introduction to absorbing gases in radiation. As discussed in this section, solids and liquids emit radiation over a continuous spectrum. However, most gases that are monoatomic or diatomic, such as He, Ar, H_2, O_2, and N_2, are virtually transparent to thermal radiation; i.e., they emit practically no radiation and also do not absorb radiation. Gases with a dipole moment and higher polyatomic gases emit significant amounts of radiation and also absorb radiant energy within the same bands in which they emit radiation. These gases include CO_2, H_2O, CO, SO_2, NH_3, and organic vapors.

For a particular gas, the width of the absorption or emission bands depends on the pressure and also the temperature. If an absorbing gas is heated, it radiates energy to the cooler surroundings. The net radiation heat-transfer rate between surfaces is decreased in these cases because the gas absorbs some of the radiant energy being transported between the surfaces.

2. Absorption of radiation by a gas. The absorption of radiation in a gas layer can be described analytically since the absorption by a given gas depends on the number of molecules in the path of radiation. Increasing the partial pressure of the absorbing gas or the path length increases the amount of absorption. We define $I_{\lambda 0}$ as the intensity of radiation at a particular wavelength before it enters the gas and $I_{\lambda L}$ as the intensity at the same wavelength after having traveled a distance of L in the gas. If the beam impinges on a gas layer of thickness dL, the decrease in intensity, dI_λ, is proportional to I_λ and dL.

$$dI_\lambda = -\alpha_\lambda I_\lambda \, dL \tag{4.11-50}$$

where I_λ is in W/m². Integrating,

$$I_{\lambda L} = I_{\lambda 0}\, e^{-\alpha_\lambda L} \qquad\qquad (4.11\text{-}51)$$

The constant α_λ depends on the particular gas, its partial pressure, and the wavelength of radiation. This equation is called Beer's law. Gases frequently absorb only in narrow-wavelength bands.

3. Characteristic mean beam length of absorbing gas. The calculation methods for gas radiation are quite complicated. For the purpose of engineering calculations, Hottel (M1) has presented approximate methods to calculate radiaton and absorption when gases such as CO_2 and water vapor are present. Thick layers of a gas absorb more energy than do thin layers. Hence, in addition to specifying the pressure and temperature of a gas, we must specify a characteristic length (mean beam length) of a gas mass to determine the emissivity and absorptivity of a gas. The mean beam length L depends on the specific geometry.

For a black differential receiving surface area dA located in the center of the base of a hemisphere of radius L containing a radiating gas, the mean beam length is L. The mean beam length has been evaluated for various geometries and is given in Table 4.11-1. For other shapes, L can be approximated by

$$L = 3.6\, \frac{V}{A} \qquad\qquad (4.11\text{-}52)$$

where V is volume of the gas in m³, A the surface area of the enclosure in m², and L is in m.

4. Emissivity, absorptivity, and radiation of a gas. Gas emissivities have been correlated and Fig. 4.11-10 gives the gas emissivity ε_G of CO_2 at a total pressure of the system of 1.0 atm abs. The p_G is the partial pressure of CO_2 in atm and the mean beam length L in m. The emissivity ε_G is defined as the ratio of the rate of energy transfer from the hemispherical body of gas to a surface element at the midpoint divided by the rate of energy transfer from a black hemisphere surface of radius L and temperature T_G to the same element.

TABLE 4.11-1. *Mean Beam Length for Gas Radiation to
Entire Enclosure Surface (M1, R2, P1)*

Geometry of Enclosure	Mean Beam Length, L
Sphere, diameter D	0.65D
Infinite cylinder, diameter D	0.95D
Cylinder, length = diameter D	0.60D
Infinite parallel plates, separation distance D	1.8D
Hemisphere, radiation to element in base, radius R	R
Cube, radiation to any face, side D	0.60D
Volume surrounding bank of long tubes with centers on equilateral triangle, clearance = tube diameter D	2.8D

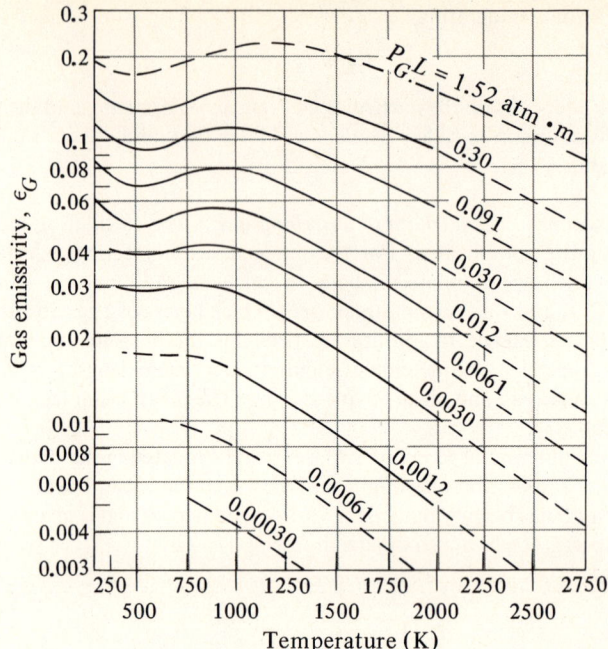

FIGURE 4.11-10. *Total emissivity of the gas carbon dioxide at a total pressure of 1.0 atm. (From W. H. McAdams, Heat Transmission, 3rd ed. New York: McGraw-Hill Book Company, 1954. With permission.)*

The rate of radiation emitted from the gas is $\sigma \varepsilon_G T_G^4$ in W/m² of receiving surface element, where ε_G is evaluated at T_G. If the surface element at the midpoint at T_1 is radiating heat back to the gas, the absorption rate of the gas will be $\sigma \alpha_G T_1^4$, where α_G is the absorptivity of the gas for blackbody radiation from the surface at T_1. The α_G of CO_2 is determined from Fig. 4.11-10 at T_1 but instead of using the parameter of $p_G L$, the parameter $p_G L(T_1/T_G)$ is used. The resulting value from the chart is then multiplied by $(T_G/T_1)^{0.65}$ to give α_G. The net rate of radiant transfer between a gas at T_G and a black surface of finite area A_1 at T_1 is then

$$q = \sigma A(\varepsilon_G T_G^4 - \alpha_G T_1^4) \tag{4.11-53}$$

When the total pressure is not 1.0 atm, a correction chart is available to correct the emissivity of CO_2. Also, charts are available for water vapor (H1, K1, M1, P1). When both CO_2 and H_2O are present the total radiation is reduced somewhat, since each gas is somewhat opaque to radiation from the other gas. Charts for these interactions are also available (H1, K1, M1, P1).

EXAMPLE 4.11-7. *Gas Radiation to a Furnace Enclosure*
A furnace is in the form of a cube 0.3 m on a side inside, and these interior walls can be approximated as black surfaces. The gas inside at 1.0 atm total pressure and 1100 K contains 10 mole % CO_2 and the rest is O_2 and N_2. The small amount of water vapor present will be neglected. The walls of the furnace are maintained at 600 K by external cooling. Calculate the total heat transfer to the walls neglecting heat transfer by convection.

Solution: From Table 4.11-1, the mean beam length for radiation to a cube face is $L = 0.60D = 0.60(0.30) = 0.180$ m. The partial pressure of CO_2 is $p_G = 0.10(100) = 0.10$ atm. Then $p_G L = 0.10(0.180) = 0.0180$ atm·m. From Fig. 4.11-10, $\varepsilon_G = 0.064$ at $T_G = 1100$ K.

To obtain α_G, we evaluate α_G at $T_1 = 600$ K and $p_G L(T_1/T_G) = (0.0180)(600/1100) = 0.00982$ atm·m. From Fig. 4.11-10, the uncorrected value of $\alpha_G = 0.048$. Multiplying this by the correction factor $(T_G/T_1)^{0.65}$, the final correction value is

$$\alpha_G = 0.048(1100/600)^{0.65} = 0.0712$$

Substituting into Eq. (4.11-53),

$$\frac{q}{A} = \sigma(\varepsilon_G T_G^4 - \alpha_G T_1^4)$$

$$= (5.676 \times 10^{-8})[0.064(1100)^4 - 0.0712(600)^4]$$

$$= 4.795 \times 10^3 \ \text{W/m}^2 = 4.795 \ \text{kW/m}^2$$

For six sides, $A = 6(0.3 \times 0.3) = 0.540 \ \text{m}^2$. Then,

$$q = 4.795(0.540) = 2.589 \ \text{kW}$$

For the case where the walls of the enclosure are not black, some of the radiation striking the walls is reflected back to the other walls and into the gas. As an approximation when the emissivity of the walls is greater than 0.7, an effective emissivity ε' can be used.

$$\varepsilon' = \frac{\varepsilon + 1.0}{2} \tag{4.11-54}$$

where ε is the actual emissivity of the enclosure walls. Then Eq. (4.11-53) is modified to give the following (M1):

$$q = \sigma A \varepsilon'(\varepsilon_G T_G^4 - \alpha_G T_1^4) \tag{4.11-55}$$

Other approximate methods are available for gases containing suspended luminous flames, clouds of nonblack particles, refractory walls and absorbing gases present, and so on (M1, P1).

4.12 (SELECTED TOPIC) HEAT TRANSFER OF NON-NEWTONIAN FLUIDS

4.12A Introduction

Most of the studies on heat transfer with fluids have been done with Newtonian fluids. However, a wide variety of non-Newtonian fluids are encountered in the industrial chemical, biological, and food processing industries. To design equipment to handle these fluids, the flow property constants (rheological constants) must be available or must be measured experimentally. Section 3.5 gave a detailed discussion of rheological constants for non-Newtonian fluids. Since many non-Newtonian fluids have high effective viscosities, they are often in laminar flow. Since the majority of non-Newtonian fluids are pseudoplastic fluids, which can usually be represented by the power law, Eq. (3.5-2), the discussion will be concerned with such fluids. For other fluids, the reader is referred to Skelland (S3).

4.12B Heat Transfer Inside Tubes

1. Laminar flow in tubes. A large portion of the experimental investigations have been concerned with heat transfer of non-Newtonian fluids in laminar flow through cylindrical tubes. The physical properties that are needed for heat transfer coefficients are density, heat capacity, thermal conductivity, and the rheological constants K' and n' or K and n.

In heat transfer in a fluid in laminar flow, the mechanism is one of primarily conduction. However, for low flow rates and low viscosities, natural convection effects can be present. Since many non-Newtonian fluids are quite "viscous," natural convection effects are reduced substantially. For laminar flow inside circular tubes of power-law fluids, the equation of Metzner and Gluck (M2) can be used with highly "viscous" non-Newtonian fluids with negligible natural convection for horizontal or vertical tubes for the Graetz number $N_{Gz} > 20$ and $n' > 0.10$.

$$(N_{Nu})_a = \frac{h_a D}{k} = 1.75 \delta^{1/3} (N_{Gz})^{1/3} \left(\frac{\gamma_b}{\gamma_w}\right)^{0.14} \tag{4.12-1}$$

where

$$\delta = \frac{3n' + 1}{4n'} \tag{4.12-2}$$

$$N_{Gz} = \frac{m c_p}{kL} = \frac{\pi}{4} \frac{Dv\rho}{\mu} \frac{c_p \mu}{k} \frac{D}{L} = \frac{\pi}{4} N_{Re} N_{Pr} \frac{D}{L} \tag{4.12-3}$$

The viscosity coefficients γ_b at temperature T_b and γ_w at T_w are defined as

$$\frac{\gamma_b}{\gamma_w} = \frac{K'_b 8^{n'-1}}{K'_w 8^{n'-1}} = \frac{K'_b}{K'_w} = \frac{K_b}{K_w} \tag{4.12-4}$$

The nomenclature is as follows: k in $W/m \cdot K$, c_p in $J/kg \cdot K$, ρ in kg/m^3, flow rate m in kg/s, length of heated section of tube L in m, inside diameter D in m, the mean coefficient h_a in $W/m^2 \cdot K$, and K and n' rheological constants (see Section 3.5). The physical properties and K_b are all evaluated at the mean bulk temperature T_b and K_w at the average wall temperature T_w.

The value of the rheological constant n' or n has been found to not vary appreciably over wide temperature ranges (S3). However, the rheological constant K' or K has been found to vary appreciably. A plot of log K' versus $1/T_{abs}$ (C1) or versus $T°C$ (S3) can often be approximated by a straight line. Often data for the temperature effect on K are not available. Since the ratio K_b/K_w is taken to the 0.14 power, this factor can sometimes be neglected without causing large errors. For a value of the ratio of 2:1, the error is only about 10%. A plot of viscosity versus $1/T$ for Newtonian fluids is also often a straight line. The value of h_a obtained from Eq. (4.12-1) is the mean value to use over the tube length L with the arithmetic temperature difference ΔT_a.

$$\Delta T_a = \frac{(T_w - T_{bi}) + (T_w - T_{bo})}{2} \tag{4.12-5}$$

when T_w is the average wall temperature for the whole tube and T_{bi} is the inlet bulk temperature and T_{bo} the outlet bulk temperature. The heat flux q is

$$q = h_a A \Delta T_a = h_a (\pi DL) \Delta T_a \tag{4.12-6}$$

EXAMPLE 4.12-1. *Heating a Non-Newtonian Fluid in Laminar Flow*
A non-Newtonian fluid flowing at a rate of 7.56×10^{-2} kg/s inside a
25.4-mm-ID tube is being heated by steam condensing outside the tube. The
fluid enters the heating section of the tube, which is 1.524 m long, at a
temperature of 37.8°C. The inside wall temperature T_w is constant at 93.3°C.
The mean physical properties of the fluid are $\rho = 1041$ kg/m³, $c_{pm} = 2.093$
kJ/kg·K, and $k = 1.212$ W/m·K. The fluid is a power-law fluid having the
following flow property (rheological) constants: $n = n' = 0.40$, which is approximately constant over the temperature range encountered, and
$K = 139.9$ N·s^{n}/m² at 37.8°C and 62.5 at 93.3°C. For this fluid a plot of log
K versus T°C is approximately a straight line. Calculate the outlet bulk
temperature of the fluid if it is in laminar flow.

Solution: The solution is trial and error since the outlet bulk temperature
T_{bo} of the fluid must be known to calculate h_a from Eq. (4.12-2). Assuming
$T_{bo} = 54.4$°C for the first trial, the mean bulk temperature T_b is
$(54.4 + 37.8)/2$, or 46.1°C.
  Plotting the two values of K given at 37.8 and 93.3°C as log K versus
T°C and drawing a straight line through these two points, a value of K_b of
123.5 at $T_b = 46.1$°C is read from the plot. At $T_w = 93.3$°C, $K_w = 62.5$.
  Next, δ is calculated using Eq. (4.12-2).

$$\delta = \frac{3n' + 1}{4n'} = \frac{3(0.40) + 1}{4(0.40)} = 1.375$$

Substituting into Eq. (4.12-3),

$$N_{Gz} = \frac{mc_p}{kL} = \frac{(7.56 \times 10^{-2})(2.093 \times 10^3)}{1.212(1.524)} = 85.7$$

From Eq. (4.12-4),

$$\frac{\gamma_b}{\gamma_w} = \frac{K_b}{K_w} = \frac{123.5}{62.5}$$

Then substituting into Eq. (4.12-1),

$$\frac{h_a D}{k} = \frac{h_a(0.0254)}{1.212} = 1.75\delta^{1/3}(N_{Gz})^{1/3}\left(\frac{\gamma_b}{\gamma_w}\right)^{0.14}$$

$$= 1.75(1.375)^{1/3}(85.7)^{1/3}\left(\frac{123.5}{62.5}\right)^{0.14} \quad \textbf{(4.12-1)}$$

Solving, $h_a = 448.3$ W/m²·K.
  By a heat balance, the value of q in W is as follows:

$$q = mc_{pm}(T_{bo} - T_{bi}) \quad \textbf{(4.12-7)}$$

This is equated to Eq. (4.12-6) to obtain

$$q = mc_{pm}(T_{bo} - T_{bi}) = h_a(\pi DL)\, \Delta T_a \quad \textbf{(4.12-8)}$$

The arithmetic mean temperature difference ΔT_a by Eq. (4.12-5) is

$$\Delta T_a = \frac{(T_w - T_{bi}) + (T_w - T_{bo})}{2}$$

$$= \frac{(93.3 - 37.8) + (93.3 - T_{bo})}{2} = 74.4 - 0.5T_{bo}$$

Substituting the known values in Eq. (4.12-8) and solving for T_{bo},

$$(7.56 \times 10^{-2})(2.093 \times 10^{3})(T_{bo} - 37.8)$$

$$= 448.3(\pi \times 0.0254 \times 1.524)(74.4 - 0.5T_{bo})$$

$$T_{bo} = 54.1°C$$

This value of 54.1°C is close enough to the assumed value of 54.5°C so that a second trial is not needed. Only the value of K_b would be affected. Known values can be substituted into Eq. (3.5-11) for the Reynolds number to show that it is less than 2100 and is laminar flow.

For less "viscous" non-Newtonian power-law fluids in laminar flow, natural convection may affect the heat-transfer rates. Metzner and Gluck (M2) recommend use of an empirical correction to Eq. (4.12-1) for horizontal tubes.

2. Turbulent flow in tubes. For turbulent flow of power-law fluids through tubes, Clapp (C4) presents the following empirical equation for heat transfer:

$$N_{\text{Nu}} = \frac{h_L D}{k} = 0.0041(N_{\text{Re, gen}})^{0.99}\left[\frac{K'c_p}{k}\left(\frac{8V}{D}\right)^{n-1}\right]^{0.4} \qquad \textbf{(4.12-9)}$$

where $N_{\text{Re, gen}}$ is defined by Eq. (3.5-11) and h_L is the heat-transfer coefficient based on the log mean temperature driving force. The fluid properties are evaluated at the bulk mean temperature. Metzner and Friend (M3) also give equations for turbulent heat transfer.

4.12C Natural Convection

Acrivos (A1, S3) gives relationships for natural convection heat transfer to power-law fluids from various geometries of surfaces such as spheres, cylinders, and plates.

4.13 (SELECTED TOPIC) SPECIAL HEAT-TRANSFER COEFFICIENTS

4.13A Heat Transfer in Agitated Vessels

1. Introduction. Many chemical and biological processes are often carried out in agitated vessels. As discussed in Section 3.4, the liquids are generally agitated in cylindrical vessels with an impeller mounted on a shaft and driven by an electric motor. Typical agitators and vessel assemblies have been shown in Figs. 3.4-1 and 3.4-3. Often it is necessary to cool or heat the contents of the vessel during the agitation. This is usually done by heat-transfer surfaces, which may be in the form of cooling or heating jackets in the wall of the vessel or coils of pipe immersed in the liquid.

2. Vessel with heating jacket. In Fig. 4.13-1a a vessel with a cooling or heating jacket is shown. When heating, the fluid entering is often steam, which condenses inside the jacket and leaves at the bottom. The vessel is equipped with an agitator and in most cases also with baffles (not shown).

Correlations for the heat-transfer coefficient from the agitated Newtonian liquid inside the vessel to the jacket walls of the vessel have the following form:

$$\frac{hD_t}{k} = a\left(\frac{D_a^2 N\rho}{\mu}\right)^{b}\left(\frac{c_p\mu}{k}\right)^{1/3}\left(\frac{\mu}{\mu_w}\right)^{m} \qquad \textbf{(4.13-1)}$$

where h is the heat-transfer coefficient for the agitated liquid to the inner wall in $W/m^2 \cdot K$, D_t is the inside diameter of the tank in m, k is thermal conductivity in $W/m \cdot K$, D_a is diameter of agitator in m, N is rotational speed in revolutions per sec, ρ is fluid density in kg/m^3, and μ is liquid viscosity in $Pa \cdot s$. All the liquid physical properties are evaluated at the bulk liquid temperature except μ_w, which is evaluated at the wall temperature T_w. Below are listed some correlations available and the Reynolds number range ($N'_{Re} = D_a^2 N \rho / \mu$).

1. Paddle agitator with no baffles (C5, U1)

$$a = 0.36, \quad b = \tfrac{2}{3}, \quad m = 0.21, \quad N'_{Re} = 300 \text{ to } 3 \times 10^5$$

2. Flat-blade turbine agitator with no baffles (B4)

$$a = 0.54, \quad b = \tfrac{2}{3}, \quad m = 0.14, \quad N'_{Re} = 30 \text{ to } 3 \times 10^5$$

3. Flat-blade turbine agitator with baffles (B4, B5)

$$a = 0.74, \quad b = \tfrac{2}{3}, \quad m = 0.14. \quad N'_{Re} = 500 \text{ to } 3 \times 10^5$$

4. Anchor agitator with no baffles (U1)

$$a = 1.0, \quad b = \tfrac{1}{2}, \quad m = 0.18, \quad N'_{Re} = 10 \text{ to } 300$$

$$a = 0.36, \quad b = \tfrac{2}{3}, \quad m = 0.18, \quad N'_{Re} = 300 \text{ to } 4 \times 10^4$$

5. Helical ribbon agitator with no baffles (G4)

$$a = 0.633, \quad b = \tfrac{1}{2}, \quad m = 0.18, \quad N'_{Re} = 8 \text{ to } 10^5$$

Some typical overall U values for jacketed vessels for various process applications are tabulated in Table 4.13-1 (P1).

EXAMPLE 4.13-1.   Heat-Transfer Coefficient in Agitated Vessel with Jacket

A jacketed 1.83-m-diameter agitated vessel with baffles is being used to heat a liquid which is at 300 K. The agitator is 0.61 m in diameter and is a

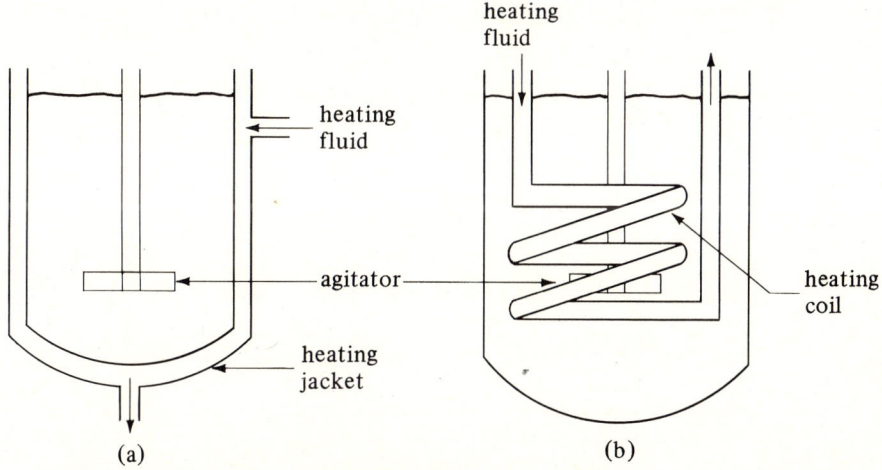

FIGURE 4.13-1.   *Heat transfer in agitated vessels: (a) vessel with heating jacket, (b) vessel with heating coils.*

TABLE 4.13-1. *Typical Overall Heat-Transfer Coefficients in Jacketed Vessels*

Fluid in Jacket	Fluid in Vessel	Wall Material	Agitation	$\dfrac{btu}{h \cdot ft^2 \cdot °F}$	$\dfrac{W}{m^2 \cdot K}$	Ref.
Steam	Water	Copper	None	150	852	(P1)
			Simple stirring	250	1420	
Steam	Paste	Cast iron	Double scrapers	125	710	(P1)
Steam	Boiling water	Copper	None	250	1420	(P1)
Steam	Milk	Enameled cast iron	None	200	1135	(P1)
			Stirring	300	1700	
Hot water	Cold water	Enameled cast iron	None	70	398	(P1)
Steam	Tomato purée	Metal	Agitation	30	170	(C1)

The heading **U** spans the two columns $\dfrac{btu}{h \cdot ft^2 \cdot °F}$ and $\dfrac{W}{m^2 \cdot K}$.

flat-blade turbine rotating at 100 rpm. Hot water is in the heating jacket. The wall surface temperature is constant at 355.4 K. The liquid has the following bulk physical properties: $\rho = 961$ kg/m^3, $c_p = 2500$ J/kg $\cdot$ K, $k = 0.173$ W/m $\cdot$ K, and $\mu = 1.00$ Pa $\cdot$ s at 300 K and 0.084 Pa $\cdot$ s at 355.4 K. Calculate the heat-transfer coefficient to the wall of the jacket.

Solution: The following are given:

$$D_t = 1.83 \text{ m} \qquad D_a = 0.61 \text{m} \qquad N = 100/60 \text{ rev/s}$$

$$\mu(300 \text{ K}) = 1.00 \text{ Pa} \cdot \text{s} = 1.00 \text{ kg/m} \cdot \text{s}$$

$$\mu_w(355.4 \text{ K}) = 0.084 \text{ Pa} \cdot \text{s} = 0.084 \text{ kg/m} \cdot \text{s}$$

First, calculating the Reynolds number at 300 K,

$$N'_{Re} = \frac{D_a^2 N \rho}{\mu} = \frac{(0.61)^2(100/60)(961)}{1.00} = 596$$

The Prandtl number is

$$N_{Pr} = \frac{c_p \mu}{k} = \frac{2500(1.00)}{0.173} = 14\,450$$

Using Eq. (4.13-1) with $a = 0.74$, $b = 2/3$, and $m = 0.14$

$$\frac{hD_t}{k} = 0.74(N'_{Re})^{2/3}(N_{Pr})^{1/3}\left(\frac{\mu}{\mu_w}\right)^{0.14} \tag{4.13-1}$$

Substituting and solving for h,

$$\frac{h(1.83)}{0.173} = 0.74(596)^{2/3}(14\,450)^{1/3}\left(\frac{1000}{84}\right)^{0.14}$$

$$h = 170.6 \text{ W/m}^2 \cdot \text{K } (30.0 \text{ btu/h} \cdot \text{ft}^2 \cdot °F)$$

A correlation to predict the heat-transfer coefficient of a power-law non-Newtonian fluid in a jacketed vessel with a turbine agitator is also available elsewhere (C6).

3. Vessel with heating coils. In Fig. 4.13-1b an agitated vessel with a helical heating or cooling coil is shown. Correlations for the heat-transfer coefficient to the outside surface of the coils in agitated vessels are listed below for various types of agitators.

For a paddle agitator with no baffles (C5),

$$\frac{hD_t}{k} = 0.87\left(\frac{D_a^2 N\rho}{\mu}\right)^{0.62}\left(\frac{c_p\mu}{k}\right)^{1/3}\left(\frac{\mu}{\mu_w}\right)^{0.14} \tag{4.13-2}$$

This holds for a Reynolds number range of 300 to 4×10^5.

For a flat-blade turbine agitator with baffles, see (O1).

When the heating or cooling coil is in the form of vertical tube baffles with a flat-blade turbine, the following correlation can be used (D1).

$$\frac{hD_o}{k} = 0.09\left(\frac{D_a^2 N\rho}{\mu}\right)^{0.65}\left(\frac{c_p\mu}{k}\right)^{1/3}\left(\frac{D_a}{D_t}\right)^{1/3}\left(\frac{2}{n_b}\right)^{0.2}\left(\frac{\mu}{\mu_f}\right)^{0.4} \tag{4.13-3}$$

where D_o is the outside diameter of the coil tube in m, n_b is the number of vertical baffle tubes, and μ_f is the viscosity at the mean film temperature.

Perry and Chilton (P1) give typical values of overall heat-transfer coefficients U for coils immersed in various liquids in agitated and nonagitated vessels.

4.13B Scraped Surface Heat Exchangers

Liquid–solid suspensions, viscous aqueous and organic solutions, and numerous food products, such as margarine and orange juice concentrate, are often cooled or heated in a scraped-surface exchanger. This consists of a double-pipe heat exchanger with a jacketed cylinder containing steam or cooling liquid and an internal shaft rotating and fitted with wiper blades, as shown in Fig. 4.13-2.

The viscous liquid product flows at low velocity through the central tube between the rotating shaft and the inner pipe. The rotating scrapers or wiper blades continually scrape the surface of liquid, preventing localized overheating and giving rapid heat transfer. This device in some cases is also called a *votator heat exchanger*.

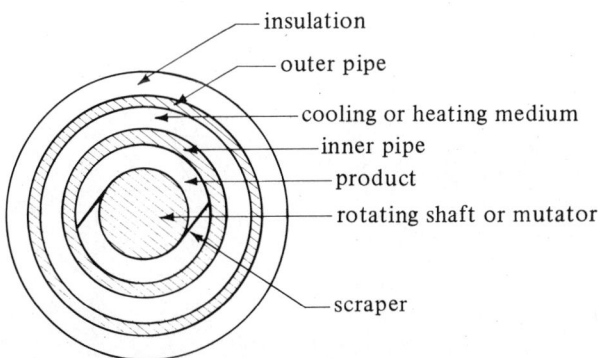

FIGURE 4.13-2. *Scraped surface heat exchanger.*

Skelland et al. (S4) give the following equation to predict the inside heat-transfer coefficient for the votator.

$$\frac{hD}{k} = \alpha \left(\frac{c_p \mu}{k}\right)^\beta \left(\frac{(D - D_S)v\rho}{\mu}\right)^{1.0} \left(\frac{DN}{v}\right)^{0.62} \left(\frac{D_S}{D}\right)^{0.55} (n_B)^{0.53} \qquad \text{(4.13-4)}$$

$$\alpha = 0.014 \quad \beta = 0.96 \qquad \text{for viscous liquids}$$

$$\alpha = 0.039 \quad \beta = 0.70 \qquad \text{for nonviscous liquids}$$

where D = diameter of vessel in m, D_S = diameter of rotating shaft in m, v = axial flow velocity of liquid in m/s, N = agitator speed in rev/s, and n_B = number of blades on agitator. Data cover a region of axial flow velocities of 0.076 to 0.38 m/min and rotational speeds of 100 to 750 rpm.

Typical overall heat-transfer coefficients in food applications are $U = 1700 \text{ W/m}^2 \cdot \text{K}$ (300 btu/h · ft² · °F) for cooling margarine with NH_3, 2270 (400) for heating applesauce with steam, 1420 (250) for chilling shortening with NH_3, and 2270 (400) for cooling cream with water (B6).

4.13C Extended Surface or Finned Exchangers

1. Introduction. The use of fins or extended surfaces on the outside of a heat exchanger pipe wall to give relatively high heat-transfer coefficients in the exchanger is quite common. An automobile radiator is such a device, where hot water passes inside through a bank of tubes and loses heat to the air. On the outside of the tubes, extended surfaces receive heat from the tube walls and transmit it to the air by forced convection.

Two common types of fins attached to the outside of a tube wall are shown in Fig. 4.13-3. In Fig. 4.13-3a there are a number of longitudinal fins spaced around the tube wall and the direction of gas flow is parallel to the axis of the tube. In Fig. 4.13-3b the gas flows normal to the tubes containing many circular or transverse fins.

The qualitative effect of using extended surfaces can be shown approximately in Eq. (4.13-5) for a fluid inside a tube having a heat-transfer coefficient of h_i and an outside coefficient of h_o.

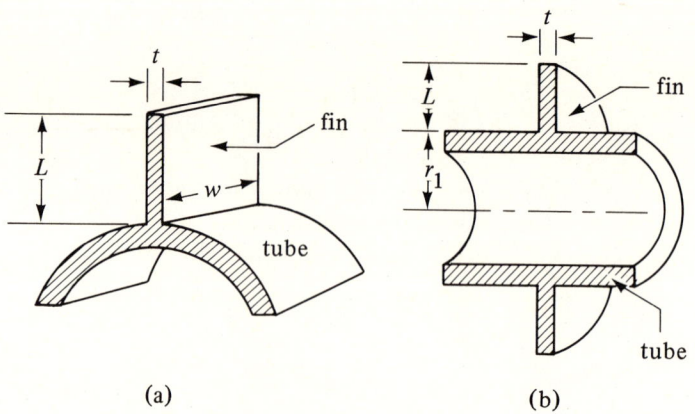

(a)   (b)

FIGURE 4.13-3. *Two common types of fins on a section of circular tube: (a) longitudinal fin, (b) circular or transverse fin.*

Transport Processes: Momentum, Heat, and Mass

$$\frac{1}{U_i A_i} = \sum R \simeq \frac{1}{h_i A_i} + R_{\text{metal}} + \frac{1}{h_o A_o} \tag{4.13-5}$$

The resistance R_{metal} of the wall can often be neglected. The presence of the fins on the outside increases A_o and hence reduces the resistance $1/h_o A_o$ of the fluid on the outside of the tube. For example, if we have h_i for condensing steam, which is very large, and h_o for air outside the tube, which is quite small, increasing A_o greatly reduces $1/h_o A_o$. This in turn greatly reduces the total resistance, which increases the heat-transfer rate. If the positions of the two fluids are reversed with air inside and steam outside, little increase in heat transfer could be obtained by using fins.

Equation (4.13-5) is only an approximation, since the temperature on the outside surface of the bare tube is not the same as that at the end of the fin because of the added resistance to heat flow by conduction from the fin tip to the base of the fin. Hence, a unit area of fin surface is not as efficient as a unit area of bare tube surface at the base of the fin. A fin efficiency η_f has been mathematically derived for various geometries of fins.

2. Derivation of equation for fin efficiency. We will consider a one-dimensional fin exposed to a surrounding fluid at temperature T_∞ as shown in Fig. 4.13-4. At the base of the fin the temperature is T_0 and at point x it is T. At steady state, the rate of heat conducted in to the element at x is $q_{x|x}$ and is equal to the rate of heat conducted out plus the rate of heat lost by convection.

$$q_{x|x} = q_{x|x+\Delta x} + q_c \tag{4.13-6}$$

Substituting Fourier's equation for conduction and the convection equation,

$$-kA \left.\frac{dT}{dx}\right|_x = -kA \left.\frac{dT}{dx}\right|_{x+\Delta x} + h(P \, \Delta x)(T - T_\infty) \tag{4.13-7}$$

where A is the cross-sectional area of the fin in m², P the perimeter of the fin in m, and $(P \, \Delta x)$ the area for convection. Rearranging Eq. (4.13-7), dividing by Δx, and letting Δx approach zero,

$$\frac{d^2 T}{dx^2} - \frac{hP}{kA}(T - T_\infty) = 0 \tag{4.13-8}$$

Letting $\theta = T - T_\infty$, Eq. (4.13-8) becomes

$$\frac{d^2 \theta}{dx^2} = \frac{hP}{kA}\theta = 0 \tag{4.13-9}$$

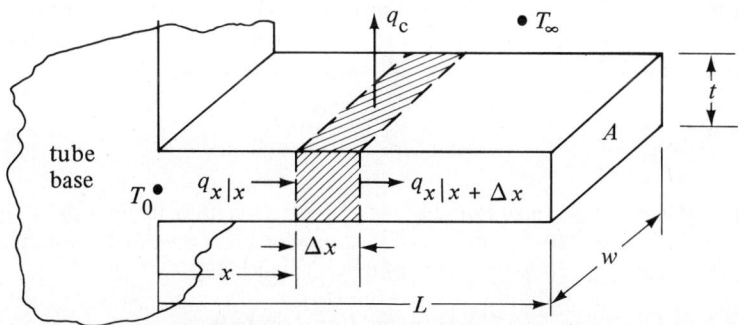

FIGURE 4.13-4. *Heat balance for one-dimensional conduction and convection in a rectangular fin with constant cross-sectional area.*

The first boundary condition is that $\theta = \theta_0 = T_0 - T_\infty$ at $x = 0$. For the second boundary condition needed to integrate Eq. (4.13-9), several cases can be considered, depending upon the physical conditions at $x = L$. In the first case, the end of the fin is insulated and $d\theta/dx = 0$ at $x = L$. In the second case the fin loses heat by convection from the tip surface so that $-k(dT/dx)_L = h(T_L - T_\infty)$. The solution using case 2 is quite involved and will not be considered here. Using the first case where the tip is insulated, integration of Eq. (4.13-9) gives

$$\frac{\theta}{\theta_0} = \frac{\cosh\,[m(L-x)]}{\cosh\,mL} \qquad (4.13\text{-}10)$$

where $m = (hP/kA)^{1/2}$.

The heat lost by the fin is expressed as

$$q = -kA\,\frac{dT}{dx}\bigg|_{x=0} \qquad (4.13\text{-}11)$$

Differentiating Eq. (4.13-10) with respect to x and combining it with Eq. (4.13-11),

$$q = (hPkA)^{1/2}(T_0 - T_\infty)\,\tanh\,mL \qquad (4.13\text{-}12)$$

In the actual fin the temperature T in the fin decreases as the tip of the fin is approached. Hence, the rate of heat transfer per unit area decreases as the distance from the tube base is increased. To indicate this effectiveness of the fin to transfer heat, the fin efficiency η_f is defined as the ratio of the actual heat transferred from the fin to the heat transferred if the entire fin were at the base temperature T_0.

$$\eta_f = \frac{(hPkA)^{1/2}(T_0 - T_\infty)\,\tanh\,mL}{h(PL)(T_0 - T_\infty)} = \frac{\tanh\,mL}{mL} \qquad (4.13\text{-}13)$$

where PL is the entire surface area of fin.

The expression for mL is

$$mL = \left(\frac{hP}{kA}\right)^{1/2} L = \left[\frac{h(2w + 2t)}{k(wt)}\right]^{1/2} L \qquad (4.13\text{-}14)$$

For fins which are thin, $2t$ is small compared to $2w$ and

$$mL = \left(\frac{2h}{kt}\right)^{1/2} L \qquad (4.13\text{-}15)$$

Equation (4.13-15) holds for a fin with an insulated tip. This equation can be modified to hold for the case where the fin loses heat from its tip. This can be done by extending the length of the fin by $t/2$, where the corrected length L_c to use in Eqs. (4.13-13) to (4.13-15) is

$$L_c = L + \frac{t}{2} \qquad (4.13\text{-}16)$$

The fin efficiency calculated from Eq. (4.13-13) for a longitudinal fin is shown in Fig. 4.13-5a. In Fig. 4.13-5b the fin efficiency for a circular fin is presented. Note that the abscissa on the curves is $L_c(h/kt)^{1/2}$ and not $L_c(2h/kt)^{1/2}$ as in Eq. (4.13-15).

EXAMPLE 4.13.-2. Fin Efficiency and Heat Loss from Fin
A circular aluminum fin as shown in Fig. 4.13-3b ($k = 222$ W/m · K) is attached to a copper tube having an outside radius of 0.04 m. The length of the fin is 0.04 m and the thickness is 2 mm. The outside wall or tube base is

at 523.2 K and the external surrounding air at 343.2 K has a convective coefficient of 30 W/m² · K. Calculate the fin efficiency and the rate of heat loss from the fin.

Solution: The given data are $T_0 = 523.2$ K, $T_\infty = 343.2$ K, $L = 0.04$ m, $r_1 = 0.04$ m, $t = 0.002$ m, $k = 222$ W/m · K, $h = 30$ W/m² · K. By Eq. (4.13-16), $L_c = L + t/2 = 0.040 + 0.002/2 = 0.041$ m. Then,

$$L_c \left(\frac{h}{kt} \right)^{1/2} = (0.041) \left[\frac{30}{222(0.002)} \right]^{1/2} = 0.337$$

Also, $(L_c + r_1)/r_1 = (0.041 + 0.040)/0.040 = 2.025$. Using Fig. 4.13-5b, $\eta_f = 0.89$. The heat transfer from the fin itself is

$$q_f = \eta_f h A_f (T_0 - T_\infty) \tag{4.13-17}$$

where A_f is the outside surface area (annulus) of the fin and is given by the following for both sides of the fin:

$$A_f = 2\pi[(L_c + r_1)^2 - (r_1)^2] \quad \text{(circular fin)}$$
$$A_f = 2\pi(L_c \, xw) \quad \text{(longitudinal fin)} \tag{4.13-18}$$

Hence,

$$A_f = 2\pi[(0.041 + 0.040)^2 - (0.040)^2] = 3.118 \times 10^{-2} \text{ m}^2$$

Substituting into Eq. (4.13-17),

$$q_f = 0.89(30)(3.118 \times 10^{-2})(523.2 - 343.2) = 149.9 \text{ W}$$

3. Overall heat-transfer coefficient for finned tubes. We consider here the general case similar to Fig. 4.3-3b, where heat transfer occurs from a fluid inside a cylinder or tube, through the cylinder metal wall A of thickness Δx_A, and then to the fluid outside the tube, where the tube has fins on the outside. The heat is transferred through a series of resistances. The total heat q leaving the outside of the tube is the sum of heat loss by

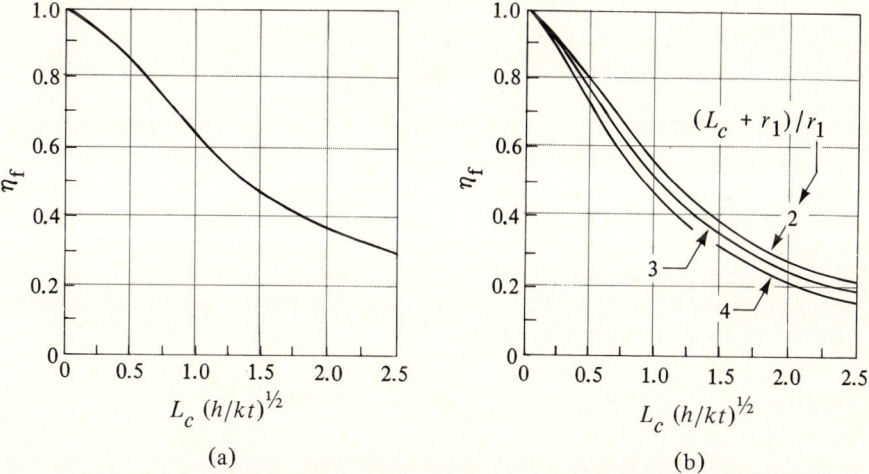

FIGURE 4.13-5. *Fin efficiency η_f for various fins : (a) longitudinal or straight fins, (b) circular or transverse fins. (See Fig. 4.13-3 for the dimensions of the fins.)*

convection from the base of the bare tube q_t and the loss by convection from the fins, q_f.

$$q = q_t + q_f = h_o A_t(T_0 - T_\infty) + h_o A_f \eta_f(T_0 - T_\infty) \tag{4.13-19}$$

This can be written as follows as a resistance since the paths are in parallel.

$$q = (h_o A_t + h_o A_f \eta_f)(T_1 - T_\infty) = \frac{T_0 - T_\infty}{\dfrac{1}{h_o(A_t + A_f \eta_f)}} = \frac{T_0 - T_\infty}{R} \tag{4.13-20}$$

where A_t is the area of the bare tube between the fins, A_f the area of the fins, and h_o the outside convective coefficient. The resistance in Eq. (4.3-20) can be substituted for the resistance $(1/h_o A_o)$ in Eq. (4.3-15) for a bare tube to give the overall equation for a finned tube exchanger.

$$q = \frac{T_4 - T_1}{1/h_i A_i + \Delta x_A/k_A A_{A\,lm} + 1/h_o(A_t + A_f \eta_f)} = \frac{T_4 - T_1}{\sum R} \tag{4.13-21}$$

where T_4 is the temperature of the fluid inside the tube and T_1 the outside fluid temperature. Writing Eq. (4.13-21) in the form of an overall heat-transfer coefficient U_i based on the inside area A_i, $q = U_i A_i(T_4 - T_1)$ and

$$U_i = \frac{1}{1/h_i + \Delta x_A A_i/k_A A_{A\,lm} + A_i/h_o(A_t + A_f \eta_f)} \tag{4.13-22}$$

The presence of fins on the outside of the tube changes the characteristics of the fluid flowing by the tube (either flowing parallel to the longitudinal finned tube or transverse to the circular finned tube). Hence, the correlations for fluid flow parallel to or transverse to bare tubes cannot be used to predict the outside convective coefficient h_o. Correlations are available in the literature (K4, M1, P1) for heat transfer to various types of fins.

4.14 (SELECTED TOPIC) DIMENSIONAL ANALYSIS IN HEAT TRANSFER

4.14A Introduction

As seen in many of the correlations for fluid flow and heat transfer, many dimensionless groups, such as the Reynolds number and Prandtl number, occur in these correlations. Dimensional analysis is often used to group the variables in a given physical situation into dimensionless parameters or numbers which can be useful in experimentation and correlating data.

An important way of obtaining these dimensionless groups is to use dimensional analysis of differential equations described in Section 3.8. Another useful method is the Buckingham method, in which the listing of the significant variables in the particular physical problem is done first. Then we determine the number of dimensionless parameters into which the variables may be combined.

4.14B Buckingham Method

1. Heat transfer inside a pipe.   The Buckingham theorem, given in Section 3.8 (Selected Topic) states that the functional relationship among q quantities or variables whose units may be given in terms of u fundamental units or dimensions may be written as $(q - u)$ dimensionless groups.

As an additional example to illustrate the use of this method, let us consider a fluid flowing in turbulent flow at velocity v inside a pipe of diameter D and undergoing heat transfer to the wall. We wish to predict the dimensionless groups relating the heat-transfer coefficient h to the variables D, ρ, μ, c_p, k, and v. The total number of variables is $q = 7$.

The fundamental units or dimensions are $u = 4$ and are mass M, length L, time t, and temperature T. The units of the variables in terms of these fundamental units are as follows:

$$h = \frac{M}{t^3 T} \qquad D = L \qquad \rho = \frac{M}{L^3} \qquad \mu = \frac{M}{Lt} \qquad c_p = \frac{L^2}{t^2 T} \qquad k = \frac{ML}{t^3 T} \qquad v = \frac{L}{t}$$

Hence, the number of dimensionless groups or π's can be assumed to be $7 - 4$, or 3. Then

$$\pi_1 = f(\pi_2, \pi_3) \tag{4.14-1}$$

We will choose the four variables D, k, μ, and v to be common to all the dimensionless groups. Then the three dimensionless groups are

$$\pi_1 = D^a k^b \mu^c v^d \rho \tag{4.14-2}$$

$$\pi_2 = D^e k^f \mu^g v^h c_p \tag{4.14-3}$$

$$\pi_3 = D^i k^j \mu^k v^l h \tag{4.14-4}$$

For π_1, substituting the actual dimensions,

$$M^0 L^0 t^0 T^0 = 1 = L^a \left(\frac{ML}{t^3 T}\right)^b \left(\frac{M}{LT}\right)^c \left(\frac{L}{T}\right)^d \left(\frac{M}{L^3}\right) \tag{4.14-5}$$

Summing for each exponent,

$$(L) \quad 0 = a + b - c + d - 3$$
$$(M) \quad 0 = b + c + 1$$
$$(t) \quad 0 = -3b - c - d \tag{4.14-6}$$
$$(T) \quad 0 = -b$$

Solving these equations simultaneously, $a = 1$, $b = 0$, $c = -1$, $d = 1$.

Substituting these values into Eq. (4.14-2),

$$\pi_1 = \frac{Dv\rho}{\mu} = N_{\text{Re}} \tag{4.14-7}$$

Repeating for π_2 and π_3 and substituting the actual dimensions,

$$\pi_2 = \frac{c_p \mu}{k} = N_{\text{Pr}} \tag{4.14-8}$$

$$\pi_3 = \frac{hD}{k} = N_{\text{Nu}} \tag{4.14-9}$$

Substituting π_1, π_2, and π_3 into Eq. (4.14-1) and rearranging,

$$\frac{hD}{k} = f\left(\frac{Dv\rho}{\mu}, \frac{c_p \mu}{k}\right) \tag{4.14-10}$$

This is in the form of the familiar equation for heat transfer inside pipes, Eq. (4.5-8).

This type of analysis is useful in empirical correlations of heat-transfer data. The importance of each dimensionless group, however, must be determined by experimentation (B1, M1).

2. *Natural convection heat transfer outside a vertical plane.* In the case of natural-convection heat transfer from a vertical plane wall of length L to an adjacent fluid, different dimensionless groups should be expected when compared to forced convection inside a pipe since velocity is not a variable. The buoyant force due to the difference in density between the cold and heated fluid should be a factor. As seen in Eqs. (4.7-1) and (4.7-2), the buoyant force depends upon the variables β, g, ρ, and ΔT. Hence, the list of variables to be considered and their fundamental units are as follows:

$$L = L \qquad \rho = \frac{M}{L^3} \qquad \mu = \frac{M}{Lt} \qquad c_p = \frac{L^2}{t^2 T} \qquad \beta = \frac{1}{T}$$

$$g = \frac{L}{t^2} \qquad \Delta T = T \qquad h = \frac{M}{t^3 T} \qquad k = \frac{ML}{t^3 T}$$

The number of variables is $q = 9$. Since $u = 4$, the number of dimensionless groups or π's is $9 - 4$, or 5. Then $\pi_1 = f(\pi_2, \pi_3, \pi_4, \pi_5)$.

We will choose the four variables, L, μ, k, and g to be common to all the dimensionless groups.

$$\pi_1 = L^a \mu^b k^c g^d \rho \qquad \pi_2 = L^e \mu^f k^g g^h c_p \qquad \pi_3 = L^i \mu^j k^k g^l \beta$$

$$\pi_4 = L^m \mu^n k^o g^p \Delta T \qquad \pi_5 = L^q \mu^r k^s g^t h$$

For π_1, substituting the dimensions,

$$1 = L^a \left(\frac{M}{Lt}\right)^b \left(\frac{ML}{t^3 T}\right)^c \left(\frac{L}{t^2}\right)^d \left(\frac{M}{L^3}\right) \tag{4.14-11}$$

Solving for the exponents as before, $a = \frac{3}{2}, b = -1, c = 0, d = \frac{1}{2}$. Then π_1 becomes

$$\pi_1 = \frac{L^{3/2} \rho g^{1/2}}{\mu} \tag{4.14-12}$$

Taking the square of both sides to eliminate fractional exponents,

$$\pi_1 = \frac{L^3 \rho^2 g}{\mu^2} \tag{4.14-13}$$

Repeating for the other π equations,

$$\pi_1 = \frac{L^3 \rho^2 g}{\mu^2} \qquad \pi_2 = \frac{c_p \mu}{k} = N_{Pr} \qquad \pi_3 = \frac{L \mu g \beta}{k}$$

$$\pi_4 = \frac{k \Delta T}{L \mu g} \qquad \pi_5 = \frac{hL}{k} = N_{Nu}$$

Combining the dimensionless groups π_1, π_3, and π_4 as follows,

$$\pi_1 \pi_3 \pi_4 = \frac{L^3 \rho^2 g}{\mu^2} \frac{L \mu g \beta}{k} \frac{k \Delta T}{L \mu g} = \frac{L^3 \rho^2 g \beta \Delta T}{\mu^2} = N_{Gr} \tag{4.14-14}$$

Equation (4.14-14) is the Grashof group given in Eq. (4.7-4). Hence,

$$N_{Nu} = f(N_{Gr}, N_{Pr}) \tag{4.14-15}$$

4.15 (SELECTED TOPIC) NUMERICAL METHODS FOR STEADY-STATE CONDUCTION IN TWO DIMENSIONS

4.15A Analytical Equation for Conduction

In Section 4.4 we discussed methods for solving two-dimensional heat-conduction problems using graphical procedures and shape factors. In this section we consider analytical and numerical methods.

The equation for conduction in the x direction is as follows:

$$q_x = -kA \frac{\partial T}{\partial x} \tag{4.15-1}$$

Now we shall derive an equation for steady-state conduction in two directions x and y. Referring to Fig. 4.15-1, a rectangular block Δx by Δy by L is shown. The total heat input to the block is equal to the output.

$$q_{x|x} + q_{y|y} = q_{x|x+\Delta x} + q_{y|y+\Delta y} \tag{4.15-2}$$

Now, from Eq. (4.15-1),

$$q_{x|x} = -k(\Delta yL) \left. \frac{\partial T}{\partial x} \right|_x \tag{4.15-3}$$

Writing similar equations for the other three terms and substituting into Eq. (4.15-2),

$$-k(\Delta yL) \left. \frac{\partial T}{\partial x} \right|_x - k(\Delta xL) \left. \frac{\partial T}{\partial y} \right|_y = -k(\Delta yL) \left. \frac{\partial T}{\partial x} \right|_{x+\Delta x} - k(\Delta xL) \left. \frac{\partial T}{\partial y} \right|_{y+\Delta y} \tag{4.15-4}$$

Dividing through by $\Delta x \, \Delta yL$ and letting Δx and Δy approach zero, we obtain the final equation for steady-state conduction in two directions.

$$\frac{\partial^2 T}{\partial x^2} + \frac{\partial^2 T}{\partial y^2} = 0 \tag{4.15-5}$$

This is called the *Laplace equation*. There are a number of analytical methods to solve this equation. In the method of separation of variables, the final solution is expressed as an infinite Fourier series (H1, G2, K1). We consider the case shown in Fig. 4.15-2. The solid is called a semi-infinite solid since one of its dimensions is ∞. The two edges or boundaries at $x = 0$ and $x = L$ are held constant at T_1 K. The edge at $y = 0$ is held at T_2. And at $y = \infty$, $T = T_1$. The solution relating T to position y and x is

$$\frac{T - T_1}{T_2 - T_1} = \frac{4}{\pi} \left[\frac{1}{1} e^{-(\pi/L)y} \sin \frac{1\pi x}{L} + \frac{1}{3} e^{-(3\pi/L)y} \sin \frac{3\pi x}{L} + \cdots \right] \tag{4.15-6}$$

FIGURE 4.15-1. *Steady-state conduction in two directions.*

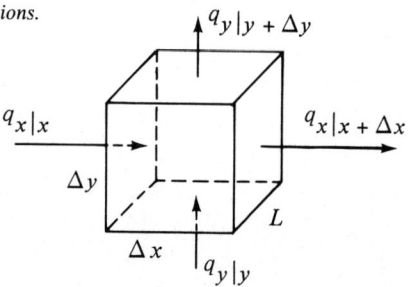

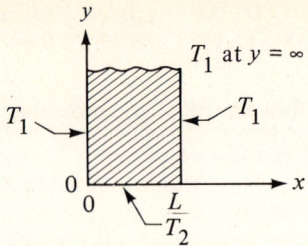

FIGURE 4.15-2. *Steady-state heat conduction in two directions in a semi-infinite plate.*

Other analytical methods are available and are discussed in many texts (C2, P1, H1, G2, K1). A large number of such analytical solutions have been given in the literature. However, there are many practical situations where the geometry or boundary conditions are too complex for analytical solutions, so that finite-difference numerical methods are used. These are discussed in the next section.

4.15B   Finite-Difference Numerical Methods

1. Derivation of the method.   Since the advent of the fast digital computers, solutions to many complex two-dimensional heat-conduction problems by numerical methods are readily possible. In deriving the equations we can start with the partial differential equation (4.15-5). Setting up the finite difference of $\partial^2 T/\partial x^2$,

$$\frac{\partial^2 T}{\partial x^2} = \frac{\partial(\partial T/\partial x)}{\partial x} = \frac{\dfrac{T_{n+1,\,m} - T_{n,\,m}}{\Delta x} - \dfrac{T_{n,\,m} - T_{n-1,\,m}}{\Delta x}}{\Delta x}$$

$$= \frac{T_{n+1,\,m} - 2T_{n,\,m} + T_{n-1,\,m}}{(\Delta x)^2} \tag{4.15-7}$$

where the index m stands for a given value of y, $m + 1$ stands for $y + 1\,\Delta y$, and n is the index indicating the position of T on the x scale. This is shown in Fig. 4.15-3. The two-dimensional solid is divided into squares. The solid inside a square is imagined to be concentrated at the center of the square, and this concentrated mass is a "node." Each node is imagined to be connected to the adjacent nodes by a small conducting rod as shown.

The finite difference of $\partial^2 T/\partial y^2$ is written in a similar manner.

$$\frac{\partial^2 T}{\partial y^2} = \frac{T_{n,\,m+1} - 2T_{n,\,m} + T_{n,\,m-1}}{(\Delta y)^2} \tag{4.15-8}$$

Substituting Eqs. (4.15-7) and (4.15-8) into Eq. (4.15-5) and setting $\Delta x = \Delta y$,

$$T_{n,\,m+1} + T_{n,\,m-1} + T_{n+1,\,m} + T_{n-1,\,m} - 4T_{n,\,m} = 0 \tag{4.15-9}$$

This equation states that the net heat flow into any point or node is zero at steady state. The shaded area in Fig. 4.15-3 represents the area on which the heat balance was made. Alternatively, Eq. (4.15-9) can be derived by making a heat balance on this shaded area.

Transport Processes: Momentum, Heat, and Mass

The total heat in for unit thickness is

$$\frac{k\,\Delta y}{\Delta x}(T_{n-1,\,m} - T_{n,\,m}) + \frac{k\,\Delta y}{\Delta x}(T_{n+1,\,m} - T_{n,\,m})$$

$$+ \frac{k\,\Delta x}{\Delta y}(T_{n,\,m+1} - T_{n,\,m}) + \frac{k\,\Delta x}{\Delta y}(T_{n,\,m-1} - T_{n,\,m}) = 0 \quad \textbf{(4.15-10)}$$

Rearranging, this becomes Eq. (4.15-9). In Fig. 4.15-3 the rods connecting the nodes act as fictitious heat conducting rods.

To use the numerical method, Eq. (4.15-9) is written for each node or point. Hence, for N unknown nodes, N linear algebraic equations must be written and the system of equations solved for the various node temperatures. For a hand calculation using a modest number of nodes, the iteration method can be used to solve the system of equations.

2. Iteration method of solution. In using the iteration method, the right-hand side of Eq. (4.15-9) is set equal to a residual $\bar{q}_{n,\,m}$.

$$\bar{q}_{n,\,m} = T_{n-1,\,m} + T_{n+1,\,m} + T_{n,\,m+1} + T_{n,\,m-1} - 4T_{n,\,m} \quad \textbf{(4.15-11)}$$

Since $\bar{q}_{n,\,m} = 0$ at steady state, solving for $T_{n,\,m}$ in Eq. (4.15-11) or (4.15-9),

$$T_{n,\,m} = \frac{T_{n-1,\,m} + T_{n+1,\,m} + T_{n,\,m+1} + T_{n,\,m-1}}{4} \quad \textbf{(4.15-12)}$$

Equations (4.15-11) and (4.15-12) are the final equations to be used. Their use is illustrated in the following example.

EXAMPLE 4.15-1.   *Steady-State Heat Conduction in Two Directions*
Figure 4.15-4 shows a cross section of a hollow rectangular chamber with the inside dimensions 4 × 2 m and outside dimensions 8 × 8 m. The chamber is 20 m long. The inside walls are held at 600 K and the outside at

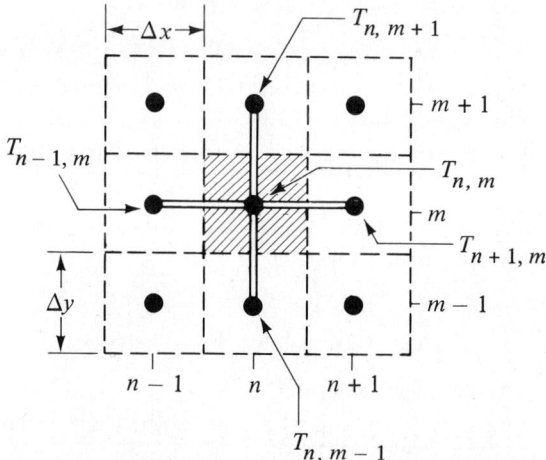

FIGURE 4.15-3.   *Temperatures and arrangement of nodes for two-dimensional steady-state heat conduction.*

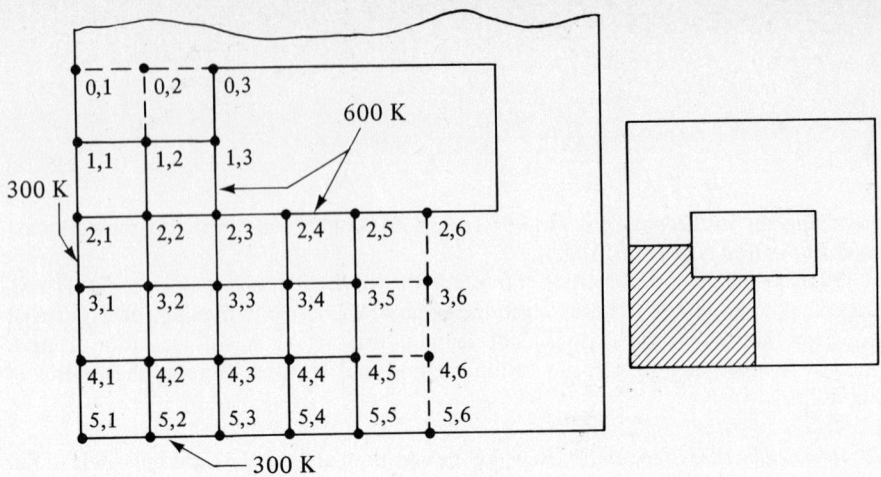

FIGURE 4.15-4. *Square grid pattern for Example 4.15-1.*

300 K. The k is 1.5 W/m · K. For steady-state conditions find the heat loss per unit chamber length. Use grids 1×1 m.

Solution: Since the chamber is symmetrical, one-fourth of the chamber (shaded part) will be used. Preliminary estimates will be made for the first approximation. These are for node $T_{1,2} = 450$ K, $T_{2,2} = 400$, $T_{3,2} = 400$, $T_{3,3} = 400$, $T_{3,4} = 450$, $T_{3,5} = 500$, $T_{4,2} = 325$, $T_{4,3} = 350$, $T_{4,4} = 375$, and $T_{4,5} = 400$. Note that $T_{0,2} = T_{2,2}$, $T_{3,6} = T_{3,4}$, and $T_{4,6} = T_{4,4}$ by symmetry.

To start the calculation, one can select any interior point, but it is usually better to start near a boundary. Using $T_{1,2}$, we calculate the residual $\bar{q}_{1,2}$ by Eq. (4.15-11).

$$\bar{q}_{1,2} = T_{1,1} + T_{1,3} + T_{0,2} + T_{2,2} - 4T_{1,2}$$
$$= 300 + 600 + 400 + 400 - 4(450) = -100$$

Hence, $T_{1,2}$ is not at steady state. Next, we set $\bar{q}_{1,2}$ to 0 and calculate a new value of $T_{1,2}$ by Eq. (4.15-12).

$$T_{1,2} = \frac{T_{1,1} + T_{1,3} + T_{0,2} + T_{2,2}}{4} = \frac{300 + 600 + 400 + 400}{4} = 425$$

This new value of $T_{1,2}$ of 425 K will replace the old one of 450 and be used to calculate the other nodes. Next,

$$\bar{q}_{2,2} = T_{2,1} + T_{2,3} + T_{1,2} + T_{3,2} - 4T_{2,2}$$
$$= 300 + 600 + 425 + 400 - 4(400) = 125$$

Setting $\bar{q}_{2,2}$ to zero and using Eq. (4.15-12),

$$T_{2,2} = \frac{T_{2,1} + T_{2,3} + T_{1,2} + T_{3,2}}{4} = \frac{300 + 600 + 425 + 400}{4} = 431$$

Continuing for all the rest of the interior nodes,

$$\bar{q}_{3,2} = 300 + 400 + 431 + 325 - 4(400) = -144$$

Using Eq. (4.15-12), $T_{3,2} = 364$,

$$\bar{q}_{3,3} = 364 + 450 + 600 + 350 - 4(400) = 164$$

$$T_{3,3} = 441$$

$$\bar{q}_{3,4} = 441 + 500 + 600 + 375 - 4(450) = 116$$

$$T_{3,4} = 479$$

$$\bar{q}_{3,5} = 479 + 479 + 600 + 400 - 4(500) = -42$$

$$T_{3,5} = 489$$

$$\bar{q}_{4,2} = 300 + 350 + 364 + 300 - 4(325) = 14$$

$$T_{4,2} = 329$$

$$\bar{q}_{4,3} = 329 + 375 + 441 + 300 - 4(350) = 45$$

$$T_{4,3} = 361$$

$$\bar{q}_{4,4} = 361 + 400 + 479 + 300 - 4(375) = 40$$

$$T_{4,4} = 385$$

$$\bar{q}_{4,5} = 385 + 385 + 489 + 399 - 4(400) = -41$$

$$T_{4,5} = 390$$

Having completed one sweep across the grid map, we can start a second approximation, using, of course, the new values calculated. We can start again with $T_{1,2}$ or we can select the node with the largest residual. Starting with $T_{1,2}$ again,

$$\bar{q}_{1,2} = 300 + 600 + 431 + 431 - 4(425) = 62$$

$$T_{1,2} = 440$$

$$\bar{q}_{2,2} = 300 + 600 + 440 + 364 - 4(431) = -20$$

$$T_{2,2} = 426$$

This is continued until the residuals are as small as desired. The final values are as follows:

$$T_{1,2} = 441, \quad T_{2,2} = 432, \quad T_{3,2} = 384, \quad T_{3,3} = 461, \quad T_{3,4} = 485,$$

$$T_{3,5} = 490, \quad T_{4,2} = 340, \quad T_{4,3} = 372, \quad T_{4,4} = 387, \quad T_{4,5} = 391$$

To calculate the total heat loss from the chamber per unit chamber length, we use Fig. 4.15-5. For node $T_{2,4}$ to $T_{3,4}$ with $\Delta x = \Delta y$ and 1 m

FIGURE 4.15-5.   *Drawing for calculation of total heat conduction.*

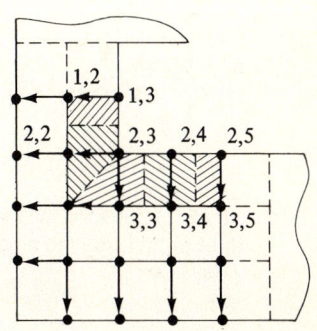

deep,

$$q = \frac{kA \, \Delta T}{\Delta x} = \frac{k[\Delta x(1)]}{\Delta x}(T_{2,4} - T_{3,4}) = k(T_{2,4} - T_{3,4}) \quad \textbf{(4.15-13)}$$

The heat flux for node $T_{2,5}$ to $T_{3,5}$ and for $T_{1,3}$ to $T_{1,2}$ should be multiplied by $\frac{1}{2}$ because of symmetry. The total heat conducted is the sum of the five paths for $\frac{1}{4}$ of the solid. For four duplicate parts,

$$
\begin{aligned}
q_I &= 4k[\tfrac{1}{2}(T_{1,3} - T_{1,2}) + (T_{2,3} - T_{2,2}) + (T_{2,3} - T_{3,3}) \\
&\quad + (T_{2,4} - T_{3,4}) + \tfrac{1}{2}(T_{2,5} - T_{3,5})] \qquad\qquad \textbf{(4.15-14)}\\
&= 4(1.5)[\tfrac{1}{2}(600 - 441) + (600 - 432) + (600 - 461) \\
&\quad + (600 - 485) + \tfrac{1}{2}(600 - 490)] \\
&= 3340 \text{ W per 1.0 m deep}
\end{aligned}
$$

Also, the total heat conducted can be calculated using the nodes at the outside, as shown in Fig. 4.15-5. This gives $q_{II} = 3430$ W. The average value is

$$q_{av} = \frac{3340 + 3430}{2} = 3385 \text{ W per 1.0 m deep}$$

If a larger number of nodes, i.e., a smaller grid size, is used, a more accurate solution can be obtained. Using a grid size of 0.5 m instead of 1.0 m for Example 4.15-1, a q_{av} of 3250 W is obtained. If a very fine grid is used, more accuracy can be obtained but a digital computer would be needed for the large number of calculations. Matrix methods are also available for solving a set of simultaneous equations on a computer. The iteration method used here is often called the Gauss–Seidel method. Conte (C7) gives an actual subroutine to solve such a system of equations. Most computers have standard subroutines for solving these equations (G2, K1).

3. Equations for other boundary conditions. In Example 4.15-1 the conditions at the boundaries were such that the node points were known and constant. For the case where there is convection at the boundary to a constant temperature T_∞, a heat balance on the node *n,m* in Fig. 4.15-6a is as follows, where heat in = heat out (K1):

$$\frac{k \, \Delta y}{\Delta x}(T_{n-1,m} - T_{n,m}) + \frac{k \, \Delta x}{2 \, \Delta y}(T_{n,m+1} - T_{n,m})$$

$$+ \frac{k \, \Delta x}{2 \, \Delta y}(T_{n,m-1} - T_{n,m}) = h \, \Delta y(T_{n,m} - T_\infty) \quad \textbf{(4.15-15)}$$

Setting $\Delta x = \Delta y$, rearranging, and setting the resultant equation $= \bar{q}_{n,m}$ residual, the following results.

(a) For convection at a boundary,

$$\frac{h \, \Delta x}{k} T_\infty + \tfrac{1}{2}(2T_{n-1,m} + T_{n,m+1} + T_{n,m-1}) - T_{n,m}\left(\frac{h \, \Delta x}{k} + 2\right) = \bar{q}_{n,m} \quad \textbf{(4.15-16)}$$

In a similar manner for the cases in Fig. 4.15-6:

(b) For an insulated boundary,

$$\tfrac{1}{2}(T_{n,m+1} + T_{n,m-1}) + T_{n-1,m} - 2T_{n,m} = \bar{q}_{n,m} \qquad\qquad \textbf{(4.15-17)}$$

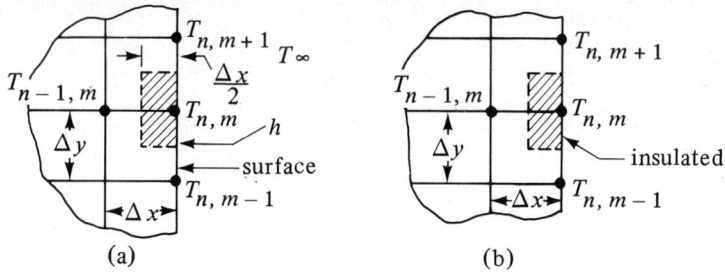

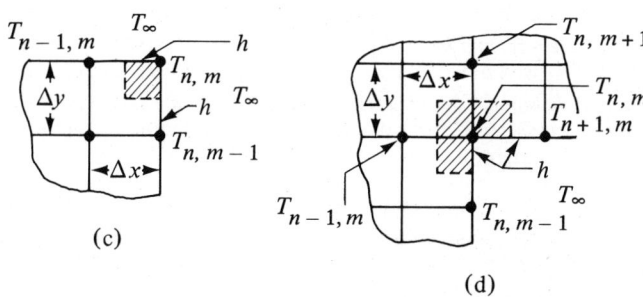

FIGURE 4.15-6. *Other types of boundary conditions: (a) convection at a boundary, (b) insulated boundary, (c) exterior corner with convective boundary, (d) interior corner with convective boundary.*

(c) For an exterior corner with convection at the boundary,

$$\frac{h\,\Delta x}{k}T_\infty + \tfrac{1}{2}(T_{n-1,m} + T_{n,m-1}) - \left(\frac{h\,\Delta x}{k} + 1\right)T_{n,m} = \bar{q}_{n,m} \qquad \textbf{(4.15-18)}$$

(d) For an interior corner with convection at the boundary,

$$\frac{h\,\Delta x}{k}T_\infty + T_{n-1,m} + T_{n,m+1} + \tfrac{1}{2}(T_{n+1,m} + T_{n,m-1}) - \left(3 + \frac{h\,\Delta x}{k}\right)T_{n,m} = \bar{q}_{n,m} \qquad \textbf{(4.15-19)}$$

For curved boundaries and other types of boundaries, see (C3, K1). To use Eqs. (4.15-16)–(4.15-19), the residual $\bar{q}_{n,m}$ is first obtained using the proper equation. Then $\bar{q}_{n,m}$ is set equal to zero and $T_{n,m}$ solved for in the resultant equation.

PROBLEMS

4.1-1. *Insulation in a Cold Room.* Calculate the heat loss per m² of surface area for a temporary insulating wall of a food cold storage room where the outside temperature is 299.9 K and the inside temperature 276.5 K. The wall is composed of 25.4 mm of corkboard having a k of 0.0433 W/m · K.

Ans. 39.9 W/m²

4.1-2. *Determination of Thermal Conductivity.* In determining the thermal conductivity of an insulating material, the temperatures were measured on both sides of a flat slab of 25 mm of the material and were 318.4 and 303.2 K. The heat flux was measured as 35.1 W/m². Calculate the thermal conductivity in btu/h · ft · °F and in W/m · K.

4.2-1. Mean Thermal Conductivity in a Cylinder. Prove that if the thermal conductivity varies linearly with temperature as in Eq. (4.1-11), the proper mean value k_m to use in the cylindrical equation is given by Eq. (4.2-3) as in a slab.

4.2-2. Heat Removal of a Cooling Coil. A cooling coil of 1.0 ft of 304 stainless steel tubing having an inside diameter of 0.25 in. and an outside diameter of 0.40 in. is being used to remove heat from a bath. The temperature at the inside surface of the tube is 40°F and 80°F on the outside. The thermal conductivity of 304 stainless steel is a function of temperature.

$$k = 7.75 + 7.78 \times 10^{-3} \ T$$

where k is in btu/h · ft · °F and T is in °F. Calculate the heat removal in btu/s and watts.

Ans. 1.225 btu/s, 1292 W

4.2-3. Removal of Heat from a Bath. Repeat Problem 4.2-2 but for a cooling coil made of 308 stainless steel having an average thermal conductivity of 15.23 W/m · K.

4.2-4. Variation of Thermal Conductivity. A flat plane of thickness Δx has one surface maintained at T_1 and the other at T_2. If the thermal conductivity varies according to temperature as

$$k = A + bT + cT^3$$

where a, b, and c are constants, derive an expression for the one-dimensional heat flux q/A.

4.2-5. Temperature Distribution in a Hollow Sphere. Derive Eq. (4.2-14) for the steady-state conduction of heat in a hollow sphere. Also, derive an equation which shows that the temperature varies hyperbolically with the radius r.

Ans. $\dfrac{T - T_1}{T_2 - T_1} = \dfrac{r_2}{r_2 - r_1}\left(1 - \dfrac{r_1}{r}\right)$

4.3-1. Insulation Needed for Food Cold Storage Room. A food cold storage room is to be constructed of an inner layer of 19.1 mm of pine wood, a middle layer of cork board, and an outer layer of 50.8 mm of concrete. The inside wall surface temperature is $-17.8°C$ and the outside surface temperature is 29.4°C at the outer concrete surface. The mean conductivities are for pine, 0.151; cork, 0.0433; and concrete, 0.762 W/m · K. The total inside surface area of the room to use in the calculation is approximately 39 m² (neglecting corner and end effects). What thickness of cork board is needed to keep the heat loss to 586 W?

Ans. 0.128 m thickness

4.3-2. Insulation of a Furnace. A wall of a furnace 0.244 m thick is constructed of material having a thermal conductivity of 1.30 W/m · K. The wall will be insulated on the outside with material having an average k of 0.346 W/m · K, so the heat loss from the furnace will be equal to or less than 1830 W/m². The inner surface temperature is 1588 K and the outer 299 K. Calculate the thickness of insulation required.

Ans. 0.179 m

4.3-3. Heat Loss Through Thermopane Double Window. A double window called thermopane is one in which two layers of glass are used separated by a layer of dry stagnant air. In a given window, each of the glass layers is 6.35 mm thick separated by a 6.35 mm space of stagnant air. The thermal conductivity of the glass is 0.869 W/m · K and that of air is 0.026 over the temperature range used. For a temperature drop of 27.8 K over the system, calculate the heat loss for a window 0.914 m × 1.83 m. (*Note:* This calculation neglects the effect of the convective coefficient on the one outside surface of one side of the window, the convective coefficient on the other outside surface, and convection inside the window.)

4.3-4. Heat Loss from Steam Pipeline. A steel pipeline, 2-in. schedule 40 pipe, contains saturated steam at 121.1°C. The line is insulated with 25.4 mm of asbestos. Assuming that the inside surface temperature of the metal wall is at 121.1°C and the outer surface of the insulation is at 26.7°C, calculate the heat loss for 30.5 m of pipe. Also, calculate the kg of steam condensed per hour in the pipe due to the heat loss. The average k for steel from Appendix A.3 is 45 W/m · K and by linear interpolation for an average temperature of $(121.1 + 26.7)/2$ or 73.9°C, the k for asbestos is 0.182.

Ans.   5384 W, 8.81 kg steam/h

4.3-5. Heat Loss with Trial-and-Error Solution. The exhaust duct from a heater has an inside diameter of 114.3 mm with ceramic walls 6.4 mm thick. The average $k = 1.52$ W/m · K. Outside this wall, an insulation of rock wool 102 mm thick is installed. The thermal conductivity of the rock wool is $k = 0.046 + 1.56 \times 10^{-4}T°$C(W/m · K). The inside surface temperature of the ceramic is $T_1 = 588.7$ K, and the outside surface temperature of the insulation is $T_3 = 311$ K. Calculate the heat loss for 1.5 m of duct and the interface temperature T_2 between the ceramic and the insulation. [*Hint :* The correct value of k_m for the insulation is that evaluated at the mean temperature of $(T_2 + T_3)/2$. Hence, for the first trial assume a mean temperature of, say, 448 K. Then calculate the heat loss and T_2. Using this new T_2, calculate a new mean temperature and proceed as before.]

4.3-6. Heat Loss by Convection and Conduction. A glass window with an area of 0.557 m² is installed in the wooden outside wall of a room. The wall dimensions are 2.44×3.05 m. The wood has a k of 0.1505 W/m · K and is 25.4 mm thick. The glass is 3.18 mm thick and has a k of 0.692. The inside room temperature is 299.9 K (26.7°C) and the outside air temperature is 266.5 K. The convection coefficient h_i on the inside wall of the glass and of the wood is estimated as 8.5 W/m² · K and the outside h_o also as 8.5 for both surfaces. Calculate the heat loss through the wooden wall, through the glass, and the total.

Ans.   569.2 W (wood) (1942 btu/h); 77.6 W (glass) (265 btu/h); 646.8 W (total) (2207 btu/h)

4.3-7. Convection, Conduction, and Overall U. A gas at 450 K is flowing inside a 2-in. steel pipe, schedule 40. The pipe is insulated with 51 mm of lagging having a mean $k = 0.0623$ W/m · K. The convective heat-transfer coefficient of the gas inside the pipe is 30.7 W/m² · K and the convective coefficient on the outside of the lagging is 10.8. The air is at a temperature of 300 K.
(a) Calculate the heat loss per unit length of 1 m of pipe using resistances.
(b) Repeat using the overall U_o based on the outside area A_o.

4.3-8. Heat Transfer in Steam Heater. Water at an average of 70°F is flowing in a 2-in. steel pipe, schedule 40. Steam at 220°F is condensing on the outside of the pipe. The convective coefficient for the water inside the pipe is $h = 500$ btu/h · ft² · °F and the condensing steam coefficient on the outside is $h = 1500$.
(a) Calculate the heat loss per unit length of 1 ft of pipe using resistances.
(b) Repeat using the overall U_i based on the inside area A_i.
(c) Repeat using U_o.

Ans.   (a) $q = 26\,710$ btu/h (7.828 kW),
(b) $U_i = 329.1$ btu/h · ft² · °F (1869 W/m² · K),
(c) $U_o = 286.4$ btu/h · ft² · °F (1626 W/m² · K)

4.3-9. Heat Loss from Temperature Measurements. A steel pipe carrying steam has an outside diameter of 89 mm. It is lagged with 76 mm of insulation having an average $k = 0.043$ W/m · K. Two thermocouples, one located at the interface between the pipe wall and the insulation and the other at the outer surface of the insulation, give temperatures of 115°C and 32°C, respectively. Calculate the heat loss in W per m of pipe.

4.3-10. *Effect of Convective Coefficients on Heat Loss in Double Window*. Repeat Problem 4.3-3 for heat loss in the double window. However, include a convective coefficient of $h = 11.35$ W/m$^2 \cdot$ K on the one outside surface of one side of the window and an h of 11.35 on the other outside surface. Also calculate the overall U.

Ans. $q = 106.7$ W, $U = 2.29$ W/m$^2 \cdot$ K

4.3-11. *Uniform Chemical Heat Generation*. Heat is being generated uniformly by a chemical reaction in a long cylinder of radius 91.4 mm. The generation rate is constant at 46.6 W/m^3. The walls of the cylinder are cooled so that the wall temperature is held at 311.0 K. The thermal conductivity is 0.865 W/m $\cdot$ K. Calculate the center-line temperature at steady state.

Ans. $T_o = 311.112$ K

4.3-12. *Heat of Respiration of a Food Product*. A fresh food product is held in cold storage at 278.0 K. It is packed in a container in the shape of a flat slab with all faces insulated except for the top flat surface, which is exposed to the air at 278.0 K. For estimation purposes the surface temperature will be assumed to be 278 K. The slab is 152.4 mm thick and the exposed surface area is 0.186 m^2. The density of the foodstuff is 641 kg/m^3. The heat of respiration is 0.070 kJ/kg $\cdot$ h and the thermal conductivity is 0.346 W/m $\cdot$ K. Calculate the maximum temperature in the food product at steady state and the total heat given off in W. (*Note*: It is assumed in this problem that there is no air circulation inside the foodstuff. Hence, the results will be conservative, since circulation during respiration will reduce the temperature.)

Ans. 278.42 K, 0.353 W (1.22 btu/h)

4.3-13. *Temperature Rise in Heating Wire*. A current of 250 A is passing through a stainless steel wire having a diameter of 5.08 mm. The wire is 2.44 m long and has a resistance of 0.0843 Ω. The outer surface is held constant at 427.6 K. The thermal conductivity is $k = 22.5$ W/m $\cdot$ K. Calculate the center-line temperature at steady state.

4.3-14. *Critical Radius for Insulation*. A metal steam pipe having an outside diameter of 30 mm has a surface temperature of 400 K and is to be insulated with an insulation having a thickness of 20 mm and a k of 0.08 W/m $\cdot$ K. The pipe is exposed to air at 300 K and a convection coefficient of 30 W/m$^2 \cdot$ K.
  (a) Calculate the critical radius and the heat loss per m of length for the bare pipe.
  (b) Calculate the heat loss for the insulated pipe assuming that the surface temperature of the pipe remains constant.

Ans. (b) $q = 54.4$ W

4.4-1. *Curvilinear-Squares Graphical Method*. Repeat Example 4.4-1 but with the following changes.
  (a) Select the number of equal temperature subdivisions between the isothermal boundaries to be five instead of four. Draw in the curvilinear squares and determine the total heat flux. Also calculate the shape factor S. Label each isotherm with the actual temperature.
  (b) Repeat part (a), but in this case the thermal conductivity is not constant but $k = 0.85 (1 + 0.00040T)$, where T is temperature in K. [*Note*: To calculate the overall q, the mean value of k at the mean temperature is used. The spacing of the isotherms is independent of how k varies with T (M1). However, the temperatures corresponding to the individual isotherms are a function of how the value of k depends upon T. Write the equation for q' for a given curvilinear section using the mean value of k over the temperature interval. Equate this to the overall value of q divided by M or q/M. Then solve for the isotherm temperature.]

4.4-2. *Heat Loss from a Furnace*. A rectangular furnace with inside dimensions of $1.0 \times 1.0 \times 2.0$ m has a wall thickness of 0.20 m. The k of the walls is

0.95 W/m · K. The inside of the furnace is held at 800 K and the outside at 350 K. Calculate the total heat loss from the furnace.

Ans. $q = 25\,081$ W

4.4-3. *Heat Loss from a Buried Pipe.* A water pipe whose wall temperature is 300 K has a diameter of 150 mm and a length of 10 m. It is buried horizontally in the ground at a depth of 0.40 m measured to the center line of the pipe. The ground surface temperature is 280 K and $k = 0.85$ W/m · K. Calculate the loss of heat from the pipe.

Ans. $q = 451.2$ W

4.5-1. *Heating Air by Condensing Steam.* Air is flowing through a tube having an inside diameter of 38.1 mm at a velocity of 6.71 m/s, average temperature of 449.9 K, and pressure of 138 kPa. The inside wall temperature is held constant at 204.4°C (477.6 K) by steam condensing outside the tube wall. Calculate the heat-transfer coefficient for a long tube and the heat-transfer flux.

Ans. $h = 33.45$ W/m^2 · K (5.89 btu/h · ft^2 · °F)

4.5-2. *Trial-and-Error Solution for Heating Water.* Water is flowing inside a horizontal $1\frac{1}{4}$-in. schedule 40 steel pipe at 37.8°C and a velocity of 1.52 m/s. Steam at 108.3°C is condensing on the outside of the pipe wall and the steam coefficient is assumed constant at 9100 W/m^2 · K.
   (a) Calculate the convective coefficient h_i for the water. (Note that this is trial and error. A wall temperature on the inside must be assumed first.)
   (b) Calculate the overall coefficient U_i based on the inside area and the heat transfer flux q/A_i in W/m^2.

4.5-3. *Heat-Transfer Area and Use of Log Mean Temperature Difference.* A reaction mixture having a $c_{pm} = 2.85$ kJ/kg · K is flowing at a rate of 7260 kg/h and is to be cooled from 377.6 K to 344.3 K. Cooling water at 288.8 K is available and the flow rate is 4536 kg/h. The overall U_o is 653 W/m^2 · K.
   (a) For counterflow, calculate the outlet water temperature and the area A_o of the exchanger.
   (b) Repeat for cocurrent flow.

Ans. (a) $T_1 = 325.2$ K, $A_o = 5.43$ m^2, (b) $A_o = 6.46$ m^2

4.5-4. *Heating Water with Hot Gases and Heat-Transfer Area.* A water flow rate of 13.85 kg/s is to be heated from 54.5 to 87.8°C in a heat exchanger by 54 430 kg/h of hot gas flowing counterflow and entering at 427°C ($c_{pm} = 1.005$ kJ/kg · K). The overall $U_o = 69.1$ W/m^2 · K. Calculate the exit-gas temperature and the heat-transfer area.

Ans. $T = 299.5$°C

4.5-5. *Cooling Oil and Overall U.* Oil flowing at the rate of 7258 kg/h with a $c_{pm} = 2.01$ kJ/kg · K is cooled from 394.3 K to 338.9 K in a counterflow heat exchanger by water entering at 294.3 K and leaving at 305.4 K. Calculate the flow rate of the water and the overall U_i if the A_i is 5.11 m^2.

Ans. 17 420 kg/h, $U_i = 686$ W/m^2 · K

4.5-6. *Laminar Flow and Heating of Oil.* A hydrocarbon oil having the same physical properties as the oil in Example 4.5-5 enters at 175°F inside a pipe having an inside diameter of 0.0303 ft and a length of 15 ft. The inside pipe surface temperature is constant at 325°F. The oil is to be heated to 250°F in the pipe. How many lb$_m$/h oil can be heated? (*Hint:* This solution is trial and error. One method is to assume a flow rate of say $m = 75$ lb mass/h. Calculate the N_{Re} and the value of h_a. Then make a heat balance to solve for q in terms of m. Equate this q to the q from the equation $q = h_a A \Delta T_a$. Solve for m. This is the new m to use for the second trial.)

Ans. $m = 84.2$ lb$_m$/h (38.2 kg/h)

4.5-7. *Heating Air by Condensing Steam.* Air at a pressure of 101.3 kPa and 288.8 K enters inside a tube having an inside diameter of 12.7 mm and a length of 1.52 m with a velocity of 24.4 m/s. Condensing steam on the outside of the tube maintains the inside wall temperature at 372.1 K. Calculate the convection coefficient of the air. (*Note :* This solution is trial and error. First assume an outlet temperature of the air.)

4.5-8. *Heat Transfer with a Liquid Metal.* The liquid metal bismuth at a flow rate of 2.00 kg/s enters a tube having an inside diameter of 35 mm at 425°C and is heated to 430°C in the tube. The tube wall is maintained at a temperature of 25°C above the liquid bulk temperature. Calculate the tube length required. The physical properties are as follows (H1): $k = 15.6$ W/m · K, $c_p = 149$ J/kg · K, $\mu = 1.34 \times 10^{-3}$ Pa · s.

4.6-1. *Heat Transfer from a Flat Plate.* Air at a pressure of 101.3 kPa and a temperature of 288.8 K is flowing over a thin, smooth flat plate at 3.05 m/s. The plate length in the direction of flow is 0.305 m and is at 333.2 K. Calculate the heat-transfer coefficient assuming laminar flow.

Ans. $h = 12.35$ W/m² · K (2.18 btu/h · ft² · °F)

4.6-2. *Chilling Frozen Meat.* Cold air at −28.9°C and 1 atm is recirculated at a velocity of 0.61 m/s over the exposed top flat surface of a piece of frozen meat. The sides and bottom of this rectangular slab of meat are insulated and the top surface is 254 mm by 254 mm square. If the surface of the meat is at −6.7°C, predict the average heat-transfer coefficient to the surface. As an approximation, assume that either Eq. (4.6-2) or (4.6-3) can be used, depending on the $N_{Re\,L}$. **Ans.** $h = 6.05$ W/m² · K

4.6-3. *Heat Transfer to an Apple.* It is desired to predict the heat-transfer coefficient for air being blown by an apple lying on a screen with large openings. The air velocity is 0.61 m/s at 101.32 kPa pressure and 316.5 K. The surface of the apple is at 277.6 K and its average diameter is 114 mm. Assume that it is a sphere.

4.6-4. *Heating Air by a Steam Heater.* A total of 13 610 kg/h of air at 1 atm abs pressure and 15.6°C is to be heated by passing over a bank of tubes in which steam at 100°C is condensing. The tubes are 12.7 mm OD, 0.61 m long, and arranged in-line in a square pattern with $S_p = S_n = 19.05$ mm. The bank of tubes contains 6 transverse rows in the direction of flow and 19 rows normal to the flow. Assume that the tube surface temperature is constant at 93.33°C. Calculate the outlet air temperature.

4.7-1. *Natural Convection from an Oven Wall.* The oven wall in Example 4.7-1 is insulated so that the surface temperature is 366.5 K instead of 505.4 K. Calculate the natural convection heat-transfer coefficient and the heat-transfer rate per m of width. Use both Eq. (4.7-4) and the simplified equation. (*Note :* Radiation is being neglected in this calculation.) Use both SI and English units.

4.7-2. *Losses by Natural Convection from a Cylinder.* A vertical cylinder 76.2 mm in diameter and 121.9 mm high is maintained at 397.1 K at its surface. It loses heat by natural convection to air at 294.3 K. Heat is lost from the cylindrical side and the flat circular end at the top. Calculate the heat loss neglecting radiation losses. Use the simplified equations of Table 4.7-2 and those equations for the lowest range of $N_{Gr}\,N_{Pr}$. The equivalent L to use for the top flat surface is 0.9 times the diameter.

Ans. $q = 26.0$ W

4.7-3. *Heat Loss from a Horizontal Tube.* A horizontal tube carrying hot water has a surface temperature of 355.4 K and an outside diameter of 25.4 mm. The tube is exposed to room air at 294.3 K. What is the natural convection heat loss for a 1-m length of pipe?

4.7-4. *Natural Convection Cooling of an Orange.* An orange 102 mm in diameter having a surface temperature of 21.1°C is placed on an open shelf in a refrigerator held at 4.4°C. Calculate the heat loss by natural convection, neglecting radiation. As an approximation, the simplified equation for vertical planes can be used with L replaced by the radius of the sphere (M1). For a more accurate correlation, see (S2).

4.7-5. *Natural Convection in Enclosed Horizontal Space.* Repeat Example 4.7-3 but for the case where the two plates are horizontal and the bottom plate is hotter than the upper plate. Compare the results.

$$\textbf{Ans.} \quad q = 12.54 \text{ W}$$

4.7-6. *Natural Convection Heat Loss in Double Window.* A vertical double plate-glass window has an enclosed air-gap space of 10 mm. The window is 2.0 m high by 1.2 m wide. One window surface is at 25°C and the other at 10°C. Calculate the free-convection heat-transfer rate through the air gap.

4.7-7. *Free Convection Heat Loss for Water in Vertical Plates.* Two vertical square metal plates having dimensions of 0.40×0.40 m are separated by a gap of 12 mm and this enclosed space is filled with water. The average surface temperature of one plate is 65.6°C and the other plate is at 37.8°C. Calculate the heat-transfer rate through this gap.

4.7-8. *Heat Loss from a Furnace.* Two horizontal metal plates having dimensions of 0.8×1.0 m comprise the top of a furnace and are separated by a distance of 15 mm. The lower plate is at 400°C and the upper at 100°C and air at 1 atm abs is enclosed in the gap. Calculate the heat-transfer rate between the plates.

4.8-1. *Boiling Coefficient in a Jacketed Kettle.* Predict the boiling heat-transfer coefficient for the vertical jacketed sides of the kettle given in Example 4.8-1. Then, using this coefficient for the sides and the coefficient from Example 4.8-1 for the bottom, predict the total heat transfer.

$$\textbf{Ans.} \quad T_w = 107.65°C, \Delta T = 7.65 \text{ K, and } h(\text{vertical}) = 3560 \text{ W/m}^2 \cdot \text{K}.$$

4.8-2. *Boiling Coefficient on a Horizontal Tube.* Predict the boiling heat-transfer coefficient for water under pressure boiling at 250°F for a horizontal surface of $\frac{1}{16}$-in.-thick stainless steel having a k of 9.4 btu/h · ft · °F. The heating medium on the other side of this surface is a hot fluid at 290°F having an h of 275 btu/h · ft^2 · °F. Use the simplified equations. Be sure and correct this h value for the effect of pressure.

4.8-3. *Condensation on a Vertical Tube.* Repeat Example 4.8-2 but for a vertical tube 1.22 m (4.0 ft) high instead of 0.305 m (1.0 ft) high. Use SI and English units.

$$\textbf{Ans.} \quad h = 9438 \text{ W/m}^2 \cdot \text{K, } 1663 \text{ btu/h} \cdot \text{ft}^2 \cdot °F; N_{\text{Re}} = 207.2 \text{ (laminar flow)}$$

4.8-4. *Condensation of Steam on Vertical Tubes.* Steam at 1 atm pressure abs and 100°C is condensing on a bank of five vertical tubes each 0.305 m high and having an OD of 25.4 mm. The tubes are arranged in a bundle spaced far enough apart so that they do not interfere with each other. The surface temperature of the tubes is 97.78°C. Calculate the average heat-transfer coefficient and the total kg condensate per hour.

$$\textbf{Ans.} \quad h = 15\,240 \text{ W/m}^2 \cdot \text{K}$$

4.8-5. *Condensation on a Bank of Horizontal Tubes.* Steam at 1 atm abs pressure and 100°C is condensing on a horizontal tube bank with five layers of tubes ($N = 5$) placed one below the other. Each layer has four tubes (total tubes = $4 \times 5 = 20$) and the OD of each tube is 19.1 mm. The tubes are each 0.61 m long and the tube surface temperature is 97.78°C. Calculate the average heat-transfer coefficient and the kg condensate per second for the whole condenser. Make a sketch of the tube bank.

4.9-1. *Radiation to a Tube from a Large Enclosure.* Repeat Example 4.9-1 but use the slightly more accurate Eq. (4.9-5) with two different emissivities.

$$\textbf{Ans.} \quad q = -2171 \text{ W} (-7410 \text{ btu/h})$$

4.9-2. *Baking a Loaf of Bread in an Oven.* A loaf of bread having a surface temperature of 373 K is being baked in an oven whose walls and the air are at 477.4 K.

The bread moves continuously through the large oven on an open chain belt conveyor. The emissivity of the bread is estimated as 0.85 and the loaf can be assumed a rectangular solid 114.3 mm high × 114.3 mm wide × 330 mm long. Calculate the radiation heat-transfer rate to the bread, assuming that it is small compared to the oven and neglecting natural convection heat transfer.

Ans. $q = 278.4$ W (950 btu/h)

4.9-3. Radiation and Convection from a Steam Pipe. A horizontal oxidized steel pipe carrying steam and having an OD of 0.1683 m has a surface temperature of 374.9 K and is exposed to air at 297.1 K in a large enclosure. Calculate the heat loss for 0.305 m of pipe from natural convection plus radiation. For the steel pipe, use an ε of 0.79.

Ans. $q = 163.3$ W (557 btu/h)

4.9-4. Radiation and Convection to a Loaf of Bread. Calculate the total heat-transfer rate to the loaf of bread in Problem 4.9-2, including the radiation plus natural convection heat transfer. For radiation first calculate a value of h_r. For natural convection, use the simplified equations for the lower $N_{Gr} N_{Pr}$ range. For the four vertical sides, the equation for vertical planes can be used with an L of 114.3 mm. For the top surface, use the equation for a cooled plate facing upward and for the bottom, a cooled plate facing downward. The characteristic L for a horizontal rectangular plate is the linear mean of the two dimensions.

4.9-5. Heat Loss from a Pipe. A bare stainless steel tube having an outside diameter of 76.2 mm and an ε of 0.55 is placed horizontally in air at 294.2 K. The pipe surface temperature is 366.4 K. Calculate the value of $h_c + h_r$ for convection plus radiation and the heat loss for 3 m of pipe.

4.10-1. Mean Temperature Difference in an Exchanger. A 1–2 exchanger with one shell pass and two tube passes is used to heat a cold fluid from 37.8°C to 121.1°C by using a hot fluid entering at 315.6°C and leaving at 148.9°C. Calculate the ΔT_{lm} and the mean temperature difference ΔT_m in K.

Ans. $\Delta T_{lm} = 148.9$ K; $\Delta T_m = 131.8$ K

4.10-2. Cooling Oil by Water in an Exchanger. Oil flowing at the rate of 5.04 kg/s ($c_{pm} = 2.09$ kJ/kg · K) is cooled in a 1–2 heat exchanger from 366.5 K to 344.3 K by 2.02 kg/s of water entering at 283.2 K. The overall heat-transfer coefficient U_o is 340 W/m² · K. Calculate the area required. (*Hint :* A heat balance must first be made to determine the outlet water temperature.)

4.10-3. Heat Exchange Between Oil and Water. Water is flowing at the rate of 1.13 kg/s in a 1–2 shell-and-tube heat exchanger and is heated from 45°C to 85°C by an oil having a heat capacity of 1.95 kJ/kg · K. The oil enters at 120°C and leaves at 85°C. Calculate the area of the exchanger if the overall heat-transfer coefficient is 300 W/m² · K.

4.10-4. Outlet Temperature and Effectiveness of an Exchanger. Hot oil at a flow rate of 3.00 kg/s ($c_p = 1.92$ kJ/kg · K) enters an existing counterflow exchanger at 400 K and is cooled by water entering at 325 K (under pressure) and flowing at a rate of 0.70 kg/s. The overall $U = 350$ W/m² · K and $A = 12.9$ m². Calculate the heat-transfer rate and the exit oil temperature.

4.11-1. (*Selected Topic*) Radiation Shielding. Two very large and parallel planes each have an emissivity of 0.7. Surface 1 is at 866.5 K and surface 2 is at 588.8 K. Use SI and English units.

(a) What is the net radiation loss of surface 1?

(b) To reduce this loss, two additional radiation shields also having an emissivity of 0.7 are placed between the original surfaces. What is the new radiation loss?

Ans. (a) 13 565 W/m², 4300 btu/h · ft²; (b) 4521 W/m², 1433 btu/h · ft²

4.11-2. (*Selected Topic*) Radiation from a Craft in Space. A space satellite in the shape of a sphere is traveling in outer space, where its surface temperature is held at

283.2 K. The sphere "sees" only outer space, which can be considered as a black body with a temperature of 0 K. The polished surface of the sphere has an emissivity of 0.1. Calculate the heat loss per m^2 by radiation.

Ans. $q_{12}/A_1 = 36.5$ W/m^2

4.11-3. *(Selected Topic) Radiation and Complex View Factor.* Find the view factor F_{12} for the configuration shown in Fig. P4.11-3. The areas A_4 and A_3 are fictitious

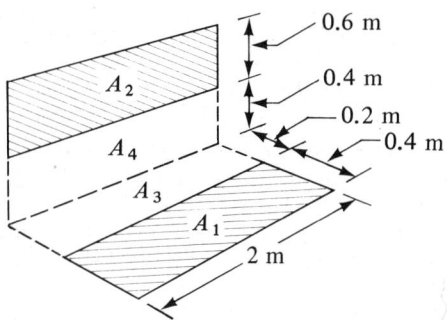

FIGURE P4.11-3. *Geometric configuration for Problem 4.11-3.*

areas (C3). The area $A_2 + A_4$ is called $A_{(24)}$ and $A_1 + A_3$ is called $A_{(13)}$. Areas $A_{(24)}$ and $A_{(13)}$ are perpendicular to each other. [*Hint :* Follow the methods in Example 4.11-5. First, write an equation similar to Eq. (4.11-48) which relates the interchange between A_3 and $A_{(24)}$. Then relate the interchange between $A_{(13)}$ and $A_{(24)}$. Finally, relate $A_{(13)}$ and A_4.]

Ans. $A_1 F_{12} = A_{(13)} F_{(13)(24)} + A_3 F_{34} - A_3 F_{3(24)} - A_{(13)} F_{(13)4}$

4.11-4. *(Selected Topic) Radiation Between Parallel Surfaces.* Two parallel surfaces each 1.83 × 1.83 m square are spaced 0.91 m apart. The surface temperature of A_1 is 811 K and that of A_2 is 533 K. Both are black surfaces.
(a) Calculate the radiant heat transfer between the two surfaces.
(b) Do the same as for part (a), but for the case where the two surfaces are connected by nonconducting reradiating walls.
(c) Repeat part (b), but A_1 has an emissivity of 0.8 and A_2 an emissivity of 0.7.

4.11-5. *(Selected Topic) Radiation Between Adjacent Perpendicular Plates.* Two adjacent rectangles are perpendicular to each other. The first rectangle is 1.52 × 2.44 m and the second is 1.83 × 2.44 m with the 2.44-m side common to both. The temperature of the first surface is 699 K and that of the second is 478 K. Both surfaces are black. Calculate the radiant heat transfer between the two surfaces.

4.11-6. *(Selected Topic) View Factor for Complex Geometry.* Using the dimensions given in Fig. P4.11-3, calculate the individual view factors and also F_{12}.

4.11-7. *(Selected Topic) Radiation from a Surface to the Sky.* A plane surface having an area of 1.0 m^2 is insulated on the bottom side and is placed on the ground exposed to the atmosphere at night. The upper surface is exposed to air at 290 K and the convective heat-transfer coefficient from the air to the plane is 12 W/m$^2 \cdot$ K. The plane radiates to the clear sky. The effective radiation temperature of the sky can be assumed as 80 K. If the plane is a black body, calculate the temperature of the plane at equilibrium.

Ans. $T = 266.5$ K $= -6.7°$C

4.11-8. *(Selected Topic) Radiation and Heating of Planes.* Two plane disks each 1.25 m in diameter are parallel and directly opposed to each other. They are separated by a distance of 0.5 m. Disk 1 is heated by electrical resistance to 833.3 K. Both disks are insulated on all faces except the two faces directly opposed to each other. Assume that the surroundings emit no radiation and that the disks are in space. Calculate the temperature of disk 2 at steady state and also the electrical energy input to disk 1. (*Hint :* The fraction of heat lost from area number 1 to space is $1 - F_{12}$.)

Ans. $F_{12} = 0.45$, $T_2 = 682.5$ K

4.11-9. (*Selected Topic*) *Radiation by Disks to Each Other and to Surroundings.* Two disks each 2.0 m in diameter are parallel and directly opposite each other and separated by a distance of 2.0 m. Disk 1 is held at 1000 K by electric heating and disk 2 at 400 K by cooling water in a jacket at the rear of the disk. The disks radiate only to each other and to the surrounding space at 300 K. Calculate the electric heat input and also the heat removed by the cooling water.

4.11-10. (*Selected Topic*) *View Factor by Integration.* A small black disk is vertical having an area of 0.002 m^2 and radiates to a vertical black plane surface that is 0.03 m wide and 2.0 m high and is opposite and parallel to the small disk. The disk source is 2.0 m away from the vertical plane and placed opposite the bottom of the plane. Determine F_{12} by integration of the view-factor equation.

Ans. $F_{12} = 0.00307$

4.11-11. (*Selected Topic*) *Gas Radiation to Gray Enclosure.* Repeat Example 4.11-7 but with the following changes.
   (a) The interior walls are not black surfaces but gray surfaces with an emissivity of 0.75.
   (b) The same conditions as part (a) with gray walls, but in addition heat is transferred by natural convection to the interior walls. Assume an average convective coefficient of 8.0 W/m$^2 \cdot$ K.

Ans. (b) q(convection + radiation) = 4.426 W

4.11-12. (*Selected Topic*) *Gas Radiation and Convection to a Stack.* A furnace discharges hot flue gas at 1000 K and 1 atm abs pressure containing 5% CO_2 into a stack having an inside diameter of 0.50 m. The inside walls of the refractory lining are at 900 K and the emissivity of the lining is 0.75. The convective heat-transfer coefficient of the gas has been estimated as 10 W/m$^2 \cdot$ K. Calculate the rate of heat transfer q/A from the gas by radiation plus convection.

4.12-1. (*Selected Topic*) *Laminar Heat Transfer of a Power-Law Fluid.* A non-Newtonian power-law fluid banana purée flowing at a rate of 300 lb$_m$/h inside a 1.0-in.-ID tube is being heated by a hot fluid flowing outside the tube. The banana purée enters the heating section of the tube, which is 5 ft long, at a temperature of 60°F. The inside wall temperature is constant at 180°F. The fluid properties as given by Charm (C1) are $\rho = 69.9$ lb$_m$/ft^3, $c_p = 0.875$ btu/lb$_m \cdot$ °F, and $k = = 0.320$ btu/h $\cdot$ ft $\cdot$ °F. The fluid has the following rheological constants: $n = n' = 0.458$, which can be assumed constant and $K = 0.146$ lb$_f \cdot$ s$^n \cdot$ ft^{-2} at 70°F and 0.0417 at 190°F. A plot of log K versus T°F can be assumed to be a straight line. Calculate the outlet bulk temperature of the fluid in laminar flow.

4.12-2. (*Selected Topic*) *Heating a Power-Law Fluid in Laminar Flow.* A non-Newtonian power-law fluid having the same physical properties and rheological constants as the fluid in Example 4.12-1 is flowing in laminar flow at a rate of 6.30×10^{-2} kg/s inside a 25.4 mm-ID tube. It is being heated by a hot fluid outside the tube. The fluid enters the heating section of the tube at 26.7°C and leaves the heating section at an outlet bulk temperature of 46.1°C. The inside wall temperature is constant at 82.2°C. Calculate the length of tube needed in m. (*Note :* In this case the unknown tube length L appears in the equation for h_a and in the heat-balance equation.)

Ans. $L = 1.722$ m

4.13-1. (*Selected Topic*) *Heat Transfer in a Jacketed Vessel with a Paddle Agitator.* A vessel with a paddle agitator and no baffles is used to heat a liquid at 37.8°C in this vessel. A steam-heated jacket furnishes the heat. The vessel inside diameter is 1.22 m and the agitator diameter is 0.406 m and is rotating at 150 rpm. The wall surface temperature is 93.3°C. The physical properties of the liquid are $\rho = 977$ kg/m³, $c_p = 2.72$ kJ/kg · K, $k = 0.346$ W/m · K, and $\mu = 0.100$ kg/m · s at 37.8°C and 7.5×10^{-3} at 93.3°C. Calculate the heat-transfer coefficient to the wall of the jacket.

4.13-2. (*Selected Topic*) *Heat Loss from Circular Fins.* Use the same data and conditions from Example 4.13-2 and calculate the fin efficiency and rate of heat loss from the following different fin materials.
   (a) Carbon steel ($k = 44$ W/m · K).
   (b) Stainless steel ($k = 17.9$ W/m · K).

Ans. (a) $\eta_f = 0.66$, $q = 111.1$ W

4.13-3. (*Selected Topic*) *Heat Loss from Longitudinal Fin.* A longitudinal aluminum fin as shown in Fig. 4.13-3a ($k = 230$ W/m · K) is attached to a copper tube having an outside radius of 0.04 m. The length of the fin is 0.080 m and the thickness is 3 mm. The tube base is held at 450 K and the external surrounding air at 300 K has a convective coefficient of 25 W/m² · K. Calculate the fin efficiency and the heat loss from the fin per 1.0 m of length.

4.13-4. (*Selected Topic*) *Heat Transfer in Finned Tube Exchanger.* Air at an average temperature of 50°C is being heated by flowing outside a steel tube ($k = 45.1$ W/m · K) having an inside diameter of 35 mm and a wall thickness of 3 mm. The outside of the tube is covered with 16 longitudinal steel fins with a length $L = 13$ mm and a thickness of $t = 1.0$ mm. Condensing steam inside at 120°C has a coefficient of 7000 W/m² · K. The outside coefficient of the air has been estimated as 30 W/m² · K. Neglecting fouling factors and using a tube 1.0 m long, calculate the overall heat-transfer coefficient U_i based on the inside area A_i.

4.14-1. (*Selected Topic*) *Dimensional Analysis for Natural Convection.* Repeat the dimensional analysis for natural convection heat transfer to a vertical plate as given in Section 4.14. However, do as follows.
   (a) Carry out all the detailed steps solving for all the exponents in the π's.
   (b) Repeat, but in this case select the four variables L, μ, c_p, and g to be common to all the dimensionless groups.

4.14-2. (*Selected Topic*) *Dimensional Analysis for Unsteady-State Conduction.* For unsteady-state conduction in a solid the following variables are involved: ρ, c_p, L (dimension of solid), t, k, and z (location in solid). Determine the dimensionless groups relating the variables.

Ans. $\pi_1 = \dfrac{kt}{\rho c_p L^2}$, $\pi_2 = \dfrac{z}{L}$

4.15-1. (*Selected Topic*) *Temperatures in a Semi-Infinite Plate.* A semi-infinite plate is similar to that in Fig. 4.15-2. At the surfaces $x = 0$ and $x = L$, the temperature is held constant at 200 K. At the surface $y = 0$, the temperature is held at 400 K. If $L = 1.0$ m, calculate the temperature at the point $y = 0.5$ m and $x = 0.5$ m at steady state.

4.15-2. (*Selected Topic*) *Heat Conduction in a Two-Dimensional Solid.* For two-dimensional heat conduction as given in Example 4.15-1, derive the equation to calculate the total heat loss from the chamber per unit length using the nodes at the outside. There should be eight paths for one-fourth of the chamber. Substitute the actual temperatures into the equation and obtain the heat loss.

Ans. $q = 3426$ W

4.15-3. (*Selected Topic*) *Steady-State Heat Loss from a Rectangular Duct.* A chamber that is in the shape of a long hollow rectangular duct has outside dimensions of 3×4 m and inside dimensions of 1×2 m. The walls are 1 m thick. The inside surface temperature is constant at 800 K and the outside constant at 200 K. The $k = 1.4$ W/m $\cdot$ K. Calculate the steady-state heat loss per unit m of length of duct. Use a grid size of $\Delta x = \Delta y = 0.5$ m. Also, use the outside nodes to calculate the total heat conduction.

Ans. $q = 7428$ W

4.15-4. (*Selected Topic*) *Two-Dimensional Heat Conduction and Different Boundary Conditions.* A very long solid piece of material 1 by 1 m square has its top face maintained at a constant temperature of 1000 K and its left face at 200 K. The bottom face and right face are exposed to an environment at 200 K and have a convection coefficient of $h = 10$ W/m$^2 \cdot$ K. The $k = 10$ W/m $\cdot$ K. Use a grid size of $\Delta x = \Delta y = \frac{1}{3}$ m and calculate the steady-state temperatures of the various nodes.

4.15-5. (*Selected Topic*) *Nodal Point at Exterior Corner Between Insulated Surfaces.* Derive the finite-difference equation for the case of the nodal point $T_{n,m}$ at an exterior corner between insulated surfaces. The diagram is similar to Fig. 4.15-6c except that the two boundaries are insulated.

Ans. $\bar{q}_{n,m} = \frac{1}{2}(T_{n-1,m} + T_{n,m-1}) - T_{n,m}$

REFERENCES

(A1)   ACRIVOS, A. *A.I.Ch.E. J.*, **6**, 584 (1960).

(B1)   BIRD, R. B., STEWART, W. E., and LIGHTFOOT, E. N. *Transport Phenomena.* New York: John Wiley & Sons, Inc., 1960.

(B2)   BADGER, W. L., and BANCHERO, J. T. *Introduction to Chemical Engineering.* New York: McGraw-Hill Book Company, 1955.

(B3)   BROMLEY, L. A. *Chem. Eng. Progr.*, **46**, 221 (1950).

(B4)   BOWMAN, R. A., MUELLER, A. C., and NAGLE, W. M. *Trans. A.S.M.E.*, **62**, 283 (1940).

(B5)   BROOKS, G., and SU, G. *Chem. Eng. Progr.*, **55**, 54 (1959).

(B6)   BOLANOWSKI, J. P., and LINEBERRY, D. D. *Ind. Eng. Chem.*, **44**, 657 (1952).

(C1)   CHARM, S. E. *The Fundamentals of Food Engineering*, 2nd ed. Westport, Conn.: Avi Publishing Co., Inc., 1971.

(C2)   CARSLAW, H. S., and JAEGER, J. E. *Conduction of Heat in Solids*, 2nd ed. New York: Oxford University Press, Inc., 1959.

(C3)   CHAPMAN, A. J. *Heat Transfer.* New York: Macmillan Publishing Co., Inc., 1960.

(C4)   CLAPP, R. M. *International Developments in Heat Transfer*, Part III. New York: American Society of Mechanical Engineers, 1961.

(C5)   CHILTON, T. H., DREW, T. B., and JEBENS, R. H. *Ind. Eng. Chem.*, **36**, 510 (1944).

(C6)   CARREAU, P., CHAREST, G., and CORNEILLE, J. L. *Can. J. Chem. Eng.*, **44**, 3 (1966).

(C7)   CLAUSING, A. M. *Int. J. Heat Mass Transfer*, **9**, 791 (1966).

(D1)   DUNLAP, I. R., and RUSHTON, J. H. *Chem. Eng. Progr. Symp.*, **49**(5), 137 (1953).

(F1)   FUJII, T., and IMURA, H. *Int. J. Heat Mass Transfer*, **15**, 755 (1972).

(G1)   GRIMISON, E. D. *Trans. A.S.M.E.*, **59**, 583 (1937).

(G2)   GEANKOPLIS, C. J. *Mass Transport Phenomena.* Columbus, Ohio: Ohio State University Bookstores, 1972.

(G3)   GUPTA, A. S., CHAUBE, R. B., and UPADHYAY, S. N. *Chem. Eng. Sci.*, **29**, 839 (1974).

(G4)   GLUZ, M. D., and PAVLUSHENKO, L. S. *J.Appl. Chem., U.S.S.R.*, **39**, 2323 (1966).

(G5)   GLOBE, S., and DROPKIN, D. *J. Heat Transfer*, **81**, 24 (1959).

(H1)   HOLMAN, J. P. *Heat Transfer*, 4th ed. New York: McGraw-Hill Book Company, 1976.

(J1)   JACOB, M. *Heat Transfer*, Vol. 1. New York: John Wiley & Sons, Inc., 1949.

(J2)   JACOB, M., and HAWKINS, G. *Elements of Heat Transfer*, 3rd ed. New York: John Wiley & Sons, Inc., 1957.

(J3)   JACOB, M. *Heat Transfer*, Vol. 2. New York: John Wiley & Sons, Inc., 1957.

(K1)   KREITH, F., and BLACK, W. Z. *Basic Heat Transfer*. New York: Harper & Row, Publishers, 1980.

(K2)   KEYES, F. G. *Trans. A.S.M.E.*, **73**, 590 (1951); **74**, 1303 (1952).

(K3)   KNUDSEN, J. G., and KATZ, D. L. *Fluid Dynamics and Heat Transfer*. New York: McGraw-Hill Book Company, 1958.

(K4)   KERN, D. Q. *Process Heat Transfer*. New York: McGraw-Hill Book Company, 1950.

(L1)   LUBARSKY, B., and KAUFMAN, S. J. *NACA Tech. Note No. 3336* (1955).

(M1)   MCADAMS, W. H. *Heat Transmission*, 3rd ed. New York: McGraw-Hill Book Company, 1954.

(M2)   METZNER, A. B., and GLUCK, D. F. *Chem. Eng. Sci.*, **12**, 185 (1960).

(M3)   METZNER, A. B., and FRIEND, P. S. *Ind. Eng. Chem.*, **51**, 879 (1959).

(N1)   NELSON, W. L. *Petroleum Refinery Engineering*, 4th ed. New York: McGraw-Hill Book Company, 1949.

(O1)   OLDSHUE, J. Y., and GRETTON, A. I. *Chem. Eng. Progr.*, **50**, 615 (1954).

(P1)   PERRY, R. H., and CHILTON, C. H. *Chemical Engineers' Handbook*, 5th ed. New York: McGraw-Hill Book Company, 1973.

(P2)   PERRY, J. H. *Chemical Engineers' Handbook*, 4th ed. New York: McGraw-Hill Book Company, 1963.

(R1)   REID, R. C., PRAUSNITZ, J. M., and SHERWOOD, T. K. *The Properties of Gases and Liquids*, 3rd ed. New York: McGraw-Hill Book Company, 1977.

(R2)   ROHESENOW, W. M., and HARTNETT, J. P., eds. *Handbook of Heat Transfer*. New York: McGraw-Hill Book Company, 1973.

(S1)   SIEDER, E. N., and TATE, G. E. *Ind. Eng. Chem.*, **28**, 1429 (1936).

(S2)   STEINBERGER, R. L., and TREYBAL, R. E. *A.I.Ch.E. J.*, **6**, 227 (1960).

(S3)   SKELLAND, A. H. P. *Non-Newtonian Flow and Heat Transfer*. New York: John Wiley & Sons, Inc., 1967.

(S4)   SKELLAND, A. H. P., OLIVER, D. R., and TOOKE, S. *Brit. Chem. Eng.*, **7**(5), 346 (1962).

(U1)   UHL, V. W. *Chem. Eng. Progr. Symp.*, **51**(17), 93 (1955).

(W1)   WELTY, J. R., WICKS, C. E., and WILSON, R. E. *Fundamentals of Momentum, Heat and Mass Transfer*, 2nd ed. New York: John Wiley & Sons, Inc., 1976.

CHAPTER 5

Principles of
Unsteady-State
Heat Transfer

5.1 DERIVATION OF BASIC EQUATION

5.1A Introduction

In Chapter 4 we considered various heat-transfer systems in which the temperature at any given point and the heat flux were always constant with time, i.e., in steady state. In the present chapter we will study processes in which the temperature at any given point in the system changes with time, i.e., heat transfer is unsteady state or transient.

Before steady-state conditions can be reached in a process, some time must elapse after the heat-transfer process is initiated to allow the unsteady-state conditions to disappear. For example, in Section 4.2A we determined the heat flux through a wall at steady state. We did not consider the period during which the one side of the wall was being heated up and the temperatures were increasing.

Unsteady-state heat transfer is important because of the large number of heating and cooling problems occurring industrially. In metallurgical processes it is necessary to predict cooling and heating rates of various geometries of metals in order to predict the time required to reach certain temperatures. In food processing, such as in the canning industry, perishable canned foods are heated by immersion in steam baths or chilled by immersion in cold water. In the paper industry wood logs are immersed in steam baths before processing. In most of these processes the material is suddenly immersed into a fluid of higher or lower temperature.

5.1B Derivation of Unsteady-State Conduction Equation

To derive the equation for unsteady-state condition in one direction in a solid, we refer to Fig. 5.1-1. Heat is being conducted in the x direction in the cube Δx, Δy, Δz in size. For conduction in the x direction, we write

$$q_x = -kA\frac{\partial T}{\partial x} \tag{5.1-1}$$

The term $\partial T/\partial x$ means the partial or derivative of T with respect to x with the other variables, y, z, and time t, being held constant. Next, making a heat balance on the cube, we can write

rate of heat input + rate of heat generation = rate of heat output

$$+ \text{ rate of heat accumulation} \qquad \text{(5.1-2)}$$

The rate of heat input to the cube is

$$\text{rate of heat input} = q_{x|x} = -k(\Delta y \ \Delta z) \left. \frac{\partial T}{\partial x} \right|_{x} \qquad \text{(5.1-3)}$$

Also,

$$\text{rate of heat output} = q_{x|x+\Delta x} = -k(\Delta y \ \Delta z) \left. \frac{\partial T}{\partial x} \right|_{x+\Delta x} \qquad \text{(5.1-4)}$$

The rate of accumulation of heat in the volume $\Delta x \ \Delta y \ \Delta z$ in time ∂t is

$$\text{rate of heat accumulation} = (\Delta x \ \Delta y \ \Delta z)\rho c_p \frac{\partial T}{\partial t} \qquad \text{(5.1-5)}$$

The rate of heat generation in volume $\Delta x \ \Delta y \ \Delta z$ is

$$\text{rate of heat generation} = (\Delta x \ \Delta y \ \Delta z)\dot{q} \qquad \text{(5.1-6)}$$

Substituting Eqs. (5.1-3)–(5.1-6) into (5.1-2) and dividing by $\Delta x \ \Delta y \ \Delta z$,

$$\dot{q} + \frac{-k\left(\left. \frac{\partial T}{\partial x} \right|_{x} - \left. \frac{\partial T}{\partial x} \right|_{x+\Delta x} \right)}{\Delta x} = \rho c_p \frac{\partial T}{\partial t} \qquad \text{(5.1-7)}$$

Letting Δx approach zero, we have the second partial of T with respect to x or $\partial^2 T/\partial x^2$ on the left side. Then, rearranging,

$$\frac{\partial T}{\partial t} = \frac{k}{\rho c_p} \frac{\partial^2 T}{\partial x^2} + \frac{\dot{q}}{\rho c_p} = \alpha \frac{\partial^2 T}{\partial x^2} + \frac{\dot{q}}{\rho c_p} \qquad \text{(5.1-8)}$$

where α is $k/\rho c_p$, thermal diffusivity. This derivation assumes constant k, ρ, and c_p. In SI units, $\alpha = \text{m}^2/\text{s}$, $T = \text{K}$, $t = \text{s}$, $k = \text{W/m} \cdot \text{K}$, $\rho = \text{kg/m}^3$, $\dot{q} = \text{W/m}^3$, and $c_p = \text{J/kg} \cdot \text{K}$. In English units, $\alpha = \text{ft}^2/\text{h}$, $T = {}^\circ\text{F}$, $t = \text{h}$, $k = \text{btu/h} \cdot \text{ft} \cdot {}^\circ\text{F}$, $\rho = \text{lb}_\text{m}/\text{ft}^3$, $\dot{q} = \text{btu/h} \cdot \text{ft}^3$, and $c_p = \text{btu/lb}_\text{m} \cdot {}^\circ\text{F}$.

For conduction in three dimensions, a similar derivation gives

$$\frac{\partial T}{\partial t} = \alpha \left(\frac{\partial^2 T}{\partial x^2} + \frac{\partial^2 T}{\partial y^2} + \frac{\partial^2 T}{\partial z^2} \right) + \frac{\dot{q}}{\rho c_p} \qquad \text{(5.1-9)}$$

FIGURE 5.1-1. *Unsteady-state conduction in one direction.*

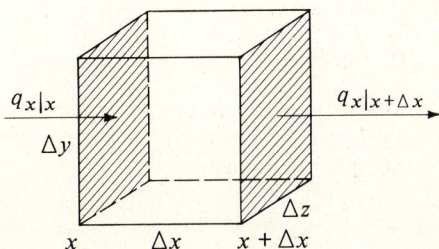

In many cases, unsteady-state heat conduction is occurring but the rate of heat generation is zero. Then Eqs. (5.1-8) and (5.1-9) become

$$\frac{\partial T}{\partial t} = \alpha \frac{\partial^2 T}{\partial x^2}$$

(5.1-10)

$$\frac{\partial T}{\partial t} = \alpha\left(\frac{\partial^2 T}{\partial x^2} + \frac{\partial^2 T}{\partial y^2} + \frac{\partial^2 T}{\partial z^2}\right)$$

(5.1-11)

Equations (5.1-10) and (5.1-11) relate the temperature T with position x, y, and z and time t. The solutions of Eqs. (5.1-10) and (5.1-11) for certain specific cases as well as for the more general cases are considered in much of the remainder of this chapter.

5.2 SIMPLIFIED CASE FOR SYSTEMS WITH NEGLIGIBLE INTERNAL RESISTANCE

5.2A Basic Equation

We begin our treatment of transient heat conduction by analyzing a simplified case. In this situation we consider a solid which has a very high thermal conductivity or very low internal conductive resistance compared to the external surface resistance, where convection occurs from the external fluid to the surface of the solid. Since the internal resistance is very small, the temperature within the solid is essentially uniform at any given time.

An example would be a small, hot cube of steel at T_0 K at time $t = 0$, suddenly immersed into a large bath of cold water at T_∞ which is held constant with time. Assume that the heat-transfer coefficient h in $W/m^2 \cdot K$ is constant with time. Making a heat balance on the solid object for a small time interval of time dt s, the heat transfer from the bath to the object must equal the change in internal energy of the object.

$$hA(T_\infty - T)\, dt = c_p \rho V\, dT$$

(5.2-1)

where A is the surface area of the object in m^2, T the average temperature of the object at time t in s, ρ the density of the object in kg/m^3, and V the volume in m^3. Rearranging the equation and integrating between the limits of $T = T_0$ when $t = 0$ and $T = T$ when $t = t$,

$$\int_{T=T_0}^{T=T} \frac{dT}{T_\infty - T} = \frac{hA}{c_p \rho V} \int_{t=0}^{t=t} dt$$

(5.2-2)

$$\frac{T - T_\infty}{T_0 - T_\infty} = e^{-(hA/c_p \rho V)t}$$

(5.2-3)

This equation describes the time–temperature history of the solid object. The term $c_p \rho V$ is often called the *lumped thermal capacitance* of the system. This type of analysis is often called the *lumped capacity method* or *Newtonian heating or cooling method*.

5.2B Equation for Different Geometries

In using Eq. (5.2-3) the surface/volume ratio of the object must be known. The basic assumption of negligible internal resistance was made in the derivation. This assumption is reasonably accurate when

$$N_{Bi} = \frac{hx_1}{k} < 0.1$$

(5.2-4)

where hx_1/k is called the Biot number N_{Bi}, which is dimensionless, and x_1 is a characteristic dimension of the body obtained from $x_1 = V/A$. The Biot number compares the relative values of internal conduction resistance and surface convective resistance to heat transfer.

For a sphere,

$$x_1 = \frac{V}{A} = \frac{4\pi r^3/3}{4\pi r^2} = \frac{r}{3} \qquad (5.2\text{-}5)$$

For a long cylinder,

$$x_1 = \frac{V}{A} = \frac{\pi D^2 L/4}{\pi DL} = \frac{D}{4} = \frac{r}{2} \qquad (5.2\text{-}6)$$

For a long square rod,

$$x_1 = \frac{V}{A} = \frac{(2x)^2 L}{4(2x)L} = \frac{x}{2} \qquad (x = \tfrac{1}{2}\text{ thickness}) \qquad (5.2\text{-}7)$$

EXAMPLE 5.2-1.   Cooling of a Steel Ball
A steel ball having a radius of 1.0 in. (25.4 mm) is at a uniform temperature of 800°F (699.9 K). It is suddenly plunged into a medium whose temperature is held constant at 250°F (394.3 K). Assuming a convective coefficient of $h = 2.0$ btu/h·ft²·°F (11.36 W/m²·K), calculate the temperature of the ball after 1 h (3600 s). The average physical properties are $k = 25$ btu/h·ft·°F (43.3 W/m·K), $\rho = 490$ lb$_m$/ft³ (7849 kg/m³), and $c_p = 0.11$ btu/lb$_m$·°F (0.4606 kJ/kg·K). Use SI and English units.

Solution: For a sphere from Eq. (5.2-5),

$$x_1 = \frac{V}{A} = \frac{r}{3} = \frac{\tfrac{1}{12}}{3} = \tfrac{1}{36}\text{ ft}$$

$$= \frac{25.4}{1000 \times 3} = 8.47 \times 10^{-3}\text{ m}$$

From Eq. (5.2-4) for the Biot number,

$$N_{Bi} = \frac{hx_1}{k} = \frac{2(\tfrac{1}{36})}{25} = 0.00222$$

$$N_{Bi} = \frac{11.36(8.47 \times 10^{-3})}{43.3} = 0.00222$$

This value is <0.1; hence, the lumped capacity method can be used. Then,

$$\frac{hA}{c_p \rho V} = \frac{2}{0.11(490)(\tfrac{1}{36})} = 1.335\text{ h}^{-1}$$

$$\frac{hA}{c_p \rho V} = \frac{11.36}{(0.4606 \times 1000)(7849)(8.47 \times 10^{-3})} = 3.71 \times 10^{-4}\text{ s}^{-1}\ (1.335\text{ h}^{-1})$$

Substituting into Eq. (5.2-3) for $t = 1.0$ h and solving for T,

$$\frac{T - T_\infty}{T_0 - T_\infty} = \frac{T - 250°F}{800 - 250} = e^{-(hA/c_p\rho V)t} = e^{-(1.335)(1.0)} \qquad T = 395°F$$

$$\frac{T - 394.3\text{ K}}{699.9 - 394.3} = e^{-(3.71 \times 10^{-4})(3600)} \qquad T = 474.9\text{ K}$$

5.2C Total Amount of Heat Transferred

The temperature of the solid at any time t can be calculated from Eq. (5.2-3). At any time t, the instantaneous rate of heat transfer $q(t)$ in W from the solid of negligible internal resistance can be calculated from

$$q(t) = hA(T - T_\infty) \tag{5.2-8}$$

Substituting the instantaneous temperature T from Eq. (5.2-3) into Eq. (5.2-8),

$$q(t) = hA(T_0 - T_\infty)e^{-(hA/c_p\rho V)t} \tag{5.2-9}$$

To determine the total amount of heat Q in W · s or J transferred from the solid from time $t = 0$ to $t = t$, we can integrate Eq. (5.2-9).

$$Q = \int_{t=0}^{t=t} q(t)dt = \int_{t=0}^{t=t} hA(t_0 - T_\infty)e^{-(hA/c_p\rho V)t}\,dt \tag{5.2-10}$$

$$Q = c_p \rho V(T_0 - T_\infty)[1 - e^{-(hA/c_p\rho V)t}] \tag{5.2-11}$$

EXAMPLE 5.2-2. Total Amount of Heat in Cooling
For the conditions in Example 5.2-1, calculate the total amount of heat removed up to time $t = 3600$ s.

Solution: From Example 5.2-1, $hA/c_p\rho V = 3.71 \times 10^{-4}$ s^{-1}. Also, $V = 4\pi r^3/3 = 4(\pi)(0.0254)^3/3 = 6.864 \times 10^{-5}$ m^3. Substituting into Eq. (5.2-11),

$$Q = (0.4606 \times 1000)(7849)(6.864 \times 10^{-5})(699.9 - 394.3)$$

$$\cdot \, [1 - e^{-(3.71 \times 10^{-4})(3600)}]$$

$$= 5.589 \times 10^4 \text{ J}$$

5.3 UNSTEADY-STATE HEAT CONDUCTION IN VARIOUS GEOMETRIES

5.3A Introduction and Analytical Methods

In Section 5.2 we considered a simplified case of negligible internal resistance where the object has a very high thermal conductivity. Now we will consider the more general situation where the internal resistance is not small, and hence the temperature is not constant in the solid. The first case that we shall consider is one where the surface convective resistance is negligible compared to the internal resistance. This could occur because of a very large heat-transfer coefficient at the surface or because of a relatively large conductive resistance in the object.

To illustrate an analytical method of solving this first case, we will derive the equation for unsteady-state conduction in the x direction only in a flat plate of thickness $2H$ as shown in Fig. 5.3-1. The initial profile of the temperature in the plate at $t = 0$ is uniform at $T = T_0$. At time $t = 0$, the ambient temperature is suddenly changed to T_1 and held there. Since there is no convection resistance, the temperature of the surface is also held constant at T_1. Since this is conduction in the x direction, Eq. (5.1-10) holds.

$$\frac{\partial T}{\partial t} = \alpha \frac{\partial^2 T}{\partial x^2} \tag{5.1-10}$$

The initial and boundary conditions are

$$T = T_0, \quad t = 0, \quad x = x$$

$$T = T_1, \quad t = t, \quad x = 0 \tag{5.3-1}$$

$$T = T_1, \quad t = t, \quad x = 2H$$

Generally, it is convenient to define a dimensionless temperature Y so that it varies between 0 and 1. Hence,

$$Y = \frac{T_1 - T}{T_1 - T_0} \tag{5.3-2}$$

Substituting Eq. (5.3-2) into (5.1-10),

$$\frac{\partial Y}{\partial t} = \alpha \frac{\partial^2 Y}{\partial x^2} \tag{5.3-3}$$

Redefining the boundary and initial conditions,

$$Y = \frac{T_1 - T_0}{T_1 - T_0} = 1, \quad t = 0, \quad x = x$$

$$Y = \frac{T_1 - T_1}{T_1 - T_0} = 0, \quad t = t, \quad x = 0 \tag{5.3-4}$$

$$Y = \frac{T_1 - T_1}{T_1 - T_0} = 0, \quad t = t, \quad x = 2H$$

A convenient procedure to use to solve Eq. (5.3-3) is the method of separation of variables, which leads to a product solution

$$Y = e^{-a^2 \alpha t}(A \cos ax + B \sin ax) \tag{5.3-5}$$

where A and B are constants and a is a parameter. Applying the boundary and initial conditions of Eq. (5.3-4) to solve for these constants in Eq. (5.3-5), the final solution is an infinite Fourier series (G1).

FIGURE 5.3-1. *Unsteady-state conduction in a flat plate with negligible surface resistance.*

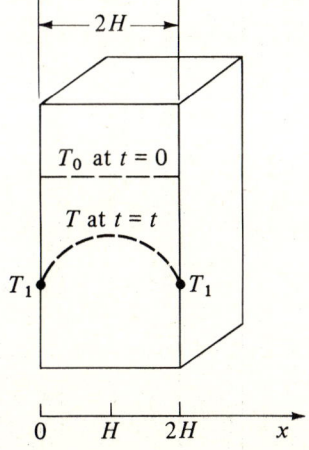

$2H$

T_0 at $t = 0$

T at $t = t$

T_1 T_1

$0 \quad H \quad 2H \qquad x$

$$\frac{T_1 - T}{T_1 - T_0} = \frac{4}{\pi}\left(\frac{1}{1}\exp\frac{-1^2\pi^2\alpha t}{4H^2}\sin\frac{1\pi x}{2H} + \frac{1}{3}\exp\frac{-3^2\pi^2\alpha t}{4H^2}\sin\frac{3\pi x}{2H}\right.$$

$$\left. + \frac{1}{5}\exp\frac{-5^2\pi^2\alpha t}{4H^2}\sin\frac{5\pi x}{2H} + \cdots\right) \qquad (5.3\text{-}6)$$

Hence from Eq. (5.3-6), the temperature T at any position x and time t can be determined. However, these types of equations are very time consuming to use, and convenient charts have been prepared which are discussed in Sections 5.3B, 5.3C, 5.3D, and 5.3E, where a surface resistance is present.

5.3B Unsteady-State Conduction in a Semiinfinite Solid

In Fig. 5.3-2 a semiinfinite solid is shown that extends to ∞ in the $+x$ direction. Heat conduction occurs only in the x direction. Originally, the temperature in the solid is uniform at T_0. At time $t = 0$, the solid is suddenly exposed to or immersed in a large mass of ambient fluid at temperature T_1, which is constant. The convection coefficient h in W/m$^2 \cdot$ K or btu/h $\cdot$ ft$^2 \cdot$ °F is present and is constant; i.e., a surface resistance is present. Hence, the temperature T_S at the surface is not the same as T_1.

The solution of Eq. (5.1-10) for these conditions has been obtained (S1) and is

$$\frac{T - T_0}{T_1 - T_0} = 1 - Y = \mathrm{erfc}\,\frac{x}{2\sqrt{\alpha t}}$$

$$- \exp\left[\frac{h\sqrt{\alpha t}}{k}\left(\frac{x}{\sqrt{\alpha t}} + \frac{h\sqrt{\alpha t}}{k}\right)\right]\mathrm{erfc}\left(\frac{x}{2\sqrt{\alpha t}} + \frac{h}{k}\sqrt{\alpha t}\right) \qquad (5.3\text{-}7)$$

where x is the distance into the solid from the surface in SI units in m, $t =$ time in s, $\alpha = k/\rho c_p$ in m^2/s. In English units, $x =$ ft, $t =$ h, and $\alpha =$ ft^2/h. The function erfc is $(1 - \mathrm{erf})$, where erf is the error function and numerical values are tabulated in standard tables and texts (G1, P1, S1), Y is fraction of unaccomplished change $(T_1 - T)/(T_1 - T_0)$, and $1 - Y$ is fraction of change.

Figure 5.3-3, calculated using Eq. (5.3-7), is a convenient plot used for unsteady-state heat conduction into a semiinfinite solid with surface convection. If conduction into the solid is slow enough or h is very large, the top line with $h\sqrt{\alpha t}/k = \infty$ is used.

> #### *EXAMPLE 5.3-1. Freezing Temperature in the Ground*
> The depth in the soil of the earth at which freezing temperatures penetrate is often of importance in agriculture and construction. During a certain fall day, the temperature in the earth is constant at 15.6°C (60°F) to a depth of several meters. A cold wave suddenly reduces the air temperature from 15.6 to -17.8°C (0°F). The convective coefficient above the soil is 11.36 W/m$^2 \cdot$ K (2 btu/h $\cdot$ ft$^2 \cdot$ °F). The soil properties can be assumed as $\alpha = 4.65$

FIGURE 5.3-2.   *Unsteady-state conduction in a semiinfinite solid.*

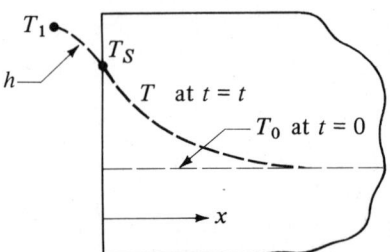

$\times 10^{-7}$ m²/s (0.018 ft²/h) and $k = 0.865$ W/m·K (0.5 btu/h·ft·°F). Neglect any latent heat effects. Use SI and English units.

    (a) What is the surface temperature after 5 h?

    (b) To what depth in the soil will the freezing temperature of 0°C (32°F) penetrate in 5 h?

Solution: This is a case of unsteady-state conduction in a semiinfinite solid. For part (a), the value of x which is the distance from the surface is $x = 0$ m. Then the value of $x/2\sqrt{\alpha t}$ is calculated as follows for $t = 5$ h, $\alpha = 4.65 \times 10^{-7}$ m²/s, $k = 0.865$ W/m·°C, and $h = 11.36$ W/m²·°C. Using SI and English units,

$$\frac{x}{2\sqrt{\alpha t}} = \frac{0}{2\sqrt{(4.65 \times 10^{-7})(5 \times 3600)}} \qquad \frac{x}{2\sqrt{\alpha t}} = \frac{0}{2\sqrt{0.018(5)}}$$

Also,

$$\frac{h\sqrt{\alpha t}}{k} = \frac{11.36\sqrt{(4.65 \times 10^{-7})(5 \times 3600)}}{0.865} \qquad \frac{h\sqrt{\alpha t}}{k} = \frac{2\sqrt{0.018(5)}}{0.5}$$

$$= 1.2 \qquad\qquad\qquad = 1.2$$

Using Fig. 5.3-3, for $x/2\sqrt{\alpha t} = 0$ and $h\sqrt{\alpha t}/k = 1.2$, the value of $1 - Y = 0.63$ is read off the curve. Converting temperatures to K, $T_0 = 15.6°C + 273.2 = 288.8$ K (60°F) and $T_1 = -17.8°C + 273.2 = 255.4$ K (0°F). Then

$$1 - Y = \frac{T - T_0}{T_1 - T_0} = 0.63 = \frac{T - 288.8}{255.4 - 288.8}$$

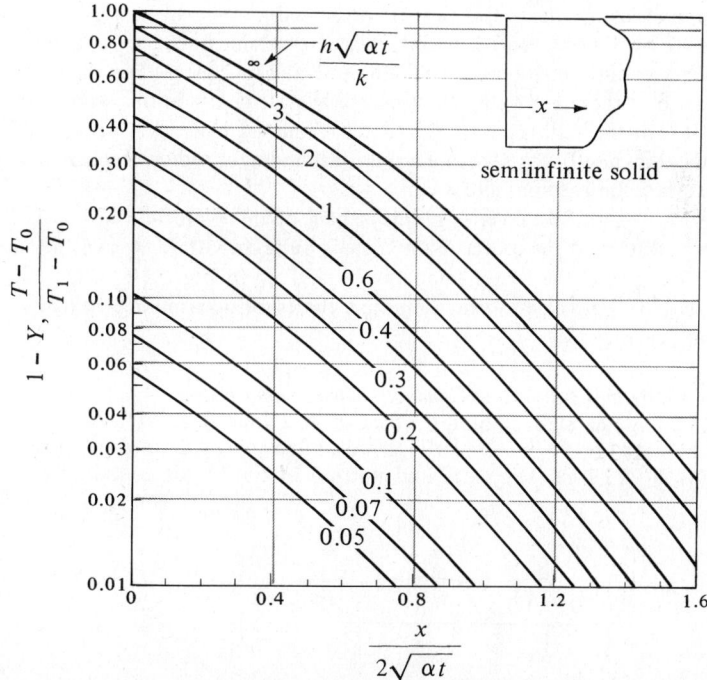

FIGURE 5.3-3. *Unsteady-state heat conducted in a semiinfinite solid with surface convection. Calculated from Eq. (5.3-7)(SI).*

Solving for T at the surface after 5 h,

$$T = 267.76 \ K \quad \text{or} \quad -5.44°C \ (22.2°F)$$

For part (b), $T = 273.2$ K or 0°C, and the distance x is unknown. Substituting the known values,

$$\frac{T - T_0}{T_1 - T_0} = \frac{273.2 - 288.8}{255.4 - 288.8} = 0.467$$

From Fig. 5.3-3 for $(T - T_0)/(T_1 - T_0) = 0.467$ and $h\sqrt{\alpha t}/k = 1.2$, a value of 0.16 is read off the curve for $x/2\sqrt{\alpha t}$. Hence,

$$\frac{x}{2\sqrt{\alpha t}} = \frac{x}{2\sqrt{(4.65 \times 10^{-7})(5 \times 3600)}} = 0.16 \qquad \frac{x}{2\sqrt{\alpha t}} = \frac{x}{2\sqrt{0.018(5)}} = 0.16$$

Solving for x, the distance the freezing temperature penetrates in 5 h,

$$x = 0.0293 \ \text{m} \ (0.096 \ \text{ft})$$

5.3C Unsteady-State Conduction in a Large Flat Plate

A geometry that often occurs in heat-conduction problems is a flat plate of thickness $2x_1$ in the x direction and having large or infinite dimensions in the y and z directions, as shown in Fig. 5.3-4. Heat is being conducted only from the two flat and parallel surfaces in the x direction. The original uniform temperature of the plate is T_0, and at time $t = 0$, the solid is exposed to an environment at temperature T_1 and unsteady-state conduction occurs. A surface resistance is present.

The numerical results of this case are presented graphically in Figs. 5.3-5 and 5.3-6. Figure 5.3-5 by Gurney and Lurie (G2) is a convenient chart for determining the temperatures at any position in the plate and at any time t. The dimensionless parameters used in these and subsequent unsteady-state charts in this section are given in Table 5.3-1 (x is the distance from the center of the flat plate, cylinder, or sphere. x_1 is one half the thickness of the flat plate radius of cylinder, or radius of sphere. $x =$ distance from the surface for a semiinfinite solid.)

When $n = 0$, the position is at the center of the plate in Fig. 5.3-5. Often the temperature history at the center of the plate is quite important. A more accurate chart for determining only the center temperature is given in Fig. 5.3-6 in the Heisler (H1) chart. Heisler (H1) has also prepared multiple charts for determining the temperatures at other positions.

EXAMPLE 5.3-2. Heat Conduction in a Slab of Butter
A rectangular slab of butter which is 46.2 mm thick at a temperature of 277.6 K (4.4°C) in a cooler is removed and placed in an environment at 297.1 K (23.9°C). The sides and bottom of the butter container can be

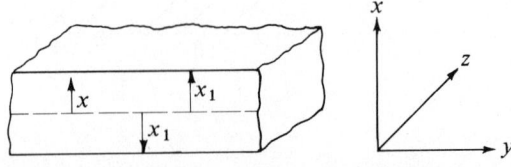

FIGURE 5.3-4. Unsteady-state conduction in a large flat plate.

considered to be insulated by the container side walls. The flat top surface of the butter is exposed to the environment. The convective coefficient is constant and is 8.52 W/m^2 · K. Calculate the temperature in the butter at the surface, at 25.4 mm below the surface, and at 46.2 mm below the surface at the insulated bottom after 5 h of exposure.

Solution: The butter can be considered as a large flat plate with conduction vertically in the x direction. Since heat is entering only at the top face and the bottom face is insulated, the 46.2 mm of butter is equivalent to a half plate with thickness $x_1 = 46.2$ mm. In a plate with two exposed surfaces as in Fig. 5.3-4, the center at $x = 0$ acts as an insulated surface and both halves are mirror images of each other.

The physical properties of butter from Appendix A.4 are $k = 0.197$ W/m · K, $c_p = 2.30$ kJ/kg · K = 2300 J/kg · K, and $\rho = 998$ kg/m^3. The thermal diffusivity is

$$\alpha = \frac{k}{\rho c_p} = \frac{0.197}{998(2300)} = 8.58 \times 10^{-8} \text{ m}^2/\text{s}$$

Also, $x_1 = 46.2/1000 = 0.0462$ m.

The parameters needed for use in Fig. 5.3-5 are

$$m = \frac{k}{hx_1} = \frac{0.197}{8.52(0.0462)} = 0.50$$

$$X = \frac{\alpha t}{x_1^2} = \frac{(8.58 \times 10^{-8})(5 \times 3600)}{(0.0462)^2} = 0.72$$

For the top surface where $x = x_1 = 0.0462$ m,

$$n = \frac{x}{x_1} = \frac{0.0462}{0.0462} = 1.0$$

Then using Fig. 5.3-5,

$$Y = 0.25 = \frac{T_1 - T}{T_1 - T_0} = \frac{297.1 - T}{297.1 - 277.6}$$

Solving, $T = 292.2$ K (19.0°C).

TABLE 5.3-1. *Dimensionless Parameters for Use in Unsteady-State Conduction Charts*

$$Y = \frac{T_1 - T}{T_1 - T_0} \qquad m = \frac{k}{hx_1}$$

$$1 - Y = \frac{T - T_0}{T_1 - T_0} \qquad n = \frac{x}{x_1}$$

$$X = \frac{\alpha t}{x_1^2}$$

SI units: $\alpha = $ m^2/s, $T = $ K, $t = $ s, $x = $ m, $x_1 = $ m, $k = $ W/m · K, $h = $ W/m^2 · K

English units: $\alpha = $ ft^2/h, $T = $ °F, $t = $ h, $x = $ ft, $x_1 = $ ft, $k = $ btu/h · ft · °F, $h = $ btu/h · ft^2 · °F

Cgs units: $\alpha = $ cm^2/s, $T = $ °C, $t = $ s, $x = $ cm, $x_1 = $ cm, $k = $ cal/s · cm · °C, $h = $ cal/s · cm^2 · °C

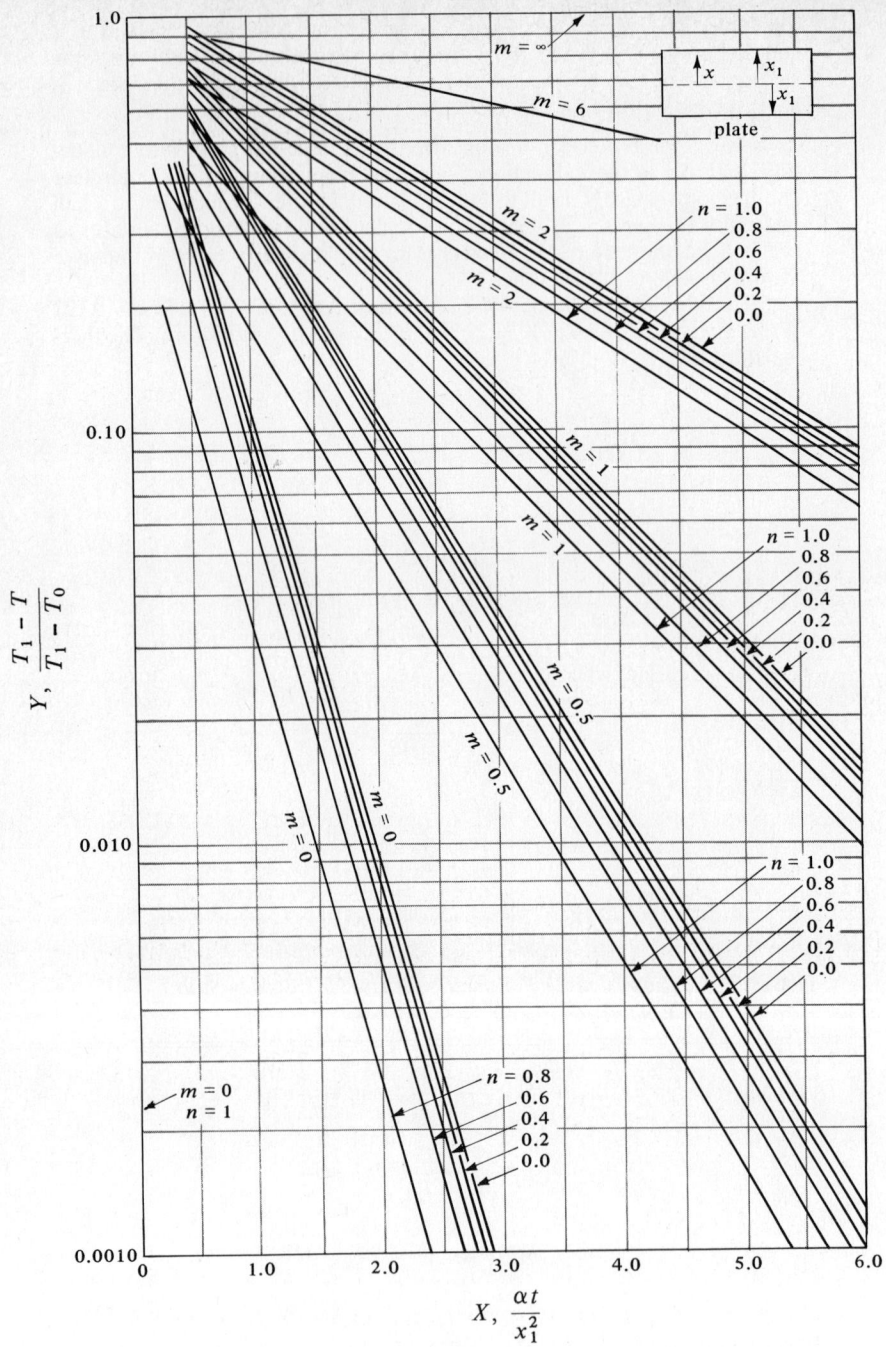

FIGURE 5.3-5. *Unsteady-state heat conduction in a large flat plate.* [*From H. P. Gurney and J. Lurie, Ind. Eng. Chem.,* **15**, *1170 (1923).*]

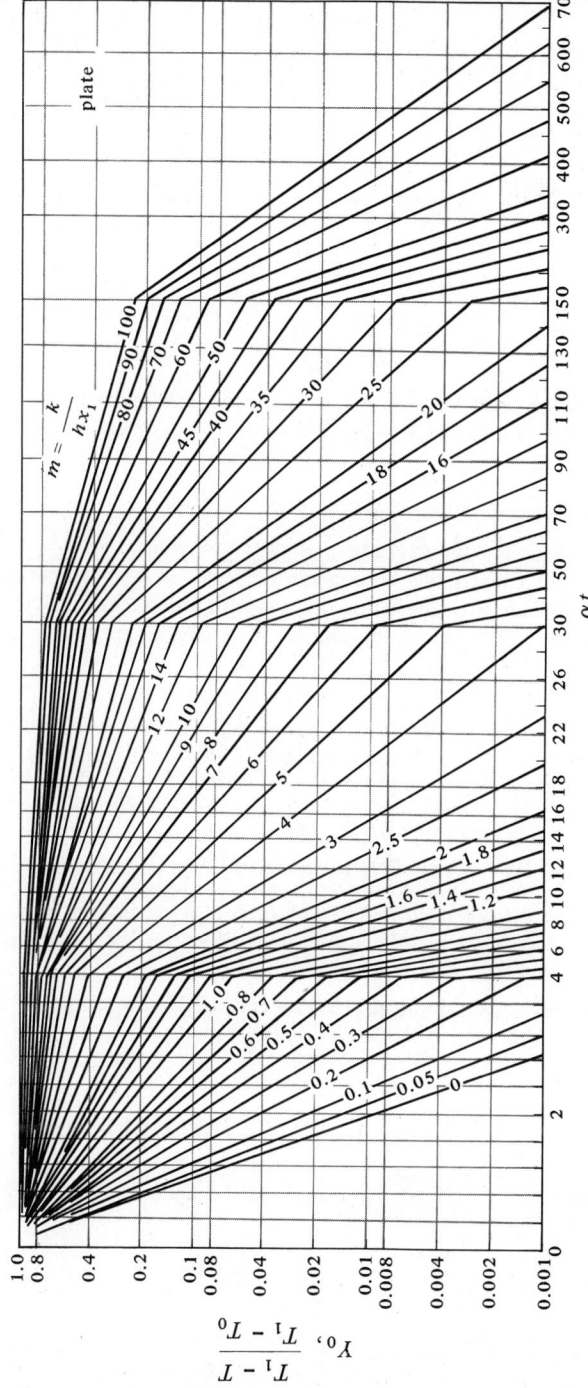

FIGURE 5.3-6. *Chart for determining temperature at the center of a large flat plate for unsteady-state heat conduction.* [*From H. P. Heisler, Trans. A.S.M.E.,* **69,** *227 (1947). With permission.*]

The chart axes and labels:

Vertical axis: $Y_0,\ \dfrac{T_1 - T}{T_1 - T_0}$

Horizontal axis: $X,\ \dfrac{\alpha t}{x_1^2}$

$m = \dfrac{k}{hx_1}$

plate

At the point 25.4 mm from the top surface or 20.8 mm from the center, $x = 0.0208$ m, and

$$n = \frac{x}{x_1} = \frac{0.0208}{0.0462} = 0.45$$

From Fig. 5.3-5,

$$Y = 0.45 = \frac{T_1 - T}{T_1 - T_0} = \frac{297.1 - T}{297.1 - 277.6}$$

Solving, $T = 288.3$ K (15.1°C).
    For the bottom point or 0.0462 m from the top, $x = 0$ and

$$n = \frac{x}{x_1} = \frac{0}{x_1} = 0$$

Then, from Fig. 5.3-5,

$$Y = 0.50 = \frac{T_1 - T}{T_1 - T_0} = \frac{297.1 - T}{297.1 - 277.6}$$

Solving, $T = 287.4$ K (14.2°C). Alternatively, using Fig. 5.3-6, which is only for the center point, $Y = 0.53$ and $T = 286.8$ K (13.6°C).

5.3D Unsteady-State Conduction in a Long Cylinder

Here we consider unsteady-state conduction in a long cylinder where conduction occurs only in the radial direction. The cylinder is long so that conduction at the ends can be neglected or the ends are insulated. Charts for this case are presented in Fig. 5.3-7 for determining the temperatures at any position and Fig. 5.3-8 for the center temperature only.

EXAMPLE 5.3-3. *Transient Heat Conduction in a Can of Pea Purée*
A cylindrical can of pea purée (C2) has a diameter of 68.1 mm and a height of 101.6 mm and is initially at a uniform temperature of 29.4°C. The cans are stacked vertically in a retort and steam at 115.6°C is admitted. For a heating time of 0.75 h at 115.6°C, calculate the temperature at the center of the can. Assume that the can is in the center of a vertical stack of cans and that it is insulated on its two ends by the other cans. The heat capacity of the metal wall of the can will be neglected. The heat-transfer coefficient of the steam is estimated as 4540 W/m² · K. Physical properties of purée are $k = 0.830$ W/m · K and $\alpha = 2.007 \times 10^{-7}$ m²/s.

Solution: Since the can is insulated at the two ends we can consider it as a long cylinder. The radius is $x_1 = 0.0681/2 = 0.03405$ m. For the center with $x = 0$,

$$n = \frac{x}{x_1} = \frac{0}{x_1} = 0$$

Also,

$$m = \frac{k}{hx_1} = \frac{0.830}{4540(0.03405)} = 0.00537$$

$$X = \frac{\alpha t}{x_1^2} = \frac{(2.007 \times 10^{-7})(0.75 \times 3600)}{(0.03405)^2} = 0.468$$

Using Fig. 5.3-8 by Heisler for the center temperature,

$$Y = 0.13 = \frac{T_1 - T}{T_1 - T_0} = \frac{115.6 - T}{115.6 - 29.4}$$

Solving, $T = 104.4$°C.

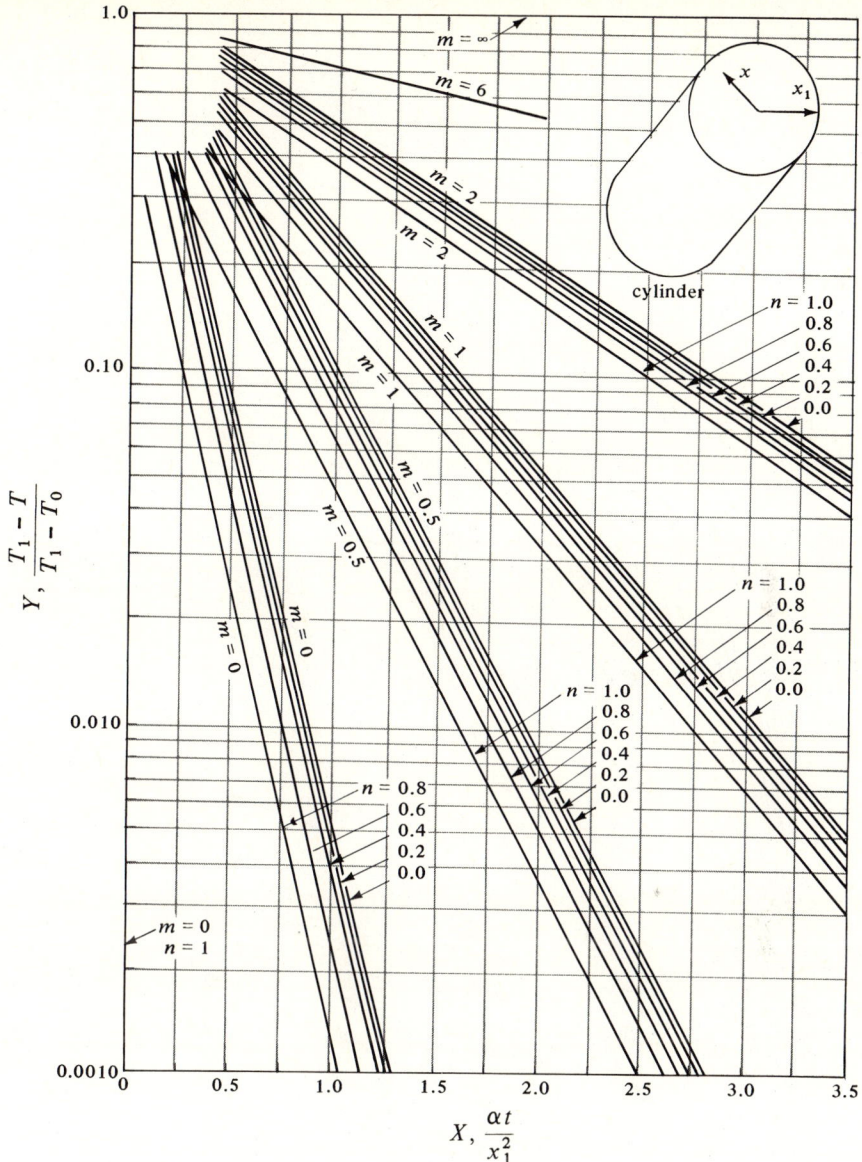

FIGURE 5.3-7. *Unsteady-state heat conduction in a long cylinder.* [*From H. P. Gurney and J. Lurie, Ind. Eng. Chem., 15, 1170 (1923).*]

5.3E Unsteady-State Conduction in a Sphere

In Fig. 5.3-9 a chart is given by Gurney and Lurie for determining the temperatures at any position in a sphere. In Fig. 5.3-10 a chart by Heisler is given for determining the center temperature only in a sphere.

Principles of Unsteady-State Heat Transfer

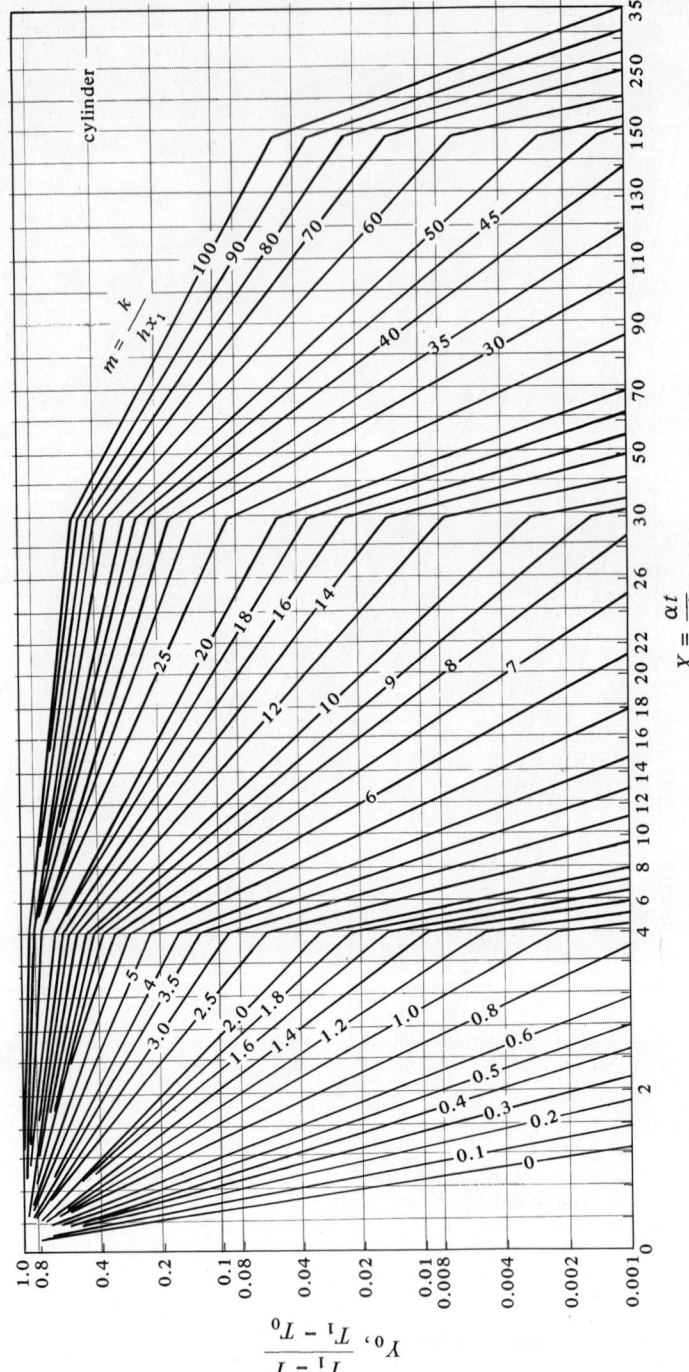

FIGURE 5.3-8.   Chart for determining temperature at the center of a long cylinder for unsteady-state heat conduction. [From H. P. Heisler, Trans. A.S.M.E., **69**, 227 (1947). With permission.]

In the figure:

$$X = \frac{\alpha t}{x_1^2}$$

$$m = \frac{k}{h x_1}$$

$$Y_0, \ \frac{T_1 - T}{T_1 - T_0}$$

cylinder

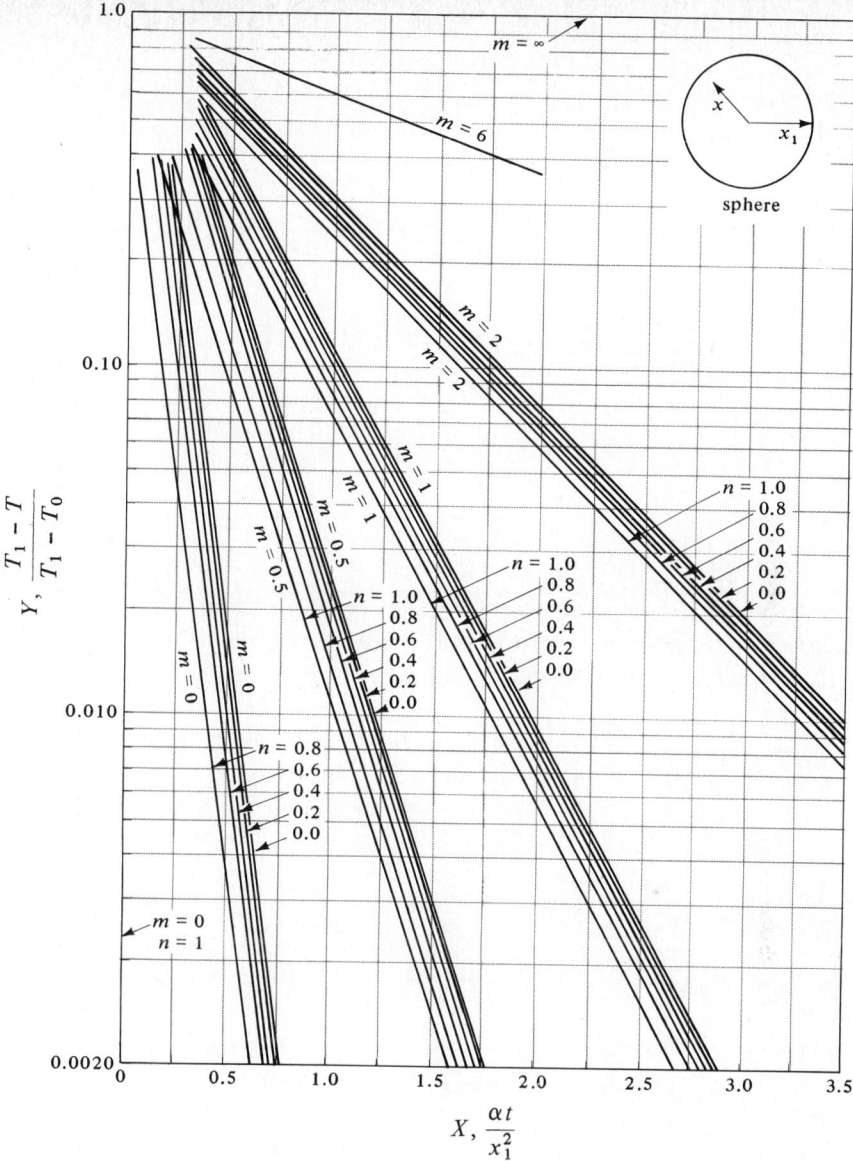

FIGURE 5.3-9.   *Unsteady-state heat conduction in a sphere.* [*From H. P. Gurney and J. Lurie, Ind. Eng. Chem.,* **15**, *1170 (1923).*]

5.3F   Unsteady-State Conduction in Two- and Three-Dimensional Systems

The heat-conduction problems considered so far have been limited to one dimension. However, many practical problems are involved with simultaneous unsteady-state conduction in two and three directions. We shall illustrate how to combine one-dimensional solutions to yield solutions for several-dimensional systems.

Newman (N1) used the principle of superposition and showed mathematically how

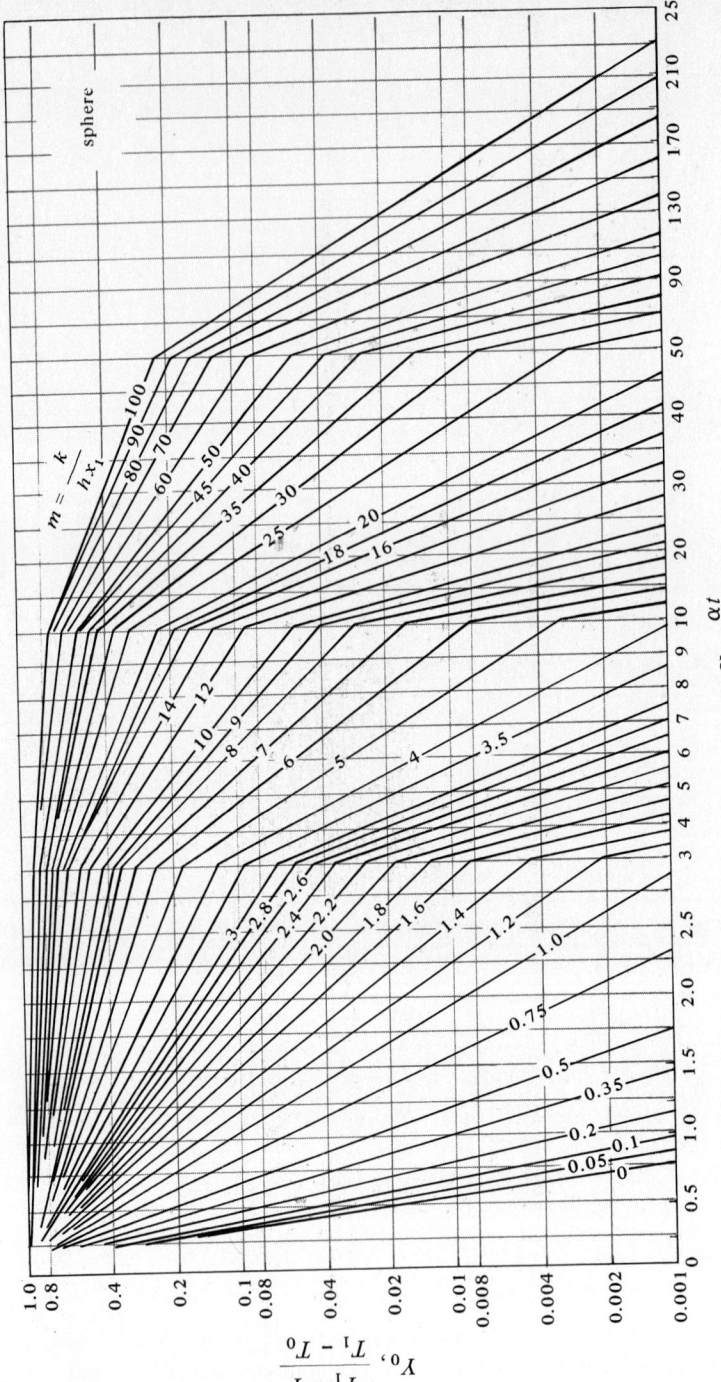

FIGURE 5.3-10. *Chart for determining the temperature at the center of a sphere for unsteady-state heat conduction.* [*From H. P. Heisler, Trans. A.S.M.E.,* **69**, *227 (1947). With permission.*]

The axes and labels shown in the figure:

$$Y_0, \quad \frac{T_1 - T}{T_1 - T_0}$$

$$X = \frac{\alpha t}{x_1^2}$$

$$m = \frac{k}{h x_1}$$

sphere

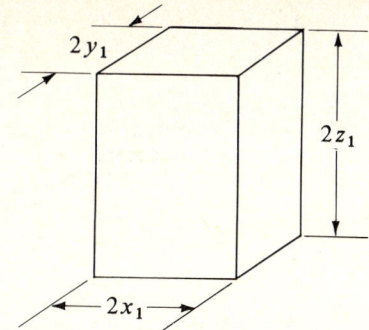

to combine the solutions for one-dimensional heat conduction in the x, the y, and the z direction into an overall solution for simultaneous conduction in all three directions. For example, a rectangular block with dimensions $2x_1$, $2y_1$, and $2z_1$ is shown in Fig. 5.3-11. For the Y value in the x direction, as before,

$$Y_x = \frac{T_1 - T_x}{T_1 - T_0} \tag{5.3-8}$$

where T_x is the temperature at time t and position x distance from the center line, as before. Also, $n = x/x_1$, $m = k/hx_1$, and $X_x = \alpha t/x_1^2$, as before. Then for the y direction,

$$Y_y = \frac{T_1 - T_y}{T_1 - T_0} \tag{5.3-9}$$

and $n = y/y_1$, $m = k/hy_1$ and $X_y = \alpha t/y_1^2$. Similarly, for the z direction,

$$Y_z = \frac{T_1 - T_z}{T_1 - T_0} \tag{5.3-10}$$

Then, for the simultaneous transfer in all three directions,

$$Y_{x, y, z} = (Y_x)(Y_y)(Y_z) = \frac{T_1 - T_{x, y, z}}{T_1 - T_0} \tag{5.3-11}$$

where $T_{x, y, z}$ is the temperature at the point x, y, z from the center of the rectangular block. The value of Y_x for the two parallel faces is obtained from Figs. 5.3-5 and 5.3-6 for conduction in a flat plate. The values of Y_y and Y_z are similarly obtained from the same charts.

For a short cylinder with radius x_1 and length $2y_1$, the following procedure is followed. First Y_x for the radial conduction is obtained from the figures for a long cylinder. Then Y_y for conduction between two parallel planes is obtained from Fig. 5.3-5 or 5.3-6 for conduction in a flat plate. Then,

$$Y_{x, y} = (Y_x)(Y_y) = \frac{T_1 - T_{x, y}}{T_1 - T_0} \tag{5.3-12}$$

EXAMPLE 5.3-4. *Two-Dimensional Conduction in a Short Cylinder*
Repeat Example 5.3-3 for transient conduction in a can of pea purée but assume that conduction also occurs from the two flat ends.

Solution: The can, which has a diameter of 68.1 mm and a height of 101.6 mm, is shown in Fig. 5.3-12. The given values from Example 5.3-3 are

FIGURE 5.3-12. *Two dimensional conduction in a short cylinder in Example 5.3-4.*

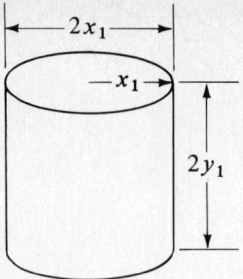

$x_1 = 0.03405$ m, $y_1 = 0.1016/2 = 0.0508$ m, $k = 0.830$ W/m·K, $\alpha = 2.007 \times 10^{-7}$ m²/s, $h = 4540$ W/m²·K and $t = 0.75(3600) = 2700$ s.

For conduction in the x (radial) direction as calculated previously,

$$n = \frac{x}{x_1} = \frac{0}{x_1} = 0, \qquad m = \frac{k}{hx_1} = \frac{0.830}{4540(0.03405)} = 0.00537$$

$$X = \frac{\alpha t}{x_1^2} = \frac{(2.007 \times 10^{-7})2700}{(0.03405)^2} = 0.468$$

From Fig. 5.3-8 for the center temperature,

$$Y_x = 0.13$$

For conduction in the y (axial) direction for the center temperature,

$$n = \frac{y}{y_1} = \frac{0}{0.0508} = 0$$

$$m = \frac{k}{hy_1} = \frac{0.830}{4540(0.0508)} = 0.00360$$

$$X = \frac{\alpha t}{y_1^2} = \frac{(2.007 \times 10^{-7})2700}{(0.0508)^2} = 0.210$$

Using Fig. 5.3-6 for the center of a large plate (two parallel opposed planes),

$$Y_y = 0.80$$

Substituting into Eq. (5.3-12),

$$Y_{x,y} = (Y_x)(Y_y) = 0.13(0.80) = 0.104$$

Then,

$$\frac{T_1 - T_{x,y}}{T_1 - T_0} = \frac{115.6 - T_{x,y}}{115.6 - 29.4} = 0.104$$

$$T_{x,y} = 106.6°C$$

This compares with 104.4°C obtained in Example 5.3-3 for only radial conduction.

5.3G Charts for Average Temperature in a Plate, Cylinder, and Sphere with Negligible Surface Resistance

If the surface resistance is negligible, the curves given in Fig. 5.3-13 will give the total fraction of unaccomplished change, E, for slabs, cylinders, or spheres for unsteady-state conduction. The value of E is

$$E = \frac{T_1 - T_{av}}{T_1 - T_0} \qquad (5.3\text{-}13)$$

where T_0 is the original uniform temperature, T_1 is the environment temperature to which the solid is suddenly subjected, and T_{av} is the average temperature of the solid after t hours.

The values of E_a, E_b, and E_c are each used for conduction between a pair of parallel faces as in a plate. For example, for conduction in the a and b directions in a rectangular bar,

$$E = E_a E_b \qquad (5.3\text{-}14)$$

For conduction from all three sets of faces,

$$E = E_a E_b E_c \qquad (5.3\text{-}15)$$

For conduction in a short cylinder $2c$ long and radius a,

$$E = E_c E_r \qquad (5.3\text{-}16)$$

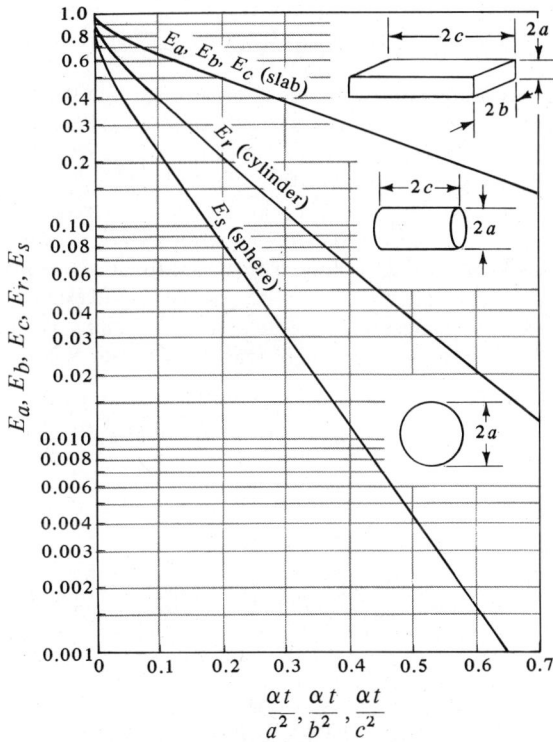

FIGURE 5.3-13. *Unsteady-state conduction and average temperatures for negligible surface resistance. (From R. E. Treybal, Mass Transfer Operations, 2nd ed. New York: McGraw-Hill Book Company, 1968. With permission.)*

5.4 NUMERICAL FINITE-DIFFERENCE METHODS FOR UNSTEADY-STATE CONDUCTION

5.4A Unsteady-State Conduction in a Slab

1. Introduction. As discussed in previous sections of this chapter, the partial differential equations for unsteady-state conduction in various simple geometries can be solved analytically if the boundary conditions are constant at $T = T_1$ with time. Also, in the solutions the initial profile of the temperature at $t = 0$ is uniform at $T = T_0$. The unsteady-state charts used also have these same boundary conditions and initial condition. However, when the boundary conditions are not constant with time and/or the initial conditions are not constant with position, numerical methods must be used.

Numerical calculation methods for unsteady-state heat conduction are similar to numerical methods for steady state discussed in Section 4.15 (Selected Topic). The solid is subdivided into sections or slabs of equal length and a fictitious node is placed at the center of each section. Then a heat balance is made for each node. This method differs from the steady-state method in that we have heat accumulation in a node for unsteady-state conduction.

2. Equations for a slab. The unsteady-state equation for conduction in the x direction in a slab is

$$\frac{\partial T}{\partial t} = \alpha \frac{\partial^2 T}{\partial x^2} \tag{5.1-10}$$

This can be set up for a numerical solution by expressing each partial derivative as an actual finite difference in ΔT, Δt, and Δx. However, an alternative method will be used to derive the final result by making a heat balance. Figure 5.4-1 shows a slab centered at position n, represented by the shaded area. The slab has a width of Δx m and a cross-sectional area of A m^2. The node at position n having a temperature of T_n is placed at the center of the shaded section and this node represents the total mass and heat capacity of the section or slab. Each node is imagined to be connected to the adjacent node by a fictitious, small conducting rod. (See Fig. 4.15-3 for an example.)

The figure shows the temperature profile at a given instant of time t s. Making a heat balance on this node or slab, the rate of heat in − the rate of heat out = the rate of heat accumulation in Δt s.

$$\frac{kA}{\Delta x}({}_tT_{n-1} - {}_tT_n) - \frac{kA}{\Delta x}({}_tT_n - {}_tT_{n+1}) = \frac{(A\,\Delta x)\rho c_p}{\Delta t}({}_{t+\Delta t}T_n - {}_tT_n) \tag{5.4-1}$$

where ${}_tT_n$ is the temperature at point n at time t and ${}_{t+\Delta t}T_n$ is the temperature at point n at time $t + 1\,\Delta t$ later. Rearranging and solving for ${}_{t+\Delta t}T_n$,

$$_{t+\Delta t}T_n = \frac{1}{M}\left[{}_tT_{n+1} + (M-2){}_tT_n + {}_tT_{n-1}\right] \tag{5.4-2}$$

where

$$M = \frac{(\Delta x)^2}{\alpha\,\Delta t} \tag{5.4-3}$$

Note that in Eq. (5.4-2) the temperature ${}_{t+\Delta t}T_n$ at position or node n and at a new time $t + \Delta t$ is calculated from the three points which are known at time t, the starting time. This is called the *explicit method*, because the temperature at a new time can be

Transport Processes: Momentum, Heat, and Mass

calculated explicitly from the temperatures at the previous time. In this method the calculation proceeds directly from one time increment to the next until the final temperature distribution is calculated at the desired final time. Of course, the temperature distribution at the initial time and the boundary conditions must be known.

Once the value of Δx has been selected, then from Eq. (5.4-3) a value of M or the time increment Δt may be picked. For a given value of M, smaller values of Δx mean smaller values of Δt. The value of M must be as follows:

$$M \geq 2 \tag{5.4-4}$$

If M is less than 2, the second law of thermodynamics is violated. It also can be shown that for stability and convergence of the finite-difference solution M must be ≥ 2.

Stability means the errors in the solution do not grow exponentially as the solution proceeds but damp out. Convergence means that the solution of the difference equation approaches the exact solution of the partial differential equation as Δt and Δx go to zero with M fixed. Using smaller sizes of Δt and Δx increases the accuracy in general but greatly increases the number of calculations required. Hence, a digital computer is often ideally suited for this type of calculation.

3. Simplified Schmidt method for a slab.   If the value of $M = 2$, then a great simplification of Eq. (5.4-2) occurs, giving the Schmidt method.

$$_{t+\Delta t}T_n = \frac{_tT_{n-1} + {}_tT_{n+1}}{2} \tag{5.4-5}$$

This means that when a time $1\ \Delta t$ has elapsed, the new temperature at a given point n at $t + \Delta t$ is the arithmetic average of the temperatures at the two adjacent nodes $n + 1$ and $n - 1$ at the original time t.

5.4B Boundary Conditions for Numerical Method for a Slab

1. Convection at the boundary.   For the case where there is a finite convective resistance at the boundary and the temperature of the environment or fluid outside is suddenly changed to T_a, we can derive the following for a slab. Referring to Fig. 5.4-1, we make a

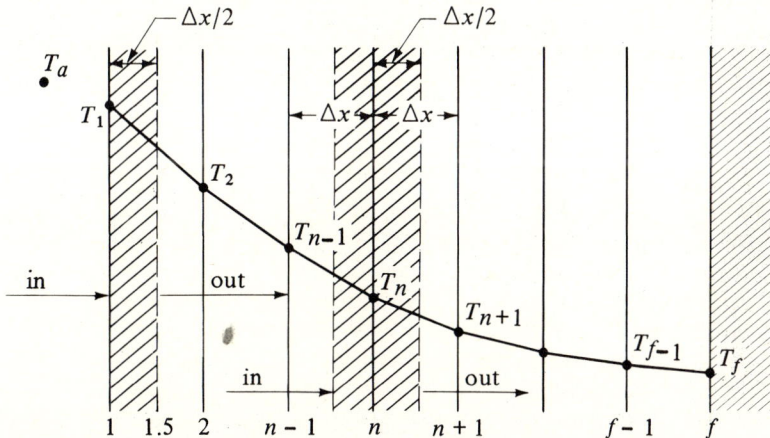

FIGURE 5.4-1.   *Unsteady-state conduction in a slab.*

heat balance on the outside $\frac{1}{2}$ element. The rate of heat in by convection — the rate of heat out by conduction = the rate of heat accumulations in Δt s.

$$hA(_tT_a - _tT_1) - \frac{kA}{\Delta x}(_tT_1 - _tT_2) = \frac{(A\,\Delta x/2)\rho c_p}{\Delta t}(_{t+\Delta t}T_{1.25} - _tT_{1.25}) \qquad (5.4\text{-}6)$$

where $_tT_{1.25}$ is the temperature at the midpoint of the $0.5\,\Delta x$ outside slab. As an approximation, the temperature T_1 at the surface can be used to replace that of $T_{1.25}$. Rearranging,

$$_{t+\Delta t}T_1 = \frac{1}{M}[2N\,_tT_a + [M - (2N + 2)]\,_tT_1 + 2\,_tT_2] \qquad (5.4\text{-}7)$$

where

$$N = \frac{h\,\Delta x}{k} \qquad (5.4\text{-}8)$$

Note that the value of M must be such that

$$M \geq 2N + 2 \qquad (5.4\text{-}9)$$

2. *Insulated boundary condition.* In the case for the boundary condition where the rear face is insulated, a heat balance is made on the rear $\frac{1}{2}\,\Delta x$ slab just as on the front $\frac{1}{2}\,\Delta x$ slab in Fig. 5.4-1. The resulting equation is the same as Eqs. (5.4-6) and (5.4-7), but $h = 0$ or $N = 0$ and $_tT_{f-1} = _tT_{f+1}$ because of symmetry.

$$_{t+\Delta t}T_f = \frac{1}{M}[(M - 2)\,_tT_f + 2\,_tT_{f-1}] \qquad (5.4\text{-}10)$$

3. *Alternative convective condition.* To use the equations above for a given problem, the same values of M, Δx, and Δt must be used. If N gets too large, so that M may be inconveniently too large, another form of Eq. (5.4-7) can be derived. By neglecting the heat accumulation in the front half-slab in Eq. (5.4-6),

$$_{t+\Delta t}T_1 = \frac{N}{N + 1}\,_{t+\Delta t}T_a + \frac{1}{N + 1}\,_{t+\Delta t}T_2 \qquad (5.4\text{-}11)$$

Here the value of M is not restricted by the N value. This approximation works fairly well when a large number of increments in Δx are used so that the amount of heat neglected is small compared to the total.

4. *Procedures for use of initial boundary temperature.* When the temperature of the environment outside is suddenly changed to T_a, the following procedures should be used.

1. When $M = 2$ and a hand calculation of a limited number of increments is used, a special procedure should be used in Eqs. (5.4-5) and (5.4-7) or (5.4-11). For the first time increment one should use an average value for $_1T_a$ of $(T_a + _0T_1)/2$, where $_0T_1$ is the initial temperature at point 1. For all succeeding Δt values, the value of T_a should be used (D1, K1). This special procedure for the value of T_a to use for the first time increment increases the accuracy of the numerical method, especially after a few time intervals. If T_a varies with time t, a new value can be used for each Δt interval.

2. When $M = 2$ and many time increments are used with a digital computer, this special procedure is not needed and the one value of T_a is used for all time increments.

3. When $M = 3$ or more and a hand calculation of a limited number of increments or a

digital computer calculation of many increments is used, only the one value of T_a is used for all time increments. Note that when $M = 3$ or more, many more calculations are needed compared to the case for $M = 2$. The most accurate results are obtained when $M = 4$, which is the preferred method, with slightly less accurate results for $M = 3$ (D1, K1, K2).

EXAMPLE 5.4-1. Unsteady-State Conduction and the Schmidt Numerical Method

A slab of material 1.00 m thick is at a uniform temperature of 100°C. The front surface is suddenly exposed to a constant environmental temperature of 0°C. The convective resistance is zero ($h = \infty$). The back surface of the slab is insulated. The thermal diffusivity $\alpha = 2.00 \times 10^{-5}$ m²/s. Using five slices each 0.20 m thick and the Schmidt numerical method with $M = 2.0$, calculate the temperature profile at $t = 6000$ s. Use the special procedure for the first time increment.

Solution: Figure 5.4-2 shows the temperature profile at $t = 0$ and the environmental temperature of $T_a = 0°C$ with five slices used. For the Schmidt method, $M = 2$. Substituting into Eq. (5.4-3) with $\alpha = 2.00 \times 10^{-5}$ and $\Delta x = 0.20$ and solving for Δt,

$$M = 2 = \frac{(\Delta x)^2}{\alpha\, \Delta t} = \frac{(0.20)^2}{(2.00 \times 10^{-5})\, \Delta t} \qquad \Delta t = 1000 \text{ s} \qquad \textbf{(5.4-3)}$$

This means that (6000 s)/(1000 s)/increment), or six time increments must be used to reach 6000 s.

For the front surface where $n = 1$, the temperature $_1T_a$ to use for the first Δt time increment, as stated previously, is

$$_1T_a = \frac{T_a + {_0}T_1}{2} = {_1}T_1 \qquad n = 1 \qquad \textbf{(5.4-12)}$$

where $_0T_1$ is the initial temperature at point 1. For the remaining time

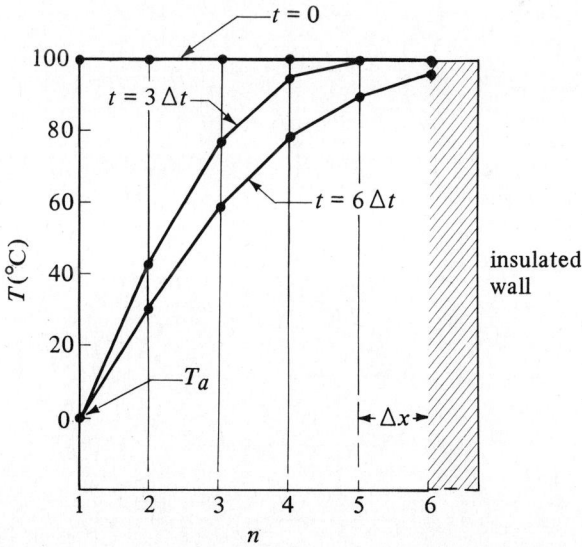

FIGURE 5.4-2. *Temperature for numerical method, Example 5.4-1.*

increments,

$$T_1 = T_a \qquad n = 1 \qquad \text{(5.4-13)}$$

To calculate the temperatures for all time increments for the slabs $n = 2$ to 5, using Eq. (5.4-5),

$$_{t+\Delta t}T_n = \frac{{}_tT_{n-1} + {}_tT_{n+1}}{2} \qquad n = 2, 3, 4, 5 \qquad \text{(5.4-14)}$$

For the insulated end for all time increments at $n = 6$, substituting $M = 2$ and $f = 6$ into Eq. (5.4-12),

$$_{t+\Delta t}T_6 = \frac{(2-2)_tT_6 + 2_tT_5}{2} = {}_tT_5 \qquad \text{(5.4-15)}$$

For the first time increment of $t + \Delta t$ and calculating the temperature at $n = 1$ by Eq. (5.4-12),

$$_{t+\Delta t}T_1 = \frac{T_a + {}_0T_1}{2} = \frac{0 + 100}{2} = 50°C = {}_1T_a$$

For $n = 2$, using Eq. (5.4-14),

$$_{t+\Delta t}T_2 = \frac{{}_tT_1 + {}_tT_3}{2} = \frac{50 + 100}{2} = 75$$

Continuing for $n = 3, 4, 5$,

$$_{t+\Delta t}T_3 = \frac{{}_tT_2 + {}_tT_4}{2} = \frac{100 + 100}{2} = 100$$

$$_{t+\Delta t}T_4 = \frac{{}_tT_3 + {}_tT_5}{2} = \frac{100 + 100}{2} = 100$$

$$_{t+\Delta t}T_5 = \frac{{}_tT_4 + {}_tT_6}{2} = \frac{100 + 100}{2} = 100$$

For $n = 6$, using Eq. (5.4-15),

$$_{t+\Delta t}T_6 = {}_tT_5 = 100$$

For $2\,\Delta t$ using Eq. (5.4-13) for $n = 1$ and continuing for $n = 2$ to 6, using Eqs. (5.4-14) and (5.4-15),

$$_{t+2\,\Delta t}T_1 = T_a = 0$$

$$_{t+2\,\Delta t}T_2 = \frac{{}_{t+\Delta t}T_1 + {}_{t+\Delta t}T_3}{2} = \frac{0 + 100}{2} = 50$$

$$_{t+2\,\Delta t}T_3 = \frac{{}_{t+\Delta t}T_2 + {}_{t+\Delta t}T_4}{2} = \frac{75 + 100}{2} = 87.5$$

$$_{t+2\,\Delta t}T_4 = \frac{{}_{t+\Delta t}T_3 + {}_{t+\Delta t}T_5}{2} = \frac{100 + 100}{2} = 100$$

$$_{t+2\,\Delta t}T_5 = \frac{{}_{t+\Delta t}T_4 + {}_{t+\Delta t}T_6}{2} = \frac{100 + 100}{2} = 100$$

$$_{t+2\,\Delta t}T_6 = {}_{t+\Delta t}T_5 = 100$$

For 3 Δt,

$$_{t+3\,\Delta t}T_1 = 0$$

$$_{t+3\,\Delta t}T_2 = \frac{0 + 87.5}{2} = 43.75$$

$$_{t+3\,\Delta t}T_3 = \frac{50 + 100}{2} = 75$$

$$_{t+3\,\Delta t}T_4 = \frac{87.5 + 100}{2} = 93.75$$

$$_{t+3\,\Delta t}T_5 = \frac{100 + 100}{2} = 100$$

$$_{t+3\,\Delta t}T_6 = 100$$

For 4 Δt,

$$_{t+4\,\Delta t}T_1 = 0$$

$$_{t+4\,\Delta t}T_2 = \frac{0 + 75}{2} = 37.5$$

$$_{t+4\,\Delta t}T_3 = \frac{43.75 + 93.75}{2} = 68.75$$

$$_{t+4\,\Delta t}T_4 = \frac{75 + 100}{2} = 87.5$$

$$_{t+4\,\Delta t}T_5 = \frac{93.75 + 100}{2} = 96.88$$

$$_{t+4\,\Delta t}T_6 = 100$$

For 5 Δt,

$$_{t+5\,\Delta t}T_1 = 0$$

$$_{t+5\,\Delta t}T_2 = \frac{0 + 68.75}{2} = 34.38$$

$$_{t+5\,\Delta t}T_3 = \frac{37.5 + 87.5}{2} = 62.50$$

$$_{t+5\,\Delta t}T_4 = \frac{68.75 + 96.88}{2} = 82.81$$

$$_{t+5\,\Delta t}T_5 = \frac{87.5 + 100}{2} = 93.75$$

$$_{t+5\,\Delta t}T_6 = 96.88$$

For 6 Δt (final time),

$$_{t+6\,\Delta t}T_1 = 0$$

$$_{t+6\,\Delta t}T_2 = \frac{0 + 62.5}{2} = 31.25$$

$$_{t+6\,\Delta t}T_3 = \frac{34.38 + 82.81}{2} = 58.59$$

$$_{t+6\,\Delta t}T_4 = \frac{62.50 + 93.75}{2} = 78.13$$

$$_{t+6\,\Delta t}T_5 = \frac{82.81 + 96.88}{2} = 89.84$$

$$_{t+6\,\Delta t}T_6 = 93.75$$

The temperature profiles for 3 Δt increments and the final time of 6 Δt increments are plotted in Fig. 5.4-2. This example shows how a hand calculation can be done. To increase the accuracy, more slab increments and more time increments are required. This, then, is ideally suited for computation using the digital computer.

EXAMPLE 5.4-2. *Unsteady-State Conduction Using the Digital Computer*
Repeat Example 5.4-1 using the digital computer. Use a $\Delta x = 0.05$ m. Write the Fortran program and compare the final temperatures with Example 5.4-1. Use the explicit method of Schmidt for $M = 2$. Although not needed for many time increments using the digital computer, use the special procedure for the value of $_1T_a$ for the first time increment. Hence, a direct comparison of the effect of the number of increments on the results can be made with Example 5.4-1.

Solution: The number of slabs to use is 1.00 m/(0.05 m/slab) or 20 slabs. Substituting into Eq. (5.4-3) with $\alpha = 2.00 \times 10^{-5}$ m^2/s, $\Delta x = 0.05$ m, and $M = 2$, and solving for Δt,

$$M = 2 = \frac{(\Delta x)^2}{\alpha\,\Delta t} = \frac{(0.05)^2}{(2.00 \times 10^{-5})(\Delta t)}$$

$$\Delta t = 62.5 \text{ s}$$

Hence, $(6000/62.5) = 96$ time increments to be used. The value of n goes from $n = 1$ to 21.

The equations to use to calculate the temperatures are again Eqs. (5.4-12)–(5.4-15). However, the only differences are that in Eq. (5.4-14) n goes from 2 to 20, and in Eq. (5.4-15) $n = 21$, so that $_{t+\Delta t}T_{21} = {}_t T_{20}$.

The Fortran program for these equations is easily written and is left up to the reader. The results are tabulated in Table 5.4-1 for comparison with Example 5.4-1, where only 5 slices were used. The table shows that the

TABLE 5.4-1. *Comparison of Results for Examples 5.4-1 and 5.4-2*

| Distance from Front Face | Results Using $\Delta x = 0.20$ m | | Results Using $\Delta x = 0.05$ m | |
| | | Temperature | | Temperature |
m	n	°C	n	°C
0	1	0.0	1	0.0
0.20	2	31.25	5	31.65
0.40	3	58.59	9	58.47
0.60	4	78.13	13	77.55
0.80	5	89.84	17	88.41
1.00	6	93.75	21	91.87

results for 5 slices are reasonably close to those for 20 slices with values from both cases deviating by 2% or less from each other.

As a rule-of-thumb guide for hand calculations, using a minimum of five slices and at least 8 to 10 time increments should give sufficient accuracy for most purposes. Only when very high accuracy is desired or several cases are to be solved is it desirable to solve the problem using a digital computer. In some cases the physical properties are not known with sufficient accuracy to justify a computer solution.

EXAMPLE 5.4-3. Unsteady-State Conduction with Convective Boundary Condition

Use the same conditions as Example 5.4-1, but a convective coefficient of $h = 25.0 \text{ W/m}^2 \cdot \text{K}$ is now present at the surface. The thermal conductivity $k = 10.0 \text{ W/m} \cdot \text{K}$.

Solution: Equations (5.4-7) and (5.4-8) can be used for convection at the surface. From Eq. (5.4-8), $N = h\,\Delta x / k = 25.0(0.20)/10.0 = 0.50$. Then $2N + 2 = 2(0.50) + 2 = 3.0$. However, by Eq. (5.4-9), the value of M must be equal to or greater than $2N + 2$. This means that a value of $M = 2$ cannot be used. We will select the preferred method where $M = 4.0$. [Another less accurate alternative is to use Eq. (5.4-11) for convection and then the value of M is not restricted by the N value.]
Substituting into Eq. (5.4-3) and solving for Δt,

$$M = 4 = \frac{(\Delta x)^2}{\alpha\,\Delta t} = \frac{(0.20)^2}{(2.00 \times 10^{-5})(\Delta t)} \qquad \Delta t = 500 \text{ s}$$

Hence, $6000/500 = 12$ time increments must be used.
For the first Δt time increment and for all time increments, the value of the environmental temperature T_a to use is $T_a = 0°\text{C}$ since $M > 3$. For convection at the node or point $n = 1$ we use Eq. (5.4-7), where $M = 4$ and $N = 0.50$.

$$_{t+\Delta t}T_1 = \tfrac{1}{4}[2(0.5)T_a + [4 - (2 \times 0.5 + 2)]_t T_1 + 2_t T_2]$$
$$= 0.25 T_a + 0.25_t T_1 + 0.50_t T_2 \qquad n = 1 \qquad \textbf{(5.4-16)}$$

For $n = 2, 3, 4, 5$, we use Eq. (5.4-2),

$$_{t+\Delta t}T_n = \tfrac{1}{4}[_t T_{n+1} + (4 - 2)_n T_t + {}_t T_{n-1}]$$
$$= 0.25_t T_{n+1} + 0.50_t T_n + 0.25_t T_{n-1} \qquad n = 2, 3, 4, 5 \; \textbf{(5.4-17)}$$

For $n = 6$ (insulated boundary), we use Eq. (5.4-10) and $f = 6$.

$$_{t+\Delta t}T_6 = \tfrac{1}{4}[(4 - 2)_t T_6 + 2_t T_5]$$
$$= 0.50_t T_6 + 0.50_t T_5 \qquad n = 6 \qquad \textbf{(5.4-18)}$$

For $1\,\Delta t$, for the first time increment of $t + \Delta t$, $T_a = 0$. Using Eq. (5.4-16) to calculate the temperature at node 1,

$$_{t+\Delta t}T_1 = 0.25(0) + 0.25(100) + 0.50(100) = 75.0$$

For $n = 2, 3, 4, 5$ using Eq. (5.4-17),

$$_{t+\Delta t}T_2 = 0.25_t T_3 + 0.50_t T_2 + 0.25_t T_1$$
$$= 0.25(100) + 0.50(100) + 0.25(100) = 100.0$$

Also, in a similar calculation, T_3, T_4, and $T_5 = 100.0$. For $n = 6$, using Eq. (5.4-18), $T_6 = 100.0$.

For 2 Δt, $T_a = 0$. Using Eq. (5.4-16),

$$_{t+2\,\Delta t}T_1 = 0.25(0) + 0.25(75.0) + 0.50(100) = 68.75$$

Using Eq. (5.4-17) for $n = 2, 3, 4, 5$,

$$_{t+2\,\Delta t}T_2 = 0.25(100) + 0.50(100) + 0.25(75.00) = 93.75$$

$$_{t+2\,\Delta t}T_3 = 0.25(100) + 0.50(100) + 0.25(100) = 100.0$$

Also, T_4 and $T_5 = 100.0$.
For $n = 6$, using Eq. (5.4-18), $T_6 = 100.0$.
    For 3 Δt, $T_a = 0$. Using Eq. (5.4-16),

$$_{t+3\,\Delta t}T_1 = 0.25(0) + 0.25(68.75) + 0.50(93.75) = 64.07$$

Using Eq. (5.4-17) for $n = 2, 3, 4, 5$,

$$_{t+3\,\Delta t}T_2 = 0.25(100) + 0.50(93.75) + 0.25(68.75) = 89.07$$

$$_{t+3\,\Delta t}T_3 = 0.25(100) + 0.50(100) + 0.25(93.75) = 98.44$$

$$_{t+3\,\Delta t}T_4 = 0.25(100) + 0.50(100) + 0.25(100) = 100.0$$

Also, $T_5 = 100.0$ and $T_6 = 100.0$.
    In a similar manner the calculations can be continued for the remaining time until a total of 12 Δt increments have been used.

5.4C. Other Numerical Methods for Unsteady-State Conduction

1. Unsteady-state conduction in a cylinder. In deriving the numerical equations for unsteady-state conduction in a flat slab, the cross-sectional area was constant throughout. In a cylinder it changes radially. To derive the equation for a cylinder, Fig. 5.4-3 is used where the cylinder is divided into concentric hollow cylinders whose walls are Δx m thick. Assuming a cylinder 1 m long and making a heat balance on the slab at point n, the rate of heat in $-$ rate of heat out = rate of heat accumulation.

$$\frac{k[2\pi(n + 1/2)\,\Delta x]}{\Delta x}(_tT_{n+1} - _tT_n) - k\frac{[2\pi(n - 1/2)\,\Delta x]}{\Delta x}(_tT_n - _tT_{n-1})$$

$$= \frac{2\pi n(\Delta x)^2 \rho c_p}{\Delta t}(_{t+\Delta t}T_n - _tT_n) \qquad (5.4\text{-}19)$$

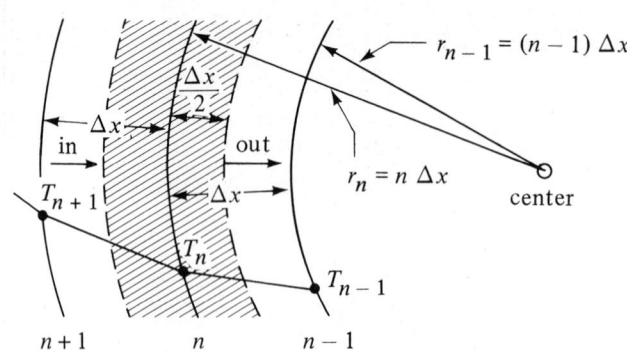

FIGURE 5.4-3. *Unsteady-state conduction in a cylinder.*

Rearranging, the final equation is

$$_{t+\Delta t}T_n = \frac{1}{M}\left[\frac{2n+1}{2n}\,_t T_{n+1} + (M-2)_t T_n + \frac{2n-1}{2n}\,_t T_{n-1}\right] \qquad (5.4\text{-}20)$$

where $M = (\Delta x)^2/(\alpha\,\Delta t)$ as before. Also, at the center where $n = 0$,

$$_{t+\Delta t}T_0 = \frac{4}{M}\,_t T_1 + \frac{M-4}{M}\,_t T_0 \qquad (5.4\text{-}21)$$

To use Equations (5.4-20) and (5.4-21),

$$M \geq 4 \qquad (5.4\text{-}22)$$

Equations for convection at the outer surface of the cylinder have been derived (D1). If the heat capacity of the outer $\frac{1}{2}$ slab is neglected,

$$_{t+\Delta t}T_n = \frac{nN}{\dfrac{2n-1}{2}+nN}\,_{t+\Delta t}T_a + \frac{(2n-1)/2}{\dfrac{2n-1}{2}+nN}\,_{t+\Delta t}T_{n-1} \qquad (5.4\text{-}23)$$

where T_n is the temperature at the surface and T_{n-1} at a position in the solid 1 Δx below the surface.

Equations for numerical methods for two-dimensional unsteady-state conduction have been derived and are available in a number of references (D1, K2).

2. *Unsteady-state conduction and implicit numerical method.* In some practical problems the restrictions imposed on the value $M \geq 2$ by stability requirements may prove inconvenient. Also, to minimize the stability problems, implicit methods using different finite-difference formulas have been developed. An important one of these formulas is the Crank–Nicolson method, which will be considered here.

In deriving Eqs. (5.4-1) and (5.4-2), the rate at which heat entered the slab in Fig. 5.4-1 was taken to be the rate at time t.

$$\text{Rate of heat in at } t = \frac{kA}{\Delta x}(_t T_{n-1} - _t T_n) \qquad (5.4\text{-}24)$$

It was then assumed that this rate could be used during the whole interval from t to $t + \Delta t$. However, this is an approximation since the rate changes during this Δt interval. A better value would be the average value of the rate at t and at $t + \Delta t$, or

$$\text{average rate of heat in} = \frac{kA}{\Delta x}\left[\frac{(_t T_{n-1} - _t T_n) + (_{t+\Delta t}T_{n-1} - _{t+\Delta t}T_n)}{2}\right] \qquad (5.4\text{-}25)$$

Also, for the heat leaving, a similar type of average is used. The final equation is

$$_{t+\Delta t}T_{n+1} - (2M+2)_{t+\Delta t}T_n + _{t+\Delta t}T_{n-1} = -_t T_{n+1} + (2-2M)_t T_n - _t T_{n-1} \qquad (5.4\text{-}26)$$

This means that now a new value of $_{t+\Delta t}T_n$ cannot be calculated only from values at time t as in Eq. (5.4-2) but that all the new values of T at $t + \Delta t$ at all points must be calculated simultaneously. To do this an equation is written similar to Eq. (5.4-26) for each of the internal points. Each of these equations and the boundary equations are linear algebraic equations. These then can be solved simultaneously by the standard methods used, such as the Gauss–Seidel iteration technique, matrix inversion technique, and so on (G1, K1).

An important advantage of Eq. (5.4-26) is that the stability and convergence criteria are satisfied for all positive values of M. This means that M can have values less than 2.0. A disadvantage of the implicit method is the larger number of calculations needed for each time step. Explicit methods are simpler to use but because of stability considerations, especially in complex situations, implicit methods are often used.

5.5 (SELECTED TOPIC) CHILLING AND FREEZING OF FOOD AND BIOLOGICAL MATERIALS

5.5A Introduction

Unlike many inorganic and organic materials which are relatively stable, food and other biological materials decay and deteriorate more or less rapidly with time at room temperature. This spoilage is due to a number of factors. Tissues of foods such as fruits and vegetables, after harvesting, continue to undergo metabolic respiration and ripen and eventually spoil. Enzymes of the dead tissues of meats and fish remain active and induce oxidation and other deteriorating effects. Microorganisms attack all types of foods by decomposing the foods so that spoilage occurs; also, chemical reactions occur, such as the oxidation of fats.

At low temperatures the growth rate of microorganisms will be slowed if the temperature is below that which is optimum for growth. Also, enzyme activity and chemical reaction rates are reduced at low temperatures. The rates of most chemical and biological reactions in storage of chilled or frozen foods and biological materials are reduced by factors of $\frac{1}{2}$ to $\frac{1}{3}$ for each 10 K (10°C) drop in temperature.

Water plays an important part in these rates of deterioration, and it is present to a substantial percentage in most biological materials. To reach a low enough temperature for most of these rates to approximately cease, most of the water must be frozen. Materials such as food do not freeze at 0°C (32°F) as pure water does but at a range of temperatures below 0°C. However, because of some of the physical effects of ice crystals and other effects, such as concentrating of solutions, chilling of biological materials is often used instead of freezing for preservation.

Chilling of materials involves removing the sensible heat and heat of metabolism and reducing the temperature usually to a range of 4.4°C (40°F) to just above freezing. Essentially no latent heat of freezing is involved. The materials can be stored for a week or so up to a few months, depending on the product stored and the gaseous atmosphere. Each material has its optimum chill storage temperature.

In the freezing of food and biological materials, the temperature is reduced so that most of the water is frozen to ice. Depending on the final storage temperatures of down to $-30°C$, the materials can be stored for up to 1 year or so. Often in the production of many frozen foods, they are first treated by blanching or scalding to destroy enzymes.

5.5B Chilling of Food and Biological Materials

In the chilling of food and biological materials, the temperature of the materials is reduced to the desired chill storage temperature, which can be about $-1.1°C$ (30°F) to 4.4°C (40°F). For example, after slaughter, beef has a temperature of 37.8°C (100°F) to 40°C (104°F), and it is often cooled to about 4.4°C (40°F). Milk from cows must be chilled quickly to temperatures just above freezing. Some fish fillets at the time of packing are at a temperature of 7.2°C (45°F) to 10°C (50°F) and are chilled to close to 0°C.

These rates of chilling or cooling are governed by the laws of unsteady-state heat conduction discussed in Sections 5.1 to 5.4. The heat is removed by convection at the surface of the material and by unsteady-state conduction in the material. The fluid outside the foodstuff or biological materials is used to remove this heat, and in many cases it is air. The air has previously been cooled by refrigeration to $-1.1°C$ to $+4.4°C$, depending on the material and other conditions. The convective heat-transfer coefficients, which usually include radiation effects, can also be predicted by the methods in Chapter 4, and for air the coefficient varies from about 8.5 to 40 $W/m^2 \cdot K$ (1.5 to 7 btu/h $\cdot ft^2 \cdot °F$), depending primarily on air velocity.

In some cases the fluid used for chilling is a liquid flowing over the surface and the values of h can vary from about 280 to 1700 $W/m^2 \cdot K$ (50–300 btu/h $\cdot ft^2 \cdot °F$). Also, in other cases, a contact or plate cooler is used where chilled plates are in direct contact with the material. Then the temperature of the surface of the material is usually assumed to be equal to or close to that of the contact plates. Contact freezers are used for freezing biological materials.

Where the food is packaged in boxes or where the material is tightly covered by a film of plastic, this additional resistance must be considered. One method to do this is to add the resistance of the package covering to that of the convective film:

$$R_T = R_P + R_C \qquad (5.5\text{-}1)$$

where R_P is the resistance of covering, R_C the resistance of the outside convective film, and R_T the total resistance. Then, for each resistance,

$$R_C = \frac{1}{h_c A} \qquad (5.5\text{-}2)$$

$$R_P = \frac{\Delta x}{kA} \qquad (5.5\text{-}3)$$

$$R_T = \frac{1}{hA} \qquad (5.5\text{-}4)$$

where h_c is the convective gas or liquid coefficient, A is the area, Δx is the thickness of the covering, k is the thermal conductivity of the covering, and h is the overall coefficient. The overall coefficient h is the one to use in the unsteady-state charts. This assumes a negligible heat capacity of the covering, which is usually the case. Also, it assumes that the covering closely touches the food material so there is no resistance between the covering and the food.

The major sources of error in using the unsteady-state charts are the inadequate data on the density, heat capacity, and thermal conductivity of the foods and the prediction of the convective coefficient. Food materials are irregular anisotropic substances, and the physical properties are often difficult to evaluate. Also, if evaporation of water occurs on chilling, latent heat losses can affect the accuracy of the results.

EXAMPLE 5.5-1.   Chilling Dressed Beef
Hodgson (H2) gives physical properties of beef carcasses during chilling of $\rho = 1073$ kg/m^3, $c_p = 3.48$ kJ/kg $\cdot$ K, and $k = 0.498$ W/m $\cdot$ K. A large slab of beef 0.203 m thick and initially at a uniform temperature of 37.8°C is to be cooled so that the center temperature is 10°C. Chilled air at 1.7°C (assumed constant) and having an $h = 39.7$ W/m$^2 \cdot$ K is used. Calculate the time needed.

Solution: The thermal diffusivity α is

$$\alpha = \frac{k}{\rho c_p} = \frac{0.498}{(1073)(3.48 \times 1000)} = 1.334 \times 10^{-7} \text{ m}^2/\text{s}$$

Then for the half-thickness x_1 of the slab,

$$x_1 = \frac{0.203}{2} = 0.1015 \text{ m}$$

For the center of the slab,

$$n = \frac{x}{x_1} = \frac{0}{x_1} = 0$$

Also,

$$m = \frac{k}{hx_1} = \frac{0.498}{(39.7)(0.1015)} = 0.123$$

$$T_1 = 1.7°C + 273.2 = 274.9 \text{ K} \qquad T_0 = 37.8 + 273.2 = 311.0 \text{ K}$$

$$T = 10 + 273.2 = 283.2 \text{ K}$$

$$Y = \frac{T_1 - T}{T_1 - T_0} = \frac{274.9 - 283.2}{274.9 - 311.0} = 0.230$$

Using Fig. 5.3-6 for the center of a large flat plate,

$$X = 0.90 = \frac{\alpha t}{x_1^2} = \frac{(1.334 \times 10^{-7})(t)}{(0.1015)^2}$$

Solving, $t = 6.95 \times 10^4$ s (19.3 h).

5.5C   Freezing of Food and Biological Materials

1. Introduction.   In the freezing of food and other biological materials, the removal of sensible heat in chilling first occurs and then the removal of the latent heat of freezing. The latent heat of freezing water of 335 kJ/kg (144 btu/lb$_m$) is a substantial portion of the total heat removed on freezing. Other slight effects, such as the heats of solution of salts, and so on, may be present but are quite small. Actually, when materials such as meats are frozen to $-29°C$, only about 90% of the water is frozen to ice, with the rest thought to be bound water (B1).

Riedel (R1) gives enthalpy–temperature–composition charts for the freezing of many different foods. These charts show that freezing does not occur at a given temperature but extends over a range of several degrees. As a consequence, there is no one freezing point with a single latent heat of freezing.

Since the latent heat of freezing is present in the unsteady-state process of freezing, the standard unsteady-state conduction equations and charts given in this chapter cannot be used for prediction of freezing times. A full analytical solution of the rate of freezing of food and biological materials is very difficult because of the variation of physical properties with temperature, the amount of freezing varying with temperature, and other factors. An approximate solution by Plank is often used.

2. Approximate solution of Plank for freezing.   Plank (P2) has derived an approximate solution for the time of freezing which is often sufficient for engineering purposes. The assumptions in the derivation are as follows. Initially, all the food is at the freezing temperature but is unfrozen. The thermal conductivity of the frozen part is constant. All the material freezes at the freezing point, with a constant latent heat. The heat transfer by conduction in the frozen layer occurs slowly enough so that it is under pseudo-steady-state conditions.

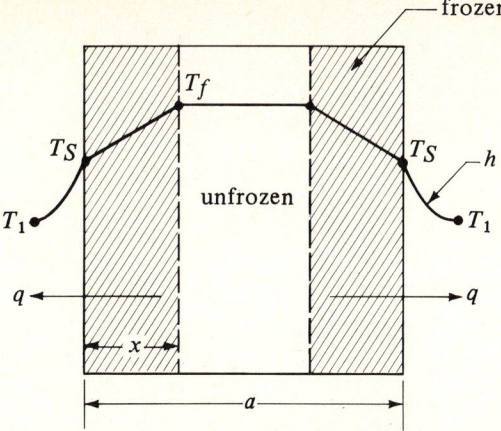

FIGURE 5.5-1.   *Temperature profile during freezing.*

In Fig. 5.5-1 a slab of thickness a m is cooled from both sides by convection. At a given time t s, a thickness of x m of frozen layer has formed on both sides. The temperature of the environment is constant at T_1 K and the freezing temperature is constant at T_f. An unfrozen layer in the center at T_f is present.

The heat leaving at time t is q W. Since we are at pseudo-steady state, at time t, the heat leaving by convection on the outside is

$$q = hA(T_S - T_1) \qquad (5.5\text{-}5)$$

where A is the surface area. Also, the heat being conducted through the frozen layer of x thickness at steady state is

$$q = \frac{kA}{x}(T_f - T_S) \qquad (5.5\text{-}6)$$

where k is the thermal conductivity of the frozen material. In a given time dt s, a layer dx thick of material freezes. Then multiplying A times dx times ρ gives the kg mass frozen. Multiplying this by the latent heat λ in J/kg and dividing by dt,

$$q = \frac{A\,dx\,\rho\lambda}{dt} = A\rho\lambda\,\frac{dx}{dt} \qquad (5.5\text{-}7)$$

where ρ is the density of the unfrozen material.

Next, to eliminate T_S from Eqs. (5.5-5) and (5.5-6), Eq. (5.5-5) is solved for T_S and substituted into Eq. (5.5-6), giving

$$q = \frac{(T_f - T_1)A}{x/k + 1/h} \qquad (5.5\text{-}8)$$

Equating Eq. (5.5-8) to (5.5-7),

$$\frac{(T_f - T_1)A}{x/k + 1/h} = A\rho\lambda\,\frac{dx}{dt} \qquad (5.5\text{-}9)$$

Rearranging and integrating between $t = 0$ and $x = 0$, to $t = t$ and $x = a/2$,

$$(T_f - T_1)\int_0^t dt = \lambda\rho \int_0^{a/2} \left(\frac{x}{k} + \frac{1}{h}\right) dx \qquad (5.5\text{-}10)$$

Integrating and solving for t,

$$t = \frac{\lambda \rho}{T_f - T_1} \left(\frac{a}{2h} + \frac{a^2}{8k} \right) \qquad (5.5\text{-}11)$$

To generalize the equation for other shapes,

$$t = \frac{\lambda \rho}{T_f - T_1} \left(\frac{Pa}{h} + \frac{Ra^2}{k} \right) \qquad (5.5\text{-}12)$$

where a is the thickness of an infinite slab (as in Fig. 5.5-1), diameter of a sphere, diameter of a long cylinder, or the smallest dimension of a rectangular block or brick. Also,

$$P = \tfrac{1}{2} \text{ for infinite slab, } \tfrac{1}{6} \text{ for sphere, } \tfrac{1}{4} \text{ for infinite cylinder}$$

$$R = \tfrac{1}{8} \text{ for infinite slab, } \tfrac{1}{24} \text{ for sphere, } \tfrac{1}{16} \text{ for infinite cylinder}$$

For a rectangular brick having dimensions a by $\beta_1 a$ by $\beta_2 a$, where a is the shortest side, Ede (B1) has prepared a chart to determine the values of P and R to be used to calculate t in Eq. (5.5-12). Equation (5.5-11) can also be used for calculation of thawing times by replacing the k of the frozen material by the k of the thawed material.

> **EXAMPLE 5.5-2. *Freezing of Meat***
> Slabs of meat 0.0635 m thick are to be frozen in an air-blast freezer at 244.3 K ($-28.9°C$). The meat is initially at the freezing temperature of 270.4 K ($-2.8°C$). The meat contains 75% moisture. The heat-transfer coefficient is $h = 17.0$ W/m$^2 \cdot$ K. The physical properties are $\rho = 1057$ kg/m^3 for the unfrozen meat and $k = 1.038$ W/m $\cdot$ K for the frozen meat. Calculate the freezing time.
>
> **Solution:** Since the latent heat of fusion of water to ice is 335 kJ/kg (144 btu/lb$_m$), for meat with 75% water,
>
> $$\lambda = 0.75(335) = 251.2 \text{ kJ/kg}$$
>
> The other given variables are $a = 0.0635$ m, $T_f = 270.4$ K, $T_1 = 244.3$ K, $\rho = 1057$ kg/m^3, $h = 17.0$ W/m$^2 \cdot$ K, $k = 1.038$ W/m $\cdot$ K. Substituting into Eq. (5.5-11),
>
> $$t = \frac{\lambda \rho}{T_f - T_1} \left(\frac{a}{2h} + \frac{a^2}{8k} \right) = \frac{(2.512 \times 10^5)1057}{270.4 - 244.3} \left[\frac{0.0635}{2(17.0)} + \frac{(0.0635)^2}{8(1.038)} \right]$$
>
> $$= 2.395 \times 10^4 \text{ s (6.65 h)}$$

3. Other methods to calculate freezing times. Neumann (C1, C2) has derived a complicated equation for freezing in a slab. He assumes the following conditions. The surface temperature is the same as the environment, i.e., no surface resistance. The temperature of freezing is constant. This method suffers from this limitation that a convection coefficient cannot be used at the surface, since it assumes no surface resistance. However, the method does include the effect of cooling from an original temperature, which may be above the freezing point.

Plank's equation does not make provision for an original temperature, which may be above the freezing point. An approximate method to calculate the additional time necessary to cool from temperature T_0 down to the freezing point T_f is as follows. Calculate by means of the unsteady-state charts the time for the average temperature in the material to reach T_f assuming that no freezing occurs using the physical properties of the unfrozen material. If there is no surface resistance, Fig. 5.3-13 can be used directly for this. If a resistance is present, the temperature at several points in the material will have to be obtained from the unsteady-state charts and the average temperature calculated from these point temperatures. This may be partial trial and error since the time is unknown, which must be assumed. If the average temperature calculated is not at the freezing point, a new time must be assumed. This is an approximate method since some material will actually freeze.

5.6 (SELECTED TOPIC) DIFFERENTIAL EQUATION OF ENERGY CHANGE

5.6A Introduction

In Section 3.6 (Selected Topic) we derived a differential equation of continuity and a differential equation of momentum transfer for a pure fluid. These equations were derived because overall mass, energy, and momentum balances made on a finite volume in the earlier parts of Chapter 2 did not tell us what goes on inside a control volume. In the overall balances performed, a new balance was made for each new system studied. However, it is often easier to start with the differential equations of continuity and momentum transfer in general form and then to simplify the equations by discarding unneeded terms for each specific problem.

In Chapter 4 on steady-state heat transfer and Chapter 5 on unsteady-state heat transfer new overall energy balances were made on a finite control volume for each new situation. To advance further in our study of heat or energy transfer in flow and nonflow systems we must use a differential volume to investigate in greater detail what goes on inside this volume. The balance will be made on a single phase and the boundary conditions at the phase boundary will be used for integration.

In the next section we derive a general differential equation of energy change: the conservation-of-energy equation. Then this equation is modified for certain special cases that occur frequently. Finally, applications of the uses of these equations are given. Cases for both steady-state and unsteady-state energy transfer are studied using this conservation-of-energy equation, which is perfectly general and holds for steady- or unsteady-state conditions.

5.6B Derivation of Differential Equation of Energy Change

As in the derivation of the differential equation of momentum transfer, we write a balance on an element of volume of size Δx, Δy, and Δz which is stationary. We then write the law of conservation of energy, which is really the first law of thermodynamics for the fluid in this volume element at any time. The following is the same as Eq. (2.7-7) for a control volume given in Section 2.7.

$$\begin{pmatrix} \text{rate of} \\ \text{energy in} \end{pmatrix} - \begin{pmatrix} \text{rate of} \\ \text{energy out} \end{pmatrix} - \begin{pmatrix} \text{rate of} \\ \text{external work done by} \\ \text{system on surroundings} \end{pmatrix}$$
$$= \begin{pmatrix} \text{rate of} \\ \text{accumulation of} \\ \text{energy} \end{pmatrix} \text{(5.6-1)}$$

As in momentum transfer, the transfer of energy into and out of the volume element is by convection and molecular transport or conduction. There are two kinds of energy being transferred. The first is internal energy U in J/kg (btu/lb$_m$) or any other set of units.

This is the energy associated with random translational and internal motions of the molecules plus molecular interactions. The second is kinetic energy $\rho v^2/2$, which is the energy associated with the bulk fluid motion, where v is the local fluid velocity, m/s (ft/s). Hence, the total energy per unit volume is $(\rho U + \rho v^2/2)$. The rate of accumulation of energy in the volume element in m³ (ft³) is then

$$\Delta x \, \Delta y \, \Delta z \, \frac{\partial}{\partial t}\left(\rho U + \frac{\rho v^2}{2}\right) \tag{5.6-2}$$

The total energy coming in by convection in the x direction at x minus that leaving at $x + \Delta x$ is

$$\Delta y \, \Delta z\left[v_x\left(\rho U + \frac{\rho v^2}{2}\right)\right]_x - \Delta y \, \Delta z\left[v_x\left(\rho U + \frac{\rho v^2}{2}\right)\right]_{x+\Delta x} \tag{5.6-3}$$

Similar equations can be written for the y and z directions using velocities v_y and v_z, respectively.

The net rate of energy into the element by conduction in the x direction is

$$\Delta y \, \Delta z[(q_x)_x - (q_x)_{x+\Delta x}] \tag{5.6-4}$$

Similar equations can be written for the y and z directions where q_x, q_y, and q_z are the components of the heat flux vector $\mathbf{q}$, which is in W/m² (btu/s·ft²) or any other convenient set of units.

The net work done by the system on its surroundings is the sum of the following three parts for the x direction. For the net work done against the gravitational force,

$$-\rho \, \Delta x \, \Delta y \, \Delta z(v_x g_x) \tag{5.6-5}$$

where g_x is gravitational force. The net work done against the static pressure p is

$$\Delta y \, \Delta z[(pv_x)_{x+\Delta x} - (pv_x)_x] \tag{5.6-6}$$

where p is N/m² (lb$_f$/ft²) or any other convenient set of units. For the net work against the viscous forces,

$$(\Delta y \, \Delta z)[(\tau_{xx}v_x + \tau_{xy}v_y + \tau_{xz}v_z)_{x+\Delta x} - (\tau_{xx}v_x + \tau_{xy}v_y + \tau_{xz}v_z)_x] \tag{5.6-7}$$

In Section 3.6 these viscous forces are discussed in more detail.

Writing equations similar to (5.6-3)–(5.6-7) in all three directions; substituting these equations and Eq. (5.6-2) into (5.6-1); dividing by Δx, Δy, and Δz; and letting Δx, Δy, and Δz approach zero, we obtain

$$\frac{\partial}{\partial t}\left(\rho U + \frac{\rho v^2}{2}\right) = -\left[\frac{\partial}{\partial x}v_x\left(\rho U + \frac{\rho v^2}{2}\right) + \frac{\partial}{\partial y}v_y\left(\rho U + \frac{\rho v^2}{2}\right) + \frac{\partial}{\partial z}v_z\left(\rho U + \frac{\rho v^2}{2}\right)\right]$$

$$-\left(\frac{\partial q_x}{\partial x} + \frac{\partial q_y}{\partial y} + \frac{\partial q_z}{\partial z}\right) + \rho(v_x g_x + v_y g_y + v_z g_z)$$

$$-\left[\frac{\partial}{\partial x}(pv_x) + \frac{\partial}{\partial y}(pv_y) + \frac{\partial}{\partial z}(pv_z)\right]$$

$$-\left[\frac{\partial}{\partial x}(\tau_{xx}v_x + \tau_{xy}v_y + \tau_{xz}v_z) + \frac{\partial}{\partial y}(\tau_{yx}v_x + \tau_{yy}v_y + \tau_{yz}v_z)\right.$$

$$\left. + \frac{\partial}{\partial z}(\tau_{zx}v_x + \tau_{zy}v_y + \tau_{zz}v_z)\right] \tag{5.6-8}$$

For further details of this derivation, see (B2).

Equation (5.6-8) is the final equation of energy change relative to a stationary point. However, it is not in a convenient form. We first combine Eq. (5.6-8) with the equation of continuity, Eq. (3.6-23), with the equation of motion, Eq. (3.6-40), and express the internal energy in terms of fluid temperature T and heat capacity. Then writing the resultant equation for a Newtonian fluid with constant thermal conductivity k, we obtain

$$\rho c_v \frac{DT}{Dt} = k\nabla^2 T - T\left(\frac{\partial p}{\partial T}\right)_\rho (\nabla \cdot \mathbf{v}) + \mu\phi \qquad (5.6\text{-}9)$$

This equation utilizes Fourier's second law in three directions, where

$$k\nabla^2 T = k\left(\frac{\partial^2 T}{\partial x^2} + \frac{\partial^2 T}{\partial y^2} + \frac{\partial^2 T}{\partial z^2}\right) \qquad (5.6\text{-}10)$$

The viscous dissipation term $\mu\phi$ is generally negligible except where extremely large velocity gradients exist. It will be omitted in the discussions to follow. Equation (5.6-9) is the equation of energy change for a Newtonian fluid with constant k in terms of the fluid temperature T.

5.6C Special Cases of the Equation of Energy Change

The following special forms of Eq. (5.6-9) for a Newtonian fluid with constant thermal conductivity are commonly encountered. First, Eq. (4.6-9) will be written in rectangular coordinates without the $\mu\phi$ term.

$$\rho c_v\left(\frac{\partial T}{\partial t} + v_x \frac{\partial T}{\partial x} + v_y \frac{\partial T}{\partial y} + v_z \frac{\partial T}{\partial z}\right)$$

$$= k\left(\frac{\partial^2 T}{\partial x^2} + \frac{\partial^2 T}{\partial y^2} + \frac{\partial^2 T}{\partial z^2}\right) - T\left(\frac{\partial p}{\partial T}\right)_\rho \left(\frac{\partial v_x}{\partial x} + \frac{\partial v_y}{\partial y} + \frac{\partial v_z}{\partial z}\right) \qquad (5.6\text{-}11)$$

1. Fluid at constant pressure. The equations below can be used for constant-density fluids as well as for constant pressure.

$$\rho c_p \frac{DT}{Dt} = k\nabla^2 T \qquad (5.6\text{-}12)$$

In rectangular coordinates,

$$\rho c_p\left(\frac{\partial T}{\partial t} + v_x \frac{\partial T}{\partial x} + v_y \frac{\partial T}{\partial y} + v_z \frac{\partial T}{\partial z}\right) = k\left(\frac{\partial^2 T}{\partial x^2} + \frac{\partial^2 T}{\partial y^2} + \frac{\partial^2 T}{\partial z^2}\right) \qquad (5.6\text{-}13)$$

In cylindrical coordinates,

$$\rho c_p\left(\frac{\partial T}{\partial t} + v_r \frac{\partial T}{\partial r} + \frac{v_\theta}{r} \frac{\partial T}{\partial \theta} + v_z \frac{\partial T}{\partial z}\right) = k\left(\frac{\partial^2 T}{\partial r^2} + \frac{1}{r} \frac{\partial T}{\partial r} + \frac{1}{r^2} \frac{\partial^2 T}{\partial \theta^2} + \frac{\partial^2 T}{\partial z^2}\right) \qquad (5.6\text{-}14)$$

In spherical coordinates,

$$\rho c_p\left(\frac{\partial T}{\partial t} + v_r \frac{\partial T}{\partial r} + \frac{v_\theta}{r} \frac{\partial T}{\partial \theta} + \frac{v_\phi}{r \sin\theta} \frac{\partial T}{\partial \phi}\right)$$

$$= k\left[\frac{1}{r^2} \frac{\partial}{\partial r}\left(r^2 \frac{\partial T}{\partial r}\right) + \frac{1}{r^2 \sin\theta} \frac{\partial}{\partial \theta}\left(\sin\theta \frac{\partial T}{\partial \theta}\right) + \frac{1}{r^2 \sin^2\theta} \frac{\partial^2 T}{\partial \phi^2}\right] \qquad (5.6\text{-}15)$$

Principles of Unsteady-State Heat Transfer

For definitions of cylindrical and spherical coordinates, see Section 3.6. If the velocity **v** is zero, DT/Dt becomes $\partial T/\partial t$.

2. Fluid at constant density

$$\rho c_p \frac{DT}{Dt} = k\nabla^2 T \qquad (5.6\text{-}16)$$

Note that this is identical to Eq. (5.6-12) for constant pressure.

3. Solid. Here we consider ρ is constant and **v** = 0.

$$\rho c_p \frac{\partial T}{\partial t} = k\nabla^2 T \qquad (5.6\text{-}17)$$

This is often referred to as *Fourier's second law* of heat conduction. This also holds for a fluid with zero velocity at constant pressure.

4. Heat generation. If there is heat generation in the fluid by electrical or chemical means, then $\dot{q}$ can be added to the right side of Eq. (5.6-17).

$$\rho c_p \frac{\partial T}{\partial t} = k\nabla^2 T + \dot{q} \qquad (5.6\text{-}18)$$

where $\dot{q}$ is the rate of heat generation in W/m³ (btu/h · ft³) or other suitable units. Viscous dissipation is also a heat source, but its inclusion greatly complicates problem solving because the equations for energy and motion are then coupled.

5. Other coordinate systems. Fourier's second law of unsteady-state heat conduction can be written as follows.

For rectangular coordinates,

$$\frac{\partial T}{\partial t} = \frac{k}{\rho c_p} \nabla^2 T = \alpha \left(\frac{\partial^2 T}{\partial x^2} + \frac{\partial^2 T}{\partial y^2} + \frac{\partial^2 T}{\partial z^2} \right) \qquad (5.6\text{-}19)$$

where α is $k/\rho c_p$ and is thermal diffusivity in m²/s (ft²/h).

For cylindrical coordinates,

$$\frac{\partial T}{\partial t} = \alpha \left(\frac{\partial^2 T}{\partial r^2} + \frac{1}{r} \frac{\partial T}{\partial r} + \frac{1}{r^2} \frac{\partial^2 T}{\partial \theta^2} + \frac{\partial^2 T}{\partial z^2} \right) \qquad (5.6\text{-}20)$$

For spherical coordinates,

$$\frac{\partial T}{\partial t} = \alpha \left[\frac{1}{r^2} \frac{\partial}{\partial r}\left(r^2 \frac{\partial T}{\partial r} \right) + \frac{1}{r^2 \sin \theta} \frac{\partial}{\partial \theta}\left(\sin \theta \frac{\partial T}{\partial \theta} \right) + \frac{1}{r^2 \sin^2 \theta} \frac{\partial^2 T}{\partial \phi^2} \right] \qquad (5.6\text{-}21)$$

5.6D Uses of Equation of Energy Change

In Section 3.6 we used the differential equations of continuity and of motion to set up fluid flow problems. We did this by discarding the terms that are zero or near zero and using the remaining equations to solve for the velocity and pressure distributions. This was done instead of making new mass and momentum balances for each new situation. In a similar manner, to solve problems of heat transfer, the differential equations of continuity, motion, and energy will be used with the unneeded terms being discarded. Several examples will be given to illustrate the general methods used.

EXAMPLE 5.6-1. Temperature Profile with Heat Generation

A solid cylinder in which heat generation is occurring uniformly as $\dot{q}$ W/m^3 is insulated on the ends. The temperature of the surface of the cylinder is held constant at T_w K. The radius of the cylinder is $r = R$ m. Heat flows only in the radial direction. Derive the equation for the temperature profile at steady state if the solid has a constant thermal conductivity.

Solution: Equation (5.6-20) will be used for cylindrical coordinates. Also the term $\dot{q}/\rho c_p$ for generation will be added to the right side, giving

$$\frac{\partial T}{\partial t} = \frac{k}{\rho c_p}\left(\frac{\partial^2 T}{\partial r^2} + \frac{1}{r}\frac{\partial T}{\partial r} + \frac{1}{r^2}\frac{\partial^2 T}{\partial \theta^2} + \frac{\partial^2 T}{\partial z^2}\right) + \frac{\dot{q}}{\rho c_p} \tag{5.6-22}$$

For steady state $\partial T/\partial t = 0$. Also, for conduction only in the radial direction $\partial^2 T/\partial z^2 = 0$ and $\partial^2 T/\partial \theta^2 = 0$. This gives the following differential equation:

$$\frac{d^2 T}{dr^2} + \frac{1}{r}\frac{dT}{dr} = -\frac{\dot{q}}{k} \tag{5.6-23}$$

This can be rewritten as

$$r\frac{d^2 T}{dr^2} + \frac{dT}{dr} = -\frac{\dot{q}r}{k} \tag{5.6-24}$$

Note that Eq. (5.6-24) can be rewritten as follows:

$$\frac{d}{dr}\left(r\frac{dT}{dr}\right) = -\frac{\dot{q}r}{k} \tag{5.6-25}$$

Integrating Eq. (5.6-25) once,

$$r\frac{dT}{dr} = -\frac{\dot{q}r^2}{2k} + K_1 \tag{5.6-26}$$

where K_1 is a constant. Integrating again,

$$T = -\frac{\dot{q}r^2}{4k} + K_1 \ln r + K_2 \tag{5.6-27}$$

where K_2 is a constant. The boundary conditions are when $r = 0$, $dT/dr = 0$ (by symmetry); and when $r = R$, $T = T_w$. The final equation is

$$T = \frac{\dot{q}(R^2 - r^2)}{4k} + T_w \tag{5.6-28}$$

This is the same as Eq. (4.3-29), which was obtained by another method.

EXAMPLE 5.6-2. Laminar Flow and Heat Transfer Using Equation of Energy Change

Using the differential equation of energy change, derive the partial differential equation and boundary conditions needed for the case of laminar flow of a constant density fluid in a horizontal tube which is being heated. The fluid is flowing at a constant velocity v_z. At the wall of the pipe where the radius $r = r_0$, the heat flux is constant at q_0. The process is at steady state and it is assumed at $z = 0$ at the inlet that the velocity profile is established. Constant physical properties will be assumed.

Solution: From Example 3.6-3, the equation of continuity gives $\partial v_z / \partial z = 0$. Solution of the equation of motion for steady state using cylindrical coordinates gives the parabolic velocity profile.

$$v_z = v_{z\,max} \left[1 - \left(\frac{r}{r_0} \right)^2 \right] \tag{5.6-29}$$

Since the fluid has a constant density, Eq. (5.6-14) in cylindrical coordinates will be used for the equation of energy change. For this case $v_r = 0$ and $v_\theta = 0$. Since this will be symmetrical $\partial T / \partial \theta$ and $\partial^2 T / \partial \theta^2$ will be zero. For steady state, $\partial T / \partial t = 0$. Hence, Eq. (5.6-14) reduces to

$$v_z \frac{\partial T}{\partial z} = \frac{k}{\rho c_p} \left(\frac{\partial^2 T}{\partial r^2} + \frac{1}{r} \frac{\partial T}{\partial r} + \frac{\partial^2 T}{\partial z^2} \right) \tag{5.6-30}$$

Usually conduction in the z direction ($\partial^2 T / \partial z^2$ term) is small compared to the convective term $v_z \, \partial T / \partial z$ and can be dropped. Finally, substituting Eq. (5.6-29) into (5.6-30), we obtain

$$v_{z\,max} \left[1 - \left(\frac{r}{r_0} \right)^2 \right] \frac{\partial T}{\partial z} = \frac{k}{\rho c_p} \left(\frac{\partial^2 T}{\partial r^2} + \frac{1}{r} \frac{\partial T}{\partial r} \right) \tag{5.6-31}$$

The boundary conditions are

At $z = 0$, $T = T_0$ (all r)

At $r = 0$, $T = $ finite

At $r = r_0$, $q_0 = -k \dfrac{\partial T}{\partial r}$ (constant)

For details on the actual solution of this equation, see Siegel et al. (S2).

5.7 (SELECTED TOPIC) BOUNDARY-LAYER FLOW AND TURBULENCE IN HEAT TRANSFER

5.7A Laminar Flow and Boundary-Layer Theory in Heat Transfer

In section 3.7C an exact solution was obtained for the velocity profile for isothermal laminar flow past a flat plate. The solution of Blasius can be extended to include the convective heat-transfer problem for the same geometry and laminar flow. In Fig. 5.7-1 the thermal boundary layer is shown. The temperature of the fluid approaching the plate is T_∞ and that of the plate is T_S at the surface.

We start by writing the differential energy balance, Eq. (5.6-13).

$$\frac{\partial T}{\partial t} + v_x \frac{\partial T}{\partial x} + v_y \frac{\partial T}{\partial y} + v_z \frac{\partial T}{\partial z} = \frac{k}{\rho c_p} \left(\frac{\partial^2 T}{\partial x^2} + \frac{\partial^2 T}{\partial y^2} + \frac{\partial^2 T}{\partial z^2} \right) \tag{5.7-1}$$

If the flow is in the x and y directions, $v_z = 0$. At steady state, $\partial T / \partial t = 0$. Conduction is neglected in the x and z directions, so $\partial^2 T / \partial x^2 = \partial^2 T / \partial z^2 = 0$. Conduction occurs in the y direction. The result is

$$v_x \frac{\partial T}{\partial x} + v_y \frac{\partial T}{\partial y} = \frac{k}{\rho c_p} \frac{\partial^2 T}{\partial y^2} \tag{5.7-2}$$

The simplified momentum balance equation used in the velocity boundary-layer derivation is very similar and is

$$v_x \frac{\partial v_x}{\partial x} + v_y \frac{\partial v_x}{\partial y} = \frac{\mu}{\rho} \frac{\partial^2 v_x}{\partial y^2} \tag{3.7-5}$$

The continuity equation used previously is

$$\frac{\partial v_x}{\partial x} + \frac{\partial v_y}{\partial y} = 0 \tag{3.7-3}$$

Equations (3.7-5) and (3.7-3) were used by Blasius for solving the case for laminar boundary-layer flow. The boundary conditions used were

$$\frac{v_x}{v_\infty} = \frac{v_y}{v_\infty} = 0 \qquad \text{at } y = 0$$

$$\frac{v_x}{v_\infty} = 1 \qquad \text{at } y = \infty \tag{5.7-3}$$

$$\frac{v_x}{v_\infty} = 1 \qquad \text{at } x = 0$$

The similarity between Eqs. (3.7-5) and (5.7-2) is obvious. Hence, the Blasius solution can be applied if $k/\rho c_p = \mu/\rho$. This means the Prandtl number $c_p \mu/k = 1$. Also, the boundary conditions must be the same. This is done by replacing the temperature T in Eq. (5.7-2) by the dimensionless variable $(T - T_S)/(T_\infty - T_S)$. The boundary conditions become

$$\frac{v_x}{v_\infty} = \frac{v_y}{v_\infty} = \frac{T - T_S}{T_\infty - T_S} = 0 \qquad \text{at } y = 0$$

$$\frac{v_x}{v_\infty} = \frac{T - T_S}{T_\infty - T_S} = 1 \qquad \text{at } y = \infty \tag{5.7-4}$$

$$\frac{v_x}{v_\infty} = \frac{T - T_S}{T_\infty - T_S} = 1 \qquad \text{at } x = 0$$

We see that the equations and boundary conditions are identical for the temperature profile and the velocity profile. Hence, for any point x, y in the flow system, the dimensionless velocity variables v_x/v_∞ and $(T - T_S)/(T_\infty - T_S)$ are equal. The velocity-profile solution is the same as the temperature-profile solution.

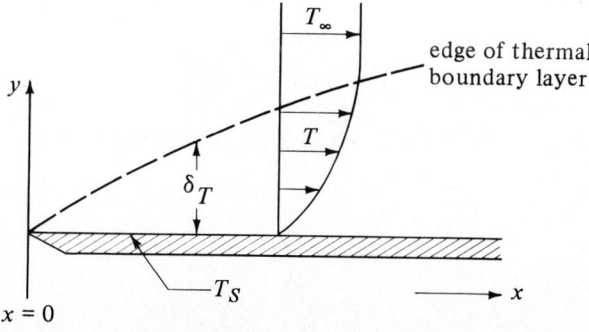

FIGURE 5.7-1. *Laminar flow of fluid past a flat plate and thermal boundary layer.*

This means that the transfer of momentum and heat are directly analogous and the boundary-layer thickness δ for the velocity profile (hydrodynamic boundary layer) and the thermal boundary-layer thickness δ_T are equal. This is important for gases, where the Prandtl numbers are close to 1.

By combining Eqs. (3.7-7) and (3.7-8), the velocity gradient at the surface is

$$\left(\frac{\partial v_x}{\partial y}\right)_{y=0} = 0.332 \frac{v_\infty}{x} N_{\text{Re}, x}^{1/2} \qquad (5.7\text{-}5)$$

where $N_{\text{Re}, x} = xv_\infty \rho/\mu$. Also,

$$\frac{v_x}{v_\infty} = \frac{T - T_S}{T_\infty - T_S} \qquad (5.7\text{-}6)$$

Combining Eqs. (5.7-5) and (5.7-6),

$$\left(\frac{\partial T}{\partial y}\right)_{y=0} = (T_\infty - T_S)\left(\frac{0.332}{x} N_{\text{Re}, x}^{1/2}\right) \qquad (5.7\text{-}7)$$

The convective equation can be related to the Fourier equation by the following, where q_y is in J/s or W (btu/h).

$$\frac{q_y}{A} = h_x(T_S - T_\infty) = -k\left(\frac{\partial T}{\partial y}\right)_{y=0} \qquad (5.7\text{-}8)$$

Combining Eqs. (5.7-7) and (5.7-8),

$$\frac{h_x x}{k} = N_{\text{Nu}, x} = 0.332 N_{\text{Re}, x}^{1/2} \qquad (5.7\text{-}9)$$

where $N_{\text{Nu}, x}$ is the dimensionless Nusselt number and h_x is the local heat-transfer coefficient at point x on the plate.

Pohlhausen (K1) was able to show that the relation between the hydrodynamic and thermal boundary layers for fluids with Prandtl number > 0.6 gives approximately

$$\frac{\delta}{\delta_T} = N_{\text{Pr}}^{1/3} \qquad (5.7\text{-}10)$$

As a result, the equation for the local heat-transfer coefficient is

$$h = 0.332 \frac{k}{x} N_{\text{Re}, x}^{1/2} N_{\text{Pr}}^{1/3} \qquad (5.7\text{-}11)$$

Also,

$$\frac{h_x x}{k} = N_{\text{Nu}, x} = 0.332 N_{\text{Re}, x}^{1/2} N_{\text{Pr}}^{1/3} \qquad (5.7\text{-}12)$$

The equation for the mean heat-transfer coefficient h from $x = 0$ to $x = L$ is for a plate of width b and area bL,

$$h = \frac{b}{A} \int_0^L h_x \, dx$$

$$= \frac{1}{L} 0.332 k \left(\frac{\rho v_\infty}{\mu}\right)^{1/2} N_{\text{Pr}}^{1/3} \int_0^L \frac{dx}{x^{1/2}} \qquad (5.7\text{-}13)$$

Integrating,

$$h = 0.644 \frac{k}{L} N_{Re, L}^{1/2} N_{Pr}^{1/3} \tag{5.7-14}$$

$$\frac{hL}{k} = N_{Nu} = 0.644 N_{Re, L}^{1/2} N_{Pr}^{1/3} \tag{5.7-15}$$

As pointed out previously, this laminar boundary layer on smooth plates holds up to a Reynolds number of about 5×10^5. In using the results above, the fluid properties are usually evaluated at the film temperature $T_f = (T_S + T_\infty)/2$.

5.7B Approximate Integral Analysis of the Thermal Boundary Layer

As discussed in the analysis of the hydrodynamic boundary layer, the Blasius solution is accurate but limited in its scope. Other more complex systems cannot be solved by this method. The approximate integral analysis was used by von Kármán to calculate the hydrodynamic boundary layer and was covered in Section 3.7. This approach can be used to analyze the thermal boundary layer.

This method will be outlined briefly. First, a control volume, as previously given in Fig. 3.7-5, is used to derive the final energy integral expression.

$$\frac{k}{\rho c_p} \left(\frac{\partial T}{\partial y} \right)_{y=0} = \frac{d}{dx} \int_0^{\delta_T} v_x (T_\infty - T) \, dy \tag{5.7-16}$$

This equation is analogous to Eq. (3.7-48) combined with Eq. (3.7-51) for the momentum analysis, giving

$$\frac{\mu}{\rho} \left(\frac{\partial v_x}{\partial y} \right)_{y=0} = \frac{d}{dx} \int_0^{\delta} v_x (v_\infty - v_x) \, dy \tag{5.7-17}$$

Equation (5.7-16) can be solved if both a velocity profile and temperature profile are known. The assumed velocity profile used is Eq. (3.7-50).

$$\frac{v_x}{v_\infty} = \frac{3}{2} \frac{y}{\delta} - \frac{1}{2} \left(\frac{y}{\delta} \right)^3 \tag{3.7-50}$$

The same form of temperature profile is assumed.

$$\frac{T - T_S}{T_\infty - T_S} = \frac{3}{2} \frac{y}{\delta_T} - \frac{1}{2} \left(\frac{y}{\delta_T} \right)^3 \tag{5.7-18}$$

Substituting Eqs. (3.7-50) and (5.7-18) into the integral expression and solving,

$$N_{Nu, x} = 0.36 N_{Re, x}^{1/2} N_{Pr}^{1/3} \tag{5.7-19}$$

This is only about 8% greater than the exact result in Eq. (5.7-11), which indicates that this approximate integral method can be used with confidence in cases where exact solutions cannot be obtained.

In a similar fashion, the integral momentum analysis method used for the turbulent hydrodynamic boundary layer in Section 3.7 can be used for the thermal boundary layer in turbulent flow. Again, the Blasius $\frac{1}{7}$-power law is used for the temperature distribution. These give results that are quite similar to the experimental equations as given in Section 4.6.

5.7C Prandtl Mixing Length and Eddy Thermal Diffusivity

1. Eddy momentum diffusivity in turbulent flow. In Section 3.7F the total shear stress τ_{yx}^t for turbulent flow was written as follows when the molecular and turbulent contributions are summed together:

$$\tau_{yx}^t = -\rho\left(\frac{\mu}{\rho} + \varepsilon_t\right)\frac{d\bar{v}_x}{dy} \tag{5.7-20}$$

The molecular momentum diffusivity μ/ρ in m²/s is a function only of the fluid molecular properties. However, the turbulent momentum eddy diffusivity ε_t depends on the fluid motion. In Eq. (3.7-29) we related ε_t to the Prandtl mixing length L as follows:

$$\varepsilon_t = L^2\left|\frac{d\bar{v}_x}{dy}\right| \tag{3.7-29}$$

2. Prandtl mixing length and eddy thermal diffusivity. We can derive in a similar manner the eddy thermal diffusivity α_t for turbulent heat transfer as follows. Eddies or clumps of fluid are transported a distance L in the y direction. At this point L the clump of fluid differs in mean velocity from the adjacent fluid by the velocity v'_x, which is the fluctuating velocity component discussed in Section 3.7F. Energy is also transported the distance L with a velocity v'_y in the y direction together with the mass being transported. The instantaneous temperature of the fluid is $T = T' + \bar{T}$, where $\bar{T}$ is the mean value and T' the deviation from the mean value. This fluctuating T' is similar to the fluctuating velocity v'_x. The mixing length is small enough so that the temperature difference can be written as

$$T' = L\frac{d\bar{T}}{dy} \tag{5.7-21}$$

The rate of energy transported per unit area is q_y/A and is equal to the mass flux in the y direction times the heat capacity times the temperature difference.

$$\frac{q_y}{A} = \frac{-v'_y \rho c_p L \, d\bar{T}}{dy} \tag{5.7-22}$$

In Section 3.7F we assumed $v'_x \cong v'_y$ and that

$$v'_x = v'_y = L\left|\frac{d\bar{v}_x}{dy}\right| \tag{5.7-23}$$

Substituting Eq. (5.7-23) into (5.7-22),

$$\frac{q_y}{A} = -\rho c_p L^2\left|\frac{d\bar{v}_x}{d_y}\right|\frac{d\bar{T}}{dy} \tag{5.7-24}$$

The term $L^2\,|d\bar{v}_x/dy|$ by Eq. (3.7-29) is the momentum eddy diffusivity ε_t. When this term is in the turbulent heat-transfer equation (5.7-24), it is called α_t, eddy thermal diffusivity. Then Eq. (5.7-24) becomes

$$\frac{q_y}{A} = -\rho c_p \alpha_t \frac{d\bar{T}}{dy} \tag{5.7-25}$$

Combining this with the Fourier equation written in terms of the molecular thermal diffusivity α,

$$\frac{q_y}{A} = -\rho c_p (\alpha + \alpha_t) \frac{d\bar{T}}{dy} \tag{5.7-26}$$

3. Similarities among momentum, heat, and mass transport. Equation (5.7-26) is similar to Eq. (5.7-20) for total momentum transport. The eddy thermal diffusivity α_t and the eddy momentum diffusivity ε_t have been assumed equal in the derivations. Experimental data show that this equality is only approximate. An eddy mass diffusivity for mass transfer has also been defined in a similar manner using the Prandtl mixing length theory and is assumed equal to α_t and ε_t.

PROBLEMS

5.2-1. *Temperature Response in Cooling a Wire.* A small copper wire with a diameter of 0.792 mm and initially at 366.5 K is suddenly immersed into a liquid held constant at 311 K. The convection coefficient $h = 85.2$ W/m$^2 \cdot$ K. The physical properties can be assumed constant and are $k = 374$ W/m $\cdot$ K, $c_p = 0.389$ kJ/kg $\cdot$ K, and $\rho = 8890$ kg/m^3.
(a) Determine the time in seconds for the average temperature of the wire to drop to 338.8 K (one half the initial temperature difference).
(b) Do the same but for $h = 11.36$ W/m$^2 \cdot$ K.
(c) For part (b), calculate the total amount of heat removed for a wire 1.0 m long.
$\quad$ **Ans.** (a) $t = 5.66$ s

5.2-2. *Quenching Lead Shot in a Bath.* Lead shot having an average diameter of 5.1 mm is at an initial temperature of 204.4°C. To quench the shot it is added to a quenching oil bath held at 32.2°C and falls to the bottom. The time of fall is 15 s. Assuming an average convection coefficient of $h = 199$ W/m$^2 \cdot$ K, what will be the temperature of the shot after the fall? For lead, $\rho = 11\,370$ kg/m^3 and $c_p = 0.138$ kJ/kg $\cdot$ K.

5.2-3. *Unsteady-State Heating of a Stirred Tank.* A vessel is filled with 0.0283 m^3 of water initially at 288.8 K. The vessel, which is well stirred, is suddenly immersed into a steam bath held at 377.6 K. The overall heat-transfer coefficient U between the steam and water is 1136 W/m$^2 \cdot$ K and the area is 0.372 m^2. Neglecting the heat capacity of the walls and agitator, calculate the time in hours to heat the water to 338.7 K. [*Hint :* Since the water is well stirred, its temperature is uniform. Show that Eq. (5.2-3) holds by starting with Eq. (5.2-1).]

5.3-1. *Temperature in a Refractory Lining.* A combustion chamber has a 2-in.-thick refractory lining to protect the outer shell. To predict the thermal stresses at startup, the temperature 0.2 in. below the surface is needed 1 min after startup. The following data are available. The initial temperature $T_0 = 100$°F, the hot gas temperature $T_1 = 3000$°F, $h = 40$ btu/h $\cdot$ ft$^2 \cdot$ °F, $k = 0.6$ btu/h $\cdot$ ft $\cdot$ °F, and $\alpha = 0.020$ ft^2/h. Calculate the temperature at a 0.2 in. depth and at a 0.6 in. depth. Use Fig. 5.3-3 and justify its use by seeing if the lining acts as a semiinfinite solid during this 1-min period.
$\quad$ **Ans.** For $x = 0.2$ in., $(T - T_0)/(T_1 - T_0) = 0.28$ and $T = 912$°F (489°C); for $x = 0.6$ in., $(T - T_0)/(T_1 - T_0) = 0.02$ and $T = 158$°F (70°C)

5.3-2. *Freezing Temperature in the Soil.* The average temperature of the soil to a considerable depth is approximately 277.6 K (40°F) during a winter day. If the outside air temperature suddenly drops to 255.4 K (0°F) and stays there, how long will it take for a pipe 3.05 m (10 ft) below the surface to reach 273.2 K (32°F)? The convective coefficient is $h = 8.52$ W/m$^2 \cdot$ K (1.5 btu/h $\cdot$ ft$^2 \cdot$ °F). The soil physical properties can be taken as 5.16×10^{-7} m^2/s (0.02 ft^2/h) for the thermal diffusivity and $1.384 \text{ W/m} \cdot \text{K}$ $(0.8 \text{ btu/h} \cdot \text{ft} \cdot °F)$ for the thermal conductivity. (*Note:* The solution is trial and error, since the unknown time appears twice in the graph for a semiinfinite solid.)

5.3-3. *Cooling a Slab of Aluminum.* A large piece of aluminum that can be considered a semiinfinite solid initially has a uniform temperature of 505.4 K. The surface is suddenly exposed to an environment at 338.8 K with a surface convection coefficient of $455 \text{ W/m}^2 \cdot \text{K}$. Calculate the time in hours for the temperature to reach 388.8 K at a depth of 25.4 mm. The average physical properties are $\alpha = 0.340 \text{ m}^2/\text{h}$ and $k = 208 \text{ W/m} \cdot \text{K}$.

5.3-4. *Transient Heating of a Concrete Wall.* A wall made of concrete 0.305 m thick is insulated on the rear side. The wall at a uniform temperature of $10°C$ (283.2 K) is exposed on the front side to a gas at $843°C$ (1116.2 K). The convection coefficient is $28.4 \text{ W/m}^2 \cdot \text{K}$, the thermal diffusivity is $1.74 \times 10^{-3} \text{ m}^2/\text{h}$, and the thermal conductivity is $0.935 \text{ W/m} \cdot \text{K}$.
(a) Calculate the time for the temperature at the insulated face to reach $232°C$ (505.2 K).
(b) Calculate the temperature at a point 0.152 m below the surface at this same time.

Ans. (a) $\alpha t/x_1^2 = 0.25, t = 13.4 \text{ h}$

5.3-5. *Cooking a Slab of Meat.* A slab of meat 25.4 mm thick originally at a uniform temperature of $10°C$ is to be cooked from both sides until the center reaches $121°C$ in an oven at $177°C$. The convection coefficient can be assumed constant at $25.6 \text{ W/m}^2 \cdot \text{K}$. Neglect any latent heat changes and calculate the time required. The thermal conductivity is $0.69 \text{ W/m} \cdot \text{K}$ and the thermal diffusivity $5.85 \times 10^{-4} \text{ m}^2/\text{h}$. Use the Heisler chart.

Ans. 0.80 h (2880 s)

5.3-6. *Unsteady-State Conduction in a Brick Wall.* A flat brick wall 1.0 ft thick is the lining on one side of a furnace. If the wall is at a uniform temperature of $100°F$ and the one side is suddenly exposed to a gas at $1100°F$, calculate the time for the furnace wall at a point 0.5 ft from the surface to reach $500°F$. The rear side of the wall is insulated. The convection coefficient is $2.6 \text{ btu/h} \cdot \text{ft}^2 \cdot °F$ and the physical properties of the brick are $k = 0.65 \text{ btu/h} \cdot \text{ft} \cdot °F$ and $\alpha = 0.02 \text{ ft}^2/\text{h}$.

5.3-7. *Cooling a Steel Rod.* A long steel rod 0.305 m in diameter is initially at a temperature of 588 K. It is immersed in an oil bath maintained at 311 K. The surface convective coefficient is $125 \text{ W/m}^2 \cdot \text{K}$. Calculate the temperature at the center of the rod after 1 h. The average physical properties of the steel are $k = 38 \text{ W/m} \cdot \text{K}$ and $\alpha = 0.0381 \text{ m}^2/\text{h}$.

Ans. $T = 391 \text{ K}$

5.3-8. *Effect of Size on Heat-Processing Meat.* An autoclave held at $121.1°C$ is being used to process sausage meat 101.6 mm in diameter and 0.61 m long which is originally at $21.1°C$. After 2 h the temperature at the center is $98.9°C$. If the diameter is increased to 139.7 mm, how long will it take for the center to reach $98.9°C$? The heat transfer coefficient to the surface is $h = 1100 \text{ W/m}^2 \cdot \text{K}$, which is very large so that the surface resistance can be considered negligible. (Show this.) Neglect the heat transfer from the ends of the cylinder. The thermal conductivity $k = 0.485 \text{ W/m} \cdot \text{K}$.

Ans. 3.78 h

5.3-9. *Temperature of Oranges on Trees During Freezing Weather.* In orange-growing areas, the freezing of the oranges on the trees during cold nights is economically important. If the oranges are initially at a temperature of $21.1°C$, calculate the center temperature of the orange if exposed to air at $-3.9°C$ for 6 h. The oranges are 102 mm in diameter and the convective coefficient is estimated as $11.4 \text{ W/m}^2 \cdot \text{K}$. The thermal conductivity k is $0.431 \text{ W/m} \cdot \text{K}$ and α is $4.65 \times 10^{-4} \text{ m}^2/\text{h}$. Neglect any latent heat effects.

Ans. $(T_1 - T)/(T_1 - T_0) = 0.05, T = -2.65°C$

5.3-10. Hardening a Steel Sphere. To harden a steel sphere having a diameter of 50.8 mm, it is heated to 1033 K and then dunked into a large water bath at 300 K. Determine the time for the center of the sphere to reach 366.5 K. The surface coefficient can be assumed as 710 W/m² · K, $k = 45$ W/m · K, and $\alpha = 0.0325$ m²/h.

5.3-11. Unsteady-State Conduction in a Short Cylinder. An aluminum cylinder is initially heated so it is at a uniform temperature of 204.4°C. Then it is plunged into a large bath held at 93.3°C, where $h = 568$ W/m² · K. The cylinder has a diameter of 50.8 mm and is 101.6 mm long. Calculate the center temperature after 60 s. The physical properties are $\alpha = 9.44 \times 10^{-5}$ m²/s and $k = 207.7$ W/m · K.

5.3-12. Conduction in Three Dimensions in a Rectangular Block. A rectangular steel block 0.305 m by 0.457 m by 0.61 m is initially at 315.6°C. It is suddenly immersed into an environment at 93.3°C. Determine the temperature at the center of the block after 1 h. The surface convection coefficient is 34 W/m² · K. The physical properties are $k = 38$ W/m · K and $\alpha = 0.0379$ m²/h.

5.4-1. Schmidt Numerical Method for Unsteady-State Conduction. A material in the form of an infinite plate 0.762 m thick is at an initial uniform temperature of 366.53 K. The rear face of the plate is insulated. The front face is suddenly exposed to a temperature of 533.2 K. The convective resistance at this face can be assumed as zero. Calculate the temperature profile after 0.875 h using the Schmidt numerical method with $M = 2$ and slabs 0.1524 m thick. The thermal diffusivity is 0.0929 m²/h.

Ans. $\Delta t = 0.125$ h, seven time increments needed

5.4-2. Unsteady-State Conduction with Nonuniform Initial Temperature Profile. Use the same conditions as Problem 5.4-1 but with the following change. The initial temperature profile is not uniform but is 366.53 K at the front face and 422.1 K at the rear face with a linear variation between the two faces.

5.4-3. Unsteady-State Conduction Using the Digital Computer. Repeat Problem 5.4-2 but use the digital computer and write the Fortran program. Use slabs 0.03048 m thick and $M = 2.0$. Calculate the temperature profile after 0.875 h.

5.4-4. Chilling Meat Using Numerical Methods. A slab of beef 45.7 mm thick and initially at a uniform temperature of 283 K is being chilled by a surface contact cooler at 274.7 K on the front face. The rear face of the meat is insulated. Assume that the convection resistance at the front surface is zero. Using five slices and $M = 2$, calculate the temperature profile after 0.54 h. The thermal diffusivity is 4.64×10^{-4} m²/h.

Ans. $\Delta t = 0.090$ h, six time increments

5.4-5. Cooling Beef with Convective Resistance. A large slab of beef is 45.7 mm thick and is at an initial uniform temperature of 37.78°C. It is being chilled at the front surface in a chilled air blast at −1.11°C with a convective heat-transfer coefficient of $h = 38.0$ W/m² · K. The rear face of the meat is insulated. The thermal conductivity of the beef is $k = 0.498$ W/m · K and $\alpha = 4.64 \times 10^{-4}$ m²/h. Using a numerical method with five slices and $M = 4.0$, calculate the temperature profile after 0.27 h. [*Hint :* Since there is a convective resistance, the value of N must be calculated. Also, Eq. (5.4-7) should be used to calculate the surface temperature $_{t+\Delta t}T_1$.]

5.4-6. Cooling Beef Using the Digital Computer. Repeat Problem 5.4-5 using the digital computer. Use 20 slices and $M = 4.0$. Write the Fortran program.

5.4-7. Convection and Unsteady-State Conduction. For the conditions of Example 5.4-3, continue the calculations for a total of 12 time increments. Plot the temperature profile.

5.4-8. Alternative Convective Boundary Condition for Numerical Method. Repeat Example 5.4-3 but instead use the alternative boundary condition, Eq. (5.4-11). Also, use $M = 4$. Calculate the profile for the full 12 time increments.

5.4-9. Numerical Method for Semiinfinite Solid and Convection. A semiinfinite solid initially at a uniform temperature of 200°C is cooled at its surface by convection. The cooling fluid at a constant temperature of 100°C has a convective coefficient of $h = 250$ W/m$^2 \cdot$ K. The physical properties of the solid are $k = 20$ W/m $\cdot$ K and $\alpha = 4 \times 10^{-5}$ m^2/s. Using a numerical method with $\Delta x = 0.040$ m and $M = 4.0$, calculate the temperature profile after 50 s total time.

Ans. $T_1 = 157.72$, $T_2 = 181.84$, $T_3 = 194.44$, $T_4 = 198.93$, $T_5 = 199.90°C$

5.5-1. (Selected Topic) Chilling Slab of Beef. Repeat Example 5.5-1, where the slab of beef is cooled to 10°C at the center but use air of 0°C at a lower value of $h = 22.7$ W/m$^2 \cdot$ K.

Ans. $(T_1 - T)/(T_1 - T_0) = 0.265$, $X = 0.92$, $t = 19.74$ h

5.5-2. (Selected Topic) Chilling Fish Fillets. Cod fish fillets originally at 10°C are packed to a thickness of 102 mm. Ice is packed on both sides of the fillets and wet-strength paper separates the ice and fillets. The surface temperature of the fish can be assumed as essentially 0°C. Calculate the time for the center of the fillets to reach 2.22°C and the temperature at this time at a distance of 25.4 mm from the surface. Also, plot temperature versus position for the slab. The physical properties are (B1) $k = 0.571$ W/m $\cdot$ K, $\rho = 1052$ kg/m^3, and $c_p = 4.02$ kJ/kg $\cdot$ K.

5.5-3. (Selected Topic) Average Temperature in Chilling Fish. Fish fillets having the same physical properties given in Problem 5.5-2 are originally at 10°C. They are packed to a thickness of 102 mm with ice on each side. Assuming that the surface temperature of the fillets is 0°C, calculate the time for the average temperature to reach 1.39°C. (*Note :* This is a case where the surface resistance is zero. Can Fig. 5.3-13 be used for this case?)

5.5-4. (Selected Topic) Time to Freeze a Slab of Meat. Repeat Example 5.5-2 using the same conditions except that a plate or contact freezer is used where the surface coefficient can be assumed as $h = 142$ W/m$^2 \cdot$ K.

Ans. $t = 2.00$ h

5.5-5. (Selected Topic) Freezing a Cylinder of Meat. A package of meat containing 75% moisture and in the form of a long cylinder 5 in. in diameter is to be frozen in an air-blast freezer at $-25°F$. The meat is initially at the freezing temperature of 27°F. The heat-transfer coefficient is $h = 3.5$ btu/h $\cdot$ ft$^2 \cdot$ °F. The physical properties are $\rho = 64$ lb$_m$/ft^3 for the unfrozen meat and $k = 0.60$ btu/h $\cdot$ ft $\cdot$ °F for the frozen meat. Calculate the freezing time.

5.6-1. (Selected Topic) Heat Generation Using Equation of Energy Change. A plane wall with uniform internal heat generation of $\dot{q}$ W/m^3 is insulated at four surfaces with heat conduction only in the x direction. The wall has a thickness of $2L$ m. The temperature at the one wall at $x = +L$ and at the other wall at $x = -L$ is held constant at T_w K. Using the differential equation of energy change, Eq. (5.6-18), derive the equation for the final temperature profile.

Ans. $T = \dfrac{\dot{q}(L^2 - x^2)}{2k} + T_w$

5.6-2. (Selected Topic) Heat Transfer in a Solid Using Equation of Energy Change. A solid of thickness L is at a uniform temperature of T_0 K. Suddenly the front surface temperature of the solid at $z = 0$ m is raised to T_1 at $t = 0$ and held there and at $z = L$ at the rear to T_2 and held constant. Heat transfer occurs only in the z direction. For constant physical properties and using the differential equation of energy change, do as follows.

(a) Derive the partial differential equation and the boundary conditions for unsteady-state energy transfer.

(b) Do the same for steady state and integrate the final equation.

Ans. (a) $\partial T/\partial t = \alpha\, \partial^2 T/\partial z^2$; B.C.(1): $t = 0$, $z = z$, $T = T_0$; B.C.(2): $t = t$, $z = 0$, $T = T_1$; B.C.(3): $t = t$, $z = L$, $T = T_2$; (b) $T = (T_2 - T_1)z/L + T_1$

5.6-3. (*Selected Topic*) *Radial Temperature Profile Using the Equation of Energy Change.* Radial heat transfer is occurring by conduction through a long hollow cylinder of length L with the ends insulated.
    (a) What is the final differential equation for steady-state conduction? Start with Fourier's second law in cylindrical coordinates, Eq. (5.6-20).
    (b) Solve the equation for the temperature profile from part (a) for the boundary conditions given as follows: $T = T_i$ for $r = r_i$, $T = T_o$ for $r = r_o$.
    (c) Using part (b), derive an expression for the heat flow q in W.

$$\textbf{Ans.}\quad \text{(b)}\quad T = T_i - \frac{T_i - T_o}{\ln(r_o/r_i)}\ln\frac{r}{r_i}$$

5.6-4. (*Selected Topic*) *Heat Conduction in a Sphere.* Radial energy flow is occurring in a hollow sphere with an inside radius of r_i and an outside radius of r_o. At steady state the inside surface temperature is constant at T_i and constant at T_o on the outside surface.
    (a) Using the differential equation of energy change, solve the equation for the temperature profile.
    (b) Using part (a), derive an expression for the heat flow in W.

5.6-5. (*Selected Topic*) *Variable Heat Generation and Equation of Energy Change.* A plane wall is insulated so that conduction occurs only in the x direction. The boundary conditions which apply at steady state are $T = T_0$ at $x = 0$ and $T = T_L$ at $x = L$. Internal heat generation per unit volume is occurring and varies as $\dot{q} = \dot{q}_0 e^{-\beta x/L}$, where $\dot{q}_0$ and β are constants. Solve the general differential equation of energy change for the temperature profile.

5.7-1. (*Selected Topic*) *Thermal and Hydrodynamic Boundary Layer Thicknesses.* Air at 294.3 K and 101.3 kPa with a free stream velocity of 12.2 m/s is flowing parallel to a smooth flat plate held at a surface temperature of 383 K. Do the following.
    (a) At the critical $N_{\text{Re},L} = 5 \times 10^5$, calculate the critical length $x = L$ of the plate, the thickness δ of the hydrodynamic boundary layer, and the thickness δ_T of the thermal boundary layer. Note that the Prandtl number is not 1.0.
    (b) Calculate the average heat-transfer coefficient over the plate covered by the laminar boundary layer.

5.7-2. (*Selected Topic*) *Boundary-Layer Thicknesses and Heat Transfer.* Air at 37.8°C and 1 atm abs flows at a velocity of 3.05 m/s parallel to a flat plate held at 93.3°C. The plate is 1 m wide. Calculate the following at a position 0.61 m from the leading edge.
    (a) The thermal boundary-layer thickness δ_T and the hydrodynamic boundary-layer thickness δ.
    (b) Total heat transfer from the plate.

REFERENCES

(B1)   BLAKEBROUGH, N. *Biochemical and Biological Engineering Science*, Vol. 2. New York: Academic Press, Inc., 1968.

(B2)   BIRD, R. B., STEWART, W. E., and LIGHTFOOT, E. N. *Transport Phenomena*. New York: John Wiley & Sons, Inc., 1960.

(C1)   CARSLAW, H. S., and JAEGER, J. C. *Conduction of Heat in Solids*. Oxford: Clarendon Press, 1959.

(C2)   CHARM, S. E. *The Fundamentals of Food Engineering*, 2nd ed. Westport, Conn.: Avi Publishing Co., Inc., 1971.

(D1)   DUSINBERRE, G. M. *Heat Transfer Calculations by Finite Differences*. Scranton, Pa.: International Textbook Co., Inc., 1961.

(G1)   GEANKOPLIS, C. J. *Mass Transport Phenomena.* Columbus, Ohio: Ohio State University Bookstores, 1972.

(G2)   GURNEY, H. P., and LURIE, J. *Ind. Eng. Chem.*, **15**, 1170 (1923).

(H1)   HEISLER, H. P. *Trans. A.S.M.E.*, **69**, 227 (1947).

(H2)   HODGSON, T. *Fd. Inds. S. Afr.*, **16**, 41 (1964); *Int. Inst. Refrig. Annexe*, **1966**, 633 (1966).

(K1)   KREITH, F. *Principles of Heat Transfer*, 2nd ed. Scranton, Pa.: International Textbook Company, 1965.

(K2)   KREITH, F., and BLACK, W. Z. *Basic Heat Transfer.* New York: Harper & Row, Publishers, 1980.

(N1)   NEWMAN, A. H. *Ind. Eng. Chem.*, **28**, 545 (1936).

(P1)   PERRY, R. H., and CHILTON, C. H. *Chemical Engineers' Handbook*, 5th ed. New York: McGraw-Hill Book Company, 1973.

(P2)   PLANK, R. *Z. Ges. Kalteind.*, **20**, 109 (1913); *Z. Ges. Kalteind. Bieh. Reih*, **10** (3), 1 (1941).

(R1)   RIEDEL, L. *Kaltetchnik*, **8**, 374 (1956); **9**, 38 (1957); **11**, 41 (1959); **12**, 4 (1960).

(S1)   SCHNEIDER, P. J. *Conduction Heat Transfer.* Reading, Mass.: Addison-Wesley Publishing Company, Inc., 1955.

(S2)   SIEGEL, R., SPARROW, E. M., and HALLMAN, T. M. *Appl. Sci. Res.*, **A7**, 386 (1958).

CHAPTER 6

Principles of
Mass Transfer

6.1 INTRODUCTION TO MASS TRANSFER AND DIFFUSION

6.1A Similarity of Mass, Heat, and Momentum Transfer Processes

1. Introduction. In Chapter 1 we noted that the various unit operations could be classified into three fundamental transfer (or "transport") processes: momentum transfer, heat transfer, and mass transfer. The fundamental process of *momentum transfer* occurs in such unit operations as fluid flow, mixing, sedimentation, and filtration. *Heat transfer* occurs in conductive and convective transfer of heat, evaporation, distillation, and drying.

The third fundamental transfer process, *mass transfer*, occurs in distillation, absorption, drying, liquid–liquid extraction, and membrane processes. When mass is being transferred from one distinct phase to another or through a single phase, the basic mechanisms are the same whether the phase is a gas, liquid, or solid. This was also shown in heat transfer, where the transfer of heat by conduction followed Fourier's law in a gas, solid, or liquid.

2. General molecular transport equation. All three of the molecular transport processes of momentum, heat, and mass are characterized by the same general type of equation given in Section 2.3A.

$$\text{rate of a transfer process} = \frac{\text{driving force}}{\text{resistance}} \qquad \text{(2.3-1)}$$

This can be written as follows for molecular diffusion of the property momentum, heat, and mass:

$$\psi_z = -\delta \frac{d\Gamma}{dz} \qquad \text{(2.3-2)}$$

3. Molecular diffusion equations for momentum, heat, and mass transfer. Newton's equation for momentum transfer for constant density can be written as follows in a manner similar to Eq. (2.3-2):

$$\tau_{zx} = -\frac{\mu}{\rho} \frac{d(v_x \rho)}{dz} \qquad \text{(6.1-1)}$$

where τ_{zx} is momentum transferred/s $\cdot$ m^2, μ/ρ is kinematic viscosity in m^2/s, z is distance in m, and $v_x \rho$ is momentum/m^3, where the momentum has units of kg $\cdot$ m/s.

Fourier's law for heat conduction can be written as follows for constant ρ and c_p:

$$\frac{q_z}{A} = -\alpha \frac{d(\rho c_p T)}{dz} \tag{6.1-2}$$

where q_z/A is heat flux in W/m^2, α is the thermal diffusivity in m^2/s, and $\rho c_p T$ is J/m^3.

The equation for molecular diffusion of mass is Fick's law and is also similar to Eq. (2.3-2). It is written as follows for constant total concentration in a fluid:

$$J_{Az}^* = -D_{AB} \frac{dc_A}{dz} \tag{6.1-3}$$

where J_{Az}^* is the molar flux of component A in the z direction due to molecular diffusion in kg mol A/s $\cdot$ m^2, D_{AB} the molecular diffusivity of the molecule A in B in m^2/s, c_A the concentration of A in kg mol/m^3, and z the distance of diffusion in m. In cgs units J_{Az}^* is g mol A/s $\cdot$ cm^2, D_{AB} is cm^2/s, and c_A is g mol A/cm^3. In English units, J_{Az}^* is lb mol/h $\cdot$ ft^2, D_{AB} is ft^2/h, and c_A is lb mol/ft^3.

The similarity of Eqs. (6.1-1), (6.1-2), and (6.1-3) for momentum, heat, and mass transfer should be obvious. All the fluxes on the left-hand side of the three equations have as units transfer of a quantity of momentum, heat, or mass per unit time per unit area. The transport properties μ/ρ, α, and D_{AB} all have units of m^2/s, and the concentrations are represented as momentum/m^3, J/m^3, or kg mol/m^3.

4. *Turbulent diffusion equations for momentum, heat, and mass transfer.* In Section 5.7C equations were given discussing the similarities among momentum, heat, and mass transfer in turbulent transfer. For turbulent momentum transfer and constant density,

$$\tau_{zx} = -\left(\frac{\mu}{\rho} + \varepsilon_t\right) \frac{d(v_x \rho)}{dz} \tag{6.1-4}$$

For turbulent heat transfer for constant ρ and c_p,

$$\frac{q_z}{A} = -(\alpha + \alpha_t) \frac{d(\rho c_p T)}{dz} \tag{6.1-5}$$

For turbulent mass transfer for constant c,

$$J_{Az}^* = -(D_{AB} + \varepsilon_M) \frac{dc_A}{dz} \tag{6.1-6}$$

In these equations ε_t is the turbulent or eddy momentum diffusivity in m^2/s, α_t the turbulent or eddy thermal diffusivity in m^2/s, and ε_M the turbulent or eddy mass diffusivity in m^2/s. Again, these equations are quite similar to each other. Many of the theoretical equations and empirical correlations for turbulent transport to various geometries are also quite similar.

6.1B   Examples of Mass-Transfer Processes

Mass transfer is important in many areas of science and engineering. Mass transfer occurs when a component in a mixture migrates in the same phase or from phase to phase because of a difference in concentration between two points. Many familiar phenomena involve mass transfer. Liquid in an open pail of water evaporates into still air because of the difference in concentration of water vapor at the water surface and the surrounding air. There is a "driving force" from the surface to the air. A piece of sugar added to a cup of coffee eventually dissolves by itself and diffuses to the surrounding solution. When newly cut and moist green timber is exposed to the atmosphere, the wood will dry partially when water in the timber diffuses through the wood, to the surface, and then to the atmosphere. In a fermentation process nutrients and oxygen dissolved in the solution diffuse to the microorganisms. In a catalytic reaction the reactants diffuse from the surrounding medium to the catalyst surface, where reaction occurs.

Many purification processes involve mass transfer. In uranium processing, a uranium salt in solution is extracted by an organic solvent. Distillation to separate alcohol from water involves mass transfer. Removal of SO_2 from flue gas is done by absorption in a basic liquid solution.

We can treat mass transfer in a manner somewhat similar to that used in heat transfer with Fourier's law of conduction. However, an important difference is that in molecular mass transfer one or more of the components of the medium is moving. In heat transfer by conduction the medium is usually stationary and only energy in the form of heat is being transported. This introduces some differences between heat and mass transfer that will be discussed in this chapter.

6.1C Fick's Law for Molecular Diffusion

Molecular diffusion or molecular transport can be defined as the transfer or movement of individual molecules through a fluid by means of the random, individual movements of the molecules. We can imagine the molecules traveling only in straight lines and changing direction by bouncing off other molecules after collisions. Since the molecules travel in a random path, molecular diffusion is often called a *random-walk process*.

In Fig. 6.1-1 the *molecular diffusion process* is shown schematically. A random path that molecule A might take in diffusing through B molecules from point (1) to (2) is shown. If there are a greater number of A molecules near point (1) than at (2), then, since molecules diffuse randomly in both directions, more A molecules will diffuse from (1) to (2) than from (2) to (1). The net diffusion of A is from high- to low-concentration regions.

As another example, a drop of blue liquid dye is added to a cup of water. The dye molecules will diffuse slowly by molecular diffusion to all parts of the water. To increase this rate of mixing of the dye, the liquid can be mechanically agitated by a spoon and *convective mass transfer* will occur. The two modes of heat transfer, conduction and convective heat transfer, are analogous to molecular diffusion and convective mass transfer.

First, we will consider the diffusion of molecules when the whole bulk fluid is not moving but is stationary. Diffusion of the molecules is due to a concentration gradient.

FIGURE 6.1-1. *Schematic diagram of molecular diffusion process.*

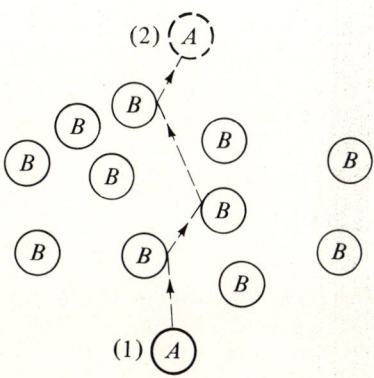

The general Fick's law equation can be written as follows for a binary mixture of A and B:

$$J_{Az}^* = -cD_{AB}\frac{dx_A}{dz} \tag{6.1-7}$$

where c is total concentration of A and B in kg mol $A + B/m^3$, and x_A is the mole fraction of A in the mixture of A and B. If c is constant, then since $c_A = cx_A$,

$$c\,dx_A = d(cx_A) = dc_A \tag{6.1-8}$$

Substituting into Eq. (6.1-7) we obtain Eq. (6.1-3) for constant total concentration.

$$J_{Az}^* = -D_{AB}\frac{dc_A}{dz} \tag{6.1-3}$$

This equation is the more commonly used one in many molecular diffusion processes. If c varies some, an average value is often used with Eq. (6.1-3).

EXAMPLE 6.1-1. *Molecular Diffusion of Helium in Nitrogen*

A mixture of He and N_2 gas is contained in a pipe at 298 K and 1 atm total pressure which is constant throughout. At one end of the pipe at point 1 the partial pressure p_{A1} of He is 0.60 atm and at the other end 0.2 m (20 cm) $p_{A2} = 0.20$ atm. Calculate the flux of He at steady state if D_{AB} of the He–N_2 mixture is 0.687×10^{-4} m²/s (0.687 cm²/s). Use SI and cgs units.

Solution: Since total pressure P is constant, then c is constant, where c is as follows for a gas from the perfect gas law.

$$PV = nRT \tag{6.1-9}$$

$$\frac{n}{V} = \frac{P}{RT} = c \tag{6.1-10}$$

where n is kg mol A plus B, V is volume in m³, T is temperature in K, R is 8314.3 m³ · Pa/kg mol · K or R is 82.057×10^{-3} m³ · atm/kg mol · K, and c is kg mol A plus B/m^3. In cgs units, R is 82.057 cm³ · atm/g mol · K.

For steady state the flux J_{Az}^* in Eq. (6.1-3) is constant. Also, D_{AB} for a gas is constant. Rearranging Eq. (6.1-3) and integrating,

$$J_{Az}^* \int_{z_1}^{z_2} dz = -D_{AB} \int_{c_{A1}}^{c_{A2}} dc_A$$

$$J_{Az}^* = \frac{D_{AB}(c_{A1} - c_{A2})}{z_2 - z_1} \tag{6.1-11}$$

Also, from the perfect gas law, $p_A V = n_A RT$, and

$$c_{A1} = \frac{p_{A1}}{RT} = \frac{n_A}{V} \tag{6.1-12}$$

Substituting Eq. (6.1-12) into (6.1-11),

$$J_{Az}^* = \frac{D_{AB}(p_{A1} - p_{A2})}{RT(z_2 - z_1)} \tag{6.1-13}$$

This is the final equation to use, which is in the form easily used for gases. Partial pressures are $p_{A1} = 0.6$ atm $= 0.6 \times 1.01325 \times 10^5 = 6.08 \times 10^4$ Pa and $p_{A2} = 0.2$ atm $= 0.2 \times 1.01325 \times 10^5 = 2.027 \times 10^4$ Pa. Then, using SI

units,
$$J_{Az}^* = \frac{(0.687 \times 10^{-4})(6.08 \times 10^4 - 2.027 \times 10^4)}{8314(298)(0.20 - 0)}$$

$$= 5.63 \times 10^{-6} \text{ kg mol } A/s \cdot m^2$$

If pressures in atm are used with SI units,

$$J_{Az}^* = \frac{(0.687 \times 10^{-4})(0.60 - 0.20)}{(82.06 \times 10^{-3})(298)(0.20 - 0)} = 5.63 \times 10^{-6} \text{ kg mol } A/s \cdot m^2$$

For cgs units, substituting into Eq. (6.1-13),

$$J_{Az}^* = \frac{0.687(0.60 - 0.20)}{82.06(298)(20 - 0)} = 5.63 \times 10^{-7} \text{ g mol } A/s \cdot cm^2$$

Other driving forces (besides concentration differences) for diffusion also occur because of temperature, pressure, electrical potential, and other gradients. Details are given elsewhere (B3).

6.1D   Convective Mass-Transfer Coefficient

When a fluid is flowing outside a solid surface in forced convection motion, we can express the rate of convective mass transfer from the surface to the fluid, or vice versa, by the following equation:

$$N_A = k_c (c_{L1} - c_{Li}) \qquad \text{(6.1-14)}$$

where k_c is a mass-transfer coefficient in m/s, c_{L1} the bulk fluid concentration in kg mol A/m^3, and c_{Li} the concentration in the fluid next to the surface of the solid. This mass-transfer coefficient is very similar to the heat-transfer coefficient h and is a function of the system geometry, fluid properties, and flow velocity. In Chapter 7 we consider convective mass transfer in detail.

6.2   MOLECULAR DIFFUSION IN GASES

6.2A   Equimolar Counterdiffusion in Gases

In Fig. 6.2-1 a diagram is given of two gases A and B at constant total pressure P in two large chambers connected by a tube where molecular diffusion at steady state is oc-

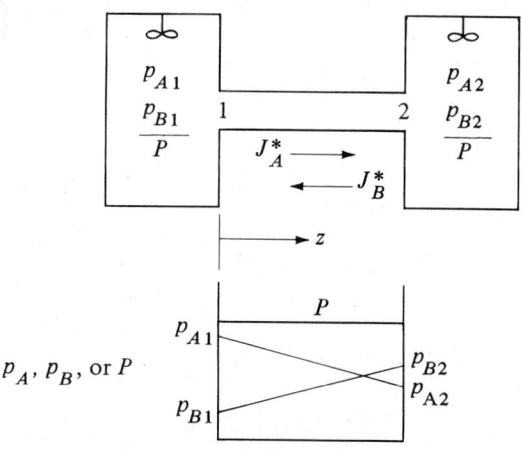

FIGURE 6.2-1.   *Equimolar counterdiffusion of gases A and B.*

curring. Stirring in each chamber keeps the concentrations in each chamber uniform. The partial pressure $p_{A1} > p_{A2}$ and $p_{B2} > p_{B1}$. Molecules of A diffuse to the right and B to the left. Since the total pressure P is constant throughout, the net moles of A diffusing to the right must equal the net moles of B to the left. If this is not so, the total pressure would not remain constant. This means that

$$J^*_{Az} = -J^*_{Bz} \tag{6.2-1}$$

The subscript z is often dropped when the direction is obvious. Writing Fick's law for B for constant c,

$$J^*_B = -D_{BA} \frac{dc_B}{dz} \tag{6.2-2}$$

Now since $P = p_A + p_B = $ constant, then

$$c = c_A + c_B \tag{6.2-3}$$

Differentiating both sides,

$$dc_A = -dc_B \tag{6.2-4}$$

Equating Eq. (6.1-3) to (6.2-2),

$$J^*_A = -D_{AB} \frac{dc_A}{dz} = -J^*_B = -(-)D_{BA} \frac{dc_B}{dz} \tag{6.2-5}$$

Substituting Eq. (6.2-4) into (6.2-5) and canceling like terms,

$$D_{AB} = D_{BA} \tag{6.2-6}$$

This shows that for a binary gas mixture of A and B the diffusivity coefficient D_{AB} for A diffusing in B is the same as D_{BA} for B diffusing into A.

EXAMPLE 6.2-1.   Equimolar Counterdiffusion
Ammonia gas (A) is diffusing through a uniform tube 0.10 m long containing N_2 gas (B) at 1.0132×10^5 Pa press and 298 K. The diagram is similar to Fig. 6.2-1. At point 1, $p_{A1} = 1.013 \times 10^4$ Pa and at point 2, $p_{A2} = 0.507 \times 10^4$ Pa. The diffusivity $D_{AB} = 0.230 \times 10^{-4}$ m²/s.
   (a) Calculate the flux J^*_A at steady state.
   (b) Repeat for J^*_B.

Solution: Equation (6.1-13) can be used where $P = 1.0132 \times 10^5$ Pa, $z_2 - z_1 = 0.10$ m, and $T = 298$ K. Substituting into Eq. (6.1-13) for part (a),

$$J^*_A = \frac{D_{AB}(p_{A1} - p_{A2})}{RT(z_2 - z_1)} = \frac{(0.23 \times 10^{-4})(1.013 \times 10^4 - 0.507 \times 10^4)}{8314(298)(0.10 - 0)}$$

$$= 4.70 \times 10^{-7} \text{ kg mol } A/\text{s} \cdot \text{m}^2$$

Rewriting Eq. (6.1-13) for component B for part (b) and noting that $p_{B1} = P - p_{A1} = 1.0132 \times 10^5 - 1.013 \times 10^4 = 9.119 \times 10^4$ Pa and $p_{B2} = P - p_{A2} = 1.0132 \times 10^5 - 0.507 \times 10^4 = 9.625 \times 10^4$ Pa,

$$J^*_B = \frac{D_{AB}(p_{B1} - p_{B2})}{RT(z_2 - z_1)} = \frac{(0.23 \times 10^{-4})(9.119 \times 10^4 - 9.625 \times 10^4)}{8314(298)(0.10 - 0)}$$

$$= -4.70 \times 10^{-7} \text{ kg mol } B/\text{s} \cdot \text{m}^2$$

The negative value for J^*_B means the flux goes from point 2 to 1.

Transport Processes: Momentum, Heat, and Mass

6.2B General Case for Diffusion of Gases A and B Plus Convection

Up to now we have considered Fick's law for diffusion in a stationary fluid; i.e., there has been no net movement or convective flow of the entire phase of the binary mixture A and B. The diffusion flux J_A^* occurred because of the concentration gradient. The rate at which moles of A passed a fixed point to the right, which will be taken as a positive flux, is J_A^* kg mol $A/\text{s} \cdot \text{m}^2$. This flux can be converted to a velocity of diffusion of A to the right by

$$J_A^* \ (\text{kg mol } A/\text{s} \cdot \text{m}^2) = v_{Ad} c_A \left(\frac{\text{m}}{\text{s}} \frac{\text{kg mol } A}{\text{m}^3} \right) \qquad (6.2\text{-}7)$$

where v_{Ad} is the diffusion velocity of A in m/s.

Now let us consider what happens when the whole fluid is moving in bulk or convective flow to the right. The molar average velocity of the whole fluid relative to a stationary point is v_M m/s. Component A is still diffusing to the right, but now its diffusion velocity v_{Ad} is measured relative to the moving fluid. To a stationary observer A is moving faster than the bulk of the phase, since its diffusion velocity v_{Ad} is added to that of the bulk phase v_M. Expressed mathematically, the velocity of A relative to the stationary point is the sum of the diffusion velocity and the average or convective velocity.

$$v_A = v_{Ad} + v_M \qquad (6.2\text{-}8)$$

where v_A is the velocity of A relative to a stationary point. Expressed pictorially,

Multiplying Eq. (6.2-8) by c_A,

$$c_A v_A = c_A v_{Ad} + c_A v_M \qquad (6.2\text{-}9)$$

Each of the three terms represents a flux. The first term, $c_A v_A$, can be represented by the flux N_A kg mol $A/\text{s} \cdot \text{m}^2$. This is the total flux of A relative to the stationary point. The second term is J_A^*, the diffusion flux relative to the moving fluid. The third term is the convective flux of A relative to the stationary point. Hence, Eq. (6.2-9) becomes

$$N_A = J_A^* + c_A v_M \qquad (6.2\text{-}10)$$

Let N be the total convective flux of the whole stream relative to the stationary point. Then,

$$N = c v_M = N_A + N_B \qquad (6.2\text{-}11)$$

Or, solving for v_M,

$$v_M = \frac{N_A + N_B}{c} \qquad (6.2\text{-}12)$$

Substituting Eq. (6.2-12) into (6.2-10),

$$N_A = J_A^* + \frac{c_A}{c} (N_A + N_B) \qquad (6.2\text{-}13)$$

Since J_A^* is Fick's law, Eq. (6.1-7),

$$N_A = -cD_{AB}\frac{dx_A}{dz} + \frac{c_A}{c}(N_A + N_B) \qquad (6.2\text{-}14)$$

Equation (6.2-14) is the final general equation for diffusion plus convection to use when the flux N_A is used, which is relative to a stationary point. A similar equation can be written for N_B.

$$N_B = -cD_{BA}\frac{dx_B}{dz} + \frac{c_B}{c}(N_A + N_B) \qquad (6.2\text{-}15)$$

To solve Eq. (6.2-14) or (6.2-15), the relation between the flux N_A and N_B must be known. Equations (6.2-14) and (6.2-15) hold for diffusion in a gas, liquid, or solid.

For equimolar counterdiffusion, $N_A = -N_B$ and the convective term in Eq. (6.2-14) becomes zero. Then, $N_A = J_A^* = -N_B = -J_B^*$.

6.2C Special Case for A Diffusing Through Stagnant, Nondiffusing B

The case of diffusion of A through stagnant or nondiffusing B at steady state often occurs. In this case one boundary at the end of the diffusion path is impermeable to component B, so it cannot pass through. One example shown in Fig. 6.2-2a is in evaporation of a pure liquid such as benzene (A) at the bottom of a narrow tube, where a large amount of inert or nondiffusing air (B) is passed over the top. The benzene vapor (A) diffuses through the air (B) in the tube. The boundary at the liquid surface at point 1 is impermeable to air, since air is insoluble in benzene liquid. Hence, air (B) cannot diffuse into or away from the surface. At point 2 the partial pressure $p_{A2} = 0$, since a large volume of air is passing by.

Another example shown in Fig. 6.2-2b occurs in the absorption of NH_3 (A) vapor which is in air (B) by water. The water surface is impermeable to the air, since air is only very slightly soluble in water. Thus, since B cannot diffuse, $N_B = 0$.

To derive the case for A diffusing in stagnant B, $N_B = 0$ is substituted into the general Eq. (6.2-14).

$$N_A = -cD_{AB}\frac{dx_A}{dz} + \frac{c_A}{c}(N_A + 0) \qquad (6.2\text{-}16)$$

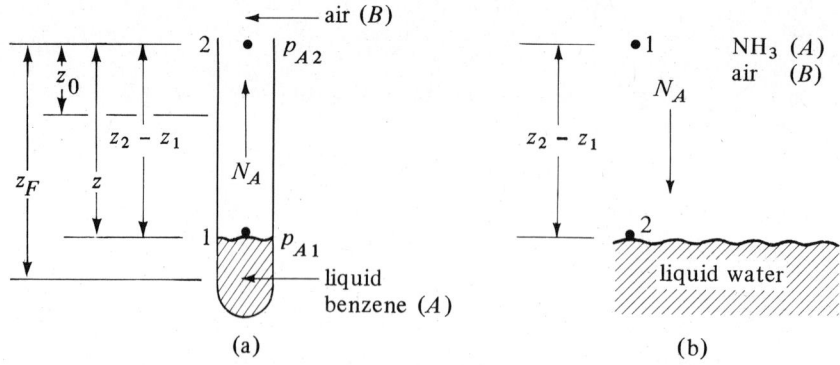

FIGURE 6.2-2. *Diffusion of A through stagnant B:* (a) *benzene evaporating into air,* (b) *ammonia in air being absorbed into water.*

Keeping the total pressure P constant, substituting $c = P/RT$, $p_A = x_A P$, and $c_A/c = p_A/P$ into Eq. (6.2-16),

$$N_A = -\frac{D_{AB}}{RT}\frac{dp_A}{dz} + \frac{p_A}{P}N_A \qquad (6.2\text{-}17)$$

Rearranging and integrating,

$$N_A\left(1 - \frac{p_A}{P}\right) = -\frac{D_{AB}}{RT}\frac{dp_A}{dz} \qquad (6.2\text{-}18)$$

$$N_A\int_{z_1}^{z_2} dz = -\frac{D_{AB}}{RT}\int_{p_{A1}}^{p_{A2}}\frac{dp_A}{1 - p_A/P} \qquad (6.2\text{-}19)$$

$$N_A = \frac{D_{AB}P}{RT(z_2 - z_1)}\ln\frac{P - p_{A2}}{P - p_{A1}} \qquad (6.2\text{-}20)$$

Equation (6.2-20) is the final equation to be used to calculate the flux of A. However, it is often written in another form as follows. A log mean value of the inert B is defined as follows. Since $P = p_{A1} + p_{B1} = p_{A2} + p_{B2}$, $p_{B1} = P - p_{A1}$, and $p_{B2} = P - p_{A2}$,

$$p_{BM} = \frac{p_{B2} - p_{B1}}{\ln(p_{B2}/p_{B1})} = \frac{p_{A1} - p_{A2}}{\ln[(P - p_{A2})/(P - p_{A1})]} \qquad (6.2\text{-}21)$$

Substituting Eq. (6.2-21) into (6.2-20),

$$N_A = \frac{D_{AB}P}{RT(z_2 - z_1)p_{BM}}(p_{A1} - p_{A2}) \qquad (6.2\text{-}22)$$

EXAMPLE 6.2-2. Diffusion of Water Through Stagnant, Nondiffusing Air
Water in the bottom of a narrow metal tube is held at a constant temperature of 293 K. The total pressure of air (assumed dry) is 1.01325×10^5 Pa (1.0 atm) and the temperature is 293 K (20°C). Water evaporates and diffuses through the air in the tube and the diffusion path $z_2 - z_1$ is 0.1524 m (0.5 ft) long. The diagram is similar to Fig. 6.2-2a. Calculate the rate of evaporation at steady state in lb mol/h·ft^2 and kg mol/s·m^2. The diffusivity of water vapor at 293 K and 1 atm pressure is 0.250×10^{-4} m^2/s. Assume that the system is isothermal. Use SI and English units.

Solution: The diffusivity is converted to ft^2/h by using the conversion factor from Appendix A.1.

$$D_{AB} = 0.250 \times 10^{-4}\,(3.875 \times 10^4) = 0.969\ \text{ft}^2/\text{h}$$

From Appendix A.2 the vapor pressure of water at 20°C is 17.54 mm or $p_{A1} = 17.54/760 = 0.0231$ atm $= 0.0231(1.01325 \times 10^5) = 2.341 \times 10^3$ Pa, $p_{A2} = 0$ (pure air). Since the temperature is 20°C (68°F), $T = 460 + 68 = 528$°R $= 293$ K. From Appendix A.1, $R = 0.730$ ft$^3 \cdot$ atm/lb mol·°R. To calculate the value of p_{BM} from Eq. (6.2-21),

$$p_{B1} = P - p_{A1} = 1.00 - 0.0231 = 0.9769\ \text{atm}$$

$$P_{B2} = P - p_{A2} = 1.00 - 0 = 1.00\ \text{atm}$$

$$p_{BM} = \frac{p_{B2} - p_{B1}}{\ln(p_{B2}/p_{B1})} = \frac{1.00 - 0.9769}{\ln(1.00/0.9769)} = 0.988\ \text{atm} = 1.001 \times 10^5\ \text{Pa}$$

Since p_{B1} is close to p_{B2}, the linear mean $(p_{B1} + p_{B2})/2$ could be used and would be very close to p_{BM}.

Substituting into Eq. (6.2-22) with $z_2 - z_1 = 0.5$ ft (0.1524 m),

$$N_A = \frac{D_{AB} P}{RT(z_2 - z_1) p_{BM}} (p_{A1} - p_{A2}) = \frac{0.969(1.0)(0.0231 - 0)}{0.730(528)(0.5)(0.988)}$$

$$= 1.175 \times 10^{-4} \text{ lb mol/h} \cdot \text{ft}^2$$

$$N_A = \frac{(0.250 \times 10^{-4})(1.01325 \times 10^5)(2.341 \times 10^3 - 0)}{8314(293)(0.1524)(1.001 \times 10^5)}$$

$$= 1.595 \times 10^{-7} \text{ kg mol/s} \cdot \text{m}^2$$

EXAMPLE 6.2-3. *Diffusion in a Tube with Change in Path Length*
Diffusion of water vapor in a narrow tube is occurring as in Example 6.2-2 under the same conditions. However, as shown in Fig. 6.2-2a, at a given time t, the level is z m from the top. As diffusion proceeds, the level drops slowly. Derive the equation for the time t_F for the level to drop from a starting point of z_0 m at $t = 0$ to z_F at $t = t_F$ s as shown.

Solution: We assume a pseudo-steady-state condition since the level drops very slowly. As time progresses, the path length z increases. At any time t, Eq. (6.2-22) holds; but the path length is z and Eq. (6.2-22) becomes as follows where N_A and z are now variables.

$$N_A = \frac{D_{AB} P}{RT z p_{BM}} (p_{A1} - p_{A2}) \tag{6.2-23}$$

Assuming a cross-sectional area of 1 m², the level drops dz m in dt s and $\rho_A (dz \cdot 1)/M_A$ is the kg mol of A that have left and diffused. Then,

$$N_A \cdot 1 = \frac{\rho_A (dz \cdot 1)}{M_A \, dt} \tag{6.2-24}$$

Equating Eq. (6.2-24) to (6.2-23), rearranging, and integrating between the limits of $z = z_0$ when $t = 0$ and $z = z_F$ when $t = t_F$,

$$\frac{\rho_A}{M_A} \int_{z_0}^{z_F} z \, dz = \frac{D_{AB} P(p_{A1} - p_{A2})}{RT p_{BM}} \int_0^{t_F} dt \tag{6.2-25}$$

Solving for t_F,

$$t_F = \frac{\rho_A (z_F^2 - z_0^2) RT p_{BM}}{2 M_A D_{AB} P(p_{A1} - p_{A2})} \tag{6.2-26}$$

The method shown in Example 6.2-3 has been used to experimentally determine the diffusivity D_{AB}. In this experiment the starting path length z_0 is measured at $t = 0$ and also the final z_F at t_F. Then Eq. (6.2-26) is used to calculate D_{AB}.

6.2D Diffusion Through a Varying Cross-Sectional Area

In the cases so far at steady state we have considered N_A and J_A^* as constants in the integrations. In these cases the cross-sectional area A m² through which the diffusion occurs has been constant with varying distance z. In some situations the area A may vary. Then it is convenient to define N_A as

$$N_A = \frac{\bar{N}_A}{A} \tag{6.2-27}$$

where $\bar{N}_A$ is kg moles of A diffusing per second or kg mol/s. At steady state, $\bar{N}_A$ will be constant but not A for a varying area.

Transport Processes: Momentum, Heat, and Mass

1. Diffusion from a sphere. To illustrate the use of Eq. (6.2-27), the important case of diffusion to or from a sphere in a gas will be considered. This situation appears often in such cases as the evaporation of a drop of liquid, the evaporation of a ball of naphthalene, and the diffusion of nutrients to a spherical-like microorganism in a liquid. In Fig. 6.2-3a is shown a sphere of fixed radius r_1 m in an infinite gas medium. Component (A) at partial pressure p_{A1} at the surface is diffusing into the surrounding stagnant medium (B), where $p_{A2} = 0$ at some large distance away. Steady-state diffusion will be assumed.

The flux N_A can be represented by Eq. (6.2-27), where A is the cross-sectional area $4\pi r^2$ at point r distance from the center of the sphere. Also, $\bar{N}_A$ is a constant at steady state.

$$N_A = \frac{\bar{N}_A}{4\pi r^2} \qquad (6.2\text{-}28)$$

Since this is a case of A diffusing through stagnant B, Eq. (6.2-18) will be used in its differential form and N_A will be equated to Eq. (6.2-28), giving

$$N_A = \frac{\bar{N}_A}{4\pi r^2} = -\frac{D_{AB}}{RT} \frac{dp_A}{(1 - p_A/P)\,dr} \qquad (6.2\text{-}29)$$

Note that dr was substituted for dz. Rearranging and integrating between r_1 and some point r_2 a large distance away,

$$\frac{\bar{N}_A}{4\pi} \int_{r_1}^{r_2} \frac{dr}{r^2} = -\frac{D_{AB}}{RT} \int_{p_{A1}}^{p_{A2}} \frac{dp_A}{(1 - p_A/P)} \qquad (6.2\text{-}30)$$

$$\frac{\bar{N}_A}{4\pi} \left(\frac{1}{r_1} - \frac{1}{r_2} \right) = \frac{D_{AB}\,P}{RT} \ln \frac{P - p_{A2}}{P - p_{A1}} \qquad (6.2\text{-}31)$$

Since $r_2 \gg r_1$, $1/r_2 \cong 0$. Substituting p_{BM} from Eq. (6.2-21) into Eq. (6.2-31),

$$\frac{\bar{N}_A}{4\pi r_1^2} = N_{A1} = \frac{D_{AB}\,P}{RT r_1} \frac{p_{A1} - p_{A2}}{p_{BM}} \qquad (6.2\text{-}32)$$

This equation can be simplified further. If p_{A1} is small compared to P (a dilute gas

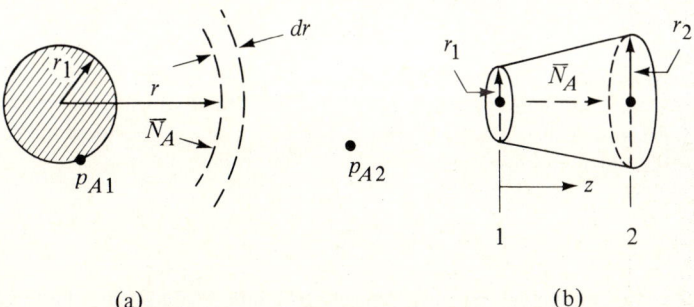

(a)   (b)

FIGURE 6.2-3. *Diffusion through a varying cross-sectional area: (a) from a sphere to a surrounding medium, (b) through a circular conduit that is tapered uniformally.*

phase), $p_{BM} \cong P$. Also, setting $2r_1 = D_1$, diameter, and $c_{A1} = p_{A1}/RT$, we obtain

$$N_{A1} = \frac{2D_{AB}}{D_1}(c_{A1} - c_{A2}) \tag{6.2-33}$$

This equation can also be used for liquids, where D_{AB} is the diffusivity of A in the liquid.

EXAMPLE 6.2-4. Evaporation of a Naphthalene Sphere

A sphere of naphthalene having a radius of 2.0 mm is suspended in a large volume of still air at 318 K and 1.01325×10^5 Pa (1 atm). The surface temperature of the naphthalene can be assumed to be at 318 K and its vapor pressure at 318 K is 0.555 mm Hg. The D_{AB} of naphthalene in air at 318 K is 6.92×10^{-6} m²/s. Calculate the rate of evaporation of naphthalene from the surface.

Solution: The flow diagram is similar to Fig. 6.2-3a. $D_{AB} = 6.92 \times 10^{-6}$ m²/s, $p_{A1} = (0.555/760)(1.01325 \times 10^5) = 74.0$ Pa, $p_{A2} = 0$, $r_1 = 2/1000$ m, $R = 8314$ m³ · Pa/kg mol · K, $p_{B1} = P - p_{A1} = 1.01325 \times 10^5 - 74.0 = 1.01251 \times 10^5$ Pa, $p_{B2} = P - p_{A2} = 1.01325 \times 10^5 - 0$. Since the values of p_{B1} and p_{B2} are close to each other,

$$p_{BM} = \frac{p_{B1} + p_{B2}}{2} = \frac{(1.0125 + 1.01325) \times 10^5}{2} = 1.0129 \times 10^5 \text{ Pa}$$

Substituting into Eq. (6.2-32),

$$N_{A1} = \frac{D_{AB}P(p_{A1} - p_{A2})}{RTr_1 p_{BM}} = \frac{(6.92 \times 10^{-6})(1.01325 \times 10^5)(74.0 - 0)}{8314(318)(2/1000)(1.0129 \times 10^5)}$$

$$= 9.68 \times 10^{-8} \text{ kg mol } A/\text{s} \cdot \text{m}^2$$

If the sphere in Fig. 6.2-3a is evaporating, the radius r of the sphere decreases slowly with time. The equation for the time for the sphere to evaporate completely can be derived by assuming pseudo-steady state and by equating the diffusion flux equation (6.2-32), where r is now a variable, to the moles of solid A evaporated per dt time and per unit area as calculated from a material balance. (See Problem 6.2-9 for this case.) The material-balance method is similar to Example 6.2-3. The final equation is

$$t_F = \frac{\rho_A r_1^2 RT p_{BM}}{2M_A D_{AB} P(p_{A1} - p_{A2})} \tag{6.2-34}$$

where r_1 is the original sphere radius, ρ_A the density of the sphere, and M_A the molecular weight.

2. Diffusion through a conduit of nonuniform cross-sectional area. In Fig. 6.2-3b component A is diffusion at steady state through a circular conduit which is tapered uniformly as shown. At point 1 the radius is r_1 and at point 2 it is r_2. At position z in the conduit, for A diffusing through stagnant B,

$$N_A = \frac{\bar{N}_A}{\pi r^2} = -\frac{D_{AB}}{RT} \frac{dp_A}{(1 - p_A/P) dz} \tag{6.2-35}$$

Using the geometry shown, the variable radius r can be related to position z in the path as follows:

$$r = \left(\frac{r_2 - r_1}{z_2 - z_1}\right)z + r_1 \tag{6.2-36}$$

Transport Processes: Momentum, Heat, and Mass

This value of r is then substituted into Eq. (6.2-35) to eliminate r and the equation integrated.

$$\frac{\bar{N}_A}{\pi} \int_{z_1}^{z_2} \frac{dz}{\left[\left(\dfrac{r_2 - r_1}{z_2 - z_1}\right)z + r_1\right]^2} = -\frac{D_{AB}}{RT} \int_{p_{A1}}^{p_{A2}} \frac{dp_A}{1 - pA/P} \tag{6.2-37}$$

A case similar to this is given in Problem 6.2-10.

6.2E   Diffusion Coefficients for Gases

1. Experimental determination of diffusion coefficients.   A number of different experimental methods have been used to determine the molecular diffusivity for binary gas mixtures. Several of the important methods are as follows. One method is to evaporate a pure liquid in a narrow tube with a gas passed over the top as shown in Fig. 6.2-2a. The fall in liquid level is measured with time and the diffusivity calculated from Eq. (6.2-26).

In another method, two pure gases having equal pressures are placed in separate sections of a long tube separated by a partition. The partition is slowly removed and diffusion proceeds. After a given time the partition is reinserted and the gas in each section analyzed. The diffusivities of the vapor of solids such as naphthalene, iodine, and benzoic acid in a gas have been obtained by measuring the rate of evaporation of a sphere. Equation (6.2-32) can be used. See Problem 6.2-9 for an example of this.

A useful method often used is the two-bulb method (N1). The apparatus consists of two glass bulbs with volumes V_1 and V_2 m^3 connected by a capillary of cross-sectional area A m^2 and length L whose volume is small compared to V_1 and V_2, as shown in Fig. 6.2-4. Pure gas A is added to V_1 and pure B to V_2 at the same pressures. The valve is opened, diffusion proceeds for a given time, and then the valve is closed and the mixed contents of each chamber are sampled separately.

The equations can be derived by neglecting the capillary volume and assuming each bulb is always of a uniform concentration. Assuming quasi-steady-state diffusion in the capillary,

$$J_A^* = -D_{AB}\frac{dc}{dz} = -\frac{D_{AB}(c_2 - c_1)}{L} \tag{6.2-38}$$

where c_2 is the concentration of A in V_2 at time t and c_1 in V_1. The rate of diffusion of A going to V_2 is equal to the rate of accumulation in V_2.

$$AJ_A^* = -\frac{D_{AB}(c_2 - c_1)A}{L} = V_2\frac{dc_2}{dt} \tag{6.2-39}$$

The average value c_{av} at equilibrium can be calculated by a material balance from the starting compositions c_1^0 and c_2^0 at $t = 0$.

$$(V_1 + V_2)c_{av} = V_1 c_1^0 + V_2 c_2^0 \tag{6.2-40}$$

FIGURE 6.2-4.   *Diffusivity   measurement   of gases by the two-bulb method.*

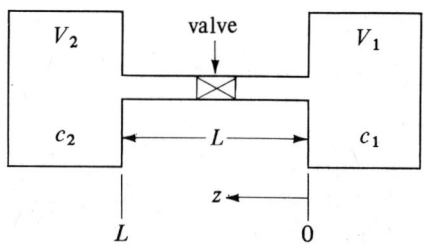

A similar balance at time t gives

$$(V_1 + V_2)c_{av} = V_1 c_1 + V_2 c_2 \qquad \text{(6.2-41)}$$

Substituting c_1 from Eq. (6.2-41) into (6.2-39), rearranging, and integrating between $t = 0$ and $t = t$, the final equation is

$$\frac{c_{av} - c_2}{c_{av} - c_2^0} = \exp\left[-\frac{D_{AB}(V_1 + V_2)}{(L/A)(V_2 V_1)} t \right] \qquad \text{(6.2-42)}$$

If c_2 is obtained by sampling at t, D_{AB} can be calculated.

2. *Experimental diffusivity data.* Some typical data are given in Table 6.2-1. Other data are tabulated in Perry and Chilton (P1) and Reid et al. (R1). The values range from about 0.05×10^{-4} m^2/s, where a large molecule is present, to about 1.0×10^{-4} m^2/s, where H_2 is present at room temperatures. The relation between diffusivity in m^2/s and ft^2/h is 1 m^2/s = 3.875×10^4 ft^2/h.

3. *Prediction of diffusivity for gases.* The diffusivity of a binary gas mixture in the dilute gas region, i.e., at low pressures near atmospheric, can be predicted using the kinetic theory of gases. The gas is assumed to consist of rigid spherical particles that are completely elastic on collision with another molecule, which implies that momentum is conserved.

In a simplified treatment it is assumed that there are no attractive or repulsive forces between the molecules. The derivation uses the mean free path λ, which is the average distance that a molecule has traveled between collisions. The final equation is

$$D_{AB} = \tfrac{1}{3}\bar{u}\lambda \qquad \text{(6.2-43)}$$

where $\bar{u}$ is the average velocity of the molecules. The final equation obtained after substituting expressions for $\bar{u}$ and λ into Eq. (6.2-43) is approximately correct, since it correctly predicts D_{AB} proportional to 1/pressure and approximately predicts the temperature effect.

A more accurate and rigorous treatment must consider the intermolecular forces of attraction and repulsion between molecules and also the different sizes of molecules A and B. Chapman and Enskog (H3) solved the Boltzmann equation, which does not utilize the mean free path λ but uses a distribution function. To solve the equation, a relation between the attractive and repulsive forces between a given pair of molecules must be used. For a pair of nonpolar molecules a reasonable approximation to the forces is the Lennard–Jones function.

The final relation to predict the diffusivity of a binary gas pair of A and B molecules is

$$D_{AB} = \frac{1.8583 \times 10^{-7} T^{3/2}}{P\sigma_{AB}^2 \Omega_{D,\,AB}} \left(\frac{1}{M_A} + \frac{1}{M_B} \right)^{1/2} \qquad \text{(6.2-44)}$$

where D_{AB} is the diffusivity in m^2/s, T temperature in K, M_A molecular weight of A in kg mass/kg mol, M_B molecular weight of B, and P absolute pressure in atm. The term σ_{AB} is an "average collision diameter" and $\Omega_{D,\,AB}$ is a collision integral based on the Lennard–Jones potential. Values of σ_A and σ_B and $\Omega_{D,\,AB}$ can be obtained from a number of sources (B3, G2, H3, P1, R1).

The collision integral $\Omega_{D,\,AB}$ is a ratio giving the deviation of a gas with interactions compared to a gas of rigid, elastic spheres. This value would be 1.0 for a gas with no interactions. Equation (6.2-44) predicts diffusivities with an average deviation of about

Transport Processes: Momentum, Heat, and Mass

TABLE 6.2-1. *Diffusion Coefficients of Gases at 101.32 kPa Pressure*

System	Temperature °C	K	Diffusivity $[(m^2/s)10^4$ or $cm^2/s]$	Ref.
Air–NH_3	0	273	0.198	(W1)
Air–H_2O	0	273	0.220	(N2)
	25	298	0.260	(L1)
	42	315	0.288	(M1)
Air–CO_2	3	276	0.142	(H1)
	44	317	0.177	
Air–H_2	0	273	0.611	(N2)
Air–C_2H_5OH	25	298	0.135	(M1)
	42	315	0.145	
Air–CH_3COOH	0	273	0.106	(N2)
Air–*n*-hexane	21	294	0.080	(C1)
Air–benzene	25	298	0.0962	(L1)
Air–toluene	25.9	298.9	0.086	(G1)
Air–*n*-butanol	0	273	0.0703	(N2)
	25.9	298.9	0.087	
H_2–CH_4	25	298	0.726	(C2)
H_2–N_2	25	298	0.784	(B1)
	85	358	1.052	
H_2–benzene	38.1	311.1	0.404	(H2)
H_2–Ar	22.4	295.4	0.83	(W2)
H_2–NH_3	25	298	0.783	(B1)
H_2–SO_2	50	323	0.61	(S1)
H_2–C_2H_5OH	67	340	0.586	(T1)
He–Ar	25	298	0.729	(S2)
He–*n*-butanol	150	423	0.587	(S2)
He–air	44	317	0.765	(H1)
He–CH_4	25	298	0.675	(C2)
He–N_2	25	298	0.687	(S2)
He–O_2	25	298	0.729	(S2)
Ar–CH_4	25	298	0.202	(C2)
CO_2–N_2	25	298	0.167	(W3)
CO_2–O_2	20	293	0.153	(W4)
N_2–*n*-butane	25	298	0.0960	(B2)
H_2O–CO_2	34.3	307.3	0.202	(S3)
CO–N_2	100	373	0.318	(A1)
CH_3Cl–SO_2	30	303	0.0693	(C3)
$(C_2H_5)_2O$–NH_3	26.5	299.5	0.1078	(S4)

8% up to about 1000 K (R1). For a polar–nonpolar gas mixture Eq. (6.2-44) can be used if the correct force constant is used for the polar gas (M1, M2). For polar–polar gas pairs the potential-energy function commonly used is the *Stockmayer potential* (M2).

The effect of concentration of *A* in *B* in Eq. (6.2-44) is not included. However, for real gases with interactions the maximum effect of concentration on diffusivity is about 4% (G2). In most cases the effect is considerably less, and hence it is usually neglected.

Equation (6.2-44) is relatively complicated to use and often some of the constants such as σ_{AB} are not available or difficult to estimate. Hence, the semiempirical method of Fuller et al. (F1), which is much more convenient to use, is often utilized. The equation was obtained by correlating many recent data and uses atomic volumes from Table 6.2-2, which are summed for each gas molecule. The equation is

$$D_{AB} = \frac{1.00 \times 10^{-7} T^{1.75} (1/M_A + 1/M_B)^{1/2}}{P[(\sum v_A)^{1/3} + (\sum v_B)^{1/3}]^2} \tag{6.2-45}$$

where $\sum v_A$ = sum of structural volume increments, Table 6.2-2, and D_{AB} = m²/s. This method can be used for mixtures of nonpolar gases or for a polar–nonpolar mixture. Its accuracy is not quite as good as that of Eq. (6.2-44).

TABLE 6.2-2. *Atomic Diffusion Volumes for Use with the Fuller, Schettler, and Giddings Method**

Atomic and structural diffusion volume increments, v

C	16.5	(Cl)	19.5
H	1.98	(S)	17.0
O	5.48	Aromatic ring	−20.2
(N)	5.69	Heterocyclic ring	−20.2

Diffusion volumes for simple molecules, $\sum v$

H_2	7.07	CO	18.9
D_2	6.70	CO_2	26.9
He	2.88	N_2O	35.9
N_2	17.9	NH_3	14.9
O_2	16.6	H_2O	12.7
Air	20.1	(CCl_2F_2)	114.8
Ar	16.1	(SF_6)	69.7
Kr	22.8	(Cl_2)	37.7
(Xe)	37.9	(Br_2)	67.2
Ne	5.59	(SO_2)	41.1

* Parentheses indicate that the value listed is based on only a few data points.
Source: Reprinted with permission from E. N. Fuller, P. D. Schettler, and J. C. Giddings, *Ind. Eng. Chem.*, **58**, 19(1966). Copyright by the American Chemical Society.

The equation shows that D_{AB} is proportional to $1/P$ and to $T^{1.75}$. If an experimental value of D_{AB} is available at a given T and P and it is desired to have a value of D_{AB} at another T and P, one should correct the experimental value to the new T and P by the relationship $D_{AB} \propto T^{1.75}/P$.

4. Schmidt number of gases. The *Schmidt number* of a gas mixture of dilute A in B is dimensionless and is defined as

$$N_{Sc} = \frac{\mu}{\rho D_{AB}} \tag{6.2-46}$$

where μ is viscosity of the gas mixture, which is viscosity of B for a dilute mixture in Pa · s or kg/m · s, D_{AB} is diffusivity in m²/s, and ρ is the density of the mixture in kg/m³. For a gas the Schmidt number can be assumed independent of temperature over moderate ranges and independent of pressure up to about 10 atm or 10×10^5 Pa.

The Schmidt number is the dimensionless ratio of the molecular momentum diffusivity μ/ρ to the molecular mass diffusivity D_{AB}. Values of the Schmidt number for gases range from about 0.5 to 2. For liquids Schmidt numbers range from about 100 to over 10 000 for viscous liquids.

EXAMPLE 6.2-5. *Estimation of Diffusivity of a Gas Mixture*

Normal butanol (A) is diffusing through air (B) at 1 atm abs. Using the Fuller et al. method, estimate the diffusivity D_{AB} for the following temperatures and compare with the experimental data.

   (a) For 0°C.
   (b) For 25.9°C.
   (c) For 0°C and 2.0 atm abs.

Solution: For part (a), $P = 1.00$ atm, $T = 273 + 0 = 273$ K, M_A (butanol) = 74.1, M_B (air) = 29. From Table 6.2-2,

$$\sum v_A = 4(16.5) + 10(1.98) + 1(5.48) = 91.28 \text{ (butanol)}$$

$$\sum v_B = 20.1 \text{ (air)}$$

Substituting into Eq. (6.2-45),

$$D_{AB} = \frac{1.0 \times 10^{-7}(273)^{1.75}(1/74.1 + 1/29)^{1/2}}{1.0[(91.28)^{1/3} + (20.1)^{1/3}]^2}$$

$$= 7.73 \times 10^{-6} \text{ m}^2/\text{s}$$

This value deviates by +10% from the experimental value of 7.03×10^{-6} m²/s from Table 6.2-1.

For part (b), $T = 273 + 25.9 = 298.9$. Substituting into Eq. (6.2-45), $D_{AB} = 9.05 \times 10^{-6}$ m²/s. This value deviates by +4% from the experimental of 8.70×10^{-6} m²/s.

For part (c), the total pressure $P = 2.0$ atm. Using the value predicted in part (a) and correcting for pressure,

$$D_{AB} = 7.73 \times 10^{-6}(1.0/2.0) = 3.865 \times 10^{-6} \text{ m}^2/\text{s}$$

6.3 MOLECULAR DIFFUSION IN LIQUIDS

6.3A Introduction

Diffusion of solutes in liquids is very important in many industrial processes, especially in such separation operations as liquid–liquid extraction or solvent extraction, gas absorption, and distillation. Diffusion in liquids also occurs in many situations in nature, such as oxygenation of rivers and lakes by the air and diffusion of salts in blood.

It should be apparent that the rate of molecular diffusion in liquids is considerably slower than in gases. The molecules in a liquid are very close together compared to a gas. Hence, the molecules of the diffusing solute A will collide with molecules of liquid B more often and diffuse more slowly than in gases. In general, the diffusion coefficient in a gas will be of the order of magnitude of about 10^5 times greater than in a liquid. However, the flux in a gas is not that much greater, being only about 100 times faster, since the concentrations in liquids are considerably higher than in gases.

6.3B Equations for Diffusion in Liquids

Since the molecules in a liquid are packed together much more closely than in gases, the density and the resistance to diffusion in a liquid are much greater. Also, because of this closer spacing of the molecules, the attractive forces between molecules play an important role in diffusion. Since the kinetic theory of liquids is only partially developed, we write the equations for diffusion in liquids similar to those for gases.

In diffusion in liquids an important difference from diffusion in gases is that the diffusivities are often quite dependent on the concentration of the diffusing components.

1. Equimolar counterdiffusion. Starting with the general equation (6.2-14), we can obtain for equimolal counterdiffusion where $N_A = -N_B$, an equation similar to Eq. (6.1-11) for gases at steady state.

$$N_A = \frac{D_{AB}(c_{A1} - c_{A2})}{z_2 - z_1} = \frac{D_{AB}c_{av}(x_{A1} - x_{A2})}{z_2 - z_1} \tag{6.3-1}$$

where N_A is the flux of A in kg mol $A/s \cdot m^2$, D_{AB} the diffusivity of A in B in m^2/s, c_{A1} the concentration of A in kg mol A/m^3 at point 1, x_{A1} the mole fraction of A at point 1, and c_{av} defined by

$$c_{av} = \left(\frac{\rho}{M}\right)_{av} = \left(\frac{\rho_1}{M_1} + \frac{\rho_2}{M_2}\right)\bigg/2 \tag{6.3-2}$$

where c_{av} is the average total concentration of $A + B$ in kg mol/m^3, M_1 the average molecular weight of the solution at point 1 in kg mass/kg mol, and ρ_1 the average density of the solution in kg/m^3 at point 1.

Equation (6.3-1) uses the average value of D_{AB}, which may vary some with concentration and the average value of c, which also may vary with concentration. Usually, the linear average of c is used as in Eq. (6.3-2). The case of equimolar counterdiffusion in Eq. (6.3-1) occurs only very infrequently in liquids.

2. Diffusion of A through nondiffusing B. The most important case of diffusion in liquids is that where solute A is diffusing and solvent B is stagnant or nondiffusing. An example is a dilute solution of propionic acid (A) in a water (B) solution being contacted with toluene. Only the propionic acid (A) diffuses through the water phase, to the boundary, and then into the toluene phase. The toluene–water interface is a barrier to diffusion of B and $N_B = 0$. Such cases often occur in industry (T2). If Eq. (6.2-22) is rewritten in terms of concentrations by substituting $c_{av} = P/RT$, $c_{A1} = p_{A1}/RT$, and $x_{BM} = p_{BM}/P$, we obtain the equation for liquids at steady state.

$$N_A = \frac{D_{AB}c_{av}}{(z_2 - z_1)x_{BM}}(x_{A1} - x_{A2}) \tag{6.3-3}$$

where

$$x_{BM} = \frac{x_{B2} - x_{B1}}{\ln(x_{B2}/x_{B1})} \tag{6.3-4}$$

Note that $x_{A1} + x_{B1} = x_{A2} + x_{B2} = 1.0$. For dilute solutions x_{BM} is close to 1.0 and c is essentially constant. Then Eq. (6.3-3) simplifies to

$$N_A = \frac{D_{AB}(c_{A1} - c_{A2})}{z_2 - z_1} \tag{6.3-5}$$

Transport Processes: Momentum, Heat, and Mass

EXAMPLE 6.3-1. Diffusion of Ethanol (A) Through Water (B)

An ethanol (A)–water (B) solution in the form of a stagnant film 2.0 mm thick at 293 K is in contact at one surface with an organic solvent in which ethanol is soluble and water is insoluble. Hence, $N_B = 0$. At point 1 the concentration of ethanol is 16.8 wt % and the solution density is $\rho_1 = 972.8$ kg/m³. At point 2 the concentration of ethanol is 6.8 wt % and $\rho_2 = 988.1$ kg/m³ (P1). The diffusivity of ethanol is 0.740×10^{-9} m²/s (T2). Calculate the steady-state flux N_A.

Solution: The diffusivity is $D_{AB} = 0.740 \times 10^{-9}$ m²/s. The molecular weights of A and B are $M_A = 46.05$ and $M_B = 18.02$. For a wt % of 6.8, the mole fraction of ethanol (A) is as follows when using 100 kg of solution.

$$x_{A2} = \frac{6.8/46.05}{6.8/46.05 + 93.2/18.02} = \frac{0.1477}{0.1477 + 5.17} = 0.0277$$

Then $x_{B2} = 1 - 0.0277 = 0.9723$. Calculating x_{A1} in a similar manner, $x_{A1} = 0.0732$ and $x_{B1} = 1 - 0.0732 = 0.9268$. To calculate the molecular weight M_2 at point 2.

$$M_2 = \frac{100 \text{ kg}}{(0.1477 + 5.17) \text{ kg mol}} = 18.75 \text{ kg/kg mol}$$

Similarly, $M_1 = 20.07$. From Eq. (6.3-2),

$$c_{av} = \frac{\rho_1/M_1 + \rho_2/M_2}{2} = \frac{972.8/20.07 + 988.1/18.75}{2} = 50.6 \text{ kg mol/m}^3$$

To calculate x_{BM} from Eq. (6.3-4), we can use the linear mean since x_{B1} and x_{B2} are close to each other.

$$x_{BM} = \frac{x_{B1} + x_{B2}}{2} = \frac{0.9268 + 0.9723}{2} = 0.949$$

Substituting into Eq. (6.3-3) and solving,

$$N_A = \frac{D_{AB} c_{av}}{(z_2 - z_1) x_{BM}} (x_{A1} - x_{A2}) = \frac{(0.740 \times 10^{-9})(50.6)(0.0732 - 0.0277)}{(2/1000)0.949}$$

$$= 8.99 \times 10^{-7} \text{ kg mol/s} \cdot \text{m}^2$$

6.3C   Diffusion Coefficients for Liquids

1. Experimental determination of diffusivities. Several different methods are used to determine diffusion coefficients experimentally in liquids. In one method unsteady-state diffusion in a long capillary tube is carried out and the diffusivity determined from the concentration profile. If the solute A is diffusing in B, the diffusion coefficient determined is D_{AB}. Also, the value of diffusivity is often very dependent upon the concentration of the diffusing solute A. Unlike gases, the diffusivity D_{AB} does not equal D_{BA} for liquids.

In a relatively common method a relatively dilute solution and a slightly more concentrated solution are placed in chambers on opposite sides of a porous membrane of sintered glass as shown in Fig. 6.3-1. Molecular diffusion takes place through the narrow passageways of the pores in the sintered glass while the two compartments are stirred. The effective diffusion length is $K_1 \delta$, where $K_1 > 1$ is a constant and corrects for the fact the path is actually greater than δ cm. In this method, discussed by Bidstrup and Geankoplis (B4), the effective diffusion length is obtained by calibrating with a solute such as KCl having a known diffusivity.

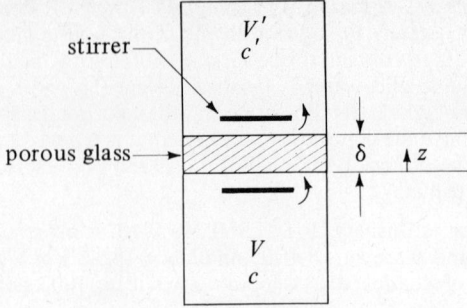

FIGURE 6.3-1.   *Diffusion cell for determination of diffusivity in a liquid.*

To derive the equation, quasi-steady-state diffusion in the membrane is assumed.

$$N_A = \varepsilon D_{AB} \frac{c - c'}{K_1 \delta} \tag{6.3-6}$$

where c is the concentration in the lower chamber at a time t, c' is the concentration in the upper, and ε is the fraction of area of the glass open to diffusion. Making a balance on solute A in the upper chamber, where the rate in = rate out + rate of accumulation, making a similar balance on the lower chamber, using volume $V = V'$, and combining and integrating the final equation is

$$\ln \frac{c_0 - c_0'}{c - c'} = \frac{2\varepsilon A}{K_1 \delta V} D_{AB} t \tag{6.3-7}$$

where $2\varepsilon A / K_1 \delta V$ is a cell constant that can be determined using a solute of known diffusivity, such as KCl. The values c_0 and c_0' are initial concentrations and c and c' final concentrations.

2. Experimental liquid diffusivity data.   Experimental diffusivity data for binary mixtures in the liquid phase are given in Table 6.3-1. All the data are for dilute solutions of the diffusing solute in the solvent. In liquids the diffusivities often vary quite markedly with concentration. Hence, the values in Table 6.3-1 should be used with some caution when outside the dilute range. Values for biological solutes in solution are given in the next section. As noted in the table, the diffusivity values are quite small and in the range of about 0.5×10^{-9} to 5×10^{-9} m^2/s for relatively nonviscous liquids. Diffusivities in gases are larger by a factor of 10^4 to 10^5.

6.3D   Prediction of Diffusivities in Liquids

The equations for predicting diffusivities of dilute solutes in liquids are by necessity semiempirical, since the theory for diffusion in liquids is not well established as yet. The Stokes–Einstein equation, one of the first theories, was derived for a very large spherical molecule (A) diffusing in a liquid solvent (B) of small molecules. Stokes' law was used to describe the drag on the moving solute molecule. Then the equation was modified by assuming that all molecules are alike and arranged in a cubic lattice and by expressing the molecular radius in terms of the molar volume (W5),

$$D_{AB} = \frac{9.96 \times 10^{-16} T}{\mu V_A^{1/3}} \tag{6.3-8}$$

TABLE 6.3-1. *Diffusion Coefficients for Dilute Liquid Solutions*

Solute	Solvent	Temperature °C	Temperature K	Diffusivity [$(m^2/s)10^9$ or $(cm^2/s)10^5$]	Ref.
NH_3	Water	12	285	1.64	(N2)
		15	288	1.77	
O_2	Water	18	291	1.98	(N2)
		25	298	2.41	(V1)
CO_2	Water	25	298	2.00	(V1)
H_2	Water	25	298	4.8	(V1)
Methyl alcohol	Water	15	288	1.26	(J1)
Ethyl alcohol	Water	10	283	0.84	(J1)
		25	298	1.24	(J1)
n-Propyl alcohol	Water	15	288	0.87	(J1)
Formic acid	Water	25	298	1.52	(B4)
Acetic acid	Water	9.7	282.7	0.769	(B4)
		25	298	1.26	(B4)
Propionic acid	Water	25	298	1.01	(B4)
HCl (9 g mol/liter)	Water	10	283	3.3	(N2)
(2.5 g mol/liter)		10	283	2.5	(N2)
Benzoic acid	Water	25	298	1.21	(C4)
Acetone	Water	25	298	1.28	(A2)
Acetic acid	Benzene	25	298	2.09	(C5)
Urea	Ethanol	12	285	0.54	(N2)
Water	Ethanol	25	298	1.13	(H4)
KCl	Water	25	298	1.870	(P2)
KCl	Ethylene glycol	25	298	0.119	(P2)

where D_{AB} is diffusivity in m^2/s, T is temperature in K, μ is viscosity of solution in Pa·s or kg/m·s, and V_A is the solute molar volume at its normal boiling point in m^3/kg mol. This equation applies very well to very large unhydrated solute molecules of about 1000 molecular weight or greater (R1), or where the V_A is above about 0.500 m^3/kg mol (W5) in aqueous solution.

For smaller solute molar volumes, Eq. (6.3-8) does not hold. Several other theoretical derivations have been attempted, but the equations do not predict diffusivities very accurately. Hence, a number of semitheoretical expressions have been developed (R1). The Wilke–Chang (T3, W5) correlation can be used for most general purposes where the solute (A) is dilute in the solvent (B).

$$D_{AB} = 1.173 \times 10^{-16}(\varphi M_B)^{1/2} \frac{T}{\mu_B V_A^{0.6}} \tag{6.3-9}$$

where M_B is the molecular weight of solvent B, μ_B is the viscosity of B in Pa·s or kg/m·s, V_A is the solute molar volume at the boiling point (L2) that can be obtained from Table 6.3-2, and φ is an "association parameter" of the solvent, where φ is 2.6 for water, 1.9 methanol, 1.5 ethanol, 1.0 benzene, 1.0 ether, 1.0 heptane, and 1.0 other unassociated solvents. When values of V_A are above 0.500 m^3/kg mol (500 cm^3/g mol), Eq. (6.3-8) should be used.

TABLE 6.3-2. *Atomic and Molar Volumes at the Normal Boiling Point*

Material	Atomic Volume (m^3/kg mol) 10^3	Material	Atomic Volume (m^3/kg mol) 10^3
C	14.8	Ring, 3-membered	−6
H	3.7	as in ethylene	
O (except as below)	7.4	oxide	
Doubly bound as	7.4	4-membered	−8.5
carbonyl		5-membered	−11.5
Coupled to two		6-membered	−15
other elements		Naphthalene ring	−30
In aldehydes, ketones	7.4	Anthracene ring	−47.5
In methyl esters	9.1		
In methyl ethers	9.9		
In ethyl esters	9.9		*Molecular Volume*
In ethyl ethers	9.9		(m^3/kg mol) 10^3
In higher esters	11.0		
In higher ethers	11.0	Air	29.9
In acids (—OH)	12.0	O_2	25.6
Joined to S, P, N	8.3	N_2	31.2
N		Br_2	53.2
Doubly bonded	15.6	Cl_2	48.4
In primary amines	10.5	CO	30.7
In secondary amines	12.0	CO_2	34.0
Br	27.0	H_2	14.3
Cl in RCHClR′	24.6	H_2O	18.8
Cl in RCl (terminal)	21.6	H_2S	32.9
F	8.7	NH_3	25.8
I	37.0	NO	23.6
S	25.6	N_2O	36.4
P	27.0	SO_2	44.8

Source : G. Le Bas, *The Molecular Volumes of Liquid Chemical Compounds.* New York: David McKay Co., Inc., 1915.

When water is the solute, values from Eq. (6.3-9) should be multiplied by a factor of 1/2.3 (R1). Equation (6.3-9) predicts diffusivities with a mean deviation of 10–15% for aqueous solutions and about 25% in nonaqueous solutions. Outside the range 278–313 K, the equation should be used with caution. For water as the diffusing solute, an equation by Reddy and Doraiswamy is preferred (R2). Skelland (S5) summarizes the correlations available for binary systems. Geankoplis (G2) discusses and gives an equation to predict diffusion in a ternary system, where a dilute solute A is diffusing in a mixture of B and C solvents. This case is often approximated in industrial processes.

EXAMPLE 6.3-2. Prediction of Liquid Diffusivity
Predict the diffusion coefficient of acetone (CH_3COCH_3) in water at 25° and 50°C using the Wilke–Chang equation. The experimental value is 1.28 × 10^{-9} m^2/s at 25°C (298 K).

Solution: From Appendix A.2 the viscosity of water at 25.0°C is $\mu_B = 0.8937 \times 10^{-3}$ Pa·s and at 50°C, 0.5494×10^{-3}. From Table 6.3-2 for

CH_3COCH_3 with 3 carbons + 6 hydrogens + 1 oxygen,

$$V_A = 3(0.0148) + 6(0.0037) + 1(0.0074) = 0.0740 \ \text{m}^3/\text{kg mol}$$

For water the association parameter $\varphi = 2.6$ and $M_B = 18.02$ kg mass/kg mol. For 25°C, $T = 298$ K. Substituting into Eq. (6.3-9),

$$D_{AB} = (1.173 \times 10^{-16})(\varphi M_B)^{1/2} \frac{T}{\mu_B V_A^{0.6}}$$

$$= \frac{(1.173 \times 10^{-16})(2.6 \times 18.02)^{1/2}(298)}{(0.8937 \times 10^{-3})(0.0740)^{0.6}}$$

$$= 1.277 \times 10^{-9} \ \text{m}^2/\text{s}$$

For 50°C or $T = 323$ K,

$$D_{AB} = \frac{(1.173 \times 10^{-16})(2.6 \times 18.02)^{1/2}(323)}{(0.5494 \times 10^{-3})(0.0740)^{0.6}}$$

$$= 2.251 \times 10^{-9} \ \text{m}^2/\text{s}$$

Electrolytes in solution such as KCl dissociate into cations and anions and diffuse more rapidly than the undissociated molecule because of their small size. Diffusion coefficients can be estimated using ionic conductance at infinite dilution in water. Equations and data are given elsewhere (P1, S5, T2). Both the negatively and positively charged ions diffuse at the same rate so that electrical neutrality is preserved.

6.4 MOLECULAR DIFFUSION IN BIOLOGICAL SOLUTIONS AND GELS

6.4A Diffusion of Biological Solutes in Liquids

1. Introduction. The diffusion of small solute molecules and especially macromolecules (e.g., proteins) in aqueous solutions are important in the processing and storing of biological systems and in the life processes of microorganisms, animals, and plants. Food processing is an important area where diffusion plays an important role. In the drying of liquid solutions of fruit juice, coffee, and tea, water and often volatile flavor or aroma constituents are removed. These constituents diffuse through the liquid during evaporation.

In fermentation processes, nutrients, sugars, oxygen, and so on, diffuse to the microorganisms and waste products and at times enzymes diffuse away. In the artificial kidney machine various waste products diffuse through the blood solution to a membrane and then through the membrane to an aqueous solution.

Macromolecules in solution having molecular weights of tens of thousands or more were often called *colloids*, but now we know they generally form true solutions. The diffusion behavior of protein macromolecules in solution is affected by their large sizes and shapes, which can be random coils, rodlike, or globular (spheres or ellipsoids). Also, interactions of the large molecules with the small solvent and/or solute molecules affect the diffusion of the macromolecules and also of the small solute molecules.

Besides the Fickian diffusion to be discussed here, mediated transport often occurs in biological systems where chemical interactions occur. This latter type of transport will not be discussed here.

2. Interaction and "binding" in diffusion. Protein macromolecules are very large compared to small solute molecules such as urea, KCl, and sodium caprylate, and often have a number of sites for interaction or "binding" of the solute or ligand molecules. An example is the binding of oxygen to hemoglobin in the blood. Human serum albumin protein binds most of the free fatty acids in the blood and increases their apparent solubility. Bovine serum albumin, which is in milk, binds 23 mol sodium caprylate/mol albumin when the albumin concentration is 30 kg/m^3 solution and the sodium caprylate is about 0.05 molar (G6). Hence, Fickian-type diffusion of macromolecules and small solute molecules can be greatly affected by the presence together of both types of molecules, even in dilute solutions.

3. Experimental methods to determine diffusivity. Methods to determine the diffusivity of biological solutes are similar to those discussed previously in Section 6.3 with some modifications. In the diaphragm diffusion cell shown in Fig. 6.3-1, the chamber is made of Lucite or Teflon instead of glass, since protein molecules bind to glass. Also, the porous membrane through which the molecular diffusion occurs is composed of cellulose acetate or other polymers (G5, G6, K1).

4. Experimental data for biological solutes. Most of the experimental data in the literature on protein diffusivities have been extrapolated to zero concentration since the diffusivity is often a function of concentration. A tabulation of diffusivities of a few proteins and also of small solutes often present in biological systems is given in Table 6.4-1.

The diffusion coefficients for the large protein molecules are of the order of magnitude of $5 \times 10^{-11} \text{ m}^2/\text{s}$ compared to the values of about $1 \times 10^{-9} \text{ m}^2/\text{s}$ for the small solutes in Table 6.4-1. This means macromolecules diffuse at a rate about 20 times as slow as small solute molecules for the same concentration differences.

TABLE 6.4-1. *Diffusion Coefficients for Dilute Biological Solutes in Aqueous Solution*

| | Temperature | | | | |
| | °C | K | Diffusivity (m^2/s) | Molecular Weight | Ref. |
Solute					
Urea	20	293	1.20×10^{-9}	60.1	(N2)
	25	298	1.378×10^{-9}		(G5)
Glycerol	20	293	0.825×10^{-9}	92.1	(G3)
Glycine	25	298	1.055×10^{-9}	75.1	(L3)
Sodium caprylate	25	298	8.78×10^{-10}	166.2	(G6)
Bovine serum albumin	25	298	6.81×10^{-11}	67 500	(C6)
Urease	25	298	4.01×10^{-11}	482 700	(C7)
	20	293	3.46×10^{-11}		(S6)
Soybean protein	20	293	2.91×10^{-11}	361 800	(S6)
Lipoxidase	20	293	5.59×10^{-11}	97 440	(S6)
Fibrinogen, human	20	293	1.98×10^{-11}	339 700	(S6)
Human serum albumin	20	293	5.93×10^{-11}	72 300	(S6)
γ-Globulin, human	20	293	4.00×10^{-11}	153 100	(S6)
Creatinine	37	310	1.08×10^{-9}	113.1	(C8)
Sucrose	37	310	0.697×10^{-9}	342.3	(C8)
	20	293	0.460×10^{-9}		(P3)

When the concentration of macromolecules such as proteins increases, the diffusion coefficient would be expected to decrease, since the diffusivity of small solute molecules decreases with increasing concentration. However, experimental data (G4, C7) show that the diffusivity of macromolecules such as proteins decreases in some cases and increases in other cases as protein concentration increases. Surface charges on the molecules appear to play a role in these phenomena.

When small solutes such as urea, KCl, and sodium caprylate, which are often present with protein macromolecules in solution diffuse through these protein solutions, the diffusivity decreases with increasing polymer concentration (C7, G5, G6, N3). Experimental data for the diffusivity of the solute sodium caprylate (A) diffusing through bovine serum albumin (P) solution show that the diffusivity D_{AP} of A through P is markedly reduced as the protein (P) concentration is increased (G5, G6). A large part of the reduction is due to the binding of A to P so that there is less free A to diffuse. The rest is due to blockage by the large molecules.

5. Prediction of diffusivities for biological solutes. For predicting the diffusivity of small solutes alone in aqueous solution with molecular weights less than about 1000 or solute molar volumes less than about 0.500 m³/kg mol, Eq. (6.3-9) should be used. For larger solutes the equations to be used are not as accurate. As an approximation the Stokes–Einstein equation (6.3-8) can be used.

$$D_{AB} = \frac{9.96 \times 10^{-16}T}{\mu V_A^{1/3}} \qquad (6.3\text{-}8)$$

Probably a better approximate equation to use is the semiempirical equation of Polson (P3), which is recommended for a molecular weight above 1000. A modification of his equation to take into account different temperatures is as follows for dilute aqueous solutions.

$$D_{AB} = \frac{9.40 \times 10^{-15}T}{\mu (M_A)^{1/3}} \qquad (6.4\text{-}1)$$

where M_A is the molecular weight of the large molecule A. When the shape of the molecule deviates greatly from a sphere, the equation should be used with caution.

> **EXAMPLE 6.4-1.** *Prediction of Diffusivity of Albumin*
> Predict the diffusivity of bovine serum albumin at 298 K in water as a dilute solution using the modified Polson equation (6.4-1) and compare with the experimental value in Table 6.4-1.
>
> *Solution:* The molecular weight of bovine serum albumin (A) from Table 6.4-1 is $M_A = 67\,500$ kg/kg mol. The viscosity of water at 25°C is 0.8937 × 10^{-3} Pa·s and $T = 298$ K. Substituting into Eq. (6.4-1),
>
> $$D_{AB} = \frac{9.40 \times 10^{-15}T}{\mu(M_A)^{1/3}} = \frac{(9.40 \times 10^{-15})298}{(0.8937 \times 10^{-3})(67\,500)^{1/3}}$$
>
> $$= 7.70 \times 10^{-11} \text{ m}^2/\text{s}$$
>
> This value is 11% higher than the experimental value of 6.81 × 10^{-11} m²/s.

6. Prediction of diffusivity of small solutes in protein solution. When a small solute (A) diffuses through a macromolecule (P) protein solution, Eq. (6.3-9) cannot be used for prediction for the small solute because of blockage to diffusion by the large molecules. The data needed to predict these effects are the diffusivity D_{AB} of solute A in water alone, the water of hydration on the protein, and an obstruction factor. A semi-theoretical equation that can be used to approximate the diffusivity D_{AP} of A in globular-type protein P solutions is as follows, where only the blockage effect is considered (C8, G5, G6) and no binding is present:

$$D_{AP} = D_{AB}(1 - 1.81 \times 10^{-3} c_p) \tag{6.4-2}$$

where $c_p = $ kg P/m^3. Then the diffusion equation is

$$N_A = \frac{D_{AP}(c_{A1} - c_{A2})}{z_2 - z_1} \tag{6.4-3}$$

where c_{A1} is concentration of A in kg mol A/m^3.

7. *Prediction of diffusivity with binding present.* When A is in a protein solution P and binds to P, the diffusion flux of A is equal to the flux of unbound solute A in the solution plus the flux of the protein–solute complex. Methods to predict this flux are available (G5, G6) when binding data have been experimentally obtained.

6.4B Diffusion in Biological Gels

Gels can be looked upon as semisolid materials which are "porous." They are composed of macromolecules which are usually in dilute aqueous solution with the gel comprising a few wt % of the water solution. The "pores" or open spaces in the gel structure are filled with water. The rates of diffusion of small solutes in the gels are somewhat less than in aqueous solution. The main effect of the gel structure is to increase the path length for diffusion, assuming no electrical-type effects (S7).

Recent studies by electron microscopy (L4) have shown that the macromolecules of the gel agarose (a major constituent of agar) exist as long and relatively straight threads. This suggests a gel structure of loosely interwoven, extensively hydrogen-bonded polysaccharide macromolecules.

Some typical gels are agarose, agar, and gelatin. A number of organic polymers exist as gels in various types of solutions. To measure the diffusivity of solutes in gels, unsteady-state methods are used. In one method the gel is melted and poured into a narrow tube open at one end. After solidification, the tube is placed in an agitated bath containing the solute for diffusion. The solute leaves the solution at the gel boundary and diffuses through the gel itself. After a period of time the amount diffusing in the gel is determined to give the diffusion coefficient of the solute in the gel.

A few typical values of diffusivities of some solutes in various gels are given in Table 6.4-2. In some cases the diffusivity of the solute in pure water is given so that the decrease in diffusivity due to the gel can be seen. For example, from Table 6.4-2 at 278 K, urea in water has a diffusivity of 0.880×10^{-9} m^2/s and in 2.9 wt % gelatin, has a value of 0.640×10^{-9} m^2/s, a decrease of 27%.

In both agar and gelatin the diffusivity of a given solute decreases approximately linearly with an increase in wt % gel. However, extrapolation to 0% gel gives a value smaller than that shown for pure water. It should be noted that in different preparations or batches of the same type of gel, the diffusivities can vary by as much as 10 to 20%.

EXAMPLE 6.4-2. *Diffusion of Urea in Agar*
A tube or bridge of a gel solution of 1.05 wt % agar in water at 278 K is 0.04 m long and connects two agitated solutions of urea in water. The urea concentration in the first solution is 0.2 g mol urea per liter solution and 0 in the other. Calculate the flux of urea in kg mol/s · m^2 at steady state.

TABLE 6.4-2. *Typical Diffusivities of Solutes in Dilute Biological Gels in Aqueous Solution*

Solute	Gel	Wt % Gel in Solution	Temperature K	Temperature °C	Diffusivity (m^2/s)	Ref.
Sucrose	Gelatin	0	278	5	0.285×10^{-9}	(F2)
		3.8	278	5	0.209×10^{-9}	(F2)
		10.35	278	5	0.107×10^{-9}	(F2)
		5.1	293	20	0.252×10^{-9}	(F3)
Urea	Gelatin	0	278	5	0.880×10^{-9}	(F2)
		2.9	278	5	0.644×10^{-9}	(F2)
		5.1	278	5	0.609×10^{-9}	(F3)
		10.0	278	5	0.542×10^{-9}	(F2)
		5.1	293	20	0.859×10^{-9}	(F3)
Methanol	Gelatin	3.8	278	5	0.626×10^{-9}	(F3)
Urea	Agar	1.05	278	5	0.727×10^{-9}	(F3)
		3.16	278	5	0.591×10^{-9}	(F3)
		5.15	278	5	0.472×10^{-9}	(F3)
Glycerin	Agar	2.06	278	5	0.297×10^{-9}	(F3)
		6.02	278	5	0.199×10^{-9}	(F3)
Dextrose	Agar	0.79	278	5	0.327×10^{-9}	(F3)
Sucrose	Agar	0.79	278	5	0.247×10^{-9}	(F3)
Ethanol	Agar	5.15	278	5	0.393×10^{-9}	(F3)
NaCl (0.05 M)	Agarose	0	298	25	1.511×10^{-9}	(S7)
		2	298	25	1.398×10^{-9}	(S7)

Solution: From Table 6.4-2 for the solute urea at 278 K, $D_{AB} = 0.727 \times 10^{-9}$ m^2/s. For urea diffusing through stagnant water in the gel, Eq. (6.3-3) can be used. However, since the value of x_{A1} is less than about 0.01, the solution is quite dilute and $x_{BM} \cong 1.00$. Hence, Eq. (6.3-5) can be used. The concentrations are $c_{A1} = 0.20/1000 = 0.0002$ g mol/cm^3 = 0.20 kg mol/m^3 and $c_{A2} = 0$. Substituting into Eq. (6.3-5),

$$N_A = \frac{D_{AB}(c_{A1} - c_{A2})}{z_2 - z_1} = \frac{0.727 \times 10^{-9}(0.20 - 0)}{0.04 - 0}$$

$$= 3.63 \times 10^{-9} \text{ kg mol/s} \cdot \text{m}^2$$

6.5 MOLECULAR DIFFUSION IN SOLIDS

6.5A Introduction and Types of Diffusion in Solids

Even though rates of diffusion of gases, liquids, and solids in solids are generally slower than rates in liquids and gases, mass transfer in solids is quite important in chemical and biological processing. Some examples are leaching of foods, such as soybeans, and of metal ores; drying of timber, salts, and foods; diffusion and catalytic reaction in solid catalysts; separation of fluids by membranes; diffusion of gases through polymer films used in packaging; and treating of metals at high temperatures by gases.

We can broadly classify transport in solids into two types of diffusion: diffusion that can be considered to follow Fick's law and does not depend primarily on the actual structure of the solid, and diffusion in porous solids where the actual structure and void channels are important. These two broad types of diffusion will be considered.

6.5B Diffusion in Solids Following Fick's Law

1. Derivation of equations. This type of diffusion in solids does not depend on the actual structure of the solid. The diffusion occurs when the fluid or solute diffusing is actually dissolved in the solid to form a more or less homogeneous solution—for example, in leaching, where the solid contains a large amount of water and a solute is diffusing through this solution, or in the diffusion of zinc through copper, where solid solutions are present. Also, the diffusion of nitrogen or hydrogen through rubber, or in some cases diffusion of water in foodstuffs can be classified here, since equations of similar type can be used.

Generally, simplified equations are used. Using the general Eq. (6.2-14) for binary diffusion,

$$N_A = -cD_{AB} \frac{dx_A}{dz} + \frac{c_A}{c} (N_A + N_B) \qquad (6.2\text{-}14)$$

the bulk flow term, $(c_A/c)(N_A + N_B)$, even if present, is usually small, since c_A/c or x_A is quite small. Hence, it is neglected. Also, c is assumed constant giving for diffusion in solids,

$$N_A = -\frac{D_{AB} \, dc_A}{dz} \qquad (6.5\text{-}1)$$

where D_{AB} is diffusivity in m^2/s of A through B and usually is assumed constant independent of pressure for solids. Note that $D_{AB} \neq D_{BA}$ in solids.

Integration of Eq. (6.5-1) for a solid slab at steady state gives

$$N_A = \frac{D_{AB}(c_{A1} - c_{A2})}{z_2 - z_1} \qquad (6.5\text{-}2)$$

For the case of diffusion radially through a cylinder wall of inner radius r_1 and outer r_2 and length L,

$$\frac{\bar{N}_A}{2\pi r L} = -D_{AB} \frac{dc_A}{dr} \qquad (6.5\text{-}3)$$

$$\bar{N}_A = D_{AB}(c_{A1} - c_{A2}) \frac{2\pi L}{\ln (r_2/r_1)} \qquad (6.5\text{-}4)$$

This case is similar to conduction heat transfer radially through a hollow cylinder in Fig. 4.3-2.

The diffusion coefficient D_{AB} in the solid as stated above is not dependent upon the pressure of the gas or liquid on the outside of the solid. For example, if CO_2 gas is outside a slab of rubber and is diffusing through the rubber, D_{AB} would be independent of p_A, the partial pressure of CO_2 at the surface. The solubility of CO_2 in the solid, however, is directly proportional to p_A. This is similar to the case of the solubility of O_2 in water being directly proportional to the partial pressure of O_2 in the air by Henry's law.

The solubility of a solute gas (A) in a solid is usually expressed as S in m^3 solute (at

STP of 0°C and 1 atm) per m^3 solid per atm partial pressure of (A). Also, $S = cm^3$ (STP)/atm · cm^3 solid in the cgs system. To convert this to c_A concentration in the solid in kg mol A/m^3 using SI units,

$$c_A = \frac{S \ m^3(STP)/m^3 \ solid \cdot atm}{22.414 \ m^3(STP)/kg \ mol \ A} \ p_A \ atm = \frac{Sp_A}{22.414} \ \frac{kg \ mol \ A}{m^3 \ solid} \qquad \textbf{(6.5-5)}$$

Using cgs units,

$$c_A = \frac{Sp_A}{22\,414} \ \frac{g \ mol \ A}{cm^3 \ solid} \qquad \textbf{(6.5-6)}$$

EXAMPLE 6.5-1. Diffusion of H_2 Through Neoprene Membrane
The gas hydrogen at 17°C and 0.010 atm partial pressure is diffusing through a membrane of vulcanized neoprene rubber 0.5 mm thick. The pressure of H_2 on the other side of the neoprene is zero. Calculate the steady-state flux, assuming that the only resistance to diffusion is in the membrane. The solubility S of H_2 gas in neoprene at 17°C is 0.051 m^3 (at STP of 0°C and 1 atm)/m^3 solid · atm and the diffusivity D_{AB} is 1.03×10^{-10} m^2/s at 17°C.

Solution: A sketch showing the concentration is shown in Fig. 6.5-1. The equilibrium concentration c_{A1} at the inside surface of the rubber is, from Eq. (6.5-5),

$$c_{A1} = \frac{S}{22.414} \ p_{A1} = \frac{0.051(0.010)}{22.414} = 2.28 \times 10^{-5} \ kg \ mol \ H_2/m^3 \ solid$$

Since p_{A2} at the other side is 0, $c_{A2} = 0$. Substituting into Eq. (6.5-2) and solving,

$$N_A = \frac{D_{AB}(c_{A1} - c_{A2})}{z_2 - z_1} = \frac{(1.03 \times 10^{-10})(2.28 \times 10^{-5} - 0)}{(0.5 - 0)/1000}$$

$$= 4.69 \times 10^{-12} \ kg \ mol \ H_2/s \cdot m^2$$

2. Permeability equations for diffusion in solids. In many cases the experimental data for diffusion of gases in solids are not given as diffusivities and solubilities but as permeabilities, P_M, in m^3 of solute gas A at STP (0°C and 1 atm press) diffusing per second per m^2 cross-sectional area through a solid 1 m thick under a pressure difference of 1 atm pressure. This can be related to Fick's equation (6.5-2) as follows.

$$N_A = \frac{D_{AB}(c_{A1} - c_{A2})}{z_2 - z_1} \qquad \textbf{(6.5-2)}$$

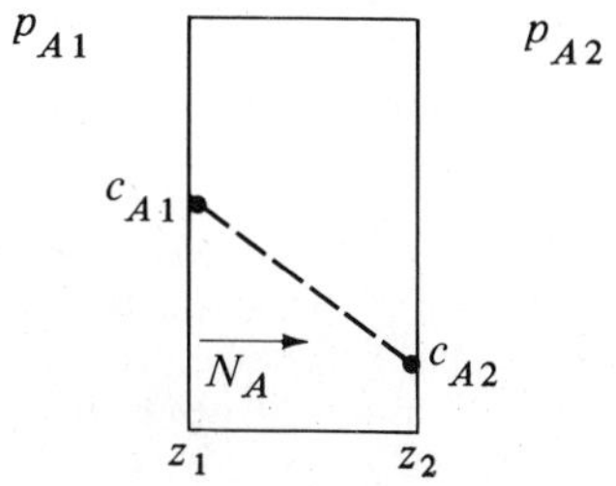

FIGURE 6.5-1. *Concentrations for Example 6.5-1.*

From Eq. (6.5-5),

$$c_{A1} = \frac{S p_{A1}}{22.414} \qquad c_{A2} = \frac{S p_{A2}}{22.414} \qquad (6.5\text{-}7)$$

Substituting Eq. (6.5-7) into (6.5-2),

$$N_A = \frac{D_{AB} S(p_{A1} - p_{A2})}{22.414(z_2 - z_1)} = \frac{P_M (p_{A1} - p_{A2})}{22.414(z_2 - z_1)} \text{ kg mol/s} \cdot \text{m}^2 \qquad (6.5\text{-}8)$$

where the permeability P_M is

$$P_M = D_{AB} S \frac{\text{m}^3 (\text{STP})}{\text{s} \cdot \text{m}^2 \text{C.S.} \cdot \text{atm/m}} \qquad (6.5\text{-}9)$$

Permeability is also given in the literature in several other ways. For the cgs system, the permeability is given as P'_M, cc(STP)/(s $\cdot$ cm^{2}C.S. $\cdot$ atm/cm). This is related to P_M by

$$P_M = 10^{-4} P'_M \qquad (6.5\text{-}10)$$

In some cases in the literature the permeability is given as P''_M, cc(STP)/(s $\cdot$ cm^{2}C.S. $\cdot$ cm Hg/mm thickness). This is related to P_M by

$$P_M = 7.60 \times 10^{-4} P''_M \qquad (6.5\text{-}11)$$

When there are several solids 1, 2, 3, ..., in series and L_1, L_2, ..., represent the thickness of each, then Eq. (6.5-8) becomes

$$N_A = \frac{p_{A1} - p_{A2}}{22.414} \frac{1}{L_1/P_{M1} + L_2/P_{M2} + \cdots} \qquad (6.5\text{-}12)$$

where $p_{A1} - p_{A2}$ is the overall partial pressure difference.

3. *Experimental diffusivities, solubilities, and permeabilities.* Accurate prediction of diffusivities in solids is generally not possible because of the lack of knowledge of the theory of the solid state. Hence, experimental values are needed. Some experimental data for diffusivities, solubilities, and permeabilities are given in Table 6.5-1 for gases diffusing in solids and solids diffusing in solids.

For the simple gases such as He, H_2, O_2, N_2, and CO_2, with gas pressures up to 1 or 2 atm, the solubility in solids such as polymers and glasses generally follows Henry's law and Eq. (6.5-5) holds. Also, for these gases the diffusivity and permeability are independent of concentration, and hence pressure. For the effect of temperature T in K, the ln P_M is approximately a linear function of $1/T$. Also, the diffusion of one gas, say H_2, is independent of the other gases present, such as O_2 and N_2.

For metals such as Ni, Cd, and Pt, where gases such as H_2 and O_2 are diffusing, it is found experimentally that the flux is approximately proportional to $(\sqrt{p_{A1}} - \sqrt{p_{A2}})$, so Eq. (6.5-8) does not hold (B5). When water is diffusing through polymers, unlike the simple gases, P_M may depend somewhat on the relative pressure difference (C9, B5). Further data are available in monographs by Crank and Park (C9) and Barrer (B5).

EXAMPLE 6.5-2. Diffusion Through a Packaging Film Using Permeability

A polyethylene film 0.00015 m (0.15 mm) thick is being considered for use in packaging a pharmaceutical product at 30°C. If the partial pressure of O_2 outside is 0.21 atm and inside the package it is 0.01 atm, calculate the diffusion flux of O_2 at steady state. Use permeability data from Table 6.5-1.

TABLE 6.5-1. *Diffusivities and Permeabilities in Solids*

Solute (A)	Solid (B)	T (K)	D_{AB}, Diffusion Coefficient $[m^2/s]$	Solubility, S $\left[\dfrac{m^3\ solute(STP)}{m^3\ solid \cdot atm}\right]$	Permeability, P_M $\left[\dfrac{m^3\ solute(STP)}{s \cdot m^2 \cdot atm/m}\right]$	Ref.
H_2	Vulcanized rubber	298	$0.85(10^{-9})$	0.040	$0.342(10^{-10})$	(B5)
O_2		298	$0.21(10^{-9})$	0.070	$0.152(10^{-10})$	(B5)
N_2		298	$0.15(10^{-9})$	0.035	$0.054(10^{-10})$	(B5)
CO_2		298	$0.11(10^{-9})$	0.90	$1.01(10^{-10})$	(B5)
H_2	Vulcanized neoprene	290	$0.103(10^{-9})$	0.051		(B5)
		300	$0.180(10^{-9})$	0.053		(B5)
H_2	Polyethylene	298			$6.53(10^{-12})$	(R3)
O_2		303			$4.17(10^{-12})$	(R3)
N_2		303			$1.52(10^{-12})$	(R3)
O_2	Nylon	303			$0.029(10^{-12})$	(R3)
N_2		303			$0.0152(10^{-12})$	(R3)
Air	English leather	298			0.15–0.68×10^{-4}	(B5)
H_2O	Wax	306			$0.16(10^{-10})$	(B5)
H_2O	Cellophane	311			0.91–$1.82(10^{-10})$	(B5)
He	Pyrex glass	293			$4.86(10^{-15})$	(B5)
		373			$20.1(10^{-15})$	(B5)
He	SiO_2	293	2.4–$5.5(10^{-14})$	0.01		(B5)
H_2	Fe	293	$2.59(10^{-13})$			(B5)
Al	Cu	293	$1.3(10^{-34})$			(B5)

Assume that the resistances to diffusion outside the film and inside are negligible compared to the resistance of the film.

Solution: From Table 6.5-1 $P_M = 4.17(10^{-12})$ m³ solute(STP)/(s·m²· atm/m). Substituting into Eq. (6.5-8),

$$N_A = \frac{P_M(p_{A1} - p_{A2})}{22.414(z_2 - z_1)} = \frac{4.17(10^{-12})(0.21 - 0.01)}{22.414(0.00015 - 0)}$$

$$= 2.480 \times 10^{-10} \text{ kg mol/s} \cdot \text{m}^2$$

Note that a film made of nylon has a much smaller value of permeability P_M for O_2 and would make a more suitable barrier.

6.5C Diffusion in Porous Solids That Depends on Structure

1. Diffusion of liquids in porous solids. In Section 6.5B we used Fick's law and treated the solid as a uniform homogeneous-like material with an experimental diffusivity D_{AB}. In this section we are concerned with porous solids that have pores or interconnected voids in the solid which affect the diffusion. A cross section of such a typical porous solid is shown in Fig. 6.5-2.

For the situation where the voids are filled completely with liquid water, the concentration of salt in water at boundary 1 is c_{A1} and at point 2 is c_{A2}. The salt in diffusing through the water in the void volume takes a tortuous path which is unknown and greater than $(z_2 - z_1)$ by a factor τ, called *tortuosity*. Diffusion does not occur in the inert solid. For a dilute solution using Eq. (6.3-5) for diffusion of salt in water at steady state,

$$N_A = \frac{\varepsilon D_{AB}(c_{A1} - c_{A2})}{\tau(z_2 - z_1)} \tag{6.5-13}$$

where ε is the open void fraction, D_{AB} is the diffusivity of salt in water, and τ is a factor which corrects for the path longer than $(z_2 - z_1)$. For inert-type solids τ can vary from about 1.5 to 5. Often the terms are combined into an effective diffusivity.

$$D_{A\,\mathrm{eff}} = \frac{\varepsilon}{\tau} D_{AB} \qquad \mathrm{m^2/s} \tag{6.5-14}$$

EXAMPLE 6.5-3. *Diffusion of KCl in Porous Silica*
A sintered solid of silica 2.0 mm thick is porous with a void fraction ε of 0.30 and a tortuosity τ of 4.0. The pores are filled with water at 298 K. At one face the concentration of KCl is held at 0.10 g mol/liter, and fresh water flows rapidly by the other face. Neglecting any other resistances but that in the porous solid, calculate the diffusion of KCl at steady state.

Solution: The diffusivity of KCl in water from Table 6.3-1 is $D_{AB} = 1.87 \times 10^{-9}$ m^2/s. Also, $c_{A1} = 0.10/1000 = 1.0 \times 10^{-4}$ g mol/cm$^3 = 0.10$ kg mol/m^3, and $c_{A2} = 0$. Substituting into Eq. (6.5-13),

$$N_A = \frac{\varepsilon D_{AB}(c_{A1} - c_{A2})}{\tau(z_2 - z_1)} = \frac{0.30(1.870 \times 10^{-9})(0.10 - 0)}{4.0(0.002 - 0)}$$

$$= 7.01 \times 10^{-9} \mathrm{~kg~mol~KCl/s \cdot m^2}$$

2. Diffusion of gases in porous solids. If the voids shown in Fig. 6.5-2 are filled with gases, then a somewhat similar situation exists. If the pores are very large so that diffusion occurs only by Fickian-type diffusion, then Eq. (6.5-13) becomes, for gases,

$$N_A = \frac{\varepsilon D_{AB}(c_{A1} - c_{A2})}{\tau(z_2 - z_1)} = \frac{\varepsilon D_{AB}(p_{A1} - p_{A2})}{\tau R T(z_2 - z_1)} \tag{6.5-15}$$

Again the value of the tortuosity must be determined experimentally. Diffusion is assumed to occur only through the voids or pores and not through the actual solid particles.

A correlation of tortuosity versus the void fraction of various unconsolidated

FIGURE 6.5-2. *Sketch of a typical porous solid.*

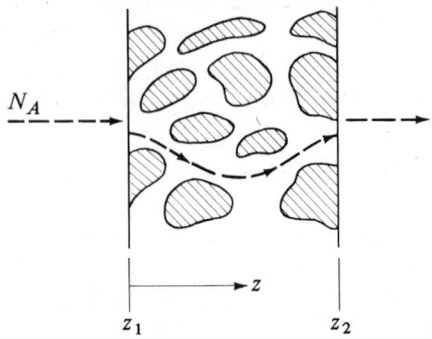

Transport Processes: Momentum, Heat, and Mass

porous media of beds of glass spheres, sand, salt, talc, and so on (S8), gives the following approximate values of τ for different values of ε: $\varepsilon = 0.2$, $\tau = 2.0$; $\varepsilon = 0.4$, $\tau = 1.75$; $\varepsilon = 0.6$, $\tau = 1.65$.

When the pores are quite small in size and of the order of magnitude of the mean free path of the gas, other types of diffusion occur, which are discussed in Section 7.6 (Selected Topic).

6.6 (SELECTED TOPIC) NUMERICAL METHODS FOR STEADY-STATE MOLECULAR DIFFUSION IN TWO DIMENSIONS

6.6A Derivation of Equations for Numerical Method

1. Derivation of method for steady state. In Fig. 6.6-1 a two-dimensional solid shown with unit thickness is divided into squares. The numerical methods for steady-state molecular diffusion are very similar to those for steady-state heat conduction discussed in Section 4.15 (Selected Topic). Hence, only a brief summary will be given here. The solid inside of a square is imagined to be concentrated at the center of the square at $c_{n,m}$ and is called a "node," which is connected to the adjacent nodes by connecting rods through which the mass diffuses.

A total mass balance is made at steady state by stating that the sum of the molecular diffusion to the shaded area for unit thickness must equal zero.

$$\frac{D_{AB}\,\Delta y}{\Delta x}(c_{n-1,m} - c_{n,m}) + \frac{D_{AB}\,\Delta y}{\Delta x}(c_{n+1,m} - c_{n,m})$$

$$+ \frac{D_{AB}\,\Delta x}{\Delta y}(c_{n,m+1} - c_{n,m}) + \frac{D_{AB}\,\Delta x}{\Delta y}(c_{n,m-1} - c_{n,m}) = 0 \quad \text{(6.6-1)}$$

where $c_{n,m}$ is concentration of A at node n,m in kg mol A/m^3. Setting $\Delta x = \Delta y$, and rearranging,

$$c_{n,m+1} + c_{n,m-1} + c_{n+1,m} + c_{n-1,m} - 4c_{n,m} = 0 \quad \text{(6.6-2)}$$

2. Iteration method of numerical solution. In order to solve Eq. (6.6-2), a separate equation is written for each unknown point giving N linear algebraic equations for N unknown points. For a hand calculation using a modest number of nodes, the iteration

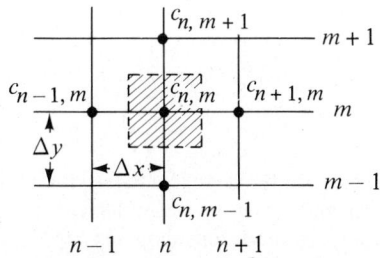

FIGURE 6.6-1. *Concentrations and spacing of nodes for two-dimensional steady-state molecular diffusion.*

method can be used to solve the equations, where the right-hand side of Eq. (6.6-2) is set equal to a residual $\bar{N}_{n, m}$.

$$c_{n, m+1} + c_{n, m-1} + c_{n+1, m} + c_{n-1, m} - 4c_{n, m} = \bar{N}_{n, m} \qquad (6.6\text{-}3)$$

Setting the equation equal to zero, $\bar{N}_{n, m} = 0$ for steady state and $c_{n, m}$ is calculated by

$$c_{n, m} = \frac{c_{n-1, m} + c_{n+1, m} + c_{n, m+1} + c_{n, m-1}}{4} \qquad (6.6\text{-}4)$$

Equations (6.6-3) and (6.6-4) are the final equations to be used to calculate all the concentrations at steady state.

Example 4.15-1 for steady-state heat conduction illustrates the detailed steps for the iteration method, which are identical to those for steady-state diffusion.

Once the concentrations have been calculated, the flux can be calculated for each element as follows. Referring to Fig. 6.6-1, the flux for the node or element $c_{n, m}$ to $c_{n, m-1}$ is ($\Delta x = \Delta y$)

$$N = \frac{A D_{AB}}{\Delta y} (c_{n, m} - c_{n, m-1}) = \frac{[(\Delta x)(1)] D_{AB}}{\Delta y} (c_{n, m} - c_{n, m-1})$$

$$= D_{AB}(c_{n, m} - c_{n, m-1}) \qquad (6.6\text{-}5)$$

where the area A is Δx times 1 m deep and N is kg mol A/s. Equations are written for the other appropriate elements and the sum of the fluxes calculated.

6.6B Equations for Special Boundary Conditions for Numerical Method

1. Equations for boundary conditions. When one of the nodal points $c_{n, m}$ is at a boundary where convective mass transfer is occurring to a constant concentration c_∞ in the bulk fluid shown in Fig. 6.6-2a a different equation must be derived. Making a mass balance on the node n, m, where mass in = mass out at steady state,

$$\frac{D_{AB} \Delta y}{\Delta x} (c_{n-1, m} - c_{n, m}) + \frac{D_{AB} \Delta x}{2 \Delta y} (c_{n, m+1} - c_{n, m})$$

$$+ \frac{D_{AB} \Delta x}{2 \Delta y} (c_{n, m-1} - c_{n, m}) = k_c \Delta y(c_{n, m} - c_\infty) \qquad (6.6\text{-}6)$$

where k_c is the convective mass-transfer coefficient in m/s defined by Eq. (6.1-14).

Setting $\Delta x = \Delta y$, rearranging, and setting the resultant equation $= \bar{N}_{n, m}$, the residual, the following results.

1. For convection at a boundary, Fig. 6.6-2a,

$$\frac{k_c \Delta x}{D_{AB}} c_\infty + \tfrac{1}{2}(2c_{n-1, m} + c_{n, m+1} + c_{n, m-1}) - c_{n, m}\left(\frac{k_c \Delta x}{D_{AB}} + 2\right) = \bar{N}_{n, m} \qquad (6.6\text{-}7)$$

This equation is similar to Eq. (4.15-16) for heat conduction and convection with $k_c \Delta x/D_{AB}$ being used in place of $h \Delta x/k$. Similarly, Eqs. (6.6-8)–(6.6-10) have been derived for the other boundary conditions shown in Fig. 6.6-2.

2. For an insulated boundary (Fig. 6.6-2b),

$$\tfrac{1}{2}(c_{n, m+1} + c_{n, m-1}) + c_{n-1, m} - 2c_{n, m} = \bar{N}_{n, m} \qquad (6.6\text{-}8)$$

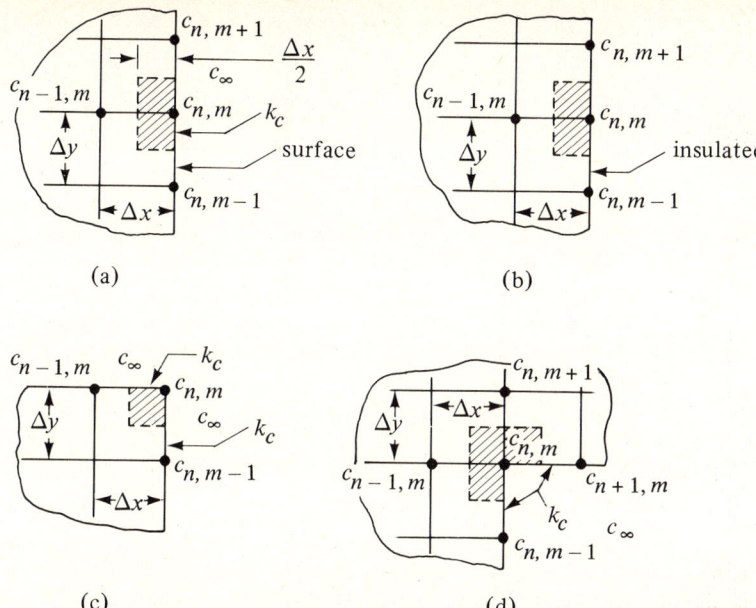

(a)   (b)   (c)   (d)

FIGURE 6.6-2. *Different boundary conditions for steady-state diffusion: (a) convection at a boundary, (b) insulated boundary, (c) exterior corner with convective boundary, (d) interior corner with convective boundary.*

3. For an exterior corner with convection at the boundary (Fig. 6.6-2c),

$$\frac{k_c \, \Delta x}{D_{AB}} c_\infty + \tfrac{1}{2}(c_{n-1,\,m} + c_{n,\,m-1}) - \left(\frac{k_c \, \Delta x}{D_{AB}} + 1\right) c_{n,\,m} = \bar{N}_{n,\,m} \qquad (6.6\text{-}9)$$

4. For an interior corner with convection at the boundary (Fig. 6.6-2d),

$$\frac{k_c \, \Delta x}{D_{AB}} c_\infty + c_{n-1,\,m} + c_{n,\,m+1} + \tfrac{1}{2}(c_{n+1,\,m} + c_{n,\,m-1}) - \left(3 + \frac{k_c \, \Delta x}{D_{AB}}\right) c_{n,\,m} = \bar{N}_{n,\,m} \qquad (6.6\text{-}10)$$

2. Boundary conditions with distribution coefficient. When Eq. (6.6-7) was derived, the distribution coefficient K between the liquid and the solid at the surface interface was 1.0. The distribution coefficient as shown in Fig. 6.6-3 is defined as

$$K = \frac{c_{n,\,mL}}{c_{n,\,m}} \qquad (6.6\text{-}11)$$

where $c_{n,\,mL}$ is the concentration in the liquid adjacent to the surface and $c_{n,\,m}$ is the concentration in the solid adjacent to the surface. Then in deriving Eq. (6.6-6), the right-hand side $k_c \, \Delta y(c_{n,\,m} - c_\infty)$ becomes

$$k_c \, \Delta y(c_{n,\,mL} - c_\infty) \qquad (6.6\text{-}12)$$

where c_∞ is the concentration in the bulk fluid. Substituting $Kc_{n,\,m}$ for $c_{n,\,mL}$ from Eq. (6.6-11) into (6.6-12) and multiplying and dividing by K,

$$Kk_c \, \Delta y\left(\frac{Kc_{n,\,m}}{K} - \frac{c_\infty}{K}\right) = Kk_c \, \Delta y\left(c_{n,\,m} - \frac{c_\infty}{K}\right) \qquad (6.6\text{-}13)$$

Principles of Mass Transfer

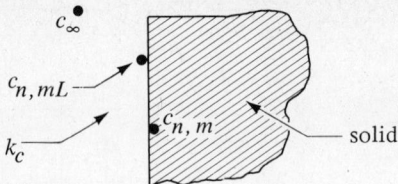

FIGURE 6.6-3. *Interface concentrations for convective mass transfer at a solid surface and an equilibrium distribution coefficient $K = c_{n,\,mL}/c_{n,\,m}$.*

Hence, whenever k_c appears as in Eq. (6.6-7), Kk_c should be substituted and when c_∞ appears, c_∞/K should be used. Then Eq. (6.6-7) becomes as follows.

1. For convection at a boundary (Fig. 6.6-2a),

$$\left(\frac{Kk_c\,\Delta x}{D_{AB}}\right)\frac{c_\infty}{K} + \tfrac{1}{2}(2c_{n-1,\,m} + c_{n,\,m+1} + c_{n,\,m-1})$$

$$- c_{n,\,m}\left(\frac{Kk_c\,\Delta x}{D_{AB}} + 2\right) = \bar{N}_{n,\,m} \quad (6.6\text{-}14)$$

Equations (6.6-9) and (6.6-10) can be rewritten in a similar manner as follows,

2. For an exterior corner with convection at the boundary (Fig. 6.6-2c),

$$\left(\frac{Kk_c\,\Delta x}{D_{AB}}\right)\frac{c_\infty}{K} + \tfrac{1}{2}(c_{n-1,\,m} + c_{n,\,m-1}) - \left(\frac{Kk_c\,\Delta x}{D_{AB}} + 1\right)c_{n,\,m} = \bar{N}_{n,\,m} \quad (6.6\text{-}15)$$

3. For an interior corner with convection at the boundary (Fig. 6.6-2d),

$$\left(\frac{Kk_c\,\Delta x}{D_{AB}}\right)\frac{c_\infty}{K} + c_{n-1,\,m} + c_{n,\,m+1}$$

$$+ \tfrac{1}{2}(c_{n+1,\,m} + c_{n,\,m-1}) - \left(3 + \frac{Kk_c\,\Delta x}{D_{AB}}\right)c_{n,\,m} = \bar{N}_{n,\,m} \quad (6.6\text{-}16)$$

EXAMPLE 6.6-1. *Numerical Method for Convection and Steady-State Diffusion*

For the two-dimensional hollow solid chamber shown in Fig. 6.6-4, determine the concentrations at the nodes as shown at steady state. At the inside surfaces the concentrations remain constant at 6.00×10^{-3} kg mol/m^3. At the outside surfaces the convection coefficient $k_c = 2 \times 10^{-7}$ m/s and $c_\infty = 2.00 \times 10^{-3}$ kg mol/m^3. The diffusivity in the solid is $D_{AB} = 1.0 \times 10^{-9}$ m^2/s. The grid size is $\Delta x = \Delta y = 0.005$ m. Also, determine the diffusion rates per 1.0-m depth. The distribution coefficient $K = 1.0$.

Solution: To simplify the calculations, all the concentrations will be multiplied by 10^3. Since the chamber is symmetrical, we do the calculations on the $\tfrac{1}{8}$ shaded portion shown. The fixed known values are $c_{1,\,3} = 6.00$, $c_{1,\,4} = 6.00$, $c_\infty = 2.00$. Because of symmetry, $c_{1,\,2} = c_{2,\,3}$, $c_{2,\,5} = c_{2,\,3}$, $c_{2,\,1} = c_{3,\,2}$, $c_{3,\,3} = c_{3,\,5}$. To speed up the calculations, we will make estimates of the unknown concentrations as follows: $c_{2,\,2} = 3.80$, $c_{2,\,3} = 4.20$, $c_{2,\,4} = 4.40$, $c_{3,\,1} = 2.50$, $c_{3,\,2} = 2.70$, $c_{3,\,3} = 3.00$, $c_{3,\,4} = 3.20$.

For the interior points $c_{2,\,2}$, $c_{2,\,3}$ and $c_{2,\,4}$ we use Eqs. (6.6-3) and (6.6-4); for the corner convection point $c_{3,\,1}$, Eq. (6.6-9); for the other

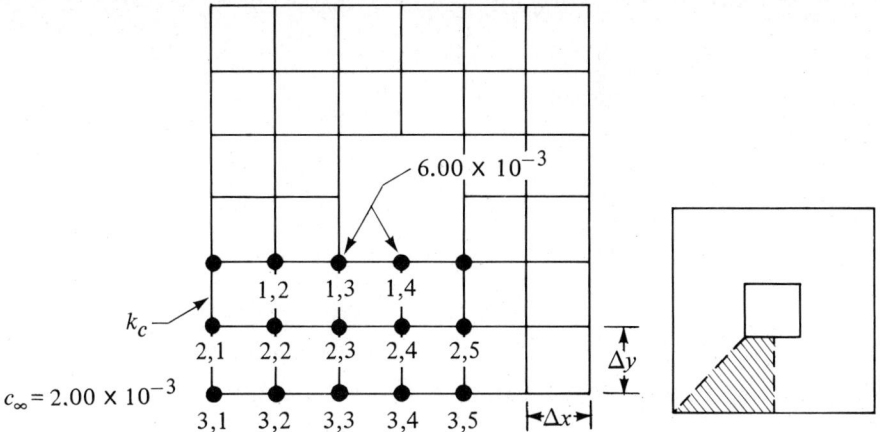

FIGURE 6.6-4. Concentrations for hollow chamber for Example 6.6-1.

convection points $c_{3,2}$, $c_{3,3}$, $c_{3,4}$, Eq. (6.6-7). The term $k_c\,\Delta x/D_{AB} =$ (2 $\times 10^{-7})(0.005)/(1.0 \times 10^{-9}) = 1.00$.

First approximation. Starting with $c_{2,2}$ we use Eq. (6.6-3) and calculate the residual $\bar{N}_{2,2}$.

$$c_{1,2} + C_{3,2} + c_{2,1} + c_{2,3} - 4c_{2,2} = \bar{N}_{2,2}$$
$$4.20 + 2.70 + 2.70 + 4.20 - 4(3.80) = -1.40$$

Hence, $c_{2,2}$ is not at steady state. Next we set $\bar{N}_{2,2}$ to zero and calculate a new value of $c_{2,2}$ from Eq. (6.6-4).

$$c_{2,2} = \frac{c_{1,2} + c_{3,2} + c_{2,1} + c_{2,3}}{4} = \frac{4.20 + 2.70 + 2.70 + 4.20}{4} = 3.45$$

This new value of $c_{2,2}$ replaces the old value.
For $c_{2,3}$,

$$c_{2,2} + c_{2,4} + c_{1,3} + c_{3,3} - 4c_{2,3} = \bar{N}_{2,3}$$
$$3.45 + 4.40 + 6.00 + 3.00 - 4(4.20) = 0.05$$

$$c_{2,3} = \frac{c_{2,2} + c_{2,4} + c_{1,3} + c_{3,3}}{4} = \frac{3.45 + 4.40 + 6.00 + 3.00}{4} = 4.21$$

For $c_{2,4}$,

$$c_{2,3} + c_{2,5} + c_{1,4} + C_{3,4} - 4c_{2,4} = \bar{N}_{2,4}$$
$$4.21 + 4.21 + 6.00 + 3.20 - 4(4.40) = 0.02$$

$$c_{2,4} = \frac{c_{2,3} + c_{2,5} + c_{1,4} + c_{3,4}}{4} = \frac{4.21 + 4.21 + 6.00 + 3.20}{4} = 4.41$$

For $c_{3,1}$, we use Eq. (6.6-9):

$$(1.0)c_\infty + \tfrac{1}{2}(c_{2,1} + c_{3,2}) - (1.0 + 1)c_{3,1} = \bar{N}_{3,1}$$
$$(1.0)2.00 + \tfrac{1}{2}(2.70 + 2.70) - 2.0(2.50) = -0.30$$

Setting Eq. (6.6-9) to zero and solving for $c_{3,1}$,

$$(1.0)2.00 + \tfrac{1}{2}(2.70 + 2.70) - (2.0)c_{3,1} = 0 \qquad c_{3,1} = 2.35$$

For $c_{3,2}$, we use Eq. (6.6-7).

$$(1.0)c_\infty + \tfrac{1}{2}(2 \times c_{2,2} + c_{3,1} + c_{3,3}) - (1.0 + 2)c_{3,2} = \bar{N}_{3,2}$$

$$(1.0)2.00 + \tfrac{1}{2}(2 \times 3.45 + 2.35 + 3.00) - (3.0)2.70 = 0.03$$

$$(1.0)2.00 + \tfrac{1}{2}(2 \times 3.45 + 2.35 + 3.00) - (3.0)c_{3,2} = 0 \qquad c_{3,2} = 2.71$$

For $c_{3,3}$,

$$(1.0)c_\infty + \tfrac{1}{2}(2 \times c_{2,3} + c_{3,2} + c_{3,4}) - (3.0)c_{3,3} = \bar{N}_{3,3}$$

$$(1.0)2.00 + \tfrac{1}{2}(2 \times 4.21 + 2.71 + 3.20) - (3.0)3.00 = 0.17$$

$$(1.0)2.00 + \tfrac{1}{2}(2 \times 4.21 + 2.71 + 3.20) - (3.0)c_{3,3} = 0 \qquad c_{3,3} = 3.06$$

For $c_{3,4}$,

$$(1.0)c_\infty + \tfrac{1}{2}(2 \times c_{2,4} + c_{3,3} + c_{3,5}) - (3.0)c_{3,4} = \bar{N}_{3,4}$$

$$(1.0)2.00 + \tfrac{1}{2}(2 \times 4.41 + 3.06 + 3.06) - (3.0)3.20 = -0.13$$

$$(1.0)2.00 + \tfrac{1}{2}(2 \times 4.41 + 3.06 + 3.06) - (3.0)c_{3,4} = 0 \qquad c_{3,4} = 3.16$$

Having completed one sweep across the grid map, we can start a second approximation using the new values calculated and starting with $c_{2,2}$ or any other node. This is continued until all the residuals are as small as desired. The final values after three approximations are $c_{2,2} = 3.47$, $c_{2,3} = 4.24$, $c_{2,4} = 4.41$, $c_{3,1} = 2.36$, $c_{3,2} = 2.72$, $c_{3,3} = 3.06$, $c_{3,4} = 3.16$.

To calculate the diffusion rates we first calculate the total convective diffusion rate, leaving the bottom surface at nodes $c_{3,1}$, $c_{3,2}$, $c_{3,3}$, and $c_{3,4}$ for 1.0-m depth.

$$N = k_c(\Delta x \cdot 1)\left[\frac{c_{3,1} - c_\infty}{2} + (c_{3,2} - c_\infty) + (c_{3,3} - c_\infty) + \frac{c_{3,4} - c_\infty}{2}\right]$$

$$= (2 \times 10^{-7})(0.005 \times 1)\left[\frac{2.36 - 2.00}{2} + (2.72\text{-}2.00)\right.$$

$$\left. + (3.06 - 2.00) + \frac{3.16 - 2.00}{2}\right] \times 10^{-3}$$

$$= 2.540 \times 10^{-12} \text{ kg mol/s}$$

Note that the first and fourth paths include only $\tfrac{1}{2}$ of a surface. Next we calculate the total diffusion rate in the solid entering the top surface inside, using an equation similar to Eq. (6.6-5).

$$N = \frac{D_{AB}(\Delta x \cdot 1)}{\Delta y}\left[(c_{1,3} - c_{2,3}) + \frac{c_{1,4} - c_{2,4}}{2}\right]$$

$$= \frac{(1.0 \times 10^{-9})(0.005 \times 1)}{0.005}\left[(6.00 - 4.24) + \frac{6.00 - 4.41}{2}\right] \times 10^{-3}$$

$$= 2.555 \times 10^{-12} \text{ kg mol/s}$$

At steady state the diffusion rate leaving by convection should equal that entering by diffusion. These results indicate a reasonable check. Using smaller grids would give even more accuracy. Note that the results for the diffusion rate should be multiplied by 8.0 for the whole chamber.

PROBLEMS

6.1-1. Diffusion of Methane Through Helium. A gas of CH_4 and He is contained in a tube at 101.32 kPa pressure and 298 K. At one point the partial pressure of methane is $p_{A1} = 60.79$ kPa and at a point 0.02 m distance away, $p_{A2} = 20.26$ kPa. If the total pressure is constant throughout the tube, calculate the flux of CH_4 (methane) at steady-state for equimolar counterdiffusion.

Ans. $J^*_{Az} = 5.52 \times 10^{-5}$ kg mol A/s·m² (5.52×10^{-6} g mol A/s·cm²)

6.1-2. Diffusion of CO_2 in a Binary Gas Mixture. The gas CO_2 is diffusing at steady state through a tube 0.20 m long having a diameter of 0.01 m and containing N_2 at 298 K. The total pressure is constant at 101.32 kPa. The partial pressure of CO_2 at one end is 456 mm Hg and 76 mm Hg at the other end. The diffusivity D_{AB} is 1.67×10^{-5} m²/s at 298 K. Calculate the flux of CO_2 in cgs and SI units for equimolar counterdiffusion.

6.2-1. Equimolar Counterdiffusion of a Binary Gas Mixture. Helium and nitrogen gas are contained in a conduit 5 mm in diameter and 0.1 m long at 298 K and a uniform constant pressure of 1.0 atm abs. The partial pressure of He at one end of the tube is 0.060 atm and 0.020 atm at the other end. The diffusivity can be obtained from Table 6.2-1. Calculate the following for steady-state equimolar counterdiffusion.
(a) Flux of He in kg mol/s·m² and g mol/s·cm².
(b) Flux of N_2.
(c) Partial Pressure of He at a point 0.05 m from either end.

6.2-2. Equimolar Counterdiffusion of NH_3 and N_2 at Steady State. Ammonia gas (A) and nitrogen gas (B) are diffusing in counterdiffusion through a straight glass tube 2.0 ft (0.610 m) long with an inside diameter of 0.080 ft (24.4 mm) at 298 K and 101.32 kPa. Both ends of the tube are connected to large mixed chambers at 101.32 kPa. The partial pressure of NH_3 in one chamber is constant at 20.0 kPa and 6.666 kPa in the other chamber. The diffusivity at 298 K and 101.32 kPa is 2.30×10^{-5} m²/s.
(a) Calculate the diffusion of NH_3 in lb mol/h and kg mol/s.
(b) Calculate the diffusion of N_2.
(c) Calculate the partial pressures at a point 1.0 ft (0.305 m) in the tube and plot p_A, p_B, and P versus distance z.

Ans. (a) Diffusion of $NH_3 = 7.52 \times 10^{-7}$ lb mol A/h,
9.48×10^{-11} kg mol A/s;
(c) $p_A = 1.333 \times 10^4$ Pa

6.2-3. Diffusion of A Through Stagnant B and Effect of Type of Boundary on Flux. Ammonia gas is diffusing through N_2 under steady-state conditions with N_2 nondiffusing since it is insoluble in one boundary. The total pressure is 1.013×10^5 Pa and the temperature is 298 K. The partial pressure of NH_3 at one point is 1.333×10^4 Pa and at the other point 20 mm away it is 6.666×10^3 Pa. The D_{AB} for the mixture at 1.013×10^5 Pa and 298 K is 2.30×10^{-5} m²/s.
(a) Calculate the flux of NH_3 in kg mol/s·m².
(b) Do the same as (a) but assume that N_2 also diffuses; i.e., both boundaries are permeable to both gases and the flux is equimolar counterdiffusion. In which case is the flux greater?

Ans. (a) $N_A = 3.44 \times 10^{-6}$ kg mol/s·m²

6.2-4. Diffusion of Methane Through Nondiffusing Helium. Methane gas is diffusing in a straight tube 0.1 m long containing helium at 298 K and a total pressure of 1.01325×10^5 Pa. The partial pressure of CH_4 at one end is 1.400×10^4 Pa and 1.333×10^3 Pa at the other end. Helium is insoluble in one boundary, and hence is nondiffusing or stagnant. The diffusivity is given in Table 6.2-1. Calculate the flux of methane in kg mol/s·m² at steady state.

6.2-5. Mass Transfer from a Naphthalene Sphere to Air. Mass transfer is occurring from a sphere of naphthalene having a radius of 10 mm. The sphere is in a large volume of still air at 52.6°C and 1 atm abs pressure. The vapor pressure of naphthalene at 52.6°C is 1.0 mm Hg. The diffusivity of naphthalene in air at 0°C is 5.16×10^{-6} m^2/s. Calculate the rate of evaporation of naphthalene from the surface in kg mol/s·m^2. [*Note :* The diffusivity can be corrected for temperature using the temperature-correction factor of the Fuller et al. Eq. (6.2-45).]

6.2-6. Estimation of Diffusivity of a Binary Gas. For a mixture of ethanol (CH$_3$CH$_2$OH) vapor and methane (CH$_4$), predict the diffusivity using the method of Fuller et al.
(a) At 1.0132×10^5 Pa and 298 and 373 K.
(b) At 2.0265×10^5 Pa and 298 K.

Ans. (a) $D_{AB} = 1.43 \times 10^{-5}$ m^2/s (298 K)

6.2-7. Diffusion Flux and Effect of Temperature and Pressure. Equimolar counterdiffusion is occurring at steady state in a tube 0.11 m long containing N$_2$ and CO gases at a total pressure of 1.0 atm abs. The partial pressure of N$_2$ is 80 mm Hg at one end and 10 mm at the other end. Predict the D_{AB} by the method of Fuller et al.
(a) Calculate the flux in kg mol/s·m^2 at 298 K for N$_2$.
(b) Repeat at 473 K. Does the flux increase?
(c) Repeat at 298 K but for a total pressure of 3.0 atm abs. The partial pressure of N$_2$ remains at 80 and 10 mm Hg, as in part (a). Does the flux change?

Ans. (a) $D_{AB} = 2.05 \times 10^{-5}$ m^2/s, $N_A = 7.02 \times 10^{-7}$ kg mol/s·m^2;
(b) $N_A = 9.92 \times 10^{-7}$ kg mol/s·m^2;
(c) $N_A = 2.34 \times 10^{-7}$ kg mol/s·m^2

6.2-8. Evaporation Losses of Water in Irrigation Ditch. Water at 25°C is flowing in a covered irrigation ditch below ground. Every 100 ft there is a vent line 1.0 in. inside diameter and 1.0 ft long to the outside atmosphere at 25°C. There are 10 vents in the 1000-ft ditch. The outside air can be assumed to be dry. Calculate the total evaporation loss of water in lb$_m$/d. Assume that the partial pressure of water vapor at the surface of the water is the vapor pressure, 23.76 mm Hg at 25°C. Use the diffusivity from Table 6.2-1.

6.2-9. Time to Completely Evaporate a Sphere. A drop of liquid toluene is kept at a uniform temperature of 25.9°C and is suspended in air by a fine wire. The initial radius $r_1 = 2.00$ mm. The vapor pressure of toluence at 25.9°C is $P_{A1} = 3.84$ kPa and the density of liquid toluene is 866 kg/m^3.
(a) Derive Eq. (6.2-34) to predict the time t_F for the drop to evaporate completely in a large volume of still air. Show all steps.
(b) Calculate the time in seconds for complete evaporation.

Ans. (b) $t_F = 1388$ s

6.2-10. Diffusion in a Nonuniform Cross-Sectional Area. The gas ammonia (A) is diffusing at steady state through N$_2$(B) by equimolar counterdiffusion in a conduit 1.22 m long at 25°C and a total pressure of 101.32 kPa abs. The partial pressure of ammonia at the left end is 25.33 kPa and 5.066 kPa at the other end. The cross section of the conduit is in the shape of an equilateral triangle, the length of each side of the triangle being 0.0610 m at the left end and tapering uniformly to 0.0305 m at the right end. Calculate the molar flux of ammonia. The diffusivity is $D_{AB} = 0.230 \times 10^{-4}$ m^2/s.

6.3-1. Diffusion of A Through Stagnant B in a Liquid. The solute HCl (A) is diffusing through a thin film of water (B) 2.0 mm thick at 283 K. The concentration of HCl at point 1 at one boundary of the film is 12.0 wt % HCl (density $\rho_1 = 1060.7$ kg/m^3), and at the other boundary at point 2 it is 6.0 wt % HCl ($\rho_2 = 1030.3$ kg/m^3). The diffusion coefficient of HCl in water is 2.5×10^{-9} m^2/s. Assuming steady state and one boundary impermeable to water, calculate the flux of HCl in kg mol/s·m^2.

Ans. $N_A = 2.370 \times 10^{-6}$ kg mol/s·m^2

6.3-2. Diffusion of Ammonia in an Aqueous Solution. An ammonia (A)–water (B) solution at 278 K and 4.0 mm thick is in contact at one surface with an organic liquid at this interface. The concentration of ammonia in the organic phase is held constant and is such that the equilibrium concentration of ammonia in the water at this surface is 2.0 wt % ammonia (density of aqueous solution is 991.7 kg/m³) and the concentration of ammonia in water at the other end of the film 4.0 mm away is 10 wt % (density of 961.7 kg/m³). Water and the organic are insoluble in each other. The diffusion coefficient of NH_3 in water is 1.24×10^{-9} m²/s.
(a) At steady state, calculate the flux N_A in kg mol/s · m².
(b) Calculate the flux N_B. Explain.

6.3-3. Estimation of Liquid Diffusivity. It is desired to predict the diffusion coefficient of dilute acetic acid (CH_3COOH) in water at 282.9 K and at 298 K using the Wilke–Chang method. Compare the predicted values with the experimental values in Table 6.3-1.

Ans. $D_{AB} = 0.897 \times 10^{-9}$ m²/s (282.9 K); $D_{AB} = 1.396 \times 10^{-9}$ m²/s (298 K)

6.3-4. Estimation of Diffusivity of Methanol in H_2O. The diffusivity of dilute methanol in water has been determined experimentally to be 1.26×10^{-9} m²/s at 288 K.
(a) Estimate the diffusivity at 293 K using the Wilke–Chang equation.
(b) Estimate the diffusivity at 293 K by correcting the experimental value at 288 K to 293 K. (*Hint*: Do this by using the relationship $D_{AB} \propto T/\mu_B$.)

6.4-1. Prediction of Diffusivity of Enzyme Urease in Solution. Predict the diffusivity of the enzyme urease in a dilute solution in water at 298 K using the modified Polson equation and compare the result with the experimental value in Table 6.4-1.

Ans. Predicted $D_{AB} = 3.995 \times 10^{-11}$ m²/s

6.4-2. Diffusion of Sucrose in Gelatin. A layer of gelatin in water 5 mm thick containing 5.1 wt % gelatin at 293 K separates two solutions of sucrose. The concentration of sucrose in the solution at one surface of the gelatin is constant at 2.0 g sucrose/100 mL solution and 0.2 g/100 mL at the other surface. Calculate the flux of sucrose in kg sucrose/s · m² through the gel at steady state.

6.4-3. Diffusivity of Oxygen in Protein Solution. Oxygen is diffusing through a solution of bovine serum albumin (BSA) at 298 K. Oxygen has been shown to not bind to BSA. Predict the diffusivity D_{AP} of oxygen in a protein solution containing 11 g protein/100 mL solution. (*Note*: See Table 6.3-1 for the diffusivity of O_2 in water.)

Ans. $D_{AP} = 1.930 \times 10^{-9}$ m²/s

6.5-1. Diffusion of CO_2 Through Rubber. A flat plug 30 mm thick having an area of 4.0×10^{-4} m² and made of vulcanized rubber is used for closing an opening in a container. The gas CO_2 at 25°C and 2.0 atm pressure is inside the container. Calculate the total leakage or diffusion of CO_2 through the plug to the outside in kg mol CO_2/s at steady state. Assume that the partial pressure of CO_2 outside is zero. From Barrer (B5) the solubility of the CO_2 gas is 0.90 m³ gas (at STP of 0°C and 1 atm) per m³ rubber per atm pressure of CO_2. The diffusivity is 0.11×10^{-9} m²/s.

Ans. 1.178×10^{-13} kg mol CO_2/s

6.5-2. Leakage of Hydrogen Through Neoprene Rubber. Pure hydrogen gas at 2.0 atm abs pressure and 27°C is flowing past a vulcanized neoprene rubber slab 5 mm thick. Using the data from Table 6.5-1, calculate the diffusion flux in kg mol/s · m² at steady state. Assume no resistance to diffusion outside the slab and zero partial pressure of H_2 on the outside.

6.5-3. Relation Between Diffusivity and Permeability. The gas hydrogen is diffusing through a sheet of vulcanized rubber 20 mm thick at 25°C. The partial pressure of H_2 inside is 1.5 atm and 0 outside. Using the data from Table 6.5-1, calculate the following.

(a) The diffusivity D_{AB} from the permeability P_M and solubility S and compare with the value in Table 6.5-1.

(b) The flux N_A of H_2 at steady state.

Ans. (b) $N_A = 1.144 \times 10^{-10}$ kg mol/s·m^2

6.5-4. *Loss from a Tube of Neoprene.* Hydrogen gas at 2.0 atm and 27°C is flowing in a neoprene tube 3.0 mm inside diameter and 11 mm outside diameter. Calculate the leakage of H_2 through a tube 1.0 m long in kg mol H_2/s at steady state.

6.5-5. *Diffusion Through Membranes in Series.* Nitrogen gas at 2.0 atm and 30°C is diffusing through a membrane of nylon 1.0 mm thick and polyethylene 8.0 mm thick in series. The partial pressure at the other side of the two films is 0 atm. Assuming no other resistances, calculate the flux N_A at steady state.

6.5-6. *Diffusion of CO_2 in a Packed Bed of Sand.* It is desired to calculate the rate of diffusion of CO_2 gas in air at steady state through a loosely packed bed of sand at 276 K and a total pressure of 1.013×10^5 Pa. The bed depth is 1.25 m and the void fraction ε is 0.30. The partial pressure of CO_2 at the top of the bed is 2.026×10^3 Pa and 0 Pa at the bottom. Use a τ of 1.87.

Ans. $N_A = 1.609 \times 10^{-9}$ kg mol CO_2/s·m^2

6.5-7. *Packaging to Keep Food Moist.* Cellophane is being used to keep food moist at 38°C. Calculate the loss of water vapor in g/d at steady state for a wrapping 0.10 mm thick and an area of 0.200 m^2 when the vapor pressure of water vapor inside is 10 mm Hg and the air outside contains water vapor at a pressure of 5 mm Hg. Use the larger permeability in Table 6.5-1.

Ans. 0.1667 g H_2O/day.

6.5-8. *Loss of Helium and Permeability.* A window of SiO_2 2.0 mm thick and 1.0×10^{-4} m^2 area is used to view the contents in a metal vessel at 20°C. Helium gas at 202.6 kPa is contained in the vessel. To be conservative use $D_{AB} = 5.5 \times 10^{-14}$ m^2/s from Table 6.5-1.

(a) Calculate the loss of He in kg mol/h at steady state.

(b) Calculate the permeability P_M and P_M''.

Ans. (a) Loss $= 8.833 \times 10^{-15}$ kg mol He/h

6.6-1. *(Selected Topic) Numerical Method for Steady-State Diffusion.* Using the results from Example 6.6-1, calculate the total diffusion rate in the solid using the bottom nodes and paths of $c_{2,2}$ to $c_{3,2}$, $c_{2,3}$ to $c_{3,3}$, and so on. Compare with the other diffusion rates in Example 6.6-1.

Ans. $N = 2.555 \times 10^{-12}$ kg mol/s

6.6-2 *(Selected Topic) Numerical Method for Steady-State Diffusion with Distribution Coefficient.* Use the conditions given in Example 6.6-1 except that the distribution coefficient defined by Eq. (6.6-11) between the concentration in the liquid adjacent to the external surface and the concentration in the solid adjacent to the external surface is $K = 1.2$. Calculate the steady-state concentrations and the diffusion rates.

6.6-3. *(Selected Topic) Digital Solution for Steady-State Diffusion.* Use the conditions given in Example 6.6-1 but instead of using $\Delta x = 0.005$ m, use $\Delta x = \Delta y = 0.001$ m. The overall dimensions of the hollow chamber remain as in Example 6.6-1. The only difference is that many more nodes will be used. Program this in Fortran and solve for the steady-state concentrations using the numerical method. Also, calculate the diffusion rates and compare with Example 6.6-1.

6.6-4. *(Selected Topic) Numerical Method with Fixed Surface Concentrations.* Steady-state diffusion is occurring in a two-dimensional solid as shown in Fig. 6.6-4. The grid $\Delta x = \Delta y = 0.010$ m. The diffusivity $D_{AB} = 2.00 \times 10^{-9}$ m^2/s. At the inside of the chamber the surface concentration is held constant at 2.00×10^{-3} kg mol/m^3. At the outside surfaces, the concentration is constant at 8.00×10^{-3}. Calculate the steady-state concentrations and the diffusion rates per m of depth.

Transport Processes: Momentum, Heat, and Mass

REFERENCES

(A1)   AMDUR, I., and SHULER, L. M. *J. Chem. Phys.*, **38**, 188 (1963).

(A2)   ANDERSON, D. K., HALL. J. R., and BABB, A. L. *J. Phys. Chem.*, **62**, 404 (1958).

(B1)   BUNDE, R. E. *Univ. Wisconsin Naval Res. Lab. Rept. No. CM-850*, August 1955.

(B2)   BOYD, C. A., STEIN, N., STEINGRIMISSON, V., and RUMPEL, W. F. *J. Chem. Phys.*, **19**, 548 (1951).

(B3)   BIRD, R. B., STEWART, W. E., and LIGHTFOOT, E. N. *Transport Phenomena*. New York: John Wiley & Sons, Inc., 1960.

(B4)   BIDSTRUP, D. E., and GEANKOPLIS, C. J. *J. Chem. Eng. Data*, **8**, 170 (1963).

(B5)   BARRER, R. M. *Diffusion in and Through Solids*. London: Cambridge University Press, 1941.

(C1)   CARMICHAEL, L. T., SAGE, B. H., and LACEY, W. N. *A.I.Ch.E.J.*, **1**, 385 (1955).

(C2)   CARSWELL, A. J., and STRYLAND, J. C. *Can. J. Phys.*, **41**, 708 (1963).

(C3)   CHAKRABORTI, P. K., and GRAY, P. *Trans. Faraday Soc.*, **62**, 3331 (1961).

(C4)   CHANG, S. Y. M. S. thesis, Massachusetts Institute of Technology, 1959.

(C5)   CHANG, PIN, and WILKE, C. R. *J. Phys. Chem.*, **59**, 592 (1955).

(C6)   CHARLWOOD, P. A. *J. Phys. Chem.*, **57**, 125 (1953).

(C7)   CAMERON, J. R. M. S. thesis, Ohio State University, 1973.

(C8)   COLTON, C. K., SMITH, K. A., MERRILL, E. W., and REECE, J. M. *Chem. Eng. Progr. Symp.*, **66**(99), 85(1970).

(C9)   CRANK, J., and PARK, G. S. *Diffusion in Polymers*. New York: Academic Press, Inc., 1968.

(F1)   FULLER, E. N., SCHETTLER, P. D., and GIDDINGS, J. C. *Ind. Eng. Chem.*, **58**, 19 (1966)

(F2)   FRIEDMAN, L., and KRAMER, E. O. *J. Am. Chem. Soc.*, **52**, 1298 (1930).

(F3)   FRIEDMAN, L. *J. Am. Chem. Soc.*, **52**, 1305, 1311 (1930).

(G1)   GILLILAND, E. R. *Ind. Eng. Chem.*, **26**, 681 (1934)

(G2)   GEANKOPLIS, C. J. *Mass Transport Phenomena*. Columbus, Ohio: Ohio State University Bookstores, 1972.

(G3)   GARNER, G. H., and MARCHANT, P. J. M. *Trans. Inst. Chem. Eng. (London)*, **39**, 397 (1961).

(G4)   GOSTING, L. S. *Advances in Protein Chemistry*, Vol. 1. New York: Academic Press, Inc., 1956.

(G5)   GEANKOPLIS, C. J., OKOS, M. R., and GRULKE, E. A. *J. Chem. Eng. Data*, **23**, 40 (1978).

(G6)   GEANKOPLIS, C. J., GRULKE, E. A., and OKOS, M. R. *Ind. Eng. Chem. Fund.*, **18**, 233 (1979).

(H1)   HOLSEN, J. N., and STRUNK, M. R. *Ind. Eng. Chem. Fund.*, **3**, 143 (1964).

(H2)   HUDSON, G. H., McCOUBREY, J. C., and UBBELOHDE, A. R. *Trans. Faraday Soc.*, **56**, 1144 (1960).

(H3)   HIRSCHFELDER, J. O., CURTISS, C. F., and BIRD, R. B. *Molecular Theory of Gases and Liquids*. New York: John Wiley & Sons, Inc., 1954.

(H4)   HAMMOND, B. R., and STOKES, R. H. *Trans. Faraday Soc.*, **49**, 890 (1953).

(J1)   JOHNSON, P. A., and BABB, A. L. *Chem. Revs.*, **56**, 387 (1956).

(K1)   KELLER, K. H., and FRIEDLANDER, S. K. *J. Gen. Physiol.*, **49**, 68 (1966).

(L1)   LEE, C. Y., and WILKE, C. R. *Ind. Eng. Chem.*, **46**, 2381 (1954).

(L2)   LE BAS, G. *The Molecular Volumes of Liquid Chemical Compounds*. New York: David McKay Co., Inc., 1915.

(L3)   LONGSWORTH, L. G. *J. Phys. Chem.*, **58**, 770 (1954).

(L4)   LANGDON, A. G., and THOMAS, H. C. *J. Phys. Chem.*, **75**, 1821 (1971).

(M1)   MASON, E. A., and MONCHICK, L. *J. Chem. Phys.*, **36**, 2746 (1962).

(M2)   MONCHICK, L., and MASON, E. A. *J. Chem. Phys.*, **35**, 1676 (1961).

(N1)   NEY, E. P., and ARMISTEAD, F. C. *Phys. Rev.*, **71**, 14 (1947).

(N2)   National Research Council, *International Critical Tables*, Vol. V. New York: McGraw-Hill Book Company, 1929.

(N3)   NARVARI, R. M., GAINER, J. L., and HALL, K. R. *A.I.Ch.E.J.*, **17**, 1028 (1971).

(P1)   PERRY, R. H., and CHILTON, C. H. *Chemical Engineers' Handbook*, 5th ed. New York: McGraw-Hill Book Company, 1973.

(P2)   PERKINS, L. R., and GEANKOPLIS, C. J. *Chem. Eng. Sci.*, **24**, 1035 (1969).

(P3)   POLSON, A. *J. Phys. Colloid Chem.*, **54**, 649 (1950).

(R1)   REID, R. C., PRAUSNITZ, J. M., and SHERWOOD, T. K. *The Properties of Gases and Liquids*, 3rd ed. New York: McGraw-Hill Book Company, 1977.

(R2)   REDDY, K. A., and DORAISWAMY, L. K. *Ind. Eng. Chem. Fund.*, **6**, 77 (1967).

(R3)   ROGERS, C. E. *Engineering Design for Plastics*. New York: Reinhold Publishing Co., Inc., 1964.

(S1)   SCHAFER, K. L. *Z. Elecktrochem.*, **63**, 111 (1959).

(S2)   SEAGER, S. L., GEERTSON, L. R., and GIDDINGS, J. C. *J. Chem. Eng. Data*, **8**, 168 (1963).

(S3)   SCHWERTZ, F. A., and BROW, J. E. *J. Chem. Phys.*, **19**, 640 (1951).

(S4)   SRIVASTAVA, B. N., and SRIVASTAVA, I. B. *J. Chem. Phys.*, **38**, 1183 (1963).

(S5)   SKELLAND, A. H. P. *Diffusional Mass Transfer*. New York: McGraw-Hill Book Company, 1974.

(S6)   SORBER, H. A. *Handbook of Biochemistry, Selected Data for Molecular Biology*. Cleveland: Chemical Rubber Co., Inc., 1968.

(S7)   SPALDING, G. E. *J. Phys. Chem.*, **73**, 3380 (1969).

(T1)   TRAUTZ, M., and MULLER, W. *Ann. Physik*, **22**, 333 (1935).

(T2)   TREYBAL, R. E. *Liquid Extraction*, 2nd ed. New York: McGraw-Hill Book Company, 1963.

(T3)   TREYBAL, R. E. *Mass Transfer Operations*, 2nd ed. New York: McGraw-Hill Book Company, 1968.

(V1)   VIVIAN, J. E., and KING, C. J. *A.I.Ch.E.J.*, **10**, 220 (1964).

(W1)   WINTERGERST, V. E. *Ann. Physik*, **4**, 323, (1930).

(W2)   WESTENBERG, A. A., and FRAZIER, G. *J. Chem. Phys.*, **36**, 3499 (1962).

(W3)   WALKER, R. E., and WESTENBERG, A. A. *J. Chem. Phys.*, **29**, 1139 (1958).

(W4)   WALKER, R. E., and WESTENBERG, A. A. *J. Chem. Phys.*, **32**, 436 (1960).

(W5)   WILKE, C. R., and CHANG, PIN. *A.I.Ch.E.J.*, **1**, 264 (1955).

CHAPTER 7

Principles of
Unsteady-State and
Convective Mass Transfer

7.1 UNSTEADY-STATE DIFFUSION

7.1A Derivation of Basic Equation

In Chapter 6 we considered various mass-transfer systems where the concentration or partial pressure at any point and the diffusion flux were constant with time, hence at steady state. Before steady state can be reached, time must elapse after the mass-transfer process is initiated for the unsteady-state conditions to disappear.

In Section 2.3 a general property balance for unsteady-state molecular diffusion was made for the properties momentum, heat, and mass. For no generation present, this was

$$\frac{\partial \Gamma}{\partial t} = \delta \frac{\partial^2 \Gamma}{\partial z^2} \qquad\qquad \textbf{(2.13-12)}$$

In Section 5.1 an unsteady-state equation for heat conduction was derived,

$$\frac{\partial T}{\partial t} = \alpha \frac{\partial^2 T}{\partial x^2} \qquad\qquad \textbf{(5.1-10)}$$

The derivation of the unsteady-state diffusion equation in one direction for mass transfer is similar to that done for heat transfer in obtaining Eq. (5.1-10). We refer to Fig. 7.1-1 where mass is diffusing in the x direction in a cube composed of a solid, stagnant gas, or stagnant liquid and having dimensions Δx, Δy, and Δz. For diffusion in the x direction we write

$$N_{Ax} = -D_{AB} \frac{\partial c_A}{\partial x} \qquad\qquad \textbf{(7.1-1)}$$

The term $\partial c_A / \partial x$ means the partial of c_A with respect to x or the rate of change of c_A with x when the other variable time t is kept constant.

Next we make a mass balance on component A in terms of moles for no generation.

$$\text{rate of input} = \text{rate of output} + \text{rate of accumulation} \qquad\qquad \textbf{(7.1-2)}$$

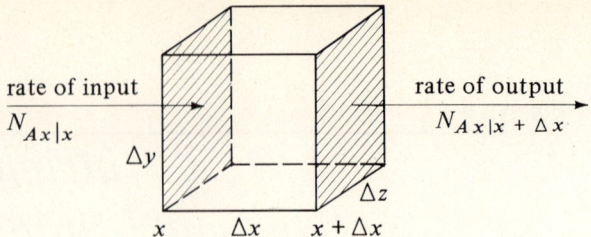

FIGURE 7.1-1.   *Unsteady-state diffusion in one direction.*

The rate of input and rate of output in kg mol A/s are

$$\text{rate of input} = N_{Ax|x} = -D_{AB} \left. \frac{\partial c_A}{\partial x} \right|_x \qquad (7.1\text{-}3)$$

$$\text{rate of output} = N_{Ax|x+\Delta x} = -D_{AB} \left. \frac{\partial c_A}{\partial x} \right|_{x+\Delta x} \qquad (7.1\text{-}4)$$

The rate of accumulation is as follows for the volume $\Delta x\, \Delta y\, \Delta z\, \text{m}^3$:

$$\text{rate of accumulation} = (\Delta x\, \Delta y\, \Delta z) \frac{\partial c_A}{\partial t} \qquad (7.1\text{-}5)$$

Substituting Eqs. (7.1-3), (7.1-4), and (7.1-5) into (7.1-2) and dividing by $\Delta x\, \Delta y\, \Delta z$,

$$-D_{AB} \frac{\left. \dfrac{\partial c_A}{\partial x} \right|_x - \left. \dfrac{\partial c_A}{\partial x} \right|_{x+\Delta x}}{\Delta x} = \frac{\partial c_A}{\partial t} \qquad (7.1\text{-}6)$$

Letting Δx approach zero,

$$\frac{\partial c_A}{\partial t} = D_{AB} \frac{\partial^2 c_A}{\partial x^2} \qquad (7.1\text{-}7)$$

The above holds for a constant diffusivity D_{AB}. If D_{AB} is a variable,

$$\frac{\partial c_A}{\partial t} = \frac{\partial(D_{AB}\, \partial c_A/\partial x)}{\partial x} \qquad (7.1\text{-}8)$$

Equation (7.1-7) relates the concentration c_A with position x and time t. For diffusion in all three directions a similar derivation gives

$$\frac{\partial c_A}{\partial t} = D_{AB}\left(\frac{\partial^2 c_A}{\partial x^2} + \frac{\partial^2 c_A}{\partial y^2} + \frac{\partial^2 c_A}{\partial z^2} \right) \qquad (7.1\text{-}9)$$

In the remainder of this section, the solutions of Eqs. (7.1-7) and (7.1-9) will be considered. Note the mathematical similarity between the equation for heat conduction,

$$\frac{\partial T}{\partial t} = \alpha \frac{\partial^2 T}{\partial x^2} \qquad (5.1\text{-}6)$$

and Eq. (7.1-7) for diffusion. Because of this similarity, the mathematical methods used for solution of the unsteady-state heat-conduction equation can be used for unsteady-state mass transfer. This is discussed more fully in Sections 7.1B, 7.1C, and 7.7.

Transport Processes: Momentum, Heat, and Mass

7.1B Diffusion in a Flat Plate with Negligible Surface Resistance

To illustrate an analytical method of solving Eq. (7.1-7), we will derive the solution for unsteady-state diffusion in the x direction for a plate of thickness $2x_1$, as shown in Fig. 7.1-2. For diffusion in one direction,

$$\frac{\partial c_A}{\partial t} = D_{AB} \frac{\partial^2 c_A}{\partial x^2} \tag{7.1-7}$$

Dropping the subscripts A and B for convenience,

$$\frac{\partial c}{\partial t} = D \frac{\partial^2 c}{\partial x^2} \tag{7.1-10}$$

The initial profile of the concentration in the plate at $t = 0$ is uniform at $c = c_0$ at all x values, as shown in Fig. 7.1-2. At time $t = 0$ the concentration of the fluid in the environment outside is suddenly changed to c_1. For a very high mass-transfer coefficient outside the surface resistance will be negligible and the concentration at the surface will be equal to that in the fluid, which is c_1.

The initial and boundary conditions are

$$c = c_0, \qquad t = 0, \qquad x = x, \qquad Y = \frac{c_1 - c_0}{c_1 - c_0} = 1$$

$$c = c_1, \qquad t = t, \qquad x = 0, \qquad Y = \frac{c_1 - c_1}{c_1 - c_0} = 0 \tag{7.1-11}$$

$$c = c_1, \qquad t = t, \qquad x = 2x_1, \qquad Y = \frac{c_1 - c_1}{c_1 - c_0} = 0$$

Redefining the concentration so it goes between 0 and 1,

$$Y = \frac{c_1 - c}{c_1 - c_0} \tag{7.1-12}$$

$$\frac{\partial Y}{\partial t} = D \frac{\partial^2 Y}{\partial x^2} \tag{7.1-13}$$

The solution of Eq. (7.1-13) is an infinite Fourier series and is identical to the solution of Eq. (5.1-6) for heat transfer.

FIGURE 7.1-2. *Unsteady-state diffusion in a flat plate with negligible surface resistance.*

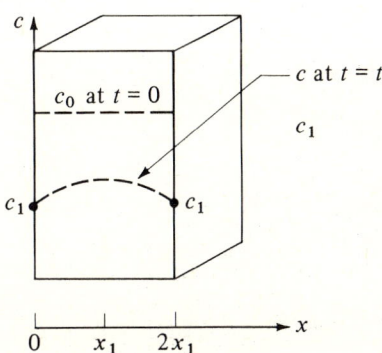

$$Y = \frac{c_1 - c}{c_1 - c_0} = \frac{4}{\pi}\left[\frac{1}{1}\exp\left(-\frac{1^2\pi^2 X}{4}\right)\sin\frac{1\pi x}{2x_1}\right.$$
$$\left. + \frac{1}{3}\exp\left(-\frac{3^2\pi^2 X}{4}\right)\sin\frac{3\pi x}{2x_1} + \frac{1}{5}\exp\left(-\frac{5^2\pi^2 X}{4}\right)\sin\frac{5\pi x}{2x_1} + \cdots\right] \quad \textbf{(7.1-14)}$$

where

$X = Dt/x_1^2$, dimensionless

c = concentration at point x and time t in slab

$Y = (c_1 - c)/(c_1 - c_0)$ = fraction of unaccomplished change, dimensionless

$1 - Y = (c - c_0)/(c_1 - c_0)$ = fraction of change

Solution of equations similar to Eq. (7.1-14) are time consuming; convenient charts for various geometries are available and will be discussed in the next section.

7.1C Unsteady-State Diffusion in Various Geometries

1. Convection and boundary conditions at the surface. In Fig. 7.1-2 there was no convective resistance at the surface. However, in many cases when a fluid is outside the solid, convective mass transfer is occurring at the surface. A convective mass-transfer coefficient k_c, similar to a convective heat-transfer coefficient, is defined as follows:

$$N_A = k_c(c_{L1} - c_{Li}) \quad \textbf{(7.1-15)}$$

where k_c is a mass-transfer coefficient in m/s, c_{L1} is the bulk fluid concentration in kg mol A/m^3, and c_{Li} is the concentration in the fluid just adjacent to the surface of the solid. The coefficient k_c is an empirical coefficient and will be discussed more fully in Section 7.2.

In Fig. 7.1-3a the case for a mass-transfer coefficient being present at the boundary is shown. The concentration drop across the fluid is $c_{L1} - c_{Li}$. The concentration in the solid c_i at the surface is in equilibrium with c_{Li}.

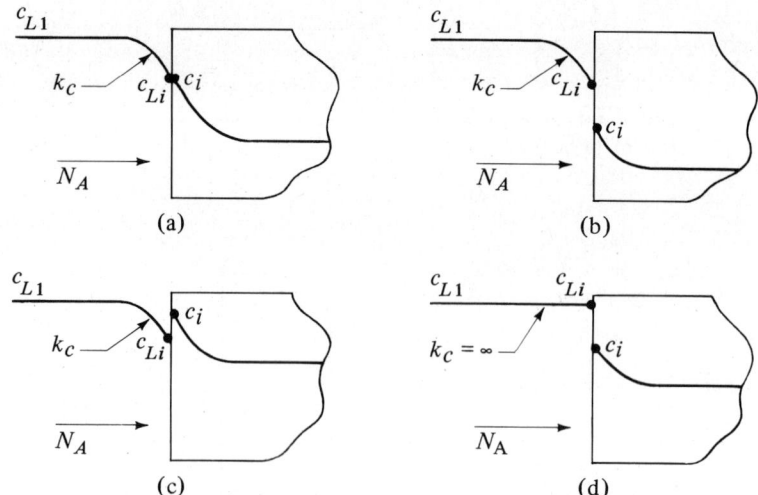

FIGURE 7.1-3. *Interface conditions for convective mass transfer and an equilibrium distribution coefficient $K = c_{Li}/c_i$: (a) $K = 1$, (b) $K > 1$, (c) $K < 1$, (d) $K > 1$ and $k_c = \infty$.*

In Fig. 7.1-3a the concentration c_{Li} in the liquid adjacent to the solid and c_i in the solid at the surface are in equilibrium and also equal. However, unlike heat transfer, where the temperatures are equal, the concentrations are in equilibrium and are related by

$$K = \frac{c_{Li}}{c_i} \tag{7.1-16}$$

where K is the equilibrium distribution coefficient (similar to Henry's law coefficient for a gas and liquid). The value of K in Fig. 7.1-3a is 1.0.

In Fig. 7.1-3b the distribution coefficient K is >1 and $c_{Li} > c_i$, even though they are in equilibrium. Other cases are shown in Fig. 7.1-3c and d. This was also discussed in Section 6.6B.

2. Relation between mass- and heat-transfer parameters. In order to use the unsteady-state heat-conduction charts in Chapter 5 for solving unsteady-state diffusion problems, the dimensionless variables or parameters for heat transfer must be related to those for mass transfer. In Table 7.1-1 the relations between these variables are tabulated. For $K \neq 1.0$, whenever k_c appears, it is given as Kk_c, and whenever c_1 appears, it is given as c_1/K.

TABLE 7.1-1. *Relation Between Mass- and Heat-Transfer Parameters for Unsteady-State Diffusion**

		Mass Transfer	
Heat Transfer		$K = c_L/c = 1.0$	$K = c_L/c \neq 1.0$
$Y,$	$\dfrac{T_1 - T}{T_1 - T_0}$	$\dfrac{c_1 - c}{c_1 - c_0}$	$\dfrac{c_1/K - c}{c_1/K - c_0}$
$1 - Y,$	$\dfrac{T - T_0}{T_1 - T_0}$	$\dfrac{c - c_0}{c_1 - c_0}$	$\dfrac{c - c_0}{c_1/K - c_0}$
$X,$	$\dfrac{\alpha t}{x_1^2}$	$\dfrac{D_{AB}t}{x_1^2}$	$\dfrac{D_{AB}t}{x_1^2}$
	$\dfrac{x}{2\sqrt{\alpha t}}$	$\dfrac{x}{2\sqrt{D_{AB}t}}$	$\dfrac{x}{2\sqrt{D_{AB}t}}$
$m,$	$\dfrac{k}{hx_1}$	$\dfrac{D_{AB}}{k_c x_1}$	$\dfrac{D_{AB}}{Kk_c x_1}$
	$\dfrac{h}{k}\sqrt{\alpha t}$	$\dfrac{k_c}{D_{AB}}\sqrt{D_{AB}t}$	$\dfrac{Kk_c}{D_{AB}}\sqrt{D_{AB}t}$
$n,$	$\dfrac{x}{x_1}$	$\dfrac{x}{x_1}$	$\dfrac{x}{x_1}$

* x is the distance from the center of the slab, cylinder, or sphere; for a semiinfinite slab, x is the distance from the surface. c_0 is the original uniform concentration in the solid, c_1 the concentration in the fluid outside the slab, and c the concentration in the solid at position x and time t.

3. *Charts for diffusion in various geometries.* The various heat-transfer charts for unsteady-state conduction can be used for unsteady-state diffusion and are as follows.

1. Semiinfinite solid, Fig. 5.3-3.
2. Flat plate, Figs. 5.3-5 and 5.3-6.
3. Long cylinder, Figs. 5.3-7 and 5.3-8.
4. Sphere, Figs. 5.3-9 and 5.3-10.
5. Average concentrations, zero convective resistance, Fig. 5.3-13.

EXAMPLE 7.1-1. *Unsteady-State Diffusion in a Slab of Agar Gel*
A solid slab of 5.15 wt % agar gel at 278 K is 10.16 mm thick and contains a uniform concentration of urea of 0.1 kg mol/m³. Diffusion is only in the x direction through two parallel flat surfaces 10.16 mm apart. The slab is suddenly immersed in pure turbulent water so that the surface resistance can be assumed to be negligible; i.e., the convective coefficient k_c is very large. The diffusivity of urea in the agar from Table 6.4-2 is 4.72×10^{-10} m²/s.
    (a) Calculate the concentration at the midpoint of the slab (5.08 mm from the surface) and 2.54 mm from the surface after 10 h.
    (b) If the thickness of the slab is halved, what would be the midpoint concentration in 10 h?

Solution: For part (a), $c_0 = 0.10$ kg mol/m³, $c_1 = 0$ for pure water, and c = concentration at distance x from center line and time t s. The equilibrium distribution coefficient K in Eq, (7.1-16) can be assumed to be 1.0, since the water in the aqueous solution in the gel and outside should be very similar in properties. From Table 7.1-1,

$$Y = \frac{c_1/K - c}{c_1/K - c_0} = \frac{0/1.0 - c}{0/1.0 - 0.10}$$

Also, $x_1 = 10.16/(1000 \times 2) = 5.08 \times 10^{-3}$ m (half-slab thickness), $x = 0$ (center), $X = D_{AB}t/x_1^2 = (4.72 \times 10^{-10})(10 \times 3600)/(5.08 \times 10^{-3})^2 = 0.658$. The relative position $n = x/x_1 = 0/5.08 \times 10^{-3} = 0$ and relative resistance $m = D_{AB}/Kk_c x_1 = 0$, since k_c is very large (zero resistance).
    From Fig. 5.3-5 for $X = 0.658$, $m = 0$, and $n = 0$,

$$Y = 0.275 = \frac{0 - c}{0 - 0.10}$$

Solving, $c = 0.0275$ kg mol/m³ for $x = 0$.
    For the point 2.54 mm from the surface or 2.54 mm from center, $x = 2.54/1000 = 2.54 \times 10^{-3}$ m, $X = 0.658$, $m = 0$, $n = x/x_1 = 2.54 \times 10^{-3}/5.08 \times 10^{-3} = 0.5$. Then from Fig. 5.3-5, $Y = 0.172$. Solving, $c = 0.0172$ kg mol/m³.
    For part (b) and half the thickness, $X = 0.658/(0.5)^2 = 2.632$, $n = 0$, and $m = 0$. Hence, $Y = 0.0020$ and $c = 2.0 \times 10^{-4}$ kg mol/m³.

EXAMPLE 7.1-2. *Unsteady-State Diffusion in a Semiinfinite Slab*
A very thick slab has a uniform concentration of solute A of $c_0 = 1.0 \times 10^{-2}$ kg mol A/m³. Suddenly, the front face of the slab is exposed to a flowing fluid having a concentration $c_1 = 0.10$ kg mol A/m³ and a convective coefficient $k_c = 2 \times 10^{-7}$ m/s. The equilibrium distribution coefficient $K = c_{Li}/c_i = 2.0$. Assuming that the slab is a semiinfinite solid, calculate the concentration in the solid at the surface ($x = 0$) and $x = 0.01$ m from the surface after $t = 3 \times 10^4$ s. The diffusivity in the solid is $D_{AB} = 4 \times 10^{-9}$ m²/s.

Solution: To use Fig. 5.3-3,

$$\frac{Kk_c}{D_{AB}}\sqrt{D_{AB}t} = \frac{2.0(2 \times 10^{-7})}{4 \times 10^{-9}}\sqrt{(4 \times 10^{-9})(3 \times 10^4)} = 1.095$$

For $x = 0.01$ m from the surface in the solid,

$$\frac{x}{2\sqrt{D_{AB}t}} = \frac{0.01}{2\sqrt{(4 \times 10^{-9})(3 \times 10^4)}} = 0.457$$

From the chart, $1 - Y = 0.26$. Then, substituting into the equation for $(1 - Y)$ from Table 7.1-1 and solving,

$$1 - Y = \frac{c - c_0}{c_1/K - c_0} = \frac{c - 1 \times 10^{-2}}{(10 \times 10^{-2})/2 - (1 \times 10^{-2})} = 0.26$$

$$c = 2.04 \times 10^{-2} \text{ kg mol/m}^3 \qquad (\text{for } x = 0.01 \text{ m})$$

For $x = 0$ m (i.e., at the surface of the solid),

$$\frac{x}{2\sqrt{D_{AB}t}} = 0$$

From the chart, $1 - Y = 0.62$. Solving, $c = 3.48 \times 10^{-2}$. This value is the same as c_i, as shown in Fig. 7.1-3b. To calculate the concentration c_{Li} in the liquid at the interface,

$$C_{Li} = Kc_i = 2.0(3.48 \times 10^{-2}) = 6.96 \times 10^{-2} \text{ kg mol/m}^3$$

A plot of these values will be similar to Fig. 7.1-3b.

4. Unsteady-state diffusion in more than one direction. In Section 5.3F a method was given for unsteady-state heat conduction to combine the one-dimensional solutions to yield solutions for several-dimensional systems. The same method can be used for unsteady-state diffusion in more than one direction. Rewriting Eq. (5.3-11) for diffusion in a rectangular block in the x, y, and z directions,

$$Y_{x,y,z} = (Y_x)(Y_y)(Y_z) = \frac{c_1/K - c_{x,y,z}}{c_1/K - c_0} \qquad (7.1\text{-}17)$$

where $c_{x,y,z}$ is the concentration at the point x, y, z from the center of the block. The value of Y_x for the two parallel faces is obtained from Fig. 5.3-5 or 5.3-6 for a flat plate in the x direction. The values of Y_y and Y_z are similarly obtained from the same charts. For a short cylinder, an equation similar to Eq. (5.3-12) is used, and for average concentrations, ones similar to Eqs. (5.3-14), (5.3-15), and (5.3-16) are used.

7.2 CONVECTIVE MASS-TRANSFER COEFFICIENTS

7.2A Introduction to Convective Mass Transfer

In the previous sections of this chapter and Chapter 6 we have emphasized molecular diffusion in stagnant fluids or fluids in laminar flow. In many cases the rate of diffusion is slow, and more rapid transfer is desired. To do this, the fluid velocity is increased until turbulent mass transfer occurs.

To have a fluid in convective flow usually requires the fluid to be flowing by another immiscible fluid or by a solid surface. An example is a fluid flowing in a pipe, where part of the pipe wall is made by a slightly dissolving solid material such as benzoic acid. The benzoic acid dissolves and is transported perpendicular to the main stream from the wall. When a fluid is in turbulent flow and is flowing past a surface, the actual velocity of small particles of fluid cannot be described clearly as in laminar flow. In laminar flow the fluid flows in streamlines and its behavior can usually be described mathematically. However, in turbulent motion there are no streamlines, but there are large eddies or "chunks" of fluid moving rapidly in seemingly random fashion.

When a solute A is dissolving from a solid surface there is a high concentration of this solute in the fluid at the surface, and its concentration, in general, decreases as the distance from the wall increases. However, minute samples of fluid adjacent to each other do not always have concentrations close to each other. This occurs because eddies having solute in them move rapidly from one part of the fluid to another, transferring relatively large amounts of solute. This turbulent diffusion or eddy transfer is quite fast in comparison to molecular transfer.

Three regions of mass transfer can be visualized. In the first, which is adjacent to the surface, a thin viscous sublayer film is present. Most of the mass transfer occurs by molecular diffusion, since few or no eddies are present. A large concentration drop occurs across this film as a result of the slow diffusion rate.

The transition or buffer region is adjacent to the first region. Some eddies are present and the mass transfer is the sum of turbulent and molecular diffusion. There is a gradual transition in this region from the transfer by mainly molecular diffusion at the one end to mainly turbulent at the other end.

In the *turbulent region* adjacent to the buffer region, most of the transfer is by turbulent diffusion, with a small amount by molecular diffusion. The concentration decrease is very small here since the eddies tend to keep the fluid concentration uniform. A more detailed discussion of these three regions is given in Section 3.7E.

A typical plot for the mass transfer of a dissolving solid from a surface to a turbulent fluid in a conduit is given in Fig. 7.2-1. The concentration drop from c_{A1} adjacent to the surface is very abrupt close to the surface and then levels off. This curve is very similar to the shapes found for heat and momentum transfer. The average or mixed concentration $\bar{c}_A$ is shown and is slightly greater than the minimum c_{A2}.

7.2B Types of Mass-Transfer Coefficients

1. Definition of mass-transfer coefficient. Since our understanding of turbulent flow is incomplete, we attempt to write the equations for turbulent diffusion in a manner similar to that for molecular diffusion. For turbulent mass transfer for constant c, Eq. (6.1-6) is

FIGURE 7.2-1. *Concentration profile in turbulent mass transfer from a surface to a fluid.*

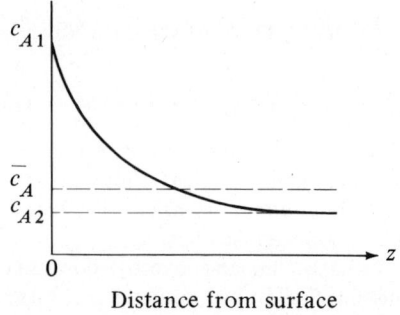

Distance from surface

Transport Processes: Momentum, Heat, and Mass

written as

$$J_A^* = -(D_{AB} + \varepsilon_M) \frac{dc_A}{dz} \tag{7.2-1}$$

where D_{AB} is the molecular diffusivity in m²/s and ε_M is the mass eddy diffusivity in m²/s. The value of ε_M is a variable and is near zero at the interface or surface and increases as the distance from the wall increases. We then use an average value $\bar{\varepsilon}_M$ since the variation of ε_M is not generally known. Integrating Eq. (7.2-1) between points 1 and 2,

$$J_{A1}^* = \frac{D_{AB} + \bar{\varepsilon}_M}{z_2 - z_1} (c_{A1} - c_{A2}) \tag{7.2-2}$$

The flux J_{A1}^* is based on the surface area A_1 since the cross-sectional area may vary. The value of $z_2 - z_1$, the distance of the path, is often not known. Hence, Eq. (7.2-2) is simplified and is written using a convective mass-transfer coefficient k_c'.

$$J_{A1}^* = k_c'(c_{A1} - c_{A2}) \tag{7.2-3}$$

where J_{A1}^* is the flux of A from the surface A_1 relative to the whole bulk phase, k_c' is $(D_{AB} + \bar{\varepsilon}_M)/(z_2 - z_1)$ an experimental mass-transfer coefficient in kg mol/s·m²·(kg mol/m³) or simplified as m/s, and c_{A2} is the concentration at point 2 in kg mol A/m³ or more usually the average bulk concentration $\bar{c}_{A2}$. This defining of a convective mass-transfer coefficient k_c' is quite similar to the convective heat-transfer coefficient h.

2. Mass-transfer coefficient for equimolar counterdiffusion. Generally, we are interested in N_A, the flux of A relative to stationary coordinates. We can start with the following, which is similar to that for molecular diffusion but the term ε_M is added.

$$N_A = -c(D_{AB} + \varepsilon_M) \frac{dx_A}{dz} + x_A(N_A + N_B) \tag{7.2-4}$$

For the case of equimolar counterdiffusion, where $N_A = -N_B$, and integrating at steady state, calling $k_c' = (D_{AB} + \bar{\varepsilon}_M)/(z_2 - z_1)$,

$$N_A = k_c'(c_{A1} - c_{A2}) \tag{7.2-5}$$

Equation (7.2-5) is the defining equation for the mass-transfer coefficient. Often, however, we define the concentration in terms of mole fraction if a liquid or gas or in terms of partial pressure if a gas. Hence, we can define the mass-transfer coefficient in several ways. If y_A is mole fraction in a gas phase and x_A in a liquid phase, then Eq. (7.2-5) can be written as follows for equimolar counterdiffusion:

Gases: $\quad N_A = k_c'(c_{A1} - c_{A2}) = k_G'(p_{A1} - p_{A2}) = k_y'(y_{A1} - y_{A2}) \tag{7.2-6}$

Liquids: $\quad N_A = k_c'(c_{A1} - c_{A2}) = k_L'(c_{A1} - c_{A2}) = k_x'(x_{A1} - x_{A2}) \tag{7.2-7}$

All of these mass-transfer coefficients can be related to each other. For example, using Eq. (7.2-6) and substituting $y_{A1} = c_{A1}/c$ and $y_{A2} = c_{A2}/c$ into the equation,

$$N_A = k_c'(c_{A1} - c_{A2}) = k_y'(y_{A1} - y_{A2}) = k_y'\left(\frac{c_{A1}}{c} - \frac{c_{A2}}{c}\right) = \frac{k_y'}{c}(c_{A1} - c_{A2}) \tag{7.2-8}$$

Hence,

$$k_c' = \frac{k_y'}{c} \tag{7.2-9}$$

These relations among mass-transfer coefficients, and the various flux equations, are given in Table 7.2-1.

TABLE 7.2-1.   *Flux Equations and Mass-Transfer Coefficients*

Flux equations for equimolar counterdiffusion

Gases: $N_A = k'_c(c_{A1} - c_{A2}) = k'_G(p_{A1} - p_{A2}) = k'_y(y_{A1} - y_{A2})$

Liquids: $N_A = k'_c(c_{A1} - c_{A2}) = k'_L(c_{A1} - c_{A2}) = k'_x(x_{A1} - x_{A2})$

Flux equations for A diffusing through stagnant B

Gases: $N_A = k_c(c_{A1} - c_{A2}) = k_G(p_{A1} - p_{A2}) = k_y(y_{A1} - y_{A2})$

Liquids: $N_A = k_c(c_{A1} - c_{A2}) = k_L(c_{A1} - c_{A2}) = k_x(x_{A1} - x_{A2})$

Conversions between mass-transfer coefficients
   Gases:

$$k'_c c = k'_c \frac{P}{RT} = k_c \frac{p_{BM}}{RT} = k'_G P = k_G p_{BM} = k_y y_{BM} = k'_y = k_c y_{BM} c = k_G y_{BM} P$$

   Liquids:

$$k'_c c = k'_L c = k_L x_{BM} c = k'_L \rho/M = k'_x = k_x x_{BM}$$

(where ρ is density of liquid and M is molecular weight)

Units of mass-transfer coefficients

	SI Units	Cgs Units	English Units
k_c, k_L, k'_c, k'_L	m/s	cm/s	ft/h
k_x, k_y, k'_x, k'_y	$\dfrac{\text{kg mol}}{\text{s} \cdot \text{m}^2 \cdot \text{mol frac}}$	$\dfrac{\text{g mol}}{\text{s} \cdot \text{cm}^2 \cdot \text{mol frac}}$	$\dfrac{\text{lb mol}}{\text{h} \cdot \text{ft}^2 \cdot \text{mol frac}}$
k_G, k'_G	$\dfrac{\text{kg mol}}{\text{s} \cdot \text{m}^2 \cdot \text{Pa}}$ $\dfrac{\text{kg mol}}{\text{s} \cdot \text{m}^2 \cdot \text{atm}}$ (preferred)	$\dfrac{\text{g mol}}{\text{s} \cdot \text{cm}^2 \cdot \text{atm}}$	$\dfrac{\text{lb mol}}{\text{h} \cdot \text{ft}^2 \cdot \text{atm}}$

---

3. Mass-transfer coefficient for A through stagnant B.   For A diffusing through stagnant nondiffusing B where $N_B = 0$, Eq. (7.2-4) gives for steady state

$$N_A = \frac{k'_c}{x_{BM}} (c_{A1} - c_{A2}) = k_c(c_{A1} - c_{A2}) \qquad (7.2\text{-}10)$$

where the x_{BM} and its counterpart y_{BM} are similar to Eq. (6.2-21) and k_c is the mass-transfer coefficient for A diffusing through stagnant B. Also,

$$x_{BM} = \frac{x_{B2} - x_{B1}}{\ln (x_{B2}/x_{B1})} \qquad y_{BM} = \frac{y_{B2} - y_{B1}}{\ln (y_{B2}/y_{B1})} \qquad (7.2\text{-}11)$$

Rewriting Eq. (7.2-10) using other units,

(Gases): $N_A = k_c(c_{A1} - c_{A2}) = k_G(p_{A1} - p_{A2}) = k_y(y_{A1} - y_{A2})$ (7.2-12)

(Liquids): $N_A = k_c(c_{A1} - c_{A2}) = k_L(c_{A1} - c_{A2}) = k_x(x_{A1} - x_{A2})$ (7.2-13)

Again all the mass-transfer coefficients can be related to each other and are given in

The viscosity μ and density ρ used are the actual flowing mixture of solute A and fluid B. If the mixture is dilute, properties of the pure fluid B can be used. The Prandtl number $c_p \mu/k$ for heat transfer is analogous to the Schmidt number for mass transfer. The Schmidt number is the ratio of the shear component for diffusivity μ/ρ to the diffusivity for mass transfer D_{AB}, and it physically relates the relative thickness of the hydrodynamic layer and mass-transfer boundary layer.

The *Sherwood number*, which is dimensionless, is

$$N_{\text{Sh}} = k_c' \frac{L}{D_{AB}} = k_c \, y_{BM} \frac{L}{D_{AB}} = \frac{k_x'}{c} \frac{L}{D_{AB}} = \cdots \qquad (7.3\text{-}3)$$

Other substitutions from Table 7.2-1 can be made for k_c' in Eq. (7.3-3).

The *Stanton number* occurs often and is

$$N_{\text{St}} = \frac{k_c'}{v} = \frac{k_y'}{G_M} = \frac{k_G' P}{G_M} = \cdots \qquad (7.3\text{-}4)$$

Again, substitution for k_c' can be made. $G_M = v\rho/M_{\text{av}} = vc$.

Often the mass-transfer coefficient is correlated as a dimensionless J_D factor which is related to k_c' and N_{Sh} as follows.

$$J_D = \frac{k_c'}{v} (N_{\text{Sc}})^{2/3} = \frac{k_G' P}{G_M} (N_{\text{Sc}})^{2/3} = \cdots = N_{\text{Sh}}/(N_{\text{Re}} \, N_{\text{Sc}}^{1/3}) \qquad (7.3\text{-}5)$$

For heat transfer a dimensionless J_H factor is as follows:

$$J_H = \frac{h}{c_p G} (N_{\text{Pr}})^{2/3} \qquad (7.3\text{-}6)$$

7.3B Analogies Among Mass, Heat, and Momentum Transfer

1. Introduction. In molecular transport of momentum, heat, or mass there are many similarities, which were pointed out in Chapters 2 to 6. The molecular diffusion equations of Newton for momentum, Fourier for heat, and Fick for mass are very similar and we can say that we have analogies among these three molecular transport processes. There are also similarities in turbulent transport, as discussed in Sections 5.7C and 6.1A, where the flux equations were written using the turbulent eddy momentum diffusivity ε_t, the turbulent eddy thermal diffusivity α_t, and the turbulent eddy mass diffusivity ε_M. However, these similarities are not as well defined mathematically or physically and are more difficult to relate to each other.

A great deal of effort has been devoted in the literature to developing analogies among these three transport processes for turbulent transfer so as to allow prediction of one from any of the others. We discuss several next.

2. Reynolds analogy. Reynolds was the first to note similarities in transport processes and relate turbulent momentum and heat transfer. Since then, mass transfer has also been related to momentum and heat transfer. We derive this analogy from Eqs. (6.1-4)–(6.1-6) for turbulent transport. For fluid flow in a pipe for heat transfer from the fluid to the wall, Eq. (6.1-5) becomes as follows, where z is distance from the wall:

$$\frac{q}{A} = -\rho c_p (\alpha + \alpha_t) \frac{dT}{dz} \qquad (7.3\text{-}7)$$

For momentum transfer, Eq. (6.1-4) becomes

$$\tau = -\rho\left(\frac{\mu}{\rho} + \varepsilon_t\right)\frac{dv}{dz} \tag{7.3-8}$$

Next we assume α and μ/ρ are negligible and that $\alpha_t = \varepsilon_t$. Then dividing Eq. (7.3-7) by (7.3-8),

$$\left(\frac{\tau}{q/A}\right)c_p\, dT = dv \tag{7.3-9}$$

If we assume that heat flux q/A in a turbulent system is analogous to momentum flux τ, the ratio $\tau/(q/A)$ must be constant for all radial positions. We now integrate between conditions at the wall where $T = T_i$ and $v = 0$ to some point in the fluid where T is the same as the bulk T and assume that the velocity at this point is the same as v_{av}, the bulk velocity. Also, q/A is understood to be the flux at the wall, as is the shear at the wall, written as τ_s. Hence,

$$\frac{\tau_s}{q/A}\, c_p(T - T_i) = v_{av} - 0 \tag{7.3-10}$$

Also, substituting $q/A = h(T - T_i)$ and $\tau_s = f v_{av}^2\, \rho/2$ from Eq. (2.10-4) into Eq. (7.3-10),

$$\frac{f}{2} = \frac{h}{c_p v_{av} \rho} = \frac{h}{c_p G} \tag{7.3-11}$$

In a similar manner using Eq. (6.1-6) for J_A^* and also $J_A^* = k_c'(c_A - c_{Ai})$, we can relate this to Eq. (7.3-8) for momentum transfer. Then, the complete Reynolds analogy is

$$\frac{f}{2} = \frac{h}{c_p G} = \frac{k_c'}{v_{av}} \tag{7.3-12}$$

Experimental data for gas streams agree approximately with Eq. (7.3-12) if the Schmidt and Prandtl numbers are near 1.0 and only skin friction is present in flow past a flat plate or inside a pipe. When liquids are present and/or form drag is present, the analogy is not valid.

3. *Other analogies.* The Reynolds analogy assumes that the turbulent diffusivities ε_t, α_t, and ε_M are all equal and that the molecular diffusivities μ/ρ, α, and D_{AB} are negligible compared to the turbulent diffusivities. When the Prandtl number $(\mu/\rho)/\alpha$ is 1.0, then $\mu/\rho = \alpha$; also, for $N_{Sc} = 1.0$, $\mu/\rho = D_{AB}$. Then, $(\mu/\rho + \varepsilon_t) = (\alpha + \alpha_t) = (D_{AB} + \varepsilon_M)$ and the Reynolds analogy can be obtained with the molecular terms present. However, the analogy breaks down when the viscous sublayer becomes important since the eddy diffusivities diminish to zero and the molecular diffusivities become important.

Prandtl modified the Reynolds analogy by writing the regular molecular diffusion equation for the viscous sublayer and a Reynolds-analogy equation for the turbulent core region. Then since these processes are in series, these equations were combined to produce an overall equation (G1). The results also are poor for fluids where the Prandtl and Schmidt numbers differ from 1.0.

Von Kármán further modified the Prandtl analogy by considering the buffer region in addition to the viscous sublayer and the turbulent core. These three regions are shown in the universal velocity profile in Fig. 3.7-4. Again, an equation is written for molecular diffusion in the viscous sublayer using only the molecular diffusivity and a Reynolds analogy equation for the turbulent core. Both the molecular and eddy diffusivity are used in an equation for the buffer layer, where the velocity in this layer is used to obtain an equation for the eddy diffusivity. These three equations are then combined to give the von Kármán analogy. Since then, numerous other analogies have appeared (P1).

4. *Chilton and Colburn J-factor analogy.* The most successful and most widely used analogy is the Chilton and Colburn *J*-factor analogy (C2). This analogy is based on experimental data for gases and liquids in both the laminar and turbulent flow regions and is written as follows:

$$\frac{f}{2} = J_H = \frac{h}{c_p G}(N_{Pr})^{2/3} = J_D = \frac{k_c'}{v_{av}}(N_{Sc})^{2/3} \qquad \text{(7.3-13)}$$

Although this is an equation based on experimental data for both laminar and turbulent flow, it can be shown to satisfy the exact solution derived from laminar flow over a flat plate in Sections 3.7 and 5.7.

Equation (7.3-13) has been shown to be quite useful in correlating momentum, heat, and mass transfer data. It permits the prediction of an unknown transfer coefficient when one of the other coefficients is known. In momentum transfer the friction factor is obtained for the total drag or friction loss, which includes form drag or momentum losses due to blunt objects and also skin friction. For flow past a flat plate or in a pipe where no form drag is present, $f/2 = J_H = J_D$. When form drag is present, such as in flow in packed beds or past other blunt objects, $f/2$ is greater than J_H or J_D and $J_H \cong J_D$.

7.3C Derivation of Mass-Transfer Coefficients in Laminar Flow

1. *Introduction.* When a fluid is flowing in laminar flow and mass transfer by molecular diffusion is occurring, the equations are very similar to those for heat transfer by conduction in laminar flow. The phenomena of heat and mass transfer are not always completely analogous since in mass transfer several components may be diffusing. Also, the flux of mass perpendicular to the direction of the flow must be small so as not to distort the laminar velocity profile.

In theory it is not necessary to have experimental mass-transfer coefficients for laminar flow, since the equations for momentum transfer and for diffusion can be solved. However, in many actual cases it is difficult to describe mathematically the laminar flow for geometries, such as flow past a cylinder or in a packed bed. Hence, experimental mass-transfer coefficients are often obtained and correlated. A simplified theoretical derivation will be given for two cases in laminar flow.

2. *Mass transfer in laminar flow in a tube.* We consider the case of mass transfer from a tube wall to a fluid inside in laminar flow, where, for example, the wall is made of solid benzoic acid which is dissolving in water. This is similar to heat transfer from a wall to the flowing fluid where natural convection is negligible. For fully developed flow, the parabolic velocity derived as Eqs. (2.6-18) and (2.6-20) is

$$v_x = v_{max}\left[1 - \left(\frac{r}{R}\right)^2\right] = 2v_{av}\left[1 - \left(\frac{r}{R}\right)^2\right] \qquad \text{(7.3-14)}$$

where v_x is the velocity in the x direction at the distance r from the center. For steady-state diffusion in a cylinder, a mass balance can be made on a differential element where the rate in by convection plus diffusion equals the rate out radially by diffusion to give

$$v_x\frac{\partial c_A}{\partial x} = D_{AB}\left(\frac{1}{r}\frac{\partial c_A}{\partial r} + \frac{\partial^2 c_A}{\partial r^2} + \frac{\partial^2 c_A}{\partial x^2}\right) \qquad \text{(7.3-15)}$$

Then, $\partial^2 c_A/\partial x^2 = 0$ if the diffusion in the x direction is negligible compared to that by convection. Combining Eqs. (7.3-14) and (7.3-15), the final solution (S1) is a complex series similar to the Graetz solution for heat transfer and a parabolic velocity profile.

If it is assumed that the velocity profile is flat as in rodlike flow, the solution is more easily obtained (S1). A third solution, called the approximate *Leveque solution*, has been obtained, where there is a linear velocity profile near the wall and the solute diffuses only a short distance from the wall into the fluid. This is similar to the parabolic velocity profile solution at high flow rates. Experimental design equations are presented in Section 7.3D for this case.

3. Diffusion in a laminar falling film.   In Section 2.9C we derived the equation for the velocity profile in a falling film shown in Fig. 7.3-1a. We will consider mass transfer of solute A into a laminar falling film, which is important in wetted-wall columns, in developing theories to explain mass transfer in stagnant pockets of fluids, and in turbulent mass transfer. The solute A in the gas is absorbed at the interface and then diffuses a distance into the liquid so that it has not penetrated the whole distance $x = \delta$ at the wall. At steady state the inlet concentration $c_A = 0$. At a point z distance from the inlet the concentration profile of c_A is shown in Fig. 7.3-1a.

A mass balance will be made on the element shown in Fig. 7.3-1b. For steady state, rate of input = rate of output.

$$N_{Ax|x}(1\ \Delta z) + N_{Az|z}(1\ \Delta x) = N_{Ax|x+\Delta x}(1\ \Delta z) + N_{Az|z+\Delta z}(1\ \Delta x) \qquad (7.3\text{-}16)$$

For a dilute solution the diffusion equation for A in the x direction is

$$N_{Ax} = -D_{AB}\frac{\partial c_A}{\partial x} + \text{zero convection} \qquad (7.3\text{-}17)$$

For the z direction the diffusion is negligible.

$$N_{Az} = 0 + c_A v_z \qquad (7.3\text{-}18)$$

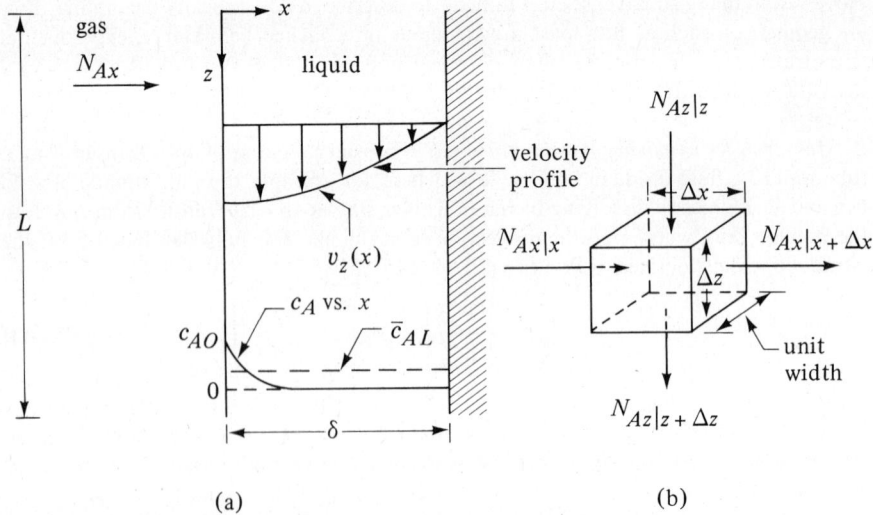

(a)   (b)

FIGURE 7.3-1.   *Diffusion of solute A in a laminar falling film: (a) velocity profile and concentration profile, (b) small element for mass balance.*

Dividing Eq. (7.3-16) by $\Delta x \, \Delta z$, letting Δx and Δz approach zero, and substituting Eqs. (7.3-17) and (7.3-18) into the result, we obtain

$$v_z \frac{\partial c_A}{\partial z} = D_{AB} \frac{\partial^2 c_A}{\partial x^2} \tag{7.3-19}$$

From Eqs. (2.9-24) and (2.9-25), the velocity profile is parabolic and is $v_z = v_{z\,max} \, [1 - (x/\delta)^2]$. Also, $v_{z\,max} = (3/2)v_{z\,av}$. If the solute has penetrated only a short distance into the fluid, i.e., short contact times of t seconds equals z/v_{max}, then the A that has diffused has been carried along at the velocity $v_{z\,max}$ or v_{max} if the subscript z is dropped. Then Eq. (7.3-19) becomes

$$\frac{\partial c_A}{\partial (z/v_{max})} = D_{AB} \frac{\partial^2 c_A}{\partial x^2} \tag{7.3-20}$$

Using the boundary conditions of $c_A = 0$ at $z = 0$, $c_A = c_{A0}$ at $x = 0$, and $c_A = 0$ at $x = \infty$, we can integrate Eq. (7.3-20) to obtain

$$\frac{c_A}{c_{A0}} = \operatorname{erfc}\left(\frac{x}{\sqrt{4D_{AB}\,z/v_{max}}}\right) \tag{7.3-21}$$

where erf y is the error function and erfc $y = 1 - \operatorname{erf} y$. Values of erf y are standard tabulated functions.

To determine the local molar flux at the surface $x = 0$ at position z from the top entrance, we write (B1)

$$N_{Ax}(z)\bigg|_{x=0} = -D_{AB} \frac{\partial c_A}{\partial x}\bigg|_{x=0} = c_{A0} \sqrt{\frac{D_{AB}\,v_{max}}{\pi z}} \tag{7.3-22}$$

The total moles of A transferred per second to the liquid over the entire length $z = 0$ to $z = L$, where the vertical surface is unit width is

$$N_A(L \cdot 1) = (1) \int_0^L (N_{Ax|x=0}) \, dz$$

$$= (1) \int_0^L c_{A0} \left(\frac{D_{AB}\,v_{max}}{\pi}\right)^{1/2} \frac{1}{z^{1/2}} \, dz$$

$$= (L \cdot 1)c_{A0} \sqrt{\frac{4D_{AB}\,v_{max}}{\pi L}} \tag{7.3-23}$$

The term L/v_{max} is t_L, time of exposure of the liquid to the solute A in the gas. This means the rate of mass transfer is proportional to $D_{AB}^{0.5}$ and $1/t_L^{0.5}$. This is the basis for the penetration theory in turbulent mass transfer where pockets of liquid are exposed to unsteady-state diffusion (penetration) for short contact times.

7.3D   Mass Transfer for Flow Inside Pipes

1. Mass transfer for laminar flow inside pipes.   When a liquid or a gas is flowing inside a pipe and the Reynolds number $Dv\rho/\mu$ is below 2100, laminar flow occurs. Experimental data obtained for mass transfer from the walls for gases (G2, L1) are plotted in Fig. 7.3-2 for values of $W/D_{AB}\rho L$ less than about 70. The ordinate is $(c_A - c_{A0})/(c_{Ai} - c_{A0})$, where c_A is the exit concentration, c_{A0} inlet concentration, and c_{Ai} concentration at the interface between the wall and the gas. The dimensionless abscissa is $W/D_{AB}\rho L$ or $N_{Re}N_{Sc}(D/L)(\pi/4)$, where W is flow in kg/s and L is length of mass-transfer section in m.

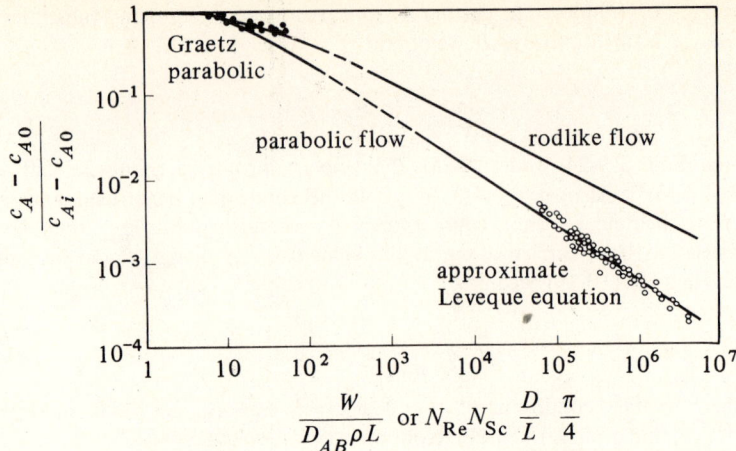

FIGURE 7.3-2. *Data for diffusion in a fluid in streamline flow inside a pipe: filled circles, vaporization data of Gilliland and Sherwood; open circles, dissolving-solids data of Linton and Sherwood. [From W. H. Linton and T. K. Sherwood, Chem. Eng. Progr., 46, 258 (1950). With permission.]*

Since the experimental data follow the rodlike plot, that line should be used. The velocity profile is assumed fully developed to parabolic form at the entrance.

For liquids that have small values of D_{AB}, data follow the parabolic flow line, which is as follows for $W/D_{AB}\rho L$ over 400.

$$\frac{c_A - c_{A0}}{c_{Ai} - c_{A0}} = 5.5\left(\frac{W}{D_{AB}\rho L}\right)^{-2/3} \tag{7.3-24}$$

2. *Mass transfer for turbulent flow inside pipes.* For turbulent flow for $Dv\rho/\mu$ above 2100 for gases or liquids flowing inside a pipe,

$$N_{Sh} = k'_c \frac{D}{D_{AB}} = \frac{k_c p_{BM}}{P} \frac{D}{D_{AB}} = 0.023\left(\frac{Dv\rho}{\mu}\right)^{0.83}\left(\frac{\mu}{\rho D_{AB}}\right)^{0.33} \tag{7.3-25}$$

The equation holds for N_{Sc} of 0.6 to 3000 (G2, L1). Note that the N_{Sc} for gases is in the range 0.5–3 and for liquids is above 100 in general. Equation (7.3-25) for mass transfer and Eq. (4.5-8) for heat transfer inside a pipe are essentially identical.

3. *Mass transfer for flow inside wetted-wall towers.* When a gas is flowing inside the core of a wetted-wall tower the same correlations that are used for mass transfer of a gas in laminar or turbulent flow in a pipe are applicable. This means that Eqs. (7.3-24) and (7.3-25) can be used to predict mass transfer for the gas. For the mass transfer in the liquid film flowing down the wetted-wall tower, Eqs. (7.3-22) and (7.3-23) can be used for Reynolds numbers of $4\Gamma/\mu$ as defined by Eq. (2.9-29) up to about 1200, and the theoretically predicted values should be multiplied by about 1.5 because of ripples and other factors. These equations hold for short contact times or Reynolds numbers above about 100 (S1).

EXAMPLE 7.3-1. Mass Transfer Inside a Tube
A tube is coated on the inside with naphthalene and has an inside diameter of 20 mm and a length of 1.10 m. Air at 318 K and an average pressure of

101.3 kPa flows through this pipe at a velocity of 0.80 m/s. Assuming that the absolute pressure remains essentially constant, calculate the concentration of naphthalene in the exit air. Use the physical properties given in Example 6.2-4.

Solution: From Example 6.2-4, $D_{AB} = 6.92 \times 10^{-6}$ m^2/s and the vapor pressure $p_{Ai} = 74.0$ Pa or $c_{Ai} = p_{Ai}/RT = 74.0/(8314.3 \times 318) = 2.799 \times 10^{-5}$ kg mol/m^3. For air from Appendix A.3, $\mu = 1.932 \times 10^{-5}$ Pa·s, $\rho = 1.114$ kg/m^3. The Schmidt number is

$$N_{Sc} = \frac{\mu}{\rho D_{AB}} = \frac{1.932 \times 10^{-5}}{1.114 \times 6.92 \times 10^{-6}} = 2.506$$

The Reynolds number is

$$N_{Re} = \frac{Dv\rho}{\mu} = \frac{0.020(0.80)(1.114)}{1.932 \times 10^{-5}} = 922.6$$

Hence, the flow is laminar. Then,

$$N_{Re}N_{Sc}\frac{D}{L}\frac{\pi}{4} = 922.6(2.506)\frac{0.020}{1.10}\frac{\pi}{4} = 33.02$$

Using Fig. 7.3-2 and the rodlike flow line, $(c_A - c_{A0})/(c_{Ai} - c_{A0}) = 0.55$. Also, c_{A0}(inlet) = 0. Then, $(c_A - 0)/(2.799 \times 10^{-5} - 0) = 0.55$. Solving, c_A(exit concentration) $= 1.539 \times 10^{-5}$ kg mol/m^3.

7.3E Mass Transfer for Flow Outside Solid Surfaces

1. Mass transfer in flow parallel to flat plates. The mass transfer and vaporization of liquids from a plate or flat surface to a flowing stream is of interest in the drying of inorganic and biological materials, in evaporation of solvents from paints, for plates in wind tunnels, and in flow channels in chemical process equipment.

When the fluid flows past a plate in a free stream in an open space the boundary layer is not fully developed. For gases or evaporation of liquids in the gas phase and for the laminar region of $N_{Re, L} = Lv\rho/\mu$ less than 15 000, the data can be represented within $\pm 25\%$ by the equation (S4)

$$J_D = 0.664N_{Re, L}^{-0.5} \tag{7.3-26}$$

Writing Eq. (7.3-26) in terms of the Sherwood number N_{Sh},

$$\frac{k_c' L}{D_{AB}} = N_{Sh} = 0.664N_{Re, L}^{0.5} N_{Sc}^{1/3} \tag{7.3-27}$$

where L is the length of plate in the direction of flow. Also, $J_D = J_H = f/2$ for this geometry. For gases and $N_{Re, L}$ of 15 000–300 000, the data are represented within $\pm 30\%$ by $J_D = J_H = f/2$ as

$$J_D = 0.036N_{Re, L}^{-0.2} \tag{7.3-28}$$

Experimental data for liquids are correlated within about $\pm 40\%$ by the following for a $N_{Re, L}$ of 600–50 000 (L2):

$$J_D = 0.99N_{Re, L}^{-0.5} \tag{7.3-29}$$

EXAMPLE 7.3-2. Mass Transfer from a Flat Plate
A large volume of pure water at 26.1°C is flowing parallel to a flat plate of solid benzoic acid, where $L = 0.244$ m in the direction of flow. The water

velocity is 0.061 m/s. The solubility of benzoic acid in water is 0.02948 kg mol/m^3. The diffusivity of benzoic acid is 1.245×10^{-9} m^2/s. Calculate the mass-transfer coefficient k_L and the flux N_A.

Solution: Since the solution is quite dilute, the physical properties of water at 26.1°C from Appendix A.2 can be used.

$$\mu = 8.71 \times 10^{-4} \text{ Pa} \cdot \text{s}$$

$$\rho = 996 \text{ kg/m}^3$$

$$D_{AB} = 1.245 \times 10^{-9} \text{ m}^2/\text{s}$$

The Schmidt number is

$$N_{Sc} = \frac{8.71 \times 10^{-4}}{996(1.245 \times 10^{-9})} = 702$$

The Reynolds number is

$$N_{Re, L} = \frac{Lv\rho}{\mu} = \frac{0.244(0.0610)(996)}{8.71 \times 10^{-4}} = 1.700 \times 10^4$$

Using Eq. (7.3-29),

$$J_D = 0.99N_{Re, L}^{-0.5} = 0.99(1.700 \times 10^4)^{-0.5} = 0.00758$$

The definition of J_D from Eq. (7.3-5) is

$$J_D = \frac{k_c'}{v}(N_{Sc})^{2/3} \qquad (7.3-5)$$

Solving for k_c', $k_c' = J_D v(N_{Sc})^{-2/3}$. Substituting known values and solving,

$$k_c' = 0.00758(0.0610)(702)^{-2/3} = 5.85 \times 10^{-6} \text{ m/s}$$

In this case, diffusion is for A through nondiffusing B, so k_c in Eq. (7.2-10) should be used.

$$N_A = \frac{k_c'}{x_{BM}}(c_{A1} - c_{A2}) = k_c(c_{A1} - c_{A2}) \qquad (7.2-10)$$

Since the solution is very dilute, $x_{BM} \cong 1.0$ and $k_c' \cong k_c$. Also, $c_{A1} = 2.948 \times 10^{-2}$ kg mol/m^3 (solubility) and $c_{A2} = 0$ (large volume of fresh water). Substituting into Eq. (7.2-10),

$$N_A = (5.85 \times 10^{-6})(0.02948 - 0) = 1.726 \times 10^{-7} \text{ kg mol/s} \cdot \text{m}^2$$

2. *Mass transfer for flow past single spheres.* For flow past single spheres and for very low $N_{Re} = D_p v\rho/\mu$, where v is the average velocity in the empty test section before the sphere, the Sherwood number, which is $k_c' D_p/D_{AB}$, should approach a value of 2.0. This can be shown from Eq. (6.2-33), which was derived for a stagnant medium. Rewriting Eq. (6.2-33) as follows, where D_p is the sphere diameter,

$$N_A = \frac{2D_{AB}}{D_p}(c_{A1} - c_{A2}) = k_c(c_{A1} - c_{A2}) \qquad (7.3-30)$$

The mass-transfer coefficient k_c, which is k_c' for a dilute solution, is then

$$k_c' = \frac{2D_{AB}}{D_p} \qquad (7.3-31)$$

Rearranging,

$$\frac{k'_c D_p}{D_{AB}} = N_{Sh} = 2.0 \tag{7.3-32}$$

Of course, natural convection effects could increase k'_c.

For gases for a Schmidt number range of 0.6–2.7 and a Reynolds number range of 1–48 000, a modified equation (G1) can be used.

$$N_{Sh} = 2 + 0.552 N_{Re}^{0.53} N_{Sc}^{1/3} \tag{7.3-33}$$

This equation also holds for heat transfer where the Prandtl number replaces the Schmidt number and the Nusselt number hD_p/k replaces the Sherwood number.

For liquids (G3) and a Reynolds number range of 2 to about 2000, the following can be used.

$$N_{Sh} = 2 + 0.95 N_{Re}^{0.50} N_{Sc}^{1/3} \tag{7.3-34}$$

For liquids and a Reynolds number of 2000–17 000, the following can be used (S5).

$$N_{Sh} = 0.347 N_{Re}^{0.62} N_{Sc}^{1/3} \tag{7.3-35}$$

EXAMPLE 7.3-3. Mass Transfer from a Sphere
Calculate the value of the mass-transfer coefficient and the flux for mass transfer from a sphere of napthalene to air at 45°C and 1 atm abs flowing at a velocity of 0.305 m/s. The diameter of the sphere is 25.4 mm. The diffusivity of naphthalene in air at 45°C is 6.92×10^{-6} m^2/s and the vapor pressure of solid naphthalene is 0.555 mm Hg. Use English and SI units.

Solution: In English units $D_{AB} = 6.92 \times 10^{-6}(3.875 \times 10^4) = 0.2682$ ft^2/h. The diameter $D_p = 0.0254$ m $= 0.0254(3.2808) = 0.0833$ ft. From Appendix A.3 the physical properties of air will be used since the concentration of naphthalene is low.

$$\mu = 1.93 \times 10^{-5} \text{ Pa} \cdot \text{s} = 1.93 \times 10^{-5}(2.4191 \times 10^3) = 0.0467 \text{ lb}_m/\text{ft} \cdot \text{h}$$

$$\rho = 1.113 \text{ kg/m}^3 = \frac{1.113}{16.0185} = 0.0695 \text{ lb}_m/\text{ft}^3$$

$$v = 0.305 \text{ m/s} = 0.305(3600 \times 3.2808) = 3600 \text{ ft/h}$$

The Schmidt number is

$$N_{Sc} = \frac{\mu}{\rho D_{AB}} = \frac{0.0467}{0.0695(0.2682)} = 2.505$$

$$N_{Sc} = \frac{1.93 \times 10^{-5}}{1.113(6.92 \times 10^{-6})} = 2.505$$

The Reynolds number is

$$N_{Re} = \frac{D_p v \rho}{\mu} = \frac{0.0833(3600)(0.0695)}{0.0467} = 446$$

$$N_{Re} = \frac{0.0254(0.3048)(1.113)}{1.93 \times 10^{-5}} = 446$$

Equation (7.3-33) for gases will be used.

$$N_{Sh} = 2 + 0.552(N_{Re})^{0.53}(N_{Sc})^{1/3} = 2 + 0.552(446)^{0.53}(2.505)^{1/3} = 21.0$$

From Eq. (7.3-3),

$$N_{Sh} = k'_c \frac{L}{D_{AB}} = k'_c \frac{D_p}{D_{AB}}$$

Substituting the knowns and solving,

$$21.0 = \frac{k'_c(0.0833)}{0.2682} \qquad k'_c = 67.6 \text{ ft/h}$$

$$21.0 = \frac{k'_c(0.0254)}{6.92 \times 10^{-6}} \qquad k'_c = 5.72 \times 10^{-3} \text{ m/s}$$

From Table 7.2-1,

$$k'_c c = k'_c \frac{P}{RT} = k'_G P$$

Hence, for $T = 45 + 273 = 318 \text{ K} = 318(1.8) = 574°\text{R}$,

$$k'_G = \frac{k'_c}{RT} = \frac{67.6}{(0.730)(573)} = 0.1616 \text{ lb mol/h} \cdot \text{ft}^2 \cdot \text{atm}$$

$$k'_G = \frac{5.72 \times 10^{-3}}{8314(318)} = 2.163 \times 10^{-9} \text{ kg mol/s} \cdot \text{m}^2 \cdot \text{Pa}$$

Since the gas is very dilute, $y_{BM} \cong 1.0$ and $k'_G \cong k_G$. Substituting into Eq. (7.2-12) for A diffusing through stagnant B and noting that $p_{A1} = 0.555/760 = 7.303 \times 10^{-4}$ atm $= 74.0$ Pa and $p_{A2} = 0$ (pure air),

$$N_A = k_G(p_{A1} - p_{A2}) = 0.1616(7.303 \times 10^{-4} - 0)$$

$$= 1.180 \times 10^{-4} \text{ lb mol/h} \cdot \text{ft}^2$$

$$= 2.163 \times 10^{-9}(74.0 - 0) = 1.599 \times 10^{-7} \text{ kg mol/s} \cdot \text{m}^2$$

The area of the sphere is

$$A = \pi D_p^2 = \pi(0.0833)^2 = 2.18 \times 10^{-2} \text{ ft}^2$$

$$= (2.18 \times 10^{-2})\left(\frac{1}{3.2808}\right)^2 = 2.025 \times 10^{-3} \text{ m}^2$$

Total amount evaporated $= N_A A = (1.18 \times 10^{-4})(2.18 \times 10^{-2}) = 2.572 \times 10^{-6}$ lb mol/h $= (1.599 \times 10^{-7})(2.025 \times 10^{-3}) = 3.238 \times 10^{-10}$ kg mol/s

3. *Mass transfer to packed beds.* Mass transfer to and from packed beds occurs often in processing operations, including drying operations, adsorption or desorption of gases or liquids by solid particles such as charcoal, and mass transfer of gases and liquids to catalyst particles. Using a packed bed a large amount of mass-transfer area can be contained in a relatively small volume.

The void fraction in a bed is ε, m^3 volume void space divided by the m^3 total volume of void space plus solid. The values range from 0.3 to 0.5 in general. Because of flow channeling, nonuniform packing, etc., accurate experimental data are difficult to obtain and data from different investigators can deviate considerably.

For a Reynolds number range of 10–10000 for gases in a packed bed of spheres (D4), the recommended correlation with an average deviation of about $\pm 20\%$ and a maximum of about $\pm 50\%$ is

$$J_D = J_H = \frac{0.4548}{\varepsilon} N_{Re}^{-0.4069} \qquad (7.3\text{-}36)$$

It has been shown (G4, G5) that J_D and J_H are approximately equal. The Reynolds number is defined as $N_{Re} = D_p v' \rho / \mu$, where D_p is diameter of the spheres and v' is the superficial mass average velocity in the empty tube without packing. For Eqs. (7.3-36)–(7.3-39) and Eqs. (7.3-5)–(7.3-6), v' is used.

For mass transfer of liquids in packed beds, the correlations of Wilson and Geankoplis (W1) should be used. For a Reynolds number $D_p v' \rho / \mu$ range of 0.0016–55 and a Schmidt number range of 165–70 600, the equation to use is

$$J_D = \frac{1.09}{\varepsilon} N_{Re}^{-2/3} \tag{7.3-37}$$

For liquids and a Reynolds number range of 55–1500 and a Schmidt number range of 165–10 690,

$$J_D = \frac{0.250}{\varepsilon} N_{Re}^{-031} \tag{7.3-38}$$

Or, as an alternate, Eq. (7.3-36) can be used for liquids for a Reynolds number range of 10–1500.

For fluidized beds of spheres, Eq. (7.3-36) can be used for gases and liquids and a Reynolds number range of 10–4000. For liquids in a fluidized bed and a Reynolds number range of 1–10 (D4),

$$\varepsilon J_D = 1.1068 \, N_{Re}^{-0.72} \tag{7.3-39}$$

If packed beds of solids other than spheres are used, approximate correction factors can be used with Eqs. (7.3-36)–(7.3-38) for spheres. This is done, for example, for a given nonspherical particle as follows. The particle diameter to use in the equations to predict J_D is the diameter of a sphere with the same surface area as the given solid particle. The flux to these particles in the bed is then calculated using the area of the given particles. An alternative approximate procedure to use is given elsewhere (G6).

4. *Calculation method for packed beds.* To calculate the total flux in a packed bed, J_D is first obtained and then k_c in m/s from the J_D. Then knowing the total volume V_b m³ of the bed (void plus solids), the total external surface area A m² of the solids for mass transfer is calculated using Eqs. (7.3-40) and (7.3-41).

$$a = \frac{6(1 - \varepsilon)}{D_p} \tag{7.3-40}$$

where a is the m² surface area/m³ total volume of bed when the solids are spheres.

$$A = a V_b \tag{7.3-41}$$

To calculate the mass-transfer rate the log mean driving force at the inlet and outlet of the bed should be used.

$$N_A A = A k_c \frac{(c_{Ai} - C_{A1}) - (c_{Ai} - c_{A2})}{\ln \dfrac{c_{Ai} - c_{A1}}{c_{Ai} - c_{A2}}} \tag{7.3-42}$$

where the final term is the log mean driving force: c_{Ai} is the concentration at the surface of the solid, in kg mol/m³; c_{A1} is the inlet bulk fluid concentration; and c_{A2} is the outlet. The material-balance equation on the bulk stream is

$$N_A A = V(c_{A2} - c_{A1}) \tag{7.3-43}$$

where V is volumetric flow rate of fluid entering in m³/s. Equations (7.3-42) and (7.3-43) must both be satisfied. The use of these two equations is similar to the use of the log mean temperature difference and heat balance in heat exchangers. These two equations can also be used for a fluid flowing in a pipe or past a flat plate, where A is the pipe wall area or plate area.

EXAMPLE 7.3-4. Mass Transfer of a Liquid in a Packed Bed
Pure water at 26.1°C flows at the rate of 5.514×10^{-7} m³/s through a packed bed of benzoic acid spheres having a diameter of 6.375 mm. The total surface area of the spheres in the bed is 0.01198 m² and the void fraction is 0.436. The tower diameter is 0.0667 m. The solubility of benzoic acid in water is 2.948×10^{-2} kg mol/m³.
   (a) Predict the mass-transfer coefficient k_c. Compare with the experimental value of 4.665×10^{-6} m/s by Wilson and Geankoplis (W1).
   (b) Using the experimental value of k_c, predict the outlet concentration of benzoic acid in the water.

Solution: Since the solution is dilute, the physical properties of water will be used at 26.1°C from Appendix A.2. At 26.1°C, $\mu = 0.8718 \times 10^{-3}$ Pa·s, $\rho = 996.7$ kg/m³. At 25.0°C, $\mu = 0.8940 \times 10^{-3}$ Pa·s and from Table 6.3-1, $D_{AB} = 1.21 \times 10^{-9}$ m²/s. To correct D_{AB} to 26.1°C using Eq. (6.3-9), $D_{AB} \propto T/\mu$. Hence,

$$D_{AB}(26.1°C) = (1.21 \times 10^{-9})\left(\frac{299.1}{298}\right)\left(\frac{0.8940 \times 10^{-3}}{0.8718 \times 10^{-3}}\right)$$

$$= 1.254 \times 10^{-9} \text{ m}^2/\text{s}$$

The tower cross-sectional area $= (\pi/4)(0.0667)^2 = 3.494 \times 10^{-3}$ m². Then $v' = (5.514 \times 10^{-7})/(3.494 \times 10^{-3}) = 1.578 \times 10^{-4}$ m/s. Then,

$$N_{Sc} = \frac{\mu}{\rho D_{AB}} = \frac{0.8718 \times 10^{-3}}{996.7(1.245 \times 10^{-9})} = 702.6$$

The Reynolds number is

$$N_{Re} = \frac{D v' \rho}{\mu} = \frac{0.006375(1.578 \times 10^{-4})(996.7)}{0.8718 \times 10^{-3}} = 1.150$$

Using Eq. (7.3-37) and assuming $k_c = k_c'$ for dilute solutions,

$$J_D = \frac{1.09}{\varepsilon}(N_{Re})^{-2/3} = \frac{1.09}{0.436}(1.150)^{-2/3} = 2.277$$

Then, using Eq. (7.3-5) and solving,

$$J_D = \frac{k_c'}{v'}(N_{Sc})^{2/3} \qquad 2.277 = \frac{k_c'}{1.578 \times 10^{-4}}(702.6)^{2/3}$$

The predicted $k_c' = 4.447 \times 10^{-6}$ m/s. This compares with the experimental value of 4.665×10^{-6} m/s.
   For part (b), using Eqs. (7.3-42) and (7.3-43),

$$A k_c \frac{(c_{Ai} - c_{A1}) - (c_{Ai} - c_{A2})}{\ln \dfrac{c_{Ai} - c_{A1}}{c_{Ai} - c_{A2}}} = V/(c_{A2} - c_{A1}) \qquad (7.3\text{-}44)$$

The values to substitute into Eq. (7.3-44) are $c_{Ai} = 2.948 \times 10^{-2}$, $c_{A1} = 0$, $A = 0.01198$, $V = 5.514 \times 10^{-7}$.

$$\frac{0.01198(4.665 \times 10^{-6})(c_{A2} - 0)}{\ln \dfrac{2.948 \times 10^{-2} - 0}{2.948 \times 10^{-2} - c_{A2}}} = (5.514 \times 10^{-7})(c_{A2} - 0)$$

Solving, $c_{A2} = 2.842 \times 10^{-3}$ kg mol/m^3.

5. Mass transfer for flow past single cylinders. Experimental data have been obtained for mass transfer from single cylinders when the flow is perpendicular to the cylinder. The cylinders are long and mass transfer to the ends of the cylinder is not considered. For the Schmidt number range of 0.6 to 2.6 for gases and 1000 to 3000 for liquids and a Reynolds number range of 50 to 50 000, data of many references (B3, L1, M1, S4, V1) have been plotted and the correlation to use is as follows:

$$J_D = 0.600(N_{Re})^{-0.487} \tag{7.3-45}$$

The data scatter considerably by up to $\pm 30\%$. This correlation can also be used for heat transfer with $J_D = J_H$.

6. Liquid metals mass transfer. In recent years several correlations for mass-transfer coefficients of liquid metals have appeared in the literature. It has been found (G1) that with moderate safety factors, the correlations for nonliquid metals mass transfer may be used for liquid metals mass transfer. Care must be taken to ensure that the solid surface is wetted. Also, if the solid is an alloy, there may exist a resistance to diffusion in the solid phase.

7.4 MASS TRANSFER TO SUSPENSIONS OF SMALL PARTICLES

7.4A Introduction

Mass transfer from or to small suspended particles in an agitated solution occurs in a number of process applications. In liquid-phase hydrogenation, hydrogen diffuses from gas bubbles, through an organic liquid, and then to small suspended catalyst particles. In fermentations, oxygen diffuses from small gas bubbles, through the aqueous medium, and then to small suspended microorganisms.

For a liquid–solid dispersion, increased agitation over and above that necessary to freely suspend very small particles has very little effect on the mass-transfer coefficient k_L to the particle (B2). When the particles in a mixing vessel are just completely suspended, turbulence forces balance those due to gravity, and the mass-transfer rates are the same as for particles freely moving under gravity. With very small particles of say a few μm or so, which is the size of many microorganisms in fermentations and some catalyst particles, their size is smaller than eddies, which are about 100 μm or so in size. Hence, increased agitation will have little effect on mass transfer except at very high agitation.

For a gas–liquid–solid dispersion, such as in fermentation, the same principles hold. However, increased agitation increases the number of gas bubbles and hence the interfacial area. The mass-transfer coefficients from the gas bubble to the liquid and from the liquid to the solid are relatively unaffected.

7.4B Equations for Mass Transfer to Small Particles

1. Mass transfer to small particles <0.6 mm. Equations to predict mass transfer to small particles in suspension have been developed which cover three size ranges of particles. The equation for particles <0.6 mm (600 μm) is discussed first.

The following equation has been shown to hold to predict mass-transfer coefficients from small gas bubbles such as oxygen or air to the liquid phase or from the liquid phase to the surface of small catalyst particles, microorganisms, other solids, or liquid drops (B2, C3).

$$k'_L = \frac{2D_{AB}}{D_p} + 0.31 N_{Sc}^{-2/3}\left(\frac{\Delta \rho \mu_c g}{\rho_c^2}\right)^{1/3} \tag{7.4-1}$$

where D_{AB} is the diffusivity of the solute A in solution in m^2/s, D_p is the diameter of the gas bubble or the solid particle in m, μ_c is the viscosity of the solution in kg/m·s, $g = 9.80665$ m/s^2, $\Delta \rho = (\rho_c - \rho_p)$ or $(\rho_p - \rho_c)$, ρ_c is the density of the continuous phase in kg/m^3, and ρ_p is the density of the gas or solid particle. The value of $\Delta \rho$ is always positive.

The first term on the right in Eq. (7.4-1) is the molecular diffusion term, and the second term is that due to free fall or rise of the sphere by gravitational forces. This equation has been experimentally checked for dispersions of low-density solids in agitated dispersions and for small gas bubbles in agitated systems.

EXAMPLE 7.4-1. Mass Transfer from Air Bubbles in Fermentation

Calculate the maximum rate of absorption of O_2 in a fermenter from air bubbles at 1 atm abs pressure having diameters of 100 μm at 37°C into water having a zero concentration of dissolved O_2. The solubility of O_2 from air in water at 37°C is 2.26×10^{-7} g mol O_2/cm^3 liquid or 2.26×10^{-4} kg mol O_2/m^3. The diffusivity of O_2 in water at 37°C is 3.25×10^{-9} m^2/s. Agitation is used to produce the air bubbles.

Solution: The mass-transfer resistance inside the gas bubble to the outside interface of the bubble can be neglected since it is negligible (B2). Hence, the mass-transfer coefficient k'_L outside the bubble is needed. The given data are

$$D_p = 100 \ \mu m = 1 \times 10^{-4} \text{ m} \qquad D_{AB} = 3.25 \times 10^{-9} \text{ m}^2/\text{s}$$

At 37°C,

$$\mu_c(\text{water}) = 6.947 \times 10^{-4} \text{ Pa·s} = 6.947 \times 10^{-4} \text{ kg/m·s}$$

$$\rho_c(\text{water}) = 994 \text{ kg/m}^3 \qquad \rho_p(\text{air}) = 1.13 \text{ kg/m}^3$$

$$N_{Sc} = \frac{\mu_c}{\rho_c D_{AB}} = \frac{6.947 \times 10^{-4}}{(994)(3.25 \times 10^{-9})} = 215$$

$$N_{Sc}^{2/3} = (215)^{2/3} = 35.9 \qquad \Delta \rho = \rho_c - \rho_p = 994 - 1.13 = 993 \text{ kg/m}^3$$

Substituting into Eq. (7.4-1),

$$k'_L = \frac{2D_{AB}}{D_p} + 0.31 N_{Sc}^{-2/3}\left(\frac{\Delta \rho \mu_c g}{\rho_c^2}\right)^{1/3}$$

$$= \frac{2(3.25 \times 10^{-9})}{1 \times 10^{-4}} + \frac{0.31}{35.9}\left[\frac{993 \times 6.947 \times 10^{-4} \times 9.806}{(994)^2}\right]^{1/3}$$

$$= 6.50 \times 10^{-5} + 16.40 \times 10^{-5} = 2.290 \times 10^{-4} \text{ m/s}$$

The flux is as follows assuming $k_L = k'_L$ for dilute solutions.

$$N_A = k_L(c_{A1} - c_{A2}) = 2.290 \times 10^{-4} (2.26 \times 10^{-4} - 0)$$

$$= 5.18 \times 10^{-8} \text{ kg mol } O_2/\text{s·m}^2$$

Knowing the total number of bubbles and their area, the maximum possible rate of transfer of O_2 to the fermentation liquid can be calculated.

In Example 7.4-1, k_L was small. For mass transfer of O_2 in a solution to a microorganism with $D_p \cong 1$ μm, the term $2D_{AB}/D_p$ would be 100 times larger. Note that at large diameters the second term in Eq. (7.4-1) becomes small and the mass-transfer coefficient k_L becomes essentially independent of size D_p. In agitated vessels with gas introduced below the agitator in aqueous solutions, or when liquids are aerated with sintered plates, the gas bubbles are often in the size range covered by Eq. (7.4-1) (B2, C3, T1).

In aerated mixing vessels the mass-transfer coefficients are essentially independent of the power input. However, as the power is increased, the bubble size decreases and the mass transfer coefficient continues to follow Eq. (7.4-1). The dispersions include those in which the solid particles are just completely suspended in mixing vessels. Increase in agitation intensity above the level needed for complete suspension of these small particles results in only a small increase in k_L (C3).

Equation (7.4-1) has also been shown to apply to heat transfer and can be written as follows (B2, C3):

$$N_{Nu} = \frac{hD_p}{k} = 2.0 + 0.31 N_{Pr}^{1/3} \left(\frac{D_p^3 \rho_c \, \Delta \rho g}{\mu_c^2} \right)^{1/3} \qquad \text{(7.4-2)}$$

2. *Mass transfer to large gas bubbles > 2.5 mm.* For large gas bubbles or liquid drops > 2.5 mm, the mass-transfer coefficient can be predicted by

$$k_L' = 0.42 N_{Sc}^{-0.5} \left(\frac{\Delta \rho \mu_c \, g}{\rho_c^2} \right)^{1/3} \qquad \text{(7.4-3)}$$

Large gas bubbles are produced when pure liquids are aerated in mixing vessels and sieve-plate columns (C1). In this case the mass-transfer coefficient k_L' or k_L is independent of the bubble size and is constant for a given set of physical properties. For the same physical properties the large bubble Eq. (7.4-3) gives values of k_L about three to four times larger than Eq. (7.4-1) for small particles. Again, Eq. (7.4-3) shows that the k_L is essentially independent of agitation intensity in an agitated vessel and gas velocity in a sieve-tray tower.

3. *Mass transfer to particles in transition region.* In mass transfer in the transition region between small and large bubbles in the size range 0.6 to 2.5 mm, the mass-transfer coefficient can be approximated by assuming that it increases linearly with bubble diameter (B2, C3).

4. *Mass transfer to particles in highly turbulent mixers.* In the preceding three regions, the density difference between phases is sufficiently large to cause the force of gravity to primarily determine the mass-transfer coefficient. This also includes solids just completely suspended in mixing vessels. When agitation power is increased beyond that needed for suspension of solid or liquid particles and the turbulence forces become larger than the gravitational forces, Eq. (7.4-1) is not followed and Eq. (7.4-4) should be used where small increases in k_L' are observed (B2, C3).

$$k_L' N_{Sc}^{2/3} = 0.13 \left(\frac{(P/V)\mu_c}{\rho_c^2} \right)^{1/4} \qquad \text{(7.4-4)}$$

where P/V is power input per unit volume defined in Section 3.4. The data deviate substantially by up to 60% from this correlation. In the case of gas–liquid dispersions it is quite impractical to exceed gravitational forces by agitation systems.

The experimental data are complicated by the fact that very small particles are easily suspended and if their size is of the order of the smallest eddies, the mass-transfer coefficient will remain constant until a large increase in power input is added above that required for suspension.

7.5 (SELECTED TOPIC) MOLECULAR DIFFUSION PLUS CONVECTION AND CHEMICAL REACTION

7.5A Different Types of Fluxes and Fick's Law

In Section 6.2B the flux J_A^* was defined as the molar flux of A in kg mol $A/\text{s} \cdot \text{m}^2$ relative to the molar average velocity v_M of the whole or bulk stream. Also, N_A was defined as the molar flux of A relative to stationary coordinates. Fluxes and velocities can also be defined in other ways. Table 7.5-1 lists the different types of fluxes and velocities often used in binary systems.

TABLE 7.5-1. *Different Types of Fluxes and Velocities in Binary Systems*

	Mass Flux (kg $A/\text{s} \cdot \text{m}^2$)	Molar Flux (kg mol $A/\text{s} \cdot \text{m}^2$)
Relative to fixed coordinates	$n_A = \rho_A v_A$	$N_A = c_A v_A$
Relative to molar average velocity v_M	$j_A^* = \rho_A(v_A - v_M)$	$J_A^* = c_A(v_A - v_M)$
Relative to mass average velocity v	$j_A = \rho_A(v_A - v)$	$J_A = c_A(v_A - v)$

Relations Between Fluxes Above

$$N_A + N_B = cv_M \qquad N_A = n_A/M_A \qquad N_A = J_A + c_A v$$

$$J_A^* + J_B^* = 0 \qquad J_A = j_A/M_A \qquad n_A + n_B = \rho v$$

$$j_A + j_B = 0 \qquad N_A = J_A^* + c_A v_M \qquad n_A = j_A + \rho_A v$$

Different Forms of Fick's Law for Diffusion Flux

$$J_A^* = -cD_{AB}\, dx_A/dz \qquad j_A = -\rho D_{AB}\, dw_A/dz$$

The velocity v is the mass average velocity of the stream relative to stationary coordinates and can be obtained by actually weighing the flow for a timed increment. It is related to the velocity v_A and v_B by

$$v = w_A v_A + w_B v_B = \frac{\rho_A}{\rho} v_A + \frac{\rho_B}{\rho} v_B \qquad (7.5\text{-}1)$$

where w_A is ρ_A/ρ, the weight fraction of A; w_B the weight fraction of B; and v_A is the velocity of A relative to stationary coordinates in m/s. The molar average velocity v_M in m/s is relative to stationary coordinates.

$$v_M = x_A v_A + x_B v_B = \frac{c_A}{c} v_A + \frac{c_B}{c} v_B \qquad (7.5\text{-}2)$$

The molar diffusion flux relative to the molar average velocity v_M defined previously is

$$J_A^* = c_A(v_A - v_M) \tag{7.5-3}$$

The molar diffusion flux J_A relative to the mass average velocity v is

$$J_A = c_A(v_A - v) \tag{7.5-4}$$

Fick's law from Table 7.5-1 as given previously is relative to v_M and is

$$J_A^* = -cD_{AB}\frac{dx_A}{dz} \tag{7.5-5}$$

Fick's law can also be defined in terms of a mass flux relative to v as given in Table 7.5-1.

$$j_A = -\rho D_{AB}\frac{dw_A}{dz} \tag{7.5-6}$$

EXAMPLE 7.5-1. *Proof of Mass Flux Equation*
Table 7.5-1 gives the following relation:

$$j_A + j_B = 0 \tag{7.5-7}$$

Prove this relationship using the definitions of the fluxes in terms of velocities.

Solution: From Table 7.5-1 substituting $\rho_A(v_A - v)$ for j_A and $\rho_B(v_B - v)$ for j_B, and rearranging,

$$\rho_A v_A - \rho_A v + \rho_B v_B - \rho_B v = 0 \tag{7.5-8}$$

$$\rho_A v_A + \rho_B v_B - v(\rho_A + \rho_B) = 0 \tag{7.5-9}$$

Substituting Eq. (7.5-1) for v and ρ for $\rho_A + \rho_B$, the identity is proved.

7.5B   Equation of Continuity for a Binary Mixture

A general equation can be derived for a binary mixture of A and B for diffusion and convection that also includes the terms for unsteady-state diffusion and chemical reaction. We shall make a mass balance on component A on an element $\Delta x\,\Delta y\,\Delta z$ fixed in space as shown in Fig. 7.5-1. The general mass balance on A is

$$\left(\begin{array}{c}\text{rate of}\\\text{mass } A \text{ in}\end{array}\right) - \left(\begin{array}{c}\text{rate of}\\\text{mass } A \text{ out}\end{array}\right)$$

$$+ \left(\begin{array}{c}\text{rate of}\\\text{generation of mass } A\end{array}\right) = \left(\begin{array}{c}\text{rate of}\\\text{accumulation of mass } A\end{array}\right) \tag{7.5-10}$$

The rate of mass A entering in the direction relative to stationary coordinates is $(n_{Ax|x})\Delta y\,\Delta z$ kg A/s and leaving is $(n_{Ax|x + \Delta x})\Delta y\,\Delta z$. Similar terms can be written for the y and z directions. The rate of chemical production of A is r_A kg A generated/s · m^3 volume and the total rate generated is $r_A(\Delta x\,\Delta y\,\Delta z)$ kg A/s. The rate of accumulation of A is $(\partial \rho_A/\partial t)\Delta x\,\Delta y\,\Delta z$. Substituting into Eq. (7.5-10) and letting Δx, Δy, and Δz approach zero,

$$\frac{\partial \rho_A}{\partial t} + \left(\frac{\partial n_{Ax}}{\partial x} + \frac{\partial n_{Ay}}{\partial y} + \frac{\partial n_{Az}}{\partial z}\right) = r_A \tag{7.5-11}$$

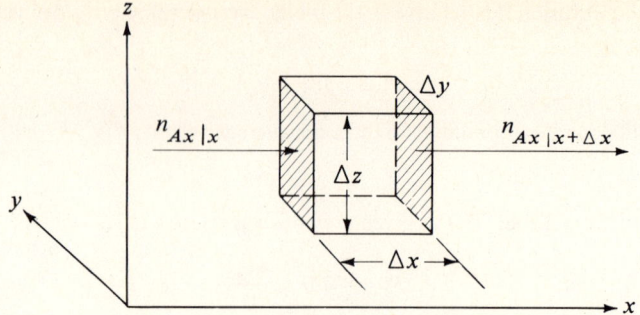

FIGURE 7.5-1. *Mass balance for A in a binary mixture.*

In vector notation,

$$\frac{\partial \rho_A}{\partial t} + (\nabla \cdot \mathbf{n}_A) = r_A \qquad (7.5\text{-}12)$$

Dividing both sides of Eq. (7.5-11) by M_A,

$$\frac{\partial c_A}{\partial t} + \left(\frac{\partial N_{Ax}}{\partial x} + \frac{\partial N_{Ay}}{\partial y} + \frac{\partial N_{Az}}{\partial z} \right) = R_A \qquad (7.5\text{-}13)$$

where R_A is kg mol A generated/s $\cdot$ m^3. Substituting N_A and Fick's law from Table 7.5-1,

$$N_A = -cD_{AB}\frac{dx_A}{dz} + c_A v_M \qquad (7.5\text{-}14)$$

and writing the equation for all three directions, Eq. (7.5-13) becomes

$$\frac{\partial c_A}{\partial t} + (\nabla \cdot c_A \mathbf{v_M}) - (\nabla \cdot cD_{AB}\nabla x_A) = R_A \qquad (7.5\text{-}15)$$

This is the final general equation.

7.5C Special Cases of the Equation of Continuity

1. Equation for constant c and D_{AB}. In diffusion with gases the total pressure P is often constant. Then, since $c = P/RT$, c is constant for constant temperature T. Starting with the general equation (7.5-15) and substituting $\nabla x_A = \nabla c_A/c$, we obtain

$$\frac{\partial c_A}{\partial t} + c_A(\nabla \cdot \mathbf{v_M}) + (\mathbf{v_M} \cdot \nabla c_A) - D_{AB}\nabla^2 c_A = R_A \qquad (7.5\text{-}16)$$

2. Equimolar counterdiffusion for gases. For the special case of equimolar counterdiffusion of gases at constant pressure and no reaction, $c = $ constant, $v_M = 0$, $D_{AB} = $ constant, $R_A = 0$, and Eq. (7.5-15) becomes

$$\frac{\partial c_A}{\partial t} = D_{AB}\left(\frac{\partial^2 c_A}{\partial x^2} + \frac{\partial^2 c_A}{\partial y^2} + \frac{\partial^2 c_A}{\partial z^2} \right) \qquad (7.1\text{-}9)$$

This equation is Eq. (7.1-9) derived previously and this equation is also used for unsteady-state diffusion of a dilute solute A in a solid or a liquid when D_{AB} is constant.

3. *Equation for constant ρ and D_{AB} (liquids).* In dilute liquid solutions the mass density ρ and D_{AB} can often be considered constant. Starting with Eq. (7.5-12) we substitute $n_A = -\rho D_{AB} \nabla w_A + c_A v$ from Table 7.5-1 into this equation. Then using the fact that for constant ρ, $\nabla w_A = \nabla \rho_A / \rho$ and also that $(\nabla \cdot v) = 0$, substituting these into the resulting equation, and dividing both sides by M_A, we obtain

$$\frac{\partial c_A}{\partial t} + (v \cdot \nabla c_A) - D_{AB} \nabla^2 c_A = R_A \qquad (7.5\text{-}17)$$

7.5D Special Cases of the General Diffusion Equation at Steady State

1. Introduction and physical conditions at the boundaries. The general equation for diffusion and convection of a binary mixture in one direction with no chemical reaction has been given previously.

$$N_A = -cD_{AB} \frac{dx_A}{dz} + \frac{c_A}{c} (N_A + N_B) \qquad (6.2\text{-}14)$$

To integrate this equation at steady state it is necessary to specify the boundary conditions at z_1 and at z_2. Often in many mass-transfer problems the molar ratio N_A/N_B is determined by the physical conditions occurring at the two boundaries.

As an example, one boundary of the diffusion path may be impermeable to species B because B is insoluble in the phase at this boundary. Diffusion of ammonia (A) and nitrogen (B) through a gas phase to a water phase at the boundary is such a case since nitrogen is essentially insoluble in water. Hence, $N_B = 0$, since at steady state N_B must have the same value at all points in the path of $z_2 - z_1$. In some cases, a heat balance in the adjacent phase at the boundary can determine the flux ratios. For example, if component A condenses at a boundary and releases its latent heat to component B, which vaporizes and diffuses back, the ratios of the latent heats determine the flux ratio.

In another example, the boundary concentration can be fixed by having a large volume of a phase flowing rapidly by with a given concentration x_{A1}. In some cases the concentration x_{A1} may be set by an equilibrium condition, whereby x_{A1} is in equilibrium with some fixed composition at the boundary. Chemical reactions can also influence the rates of diffusion and the boundary conditions.

2. Equimolar counterdiffusion. For the special case of equimolar counterdiffusion where $N_A = -N_B$, Eq. (6.2-14) becomes, as shown previously, for steady state and constant c,

$$N_A = J_A^* = -cD_{AB} \frac{dx_A}{dz} = \frac{D_{AB}(c_{A1} - c_{A2})}{z_2 - z_1} \qquad (7.5\text{-}18)$$

3. Diffusion of A through stagnant B. For gas A diffusing through stagnant nondiffusing gas B, $N_B = 0$, and integration of Eq. (6.2-14) gives Eq. (6.2-22).

$$N_A = \frac{D_{AB} P}{RT(z_2 - z_1) p_{BM}} (p_{A1} - p_{A2}) \qquad (6.2\text{-}22)$$

Several other more complicated cases of integration of Eq. (6.2-14) are considered next.

4. Diffusion and chemical reaction at a boundary. Often in catalytic reactions where A and B are diffusing to and from a catalyst surface, the relation between the fluxes N_A and

N_B at steady state is controlled by the stoichiometry of a reaction at a boundary. An example is gas A diffusing from the bulk gas phase to the catalyst surface, where it reacts instantaneously and irreversibly in a heterogeneous reaction as follows:

$$A \to 2B \tag{7.5-19}$$

Gas B then diffuses back, as is shown in Fig. 7.5-2.

At steady state 1 mol of A diffuses to the catalyst for every 2 mol of B diffusing away, or $N_B = -2N_A$. The negative sign indicates that the fluxes are in opposite directions. Rewriting Eq. (6.2-14) in terms of mole fractions,

$$N_A = -cD_{AB} \frac{dx_A}{dz} + x_A(N_A + N_B) \tag{7.5-20}$$

Next, substituting $N_B = -2N_A$ into Eq. (7.5-20),

$$N_A = -cD_{AB} \frac{dx_A}{dz} + x_A(N_A - 2N_A) \tag{7.5-21}$$

Rearranging and integrating with constant c (P = constant), we obtain the following:

1. *Instantaneous surface reaction :*

$$N_A \int_{z_1=0}^{z_2=\delta} dz = -cD_{AB} \int_{x_{A1}}^{x_{A2}} \frac{dx_A}{1 + x_A} \tag{7.5-22}$$

$$N_A = \frac{cD_{AB}}{\delta} \ln \frac{1 + x_{A1}}{1 + x_{A2}} \tag{7.5-23}$$

Since the reaction is instantaneous, $x_{A2} = 0$, because no A can exist next to the catalyst surface. Equation (7.5-23) describes the overall rate of the process of diffusion plus instantaneous chemical reaction.

2. *Slow surface reaction.* If the heterogeneous reaction at the surface is not instantaneous but slow for the reaction $A \to 2B$, and the reaction is first order,

$$N_{Az=\delta} = k_1' c_A = k_1' c x_A \tag{7.5-24}$$

where k_1' is the first-order heterogeneous reaction velocity constant in m/s. Equation (7.5-23) still holds for this case, but the boundary condition x_{A2} at $z = \delta$ is obtained by

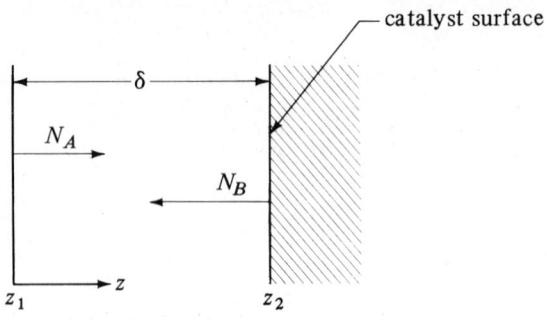

FIGURE 7.5-2. *Diffusion of A and heterogeneous reaction at a surface.*

solving for x_A in Eq. (7.5-24),

$$x_A = x_{A2} = \frac{N_{Az=\delta}}{k_1'c} = \frac{N_A}{k_1'c} \qquad (7.5-25)$$

For steady state, $N_{Az=\delta} = N_A$. Substituting Eq. (7.5-25) into (7.5-23),

$$N_A = \frac{cD_{AB}}{\delta} \ln \frac{1 + x_{A1}}{1 + N_A/k_1'c} \qquad (7.5-26)$$

The rate in Eq. (7.5-26) is less than in Eq. (7.5-23), since the denominator in the latter equation is $1 + x_{A2} = 1 + 0$ and in the former is $1 + N_A/k_1'c$.

EXAMPLE 7.5-2. Diffusion and Chemical Reaction at a Boundary

Pure gas A diffuses from point 1 at a partial pressure of 101.32 kPa to point 2 a distance 2.00 mm away. At point 2 it undergoes a chemical reaction at the catalyst surface and $A \rightarrow 2B$. Component B diffuses back at steady state. The total pressure is $P = 101.32$ kPa. The temperature is 300 K and $D_{AB} = 0.15 \times 10^{-4}$ m²s.
(a) For instantaneous rate of reaction, calculate x_{A2} and N_A.
(b) For a slow reaction where $k_1' = 5.63 \times 10^{-3}$ m/s, calculate x_{A2} and N_A.

Solution: For part (a), $p_{A2} = x_{A2} = 0$ since no A can exist next to the catalyst surface. Since $N_B = -2N_A$, Eq. (7.5-23) will be used as follows: $\delta = 2.00 \times 10^{-3}$ m, $T = 300$ K, $c = P/RT = 101.32 \times 10^3/(8314 \times 300) = 4.062 \times 10^{-2}$ kg mol/m³, $x_{A1} = p_{A1}/P = 101.32 \times 10^3/101.32 \times 10^3 = 1.00$.

$$N_A = \frac{cD_{AB}}{\delta} \ln \frac{1 + x_{A1}}{1 + x_{A2}} = \frac{(4.062 \times 10^{-2})(0.15 \times 10^{-4})}{2.00 \times 10^{-3}} \ln \frac{1 + 1.00}{1 + 0}$$

$$= 2.112 \times 10^{-4} \text{ kg mol } A/\text{s} \cdot \text{m}^2$$

For part (b), from Eq. (7.5-25), $x_{A2} = N_A/k_1'c = N_A/(5.63 \times 10^{-3} \times 4.062 \times 10^{-2})$. Substituting into Eq. (7.5-26),

$$N_A = \frac{(4.062 \times 10^{-2})(0.15 \times 10^{-4})}{2.00 \times 10^{-3}} \ln$$

$$\frac{1 + 1.00}{1 + N_A/(5.63 \times 10^{-3} \times 4.062 \times 10^{-2})}$$

Solving by trial and error, $N_A = 1.004 \times 10^{-4}$ kg mol $A/\text{s} \cdot \text{m}^2$. Then, $x_{A2} = (1.004 \times 10^{-4})/(5.63 \times 10^{-3} \times 4.062 \times 10^{-2}) = 0.4390$.

Even though in part (a) of Example 7.5-2 the rate of reaction is instantaneous, the flux N_A is diffusion controlled. As the reaction rate slows, the flux N_A is decreased also.

5. Diffusion and homogeneous reaction in a phase. Equation (7.5-23) was derived for the case of chemical reaction of A at the boundary on a catalyst surface. In some cases component A undergoes an irreversible chemical reaction in the homogeneous phase B while diffusing as follows, $A \rightarrow C$. Assume that component A is very dilute in phase B, which can be a gas or a liquid. Then at steady state the equation for diffusion of A is as follows where the bulk-flow term is dropped.

$$N_{Az} = -D_{AB} \frac{dc_A}{dz} + 0 \qquad (7.5-27)$$

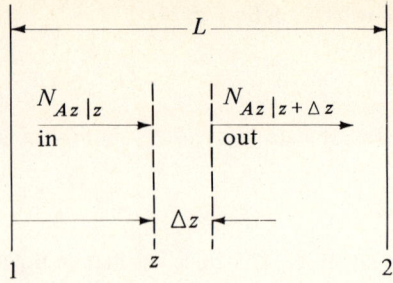

FIGURE 7.5-3. *Homogeneous chemical reaction and diffusion in a fluid.*

Writing a material balance on A shown in Fig. 7.5-3 for the Δz element for steady state,

$$\begin{pmatrix} \text{rate of} \\ A \text{ in} \end{pmatrix} + \begin{pmatrix} \text{rate of} \\ \text{generation of } A \end{pmatrix} = \begin{pmatrix} \text{rate of} \\ A \text{ out} \end{pmatrix} + \begin{pmatrix} \text{rate of} \\ \text{accumulation of } A \end{pmatrix} \qquad (7.5\text{-}28)$$

The first-order reaction rate of A per m^3 volume is

$$\text{rate of generation} = -k'c_A \qquad (7.5\text{-}29)$$

where k' is the reaction velocity constant in s^{-1}. Substituting into Eq. (7.5-28) for a cross-sectional area of $1\ m^2$ with the rate of accumulation being 0 at steady state,

$$N_{Az|z}(1) - k'c_A(1)(\Delta z) = N_{Az|z+\Delta z}(1) + 0 \qquad (7.5\text{-}30)$$

Next we divide through by Δz and let Δz approach zero.

$$-\frac{dN_{Az}}{dz} = k'c_A \qquad (7.5\text{-}31)$$

Substituting Eq. (7.5-27) into (7.5-31),

$$\frac{d^2c_A}{dz^2} = \frac{k'}{D_{AB}} c_A \qquad (7.5\text{-}32)$$

The boundary conditions are $c_A = c_{A1}$ for $z = 0$ and $c_A = c_{A2}$ for $z = L$. Solving,

$$c_A = \frac{c_{A2} \sinh\left(\sqrt{\dfrac{k'}{D_{AB}}}\, z\right) + c_{A1} \sinh\left[\sqrt{\dfrac{k'}{D_{AB}}}\, (L - z)\right]}{\sinh\left(\sqrt{\dfrac{k'}{D_{AB}}}\, L\right)} \qquad (7.5\text{-}33)$$

This equation can be used at steady state to calculate c_A at any z and can be used for reaction in gases, liquids, or even solids, where the solute A is dilute.

As an alternative derivation of Eq. (7.5-32), we can use Eq. (7.5-17) for constant ρ and D_{AB}.

$$\frac{\partial c_A}{\partial t} + (\mathbf{v} \cdot \nabla c_A) - D_{AB}\nabla^2 c_A = R_A \qquad (7.5\text{-}17)$$

We set the first term $\partial c_A/\partial t = 0$ for steady state. Since we are assuming dilute solutions and neglecting the bulk flow term, $\mathbf{v} = 0$, making the second term in Eq. (7.5-17) zero. For a first-order reaction of A where A disappears, $R_A = -k'c_A$ kg mol A generated/s $\cdot\ m^3$.

Writing the diffusion term $-D_{AB}\nabla^2 c_A$ for only the z direction, we obtain

$$D_{AB}\frac{d^2 c_A}{dz^2} = k'c_A \tag{7.5-34}$$

which is, of course, identical to Eq. (7.5-32).

7.5E Unsteady-State Diffusion and Reaction in a Semiinfinite Medium

Here we consider a case where dilute A is absorbed at the surface of a solid or stagnant fluid phase and then unsteady-state diffusion and reaction occur in the phase. The fluid or solid phase B is considered semiinfinite. At the surface where $z = 0$, the concentration of c_A is kept constant at c_{A0}. The dilute solute A reacts by a first-order mechanism

$$A + B \rightarrow C \tag{7.5-35}$$

and the rate of generation is $-k'c_A$. The same diagram as in Fig. 7.5-3 holds. Using Eq. (7.5-30) but substituting $(\partial c_A/\partial t)(\Delta z)(1)$ for the rate of accumulation,

$$N_{Az|z}(1) - k'c_A(1)(\Delta z) = N_{Az|z+\Delta z}(1) + \left(\frac{\partial c_A}{\partial t}\right)(\Delta z)(1) \tag{7.5-36}$$

This becomes

$$\frac{\partial c_A}{\partial t} = D_{AB}\frac{\partial^2 c_A}{\partial z^2} - k'c_A \tag{7.5-37}$$

The initial and boundary conditions are

$$
\begin{array}{lll}
t = 0, & c_A = 0 & \text{for } z > 0 \\[4pt]
z = 0, & c_A = c_{A0} & \text{for } t > 0 \\[4pt]
z = \infty, & c_A = 0 & \text{for } t > 0
\end{array}
\tag{7.5-38}
$$

The solution by Danckwerts (D1) is

$$\frac{c_A}{c_{A0}} = \tfrac{1}{2}\exp(-z\sqrt{k'/D_{AB}}) \cdot \text{erfc}\left(\frac{z}{2\sqrt{tD_{AB}}} - \sqrt{k't}\right)$$

$$+ \tfrac{1}{2}\exp(z\sqrt{k'/D_{AB}}) \cdot \text{erfc}\left(\frac{z}{2\sqrt{tD_{AB}}} + \sqrt{k't}\right) \tag{7.5-39}$$

The total amount Q of A absorbed up to time t is

$$Q = c_{A0}\sqrt{D_{AB}/k'}[(k't + \tfrac{1}{2})\text{erf}\sqrt{k't} + \sqrt{k't/\pi}\,e^{-k't}] \tag{7.5-40}$$

where Q is kg mol A absorbed/m^2. Many actual cases are approximated by this case. The equation is useful where absorption occurs at the surface of a stagnant fluid or a solid and unsteady-state diffusion and reaction occurs in the solid or fluid. The results can be used to measure the diffusivity of a gas in a solution, to determine reaction rate constants k' of dissolved gases, and to determine solubilities of gas in liquids with which they react. Details are given elsewhere (D3).

EXAMPLE 7.5-3. Reaction and Unsteady-State Diffusion
Pure CO_2 gas at 101.32 kPa pressure is absorbed into a dilute alkaline buffer solution containing a catalyst. The dilute, absorbed solute CO_2 undergoes a first-order reaction with $k' = 35$ s^{-1} and $D_{AB} = 1.5 \times 10^{-9}$

m^2/s. The solubility of CO_2 is 2.961×10^{-7} kg mol/m³ · Pa (D3). The surface is exposed to the gas for 0.010 s. Calculate the kg mol CO_2 absorbed/m² surface.

Solution: For use in Eq. (7.5-40), $k't = 35(0.01) = 0.350$. Also, $c_{A0} = 2.961 \times 10^{-7}$ (kg mol/m³ · Pa)(101.32 × 10³ Pa) = 3.00 × 10⁻² kg mol SO_2/m^3.

$$Q = (3.00 \times 10^{-2})\sqrt{1.5 \times 10^{-9}/35}[(0.35 + \tfrac{1}{2})\text{erf} \sqrt{0.35} + \sqrt{0.35/\pi}e^{-0.35}]$$

$$= 1.458 \times 10^{-7} \text{ kg mol } CO_2/m^2$$

7.5F Multicomponent Diffusion of Gases

The equations derived in this chapter have been for a binary system of A and B, which is probably the most important and most useful one. However, multicomponent diffusion sometimes occurs where three or more components A, B, C, ..., are present. The simplest case is for diffusion of A in a gas through a stagnant nondiffusing mixture of B, C, D, ..., at constant total pressure. Hence, $N_B = 0$, $N_c = 0$, ... The final equation derived using the Stefan–Maxwell method (G1) for steady-state diffusion is

$$N_A = \frac{D_{Am}P}{RT(z_2 - z_1)p_{iM}}(p_{A1} - p_{A2}) \qquad (7.5\text{-}41)$$

where p_{iM} is the log mean of $p_{i1} = P - p_{A1}$ and $p_{i2} = P - p_{A2}$. Also,

$$D_{Am} = \frac{1}{x'_B/D_{AB} + x'_C/D_{AC} + \cdots} \qquad (7.5\text{-}42)$$

where x'_B = mol B/mol inerts = $x_B/(1 - x_A)$, $x'_C = x_C/(1 - x_A)$, ...

> **EXAMPLE 7.5-4. Diffusion of A Through Nondiffusing B and C**
> At 298 K and 1 atm total pressure, methane (A) is diffusing at steady state through nondiffusing argon (B) and helium (C). At $z_1 = 0$, the partial pressures in atm are $p_{A1} = 0.4$, $p_{B1} = 0.4$, $p_{C1} = 0.2$ and at $z_2 = 0.005$ m, $p_{A2} = 0.1$, $p_{B2} = 0.6$, and $p_{C2} = 0.3$. The binary diffusivities from Table 6.2-1 are $D_{AB} = 2.02 \times 10^{-5}$ m²/s, $D_{AC} = 6.75 \times 10^{-5}$ m²/s, and $D_{BC} = 7.29 \times 10^{-5}$ m²/s. Calculate N_A.
>
> **Solution:** At point 1, $x'_B = x_B/(1 - x_A) = 0.4/(1 - 0.4) = 0.667$. At point 2, $x'_B = 0.6/(1 - 0.1) = 0.667$. The value of x'_B is constant throughout the path. Also, $x'_C = x_C/(1 - x_A) = 0.2/(1 - 0.4) = 0.333$.
>   Substituting into Eq. (7.5-42),
>
> $$D_{Am} = \frac{1}{x'_B/D_{AB} + x'_C/D_{AC}} = \frac{1}{0.667/2.02 \times 10^{-5} + 0.333/6.75 \times 10^{-5}}$$
>
> $$= 2.635 \times 10^{-5} \text{ m}^2/\text{s}$$
>
> For calculating p_{iM}, $p_{i1} = P - p_{A1} = 1.0 - 0.4 = 0.6$ atm, $p_{i2} = P - p_{A2} = 1.0 - 0.1 = 0.90$. Then,
>
> $$p_{iM} = \frac{p_{i2} - p_{i1}}{\ln(p_{i2}/p_{i1})} = \frac{0.90 - 0.60}{\ln(0.90/0.60)} = 0.740 \text{ atm} = 7.496 \times 10^4 \text{ Pa}$$
>
> $$p_{A1} = 0.4(1.01325 \times 10^5) = 4.053 \times 10^4 \text{ Pa}$$
>
> $$p_{A2} = 0.1(1.01325 \times 10^5) = 1.013 \times 10^4 \text{ Pa}$$

Substituting into Eq. (7.5-41),

$$N_A = \frac{D_{Am} P}{RT(z_2 - z_1)p_{iM}} (p_{A1} - p_{A2})$$

$$= \frac{(2.635 \times 10^{-5})(1.01325 \times 10^5)(4.053 - 1.013)(10^4)}{(8314)(298)(0.005 - 0)(7.496 \times 10^4)}$$

$$= 8.74 \times 10^{-5} \text{ kg mol } A/s \cdot m^2$$

Using atm pressure units,

$$N_A = \frac{D_{Am} P}{RT(z_2 - z_1)p_{iM}} (p_{A1} - p_{A2}) = \frac{(2.635 \times 10^{-5})(1.0)(0.4 - 0.1)}{(82.06 \times 10^{-3})(298)(0.005 - 0)(0.740)}$$

$$= 8.74 \times 10^{-5} \text{ kg mol } A/s \cdot m^2$$

A number of analytical solutions have been obtained for other cases such as for equimolar diffusion of three components, diffusion of components A and B through stagnant C, and the general case of two or more components diffusing in a multicomponent mixture. These are discussed in detail with examples by Geankoplis (G1) and the reader is referred there for further details.

7.6 (SELECTED TOPIC) DIFFUSION OF GASES IN POROUS SOLIDS AND CAPILLARIES

7.6A Introduction

In Section 6.5C diffusion in porous solids that depends on structure was discussed for liquids and for gases. For gases it was assumed that the pores were very large and Fickian-type diffusion occured. However, often the pores are small in diameter and the mechanism of diffusion is basically changed.

Diffusion of gases in small pores occurs often in heterogeneous catalysis where gases diffuse through very small pores to react on the surface of the catalyst. In freeze drying of foods such as turkey meat, gaseous H_2O diffuses through very fine pores of the porous structure.

Since the pores or capillaries of porous solids are often small, the diffusion of gases may depend upon the diameter of the pores. We first define a mean free path λ, which is the average distance a gas molecule travels before it collides with another gas molecule.

$$\lambda = \frac{3.2\mu}{P} \sqrt{\frac{RT}{2\pi M}} \tag{7.6-1}$$

where λ is in m, μ is viscosity in Pa·s, P is pressure in N/m^2, T is temperature in K, M = molecular weight in kg/kg mol, and $R = 8.3143 \times 10^3$ N·m/kg mol·K. Note that low pressures give large values of λ. For liquids, since λ is so small, diffusion follows Fick's law.

In the next sections we shall consider what happens to the basic mechanisms of diffusion in gases as the relative value of the mean free path compared to the pore diameter varies. The total pressure P in the system will be constant, but partial pressures of A and B may be different.

Principles of Unsteady-State and Convective Mass Transfer

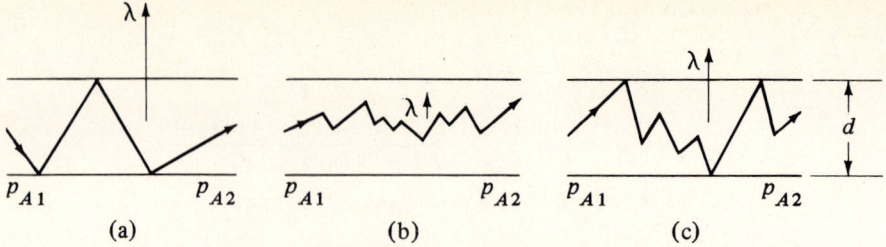

FIGURE 7.6-1. *Types of diffusion gases in small capillary tubes: (a) Knudsen gas diffusion, (b) molecular or Fick's gas diffusion, (c) transition gas diffusion.*

7.6B Knudsen Diffusion of Gases

In Fig. 7.6-1a a gas molecule A at partial pressure p_{A1} at the entrance to a capillary is diffusing through the capillary having a diameter of d m. The total pressure P is constant throughout. The mean free path λ is large compared to the diameter d. As a result, the molecule collides with the wall and molecule–wall collisions are important. This type is called *Knudsen diffusion*.

The Knudsen diffusivity is independent of pressure P and is calculated from

$$D_{KA} = \tfrac{2}{3}\bar{r}\bar{v}_A \tag{7.6-2}$$

where D_{KA} is diffusivity in m^2/s, $\bar{r}$ is average pore radius in m, and $\bar{v}_A$ is the average molecular velocity for component A in m/s. Using the kinetic theory of gases to evaluate $\bar{v}_A$ the final equation for D_{KA} is

$$D_{KA} = 97.0\bar{r}\left(\frac{T}{M_A}\right)^{1/2} \tag{7.6-3}$$

where M_A is molecular weight of A in kg/kg mol and T is temperature in K.

EXAMPLE 7.6-1. Knudsen Diffusivity of Hydrogen
A $H_2(A)$–$C_2H_6(B)$ gas mixture is diffusing in a pore of a nickel catalyst used for hydrogenation at 1.01325×10^5 Pa pressure and 373 K. The pore radius is 60 Å (angstrom). Calculate the Knudsen diffusivity D_{KA} of H_2.

Solution: Substituting into Eq. (7.6-3) for $\bar{r} = 6.0 \times 10^{-9}$ m, $M_A = 2.016$, and $T = 373$ K,

$$D_{KA} = 97.0\bar{r}\left(\frac{T}{M_A}\right)^{1/2} = 97.0(6.0 \times 10^{-9})\left(\frac{373}{2.016}\right)^{1/2}$$

$$= 7.92 \times 10^{-6} \ m^2/s$$

The flux equation for Knudsen diffusion in a pore is

$$N_A = -D_{KA}\frac{dc_A}{dz} = -\frac{D_{KA}}{RT}\frac{dp_A}{dz} \tag{7.6-4}$$

Integrating between $z_1 = 0$, $p_A = p_{A1}$ and $z_2 = L$, $p_A = p_{A2}$,

$$N_A = \frac{D_{KA}P}{RTL}(x_{A1} - x_{A2}) = \frac{D_{KA}}{RTL}(p_{A1} - p_{A2}) \tag{7.6-5}$$

The diffusion of A for Knudsen diffusion is completely independent of B, since A collides with the walls of the pore and not with B. A similar equation can be written for component B.

When the Knudsen number N_{Kn} defined as

$$N_{Kn} = \frac{\lambda}{2\bar{r}} \tag{7.6-6}$$

is $\geq 10/1$, the diffusion is primarily Knudsen and Eq. (7.6-5) predicts the flux to within about a 10% error. As N_{Kn} gets larger, this error decreases, since the diffusion approaches the Knudsen type.

7.6C Molecular Diffusion of Gases

As shown in Fig. 7.6-1b when the mean free path λ is small compared to the pore diameter d or where $N_{Kn} \leq 1/100$, molecule–molecule collisions predominate and molecule–wall collisions are few. Ordinary molecular or Fickian diffusion holds and Fick's law predicts the diffusion to within about 10%. The error diminishes as N_{Kn} gets smaller since the diffusion approaches more closely the Fickian type.

The equation for molecular diffusion given in previous sections is

$$N_A = -\frac{D_{AB}P}{RT}\frac{dx_A}{dz} + x_A(N_A + N_B) \tag{7.6-7}$$

A flux ratio factor α can be defined as

$$\alpha = 1 + \frac{N_B}{N_A} \tag{7.6-8}$$

Combining Eqs. (7.6-7) and (7.6-8) and integrating for a path length of L cm,

$$N_A = \frac{D_{AB}P}{\alpha RTL}\ln\frac{1 - \alpha x_{A2}}{1 - \alpha x_{A1}} \tag{7.6-9}$$

If the diffusion is equimolar, $N_A = -N_B$ and Eq. (7.6-7) becomes Fick's law. The molecular diffusivity D_{AB} is inversely proportional to the total pressure P.

7.6D Transition-Region Diffusion of Gases

As shown in Fig. 7.6-1c, when the mean free path λ and pore diameter are intermediate in size between the two limits given for Knudsen and molecular diffusion, transition-type diffusion occurs where molecule–molecule and molecule–wall collisions are important in diffusion.

The transition-region diffusion equation can be derived by adding the momentum loss due to molecule–wall collisions in Eq. (7.6-4) and that due to molecule–molecule collisions in Eq. (7.6-7) on a slice of capillary. No chemical reactions are occurring. The final differential equation is (G1)

$$N_A = -\frac{D_{NA}P}{RT}\frac{dx_A}{dz} \tag{7.6-10}$$

where

$$D_{NA} = \frac{1}{(1 - \alpha x_A)/D_{AB} + 1/D_{KA}} \tag{7.6-11}$$

Principles of Unsteady-State and Convective Mass Transfer

This transition region diffusivity D_{NA} depends slightly on concentration x_A.

Integrating Eq. (7.6-10),

$$N_A = \frac{D_{AB} P}{\alpha RTL} \ln \frac{1 - \alpha x_{A2} + D_{AB}/D_{KA}}{1 - \alpha x_{A1} + D_{AB}/D_{KA}} \qquad (7.6\text{-}12)$$

This equation has been shown experimentally to be valid over the entire transition region (R1). It reduces to the Knudsen equation at low pressures and to the molecular diffusion equation at high pressures. An equation similar to Eq. (7.6-12) can also be written for component B.

The term D_{AB}/D_{KA} is proportional to $1/P$. Hence, as the total pressure P increases, the term D_{AB}/D_{KA} becomes very small and N_A in Eq. (7.6-12) becomes independent of total pressure since $D_{AB} P$ is independent of P. At low total pressures Eq. (7.6-12) becomes the Knudsen diffusion equation (7.6-5) and the flux N_A becomes directly proportional to P.

This is illustrated in Fig. 7.6-2 for a fixed capillary diameter where the flux increases as total pressure increases and then levels off at high pressure. The relative position of the curve depends, of course, on the capillary diameter and the molecular and Knudsen diffusivities. Using only a smaller diameter, D_{KA} would be smaller, and the Knudsen flux line would be parallel to the existing line at low pressures. At high pressures the flux line would asymptotically approach the existing horizontal line since molecular diffusion is independent of capillary diameter.

If A is diffusing in a catalytic pore and reacts at the surface at the end of the pore so that $A \rightarrow B$, then at steady state, equimolar counterdiffusion occurs or $N_A = -N_B$. Then from Eq. (7.6-8), $\alpha = 1 - 1 = 0$. The effective diffusivity D_{NA} from Eq. (7.6-11) becomes

$$D'_{NA} = \frac{1}{1/D_{AB} + 1/D_{KA}} \qquad (7.6\text{-}13)$$

The diffusivity is then independent of concentration and is constant. Integration of

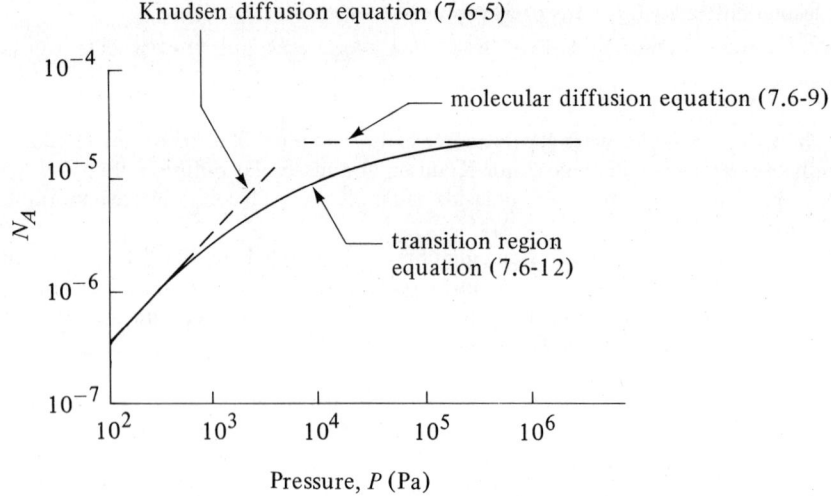

FIGURE 7.6-2. *Effect of total pressure P on the diffusion flux N_A in the transition region.*

(7.6-10) then gives

$$N_A = \frac{D'_{NA}P}{RTL}(x_{A1} - x_{A2}) = \frac{D'_{NA}}{RTL}(p_{A1} - p_{A2}) \qquad (7.6\text{-}14)$$

This simplified diffusivity D'_{NA} is often used in diffusion in porous catalysts even when equimolar counterdiffusion is not occurring. This greatly simplifies the equations for diffusion and reaction by using this simplified diffusivity.

An alternative simplified diffusivity to use is to use an average value of x_A in Eq. (7.6-11), to give

$$D''_{NA} = \frac{1}{(1 - \alpha x_{A\,av})/D_{AB} + 1/D_{KA}} \qquad (7.6\text{-}15)$$

where $x_{A\,av} = (x_{A1} + x_{A2})/2$. This diffusivity is more accurate than D'_{NA}. Integration of Eq. (7.6-10) gives

$$N_A = \frac{D''_{NA}P}{RTL}(x_{A1} - x_{A2}) = \frac{D''_{NA}}{RTL}(p_{A1} - p_{A2}) \qquad (7.6\text{-}16)$$

7.6E Flux Ratios for Diffusion of Gases in Capillaries

1. Diffusion in open system.   If diffusion in porous solids or channels with no chemical reaction is occurring where the total pressure P remains constant, then for an open binary counterdiffusing system, the ratio of N_A/N_B is constant in all of the three diffusion regimes and is (G1)

$$\frac{N_B}{N_A} = -\sqrt{\frac{M_A}{M_B}} \qquad (7.6\text{-}17)$$

Hence,

$$\alpha = 1 - \sqrt{\frac{M_A}{M_B}} \qquad (7.6\text{-}18)$$

In this case, gas flows by the two open ends of the system. However, when chemical reaction occurs, stoichiometry determines the ratio N_B/N_A and not Eq. (7.6-17).

2. Diffusion in closed system.   When molecular diffusion is occurring in a closed system shown in Fig. 6.2-1 at constant total pressure P, equimolar counterdiffusion occurs.

> **EXAMPLE 7.6-2.   Transition-Region Diffusion of He and N_2**
> A gas mixture at a total pressure of 0.10 atm abs and 298 K is composed of N_2 (A) and He (B). The mixture is diffusing through an open capillary 0.010 m long having a diameter of 5×10^{-6} m. The mole fraction of N_2 at one end is $x_{A1} = 0.8$ and at the other end is $x_{A2} = 0.2$. The molecular diffusivity D_{AB} is 6.98×10^{-5} m²/s at 1 atm, which is an average value by several investigators.
> (a) Calculate the flux N_A at steady state.
> (b) Use the approximate equations (7.6-14) and (7.6-16), for this case.
>
> *Solution:*   The given values are $T = 273 + 25 = 298$ K, $\bar{r} = 5 \times 10^{-6}/2 = 2.5 \times 10^{-6}$ m, $L = 0.01$ m, $P = 0.1(1.01325 \times 10^5) = 1.013 \times 10^4$ Pa, $x_{A1} = 0.8, x_{A2} = 0.2, D_{AB} = 6.98 \times 10^{-5}$ m²/s at 1 atm. Other values needed are $M_A = 28.02$ kg/kg mol, $M_B = 4.003$.

The molecular diffusivity at 0.1 atm is $D_{AB} = 6.98 \times 10^{-5}/0.1 = 6.98 \times 10^{-4}$ m²/s. Substituting into Eq. (7.6-3) for the Knudsen diffusivity,

$$D_{KA} = 97.0(2.5 \times 10^{-6})\sqrt{298/28.02} = 7.91 \times 10^{-4} \text{ m}^2/\text{s}$$

From Eq. (7.6-17),

$$\frac{N_B}{N_A} = -\sqrt{\frac{M_A}{M_B}} = -\sqrt{\frac{28.02}{4.003}} = -2.645$$

From Eq. (7.6-8),

$$\alpha = 1 + \frac{N_B}{N_A} = 1 - 2.645 = -1.645$$

Substituting into Eq. (7.6-12) for part (a),

$$N_A = \frac{(6.98 \times 10^{-4})(1.013 \times 10^4)}{(-1.645)(8314)(298)(0.01)} \ln \frac{1 + 1.645(0.2) + 6.98/7.91}{1 + 1.645(0.8) + 6.98/7.91}$$

$$= 6.40 \times 10^{-5} \text{ kg mol/s} \cdot \text{m}^2$$

For part (b), the approximate equation (7.6-13) is used.

$$D'_{NA} = \frac{1}{1/D_{AB} + 1/D_{KA}} = \frac{1}{1/6.98 \times 10^{-4} + 1/7.91 \times 10^{-4}}$$

$$= 3.708 \times 10^{-4} \text{ m}^2/\text{s}$$

Substituting into Eq. (7.6-14), the approximate flux is

$$N_A = \frac{D'_{NA}P}{RTL}(x_{A1} - x_{A2}) = \frac{(3.708 \times 10^{-4})(1.013 \times 10^4)}{8314(298)(0.01)}(0.8 - 0.2)$$

$$= 9.10 \times 10^{-5} \text{ kg mol/s} \cdot \text{m}^2$$

Hence, the calculated flux is approximately 40% high when using the approximation of equimolar counterdiffusion ($\alpha = 0$).

The more accurate approximate equation (7.6-15) is used next. The average concentration is $x_{A \text{ av}} = (x_{A1} + x_{A2})/2 = (0.8 + 0.2)/2 = 0.50$.

$$D''_{NA} = \frac{1}{(1 - \alpha x_{A \text{ av}})/D_{AB} + 1/D_{KA}}$$

$$= \frac{1}{(1 + 1.645 \times 0.5)/(6.98 \times 10^{-4}) + 1/(7.91 \times 10^{-4})}$$

$$= 2.581 \times 10^{-4} \text{ m}^2/\text{s}$$

Substituting into Eq. (7.6-16),

$$N_A = \frac{D''_{NA}}{RTL}(x_{A1} - x_{A2})$$

$$= \frac{(2.581 \times 10^{-4})(1.013 \times 10^4)}{8314(298)(0.01)}(0.8 - 0.2)$$

$$= 6.33 \times 10^{-5} \text{ kg mol/s} \cdot \text{m}^2$$

In this case the flux is only -1.1% low.

7.6F Diffusion of Gases in Porous Solids

In actual diffusion in porous solids the pores are not straight and cylindrical but are irregular. Hence, the equations for diffusion in pores must be modified somewhat for actual porous solids. The problem is further complicated by the fact that the pore diameters vary and the Knudsen diffusivity is a function of pore diameter.

As a result of these complications, investigators often measure effective diffusivities $D_{A \, eff}$ in porous media, where

$$N_A = \frac{D_{A \, eff} \, P}{RTL} (x_{A1} - x_{A2}) \tag{7.6-19}$$

If a tortuosity factor τ is used to correct the length L in Eq. (7.6-16), and the right-hand side is multiplied by the void fraction ε, Eq. (7.6-16) becomes

$$N_A = \frac{\varepsilon D''_{NA}}{\tau} \frac{P}{RTL} (x_{A1} - x_{A2}) \tag{7.6-20}$$

Comparing Eqs. (7.6-19) and (7.6-20),

$$D_{A \, eff} = \frac{\varepsilon D''_{NA}}{\tau} \tag{7.6-21}$$

In some cases investigators measure $D_{A \, eff}$ but use D'_{NA} instead of the more accurate D''_{NA} in Eq. (7.6-21).

Experimental data (C4, S2, S6) show that τ varies from about 1.5 to over 10. A reasonable range for many commercial porous solids is about 2–6 (S2). If the porous solid consists of a bidispersed system of micropores and macropores instead of a monodispersed pore system, the approach above should be modified (C4, S6).

Discussions and references for diffusion in porous inorganic-type solids, organic solids, and freeze-dried foods such as meat and fruit are given elsewhere (S2, S6).

Another type of diffusion that may occur is surface diffusion. When a molecular layer of absorption occurs on the solid, the molecules can migrate on the surface. Details are given elsewhere (S2, S6).

7.7 (SELECTED TOPIC) NUMERICAL METHODS FOR UNSTEADY-STATE MOLECULAR DIFFUSION

7.7A Introduction

Unsteady-state diffusion often occurs in inorganic, organic, and biological solid materials. If the boundary conditions are constant with time, if they are the same on all sides or surfaces of the solid, and if the initial concentration profile is uniform throughout the solid, the methods described in Section 7.1 can be used. However, these conditions are not always fulfilled. Hence, numerical methods must be used.

7.7B Unsteady-State Numerical Methods for Diffusion

1. Derivation for unsteady state for a slab.   For unsteady-state diffusion in one direction, Eq. (7.1-9) becomes

$$\frac{\partial c_A}{\partial t} = D_{AB} \frac{\partial^2 c_A}{\partial x^2} \tag{7.7-1}$$

Since this equation is identical mathematically to the unsteady-state heat-conduction Eq. (5.10-10),

$$\frac{\partial T}{\partial t} = \alpha \frac{\partial^2 T}{\partial x^2} \qquad (5.10\text{-}10)$$

identical mathematical methods can be used for solving both diffusion and conduction numerically.

Figure 7.7-1 shows a slab with width Δx centered at point n represented by the shaded area. Making a mole balance of A on this slab at the time t when the rate in − rate out = rate of accumulation in Δt s,

$$\frac{D_{AB} A}{\Delta x}(_tc_{n-1} - {}_tc_n) - \frac{D_{AB} A}{\Delta x}(_tc_n - {}_tc_{n+1}) = \frac{(A\,\Delta x)}{\Delta t}(_{t+\Delta t}c_n - {}_tc_n) \qquad (7.7\text{-}2)$$

where A is cross-sectional area and $_{t+\Delta t}c_n$ is concentration at point n one Δt later. Rearranging,

$$_{t+\Delta t}c_n = \frac{1}{M}\left[{}_tc_{n+1} + (M-2)_tc_n + {}_tc_{n-1} \right] \qquad (7.7\text{-}3)$$

where M is a constant.

$$M = \frac{(\Delta x)^2}{D_{AB}\,\Delta t} \qquad (7.7\text{-}4)$$

As in heat conduction, $M \geq 2$.

In using Eq. (7.7-3), the concentration $_{t+\Delta t}c_n$ at position n and the new time $t + \Delta t$ is calculated explicitly from the known three points at t. In this calculation method, starting with the known concentrations at $t = 0$, the calculations proceed directly from one time increment to the next until the final time is reached.

2. Simplified Schmidt method for a slab. If the value of $M = 2$, a simplification of Eq. (7.7-3) occurs, giving the Schmidt method.

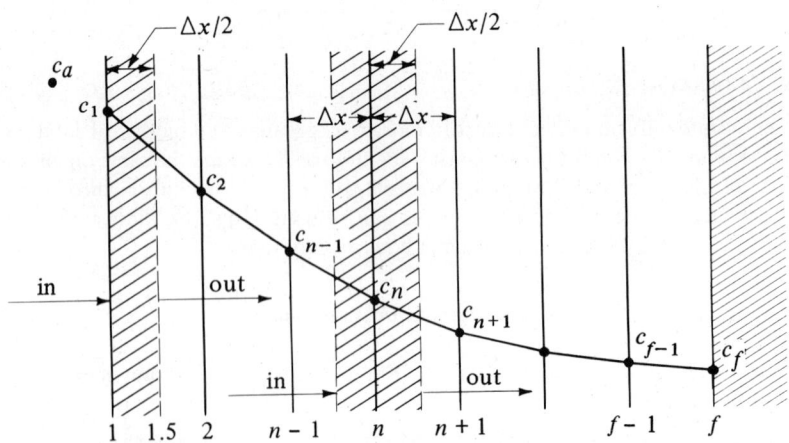

FIGURE 7.7-1. *Unsteady-state diffusion in a slab.*

$$t+\Delta t c_n = \frac{t c_{n-1} + t c_{n+1}}{2} \qquad (7.7\text{-}5)$$

7.7C Boundary Conditions for Numerical Method for a Slab

1. Convection at a boundary.   For the case where convection occurs outside in the fluid and the concentration of the fluid outside is suddenly changed to c_a, we can make a mass balance on the outside $\frac{1}{2}$ slab in Fig. 7.7-1. Following the methods used for heat transfer to derive Eq. (5.4-7), we write rate of mass entering by convection − rate of mass leaving by diffusion = rate of mass accumulation in Δt hours.

$$k_c A(t c_a - t c_1) - \frac{D_{AB} A}{\Delta x}(t c_1 - t c_2) = \frac{(A\,\Delta x/2)}{\Delta t}(t+\Delta t c_{1.25} - t c_{1.25}) \qquad (7.7\text{-}6)$$

where $t c_{1.25}$ is the concentration at the midpoint of the 0.5 Δx outside slab. As an approximation using $t c_1$ for $t c_{1.25}$ and rearranging Eq. (7.7-6),

$$t+\Delta t c_1 = \frac{1}{M}\left[2N t c_a + [M - (2N + 2)]_t c_1 + 2_t c_2\right] \qquad (7.7\text{-}7)$$

$$N = \frac{k_c\,\Delta x}{D_{AB}} \qquad (7.7\text{-}8)$$

where k_c is the convective mass-transfer coefficient in m/s. Again, note that $M \geq (2N + 2)$.

2. Insulated boundary condition.   For the insulated boundary at f in Fig. 7.7-1, setting $k_c = 0$ ($N = 0$) in Eq. (7.7-7), we obtain

$$t+\Delta t c_f = \frac{1}{M}\left[(M - 2)_t c_f + 2_t c_{f-1}\right] \qquad (7.7\text{-}9)$$

3. Alternative convective equation at the boundary.   Another form of Eq. (7.7-7) to use if N gets too large can be obtained by neglecting the accumulation in the front half-slab of Eq. (7.7-6) to give

$$t+\Delta t c_1 = \frac{N}{N+1} t+\Delta t c_a + \frac{1}{N+1} t+\Delta t c_2 \qquad (7.7\text{-}10)$$

The value of M is not restricted by the N value in this equation. When a large number of increments in Δx are used, the amount of mass neglected is small compared to the total.

4. Procedure for use of initial boundary concentration.   For the first time increment we should use an average value for $_1 c_a$ of $(c_a + _0 c_1)/2$, where $_0 c_1$ is the initial concentration at point 1. For succeeding times, the full value of c_a should be used. This special procedure for the value of c_a increases the accuracy of the numerical method, especially after a few time intervals.

In Section 5.4B for heat-transfer numerical methods, a detailed discussion is given on the best value of M to use in Eq. (7.7-3). The most accurate results are obtained for $M = 4$.

5. Boundary conditions with distribution coefficient.   Equations for boundary conditions in Eqs. (7.7-7) and (7.7-10) were derived for the distribution coefficient K given in Eq.

(7.7-7) being 1.0. When K is not 1.0, as in the boundary conditions for steady state, Kk_c should be substituted for k_c in Eq. (7.7-8) to become as follows. (See also Sections 6.6B and 7.1C.)

$$N = \frac{Kk_c \, \Delta x}{D_{AB}} \qquad (7.7\text{-}11)$$

Also, in Eqs. (7.7-7) and (7.7-10), the term c_a/K should be substituted for c_a.

Other cases such as for diffusion between dissimilar slabs in series, resistance between slabs in series, and so on, are covered in detail elsewhere (G1), with actual numerical examples being given. Also, in reference (G1) the implicit numerical method is discussed.

EXAMPLE 7.7-1. *Numerical Solution for Unsteady-State Diffusion with a Distribution Coefficient*

A slab of material 0.004 m thick has an initial concentration profile of solute A as follows, where x is distance in m from the exposed surface:

$x(m)$	Concentration $(kg \ mol \ A/m^3)$	Position, n
0	1.0×10^{-3}	1 (exposed)
0.001	1.25×10^{-3}	2
0.002	1.5×10^{-3}	3
0.003	1.75×10^{-3}	4
0.004	2.0×10^{-3}	5 (insulated)

The diffusivity $D_{AB} = 1.0 \times 10^{-9} \ m^2/s$. Suddenly, the top surface is exposed to a fluid having a constant concentration $c_a = 6 \times 10^{-3} \ kg \ mol \ A/m^3$. The distribution coefficient $K = c_a/c_n = 1.50$. The rear surface is insulated and unsteady-state diffusion is occurring only in the x direction. Calculate the concentration profile after 2500 s. The convective mass-transfer coefficient k_c can be assumed as infinite. Use $\Delta x = 0.001$ m and $M = 2.0$.

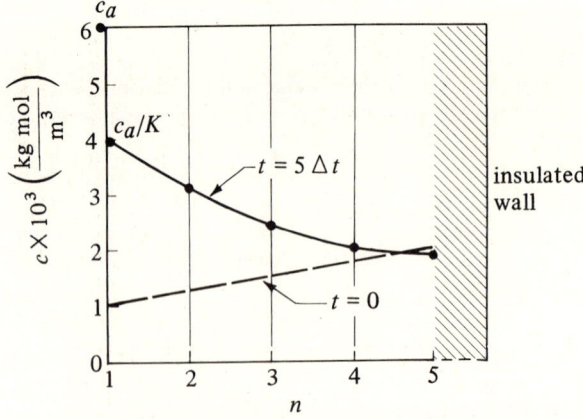

FIGURE 7.7-2.   *Concentrations for numerical method for unsteady-state diffusion, Example 7.7-1.*

Solution: Figure 7.7-2 shows the initial concentration profile for four slices and $c_a = 6 \times 10^{-3}$. Since $M = 2$, substituting into Eq. (7.7-4) with $\Delta x = 0.001$ m, and solving for Δt,

$$M = 2 = \frac{(\Delta x)^2}{D_{AB}\,\Delta t} = \frac{(0.001)^2}{(1 \times 10^{-9})(\Delta t)}$$

$$\Delta t = 500 \text{ s}$$

Hence, 2500 s/(500 s/increment) or five time increments are needed.

For the front surface where $n = 1$, the concentration to use for the first time increment, as stated previously, is

$$_1 c_a = \frac{c_a/K + {}_0 c_1}{2} = {}_1 c_1 \qquad (n = 1) \qquad \text{(7.7-12)}$$

where $_0 c_1$ is the initial concentration at $n = 1$. For the remaining time increments,

$$c_1 = \frac{c_a}{K} \qquad (n = 1) \qquad \text{(7.7-13)}$$

To calculate the concentrations for all time increments for slabs $n = 2, 3, 4$, using Eq. (7.7-5) for $M = 2$,

$$_{t + \Delta t} c_n = \frac{{}_t c_{n-1} + {}_t c_{n+1}}{2} \qquad (n = 2, 3, 4) \qquad \text{(7.7-14)}$$

For the insulated end at $n = 5$, substituting $M = 2$ and $f = n = 5$ into Eq. (7.7-9),

$$_{t + \Delta t} c_5 = \frac{(2 - 2)_t c_5 + 2_t c_4}{2} = {}_t c_4 \qquad (n = 5) \qquad \text{(7.7-15)}$$

For 1 Δt or $t + 1\ \Delta t$, the first time increment, calculating the concentration for $n = 1$ by Eq. (7.7-12),

$$_{t + \Delta t} c_1 = \frac{c_a/K + {}_0 c_1}{2} = \frac{6 \times 10^{-3}/1.5 + 1 \times 10^{-3}}{2} = 2.5 \times 10^{-3}$$

For $n = 2, 3$, and 4, using Eq. (7.7-14),

$$_{t + \Delta t} c_2 = \frac{{}_t c_{n-1} + {}_t c_{n+1}}{2} = \frac{{}_t c_1 + {}_t c_3}{2} = \frac{2.5 \times 10^{-3} + 1.5 \times 10^{-3}}{2} = 2 \times 10^{-3}$$

$$_{t + \Delta t} c_3 = \frac{{}_t c_2 + {}_t c_4}{2} = \frac{1.25 \times 10^{-3} + 1.75 \times 10^{-3}}{2} = 1.5 \times 10^{-3}$$

$$_{t + \Delta t} c_4 = \frac{{}_t c_3 + {}_t c_5}{2} = \frac{1.5 \times 10^{-3} + 2 \times 10^{-3}}{2} = 1.75 \times 10^{-3}$$

For $n = 5$, using Eq. (7.7-15),

$$_{t + \Delta t} c_5 = {}_t c_4 = 1.75 \times 10^{-3}$$

For 2 Δt using Eq. (7.7-13) for $n = 1$, using Eq. (7.7-14) for $n = 2$–4, and using Eq. (7.7-15) for $n = 5$,

$$_{t + 2\ \Delta t} c_1 = \frac{c_a}{K} = \frac{6 \times 10^{-3}}{1.5} = 4.0 \times 10^{-3} \qquad \text{(constant for rest of time)}$$

Principles of Unsteady-State and Convective Mass Transfer

$$_{t+2\,\Delta t}c_2 = \frac{_{t+\Delta t}c_1 + _{t+\Delta t}c_3}{2} = \frac{4 \times 10^{-3} + 1.5 \times 10^{-3}}{2} = 2.75 \times 10^{-3}$$

$$_{t+2\,\Delta t}c_3 = \frac{_{t+\Delta t}c_2 + _{t+\Delta t}c_4}{2} = \frac{2 \times 10^{-3} + 1.75 \times 10^{-3}}{2} = 1.875 \times 10^{-3}$$

$$_{t+2\,\Delta t}c_4 = \frac{_{t+\Delta t}c_3 + _{t+\Delta t}c_5}{2} = \frac{1.5 \times 10^{-3} + 1.75 \times 10^{-3}}{2} = 1.625 \times 10^{-3}$$

$$_{t+2\,\Delta t}c_5 = {_{t+\Delta t}c_4} = 1.75 \times 10^{-3}$$

For 3 Δt,

$$_{t+3\,\Delta t}c_1 = 4 \times 10^{-3}$$

$$_{t+3\,\Delta t}c_2 = \frac{4 \times 10^{-3} + 1.875 \times 10^{-3}}{2} = 2.938 \times 10^{-3}$$

$$_{t+3\,\Delta t}c_3 = \frac{2.75 \times 10^{-3} + 1.625 \times 10^{-3}}{2} = 2.188 \times 10^{-3}$$

$$_{t+3\,\Delta t}c_4 = \frac{1.875 \times 10^{-3} + 1.75 \times 10^{-3}}{2} = 1.813 \times 10^{-3}$$

$$_{t+3\,\Delta t}c_5 = 1.625 \times 10^{-3}$$

For 4 Δt,

$$_{t+4\,\Delta t}c_1 = 4 \times 10^{-3}$$

$$_{t+4\,\Delta t}c_2 = \frac{4 \times 10^{-3} + 2.188 \times 10^{-3}}{2} = 3.094 \times 10^{-3}$$

$$_{t+4\,\Delta t}c_3 = \frac{2.938 \times 10^{-3} + 1.813 \times 10^{-3}}{2} = 2.376 \times 10^{-3}$$

$$_{t+4\,\Delta t}c_4 = \frac{2.188 \times 10^{-3} + 1.625 \times 10^{-3}}{2} = 1.906 \times 10^{-3}$$

$$_{t+4\,\Delta t}c_5 = 1.813 \times 10^{-3}$$

For 5 Δt,

$$_{t+5\,\Delta t}c_1 = 4 \times 10^{-3}$$

$$_{t+5\,\Delta t}c_2 = \frac{4 \times 10^{-3} + 2.376 \times 10^{-3}}{2} = 3.188 \times 10^{-3}$$

$$_{t+5\,\Delta t}c_3 = \frac{3.094 \times 10^{-3} + 1.906 \times 10^{-3}}{2} = 2.500 \times 10^{-3}$$

$$_{t+5\,\Delta t}c_4 = \frac{2.376 \times 10^{-3} + 1.813 \times 10^{-3}}{2} = 2.095 \times 10^{-3}$$

$$_{t+5\,\Delta t}c_5 = 1.906 \times 10^{-3}$$

The final concentration profile is plotted in Fig. 7.7-2. To increase the accuracy, more slab increments and more time increments are needed. This type of calculation is suitable for a digital computer.

7.8 (SELECTED TOPIC) DIMENSIONAL ANALYSIS IN MASS TRANSFER

7.8A Introduction

The use of dimensional analysis enables us to predict the various dimensional groups which are very helpful in correlating experimental mass-transfer data. As we saw in fluid flow and in heat transfer, the Reynolds number, the Prandtl number, the Grashof number, and the Nusselt number were often used in correlating experimental data. The Buckingham theorem discussed in Section 3.8 (Selected Topic) and Section 4.14 (Selected Topic) states that the functional relationship among q quantities or variables whose units may be given in terms of u fundamental units or dimensions may be written as $(q - u)$ dimensionless groups.

7.8B Dimensional Analysis for Convective Mass Transfer

We consider a case of convective mass transfer where a fluid is flowing by forced convection in a pipe and mass transfer is occurring from the wall to the fluid. The fluid flows at a velocity v inside a pipe of diameter D and we wish to relate the mass-transfer coefficient k_c' to the variables D, ρ, μ, v, and D_{AB}. The total number of variables is $q = 6$. The fundamental units or dimensions are $u = 3$ and are mass M, length L, and time t. The units of the variables are

$$k_c' = \frac{L}{t} \qquad \rho = \frac{M}{L^3} \qquad \mu = \frac{M}{Lt} \qquad v = \frac{L}{t} \qquad D_{AB} = \frac{L^2}{t} \qquad D = L$$

The number of dimensionless groups or π' are then $6 - 3$, or 3. Then,

$$\pi_1 = f(\pi_2, \pi_3) \tag{7.8-1}$$

We choose the variables D_{AB}, ρ, and D to be the variables common to all the dimensionless groups, which are

$$\pi_1 = D_{AB}^a \rho^b D^c k_c' \tag{7.8-2}$$

$$\pi_2 = D_{AB}^d \rho^e D^f v \tag{7.8-3}$$

$$\pi_3 = D_{AB}^g \rho^h D^i \mu \tag{7.8-4}$$

For π_1 we substitute the actual dimensions as follows:

$$1 = \left(\frac{L^2}{t}\right)^a \left(\frac{M}{L^3}\right)^b (L)^c \left(\frac{L}{t}\right) \tag{7.8-5}$$

Summing for each exponent,

$$(L) \quad 0 = 2a - 3b + c + 1$$

$$(M) \quad 0 = b \tag{7.8-6}$$

$$(t) \quad 0 = a - 1$$

Solving these equation simultaneously, $a = -1$, $b = 0$, $c = 1$. Substituting these values into Eq. (7.8-2),

$$\pi_1 = \frac{k_c' D}{D_{AB}} = N_{Sh} \tag{7.8-7}$$

Repeating for π_2 and π_3,

$$\pi_2 = \frac{vD}{D_{AB}}$$

(7.8-8)

$$\pi_3 = \frac{\mu}{\rho D_{AB}} = N_{Sc}$$

(7.8-9)

If we divide π_2 by π_3 we obtain the Reynolds number.

$$\frac{\pi_2}{\pi_3} = \frac{vD}{D_{AB}} \bigg/ \left(\frac{\mu}{\rho D_{AB}}\right) = \frac{Dv\rho}{\mu} = N_{Re}$$

(7.8-10)

Hence, substituting into Eq. (7.8-1),

$$N_{Sh} = f(N_{Re}, N_{Sc})$$

(7.8-11)

7.9 (SELECTED TOPIC) BOUNDARY-LAYER FLOW AND TURBULENCE IN MASS TRANSFER

7.9A Laminar Flow and Boundary-Layer Theory in Mass Transfer

In Section 3.7C an exact solution was obtained for the hydrodynamic boundary layer for isothermal laminar flow past a plate and in Section 5.7A an extension of the Blasius solution was also used to derive an expression for convective heat transfer. In an analogous manner we use the Blasius solution for convective mass transfer for the same geometry and laminar flow. In Fig. 7.9-1 the concentration boundary layer is shown where the concentration of the fluid approaching the plate is $c_{A\infty}$ and c_{AS} in the fluid adjacent to the surface.

We start by using the differential mass balance, Eq. (7.5-17), and simplifying it for steady state where $\partial c_A / \partial t = 0$, $R_A = 0$, flow only in the x and y directions, so $v_z = 0$, and neglecting diffusion in the x and z directions to give

$$v_x \frac{\partial c_A}{\partial x} + v_y \frac{\partial c_A}{\partial y} = D_{AB} \frac{\partial^2 c_A}{\partial y^2}$$

(7.9-1)

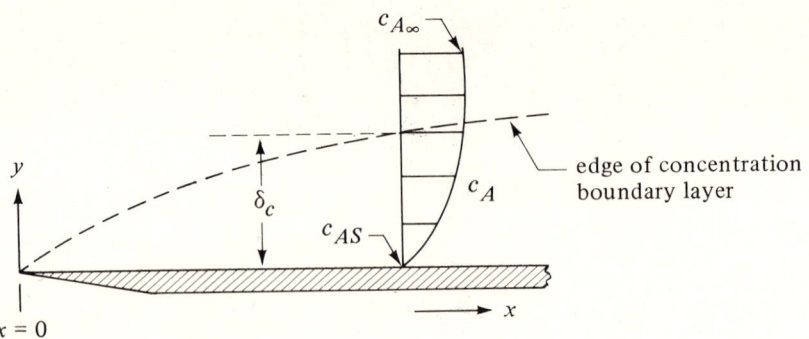

FIGURE 7.9-1. *Laminar flow of fluid past a flat plate and concentration boundary layer.*

The momentum boundary-layer equation is very similar.

$$v_x \frac{\partial v_x}{\partial x} + v_y \frac{\partial v_x}{\partial y} = \frac{\mu}{\rho} \frac{\partial^2 v_x}{\partial y^2} \qquad (3.7\text{-}5)$$

The thermal boundary-layer equation is also similar.

$$v_x \frac{\partial T}{\partial x} + v_y \frac{\partial T}{\partial y} = \frac{k}{\rho c_p} \frac{\partial^2 T}{\partial y^2} \qquad (5.7\text{-}2)$$

The continuity equation used previously is

$$\frac{\partial v_x}{\partial x} + \frac{\partial v_y}{\partial y} = 0 \qquad (3.7\text{-}3)$$

The dimensionless concentration boundary conditions are

$$\frac{v_x}{v_\infty} = \frac{T - T_S}{T_\infty - T_S} = \frac{c_A - c_{AS}}{c_{A\infty} - c_{AS}} = 0 \qquad \text{at } y = 0$$

$$\frac{v_x}{v_\infty} = \frac{T - T_S}{T_\infty - T_S} = \frac{c_A - c_{AS}}{c_{A\infty} - c_{AS}} = 1 \qquad \text{at } y = \infty$$

$$(7.9\text{-}2)$$

The similarity between the three differential equations (7.9-1), (3.7-5), and (5.7-2) is obvious, as is the similarity between the three sets of boundary conditions in Eq. (7.9-2). In Section 5.7A the Blasius solution was applied to convective heat transfer when $(\mu/\rho)/\alpha = N_{Pr} = 1.0$. We use the same type of solution for laminar convective mass transfer when $(\mu/\rho)/D_{AB} = N_{Sc} = 1.0$.

The velocity gradient at the surface was derived previously.

$$\left(\frac{\partial v_x}{\partial y}\right)_{y=0} = 0.332 \frac{v_\infty}{x} N_{Re, x}^{1/2} \qquad (5.7\text{-}5)$$

where $N_{Re, x} = x v_\infty \rho / \mu$. Also, from Eq. (7.9-2),

$$\frac{v_x}{v_\infty} = \frac{c_A - c_{AS}}{c_{A\infty} - c_{AS}} \qquad (7.9\text{-}3)$$

Differentiating Eq. (7.9-3) and combining the result with Eq. (5.7-5),

$$\left(\frac{\partial c_A}{\partial y}\right)_{y=0} = (c_{A\infty} - c_{AS})\left(\frac{0.332}{x} N_{Re, x}^{1/2}\right) \qquad (7.9\text{-}4)$$

The convective mass-transfer equation can be written as follows and also related with Fick's equation for dilute solutions:

$$N_{Ay} = k_c'(c_{AS} - c_{A\infty}) = -D_{AB}\left(\frac{\partial c_A}{\partial y}\right)_{y=0} \qquad (7.9\text{-}5)$$

Combining Eqs. (7.9-4) and (7.9-5),

$$\frac{k_c' x}{D_{AB}} = N_{Sh, x} = 0.332 N_{Re, x}^{1/2} \qquad (7.9\text{-}6)$$

This relationship is restricted to gases with a $N_{Sc} = 1.0$.

The relationship between the thickness δ of the hydrodynamic and δ_c of the concentration boundary layers where the Schmidt number is not 1.0 is

$$\frac{\delta}{\delta_c} = N_{Sc}^{1/3} \tag{7.9-7}$$

As a result, the equation for the local convective mass-transfer coefficient is

$$\frac{k'_c x}{D_{AB}} = N_{Sh, x} = 0.332 N_{Re, x}^{1/2} N_{Sc}^{1/3} \tag{7.9-8}$$

We can obtain the equation for the mean mass-transfer coefficient k'_c from $x = 0$ to $x = L$ for a plate of width b by integrating as follows:

$$k'_c = \frac{b}{bL} \int_0^L k'_c \, dx \tag{7.9-9}$$

The result is

$$\frac{k'_c L}{D_{AB}} = N_{Sh} = 0.664 N_{Re, L}^{1/2} N_{Sc}^{1/3} \tag{7.9-10}$$

This is similar to the heat-transfer equation for a flat plate, Eq. (5.7-15), and also checks the experimental mass-transfer equation (7.3-27) for a flat plate.

In Section 3.7 an approximate integral analysis was made for the laminar hydrodynamic and also for the turbulent hydrodynamic boundary layer. This was also done in Section 5.7 for the thermal boundary layer. This approximate integral analysis can also be done in exactly the same manner for the laminar and turbulent concentration boundary layers.

7.9B Prandtl Mixing Length and Turbulent Eddy Mass Diffusivity

In many applications the flow in mass transfer is turbulent and not laminar. The turbulent flow of a fluid is quite complex and the fluid undergoes a series of random eddy movements throughout the turbulent core. When mass transfer is occurring, we refer to this as eddy mass diffusion. In Sections 3.7 and 5.7 we derived equations for turbulent eddy thermal diffusivity and momentum diffusivity using the Prandtl mixing length theory.

In a similar manner we can derive a relation for the turbulent eddy mass diffusivity, ε_M. Eddies are transported a distance L, called the Prandtl mixing length, in the y direction. At this point L the fluid eddy differs in velocity from the adjacent fluid by the velocity v'_x, which is the fluctuating velocity component given in Section 3.7F. The instantaneous rate of mass transfer of A at a velocity v'_y for a distance L in the y direction is

$$J^*_{Ay} = c'_A v'_y \tag{7.9-11}$$

where c'_A is the instantaneous fluctuating concentration. The instantaneous concentration of the fluid is $c_A = c'_A + \bar{c}_A$ where $\bar{c}_A$ is the mean value and c'_A the deviation from the mean value. The mixing length L is small enough so the concentration difference is

$$c'_A = L \frac{d\bar{c}_A}{dy} \tag{7.9-12}$$

The rate of mass transported per unit area is J_{Ay}^*. Combining Eqs. (7.9-11) and (7.9-12),

$$J_{Ay}^* = -v_y' L \frac{d\bar{c}_A}{dy} \tag{7.9-13}$$

From Eq. (5.7-23),

$$v_y' = v_x' = L \left| \frac{d\bar{v}_x}{dy} \right| \tag{7.9-14}$$

Substituting Eq. (7.9-14) into (7.9-13),

$$J_{Ay}^* = -L^2 \left| \frac{d\bar{v}_x}{dy} \right| \frac{d\bar{c}_A}{dy} \tag{7.9-15}$$

The term $L^2 |d\bar{v}_x/dy|$ is called the turbulent eddy mass diffusivity ε_M. Combining Eq. (7.9-15) with the diffusion equation in terms of D_{AB}, the total flux is

$$J_{Ay}^* = -(D_{AB} + \varepsilon_M) \frac{d\bar{c}_A}{dy} \tag{7.9-16}$$

The similarities between Eq. (7.9-16) for mass transfer and heat and momentum transfer have been pointed out in detail in Section 6.1A.

7.9C Models for Mass-Transfer Coefficients

1. *Introduction.* For many years mass-transfer coefficients, which were based primarily on empirical correlations, have been used in the design of process equipment. A better understanding of the mechanisms of turbulence is needed before we can give a theoretical explanation of convective-mass-transfer coefficients. Some theories of convective mass transfer, such as the eddy diffusivity theory, have been presented in this chapter. In the following sections we present briefly some of these theories and also discuss how they can be used to extend empirical correlations.

2. *Film mass-transfer theory.* The film theory, which is the simplest and most elementary theory, assumes the presence of a fictitious laminar film next to the boundary. This film, where only molecular diffusion is assumed to be occurring, has the same resistance to mass transfer as actually exists in the viscous, transition, and turbulent core regions. Then the actual mass transfer coefficient k_c' is related to this film thickness δ_f by

$$J_A^* = k_c'(c_{A1} - c_{A2}) = \frac{D_{AB}}{\delta_f}(c_{A1} - c_{A2}) \tag{7.9-17}$$

$$k_c' = \frac{D_{AB}}{\delta_f} \tag{7.9-18}$$

The mass-transfer coefficient is proportional to $D_{AB}^{1.0}$. However, since we have shown that in Eq. (7.3-13), J_D is proportional to $(\mu/\rho D_{AB})^{2/3}$, then $k_c' \propto D_{AB}^{2/3}$. Hence, the film theory is not correct. The great advantage of the film theory is its simplicity where it can be used in complex situations such as simultaneous diffusion and chemical reaction.

3. *Penetration theory.* The penetration theory derived by Higbie and modified by Danckwerts (D3) was derived for diffusion or penetration into a laminar falling film for short contact times in Eq. (7.3-23) and is as follows:

$$k_c' = \sqrt{\frac{4 D_{AB}}{\pi t_L}} \tag{7.9-19}$$

Principles of Unsteady-State and Convective Mass Transfer

where t_L is the time of penetration of the solute in seconds. This was extended by Danckwerts. He modified this for turbulent mass transfer and postulated that a fluid eddy has a uniform concentration in the turbulent core and is swept to the surface and undergoes unsteady-state diffusion. Then the eddy is swept away to the eddy core and other eddies are swept to the surface and stay for a random amount of time. A mean surface renewal factor s in s^{-1} is defined as follows:

$$k_c' = \sqrt{D_{AB}s} \tag{7.9-20}$$

The mass-transfer coefficient k_c' is proportional to $D_{AB}^{0.5}$. In some systems, such as where liquid flows over packing and semistagnant pockets occur where the surface is being renewed, the results approximately follow Eq. (7.9-20). The value of s must be obtained experimentally. Others (D3, T2) have derived more complex combination film-surface renewal theories predicting a gradual change of the exponent on D_{AB} from 0.5 to 1.0 depending on turbulence and other factors. Penetration theories have been used in cases where diffusion and chemical reaction are occurring (D3).

4. *Boundary-layer theory.* The boundary-layer theory has been discussed in detail in Section 7.9 and is useful in predicting and correlating data for fluids flowing past solid surfaces. For laminar flow and turbulent flow the mass-transfer coefficient $k_c' \propto D_{AB}^{2/3}$. This has been experimentally verified for many cases.

PROBLEMS

7.1-1. *Unsteady-State Diffusion in a Thick Slab.* Repeat Example 7.1-2 but use a distribution coefficient $K = 0.50$ instead of 2.0. Plot the data.
  Ans. $c = c_i = 5.75 \times 10^{-2}$ ($x = 0$), $c = 2.78 \times 10^{-2}$ ($x = 0.01$ m), $c_{Li} = 2.87 \times 10^{-2}$ kg mol/m^3

7.1-2. *Plot of Concentration Profile in Unsteady-State Diffusion.* Using the same conditions as in Example 7.1-2, calculate the concentration at the points $x = 0$, 0.005, 0.01, 0.015, and 0.02 m from the surface. Also calculate c_{Li} in the liquid at the interface. Plot the concentrations in a manner similar to Fig. 7.1-3b, showing interface concentrations.

7.1-3. *Unsteady-State Diffusion in Several Directions.* Use the same conditions as in Example 7.1-1 except that the solid is a rectangular block 10.16 mm thick in the x direction, 7.62 mm thick in the y direction, and 10.16 mm thick in the z direction, and diffusion occurs at all six faces. Calculate the concentration at the midpoint after 10 h.
  Ans. $c = 6.20 \times 10^{-4}$ kg mol/m^3

7.1-4 *Drying of Moist Clay.* A very thick slab of clay has an initial moisture content of $c_0 = 14$ wt %. Air is passed over the top surface to dry the clay. Assume a relative resistance of the gas at the surface of zero. The equilibrium moisture content at the surface is constant at $c_1 = 3.0$ wt %. The diffusion of the moisture in the clay can be approximated by a diffusivity of $D_{AB} = 1.29 \times 10^{-8}$ m^2/s. After 1.0 h of drying, calculate the concentration of water at points 0.005, 0.01, and 0.02 m below the surface. Assume that the clay is a semiinfinite solid and that the Y value can be represented using concentrations of wt % rather than kg mol/m^3. Plot the values versus x.

7.1-5. *Unsteady-State Diffusion in a Cylinder of Agar Gel.* A wet cylinder of agar gel at 278 K containing a uniform concentration of urea of 0.1 kg mol/m^3 has a diameter of 30.48 mm and is 38.1 mm long with flat parallel ends. The diffusivity is 4.72×10^{-10} m^2/s. Calculate the concentration at the midpoint of the cylinder after 100 h for the following cases if the cylinder is suddenly immersed in turbulent pure water.

(a) For radial diffusion only.

(b) Diffusion occurs radially and axially.

7.1-6. *Drying of Wood.* A flat slab of Douglas fir wood 50.8 mm thick containing 30 wt % moisture is being dried from both sides (neglecting ends and edges). The equilibrium moisture content at the surface of the wood due to the drying air blown over it is held at 5 wt % moisture. The drying can be assumed to be represented by a diffusivity of 3.72×10^{-6} m²/h. Calculate the time for the center to reach 10% moisture.

7.2-1. *Flux and Conversion of Mass-Transfer Coefficient.* A value of k_G was experimentally determined to be 1.08 lb mol/h · ft² · atm for A diffusing through stagnant B. For the same flow and concentrations it is desired to predict k'_G and the flux of A for equimolar counterdiffusion. The partial pressures are $p_{A1} = 0.20$ atm, $p_{A2} = 0.05$ atm, and $P = 1.0$ atm abs total. Use English and SI units.

> **Ans.** $k'_G = 0.942$ lb mol/h · ft² · atm, 1.261×10^{-8} kg mol/s · m² · Pa,
> $N_A = 0.141$ lb mol A/h · ft², 1.912×10^{-4} kg mol/s · m²

7.2-2. *Conversion of Mass-Transfer Coefficients.* Prove or show the following relationships starting with the flux equations.

(a) Convert k'_c to k_y and k_G.

(b) Convert k_L to k_x and k'_x.

(c) Convert k_G to k_y and k_c.

7.2-3. *Absorption of H_2S by Water.* In a wetted-wall tower an air-H_2S mixture is flowing by a film of water which is flowing as a thin film down a vertical plate. The H_2S is being absorbed from the air to the water at a total pressure of 1.50 atm abs and 30°C. The value of k'_c of 9.567×10^{-4} m/s has been predicted for the gas-phase mass-transfer coefficient. At a given point the mole fraction of H_2S in the liquid at the liquid–gas interface is $2.0(10^{-5})$ and p_A of H_2S in the gas is 0.05 atm. The Henry's law equilibrium relation is p_A (atm) $= 609x_A$ (mole fraction in liquid). Calculate the rate of absorption of H_2S. (*Hint:* Call point 1 the interface and point 2 the gas phase. Then calculate p_{A1} from Henry's law and the given x_A. The value of p_{A2} is 0.05 atm.)

> **Ans.** $N_A = -1.485 \times 10^{-6}$ kg mol/s · m²

7.3-1. *Mass Transfer from a Flat Plate to a Liquid.* Using the data and physical properties of Example 7.3-2 calculate the flux for a water velocity of 0.152 m/s and a plate length of $L = 0.137$ m. Do not assume that $x_{BM} = 1.0$, but actually calculate its value.

7.3-2. *Mass Transfer from a Pipe Wall.* Pure water at 26.1°C is flowing at a velocity of 0.0305 m/s in a tube having an inside diameter of 6.35 mm. The tube is 1.829 m long with the last 1.22 m having the walls coated with benzoic acid. Assuming that the velocity profile is fully developed, calculate the average concentration of benzoic acid at the outlet. Use the physical property data of Example 7.3-2. [*Hint:* First calculate the Reynolds number Dvp/μ. Then calculate $N_{Re} N_{Sc} (D/L)(\pi/4)$, which is the same as $W/D_{AB}\rho L$.]

> **Ans.** $(c_A - c_{A0})/(c_{Ai} - c_{A0}) = 0.0744$, $c_A = 2.193 \times 10^{-3}$ kg mol/m³

7.3-3. *Mass-Transfer Coefficient for Various Geometries.* It is desired to estimate the mass-transfer coefficient k_G in kg mol/s · m² · Pa for water vapor in air at 338.6 K and 101.32 kPa flowing in a large duct past different geometry solids. The velocity in the duct is 3.66 m/s. The water vapor concentration in the air is small, so the physical properties of air can be used. Water vapor is being transferred to the solids. Do this for the following geometries.

(a) A single 25.4-mm-diameter sphere.

(b) A packed bed of 25.4-mm spheres with $\varepsilon = 0.35$.

> **Ans.** (a) $k_G = 1.98 \times 10^{-8}$ kg mol/s · m² · Pa (1.48 lb mol/h · ft² · atm)

7.3-4 Mass Transfer to Definite Shapes. Estimate the value of the mass-transfer coefficient in a stream of air at 325.6 K flowing in a duct by the following shapes made of solid naphthalene. The velocity of the air is 1.524 m/s at 325.6 K and 202.6 kPa. The D_{AB} of naphthalene in air is 5.16×10^{-6} m^2/s at 273 K and 101.3 kPa.
(a) For air flowing parallel to a flat plate 0.152 m in length.
(b) For air flowing by a single sphere 12.7 mm in diameter.

7.3-5. Mass Transfer to Packed Bed and Driving Force. Pure water at 26.1°C is flowing at a rate of 0.0701 ft^3/h through a packed bed of 0.251-in. benzoic acid spheres having a total surface area of 0.129 ft^2. The solubility of benzoic acid in water is 0.00184 lb mol benzoic acid/ft^3 solution. The outlet concentration c_{A2} is 1.80×10^{-4} lb mol/ft^3. Calculate the mass-transfer coefficient k_c.

7.3-6. Mass Transfer in Liquid Metals. Mercury at 26.5°C is flowing through a packed bed of lead spheres having a diameter of 2.096 mm with a void fraction of 0.499. The superficial velocity is 0.02198 m/s. The solubility of lead in mercury is 1.721 wt %, the Schmidt number is 124.1, the viscosity of the solution is 1.577×10^{-3} Pa · s, and the density is 13 530 kg/m^3.
(a) Predict the value of J_D. Use Eq. (7.3-38) if applicable. Compare with the experimental of $J_D = 0.076$ (D2).
(b) Predict the value of k_c for the case of A diffusing through nondiffusing B.

Ans. (a) $J_D = 0.0784$, (b) $k_c = 6.986 \times 10^{-5}$ m/s

7.3-7. Mass Transfer from a Pipe and Log Mean Driving Force. Use the same physical conditions as Problem 7.3-2, but the velocity in the pipe is now 3.05 m/s. Do as follows.
(a) Predict the mass-transfer coefficient k_c'. (Is this turbulent flow?)
(b) Calculate the average benzoic acid concentration at the outlet. [*Note*: In this case, Eqs. (7.3-42) and (7.3-43) must be used with the log mean driving force, where A is the surface area of the pipe.]
(c) Calculate the total kg mol of benzoic acid dissolved per second.

7.3-8. Derivation of Relation Between J_D and N_{Sh}. Equation (7.3-3) defines the Sherwood number and Eq. (7.3-5) defines the J_D factor. Derive the relation between N_{Sh} and J_D in terms of N_{Re} and N_{Sc}.

Ans. $N_{Sh} = J_D N_{Re} N_{Sc}^{1/3}$

7.3-9. Driving Force to Use in Mass Transfer. Derive Eq. (7.3-42) for the log mean driving force to use for a fluid flowing in a packed bed or in a tube. (*Hint* : Start by making a mass balance and a diffusion rate balance over a differential area dA as follows:

$$ N_A \, dA = k_c(c_{Ai} - c_A) \, dA = V \, dc_A $$

where $V = m^3/s$ flow rate. Assume dilute solutions.)

7.4-1. Maximum Oxygen Uptake of a Microorganism. Calculate the maximum possible rate of oxygen uptake at 37°C of microorganisms having a diameter of $\frac{2}{3}$ μm suspended in an agitated aqueous solution. It is assumed that the surrounding liquid is saturated with O_2 from air at 1 atm abs pressure. It will be assumed that the microorganism can utilize the oxygen much faster than it can diffuse to it. The microorganism has a density very close to that of water. Use physical property data from Example 7.4-1. (*Hint* : Since the oxygen is consumed faster than it is supplied, the concentration c_{A2} at the surface is zero. The concentration c_{A1} in the solution is at saturation.)

Ans. $k_c = 9.75 \times 10^{-3}$ m/s, $N_A = 2.20 \times 10^{-6}$ kg mol O_2/s · m^2

7.4-2. Mass Transfer of O_2 in Fermentation Process. A total of 5.0 g of wet microorganisms having a density of 1100 kg/m^3 and a diameter of 0.667 μm are added to 0.100 L of aqueous solution at 37°C in a shaker flask for a fermentation. Air can enter through a porous stopper. Use physical property data from Example 7.4-1.

(a) Calculate the maximum rate possible of mass transfer of oxygen in kg mol O_2/s to the surface of the microorganisms assuming that the solution is saturated with air at 101.32 kPa abs pressure.

(b) By material balances on other nutrients, the actual utilization of O_2 by the microorganisms is 6.30×10^{-6} kg mol O_2/s. What would be the actual concentration of O_2 in the solution as percent saturation during the fermentation?

Ans. (a) $k'_L = 9.82 \times 10^{-3}$ m/s, $N_A A = 9.07 \times 10^{-5}$ kg mol O_2/s, (b) 6.95% saturation

7.5-1. (*Selected Topic*) *Sum of Molar Fluxes.* Prove the following equation using the definitions in Table 7.5-1.

$$N_A + N_B = c v_M$$

7.5-2. (*Selected Topic*) *Proof of Derived Relation.* Using the definitions from Table 7.5-1, prove the following:

$$j_A = n_A - w_A(n_A + n_B)$$

7.5-3. (*Selected Topic*) *Different Forms of Fick's Law.* Using Eq. (1), prove Eq. (2).

$$j_A = -\rho D_{AB} \frac{dw_A}{dz} \tag{1}$$

$$j_A = -\frac{c^2}{\rho} M_A M_B D_{AB} \frac{dx_A}{dz} \tag{2}$$

(*Hint :* First relate w_A to x_A. Then differentiate this equation to relate dw_A and dx_A. Finally, use $M = x_A M_A + x_B M_B$ to simplify.)

7.5-4. (*Selected Topic*) *Other Form of Fick's Law.* Show that the following form of Fick's law is valid:

$$c(v_A - v_B) = -\frac{c D_{AB}}{x_A x_B} \frac{dx_A}{dz}$$

(*Hint :* Start with $N_A = J_A^* + c_A v_M$. Substitute the expression for J_A^* from Table 7.5-1 and simplify.)

7.5-5. (*Selected Topic*) *Different Form of Equation of Continuity.* Starting with Eq. (7.5-12),

$$\frac{\partial \rho_A}{\partial t} + (\nabla \cdot \mathbf{n}_A) = r_A \tag{7.5-12}$$

convert this to the following for constant ρ:

$$\frac{\partial \rho_A}{\partial t} + (\mathbf{v} \cdot \nabla \rho_A) - (\nabla \cdot D_{AB} \nabla \rho_A) = r_A \tag{1}$$

[*Hint :* From Table 7.5-1, substitute $\mathbf{n}_A = \mathbf{j}_A + \rho_A \mathbf{v}$ into Eq. (7.5-12). Note that $(\nabla \cdot \mathbf{v}) = 0$ for constant ρ. Then substitute Fick's law in terms of $\mathbf{j}_A$.]

7.5-6. (*Selected Topic*) *Diffusion and Reaction at a Surface.* Gas A is diffusing from a gas stream at point 1 to a catalyst surface at point 2 and reacts instantaneously and irreversibly as follows:

$$2A \rightarrow B$$

Gas B diffuses back to the gas stream. Derive the final equation for N_A at constant pressure P and steady state in terms of partial pressures.

Ans. $N_A = \dfrac{2 D_{AB} P}{RT(z_2 - z_1)} \ln \dfrac{1 - p_{A2}/2P}{1 - p_{A1}/2P}$

7.5-7. (Selected Topic) Unsteady-State Diffusion and Reaction. Solute A is diffusing at unsteady state into a semiinfinite medium of pure B and undergoes a first order reaction with B. Solute A is dilute. Calculate the concentration c_A at points $z = 0$, 4, and 10 mm from the surface for $t = 1 \times 10^5$ s. Physical property data are $D_{AB} = 1 \times 10^{-9}$ m^2/s, $k' = 1 \times 10^{-4}$ s^{-1}, $c_{A0} = 1.0$ kg mol/m^3. Also calculate the kg mol absorbed/m^2.

7.5-8. (Selected Topic) Multicomponent Diffusion. At a total pressure of 202.6 kPa and 358 K, ammonia gas (A) is diffusing at steady state through an inert nondiffusing mixture of nitrogen (B) and hydrogen (C). The mole fractions at $z_1 = 0$ are $x_{A1} = 0.8$, $x_{B1} = 0.15$, and $x_{C1} = 0.05$; and at $z_2 = 4.0$ mm, $x_{A2} = 0.2$, $x_{B2} = 0.6$, and $x_{C2} = 0.2$. The diffusivities at 358 K and 101.3 kPa are $D_{AB} = 3.28 \times 10^{-5}$ m^2/s and $D_{AC} = 1.093 \times 10^{-4}$ m^2/s. Calculate the flux of ammonia.

Ans. $N_A = 4.69 \times 10^{-4}$ kg mol A/s·m^2

7.5-9. (Selected Topic) Diffusion in Liquid Metals and Variable Diffusivity. The diffusion of tin (A) in liquid lead (B) at 510°C was carried out by using a 10.0-mm-long capillary tube and maintaining the mole fraction of tin at x_{A1} at the left end and x_{A2} at the right end of the tube. In the range of concentrations of $0.2 \leq x_A \leq 0.4$ the diffusivity of tin in lead has been found to be a linear function of x_A (S7).

$$D_{AB} = A + Bx_A$$

where A and B are constants and D_{AB} is in m^2/s.
(a) Assuming the molar density to be constant at $c = c_A + c_B = c_{av}$, derive the final integrated equation for the flux N_A assuming steady state and that A diffuses through stagnant B.
(b) For this experiment, $A = 4.8 \times 10^{-9}$, $B = -6.5 \times 10^{-9}$, $c_{av} = 50$ kg mol/m^3, $x_{A1} = 0.4$, $x_{A2} = 0.2$. Calculate N_A.

Ans. (b) $N_A = 4.055 \times 10^{-6}$ kg mol A/s·m^2

7.5-10. (Selected Topic) Diffusion and Chemical Reaction of Molten Iron in Process Metallurgy. In a steelmaking process using molten pig iron containing carbon, a spray of molten iron particles containing 4.0 wt % carbon fall through a pure oxygen atmosphere. The carbon diffuses through the molten iron to the surface of the drop, where it is assumed that it reacts instantly at the surface, because of the high temperature, as follows by a first-order reaction:

$$C + \tfrac{1}{2} O_2(g) \rightarrow CO(g)$$

Calculate the maximum drop size allowable so that the final drop after a 2.0-s fall contains on an average 0.1 wt % carbon. Assume that the mass transfer rate of gases at the surface is very great, so there is no outside resistance. Assume no internal circulation of the liquid. Hence, the decarburization rate is controlled by the rate of diffusion of carbon to the surface of the droplet. The diffusivity of carbon in iron is 7.5×10^{-9} m^2/s (S7). (*Hint :* Can Fig. 5.3-13 be used for this case?)

Ans. radius = 0.217 mm

7.5-11. (Selected Topic) Effect of Slow Reaction Rate on Diffusion. Gas A diffuses from point 1 to a catalyst surface at point 2, where it reacts as follows: $2A \rightarrow B$. Gas B diffuses back a distance δ to point 1.
(a) Derive the equation for N_A for a very fact reaction using mole fraction units x_{A1}, and so on.
(b) For $D_{AB} = 0.2 \times 10^{-4}$ m^2/s, $x_{A1} = 0.97$, $P = 101.32$ kPa, $\delta = 1.30$ mm, and $T = 298$ K, solve for N_A.
(c) Do the same as part (a) but for a slow first-order reaction where k'_1 is the reaction velocity constant.
(d) Calculate N_A and x_{A2} for part (c) where $k'_1 = 0.53 \times 10^{-2}$ m/s.

Ans. (b) $N_A = 8.35 \times 10^{-4}$ kg mol/s·m^2

7.5-12. (*Selected Topic*) Diffusion and Heterogeneous Reaction on Surface. In a tube of radius R m filled with a liquid dilute component A is diffusing in the nonflowing liquid phase represented by

$$N_A = -D_{AB} \frac{dc_A}{dz}$$

where z is distance along the tube axis. The inside wall of the tube exerts a catalytic effect and decomposes A so that the heterogeneous rate of decomposition on the wall in kg mol A/s is equal to $kc_A A_w$, where k is a first-order constant and A_w is the wall area in m^2. Neglect any radial gradients (this means a uniform radial concentration).

Derive the differential equation for unsteady state for diffusion and reaction for this system. [*Hint :* First make a mass balance for A for a Δz length of tube as follows: rate of input (diffusion) + rate of generation (heterogeneous) = rate of output (diffusion) + rate of accumulation.]

Ans. $\dfrac{\partial c_A}{\partial t} = D_{AB} \dfrac{\partial^2 c_A}{\partial z^2} - \dfrac{2k}{R} c_A$

7.6-1. (*Selected Topic*) Knudsen Diffusivities. A mixture of He(A) and Ar(B) is diffusing at 1.013×10^5 Pa total pressure and 298 K through a capillary having a radius of 100 Å.
(a) Calculate the Knudsen diffusivity of He (A).
(b) Calculate the Knudsen diffusivity of Ar (B).
(c) Compare with the molecular diffusivity D_{AB}.
Ans. (a) $D_{KA} = 8.37 \times 10^{-6} \, m^2/s$; (c) $D_{AB} = 7.29 \times 10^{-5} \, m^2/s$

7.6-2. (*Selected Topic*) Transition-Region Diffusion. A mixture of He (A) and Ar (B) at 298 K is diffusing through an open capillary 15 mm long with a radius of 1000 Å. The total pressure is 1.013×10^5 Pa. The molecular diffusivity D_{AB} at 1.013 $\times 10^5$ Pa is $7.29 \times 10^{-5} \, m^2/s$.
(a) Calculate the Knudsen diffusivity of He (A).
(b) Predict the flux N_A using Eq. (7.6-18) and Eq. (7.6-12) if $x_{A1} = 0.8$ and $x_{A2} = 0.2$. Assume steady state.
(c) Predict the flux N_A using the approximate Eqs. (7.6-14) and (7.6-16).

7.6-3. (*Selected Topic*) Diffusion in a Pore in the Transition Region. Pure H_2 gas (A) at one end of a noncatalytic pore of radius 50 Å and length 1.0 mm ($x_{A1} = 1.0$) is diffusing through this pore with pure C_2H_6 gas (B) at the other end at $x_{A2} = 0$. The total pressure is constant at 1013.2 kPa. The predicted molecular diffusivity of H_2–C_2H_6 is $8.60 \times 10^{-5} \, m^2/s$ at 101.32 kPa and 373 K. Calculate the Knudsen diffusivity of H_2 and flux N_A of H_2 in the mixture at 373 K and steady state.
Ans. $D_{KA} = 6.60 \times 10^{-6} \, m^2/s, \, N_A = 1.472 \times 10^{-3} \, kg \, mol \, A/s \cdot m^2$

7.6-4. (*Selected Topic*) Transition Region Diffusion in Capillary. A mixture of nitrogen gas (A) and helium (B) at 298 K is diffusing through a capillary 0.10 m long in an open system with a diameter of 10 μm. The mole fractions are constant at $x_{A1} = 1.0$ and $x_{A2} = 0$. See Example 7.6-2 for physical properties.
(a) Calculate the Knudsen diffusivity D_{KA} and D_{KB} at the total pressures of 0.001, 0.1, and 10.0 atm.
(b) Calculate the flux N_A at steady state at these pressures.
(c) Plot N_A versus P on log–log paper. What are the limiting lines at lower pressures and very high pressures? Calculate and plot these lines.

7.7-1. (*Selected Topic*) Numerical Method for Unsteady-State Diffusion. A solid slab 0.01 m thick has an initial uniform concentration of solute A of 1.00 kg mol/m^3. The diffusivity of A in the solid is $D_{AB} = 1.0 \times 10^{-10} \, m^2/s$. All surfaces of the slab are insulated except the top surface. The surface concentration is suddenly dropped to zero concentration and held there. Unsteady-state diffusion occurs in the one x direction with the rear surface insulated. Using a numerical method, determine the concentrations after 12×10^4 s. Use $\Delta x = 0.002$ m and $M = 2.0$. The value of K is 1.0.

$$\text{Ans.} \quad \begin{aligned} c_1 &= 0 \text{(front surface, } x = 0 \text{ m)}, \\ c_2 &= 0.3125 \text{ kg mol/m}^3 \ (x = 0.002 \text{ m}) \\ c_3 &= 0.5859 \ (x = 0.004 \text{ m}), \\ c_4 &= 0.7813 \ (x = 0.006 \text{ m}) \\ c_5 &= 0.8984 \ (x = 0.008 \text{ m}), \\ c_6 &= 0.9375 \text{ (insulated surface, } x = 0.01 \text{ m)} \end{aligned}$$

7.7-2. (*Selected Topic*) *Digital Computer and Unsteady-State Diffusion.* Using the conditions of Problem 7.7-1, solve that problem by the digital computer. Use $\Delta x = 0.0005$ m. Write the Fortran program and plot the final concentrations. Use the explicit method, $M = 2$.

7.7-3. (*Selected Topic*) *Numerical Method and Different Boundary Condition.* Use the same conditions as in Example 7.7-1, but in this new case the rear surface is not insulated. At time $t = 0$ the concentration at the rear surface is also suddenly changed to $c_5 = 0$ and held there. Calculate the concentration profile after 2500 s. Plot the initial and final concentration profiles and compare with the final profile of Example 7.7-1.

7.8-1. (*Selected Topic*) *Dimensional Analysis in Mass Transfer.* A fluid is flowing in a vertical pipe and mass transfer is occurring from the pipe wall to the fluid. Relate the convective mass-transfer coefficient k'_c to the variables D, ρ, μ, v, D_{AB}, g, and $\Delta\rho$, where D is pipe diameter, L is pipe length, and $\Delta\rho$ is the density difference.

$$\text{Ans.} \quad \frac{k'_c D}{D_{AB}} = f\left(\frac{gL^3 \rho \, \Delta\rho}{\mu^2}, \frac{Dv\rho}{\mu}, \frac{\mu}{\rho D_{AB}}\right)$$

7.9-1. (*Selected Topic*) *Mass Transfer and Turbulence Models.* Pure water at a velocity of 0.11 m/s is flowing at 26.1°C past a flat plate of solid benzoic acid where $L = 0.40$ m. Do as follows.
(a) Assuming dilute solutions, calculate the mass-transfer coefficient k_c. Use physical property data from Example 7.3-2.
(b) Using the film model, calculate the equivalent film thickness.
(c) Using the penetration model, calculate the time of penetration.
(d) Calculate the mean surface renewal factor using the modified penetration model.

$$\text{Ans.} \quad \text{(b)} \ \delta_f = 0.2031 \text{ mm}, \text{(d)} \ s = 3.019 \times 10^{-2} \text{ s}^{-1}$$

REFERENCES

(B1) BIRD, R. B., STEWART, W. E., and LIGHTFOOT, E. N. *Transport Phenomena.* New York: John Wiley & Sons, Inc., 1960.

(B2) BLAKEBROUGH, N. *Biochemical and Biological Engineering Science*, Vol. 1. New York: Academic Press, Inc., 1967.

(B3) BEDDINGTON, C. H., Jr., and DREW, T. B. *Ind. Eng. Chem.*, **42**, 1164 (1950).

(C1) CARMICHAEL, L. T., SAGE, B. H., and LACEY, W. N. *A.I.Ch.E. J.*, **1**, 385 (1955).

(C2) CHILTON, T. H., and COLBURN, A. P. *Ind. Eng. Chem.*, **26**, 1183 (1934).

(C3) CALDERBANK, P. H., and MOO-YOUNG, M. B. *Chem. Eng. Sci.*, **16**, 39 (1961).

(C4) CUNNINGHAM, R. S., and GEANKOPLIS, C. J. *Ind. Eng. Chem. Fund.*, **7**, 535 (1968).

(D1) DANCKWERTS, P. V. *Trans. Faraday Soc.*, **46**, 300 (1950).

(D2) DUNN, W. E., BONILLA, C. F., FERSTENBERG, C., and GROSS, B. *A.I.Ch.E. J.*, **2**, 184 (1956).

(D3)   DANCKWERTS, P. V. *Gas–Liquid Reactions.* New York: McGraw-Hill Book Company, 1970.

(D4)   DWIVEDI, P. N., and UPADHYAY, S. N. *Ind. Eng. Chem., Proc. Des. Dev.,* **16**, 157 (1977).

(G1)   GEANKOPLIS, C. J. *Mass Transport Phenomena.* Columbus, Ohio: Ohio State University Bookstores, 1972.

(G2)   GILLILAND, E. R., and SHERWOOD, T. K. *Ind. Eng. Chem.,* **26**, 516 (1934).

(G3)   GARNER, F. H., and SUCKLING, R. D. *A.I.Ch.E. J.,* **4**, 114 (1958).

(G4)   GUPTA, A. S., and THODOS, G. *Ind. Chem. Eng. Fund.,* **3**, 218 (1964).

(G5)   GUPTA, A. S., and THODOS, G. *A.I.Ch.E. J.,* **8**, 609 (1962).

(G6)   GUPTA, A. S., and THODOS, G. *Chem. Eng. Progr.,* **58** (7), 58 (1962).

(L1)   LINTON, W. H., Jr., and SHERWOOD, T. K. *Chem. Eng. Progr.,* **46**, 258 (1950).

(L2)   LITT, M., and FRIEDLANDER, S. K. *A.I.Ch.E. J.,* **5**, 483 (1959).

(M1)   MCADAMS, W. H. *Heat Transmission,* 3rd ed. New York: McGraw-Hill Book Company, 1954.

(P1)   PERRY, R. H., and CHILTON, C. H. *Chemical Engineers' Handbook,* 5th ed. New York: McGraw-Hill Book Company, 1973.

(R1)   REMICK, R. R., and GEANKOPLIS, C. J. *Ind. Eng. Chem. Fund.,* **12**, 214 (1973).

(S1)   SEAGER, S. L., GEERTSON, L. R., and GIDDINGS, J. C. *J. Chem. Eng. Data,* **8**, 168 (1963).

(S2)   SATTERFIELD, C. N. *Mass Transfer in Heterogeneous Catalysis.* Cambridge, Mass.: The MIT Press, 1970.

(S3)   STEELE, L. R., and GEANKOPLIS, C. J. *A.I.Ch.E. J.,* **5**, 178 (1959).

(S4)   SHERWOOD, T. K., PIGFORD, R. L., and WILKE, C. R. *Mass Transfer.* New York: McGraw-Hill Book Company, 1975.

(S5)   STEINBERGER, W. L., and TREYBAL, R. E. *A.I.Ch.E. J.,* **6**, 227 (1960).

(S6)   SMITH, J. M. *Chemical Engineering Kinetics,* 2nd ed. New York: McGraw-Hill Book Company, 1970.

(S7)   SZEKELY, J., and THEMELIS, N. *Rate Phenomena in Process Metallurgy.* New York: Wiley-Interscience, 1971.

(T1)   TREYBAL, R. E. *Mass Transfer Operations,* 3rd ed. New York: McGraw-Hill Book Company, 1980.

(T2)   TOOR, H. L., and MARCHELLO, J. M. *A.I.Ch.E. J.,* **1**, 97 (1958).

(V1)   VOGTLANDER, P. H., and BAKKER, C. A. P. *Chem. Eng. Sci.,* **18**, 583 (1963).

(W1)   WILSON, E. J., and GEANKOPLIS, C. J. *Ind. Eng. Chem. Fund.,* **5**, 9 (1966).

APPENDIX A.1

Fundamental Constants and Conversion Factors

A.1-1 Gas Law Constant R

Numerical Value	Units
1.9872	g cal/g mol $\cdot$ K
1.9872	btu/lb mol $\cdot$ °R
82.057	cm$^3 \cdot$ atm/g mol $\cdot$ K
8314.34	J/kg mol $\cdot$ K
82.057 $\times$ 10^{-3}	m$^3 \cdot$ atm/kg mol $\cdot$ K
8314.34	kg $\cdot$ m^2/s$^2 \cdot$ kg mol $\cdot$ K
10.731	ft$^3 \cdot$ lb$_f$/in.$^2 \cdot$ lb mol $\cdot$ °R
0.7302	ft$^3 \cdot$ atm/lb mol $\cdot$ °R
1545.3	ft $\cdot$ lb$_f$/lb mol $\cdot$ °R
8314.34	m$^3 \cdot$ Pa/kg mol $\cdot$ K

A.1-2 Volume and Density

1 g mol ideal gas at 0°C, 760 mm Hg = 22.4140 liters = 22414 cm^3
1 lb mol ideal gas at 0°C, 760 mm Hg = 359.05 ft^3
1 kg mol ideal gas at 0°C, 760 mm Hg = 22.414 m^3
Density of dry air at 0°C, 760 mm Hg = 1.2929 g/liter
$\qquad\qquad\qquad\qquad\qquad\qquad = 0.080711$ lb$_m$/ft^3
Molecular weight of air = 28.97 lb$_m$/lb mol
1 g/cm^3 = 62.43 lb$_m$/ft^3 = 1000 kg/m^3
1 g/cm^3 = 8.345 lb$_m$/U.S. gal
1 lb$_m$/ft^3 = 16.0185 kg/m^3

A.1-3 Length

1 in. = 2.540 cm
100 cm = 1 m (meter)

1 micron $= 10^{-6}$ m $= 10^{-4}$ cm $= 10^{-3}$ mm $= 1 \mu$m (micrometer)

1 Å (angstrom) $= 10^{-10}$ m $= 10^{-4} \mu$m

1 mile $= 5280$ ft

1 m $= 3.2808$ ft $= 39.37$ in.

A.1-4 Mass

1 $lb_m = 453.59$ g $= 0.45359$ kg

1 $lb_m = 16$ oz $= 7000$ grains

1 kg $= 1000$ g $= 2.2046$ lb_m

1 ton (short) $= 2000$ lb_m

1 ton (long) $= 2240$ lb_m

1 ton (metric) $= 1000$ kg

A.1-5 Standard Acceleration of Gravity

$g = 9.80665$ m/s^2

$g = 980.665$ cm/s^2

$g = 32.174$ ft/s^2

g_c (gravitational conversion factor) $= 32.1740$ $lb_m \cdot$ ft/$lb_f \cdot$ s^2

$\qquad\qquad\qquad\qquad\qquad\qquad\qquad = 980.665$ $g_m \cdot$ cm/$g_f \cdot$ s^2

A.1-6 Volume

1 L (liter) $= 1000$ cm^3	1 m^3 $= 1000$ L (liter)
1 in.^{3} $= 16.387$ cm^3	1 U.S. gal $= 4$ qt
1 ft^3 $= 28.317$ L (liter)	1 U.S. gal $= 3.7854$ L (liter)
1 ft^3 $= 0.028317$ m^3	1 U.S. gal $= 3785.4$ cm^3
1 ft^3 $= 7.481$ U.S. gal	1 British gal $= 1.20094$ U.S. gal
1 m^3 $= 264.17$ U.S. gal	

A.1-7 Force

1 g $\cdot$ cm/s^2 (dyn) $= 10^{-5}$ kg $\cdot$ m/s^2 $= 10^{-5}$ N (newton)

1 g $\cdot$ cm/s^2 $= 7.2330 \times 10^{-5}$ $lb_m \cdot$ ft/s^2 (poundal)

1 kg $\cdot$ m/s^2 $= 1$ N (newton)

1 $lb_f = 4.4482$ N

1g $\cdot$ cm/s^2 $= 2.2481 \times 10^{-6}$ lb_f

A.1-8 Pressure

1 bar $= 1 \times 10^5$ Pa (pascal) $= 1 \times 10^5$ N/m^2

1 psia $= 1$ lb_f/in.^{2}

1 psia $= 2.0360$ in. Hg at 0°C

1 psia $= 2.311$ ft H_2O at 70°F

1 psia $= 51.715$ mm Hg at 0°C ($\rho_{Hg} = 13.5955$ g/cm^3)

1 atm $= 14.696$ psia $= 1.01325 \times 10^5$ N/m^2 $= 1.01325$ bar

1 atm $= 760$ mm Hg at 0°C $= 1.01325 \times 10^5$ Pa

1 atm $= 29.921$ in. Hg at 0°C

1 atm $= 33.90$ ft H_2O at 4°C

1 psia $= 6.89476 \times 10^4$ g/cm $\cdot$ s^2
1 psia $= 6.89476 \times 10^4$ dyn/cm^2
1 dyn/cm$^2 = 2.0886 \times 10^{-3}$ lb$_f$/ft^2
1 psia $= 6.89476 \times 10^3$ N/m^2
1 lb$_f$/ft$^2 = 4.7880 \times 10^2$ dyn/cm$^2 = 47.880$ N/m^2
1 mm Hg (0°C) $= 1.333224 \times 10^2$ N/m$^2 = 0.1333224$ kPa

A.1-9 Power

1 hp $= 0.74570$ kW   1 watt (W) $= 14.340$ cal/min
1 hp $= 550$ ft $\cdot$ lb$_f$/s   1 btu/h $= 0.29307$ W (watt)
1 hp $= 0.7068$ btu/s   1 J/s (joule/s) $= 1$ W

A.1-10 Heat, Energy, Work

1 J $= 1$ N $\cdot$ m $= 1$ kg $\cdot$ m^2/s^2
1 kg $\cdot$ m^2/s$^2 = 1$ J (joule) $= 10^7$ g $\cdot$ cm^2/s^2 (erg)
1 btu $= 1055.06$ J $= 1.05506$ kJ
1 btu $= 252.16$ cal (thermochemical)
1 kcal (thermochemical) $= 1000$ cal $= 4.1840$ kJ
1 cal (thermochemical) $= 4.1840$ J
1 cal (IT) $= 4.1868$ J
1 btu $= 251.996$ cal (IT)
1 btu $= 778.17$ ft $\cdot$ lb$_f$
1 hp $\cdot$ h $= 0.7457$ kW $\cdot$ h
1 hp $\cdot$ h $= 2544.5$ btu
1 ft $\cdot$ lb$_f = 1.35582$ J
1 ft $\cdot$ lb$_f$/lb$_m = 2.9890$ J/kg

A.1-11 Thermal Conductivity

1 btu/h $\cdot$ ft $\cdot$ °F $= 4.1365 \times 10^{-3}$ cal/s $\cdot$ cm $\cdot$ °C
1 btu/h $\cdot$ ft $\cdot$ °F $= 1.73073$ W/m $\cdot$ K

A.1-12 Heat-Transfer Coefficient

1 btu/h $\cdot$ ft$^2 \cdot$ °F $= 1.3571 \times 10^{-4}$ cal/s $\cdot$ cm$^2 \cdot$ °C
1 btu/h $\cdot$ ft$^2 \cdot$ °F $= 5.6783 \times 10^{-4}$ W/cm$^2 \cdot$ °C
1 btu/h $\cdot$ ft$^2 \cdot$ °F $= 5.6783$ W/m$^2 \cdot$ K
1 kcal/h $\cdot$ m$^2 \cdot$ °F $= 0.2048$ btu/h $\cdot$ ft$^2 \cdot$ °F

A.1-13 Viscosity

1 cp $= 10^{-2}$ g/cm $\cdot$ s (poise)
1 cp $= 2.4191$ lb$_m$/ft $\cdot$ h
1 cp $= 6.7197 \times 10^{-4}$ lb$_m$/ft $\cdot$ s
1 cp $= 10^{-3}$ Pa $\cdot$ s $= 10^{-3}$ kg/m $\cdot$ s $= 10^{-3}$ N $\cdot$ s/m^2
1 cp $= 2.0886 \times 10^{-5}$ lb$_f \cdot$ s/ft^2
1 Pa $\cdot$ s $= 1$ N $\cdot$ s/m$^2 = 1$ kg/m $\cdot$ s $= 1000$ cp

A.1-14 Diffusivity

$1 \text{ cm}^2/\text{s} = 3.875 \text{ ft}^2/\text{h}$ $1 \text{ m}^2/\text{s} = 3.875 \times 10^4 \text{ ft}^2/\text{h}$
$1 \text{ cm}^2/\text{s} = 10^{-4} \text{ m}^2/\text{s}$ $1 \text{ centistoke} = 10^{-2} \text{ cm}^2/\text{s}$
$1 \text{ m}^2/\text{h} = 10.764 \text{ ft}^2/\text{h}$

A.1-15 Mass Flux and Molar Flux

$1 \text{ g/s} \cdot \text{cm}^2 = 7.3734 \times 10^3 \text{ lb}_m/\text{h} \cdot \text{ft}^2$
$1 \text{ g mol/s} \cdot \text{cm}^2 = 7.3734 \times 10^3 \text{ lb mol/h} \cdot \text{ft}^2$
$1 \text{ g mol/s} \cdot \text{cm}^2 = 10 \text{ kg mol/s} \cdot \text{m}^2 = 1 \times 10^4 \text{ g mol/s} \cdot \text{m}^2$
$1 \text{ lb mol/h} \cdot \text{ft}^2 = 1.3562 \times 10^{-3} \text{ kg mol/s} \cdot \text{m}^2$

A.1-16 Heat Flux and Heat Flow

$1 \text{ btu/h} \cdot \text{ft}^2 = 3.1546 \text{ W/m}^2$
$1 \text{ btu/h} = 0.29307 \text{ W}$
$1 \text{ cal/h} = 1.1622 \times 10^{-3} \text{ W}$

A.1-17 Heat Capacity and Enthalpy

$1 \text{ btu/lb}_m \cdot {}^\circ\text{F} = 4.1868 \text{ kJ/kg} \cdot \text{K}$
$1 \text{ btu/lb}_m \cdot {}^\circ\text{F} = 1.000 \text{ cal/g} \cdot {}^\circ\text{C}$
$1 \text{ btu/lb}_m = 2326.0 \text{ J/kg}$
$1 \text{ ft} \cdot \text{lb}_f/\text{lb}_m = 2.9890 \text{ J/kg}$
$1 \text{ cal (IT)/g} \cdot {}^\circ\text{C} = 4.1868 \text{ kJ/kg} \cdot \text{K}$
$1 \text{ kcal/g mol} = 4.1840 \times 10^3 \text{ kJ/kg mol}$

A.1-18 Mass-Transfer Coefficient

$1 \text{ } k_c \text{ cm/s} = 10^{-2} \text{ m/s}$
$1 \text{ } k_c \text{ ft/h} = 8.4668 \times 10^{-5} \text{ m/s}$
$1 \text{ } k_x \text{ g mol/s} \cdot \text{cm}^2 \cdot \text{mol frac} = 10 \text{ kg mol/s} \cdot \text{m}^2 \cdot \text{mol frac}$
$1 \text{ } k_x \text{ g mol/s} \cdot \text{cm}^2 \cdot \text{mol frac} = 1 \times 10^4 \text{ g mol/s} \cdot \text{m}^2 \cdot \text{mol frac}$
$1 \text{ } k_x \text{ lb mol/h} \cdot \text{ft}^2 \cdot \text{mol frac} = 1.3562 \times 10^{-3} \text{ kg mol/s} \cdot \text{m}^2 \cdot \text{mol frac}$
$1 \text{ } k_x a \text{ lb mol/h} \cdot \text{ft}^3 \cdot \text{mol frac} = 4.449 \times 10^{-3} \text{ kg mol/s} \cdot \text{m}^3 \cdot \text{mol frac}$
$1 \text{ } k_G \text{ kg mol/s} \cdot \text{m}^2 \cdot \text{atm} = 0.98692 \times 10^{-5} \text{ kg mol/s} \cdot \text{m}^2 \cdot \text{Pa}$
$1 \text{ } k_G a \text{ kg mol/s} \cdot \text{m}^3 \cdot \text{atm} = 0.98692 \times 10^{-5} \text{ kg mol/s} \cdot \text{m}^3 \cdot \text{Pa}$

APPENDIX A.2

Physical Properties
of Water

A.2-1   Latent Heat of Water at 273.15 K (0°C)

$$\text{Latent heat of fusion} = 1436.3 \text{ cal/g mol}$$

$$= 79.724 \text{ cal/g}$$

$$= 2585.3 \text{ btu/lb mol}$$

$$= 6013.4 \text{ kJ/kg mol}$$

Source: O. A. Hougen, K. M. Watson, and R. A. Ragatz, *Chemical Process Principles*, Part I, 2nd ed. New York: John Wiley & Sons, Inc., 1954.

Latent heat of vaporaization at 298.15 K (25°C)

Pressure (mm Hg)	Latent Heat
23.75	44 020 kJ/kg mol, 10.514 kcal/g mol, 18 925 btu/lb mol
760	44 045 kJ/kg mol, 10.520 kcal/g mol, 18 936 btu/lb mol

Source: National Bureau of Standards, *Circular 500.*

A.2-2   Vapor Pressure of Water

Temperature		Vapor Pressure		Temperature		Vapor Pressure	
K	°C	kPa	mm Hg	K	°C	kPa	mm Hg
273.15	0	0.611	4.58	323.15	50	12.333	92.51
283.15	10	1.228	9.21	333.15	60	19.92	149.4
293.15	20	2.338	17.54	343.15	70	31.16	233.7
298.15	25	3.168	23.76	353.15	80	47.34	355.1
303.15	30	4.242	31.82	363.15	90	70.10	525.8
313.15	40	7.375	55.32	373.15	100	101.325	760.0

Source: Physikalish-technishe, Reichsansalt, Holborn, Scheel, and Henning, *Warmetabellen.* Brunswick, Germany: Friedrich Viewig and Son, 1909.

A.2-3 Density of Liquid Water

Temperature		Density		Temperature		Density	
K	°C	g/cm³	kg/m³	K	°C	g/cm³	kg/m³
273.15	0	0.99987	999.87	323.15	50	0.98807	988.07
277.15	4	1.00000	1000.00	333.15	60	0.98324	983.24
283.15	10	0.99973	999.73	343.15	70	0.97781	977.81
293.15	20	0.99823	998.23	353.15	80	0.97183	971.83
298.15	25	0.99708	997.08	363.15	90	0.96534	965.34
303.15	30	0.99568	995.68	373.15	100	0.95838	958.38
313.15	40	0.99225	992.25				

Source: R. H. Perry and C. H. Chilton, *Chemical Engineers' Handbook*, 5th ed. New York: McGraw-Hill Book Company, 1973. With permission.

A.2-4 Viscosity of Liquid Water

Temperature		Viscosity [(Pa · s) 10⁵, (kg/m · s) 10³, or cp]	Temperature		Viscosity [(Pa · s) 10⁵, (kg/m · s) 10³, or cp]
K	°C		K	°C	
273.15	0	1.7921	323.15	50	0.5494
275.15	2	1.6728	325.15	52	0.5315
277.15	4	1.5674	327.15	54	0.5146
279.15	6	1.4728	329.15	56	0.4985
281.15	8	1.3860	331.15	58	0.4832
283.15	10	1.3077	333.15	60	0.4688
285.15	12	1.2363	335.15	62	0.4550
287.15	14	1.1709	337.15	64	0.4418
289.15	16	1.1111	339.15	66	0.4293
291.15	18	1.0559	341.15	68	0.4174
293.15	20	1.0050	343.15	70	0.4061
293.35	20.2	1.0000	345.15	72	0.3952
295.15	22	0.9579	347.15	74	0.3849
297.15	24	0.9142	349.15	76	0.3750
298.15	25	0.8937	351.15	78	0.3655
299.15	26	0.8737	353.15	80	0.3565
301.15	28	0.8360	355.15	82	0.3478
303.15	30	0.8007	357.15	84	0.3395
305.15	32	0.7679	359.15	86	0.3315
307.15	34	0.7371	361.15	88	0.3239
309.15	36	0.7085	363.15	90	0.3165
311.15	38	0.6814	365.15	92	0.3095
313.15	40	0.6560	367.15	94	0.3027
315.15	42	0.6321	369.15	96	0.2962
317.15	44	0.6097	371.15	98	0.2899
319.15	46	0.5883	373.15	100	0.2838
321.15	48	0.5683			

Source: Bingham, *Fluidity and Plasticity.* New York: McGraw-Hill Book Company, 1922. With permission.

A.2-5 Heat Capacity of Liquid Water at 101.325 kPa (1 Atm)

Temperature		Heat Capacity, c_p		Temperature		Heat Capacity, c_p	
°C	K	cal/g·°C	kJ/kg·K	°C	K	cal/g·°C	kJ/kg·K
0	273.15	1.0080	4.220	50	323.15	0.9992	4.183
10	283.15	1.0019	4.195	60	333.15	1.0001	4.187
20	293.15	0.9995	4.185	70	343.15	1.0013	4.192
25	298.15	0.9989	4.182	80	353.15	1.0029	4.199
30	303.15	0.9987	4.181	90	363.15	1.0050	4.208
40	313.15	0.9987	4.181	100	373.15	1.0076	4.219

Source: N. S. Osborne, H. F. Stimson, and D. C. Ginnings, *Bur. Standards J. Res.*, **23**, 197 (1939).

A.2-6 Thermal Conductivity of Liquid Water

Temperature			Thermal Conductivity	
°C	°F	K	btu/h·ft·°F	W/m·K
0	32	273.15	0.329	0.569
37.8	100	311.0	0.363	0.628
93.3	200	366.5	0.393	0.680
148.9	300	422.1	0.395	0.684
215.6	420	588.8	0.376	0.651
326.7	620	599.9	0.275	0.476

Source: D. L. Timrot and N. B. Vargaftik, *J. Tech. Phys. (U.S.S.R.)*, **10**, 1063 (1940); 6th International Conference on the Properties of Steam, Paris, 1964.

A.2-7 Vapor Pressure of Saturated Ice–Water Vapor and Heat of Sublimation

Temperature			Vapor Pressure			Heat of Sublimation	
K	°F	°C	kPa	psia	mm Hg	btu/lb$_m$	kJ/kg
273.2	32	0	6.107×10^{-1}	8.858×10^{-2}	4.581	1218.6	2834.5
266.5	20	−6.7	3.478×10^{-1}	5.045×10^{-2}	2.609	1219.3	2836.1
261.0	10	−12.2	2.128×10^{-1}	3.087×10^{-2}	1.596	1219.7	2837.0
255.4	0	−17.8	1.275×10^{-1}	1.849×10^{-2}	0.9562	1220.1	2838.0
249.9	−10	−23.3	7.411×10^{-2}	1.082×10^{-2}	0.5596	1220.3	2838.4
244.3	−20	−28.9	3.820×10^{-2}	6.181×10^{-3}	0.3197	1220.5	2838.9
238.8	−30	−34.4	2.372×10^{-2}	3.440×10^{-3}	0.1779	1220.5	2838.9
233.2	−40	−40.0	1.283×10^{-2}	1.861×10^{-3}	0.09624	1220.5	2838.9

Source: ASHRAE, *Handbook of Fundamentals.* New York: ASHRAE, 1972.

A.2-8 Heat Capacity of Ice

Temperature		c_p		Temperature		c_p	
°F	K	$btu/lb_m \cdot °F$	$kJ/kg \cdot K$	°F	K	$btu/lb_m \cdot °F$	$kJ/kg \cdot K$
32	273.15	0.500	2.093	−10	249.85	0.461	1.930
20	266.45	0.490	2.052	−20	244.25	0.452	1.892
10	260.95	0.481	2.014	−30	238.75	0.442	1.850
0	255.35	0.472	1.976	−40	233.15	0.433	1.813

Source : Adapted from ASHRAE, *Handbook of Fundamentals.* New York: ASHRAE, 1972.

A.2-9 Properties of Saturated Steam and Water (Steam Table), SI Units

Temperature (°C)	Vapor Pressure (kPa)	Specific Volume (m^3/kg)		Enthalpy (kJ/kg)		Entropy ($kJ/kg \cdot K$)	
		Liquid	Sat'd Vapor	Liquid	Sat'd Vapor	Liquid	Sat'd Vapor
0.01	0.6113	0.0010002	206.136	0.00	2501.4	0.0000	9.1562
3	0.7577	0.0010001	168.132	12.57	2506.9	0.0457	9.0773
6	0.9349	0.0010001	137.734	25.20	2512.4	0.0912	9.0003
9	1.1477	0.0010003	113.386	37.80	2517.9	0.1362	8.9253
12	1.4022	0.0010005	93.784	50.41	2523.4	0.1806	8.8524
15	1.7051	0.0010009	77.926	62.99	2528.9	0.2245	8.7814
18	2.0640	0.0010014	65.038	75.58	2534.4	0.2679	8.7123
21	2.487	0.0010020	54.514	88.14	2539.9	0.3109	8.6450
24	2.985	0.0010027	45.883	100.70	2545.4	0.3534	8.5794
25	3.169	0.0010029	43.360	104.89	2547.2	0.3674	8.5580
27	3.567	0.0010035	38.774	113.25	2550.8	0.3954	8.5156
30	4.246	0.0010043	32.894	125.79	2556.3	0.4369	8.4533
33	5.034	0.0010053	28.011	138.33	2561.7	0.4781	8.3927
36	5.947	0.0010063	23.940	150.86	2567.1	0.5188	8.3336
40	7.384	0.0010078	19.523	167.57	2574.3	0.5725	8.2570
45	9.593	0.0010099	15.258	188.45	2583.2	0.6387	8.1648
50	12.349	0.0010121	12.032	209.33	2592.1	0.7038	8.0763
55	15.758	0.0010146	9.568	230.23	2600.9	0.7679	7.9913
60	19.940	0.0010172	7.671	251.13	2609.6	0.8312	7.9096
65	25.03	0.0010199	6.197	272.06	2618.3	0.8935	7.8310
70	31.19	0.0010228	5.042	292.98	2626.8	0.9549	7.7553
75	38.58	0.0010259	4.131	313.93	2635.3	1.0155	7.6824
80	47.39	0.0010291	3.407	334.91	2643.7	1.0753	7.6122
85	57.83	0.0010325	2.828	355.90	2651.9	1.1343	7.5445
90	70.14	0.0010360	2.361	376.92	2660.1	1.1925	7.4791
95	84.55	0.0010397	1.9819	397.96	2668.1	1.2500	7.4159
100	101.35	0.0010435	1.6729	419.04	2676.1	1.3069	7.3549

Temper-ature (°C)	Vapor Pressure (kPa)	Specific Volume (m³/kg)		Enthalpy (kJ/kg)		Entropy (kJ/kg · K)	
		Liquid	Sat'd Vapor	Liquid	Sat'd Vapor	Liquid	Sat'd Vapor
105	120.82	0.0010475	1.4194	440.15	2683.8	1.3630	7.2958
110	143.27	0.0010516	1.2102	461.30	2691.5	1.4185	7.2387
115	169.06	0.0010559	1.0366	482.48	2699.0	1.4734	7.1833
120	198.53	0.0010603	0.8919	503.71	2706.3	1.5276	7.1296
125	232.1	0.0010649	0.7706	524.99	2713.5	1.5813	7.0775
130	270.1	0.0010697	0.6685	546.31	2720.5	1.6344	7.0269
135	313.0	0.0010746	0.5822	567.69	2727.3	1.6870	6.9777
140	316.3	0.0010797	0.5089	589.13	2733.9	1.7391	6.9299
145	415.4	0.0010850	0.4463	610.63	2740.3	1.7907	6.8833
150	475.8	0.0010905	0.3928	632.20	2746.5	1.8418	6.8379
155	543.1	0.0010961	0.3468	653.84	2752.4	1.8925	6.7935
160	617.8	0.0011020	0.3071	675.55	2758.1	1.9427	6.7502
165	700.5	0.0011080	0.2727	697.34	2763.5	1.9925	6.7078
170	791.7	0.0011143	0.2428	719.21	2768.7	2.0419	6.6663
175	892.0	0.0011207	0.2168	741.17	2773.6	2.0909	6.6256
180	1002.1	0.0011274	0.19405	763.22	2778.2	2.1396	6.5857
190	1254.4	0.0011414	0.15654	807.62	2786.4	2.2359	6.5079
200	1553.8	0.0011565	0.12736	852.45	2793.2	2.3309	6.4323
225	2548	0.0011992	0.07849	966.78	2803.3	2.5639	6.2503
250	3973	0.0012512	0.05013	1085.36	2801.5	2.7927	6.0730
275	5942	0.0013168	0.03279	1210.07	2785.0	3.0208	5.8938
300	8581	0.0010436	0.02167	1344.0	2749.0	3.2534	5.7045

Source : Abridged from J. H. Keenan, F. G. Keyes, P. G. Hill, and J. G. Moore, *Steam Tables—Metric Units.* New York: John Wiley & Sons, Inc., 1969. With permission of the authors and publishers.

A.2-9 Properties of Saturated Steam and Water (Steam Table), English Units

Temper-ature (°F)	Vapor Pressure (psia)	Specific Volume (ft³/lbₘ)		Enthalpy (btu/lbₘ)		Entropy (btu/lbₘ · °F)	
		Liquid	Sat'd Vapor	Liquid	Sat'd Vapor	Liquid	Sat'd Vapor
32.02	0.08866	0.016022	3302	0.00	1075.4	0.000	2.1869
35	0.09992	0.016021	2948	3.00	1076.7	0.00607	2.1764
40	0.12166	0.016020	2445	8.02	1078.9	0.01617	2.1592
45	0.14748	0.016021	2037	13.04	1081.1	0.02618	2.1423
50	0.17803	0.016024	1704.2	18.06	1083.3	0.03607	2.1259
55	0.2140	0.016029	1431.4	23.07	1085.5	0.04586	2.1099

Physical Properties of Water

Temper-ature (°F)	Vapor Pressure (psia)	Specific Volume (ft^3/lb_m)		Enthalpy (btu/lb_m)		Entropy ($btu/lb_m \cdot °F$)	
		Liquid	Sat'd Vapor	Liquid	Sat'd Vapor	Liquid	Sat'd Vapor
60	0.2563	0.016035	1206.9	28.08	1087.7	0.05555	2.0943
65	0.3057	0.016042	1021.5	33.09	1089.9	0.06514	2.0791
70	0.3622	0.016051	867.7	38.09	1092.0	0.07463	2.0642
75	0.4300	0.016061	739.7	43.09	1094.2	0.08402	2.0497
80	0.5073	0.016073	632.8	48.09	1096.4	0.09332	2.0356
85	0.5964	0.016085	543.1	53.08	1098.6	0.10252	2.0218
90	0.6988	0.016099	467.7	58.07	1100.7	0.11165	2.0083
95	0.8162	0.016114	404.0	63.06	1102.9	0.12068	1.9951
100	0.9503	0.016130	350.0	68.05	1105.0	0.12963	1.9822
110	1.2763	0.016166	265.1	78.02	1109.3	0.14730	1.9574
120	1.6945	0.016205	203.0	88.00	1113.5	0.16465	1.9336
130	2.225	0.016247	157.17	97.98	1117.8	0.18172	1.9109
140	2.892	0.016293	122.88	107.96	1121.9	0.19851	1.8892
150	3.722	0.016343	96.99	117.96	1126.1	0.21503	1.8684
160	4.745	0.016395	77.23	127.96	1130.1	0.23130	1.8484
170	5.996	0.016450	62.02	137.97	1134.2	0.24732	1.8293
180	7.515	0.016509	50.20	147.99	1138.2	0.26311	1.8109
190	9.343	0.016570	40.95	158.03	1142.1	0.27866	1.7932
200	11.529	0.016634	33.63	168.07	1145.9	0.29400	1.7762
210	14.125	0.016702	27.82	178.14	1149.7	0.30913	1.7599
212	14.698	0.016716	26.80	180.16	1150.5	0.31213	1.7567
220	17.188	0.016772	23.15	188.22	1153.5	0.32406	1.7441
230	20.78	0.016845	19.386	198.32	1157.1	0.33880	1.7289
240	24.97	0.016922	16.327	208.44	1160.7	0.35335	1.7143
250	29.82	0.017001	13.826	218.59	1164.2	0.36772	1.7001
260	35.42	0.017084	11.768	228.76	1167.6	0.38193	1.6864
270	41.85	0.017170	10.066	238.95	1170.9	0.39597	1.6731
280	49.18	0.017259	8.650	249.18	1174.1	0.40986	1.6602
290	57.33	0.017352	7.467	259.44	1177.2	0.42360	1.6477
300	66.98	0.017448	6.472	269.73	1180.2	0.43720	1.6356
310	77.64	0.017548	5.632	280.06	1183.0	0.45067	1.6238
320	89.60	0.017652	4.919	290.43	1185.8	0.46400	1.6123
330	103.00	0.017760	4.312	300.84	1188.4	0.47722	1.6010
340	117.93	0.017872	3.792	311.30	1190.8	0.49031	1.5901
350	134.53	0.017988	3.346	321.80	1193.1	0.50329	1.5793
360	152.92	0.018108	2.961	332.35	1195.2	0.51617	1.5688
370	173.23	0.018233	2.628	342.96	1197.2	0.52894	1.5585
380	195.60	0.018363	2.339	353.62	1199.0	0.54163	1.5483
390	220.2	0.018498	2.087	364.34	1200.6	0.55422	1.5383
400	247.1	0.018638	1.8661	375.12	1202.0	0.56672	1.5284
410	276.5	0.018784	1.6726	385.97	1203.1	0.57916	1.5187
450	422.1	0.019433	1.1011	430.2	1205.6	0.6282	1.4806

Source: Abridged from J. H. Keenan, F. G. Keyes, P. G. Hill, and J. G. Moore, *Steam Tables—English Units.* New York: John Wiley & Sons, Inc., 1969. With permission of the authors and publishers.

A.2-10 Properties of Superheated Steam (Steam Table), SI Units (v, specific volume, m^3/kg; H, enthalpy, kJ/kg; s, entropy, $kJ/kg \cdot K$)

Absolute Pressure, kPa (Sat. Temp., °C)		Temperature (°C)							
		100	150	200	250	300	360	420	500
10 (45.81)	v	17.196	19.512	21.825	24.136	26.445	29.216	31.986	35.679
	H	2687.5	2783.0	2879.5	2977.3	3076.5	3197.6	3320.9	3489.1
	s	8.4479	8.6882	8.9038	9.1002	9.2813	9.4821	9.6682	9.8978
50 (81.33)	v	3.418	3.889	4.356	4.820	5.284	5.839	6.394	7.134
	H	2682.5	2780.1	2877.7	2976.0	3075.5	3196.8	3320.4	3488.7
	s	7.6947	7.9401	8.1580	8.3556	8.5373	8.7385	8.9249	9.1546
75 (91.78)	v	2.270	2.587	2.900	3.211	3.520	3.891	4.262	4.755
	H	2679.4	2778.2	2876.5	2975.2	3074.9	3196.4	3320.0	3488.4
	s	7.5009	7.7496	7.9690	8.1673	8.3493	8.5508	8.7374	8.9672
100 (99.63)	v	1.6958	1.9364	2.172	2.406	2.639	2.917	3.195	3.565
	H	2672.2	2776.4	2875.3	2974.3	3074.3	3195.9	3319.6	3488.1
	s	7.3614	7.6134	7.8343	8.0333	8.2158	8.4175	8.6042	8.8342
150 (111.37)	v		1.2853	1.4443	1.6012	1.7570	1.9432	2.129	2.376
	H		2772.6	2872.9	2972.7	3073.1	3195.0	3318.9	3487.6
	s		7.4193	7.6433	7.8438	8.0720	8.2293	8.4163	8.6466
400 (143.63)	v		0.4708	0.5342	0.5951	0.6548	0.7257	0.7960	0.8893
	H		2752.8	2860.5	2964.2	3066.8	3190.3	3315.3	3484.9
	s		6.9299	7.1706	7.3789	7.5662	7.7712	7.9598	8.1913
700 (164.97)	v			0.2999	0.3363	0.3714	0.4126	0.4533	0.5070
	H			2844.8	2953.6	3059.1	3184.7	3310.9	3481.7
	s			6.8865	7.1053	7.2979	7.5063	7.6968	7.9299
1000 (179.91)	v			0.2060	0.2327	0.2579	0.2873	0.3162	0.3541
	H			2827.9	2942.6	3051.2	3178.9	3306.5	3478.5
	s			6.6940	6.9247	7.1229	7.3349	7.5275	7.7622
1500 (198.32)	v			0.13248	0.15195	0.16966	0.18988	0.2095	0.2352
	H			2796.8	2923.3	3037.6	3.1692	3299.1	3473.1
	s			6.4546	6.7090	6.9179	7.1363	7.3323	7.5698
2000 (212.42)	v				0.11144	0.12547	0.14113	0.15616	0.17568
	H				2902.5	3023.5	3159.3	3291.6	3467.6
	s				6.5453	6.7664	6.9917	7.1915	7.4317
2500 (223.99)	v				0.08700	0.09890	0.11186	0.12414	0.13998
	H				2880.1	3008.8	3149.1	3284.0	3462.1
	s				6.4085	6.6438	6.8767	7.0803	7.3234
3000 (233.90)	v				0.07058	0.08114	0.09233	0.10279	0.11619
	H				2855.8	2993.5	3138.7	3276.3	3456.5
	s				6.2872	6.5390	6.7801	6.9878	7.2338

Source: Abridged from J. Keenan, F. G. Keyes, P. G. Hill, and J. G. Moore, *Steam Tables—Metric Units.* New York: John Wiley & Sons, Inc., 1969. With permission of the authors and publishers.

A.2-10 Properties of Superheated Steam (Steam Table), English Units (v, specific volume, ft^3/lb$_m$; H, enthalpy, btu/lb$_m$; s, entropy, btu/lb$_m \cdot$ °F)

Absolute Pressure, psia (Sat. Temp., °F)		Temperature (°F)								
		200	300	400	500	600	700	800	900	1000
1.0 (101.70)	v	392.5	452.3	511.9	571.5	631.1	690.7	750.3	809.9	869.5
	H	1150.1	1195.7	1241.8	1288.5	1336.1	1384.5	1433.7	1483.8	1534.8
	s	2.0508	2.1150	2.1720	2.2235	2.2706	2.3142	2.3550	2.3932	2.4294
5.0 (162.21)	v	78.15	90.24	102.24	114.20	126.15	138.08	150.01	161.94	173.86
	H	1148.6	1194.8	1241.2	1288.2	1335.8	1384.3	1433.5	1483.7	1534.7
	s	1.8715	1.9367	1.9941	2.0458	2.0930	2.1367	2.1775	2.2158	2.2520
10.0 (193.19)	v	38.85	44.99	51.03	57.04	63.03	69.01	74.98	80.95	86.91
	H	1146.6	1193.7	1240.5	1287.7	1335.5	1384.0	1433.3	1483.5	1534.6
	s	1.7927	1.8592	1.9171	1.9690	2.0164	2.0601	2.1009	2.1393	2.1755
14.696 (211.99)	v		30.52	34.67	38.77	42.86	46.93	51.00	55.07	59.13
	H		1192.6	1239.9	1287.3	1335.2	1383.8	1433.1	1483.4	1534.5
	s		1.8157	1.8741	1.9263	1.9737	2.0175	2.0584	2.0967	2.1330
20.0 (227.96)	v		22.36	25.43	28.46	31.47	34.77	37.46	40.45	43.44
	H		1191.5	1239.2	1286.8	1334.8	1383.5	1432.9	1483.2	1534.3
	s		1.7805	1.8395	1.8919	1.9395	1.9834	2.0243	2.0627	2.0989
60.0 (292.73)	v		7.260	8.353	9.399	10.425	11.440	12.448	13.452	14.454
	H		1181.9	1233.5	1283.0	1332.1	1381.4	1431.2	1481.8	1533.2
	s		1.6496	1.7134	1.7678	1.8165	1.8609	1.9022	1.9408	1.9773
100.0 (327.86)	v			4.934	5.587	6.216	6.834	7.445	8.053	8.657
	H			1227.5	1279.1	1329.3	1379.2	1429.6	1480.5	1532.1
	s			1.6517	1.7085	1.7582	1.8033	1.8449	1.8838	1.9204
150.0 (358.48)	v			3.221	3.679	4.111	4.531	4.944	5.353	5.759
	H			1219.5	1274.1	1325.7	1376.6	1427.5	1478.8	1530.7
	s			1.5997	1.6598	1.7110	1.7568	1.7989	1.8381	1.8750
200.0 (381.86)	v			2.361	2.724	3.058	3.379	3.693	4.003	4.310
	H			1210.8	1268.8	1322.1	1373.8	1425.3	1477.1	1529.3
	s			1.5600	1.6239	1.6767	1.7234	1.7660	1.8055	1.8425
250.0 (401.04)	v				2.150	2.426	2.688	2.943	3.193	3.440
	H				1263.3	1318.3	1371.1	1423.2	1475.3	1527.9
	s				1.5948	1.6494	1.6970	1.7401	1.7799	1.8172
300.0 (417.43)	v				1.766	2.004	2.227	2.442	2.653	2.860
	H				1257.5	1314.5	1368.3	1421.0	1473.6	1526.5
	s				1.5701	1.6266	1.6751	1.7187	1.7589	1.7964
400 (444.70)	v				1.2843	1.4760	1.6503	1.8163	1.9776	2.136
	H				1245.2	1306.6	1362.5	1416.6	1470.1	1523.6
	s				1.5282	1.5892	1.6397	1.6884	1.7252	1.7632

Source: Abridged from J. H. Keenan, F. G. Keyes, P. G. Hill, and J. G. Moore, *Steam Tables—Metric Units.* New York: John Wiley & Sons, Inc., 1969. With permission of the authors and publishers.

A.2-11 Heat-Transfer Properties of Liquid Water, SI Units

T ($°C$)	T (K)	ρ (kg/m^3)	c_p ($kJ/kg \cdot K$)	$\mu \times 10^3$ ($Pa \cdot s$, or $kg/m \cdot s$)	k ($W/m \cdot K$)	N_{Pr}	$\beta \times 10^4$ ($1/K$)	$(g\beta\rho^2/\mu^2) \times 10^{-8}$ ($1/K \cdot m^3$)
0	273.2	999.6	4.229	1.786	0.5694	13.3	−0.630	
15.6	288.8	998.0	4.187	1.131	0.5884	8.07	1.44	10.93
26.7	299.9	996.4	4.183	0.860	0.6109	5.89	2.34	30.70
37.8	311.0	994.7	4.183	0.682	0.6283	4.51	3.24	68.0
65.6	338.8	981.9	4.187	0.432	0.6629	2.72	5.04	256.2
93.3	366.5	962.7	4.229	0.3066	0.6802	1.91	6.66	642
121.1	394.3	943.5	4.271	0.2381	0.6836	1.49	8.46	1300
148.9	422.1	917.9	4.312	0.1935	0.6836	1.22	10.08	2231
204.4	477.6	858.6	4.522	0.1384	0.6611	0.950	14.04	5308
260.0	533.2	784.9	4.982	0.1042	0.6040	0.859	19.8	11 030
315.6	588.8	679.2	6.322	0.0862	0.5071	1.07	31.5	19 260

A.2-11 Heat-Transfer Properties of Liquid Water, English Units

T ($°F$)	ρ $\left(\dfrac{lb_m}{ft^3}\right)$	c_p $\left(\dfrac{btu}{lb_m \cdot °F}\right)$	$\mu \times 10^3$ $\left(\dfrac{lb_m}{ft \cdot s}\right)$	k $\left(\dfrac{btu}{h \cdot ft \cdot °F}\right)$	N_{Pr}	$\beta \times 10^4$ ($1/°R$)	$(g\beta\rho^2/\mu^2) \times 10^{-6}$ ($1/°R \cdot ft^3$)
32	62.4	1.01	1.20	0.329	13.3	−0.350	
60	62.3	1.00	0.760	0.340	8.07	0.800	17.2
80	62.2	0.999	0.578	0.353	5.89	1.30	48.3
100	62.1	0.999	0.458	0.363	4.51	1.80	107
150	61.3	1.00	0.290	0.383	2.72	2.80	403
200	60.1	1.01	0.206	0.393	1.91	3.70	1010
250	58.9	1.02	0.160	0.395	1.49	4.70	2045
300	57.3	1.03	0.130	0.395	1.22	5.60	3510
400	53.6	1.08	0.0930	0.382	0.950	7.80	8350
500	49.0	1.19	0.0700	0.349	0.859	11.0	17 350
600	42.4	1.51	0.0579	0.293	1.07	17.5	30 300

A.2-12 Heat-Transfer Properties of Water Vapor (Steam) at 101.32 kPa (1 Atm Abs), SI Units

T ($°C$)	T (K)	ρ (kg/m^3)	c_p ($kJ/kg \cdot K$)	$\mu \times 10^5$ ($Pa \cdot s$, or $kg/m \cdot s$)	k ($W/m \cdot K$)	N_{Pr}	$\beta \times 10^3$ ($1/K$)	$g\beta\rho^2/\mu^2$ ($1/K \cdot m^3$)
100.0	373.2	0.596	1.888	1.295	0.02510	0.96	2.68	0.557×10^8
148.9	422.1	0.525	1.909	1.488	0.02960	0.95	2.38	0.292×10^8
204.4	477.6	0.461	1.934	1.682	0.03462	0.94	2.09	0.154×10^8
260.0	533.2	0.413	1.968	1.883	0.03946	0.94	1.87	0.0883×10^8
315.6	588.8	0.373	1.997	2.113	0.04448	0.94	1.70	52.1×10^5
371.1	644.3	0.341	2.030	2.314	0.04985	0.93	1.55	33.1×10^5
426.7	699.9	0.314	2.068	2.529	0.05556	0.92	1.43	21.6×10^5

A.2-12 Heat-Transfer Properties of Water Vapor (Steam) at 101.32 kPa (1 Atm Abs), English Units

T ($°F$)	ρ $\left(\dfrac{lb_m}{ft^3}\right)$	c_p $\left(\dfrac{btu}{lb_m \cdot °F}\right)$	$\mu \times 10^3$ $\left(\dfrac{lb_m}{ft \cdot s}\right)$	k $\left(\dfrac{btu}{h \cdot ft \cdot °F}\right)$	N_{Pr}	$\beta \times 10^3$ ($1/°R$)	$g\beta\rho^2/\mu^2$ ($1/°R \cdot ft^3$)
212	0.0372	0.451	0.870	0.0145	0.96	1.49	0.877×10^6
300	0.0328	0.456	1.000	0.0171	0.95	1.32	0.459×10^6
400	0.0288	0.462	1.130	0.0200	0.94	1.16	0.243×10^6
500	0.0258	0.470	1.265	0.0228	0.94	1.04	0.139×10^6
600	0.0233	0.477	1.420	0.0257	0.94	0.943	82×10^3
700	0.0213	0.485	1.555	0.0288	0.93	0.862	52.1×10^3
800	0.0196	0.494	1.700	0.0321	0.92	0.794	34.0×10^3

Source: D. L. Timrot and N. B. Vargaftik, *J. Tech. Phys. (U.S.S.R.)*, **10**, 1063 (1940); R. H. Perry and C. H. Chilton, *Chemical Engineers' Handbook*, 5th ed. New York: McGraw-Hill Book Company, 1973; J. H. Keenan, F. G. Keyes, P. G. Hill, and J. G. Moore, *Steam Tables*. New York: John Wiley & Sons, Inc., 1969; National Research Council, *International Critical Tables*. New York: McGraw-Hill Book Company, 1929; L. S. Marks, *Mechanical Engineers' Handbook*, 5th ed. New York: McGraw-Hill Book Company, 1951.

APPENDIX A.3

Physical Properties of Inorganic and Organic Compounds

A.3-1   Standard Heats of Formation at 298.15 K (25°C) and 101.325 kPa (1 Atm Abs), (*c*) = crystalline, (*g*) = gas, (*l*) = liquid

Compound	ΔH_f°		Compound	ΔH_f°	
	(kJ/kg mol)10⁻³	kcal/g mol		(kJ/kg mol)10⁻³	kcal/g mol
$NH_3(g)$	-46.19	-11.04	$CaCO_3(c)$	-1206.87	-288.45
$NO(g)$	$+90.374$	$+21.600$	$CaO(c)$	-635.5	-151.9
$H_2O(l)$	-285.840	-68.3174	$CO(g)$	-110.523	-26.4157
$H_2O(g)$	-241.826	-57.7979	$CO_2(g)$	-393.513	-94.0518
$HCN(g)$	$+130.1$	$+31.1$	$CH_4(g)$	-74.848	-17.889
$HCl(g)$	-92.312	-22.063	$C_2H_6(g)$	-84.667	-20.236
$H_2SO_4(l)$	-811.32	-193.91	$C_3H_8(g)$	-103.847	-24.820
$H_3PO_4(c)$	-1281.1	-306.2	$CH_3OH(l)$	-238.66	-57.04
$NaCl(c)$	-411.003	-98.232	$CH_3CH_3OH(l)$	-277.61	-66.35
$NH_4Cl(c)$	-315.39	-75.38			

Source: J. H. Perry and C. H. Chilton, *Chemical Engineers' Handbook*, 5th ed. New York: McGraw-Hill Book Company, 1973; and O. A. Hougen, K. M. Watson, and R. A. Ragatz, *Chemical Process Principles*, Part I, 2nd ed. New York: John Wiley & Sons, Inc., 1954.

A.3-2   Standard Heats of Combustion at 298.15 K (25°C) and 101.325 kPa (1 Atm Abs)
(g) = gas, (l) = liquid, (s) = solid

Compound	Combustion Reaction	ΔH_c° kcal/g mol	ΔH_c° (kJ/kg mol)10^{-3}
C(s)	$C(s) + \frac{1}{2}O_2(g) \to CO(g)$	−26.4157	−110.523
CO(g)	$CO(g) + \frac{1}{2}O_2(g) \to CO_2(g)$	−67.6361	−282.989
C(s)	$C(s) + O_2(g) \to CO_2(g)$	−94.0518	−393.513
$H_2(g)$	$H_2(g) + \frac{1}{2}O_2(g) \to H_2O(l)$	−68.3174	−285.840
$H_2(g)$	$H_2(g) + \frac{1}{2}O_2(g) \to H_2O(g)$	−57.7979	−241.826
$CH_4(g)$	$CH_4(g) + 2O_2(g) \to CO_2(g) + 2H_2O(l)$	−212.798	−890.346
$C_2H_6(g)$	$C_2H_6(g) + \frac{7}{2}O_2(g) \to 2CO_2(g) + 3H_2O(l)$	−372.820	−1559.879
$C_3H_8(g)$	$C_3H_8(g) + 5O_2(g) \to 3CO_2(g) + 4H_2O(l)$	−530.605	−2220.051
d-Glucose (dextrose) $C_6H_{12}O_6(s)$	$C_6H_{12}O_6(s) + 6O_2(g) \to 6CO_2(g) + 6H_2O(l)$	−673	−2816
Lactose (anhydrous) $C_{12}H_{22}O_{11}(s)$	$C_{12}H_{22}O_{11}(s) + 12O_2(g) \to 12CO_2(g) + 11H_2O(l)$	−1350.1	−5648.8
Sucrose $C_{12}H_{22}O_{11}(s)$	$C_{12}H_{22}O_{11}(s) + 12O_2(g) \to 12CO_2(g) + 11H_2O(l)$	−1348.9	−5643.8

Source: R. H. Perry and C. H. Chilton, *Chemical Engineers' Handbook*, 5th ed. New York: McGraw-Hill Book Company, 1973; and O. A. Hougen, K. M. Watson, and R. A. Ragatz, *Chemical Process Principles*, Part I, 2nd ed. New York: John Wiley & Sons, Inc., 1954.

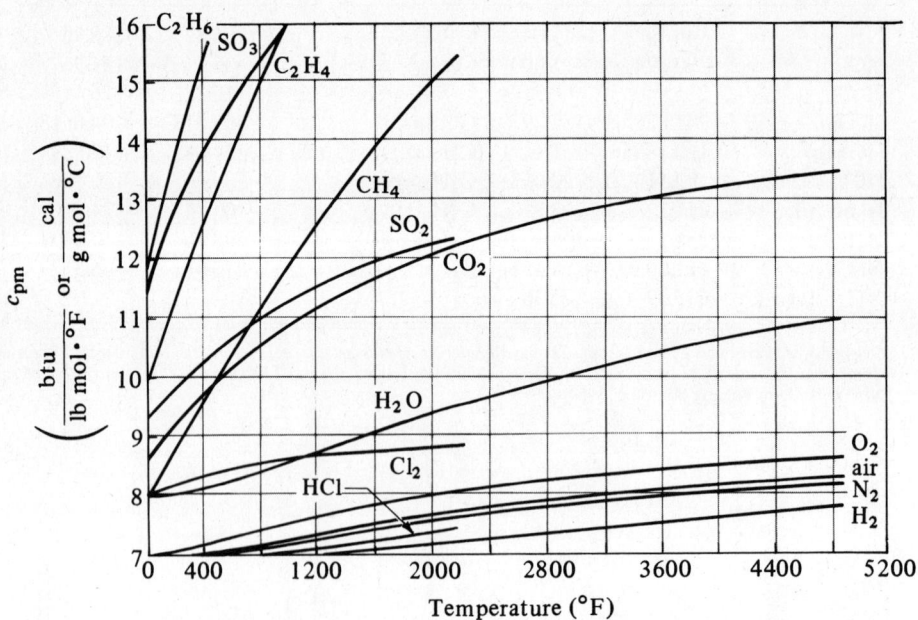

FIGURE A.3-1. *Mean molar heat capacities from 77°F (25°C) to t°F at constant pressure of 101.325 kPa (1 atm abs). (From O. A. Hougen, K. M. Watson, and R. A. Ragatz, Chemical Process Principles, Part I, 2nd ed. New York: John Wiley & Sons, Inc., 1954. With permission.)*

A.3-3 Physical Properties of Air at 101.325 kPa (1 Atm Abs), SI Units

T ($°C$)	T (K)	ρ (kg/m^3)	c_p ($kJ/kg \cdot K$)	$\mu \times 10^5$ ($Pa \cdot s$, or $kg/m \cdot s$)	k ($W/m \cdot K$)	N_{Pr}	$\beta \times 10^3$ ($1/K$)	$g\beta\rho^2/\mu^2$ ($1/K \cdot m^3$)
-17.8	255.4	1.379	1.0048	1.62	0.02250	0.720	3.92	2.79×10^8
0	273.2	1.293	1.0048	1.72	0.02423	0.715	3.65	2.04×10^8
10.0	283.2	1.246	1.0048	1.78	0.02492	0.713	3.53	1.72×10^8
37.8	311.0	1.137	1.0048	1.90	0.02700	0.705	3.22	1.12×10^8
65.6	338.8	1.043	1.0090	2.03	0.02925	0.702	2.95	0.775×10^8
93.3	366.5	0.964	1.0090	2.15	0.03115	0.694	2.74	0.534×10^8
121.1	394.3	0.895	1.0132	2.27	0.03323	0.692	2.54	0.386×10^8
148.9	422.1	0.838	1.0174	2.37	0.03531	0.689	2.38	0.289×10^8
176.7	449.9	0.785	1.0216	2.50	0.03721	0.687	2.21	0.214×10^8
204.4	477.6	0.740	1.0258	2.60	0.03894	0.686	2.09	0.168×10^8
232.2	505.4	0.700	1.0300	2.71	0.04084	0.684	1.98	0.130×10^8
260.0	533.2	0.662	1.0341	2.80	0.04258	0.680	1.87	0.104×10^8

A.3-3 Physical Properties of Air at 101.325 kPa (1 Atm Abs), English Units

T ($°F$)	ρ $\left(\dfrac{lb_m}{ft^3}\right)$	c_p $\left(\dfrac{btu}{lb_m \cdot °F}\right)$	μ (centipoise)	k $\left(\dfrac{btu}{h \cdot ft \cdot °F}\right)$	N_{pr}	$\beta \times 10^3$ ($1/°R$)	$g\beta\rho^2/\mu^2$ ($1/°R \cdot ft^3$)
0	0.0861	0.240	0.0162	0.0130	0.720	2.18	4.39×10^6
32	0.0807	0.240	0.0172	0.0140	0.715	2.03	3.21×10^6
50	0.0778	0.240	0.0178	0.0144	0.713	1.96	2.70×10^6
100	0.0710	0.240	0.0190	0.0156	0.705	1.79	1.76×10^6
150	0.0651	0.241	0.0203	0.0169	0.702	1.64	1.22×10^6
200	0.0602	0.241	0.0215	0.0180	0.694	1.52	0.840×10^6
250	0.0559	0.242	0.0227	0.0192	0.692	1.41	0.607×10^6
300	0.0523	0.243	0.0237	0.0204	0.689	1.32	0.454×10^6
350	0.0490	0.244	0.0250	0.0215	0.687	1.23	0.336×10^6
400	0.0462	0.245	0.0260	0.0225	0.686	1.16	0.264×10^6
450	0.0437	0.246	0.0271	0.0236	0.674	1.10	0.204×10^6
500	0.0413	0.247	0.0280	0.0246	0.680	1.04	0.163×10^6

Source: National Bureau of Standards, *Circular* **461C**, 1947; **564**, 1955; NBS–NACA, *Tables of Thermal Properties of Gases*, 1949; F. G. Keyes, *Trans. A.S.M.E.*, **73**, 590, 597 (1951); **74**, 1303 (1952); D. D. Wagman, *Selected Values of Chemical Thermodynamic Properties*. Washington, D.C.: National Bureau of Standards, 1953.

Physical Properties of Inorganic and Organic Compounds

A.3-4 Viscosity of Gases at 101.325 kPa (1 Atm Abs) [Viscosity in (Pa · s) 10^3, (kg/m · s) 10^3, or cp]

Temperature							
K	$°F$	$°C$	H_2	O_2	N_2	CO	CO_2
255.4	0	−17.8	0.00800	0.0181	0.0158	0.0156	0.0128
273.2	32	0	0.00840	0.0192	0.0166	0.0165	0.0137
283.2	50	10.0	0.00862	0.0197	0.0171	0.0169	0.0141
311.0	100	37.8	0.00915	0.0213	0.0183	0.0183	0.0154
338.8	150	65.6	0.00960	0.0228	0.0196	0.0195	0.0167
366.5	200	93.3	0.0101	0.0241	0.0208	0.0208	0.0179
394.3	250	121.1	0.0106	0.0256	0.0220	0.0220	0.0191
422.1	300	148.9	0.0111	0.0267	0.0230	0.0231	0.0203
449.9	350	176.7	0.0115	0.0282	0.0240	0.0242	0.0215
477.6	400	204.4	0.0119	0.0293	0.0250	0.0251	0.0225
505.4	450	232.2	0.0124	0.0307	0.0260	0.0264	0.0236
533.2	500	260.0	0.0128	0.0315	0.0273	0.0276	0.0247

Source : National Bureau of Standards, *Circular* **461C**, 1947; **564**, 1955; NBS–NACA, *Tables of Thermal Properties of Gases*, 1949; F. G. Keyes, *Trans. A.S.M.E.*, **73**, 590, 597 (1951); **74**, 1303 (1952); D. D. Wagman, *Selected Values of Chemical Thermodynamic Properties.* Washington, D.C.: National Bureau of Standards, 1953.

A.3-5 Thermal Conductivity of Gases at 101.325 kPa (1 Atm Abs)

Temperature			H₂		O₂		N₂		CO		CO₂	
K	°C	°F	$W/m \cdot K$	$btu/h \cdot ft \cdot °F$	$W/m \cdot K$	$btu/h \cdot ft \cdot °F$	$W/m \cdot K$	$btu/h \cdot ft \cdot °F$	$W/m \cdot K$	$btu/h \cdot ft \cdot °F$	$W/m \cdot K$	$btu/h \cdot ft \cdot °F$
255.4	−17.8	0	0.1592	0.0920	0.0228	0.0132	0.0228	0.0132	0.0222	0.0128	0.0132	0.0076
273.2	0	32	0.1667	0.0963	0.0246	0.0142	0.0239	0.0138	0.0233	0.0135	0.0145	0.0084
283.2	10.0	50	0.1720	0.0994	0.0253	0.0146	0.0248	0.0143	0.0239	0.0138	0.0152	0.0088
311.0	37.8	100	0.1852	0.107	0.0277	0.0160	0.0267	0.0154	0.0260	0.0150	0.0173	0.0100
338.8	65.6	150	0.1990	0.115	0.0299	0.0173	0.0287	0.0166	0.0279	0.0161	0.0190	0.0110
366.5	93.3	200	0.2111	0.122	0.0320	0.0185	0.0303	0.0175	0.0296	0.0171	0.0216	0.0125
394.3	121.1	250	0.2233	0.129	0.0343	0.0198	0.0329	0.0190	0.0318	0.0184	0.0239	0.0138
422.1	148.9	300	0.2353	0.136	0.0363	0.0210	0.0348	0.0201	0.0338	0.0195	0.0260	0.0150
449.9	176.7	350	0.2458	0.142	0.0382	0.0221	0.0365	0.0211	0.0355	0.0205	0.0286	0.0165
477.6	204.4	400	0.2579	0.149	0.0398	0.0230	0.0382	0.0221	0.0369	0.0213	0.0308	0.0178
505.4	232.2	450	0.2683	0.155	0.0422	0.0244	0.0400	0.0231	0.0384	0.0222	0.0334	0.0193
533.2	260.0	500	0.2786	0.161	0.0438	0.0253	0.0419	0.0242	0.0407	0.0235	0.0355	0.0205

Source: National Bureau of Standards, *Circular 461C*, 1947; **564**, 1955; NBS–NACA, *Table of Thermal Properties of Gases*, 1949; F. G. Keyes, *Trans. A.S.M.E.*, **73**, 590, 597 (1951); **74**, 1303 (1952); D. D. Wagman, *Selected Values of Chemical Thermodynamic Properties*. Washington, D.C.: National Bureau of Standards, 1953.

A.3-6 Heat Capacity of Gases at Constant Pressure at 101.325 kPa (1 Atm Abs)

Temperature			H_2		O_2		N_2		CO		CO_2	
K	°C	°F	$\frac{kJ}{kg \cdot K}$	$\frac{btu}{lb_m \cdot °F}$	$\frac{kJ}{kg \cdot K}$	$\frac{btu}{lb_m \cdot °F}$	$\frac{kJ}{kg \cdot K}$	$\frac{btu}{lb_m \cdot °F}$	$\frac{kJ}{kg \cdot K}$	$\frac{btu}{lb_m \cdot °F}$	$\frac{kJ}{kg \cdot K}$	$\frac{btu}{lb_m \cdot °F}$
255.4	−17.8	0	14.07	3.36	0.909	0.217	1.034	0.247	1.034	0.247	0.800	0.191
273.2	0	32	14.19	3.39	0.913	0.218	1.038	0.248	1.038	0.248	0.816	0.195
283.2	10.0	50	14.19	3.39	0.917	0.219	1.038	0.248	1.038	0.248	0.825	0.197
311.0	37.8	100	14.32	3.42	0.921	0.220	1.038	0.248	1.043	0.249	0.854	0.204
338.8	65.6	150	14.36	3.43	0.925	0.221	1.038	0.248	1.043	0.249	0.883	0.211
366.5	93.3	200	14.40	3.44	0.929	0.222	1.043	0.249	1.047	0.250	0.904	0.216
394.3	121.1	250	14.44	3.45	0.938	0.224	1.043	0.249	1.047	0.250	0.929	0.222
422.1	148.9	300	14.49	3.46	0.946	0.226	1.047	0.250	1.051	0.251	0.950	0.227
449.9	176.7	350	14.49	3.46	0.955	0.228	1.047	0.250	1.055	0.252	0.976	0.233
477.6	204.4	400	14.49	3.46	0.963	0.230	1.051	0.251	1.059	0.253	0.996	0.238
505.4	232.2	450	14.52	3.47	0.971	0.232	1.055	0.252	1.063	0.254	1.017	0.243
533.2	260.0	500	14.52	3.47	0.976	0.233	1.059	0.253	1.068	0.255	1.030	0.246

Source: National Bureau of Standards, *Circular* **461C**, 1947; **564**, 1955; NBS–NACA, *Tables of Thermal Properties of Gases*, 1949; F. G. Keyes, *Trans. A.S.M.E.*, **73**, 590, 597 (1951); **74**, 1303 (1952); D. D. Wagman, *Selected Values of Chemical Thermodynamic Properties*. Washington, D.C.: National Bureau of Standards, 1953.

A.3-7 Prandtl Number of Gases at 101.325 kPa (1 Atm Abs)

Temperature			H_2	O_2	N_2	CO	CO_2
°C	°F	K					
−17.8	0	255.4	0.720	0.720	0.720	0.740	0.775
0	32	273.2	0.715	0.711	0.720	0.738	0.770
10.0	50	283.2	0.710	0.710	0.717	0.735	0.769
37.8	100	311.0	0.700	0.707	0.710	0.731	0.764
65.6	150	338.8	0.700	0.706	0.700	0.727	0.755
93.3	200	366.5	0.694	0.703	0.700	0.724	0.752
121.1	250	394.3	0.688	0.703	0.696	0.720	0.746
148.9	300	422.1	0.683	0.703	0.690	0.720	0.738
176.6	350	449.9	0.677	0.704	0.689	0.720	0.734
204.4	400	477.6	0.670	0.706	0.688	0.720	0.725
232.2	450	505.4	0.668	0.702	0.688	0.720	0.716
260.0	500	533.2	0.666	0.700	0.688	0.720	0.702

Source : National Bureau of Standards, *Circular* **461C**, 1947; **564**, 1955; NBS–NACA, *Tables of Thermal Properties of Gases*, 1949; F. G. Keyes, *Trans. A.S.M.E.*, **73**, 590, 597 (1951); **74**, 1303 (1952); D. D. Wagman, *Selected Values of Chemical Thermodynamic Properties*. Washington, D.C.: National Bureau of Standards, 1953.

Temperature
(°C) (°F)

Viscosity
[(kg/m·s) 10^3 or c_p]

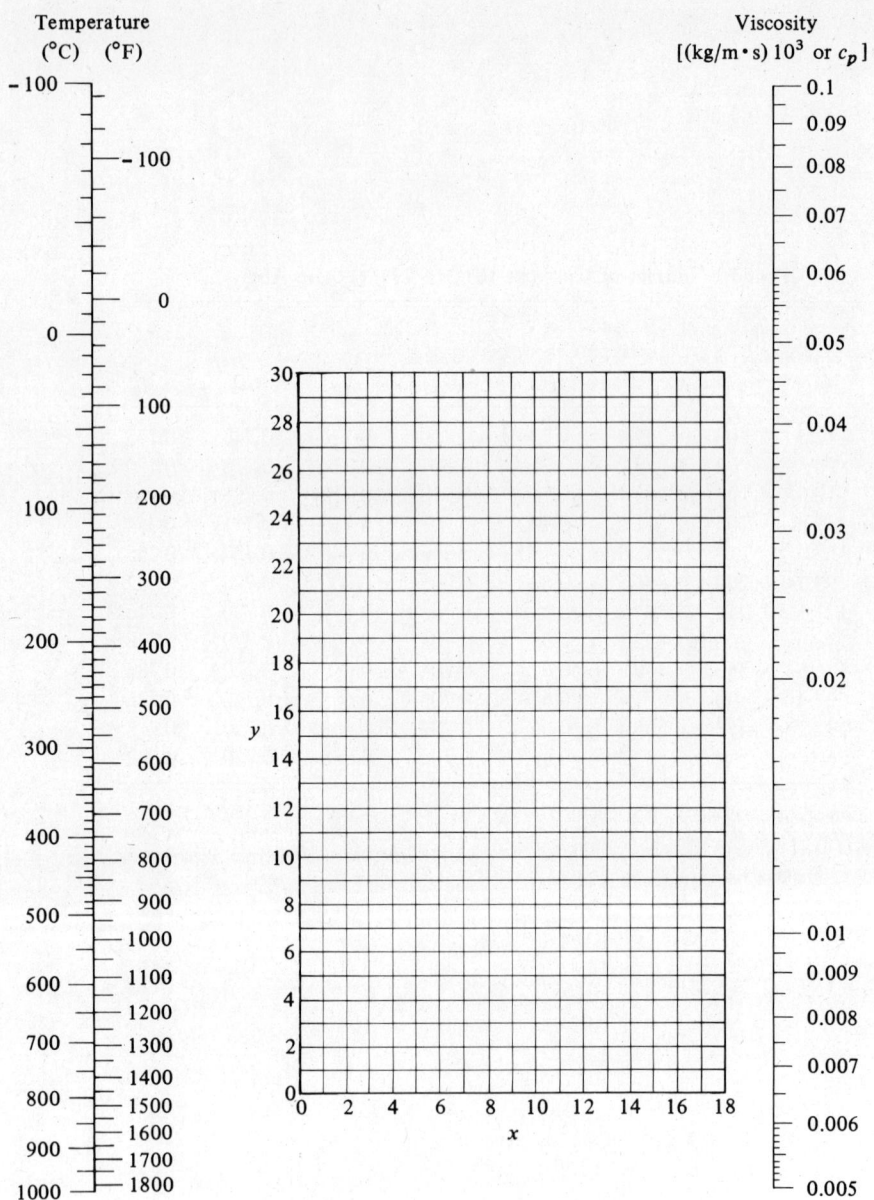

FIGURE A.3-2. *Viscosities of gases at 101.325 kPa (1 atm abs). (From R. H. Perry and C. H. Chilton, Chemical Engineers' Handbook, 5th ed. New York: McGraw-Hill Book Company, 1973. With permission.)*

Appendix

A.3-8 Viscosities of Gases (Coordinates for Use with Fig. A.3-2)

No.	Gas	X	Y	No.	Gas	X	Y
1	Acetic acid	7.7	14.3	29	Freon-113	11.3	14.0
2	Acetone	8.9	13.0	30	Helium	10.9	20.5
3	Acetylene	9.8	14.9	31	Hexane	8.6	11.8
4	Air	11.0	20.0	32	Hydrogen	11.2	12.4
5	Ammonia	8.4	16.0	33	$3H_2 + 1N_2$	11.2	17.2
6	Argon	10.5	22.4	34	Hydrogen bromide	8.8	20.9
7	Benzene	8.5	13.2	35	Hydrogen chloride	8.8	18.7
8	Bromine	8.9	19.2	36	Hydrogen cyanide	9.8	14.9
9	Butene	9.2	13.7	37	Hydrogen iodide	9.0	21.3
10	Butylene	8.9	13.0	38	Hydrogen sulfide	8.6	18.0
11	Carbon dioxide	9.5	18.7	39	Iodine	9.0	18.4
12	Carbon disulfide	8.0	16.0	40	Mercury	5.3	22.9
13	Carbon monoxide	11.0	20.0	41	Methane	9.9	15.5
14	Chlorine	9.0	18.4	42	Methyl alcohol	8.5	15.6
15	Chloroform	8.9	15.7	43	Nitric oxide	10.9	20.5
16	Cyanogen	9.2	15.2	44	Nitrogen	10.6	20.0
17	Cyclohexane	9.2	12.0	45	Nitrosyl chloride	8.0	17.6
18	Ethane	9.1	14.5	46	Nitrous oxide	8.8	19.0
19	Ethyl acetate	8.5	13.2	47	Oxygen	11.0	21.3
20	Ethyl alcohol	9.2	14.2	48	Pentane	7.0	12.8
21	Ethyl chloride	8.5	15.6	49	Propane	9.7	12.9
22	Ethyl ether	8.9	13.0	50	Propyl alcohol	8.4	13.4
23	Ethylene	9.5	15.1	51	Propylene	9.0	13.8
24	Fluorine	7.3	23.8	52	Sulfur dioxide	9.6	17.0
25	Freon-11	10.6	15.1	53	Toluene	8.6	12.4
26	Freon-12	11.1	16.0	54	2,3,3-Trimethylbutane	9.5	10.5
27	Freon-21	10.8	15.3	55	Water	8.0	16.0
28	Freon-22	10.1	17.0	56	Xenon	9.3	23.0

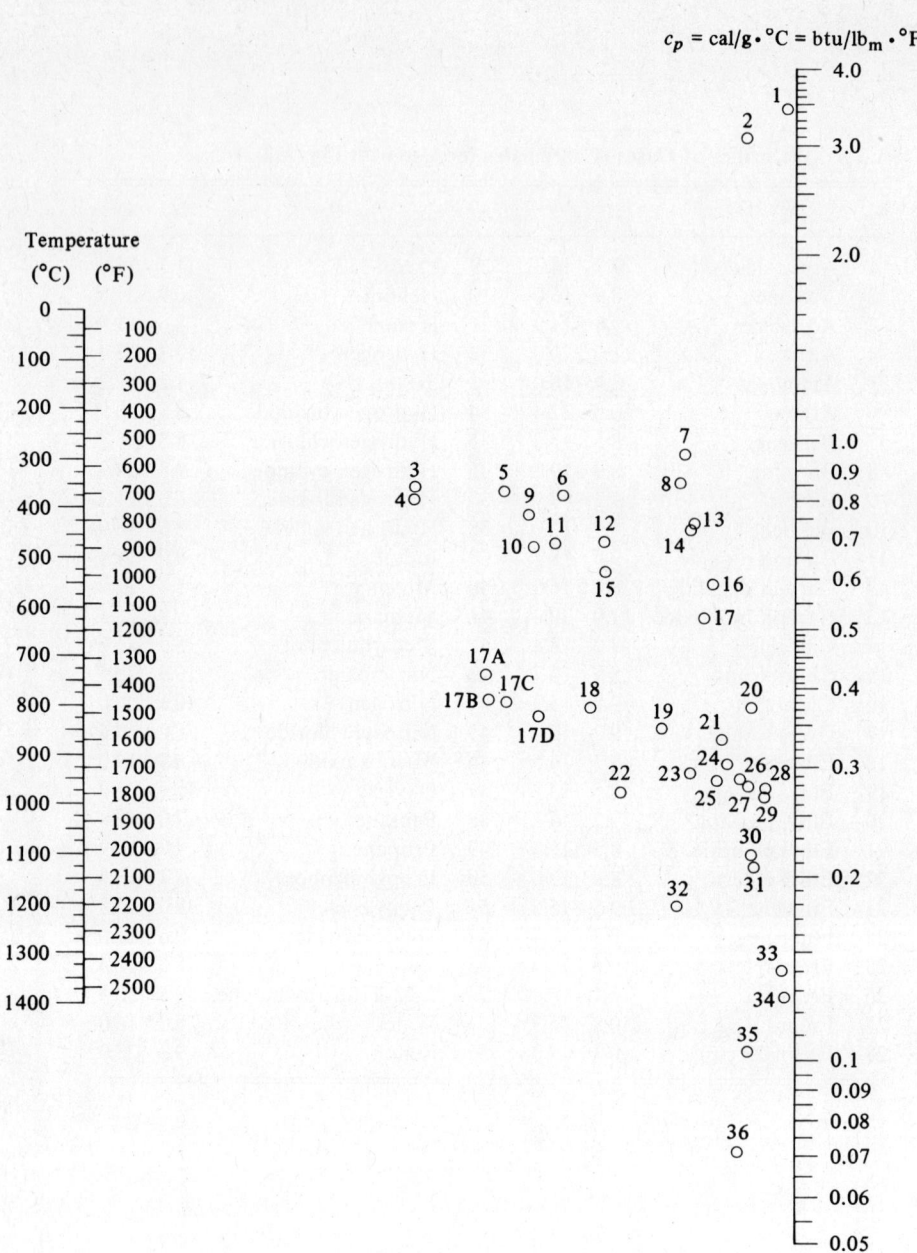

$c_p = cal/g \cdot {}^{\circ}C = btu/lb_m \cdot {}^{\circ}F$

FIGURE A.3-3.   *Heat capacity of gases at constant pressure at 101.325 kPa (1 atm abs). (From R. H. Perry and C. H. Chilton, Chemical Engineers' Handbook, 5th ed. New York: McGraw-Hill Book Company, 1973. With permission.)*

Heat Capacity of Gases at Constant Pressure (for Use with Fig. A.3-3)

No.	Gas	Range (°C)
10	Acetylene	0–200
15	Acetylene	200–400
16	Acetylene	400–1400
27	Air	0–1400
12	Ammonia	0–600
14	Ammonia	600–1400
18	Carbon dioxide	0–400
24	Carbon dioxide	400–1400
26	Carbon monoxide	0–1400
32	Chlorine	0–200
34	Chlorine	200–1400
3	Ethane	0–200
9	Ethane	200–600
8	Ethane	600–1400
4	Ethylene	0–200
11	Ethylene	200–600
13	Ethylene	600–1400
17B	Freon-11 (CCl_3F)	0–150
17C	Freon-21 ($CHCl_2F$)	0–150
17A	Freon-22 ($CHClF_2$)	0–150
17D	Freon-113 ($CCl_2F–CClF_2$)	0–150
1	Hydrogen	0–600
2	Hydrogen	600–1400
35	Hydrogen bromide	0–1400
30	Hydrogen chloride	0–1400
20	Hydrogen fluoride	0–1400
36	Hydrogen iodide	0–1400
19	Hydrogen sulfide	0–700
21	Hydrogen sulfide	700–1400
5	Methane	0–300
6	Methane	300–700
7	Methane	700–1400
25	Nitric oxide	0–700
28	Nitric oxide	700–1400
26	Nitrogen	0–1400
23	Oxygen	0–500
29	Oxygen	500–1400
33	Sulfur	300–1400
22	Sulfur dioxide	0–400
31	Sulfur dioxide	400–1400
17	Water	0–1400

A.3-9 Thermal Conductivities of Gases and Vapors at 101.325 kPa (1 Atm Abs); $k = W/m \cdot K$

Gas or Vapor	K	k	Gas or Vapor	K	k
Acetone[1]	273	0.0099	Ethane[5, 6]	239	0.0149
	319	0.0130		273	0.0183
	373	0.0171		373	0.0303
	457	0.0254	Ethyl alcohol[1]	293	0.0154
Ammonia[2]	273	0.0218		373	0.0215
	373	0.0332	Ethyl ether[1]	273	0.0133
	473	0.0484		319	0.0171
Butane[3]	273	0.0135		373	0.0227
	373	0.0234	Ethylene[6]	273	0.0175
Carbon monoxide[2]	173	0.0152		323	0.0227
	273	0.0232		373	0.0279
	373	0.0305	n-Hexane[3]	273	0.0125
Chlorine[4]	273	0.00744		293	0.0138
			Sulfur dioxide[7]	273	0.0087
				373	0.0119

Source: (1) Moser, dissertation, Berlin, 1913; (2) F. G. Keyes, *Tech. Rept.* 37, Project Squid, Apr. 1, 1952; (3) W. B. Mann and B. G. Dickens, *Proc. Roy. Soc. (London)*, **A134**, 77 (1931); (4) *International Critical Tables.* New York: McGraw-Hill Book Company, 1929; (5) T. H. Chilton and R. P. Genereaux, personal communication, 1946; (6) A. Eucken, *Physik, Z.*, **12**, 1101 (1911); **14**, 324 (1913); (7) B. G. Dickens, *Proc. Roy. Soc. (London)*, **A143**, 517 (1934).

A.3-10 Heat Capacities of Liquids ($c_p = kJ/kg \cdot K$)

Liquid	K	c_p	Liquid	K	c_p
Acetic acid	273	1.959	Hydrochloric acid (20 mol %)	273	2.43
	311	2.240		293	2.474
Acetone	273	2.119	Mercury	293	0.01390
	293	2.210	Methyl alcohol	293	2.512
Aniline	273	2.001		313	2.583
	323	2.181	Nitrobenzene	283	1.499
Benzene	293	1.700		303	1.419
	333	1.859		363	1.436
Butane	273	2.300	Sodium chloride (9.1 mol %)	293	3.39
i-Butyl alcohol	303	2.525		330	3.43
Ethyl alcohol	273	2.240	Sulfuric acid (100%)	293	1.403
	298	2.433	Toluene	273	1.616
Formic acid	273	1.825		323	1.763
	289	2.131	o-Xylene	303	1.721
Glycerol	288	2.324			
	305	2.412			

Source: N. A. Lange, *Handbook of Chemistry*, 10th ed. New York: McGraw-Hill Book Company, 1967; National Research Council, *International Critical Tables*, Vol. V. New York: McGraw-Hill Book Company, 1929; R. H. Perry and C. H. Chilton, *Chemical Engineers' Handbook*, 5th ed. New York: McGraw-Hill Book Company, 1973.

Temperature
(°C) (°F)

Viscosity
$[(kg/m \cdot s) \, 10^3 \text{ or } c_p]$

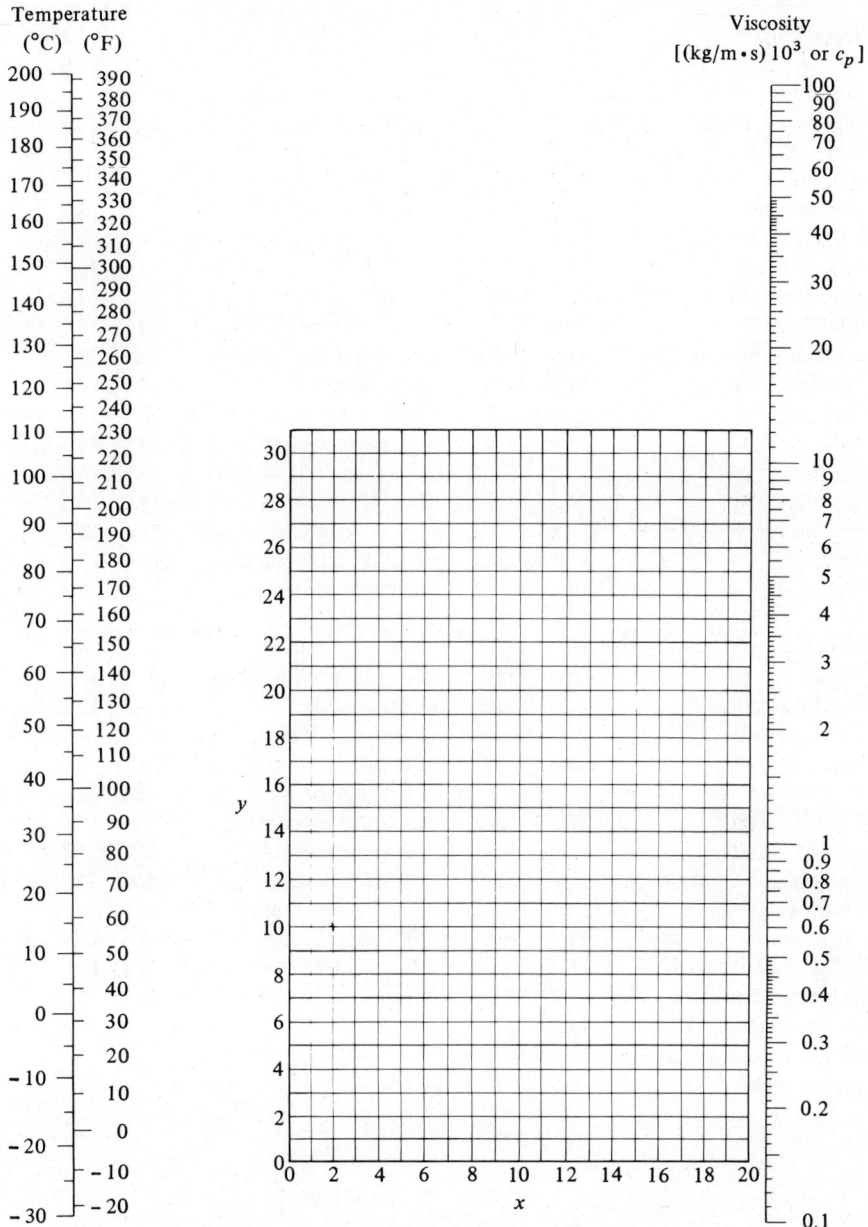

FIGURE A.3-4. *Viscosities of liquids. (From R. H. Perry and C. H. Chilton, Chemical Engineers'*
Handbook, 5th ed. New York : McGraw-Hill Book Company, 1973. With permission.)

Liquid	X	Y	Liquid	X	Y
Acetaldehyde	15.2	4.8	Cyclohexanol	2.9	24.3
Acetic acid, 100%	12.1	14.2	Cyclohexane	9.8	12.9
Acetic acid, 70%	9.5	17.0	Dibromomethane	12.7	15.8
Acetic anhydride	12.7	12.8	Dichloroethane	13.2	12.2
Acetone, 100%	14.5	7.2	Dichloromethane	14.6	8.9
Acetone, 35%	7.9	15.0	Diethyl ketone	13.5	9.2
Acetonitrile	14.4	7.4	Diethyl oxalate	11.0	16.4
Acrylic acid	12.3	13.9	Diethylene glycol	5.0	24.7
Allyl alcohol	10.2	14.3	Diphenyl	12.0	18.3
Allyl bromide	14.4	9.6	Dipropyl ether	13.2	8.6
Allyl iodide	14.0	11.7	Dipropyl oxalate	10.3	17.7
Ammonia, 100%	12.6	2.0	Ethyl acetate	13.7	9.1
Ammonia, 26%	10.1	13.9	Ethyl acrylate	12.7	10.4
Amyl acetate	11.8	12.5	Ethyl alcohol, 100%	10.5	13.8
Amyl alcohol	7.5	18.4	Ethyl alcohol, 95%	9.8	14.3
Aniline	8.1	18.7	Ethyl alcohol, 40%	6.5	16.6
Anisole	12.3	13.5	Ethyl benzene	13.2	11.5
Arsenic trichloride	13.9	14.5	Ethyl bromide	14.5	8.1
Benzene	12.5	10.9	2-Ethyl butyl acrylate	11.2	14.0
Brine, $CaCL_2$, 25%	6.6	15.9	Ethyl chloride	14.8	6.0
Brine, NaCl, 25%	10.2	16.6	Ethyl ether	14.5	5.3
Bromine	14.2	13.2	Ethyl formate	14.2	8.4
Bromotoluene	20.0	15.9	2-Ethyl hexyl acrylate	9.0	15.0
Butyl acetate	12.3	11.0	Ethyl iodide	14.7	10.3
Butyl acrylate	11.5	12.6	Ethyl propionate	13.2	9.9
Butyl alcohol	8.6	17.2	Ethyl propyl ether	14.0	7.0
Butyric acid	12.1	15.3	Ethyl sulfide	13.8	8.9
Carbon dioxide	11.6	0.3	Ethylene bromide	11.9	15.7
Carbon disulfide	16.1	7.5	Ethylene chloride	12.7	12.2
Carbon tetrachloride	12.7	13.1	Ethylene glycol	6.0	23.6
Chlorobenzene	12.3	12.4	Ethylidene chloride	14.1	8.7
Chloroform	14.4	10.2	Fluorobenzene	13.7	10.4
Chlorosulfonic acid	11.2	18.1	Formic acid	10.7	15.8
Chlorotoluene, ortho	13.0	13.3	Freon-11	14.4	9.0
Chlorotoluene, meta	13.3	12.5	Freon-12	16.8	15.6
Chlorotoluene, para	13.3	12.5	Freon-21	15.7	7.5
Cresol, meta	2.5	20.8	Freon-22	17.2	4.7

Liquid	X	Y	Liquid	X	Y
Freon-113	12.5	11.4	Octyl alcohol	6.6	21.1
Glycerol, 100%	2.0	30.0	Pentachloroethane	10.9	17.3
Glycerol, 50%	6.9	19.6	Pentane	14.9	5.2
Heptane	14.1	8.4	Phenol	6.9	20.8
Hexane	14.7	7.0	Phosphorus tribromide	13.8	16.7
Hydrochloric acid, 31.5%	13.0	16.6	Phosphorus trichloride	16.2	10.9
Iodobenzene	12.8	15.9	Propionic acid	12.8	13.8
Isobutyl alcohol	7.1	18.0	Propyl acetate	13.1	10.3
Isobutyric acid	12.2	14.4	Propyl alcohol	9.1	16.5
Isopropyl alcohol	8.2	16.0	Propyl bromide	14.5	9.6
Isopropyl bromide	14.1	9.2	Propyl chloride	14.4	7.5
Isopropyl chloride	13.9	7.1	Propyl formate	13.1	9.7
Isopropyl iodide	13.7	11.2	Propyl iodide	14.1	11.6
Kerosene	10.2	16.9	Sodium	16.4	13.9
Linseed oil, raw	7.5	27.2	Sodium hydroxide, 50%	3.2	25.8
Mercury	18.4	16.4	Stannic chloride	13.5	12.8
Methanol, 100%	12.4	10.5	Succinonitrile	10.1	20.8
Methanol, 90%	12.3	11.8	Sulfur dioxide	15.2	7.1
Methanol, 40%	7.8	15.5	Sulfuric acid, 110%	7.2	27.4
Methyl acetate	14.2	8.2	Sulfuric acid, 100%	8.0	25.1
Methyl acrylate	13.0	9.5	Sulfuric acid, 98%	7.0	24.8
Methyl *i*-butyrate	12.3	9.7	Sulfuric acid, 60%	10.2	21.3
Methyl *n*-butyrate	13.2	10.3	Sulfuryl chloride	15.2	12.4
Methyl chloride	15.0	3.8	Tetrachloroethane	11.9	15.7
Methyl ethyl ketone	13.9	8.6	Thiophene	13.2	11.0
Methyl formate	14.2	7.5	Titanium tetrachloride	14.4	12.3
Methyl iodide	14.3	9.3	Toluene	13.7	10.4
Methyl propionate	13.5	9.0	Trichloroethylene	14.8	10.5
Methyl propyl ketone	14.3	9.5	Triethylene glycol	4.7	24.8
Methyl sulfide	15.3	6.4	Turpentine	11.5	14.9
Naphthalene	7.9	18.1	Vinyl acetate	14.0	8.8
Nitric acid, 95%	12.8	13.8	Vinyl toluene	13.4	12.0
Nitric acid, 60%	10.8	17.0	Water	10.2	13.0
Nitrobenzene	10.6	16.2	Xylene, ortho	13.5	12.1
Nitrogen dioxide	12.9	8.6	Xylene, meta	13.9	10.6
Nitrotoluene	11.0	17.0	Xylene, para	13.9	10.9
Octane	13.7	10.0			

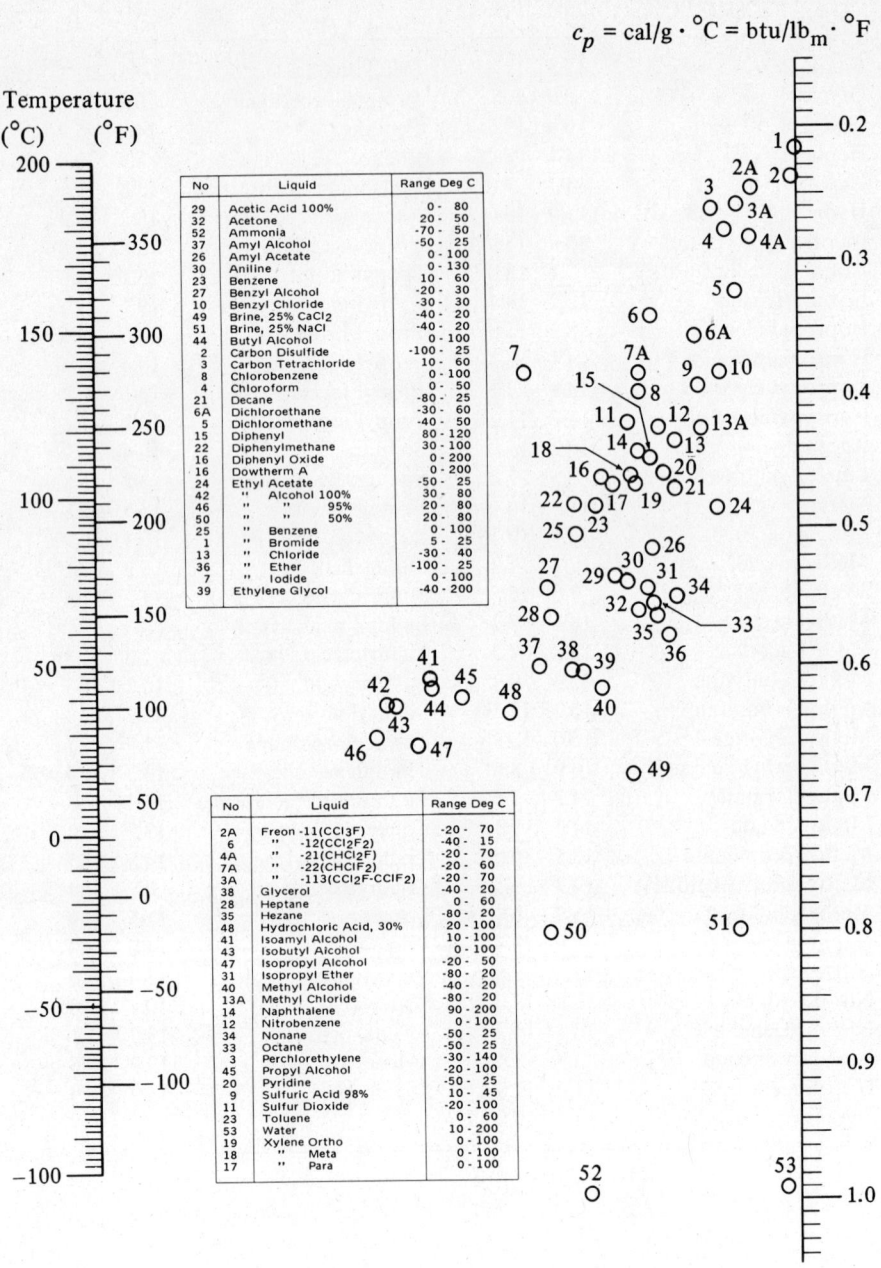

$$c_p = \text{cal/g} \cdot {}^\circ\text{C} = \text{btu/lb}_m \cdot {}^\circ\text{F}$$

No	Liquid	Range Deg C
29	Acetic Acid 100%	0 - 80
32	Acetone	20 - 50
52	Ammonia	-70 - 50
37	Amyl Alcohol	-50 - 25
26	Amyl Acetate	0 - 100
30	Aniline	0 - 130
23	Benzene	10 - 60
27	Benzyl Alcohol	-20 - 30
10	Benzyl Chloride	-30 - 30
49	Brine, 25% CaCl₂	-40 - 20
51	Brine, 25% NaCl	-40 - 20
44	Butyl Alcohol	0 - 100
2	Carbon Disulfide	-100 - 25
3	Carbon Tetrachloride	10 - 60
8	Chlorobenzene	0 - 100
4	Chloroform	0 - 50
21	Decane	-80 - 25
6A	Dichloroethane	-30 - 60
5	Dichloromethane	-40 - 50
15	Diphenyl	80 - 120
22	Diphenylmethane	30 - 100
16	Diphenyl Oxide	0 - 200
16	Dowtherm A	0 - 200
24	Ethyl Acetate	-50 - 25
42	" Alcohol 100%	30 - 80
46	" " 95%	20 - 80
50	" " 50%	20 - 80
25	" Benzene	0 - 100
1	" Bromide	5 - 25
13	" Chloride	-30 - 40
36	" Ether	-100 - 25
7	" Iodide	0 - 100
39	Ethylene Glycol	-40 - 200

No	Liquid	Range Deg C
2A	Freon -11(CCl₃F)	-20 - 70
6	" -12(CCl₂F₂)	-40 - 15
4A	" -21(CHCl₂F)	-20 - 70
7A	" -22(CHClF₂)	-20 - 60
3A	" -113(CCl₂F-CClF₂)	-20 - 70
38	Glycerol	-40 - 20
28	Heptane	0 - 60
35	Hexane	-80 - 20
48	Hydrochloric Acid, 30%	20 - 100
41	Isoamyl Alcohol	10 - 100
43	Isobutyl Alcohol	0 - 100
47	Isopropyl Alcohol	-20 - 50
31	Isopropyl Ether	-80 - 20
40	Methyl Alcohol	-40 - 20
13A	Methyl Chloride	-80 - 20
14	Naphthalene	90 - 200
12	Nitrobenzene	0 - 100
34	Nonane	-50 - 25
33	Octane	-50 - 25
3	Perchlorethylene	-30 - 140
45	Propyl Alcohol	-20 - 100
20	Pyridine	-50 - 25
9	Sulfuric Acid 98%	10 - 45
11	Sulfur Dioxide	-20 - 100
23	Toluene	0 - 60
53	Water	10 - 200
19	Xylene Ortho	0 - 100
18	" Meta	0 - 100
17	" Para	0 - 100

FIGURE A.3-5. *Heat capacity of liquids. (From R. H. Perry and C. H. Chilton, Chemical Engineers' Handbook, 5th ed. New York: McGraw-Hill Book Company, 1973. With permission.)*

A.3-12 Thermal Conductivities of Liquids (k = W/m · K)*

Liquid	K	k	Liquid	K	k
Acetic acid			Ethylene glycol	273	0.265
100%	293	0.171	Glycerol, 100%	293	0.284
50%	293	0.346	n-Hexane	303	0.138
Ammonia	243–258	0.502		333	0.135
n-Amyl alcohol	303	0.163	Kerosene	293	0.149
	373	0.154		348	0.140
Benzene	303	0.159	Methyl alcohol		
	333	0.151	100%	293	0.215
Carbon tetrachloride	273	0.185	60%	293	0.329
	341	0.163	20%	293	0.492
n-Decane	303	0.147	100%	323	0.197
	333	0.144	n-Octane	303	0.144
Ethyl acetate	293	0.175		333	0.140
Ethyl alcohol			NaCl brine		
100%	293	0.182	25%	303	0.571
60%	293	0.305	12.5%	303	0.589
20%	293	0.486	Sulfuric acid		
100%	323	0.151	90%	303	0.364
			60%	303	0.433
			Vaseline	332	0.183

* A linear variation with temperature may be assumed between the temperature limits given.
Source: R. H. Perry and C. H. Chilton, *Chemical Engineers' Handbook*, 5th ed. New York: McGraw-Hill Book Company, 1973. With permission.

A.3-13 Heat Capacities of Solids ($c_p = kJ/kg \cdot K$)

Solid	K	c_p	Solid	K	c_p
Alumina	373	0.84	Benzene	273	1.570
	1773	1.147	Benzoic acid	293	1.243
Asbestos		1.05	Camphene	308	1.591
Asphalt		0.92	Caprylic acid	271	2.629
Brick, fireclay	373	0.829	Dextrin	273	1.218
	1773	1.248	Formic acid	273	1.800
Cement, portland		0.779	Glycerol	273	1.382
Clay		0.938	Lactose	293	1.202
Concrete		0.63	Oxalic acid	323	1.612
Corkboard	303	0.167	Tartaric acid	309	1.202
Glass		0.84	Urea	293	1.340
Magnesia	373	0.980			
	1773	0.787			
Oak		2.39			
Pine, yellow	298	2.81			
Porcelain	293–373	0.775			
Rubber, vulcanized		2.01			
Steel		0.50			
Wool		1.361			

Source: R. H. Perry and C. H. Chilton, *Chemical Engineers' Handbook*, 5th ed. New York: McGraw-Hill Book Company, 1973; National Research Council, *International Critical Tables*, Vol. V. New York: McGraw-Hill Book Company, 1929; L. S. Marks, *Mechanical Engineers' Handbook*, 5th ed. New York: McGraw-Hill Book Company, 1951; F. Kreith, *Principles of Heat Transfer*, 2nd ed. Scranton, Pa.: International Textbook Co., 1965.

A.3-14 Thermal Conductivities of Building and Insulating Materials

Material	$\rho \left(\dfrac{kg}{m^3}\right)$	t^* (°C)	$k(W/m \cdot K)$		
Asbestos	577		0.151 (0°C)	0.168 (37.8°C)	0.190 (93.3°C)
Asbestos sheets	889	51	0.166		
Brick, building		20	0.69		
Brick, fireclay			1.00 (200°C)	1.47 (600°C)	1.64 (1000°C)
Clay soil, 4% H_2O	1666	4.5	0.57		
Concrete, 1:4 dry			0.762		
Corkboard	160.2	30	0.0433		
Cotton	80.1		0.055 (0°C)	0.061 (37.8°C)	0.068 (93.3°C)
Felt, wool	330	30	0.052		
Fiber insulation board	237	21	0.048		
Glass, window			0.52–1.06		
Glass wool	64.1	30	0.0310 (−6.7°C)	0.0414 (37.8°C)	0.0549 (93.3°C)
Ice	921	0	2.25		
Magnesia, 85%	271		0.068 (37.8°C)	0.071 (93.3°C)	0.080 (204.4°C)
	208		0.059 (37.8°C)	0.062 (93.3°C)	0.066 (148.9°C)
Oak, across grain	825	15	0.208		
Pine, across grain	545	15	0.151		
Paper			0.130		
Rock wool	192		0.0317 (−6.7°C)	0.0391 (37.8°C)	0.0486 (93.3°C)
	128		0.0296 (−6.7°C)	0.0395 (37.8°C)	0.0518 (93.3°C)
Rubber, hard	1198	0	0.151		
Sand soil					
4% H_2O	1826	4.5	1.51		
10% H_2O	1922	4.5	2.16		
Sandstone	2243	40	1.83		
Snow	559	0	0.47		
Wool	110.5	30	0.036		

* Room temperature when none is noted.

Source: L. S. Marks, *Mechanical Engineers' Handbook,* 5th ed. New York: McGraw-Hill Book Company, 1951; W. H. McAdams, *Heat Transmission,* 3rd ed. New York: McGraw-Hill Book Company, 1954; F. H. Norton, *Refractories.* New York: McGraw-Hill Book Company, 1949; National Research Council, *International Critical Tables.* New York: McGraw-Hill Book Company, 1929; M. S. Kersten, *Univ. Minn. Eng. Ex. Sta., Bull. 28,* June 1949; R. H. Heilman, *Ind. Eng. Chem.,* 28, 782 (1936).

A.3-15 Thermal Conductivities, Densities, and Heat Capacities of Metals

Material	t (°C)	ρ $\left(\dfrac{kg}{m^3}\right)$	c_p $\left(\dfrac{kJ}{kg \cdot K}\right)$		$k(W/m \cdot K)$	
Aluminum	20	2707	0.896	202 (0°C)	206 (100°C)	215 (200°C)
				230 (300°C)		
Brass (70–30)	20	8522	0.385	97 (0°C)	104 (100°C)	109 (200°C)
Cast iron	20	7593	0.465	55 (0°C)	52 (100°C)	48 (200°C)
Copper	20	8954	0.383	388 (0°C)	377 (100°C)	372 (200°C)
Lead	20	11 370	0.130	35 (0°C)	33 (100°C)	31 (200°C)
Steel 1%C	20	7801	0.473	45.3 (18°C)	45 (100°C)	45 (200°C)
				43 (300°C)		
308 stainless	20	7849	0.461	15.2 (100°C)	21.6 (500°C)	
304 stainless	0	7817	0.461	13.8 (0°C)	16.3 (100°C)	18.9 (300°C)
Tin	20	7304	0.227	62 (0°C)	59 (100°C)	57 (200°C)

Source: L. S. Marks, *Mechanical Engineers' Handbook*, 5th ed. New York: McGraw-Hill Book Company, 1951; E. R. G. Eckert and R. M. Drake, *Heat and Mass Transfer*, 2nd ed. New York: McGraw-Hill Book Company, 1959; R. H. Perry and C. H. Chilton, *Chemical Engineers' Handbook*, 5th ed. New York: McGraw-Hill Book Company, 1973; National Research Council, *International Critical Tables.* New York: McGraw-Hill Book Company, 1929.

A.3-16   Normal Total Emmissivities of Surfaces

Surface	K	ε	Surface	K	ε
Aluminum			Lead, unoxidized	400	0.057
highly oxidized	366	0.20	Nickel, polished	373	0.072
highly polished	500	0.039	Nickel oxide	922	0.59
	850	0.057	Oak, planed	294	0.90
Aluminum oxide	550	0.63	Paint		
Asbestos board	296	0.96	aluminum	373	0.52
Brass, highly	520	0.028	oil (16 different,		
			all colors)	373	0.92–0.96
polished	630	0.031	Paper	292	0.924
Chromium,					
polished	373	0.075	Roofing paper	294	0.91
Copper			Rubber (hard, glossy)	296	0.94
oxidized	298	0.78	Steel		
polished	390	0.023	oxidized at 867 K	472	0.79
Glass, smooth	295	0.94	polished stainless	373	0.074
Iron			304 stainless	489	0.44
oxidized	373	0.74	Water	273	0.95
tin-plated	373	0.07		373	0.963
Iron oxide	772	0.85			

Source: R. H. Perry and C. H. Chilton, *Chemical Engineers' Handbook*, 5th ed. New York: McGraw-Hill Book Company, 1973; W. H. McAdams, *Heat Transmission*, 3rd ed. New York: McGraw-Hill Book Company, 1954; E. Schmidt, *Gesundh.-Ing. Beiheft*, **20**, Reihe 1, 1 (1927).

A.3-17   Henry's Law Constants for Gases in Water ($H \times 10^{-4}$)*

T											
K	°C	CO_2	CO	C_2H_6	C_2H_4	He	H_2	H_2S	CH_4	N_2	O_2
273.2	0	0.0728	3.52	1.26	0.552	12.9	5.79	0.0268	2.24	5.29	2.55
283.2	10	0.104	4.42	1.89	0.768	12.6	6.36	0.0367	2.97	6.68	3.27
293.2	20	0.142	5.36	2.63	1.02	12.5	6.83	0.0483	3.76	8.04	4.01
303.2	30	0.186	6.20	3.42	1.27	12.4	7.29	0.0609	4.49	9.24	4.75
313.2	40	0.233	6.96	4.23		12.1	7.51	0.0745	5.20	10.4	5.35

* $p_A = Hx_A$, p_A = partial pressure of A in gas in atm, x_A = mole fraction of A in liquid, H = Henry's law constant in atm/mole frac.
Source: National Research Council, *International Critical Tables*, Vol. III. New York: McGraw-Hill Book Company, 1929.

A.3-18 Equilibrium Data for SO$_2$–Water System

Mole Fraction SO$_2$ in Liquid, x_A	Partial Pressure of SO$_2$ in Vapor, p_A (mm Hg)		Mole Fraction SO$_2$ in Vapor, y_A; P = 1 Atm	
	20°C (293 K)	30°C (303 K)	20°C	30°C
0	0	0	0	0
0.0000562	0.5	0.6	0.000658	0.000790
0.0001403	1.2	1.7	0.00158	0.00223
0.000280	3.2	4.7	0.00421	0.00619
0.000422	5.8	8.1	0.00763	0.01065
0.000564	8.5	11.8	0.01120	0.0155
0.000842	14.1	19.7	0.01855	0.0259
0.001403	26.0	36	0.0342	0.0473
0.001965	39.0	52	0.0513	0.0685
0.00279	59	79	0.0775	0.1040
0.00420	92	125	0.121	0.1645
0.00698	161	216	0.212	0.284
0.01385	336	452	0.443	0.594
0.0206	517	688	0.682	0.905
0.0273	698		0.917	

Source : T. K. Sherwood, *Ind. Eng. Chem.*, **17**, 745 (1925).

A.3-19 Equilibrium Data for Methanol–Water System

Mole Fraction Methanol in Liquid, x_A	Partial Pressure of Methanol in Vapor, p_A (mm Hg)	
	39.9°C (313.1 K)	59.4°C (332.6K)
0	0	0
0.05	25.0	50
0.10	46.0	102
0.15	66.5	151

Source : National Research Council, *International Critical Tables*, Vol. III. New York: McGraw-Hill Book Company, 1929.

A.3-20 Equilibrium Data for Acetone–Water System at 20°C (293 K)

Mole Fraction Acetone in Liquid, x_A	Partial Pressure of Acetone in Vapor, p_A (mm Hg)
0	0
0.0333	30.0
0.0720	62.8
0.117	85.4
0.171	103

Source: T. K. Sherwood, *Absorption and Extraction.* New York: McGraw-Hill Book Company, 1937. With permission.

A.3-21 Equilibrium Data for Ammonia–Water System

Mole Fraction NH_3 in Liquid, x_A	Partial Pressure of NH_3 in Vapor, p_A (mm Hg)		Mole Fraction NH_3 in Vapor, y_A; $P = 1$ Atm	
	20°C (293 K)	30°C (303 K)	20°C	30°C
0	0	0	0	0
0.0126		11.5		0.0151
0.0167		15.3		0.0201
0.0208	12	19.3	0.0158	0.0254
0.0258	15	24.4	0.0197	0.0321
0.0309	18.2	29.6	0.0239	0.0390
0.0405	24.9	40.1	0.0328	0.0527
0.0503	31.7	51.0	0.0416	0.0671
0.0737	50.0	79.7	0.0657	0.105
0.0960	69.6	110	0.0915	0.145
0.137	114	179	0.150	0.235
0.175	166	260	0.218	0.342
0.210	227	352	0.298	0.463
0.241	298	454	0.392	0.597
0.297	470	719	0.618	0.945

Source: J. H. Perry, *Chemical Engineers' Handbook*, 4th ed. New York: McGraw-Hill Book Company, 1963. With permission.

Physical Properties of Inorganic and Organic Compounds

A.3-22 Equilibrium Data for Ethanol–Water System at 101.325 kPa (1 Atm)*

Vapor–Liquid Equilibria

Temperature °C	Temperature °F	Mass Fraction Ethanol x_A	Mass Fraction Ethanol y_A
100.0	212	0	0
98.1	208.5	0.020	0.192
95.2	203.4	0.050	0.377
91.8	197.2	0.100	0.527
87.3	189.2	0.200	0.656
84.7	184.5	0.300	0.713
83.2	181.7	0.400	0.746
82.0	179.6	0.500	0.771
81.0	177.8	0.600	0.794
80.1	176.2	0.700	0.822
79.1	174.3	0.800	0.858
78.3	173.0	0.900	0.912
78.2	172.8	0.940	0.942
78.1	172.7	0.960	0.959
78.2	172.8	0.980	0.978
78.3	173.0	1.00	1.00

Enthalpy

Temperature °C	Temperature °F	Mass Fraction	Enthalpy (btu/lb$_m$ of mixture) Liquid	Enthalpy (btu/lb$_m$ of mixture) Vapor	Enthalpy (kJ/kg of mixture) Liquid	Enthalpy (kJ/kg of mixture) Vapor
100.0	212	0	180.1	1150	418.9	2675
91.8	197.2	0.1	159.8	1082	371.7	2517
84.7	184.5	0.3	135.0	943	314.0	2193
82.0	179.6	0.5	122.9	804	285.9	1870
80.1	176.2	0.7	111.1	664	258.4	1544
78.3	173.0	0.9	96.6	526	224.7	1223
78.3	173.0	1.0	89.0	457.5	207.0	1064

* Reference state for enthalpy is pure liquid at 273 K or 0°C.
Source: Data from L. W. Cornell and R. E. Montonna, *Ind. Eng. Chem.*, **25**, 1331 (1933); and W. A. Noyes and R. R. Warfel, *J. Am. Chem. Soc.*, **23**, 463 (1901), as given by G. G. Brown, *Unit Operations*. New York: John Wiley & Sons, Inc., 1950. With permission.

A.3-23 Acetic Acid–Water–Isopropyl Ether System, Liquid–Liquid Equilibria at 293 K or 20°C

Water Layer (wt %)			Isopropyl Ether Layer (wt %)		
Acetic Acid	Water	Isopropyl Ether	Acetic Acid	Water	Isopropyl Ether
0	98.8	1.2	0	0.6	99.4
0.69	98.1	1.2	0.18	0.5	99.3
1.41	97.1	1.5	0.37	0.7	98.9
2.89	95.5	1.6	0.79	0.8	98.4
6.42	91.7	1.9	1.93	1.0	97.1
13.30	84.4	2.3	4.82	1.9	93.3
25.50	71.1	3.4	11.40	3.9	84.7
36.70	58.9	4.4	21.60	6.9	71.5
44.30	45.1	10.6	31.10	10.8	58.1
46.40	37.1	16.5	36.20	15.1	48.7

Source : *Trans. A.I.Ch.E.*, **36**, 601, 628 (1940). With permission.

A.3-24 Liquid–Liquid Equilibrium Data for Acetone–Water–Methyl Isobutyl Ketone (MIK) System at 298–299 K or 25–26°C

Composition Data (wt %)			Acetone Distribution Data (wt %)	
MIK	Acetone	Water	Water Phase	MIK Phase
98.0	0	2.00	2.5	4.5
93.2	4.6	2.33	5.5	10.0
77.3	18.95	3.86	7.5	13.5
71.0	24.4	4.66	10.0	17.5
65.5	28.9	5.53	12.5	21.3
54.7	37.6	7.82	15.5	25.5
46.2	43.2	10.7	17.5	28.2
12.4	42.7	45.0	20.0	31.2
5.01	30.9	64.2	22.5	34.0
3.23	20.9	75.8	25.0	36.5
2.12	3.73	94.2	26.0	37.5
2.20	0	97.8		

Source: Reprinted with permission from D. F. Othmer, R. E. White, and E. Trueger, *Ind. Eng. Chem.*, **33**, 1240 (1941). Copyright by the American Chemical Society.

Physical Properties of Inorganic and Organic Compounds

Physical Properties
of Foods and
Biological Materials

A.4-1 Heat Capacities of Foods (Average c_p 273–373 K or 0–100°C)

Material	H_2O (wt %)	c_p (kJ/kg·K)
Apples	75–85	3.73–4.02
Apple sauce		4.02*
Asparagus		
Fresh	93	3.94†
Frozen	93	2.01‡
Bacon, lean	51	3.43
Banana purée		3.66§
Beef, lean	72	3.43
Bread, white	44–45	2.72–2.85
Butter	15	2.30¶
Cantaloupe	92.7	3.94†
Cheese, Swiss	55	2.68†
Corn, sweet		
Fresh		3.32†
Frozen		1.77‡
Cream, 45–60% fat	57–73	3.06–3.27
Cucumber	97	4.10
Eggs		
Fresh		3.18†
Frozen		1.68‡
Fish, cod		
Fresh	70	3.18
Frozen	70	1.72‡
Flour	12–13.5	1.80–1.88
Ice	100	1.958‖

Material	H_2O (wt %)	c_p ($kJ/kg \cdot K$)
Ice cream		
Fresh	58–66	3.27†
Frozen	58–66	1.88‡
Lamb	70	3.18*
Macaroni	12.5–13.5	1.84–1.88
Milk, cows'		
Whole	87.5	3.85
Skim	91	3.98–4.02
Olive oil		2.01**
Oranges		
Fresh	87.2	3.77†
Frozen	87.2	1.93‡
Peas, air-dried	14	1.84
Peas, green		
Fresh	74.3	3.31†
Frozen	74.3	1.76‡
Pea soup		4.10
Plums	75–78	3.52
Pork		
Fresh	60	2.85†
Frozen	60	1.34‡
Potatoes	75	3.52
Poultry		
Fresh	74	3.31†
Frozen	74	1.55‡
Sausage, franks		
Fresh	60	3.60†
Frozen	60	2.35‡
String beans		
Fresh	88.9	3.81†
Frozen	88.9	1.97‡
Tomatoes	95	3.98†
Veal	63	3.22
Water	100	4.185**

* 32.8°C.
† Above freezing.
‡ Below freezing.
§ 24.4°C.
¶ 4.4°C.
‖ −20°C.
** 20°C.

Source: W. O. Ordinanz, *Food Ind.,* **18,** 101 (1946); G. A. Reidy, Department of Food Science, Michigan State University, 1968; S. E. Charm, *The Fundamentals of Food Engineering,* 2nd ed. Westport, Conn.: Avi Publishing Co., Inc., 1971; R. L. Earle, *Unit Operations in Food Processing.* Oxford: Pergamon Press, Inc., 1966; ASHRAE, *Handbook of Fundamentals.* New York: ASHRAE, 1972, 1967; H. C. Mannheim, M. P. Steinberg, and A. I. Nelson, *Food Technol.,* **9,** 556 (1955).

A.4-2 Thermal Conductivities, Densities, and Viscosities of Foods

Material	H_2O (wt %)	Temperature (K)	k (W/m·K)	ρ (kg/m³)	μ [(Pa·s)10³, or c_p]
Apple sauce		295.7	0.692		
Butter	15	277.6	0.197	998	
Cantaloupe			0.571		
Fish					
Fresh		273.2	0.431		
Frozen		263.2	1.22		
Flour, wheat	8.8		0.450		
Honey	12.6	275.4	0.50		
Ice	100	273.2	2.25		
	100	253.2	2.42		
Lamb	71	278.8	0.415		
Milk					
Whole		293.2		1030	2.12
Skim		274.7	0.538		
		298.2		1041	1.4
Oil					
Cod liver		298.2		924	
Corn		288.2		921	
Olive		293.2	0.168	919	84
Peanut		277.1	0.168		
Soybean		303.2		919	40
Oranges	61.2	303.5	0.431		
Pears		281.9	0.595		
Pork, lean					
Fresh	74	275.4	0.460		
Frozen		258.2	1.109		
Potatoes					
Raw			0.554		
Frozen		260.4	1.09	977	
Salmon					
Fresh	67	277.1	0.50		
Frozen	67	248.2	1.30		
Sucrose solution	80	294.3		1073	1.92
Turkey					
Fresh	74	276.0	0.502		
Frozen		248.2	1.675		
Veal					
Fresh	75	335.4	0.485		
Frozen	75	263.6	1.30		
Water	100	293.2	0.602		
	100	273.2	0.569		

Source: R. C. Weast, *Handbook of Chemistry and Physics*, 48th ed. Cleveland: Chemical Rubber Co., Inc., 1967; C. P. Lentz, *Food Technol.*, **15**, 243 (1961); G. A. Reidy, Department of Food Science, Michigan State University, 1968; S. E. Charm, *The Fundamentals of Food Engineering*, 2nd ed. Westport, Conn.: Avi Publishing Co., Inc., 1971; R. Earle, *Unit Operations in Food Processing*, Oxford: Pergamon Press, 1966; R. H. Perry and C. H. Chilton, *Chemical Engineers' Handbook*, 5th ed. New York: McGraw-Hill Book Company, 1973; V. E. Sweat, *J. Food Sci.*, **39**, 1080 (1974).

Properties of Pipes, Tubes, and Screens

A.5-1 Dimensions of Standard Steel Pipe

Nominal Pipe Size (in.)	Outside Diameter		Sched- ule Number	Wall Thickness		Inside Diameter		Inside Cross- Sectional Area	
	in.	mm		in.	mm	in.	mm	ft^2	$m^2 \times 10^4$
$\frac{1}{8}$	0.405	10.29	40	0.068	1.73	0.269	6.83	0.00040	0.3664
			80	0.095	2.41	0.215	5.46	0.00025	0.2341
$\frac{1}{4}$	0.540	13.72	40	0.088	2.24	0.364	9.25	0.00072	0.6720
			80	0.119	3.02	0.302	7.67	0.00050	0.4620
$\frac{3}{8}$	0.675	17.15	40	0.091	2.31	0.493	12.52	0.00133	1.231
			80	0.126	3.20	0.423	10.74	0.00098	0.9059
$\frac{1}{2}$	0.840	21.34	40	0.109	2.77	0.622	15.80	0.00211	1.961
			80	0.147	3.73	0.546	13.87	0.00163	1.511
$\frac{3}{4}$	1.050	26.67	40	0.113	2.87	0.824	20.93	0.00371	3.441
			80	0.154	3.91	0.742	18.85	0.00300	2.791
1	1.315	33.40	40	0.133	3.38	1.049	26.64	0.00600	5.574
			80	0.179	4.45	0.957	24.31	0.00499	4.641
$1\frac{1}{4}$	1.660	42.16	40	0.140	3.56	1.380	35.05	0.01040	9.648
			80	0.191	4.85	1.278	32.46	0.00891	8.275
$1\frac{1}{2}$	1.900	48.26	40	0.145	3.68	1.610	40.89	0.01414	13.13
			80	0.200	5.08	1.500	38.10	0.01225	11.40
2	2.375	60.33	40	0.154	3.91	2.067	52.50	0.02330	21.65
			80	0.218	5.54	1.939	49.25	0.02050	19.05
$2\frac{1}{2}$	2.875	73.03	40	0.203	5.16	2.469	62.71	0.03322	30.89
			80	0.276	7.01	2.323	59.00	0.02942	27.30
3	3.500	88.90	40	0.216	5.49	3.068	77.92	0.05130	47.69
				0.300	7.62	2.900	73.66	0.04587	42.61
$3\frac{1}{2}$	4.000	101.6	40	0.226	5.74	3.548	90.12	0.06870	63.79
			80	0.318	8.08	3.364	85.45	0.06170	57.35
4	4.500	114.3	40	0.237	6.02	4.026	102.3	0.08840	82.19
			80	0.337	8.56	3.826	97.18	0.07986	74.17
5	5.563	141.3	40	0.258	6.55	5.047	128.2	0.1390	129.1
			80	0.375	9.53	4.813	122.3	0.1263	117.5
6	6.625	168.3	40	0.280	7.11	6.065	154.1	0.2006	186.5
			80	0.432	10.97	5.761	146.3	0.1810	168.1
8	8.625	219.1	40	0.322	8.18	7.981	202.7	0.3474	322.7
			80	0.500	12.70	7.625	193.7	0.3171	294.7

A.5-2   Dimensions of Heat-Exchanger Tubes

Outside Diameter		BWG Number	Wall Thickness		Inside Diameter		Inside Cross-Sectional Area	
in.	mm		in.	mm	in.	mm	ft^2	$m^2 \times 10^4$
$\frac{5}{8}$	15.88	12	0.109	2.77	0.407	10.33	0.000903	0.8381
		14	0.083	2.11	0.459	11.66	0.00115	1.068
		16	0.065	1.65	0.495	12.57	0.00134	1.241
		18	0.049	1.25	0.527	13.39	0.00151	1.408
$\frac{3}{4}$	19.05	12	0.109	2.77	0.532	13.51	0.00154	1.434
		14	0.083	2.11	0.584	14.83	0.00186	1.727
		16	0.065	1.65	0.620	15.75	0.00210	1.948
		18	0.049	1.25	0.652	16.56	0.00232	2.154
$\frac{7}{8}$	22.23	12	0.109	2.77	0.657	16.69	0.00235	2.188
		14	0.083	2.11	0.709	18.01	0.00274	2.548
		16	0.065	1.65	0.745	18.92	0.00303	2.811
		18	0.049	1.25	0.777	19.74	0.00329	3.060
1	25.40	10	0.134	3.40	0.732	18.59	0.00292	2.714
		12	0.109	2.77	0.782	19.86	0.00334	3.098
		14	0.083	2.11	0.834	21.18	0.00379	3.523
		16	0.065	1.65	0.870	22.10	0.00413	3.836
$1\frac{1}{4}$	31.75	10	0.134	3.40	0.982	24.94	0.00526	4.885
		12	0.109	2.77	1.032	26.21	0.00581	5.395
		14	0.083	2.11	1.084	27.53	0.00641	5.953
		16	0.065	1.65	1.120	28.45	0.00684	6.357
$1\frac{1}{2}$	38.10	10	0.134	3.40	1.232	31.29	0.00828	7.690
		12	0.109	2.77	1.282	32.56	0.00896	8.326
		14	0.083	2.11	1.334	33.88	0.00971	9.015
2	50.80	10	0.134	3.40	1.732	43.99	0.0164	15.20
		12	0.109	2.77	1.782	45.26	0.0173	16.09

A.5-3 Tyler Standard Screen Scale

Sieve Opening		Nominal Wire Diameter		
mm	in. (approx. equivalents)	mm	in. (approx. equivalents)	Tyler Equivalent Designation
26.9	1.06	3.90	0.1535	1.050 in.
25.4	1.00	3.80	0.1496	
22.6	0.875	3.50	0.1378	0.883 in.
19.0	0.750	3.30	0.1299	0.742 in.
16.0	0.625	3.00	0.1181	0.624 in.
13.5	0.530	2.75	0.1083	0.525 in.
12.7	0.500	2.67	0.1051	
11.2	0.438	2.45	0.0965	0.441 in.
9.51	0.375	2.27	0.0894	0.371 in.
8.00	0.312	2.07	0.0815	$2\frac{1}{2}$ mesh
6.73	0.265	1.87	0.0736	3 mesh
6.35	0.250	1.82	0.0717	
5.66	0.223	1.68	0.0661	$3\frac{1}{2}$ mesh
4.76	0.187	1.54	0.0606	4 mesh
4.00	0.157	1.37	0.0539	5 mesh
3.36	0.132	1.23	0.0484	6 mesh
2.83	0.111	1.10	0.0430	7 mesh
2.38	0.0937	1.00	0.0394	8 mesh
2.00	0.0787	0.900	0.0354	9 mesh
1.68	0.0661	0.810	0.0319	10 mesh
1.41	0.0555	0.725	0.0285	12 mesh
1.19	0.0469	0.650	0.0256	14 mesh
1.00	0.0394	0.580	0.0228	16 mesh
0.841	0.0331	0.510	0.0201	20 mesh
0.707	0.0278	0.450	0.0177	24 mesh
0.595	0.0234	0.390	0.0154	28 mesh
0.500	0.0197	0.340	0.0134	32 mesh
0.420	0.0165	0.290	0.0114	35 mesh
0.354	0.0139	0.247	0.0097	42 mesh
0.297	0.0117	0.215	0.0085	48 mesh
0.250	0.0098	0.180	0.0071	60 mesh
0.210	0.0083	0.152	0.0060	65 mesh
0.177	0.0070	0.131	0.0052	80 mesh
0.149	0.0059	0.110	0.0043	100 mesh
0.125	0.0049	0.091	0.0036	115 mesh
0.105	0.0041	0.076	0.0030	150 mesh
0.088	0.0035	0.064	0.0025	170 mesh
0.074	0.0029	0.053	0.0021	200 mesh
0.063	0.0025	0.044	0.0017	250 mesh
0.053	0.0021	0.037	0.0015	270 mesh
0.044	0.0017	0.030	0.0012	325 mesh
0.037	0.0015	0.025	0.0010	400 mesh

Notation

SI units are given first (followed by English and/or cgs units).

a	particle radius, m (ft)
a	area, m^2 (ft^2); also m^2 area/m^3 volume bed or packing (ft^2/ft^3)
a_v	specific surface area of particle, m^{-1} (ft^{-1})
A	cross-sectional area, m^2 (ft^2, cm^2); also area, m^2 (ft^2)
B	flow rate of dry solid, kg/h (lb_m/h)
c	concentration, kg/m^3, kg mol/m^3 (lb_m/ft^3, lb mol/ft^3, g mol/cm^3)
c_A	concentration of A, kg mol/m^3 (lb mol/ft^3, g mol/cm^3)
c_p	heat capacity at constant pressure, $J/kg \cdot K$, $kJ/kg \cdot K$, kJ/kg mol $\cdot$ K (btu/$lb_m \cdot$ °F, cal/g $\cdot$ °C)
c'_A	deviation of concentration from mean concentration $\bar{c}_A$, kg mol/m^3
c_P	concentration of P, kg P/m^3
c_v	heat capacity at constant volume, $J/kg \cdot K$
C	fluid heat capacity, W/K (btu/h $\cdot$ °F)
C	height of bottom of agitator above tank bottom, m (ft)
C_p	pitot tube coefficient, dimensionless
C_D	drag coefficient, dimensionless
C_v, C_0	Venturi coefficient, orifice coefficient, dimensionless
D	molecular diffusivity, m^2/s (ft^2/h, cm^2/s); also diameter, m (ft)
D_{AB}	molecular diffusivity, m^2/s (ft^2/h, cm^2/s)
D_p, D_P	particle diameter, m (ft)
D_{KA}	Knudsen diffusivity m^2/s (ft^2/h, cm^2/s)
D_{NA}	transition-region diffusivity, m^2/s (ft^2/h, cm^2/s)
$D_{A\text{eff}}$	effective diffusivity, m^2/s (ft^2/h, cm^2/s)
$D_{p,m}$	effective mean diameter for mixture, m (ft)
D_a	diameter of agitator, m (ft)
D_t	diameter of tank, m (ft)
D_{AP}	diffusivity of A in protein solution, m^2/s
E	total energy, J/kg (ft $\cdot$ lb_f/lb_m)

E	fraction unaccomplished change, dimensionless
E	emitted radiation energy, W/m^2 (btu/h · ft²)
$E_{B\lambda}$	monochromatic emissive power, W/m^3 (btu/h · ft³)
f	Fanning friction factor, dimensionless
F	frictional loss, J/kg (ft · lb$_f$/lb$_m$)
F	force, N (lb$_f$, dyn)
F_T	correction factor for temperature difference, dimensionless
$\mathscr{F}_{12}$	geometric view factor for gray surfaces, dimensionless
F_{12}	geometric view factor, dimensionless
g	standard acceleration of gravity (see Appendix A.1)
g_c	gravitational conversion factor (see Appendix A.1)
G	mass velocity $= v\rho$, $kg/s \cdot m^2$, $kg/h \cdot m^2$ (lb$_m$/h · ft²)
G'	mass velocity $= v'\rho$, $kg/s \cdot m^2$, $kg/h \cdot m^2$ (lb$_m$/h · ft²)
G	irradiation on a body, W/m^2 (btu/h · ft²)
h	constant spacing in x for Simpson's rule
h	head, J/kg (ft · lb$_f$/lb$_m$); also height of fluid, m (ft)
h	heat-transfer coefficient, $W/m^2 \cdot K$ (btu/h · ft² · °F)
h	enthalpy of liquid, J/kg, kJ/kg (btu/lb$_m$)
h_{fg}	latent heat of vaporization, J/kg, kJ/kg (btu/lb$_m$)
h_c	contact resistance coefficient, $W/m^2 \cdot K$ (btu/h · ft² · °F)
H	distance, m (ft)
H	head, J/kg (ft · lb$_f$/lb$_m$); also height of fluid, m (ft)
H	enthalpy, J/kg, kJ/kg, kJ/kg mol (btu/lb$_m$)
$\mathbf{i}$	unit vector along x axis
I	intensity of turbulence, dimensionless; also current, amp
I_B	radiation intensity of black body, $W/m^2 \cdot sr$ (btu/h · ft² · sr)
I_λ	intensity of radiation, W/m^2 (btu/h · ft²)
$\mathbf{j}$	unit vector along y axis
J	width of baffle, m (ft)
j_A	mass flux of A relative to mass average velocity, $kg/s \cdot m^2$
j_A^*	mass flux of A relative to molar average velocity, $kg/s \cdot m^2$
J_A	molar flux of A relative to mass average velocity, kg mol/s · m^2
$\mathbf{J}_A^*$	molar flux vector of A relative to molar average velocity, kg mol/s · m^2 (lb mol/h · ft², g mol/s · cm²)
J_D, J_H	mass-transfer and heat-transfer factors, dimensionless
$\mathbf{k}$	unit vector along z axis
k, k'	reaction velocity constant, h^{-1}, min^{-1}, or s^{-1}
k	thermal conductivity, $W/m \cdot K$ (btu/h · ft · °F)
k_1'	first-order heterogeneous reaction velocity constant, m/s (cm/s)
$k_c', k_c, k_G, k_x, \dots$	mass-transfer coefficient, kg mol/s · m^2 · conc diff (lb mol/h · ft² · conc diff, g mol/s · cm² · conc diff); kg mol/s · m^2 · Pa, kg mol/s · m^2 atm (lb mol/h · ft² · atm). See Table 7.2-1.
K	consistency index, $N \cdot s^n/m^2$ (lb$_f$ · s^n/ft²)
K_1	constant in Eq. (3.1-39), m/s (ft/s)
K'	consistency index, $N \cdot s^{n'}/m^2$ (lb$_f$ · s$^{n'}$/ft²)
K	equilibrium distribution coefficient, dimensionless
K_c, K_{ex}, K_f	contraction, expansion, fitting loss coefficient, dimensionless
L	length, m (ft); also amount, kg, kg mol (lb$_m$); also kg/h · m^2
L	Prandtl mixing length, m (ft)
L	mean beam length, m (ft)
m	flow rate, kg/s, kg/h (lb$_m$/s)
m	dimensionless ratio $= k/hx_1$; also, position m

m	parameter in Table 7.1-1, dimensionless; also position parameter
M	molecular weight, kg/kg mol (lb_m/lb mol)
M	total mass, kg (lb_m); also parameter $= (\Delta x)^2/\alpha\Delta t$, dimensionless
M	modulus $= (\Delta x)^2/D_{AB}\Delta t$, dimensionless
n	exponent, dimensionless; also flow behavior index, dimensionless
n'	slope of line for power-law fluid, dimensionless
n	position parameter; also dimensionless ratio $= x/x_1$
n	total amount, kg mol (lb mol, g mol)
n_A	flux of A relative to stationary coordinates, kg/s $\cdot$ m^2
N	rpm or rps; also number of radiation shields
N	parameter $= h\Delta x/k$, dimensionless
N	total flux relative to stationary coordinates, kg mol/s $\cdot$ m^2 (lb mol/ h $\cdot$ ft^2, g mol/s $\cdot$ cm^2)
N	modulus $= k_c\Delta x/D_{AB}$, dimensionless; also number of stages
N	number of equal temperature subdivisions, dimensionless
$\bar{N}_{n,m}$	residual defined by Eq. (6.6-3), kg mol/m^3 (lb mol/ft^3)
$\mathbf{N}_A$	molar flux vector of A relative to stationary coordinates, kg mol/ s $\cdot$ m^2 (lb mol/h $\cdot$ ft^2, g mol/s $\cdot$ cm^2)
$\bar{N}_A$	mass transfer of A relative to stationary coordinates, kg mol/s (lb mol/h, g mol/s)
N_{Gr}	Grashof number defined in Eq. (4.7-4), dimensionless
N_{Bi}	Biot number $= hx_1/k$, dimensionless
N_{Gz}	Graetz number defined in Eq. (4.12-3), dimensionless
$N_{Gr,\delta}$	Grashof number defined in Eq. (4.7-5), dimensionless
N_{Ma}	Mach number defined by Eq. (2.11-15), dimensionless
N_{Pe}	Peclet number $= N_{Re}N_{Pr}$, dimensionless
$N_{Nu,\delta}$	Nusselt number $= h\delta/k$, dimensionless
N_{Fr}	Froude number $= v^2/gL$, dimensionless
N_{Kn}	Knudsen number $= \lambda/2\bar{r}$, dimensionless
N_{Nu}	Nusselt number $= hL/k$, dimensionless
$N_{Nu,x}$	Nusselt number $= h_x x/k$, dimensionless
N_{Pr}	Prandtl number $= c_p\mu/k$, dimensionless
N_{Re}	Reynolds number $= Dv\rho/\mu$, dimensionless
N'_{Re}	Reynolds number $= D_a^2 N\rho/\mu$, dimensionless
$N_{Re,gen}$	Reynolds number defined by Eq. (3.5-11), dimensionless
$N_{Re,L}$	Reynolds number $= Lv_\infty\rho/\mu$, dimensionless
$N_{Re,x}$	Reynolds number $= xv_\infty\rho/\mu$, dimensionless
$N'_{Re,n}$	Reynolds number defined by Eq. (3.5-20), dimensionless
$N_{Re,p}$	Reynolds number defined by Eq. (3.1-15), dimensionless
$N_{Re,mf}$	Reynolds number at minimum fluidization defined by Eq. (3.1-35), dimensionless
N_{Eu}	Euler number $= p/\rho v^2$, dimensionless
N_{Sc}	Schmidt number $= \mu/\rho D_{AB}$, dimensionless
N_{Sh}	Sherwood number $= k'_c D/D_{AB}$, dimensionless
N_P	power number defined by Eq. (3.4-2), dimensionless
N_{St}	Stanton number $= k'_c/v$, dimensionless
NTU	number of transfer units, dimensionless
p	pressure, N/m^2, Pa (lb_f/ft^2, atm, psia, mm Hg)
p_A	partial pressure of A, N/m^2, Pa (lb_f/ft^2, atm, psia, mm Hg)
p_{BM}	log mean inert partial pressure of B in Eq. (6.2-21), N/m^2, Pa (lb_f/ft^2, atm, psia, mm Hg)
P	parameter in Eq. (5.5-12), dimensionless

P	total pressure, N/m^2, Pa (lb_f/ft^2, atm, psia, mm Hg)
P	power, W (ft $\cdot lb_f/s$, hp)
P	flow rate, kg/h, kg/min; also number of phases at equilibrium
$\mathbf{P}$	momentum vector, kg $\cdot$ m/s ($lb_m \cdot$ ft/s)
P_A	vapor pressure of pure A, N/m^2, Pa (lb_f/ft^2, atm, psia, mm Hg)
P_M	permeability, $m^3(STP)/(s \cdot m^2$ C.S. $\cdot$ atm/m)
P'_M	permeability, $cm^3(STP)/(s \cdot cm^2$ C.S. $\cdot$ atm/cm)
P''_M	permeability, $cm^3(STP)/(s \cdot cm^2$ C.S. $\cdot$ cm Hg/mm)
q	heat-transfer rate, W (btu/h); also net energy added to system, W (btu/h); also J (btu)
q	flow rate, m^3/s (ft^3/s)
$\mathbf{q}$	heat flux vector, W/m^2 (btu/h $\cdot ft^2$)
$\dot{q}$	rate of heat generation, W/m^3 (btu/h $\cdot ft^3$)
q'	flow rate in Darcy's law, cm^3/s
$\bar{q}_{n,m}$	residual defined by Eq. (4.5-11), K (°F)
Q	amount absorbed, kg mol/m^2; also heat loss, W (btu/h); also heat absorbed, J/kg (btu/lb_m, ft $\cdot lb_f/lb_m$)
r	radius, m (ft)
r_A	rate of generation, kg $A/s \cdot m^3$ ($lb_m A/h \cdot ft^3$)
r_H	hydraulic radius, m (ft)
$(r_2)_{cr}$	critical value of radius, m (ft)
R	scaleup ratio, dimensionless
R	radius, m (ft); also resistance, K/W (h $\cdot$ °F/btu)
R	parameter in Eq. (5.5-12), dimensionless; also resistance, ohms
R	gas constant (see Appendix A.1); also reflux ratio = L_n/D, dimensionless
R_A	rate of generation, kg mol $A/s \cdot m^3$ (lb mol $A/h \cdot ft^3$)
R_i	rate of generation of i, kg/s (lb_m/h)
R_c	contact resistance, K/W (h $\cdot$ °F/btu)
R_x	x component of force, N (lb_f, dyn)
s	mean surface renewal factor, s^{-1}
S	conduction shape factor, m (ft)
S	solubility of a gas, m^3 solute(STP)/m^3 solid $\cdot$ atm [cc solute(STP)/cc solid $\cdot$ atm]
S	distance between centers, m (ft); also steam flow rate, kg/h, kg mol/h (lb_m/h)
S	cross-sectional area of tower, m^2 (ft^2)
S_p	surface area of particle, m^2 (ft^2)
t	fin thickness, m (ft)
t	time, s, min, or h
t	temperature, K, °C (°F)
T	temperature, K, °C (°F, °R)
T'	deviation of temperature from mean temperature $\bar{T}$, K (°F)
$\bar{u}$	average velocity, m/s (ft/s)
U	overall heat-transfer coefficient, $W/m^2 \cdot$ K (btu/h $\cdot ft^2 \cdot$ °F)
U	internal energy, J/kg (btu/lb_m)
v	velocity, m/s (ft/s)
$\mathbf{v}$	velocity vector, m/s (ft/s)
v_A	velocity of A relative to stationary coordinates, m/s (ft/s, cm/s)
v_{Ad}	diffusion velocity of A relative to molar average velocity, m/s (ft/s, cm/s)

v_M	molar average velocity of stream relative to stationary coordinates, m/s (ft/s, cm/s)
v_t	terminal settling velocity, m/s (ft/s)
v^+	dimensionless velocity defined by Eq. (3.7-34)
v^*	velocity defined by Eq. (3.7-42), dimensionless
v'_x	deviation of velocity in x direction from mean velocity $\bar{v}_x$, m/s (ft/s)
v'	superficial velocity based on cross section of empty tube, m/s (ft/s, cm/s)
v'_{mf}	velocity at minimum fluidization, m/s (ft/s)
v'_t	terminal settling velocity, m/s (ft/s)
V_A	solute molar volume, m³/kg mol
V	flow rate, kg/h, kg mol/h, m³/s (lb$_m$/h, ft³/s)
V	volume, m³ (ft³, cm³); also specific volume, m³/kg (ft³/lb$_m$)
V	velocity, m/s (ft/s); also total amount, kg, kg mol (lb$_m$)
w_A	mass fraction of A
$\dot{W}$	work done on surroundings, W (ft · lb$_f$/s)
$\dot{W}_S$	mechanical shaft work done on surroundings, W (ft · lb$_f$/s)
W	flow rate, kg/h, kg mol/h (lb$_m$/h)
W	work done on surroundings, J/kg (ft · lb$_f$/lb$_m$)
W	height or width, m (ft); also power, W (hp)
W	width of paddle, m (ft)
W_S	mechanical shaft work done on surroundings, J/kg (ft · lb$_f$/lb$_m$)
W_p	shaft work delivered to pump, J/kg (ft · lb$_f$/lb$_m$)
x	distance in x direction, m (ft)
x_A	mole fraction of A
x	mass fraction or mole fraction
x'_B	inert mole fraction, mol B/mol inert
x_{BM}	log mean mole fraction of inert or stagnant B given in Eq. (6.3-4)
X	particle size, m (ft); also parameter $= \alpha t/x_1^2$, dimensionless
X	parameter $= Dt/x^2$, dimensionless
y	distance in y direction, m (ft); also mole fraction
y_A	mass fraction of A or mole fraction of A; also kg A/kg solution (lb A/lb solution)
y_{BM}	log mean mole fraction of inert or stagnant B given in Eq. (7.2-11)
y^+	dimensionless number defined by Eq. (3.7-35)
Y	temperature ratio defined by Eq. (4.10-2), dimensionless
Y	fraction of unaccomplished change in Table 5.3-1 or Eq. (7.1-12)
Y	expansion correction factor defined in Eqs. (3.2-9) and (3.2-11)
z	distance in z direction, m (ft)
Z	height, m (ft); also temperature ratio in Eq. (4.10-2)

Greek letters

α	correction factor $= 1.0$, turbulent flow and $\frac{1}{2}$, laminar flow
α	absorptivity, dimensionless; also flux ratio $= 1 + N_B/N_A$
α	thermal diffusivity $= k/\rho c_p$, m²/s (ft²/h, cm²/s)
α	angle, rad
α	kinetic-energy velocity correction factor, dimensionless
α_G	gas absorptivity, dimensionless
α_T	eddy thermal diffusivity, m²/s (ft²/h)

β	momentum velocity correction factor, dimensionless
β	volumetric coefficient of expansion, $1/\text{K}$ $(1/^\circ\text{R})$
γ	viscosity coefficient, $\text{N} \cdot \text{s}^{n'}/\text{m}^2$ $(\text{lb}_m/\text{ft} \cdot \text{s}^{2-n'})$
γ	ratio of heat capacities $= c_p/c_v$, dimensionless
Γ	flow rate, $\text{kg/s} \cdot \text{m}$ $(\text{lb}_m/\text{h} \cdot \text{ft})$
Γ	concentration of property, amount of property/m^3
δ	molecular diffusivity, m^2/s
δ	boundary-layer thickness, m (ft); also distance, m (ft)
δ	constant in Eq. (4.12-2), dimensionless
Δ	difference
ΔT_a	arithmetic temperature drop, K, $^\circ\text{C}$ $(^\circ\text{F})$
ΔT_{1m}	log mean temperature driving force, K, $^\circ\text{C}$ $(^\circ\text{F})$
ΔH	enthalpy change, J/kg, kJ/kg, kJ/kg mol $(\text{btu/lb}_m, \text{btu/lb mol})$
Δp	pressure drop, N/m^2, Pa $(\text{lb}_f/\text{ft}^2)$
ε	roughness parameter, m (ft); or void fraction, dimensionless
ε	emissivity, dimensionless; also volume fraction, dimensionless
ε_M	mass eddy diffusivity m^2/s $(\text{ft}^2/\text{h}, \text{cm}^2/\text{s})$
ε	heat-exchanger effectiveness, dimensionless
ε_G	gas emissivity, dimensionless
ε_t	momentum eddy diffusivity, m^2/s (ft^2/s)
ε_{mf}	void fraction at minimum fluidization, dimensionless
η_f	fin efficiency, dimensionless
η_t	turbulent eddy viscosity, $\text{Pa} \cdot \text{s}$, $\text{kg/m} \cdot \text{s}$ $(\text{lb}_m/\text{ft} \cdot \text{s})$
η	efficiency, dimensionless
θ	angle, rad
λ	wavelength, m (ft)
λ	latent heat, J/kg, kJ/kg (btu/lb_m); also mean free path, m (ft)
μ	viscosity, $\text{Pa} \cdot \text{s}$, $\text{kg/m} \cdot \text{s}$, $\text{N} \cdot \text{s/m}^2$ $(\text{lb}_m/\text{ft} \cdot \text{s}, \text{lb}_m/\text{ft} \cdot \text{h}, \text{cp})$
μ_a	apparent viscosity, $\text{Pa} \cdot \text{s}$, $\text{kg/m} \cdot \text{s}$ $(\text{lb}_m/\text{ft} \cdot \text{s})$
ν	momentum diffusivity μ/ρ, m^2/s $(\text{ft}^2/\text{s}, \text{cm}^2/\text{s})$
π_1	dimensionless group, dimensionless
ρ	density, kg/m^3 $(\text{lb}_m/\text{ft}^3)$; also reflectivity, dimensionless
σ	constant, 5.676×10^{-8} $\text{W/m}^2 \cdot \text{K}^4$ $(0.1714 \times 10^{-8}$ $\text{btu/h} \cdot \text{ft}^2 \cdot ^\circ\text{R}^4)$; also collision diameter, Å
τ_{zx}	flux of x-directed momentum in z direction, $(\text{kg} \cdot \text{m/s})/\text{s} \cdot \text{m}^2$ or N/m^2 $(\text{lb}_f/\text{ft}^2, \text{dyn/cm}^2)$
τ	shear stress, N/m^2 $(\text{lb}_f/\text{ft}^2, \text{dyn/cm}^2)$
τ	tortuosity, dimensionless
ϕ	angle, rad; also association parameter, dimensionless
ϕ_S	shape factor of particle, dimensionless
Ψ_z	flux of property, amount of property/$\text{s} \cdot \text{m}^2$
ψ	correction factor, dimensionless
ω	solid angle, sr
$\Omega_{D, AB}$	collision integral, dimensionless

Index

C

Capillaries (*see* Porous solids)
Centrifugal pumps, 145–146
Cgs system of units, 4
Chapman-Enskog theory, 384–386
Chemical reaction and diffusion,
    445–449
Chilling of food and biological
    materials, 350–351
Chilton-Colburn analogies, 429
Circular pipes and tubes
  compressible flow in, 112–115
  dimensions of, 519–520
  friction factors in, 97–103
  heat transfer coefficients in, 229–232
  laminar flow in, 89–91, 94–98
  mass transfer in, 429–432
  turbulent flow in, 94–95, 97–102
  universal velocity distribution in,
      190–192
Compressible flow of gases
  adiabatic conditions, 114–115
  basic differential equation, 112
  isothermal conditions, 112–114
  Mach number for, 115
  maximum flow conditions, 113–115
Compressors
  equations for, 149–150
  equipment, 149
Concentration units, 7–8
Condensation
  derivation of equation for, 254–258
  mechanisms of, 254
  outside horizontal cylinders, 257–258
  outside vertical surfaces, 254–256
Conduction heat transfer (*see also*
    Unsteady-state heat transfer)
  combined conduction and convection,
      218–219
  contact resistance at interface, 224
  critical thickness of insulation,
      222–223
  in cylinders, 212, 215–216
  effect of variable thermal
      conductivity, 211
  equations for
    in cylindrical coordinates, 358
    in rectangular coordinates, 358
    in spherical coordinates, 358
  Fourier's law, 54, 207–208

graphical curvilinear square method,
    224–226
with heat generation, 220–222
in hollow sphere, 213–214
through materials in parallel, 217–218
mechanism of, 206
shape factors in, 226–227
steady-state in two dimensions
    Laplace equation, 301
    numerical method, 302–307
    with other boundary conditions,
        306–307
through a wall, 211
through walls in series, 214–216
Continuity equation
  for binary mixture, 443–445
  for boundary layer, 185–186
  for pure fluid, 61–65, 170–171,
      173–175, 178
Control volume, 63–65
Convection heat transfer (*see also* Heat
    transfer coefficients; Natural
    convection)
  general discussion of, 206–207, 210,
      218–219
  physical mechanism of, 227–229
Convective heat transfer coefficients
    (*see* Heat transfer coefficients)
Convective mass transfer coefficients
    (*see* Mass transfer coefficients)
Conversion factors, 477–480
Countercurrent heat transfer, 235–237
Creeping flow, 182–183
Critical thickness of insulation, 222–223
Cubes, mass transfer to, 437
Curvilinear squares method, 224–226
Cylinders
  drag coefficient for, 127–128
  flow across banks of, 241–242
  heat transfer coefficients for, 240–242
  mass transfer coefficients for, 439
Cylindrical coordinates, 175, 180, 358

D

Dalton's law, 10
Darcy's law, 133–134
Density
  of foods, 518
  of metals, 510

of solids, 509
of water, 482
Differential operations
time derivatives, 171
with scalars, 172–173
with vectors, 172–173
Diffusion (*see also* Steady-state
diffusion; Unsteady-state
diffusion)
of A through stagnant B, 378–380,
445–446
in biological gels, 396–397
of biological solutes, 393–396
and blockage by proteins, 395–396
in capillaries, 451–457
with chemical reaction, 445–449
and convection, 377–379
and equation of continuity, 443–445
equimolar counterdiffusion, 375–376,
445
Fick's law, 54, 373–374, 442–443, 453
in gases, 375–387
general case for A and B, 377–378
introduction to, 53–54, 371–373
Knudsen, 452–453
in liquids, 387–389
molecular, 453
multicomponent
gases, 450–451
liquids, 392
in porous solids, 401–403, 451, 457
similarity of mass, heat, and
momentum transfer, 50–54,
371–372
in solids
Fick's law, 398–399
classification of, 397–398
permeability equations, 399–401
to a sphere, 381–382
steady-state, in two dimensions,
403–406
transition, 453–455
with variable cross-sectional area,
380–383, 398
velocities in, 377
Diffusivity (mass)
in biological gels, 396–397
of biological solutes, 394–396
effective
in capillaries, 453–455
in porous solids, 402, 457

estimation of
for biological solutes, 395–396
in gases, 384–386
in liquids, 390–393
experimental determination
in biological gels, 396
in biological solutions, 394
in gases, 380, 383–384
in liquids, 389–390
experimental values
for biological gels, 396–397
for biological solutes, 394–395
for gases, 384–385
for liquids, 390–391
for solids, 400–401
Knudsen, 452–453
molecular and eddy, 364–365
transition, 453–455
Diffusivity (momentum), 364–365,
371–372
Diffusivity (thermal), 321, 364–365,
371–372
Dilitant fluids, 163
Dimensional analysis
Buckingham pi theorem, 196–197,
298–300, 463
in differential equations, 195–196
in heat transfer, 298–300
in mass transfer, 463–464
in momentum transfer, 195–197
Dimensional homogeneity, 5
Disk, drag coefficient for, 127–128
Distribution coefficient, in mass
transfer, 418–419, 459–460
Drag coefficient
definition of, 126–127, 183
for disk, 127–128
for flat plate, 125, 185–186
form drag, 125–127, 183
for long cylinder, 127–128
skin drag, 125–127
for sphere, 125–128, 183

E

Emissivity
definition of, 259–260, 273–274
values of, 511
Energy balances
for boundary layer, 360–363

and mass, 50–54, 371–372
  for steady state, 50–51
  for unsteady state, 52–53
General property balance, 50–53
Graphical methods
  integration by, 29
  linear plot, 31
  log-log plot, 33
  semilog plot, 32–33
  two-dimensional conduction, 224–226
Grashof number, 245
Gravitational constant, 478
Gravity separator, 49–50
Gurney-Lurie charts, 330, 333, 335

H

Hagen-Poiseuille equation, 91, 98, 181
Heat balances, principles of, 23–24, 26
Heat capacity
  data for foods, 516–517
  data for gases, 20, 493, 496, 500–501
  data for liquids, 502, 506
  data for solids, 508, 510
  data for water, 483–484
  discussion of, 18–20
Heat exchangers (*see also* Heat transfer
      coefficients)
  cross-flow, 263–265, 267
  double-pipe, 263
  effectiveness of, 267–270
  extended-surface, 294–298
  fouling factors for, 270–271
  log mean temperature difference,
      235–236, 265
  scraped-surface, 293–294
  shell and tube, 263
  temperature correction factors,
      265–267
Heat of reaction, 22–24
Heat transfer coefficients
  for agitated vessels, 290–291
  approximate values of, 210
  average coefficient, 229
  to banks of tubes, 241–242
  for condensation, 254–258
  definition of, 210, 228
  entrance region effect on, 233–234
  for film boiling, 252
  for finned-surface exchangers,
      294–298

for flow parallel to flat plate, 239
for flow past a cylinder, 240
for fluidized bed, 244
fouling factors, 270–271
for laminar flow in pipes, 229
for liquid metals, 234
for natural convection, 244–250
in noncircular conducts, 232
for non-Newtonian fluids, 287–290
for nucleate boiling, 251–252
for other geometries, 244
overall, 218–219, 270–271
for packed beds, 243–244, 436–437
for radiation and convection, 260–262
for scraped-surface exchangers,
    293–294
for spheres, 240
for transition flow in pipes, 231–232
for turbulent flow in pipes, 230–231
Heat transfer mechanisms, 206–207
Heisler charts, 331, 334, 336
Henry's law, data for gases, 511

I

Ice, properties of, 484, 516
Ideal fluid, 182
Ideal gas volume, 477
Integral analysis of boundary layers
    for energy balance, 363
    for momentum balance, 192–194
Intensity of turbulence, 187–188
Interface contact resistance, 224
Internal energy, 21, 68
Isothermal compressible flow, 112–114

J

J-factor, 427, 429

K

Kinetic energy
    velocity correction factor for, 70–71,
        166–167
    definition of, 68
Kirchhoff's law, 259, 273–274
Knudsen diffusion, 452–453
Knudsen number, 453

V

Vapor pressure
  data for water, 481, 483–486
  discussion of, 11
Vectors, 171–173
Velocity
  average, 66, 91, 93
  interstitial, 426
  maximum, 94–95
  profile in laminar flow, 91, 93–95
  profile in turbulent flow, 94–95
  relation between velocities, 94–95
  representative values in pipes, 111
  superficial, 130, 426
  types of, in mass transfer, 377,
    442–443
  universal, in pipes, 190–192
Venturi meter, 140–141
View factors in radiation, 274–284

Viscoelastic fluids, 163
Viscosity
  discussion of, 54–58
  of foods, 518
  of gases, 493–494, 498–499
  of liquids, 503–505
  Newton's law of, 55–56, 371–372
  of water, 482, 490
Void fraction (packed and fluidized
    beds), 129–130, 134–135,
    436–437
Von Kármán analogy, 428–429

W

Water, properties of, 481–490
Weirs, 143
Wilke-Change correlation, 391–392
Work, 68–69, 74